T0133798

Gaseous Electronics

Theory and Practice

ELECTRICAL AND COMPUTER ENGINEERING

A Series of Reference Books and Textbooks

FOUNDING EDITOR

Marlin O. Thurston
Department of Electrical Engineering
The Ohio State University
Columbus, Ohio

1. Rational Fault Analysis, *edited by Richard Saeks and S. R. Liberty*
2. Nonparametric Methods in Communications, *edited by P. Papantoni-Kazakos and Dimitri Kazakos*
3. Interactive Pattern Recognition, *Yi-tzuu Chien*
4. Solid-State Electronics, *Lawrence E. Murr*
5. Electronic, Magnetic, and Thermal Properties of Solid Materials, *Klaus Schröder*
6. Magnetic-Bubble Memory Technology, *Hsu Chang*
7. Transformer and Inductor Design Handbook, *Colonel Wm. T. McLyman*
8. Electromagnetics: Classical and Modern Theory and Applications, *Samuel Seely and Alexander D. Poularikas*
9. One-Dimensional Digital Signal Processing, *Chi-Tsong Chen*
10. Interconnected Dynamical Systems, *Raymond A. DeCarlo and Richard Saeks*
11. Modern Digital Control Systems, *Raymond G. Jacquot*
12. Hybrid Circuit Design and Manufacture, *Roydn D. Jones*
13. Magnetic Core Selection for Transformers and Inductors: A User's Guide to Practice and Specification, *Colonel Wm. T. McLyman*
14. Static and Rotating Electromagnetic Devices, *Richard H. Engelmann*
15. Energy-Efficient Electric Motors: Selection and Application, *John C. Andreas*
16. Electromagnetic Compossibility, *Heinz M. Schlicke*
17. Electronics: Models, Analysis, and Systems, *James G. Gottling*
18. Digital Filter Design Handbook, *Fred J. Taylor*
19. Multivariable Control: An Introduction, *P. K. Sinha*
20. Flexible Circuits: Design and Applications, *Steve Gurley, with contributions by Carl A. Edstrom, Jr., Ray D. Greenway, and William P. Kelly*
21. Circuit Interruption: Theory and Techniques, *Thomas E. Browne, Jr.*
22. Switch Mode Power Conversion: Basic Theory and Design, *K. Kit Sum*
23. Pattern Recognition: Applications to Large Data-Set Problems, *Sing-Tze Bow*
24. Custom-Specific Integrated Circuits: Design and Fabrication, *Stanley L. Hurst*
25. Digital Circuits: Logic and Design, *Ronald C. Emery*
26. Large-Scale Control Systems: Theories and Techniques, *Magdi S. Mahmoud, Mohamed F. Hassan, and Mohamed G. Darwish*
27. Microprocessor Software Project Management, *Eli T. Fathi and Cedric V. W. Armstrong (Sponsored by Ontario Centre for Microelectronics)*
28. Low Frequency Electromagnetic Design, *Michael P. Perry*
29. Multidimensional Systems: Techniques and Applications, *edited by Spyros G. Tzafestas*

Gaseous Electronics

Theory and Practice

Gorur Govinda Raju

University of Windsor
Ontario, Canada

CRC Press
Taylor & Francis Group
Boca Raton London New York

CRC Press is an imprint of the
Taylor & Francis Group, an **informa** business

A TAYLOR & FRANCIS BOOK

CRC Press
Taylor & Francis Group
6000 Broken Sound Parkway NW, Suite 300
Boca Raton, FL 33487-2742

First issued in paperback 2019

ISBN-13: 978-0-8493-3763-5 (hbk)
ISBN-13: 978-0-367-39177-5 (pbk)
Library of Congress Card Number 2005044020

Library of Congress Cataloging-in-Publication Data

Raju, Gorur G., 1937-
  Gaseous electronics : theory and practice / Gorur G. Raju.
    p. cm. – (Electical and computer engineering ; 127)
  Includes bibliographical references and index.
  ISBN 0-8493-3763-1 (alk. paper)
  Electric discharges through gases – Textbooks. 2. Ionization of gases – Textbooks. 3. Gas dynamics – Textbooks. I. Title. II. Series.

QC711.R235 2005
537.5'32 – dc22   2005044020

Visit the Taylor & Francis Web site at
http://www.taylorandfrancis.com

and the CRC Press Web site at
http://www.crcpress.com

Dedicated to

———

Gorur *(India), a place I left long years ago as an aspiring youth*
and
Windsor *(Ontario, Canada) where I have been fortunate enough to*
live longer than anywhere else.

Preface

Studies of the interaction of electrons with gas neutrals, excited states, and other charge carriers are generally defined as gaseous electronics. While the ordinary fluorescent bulb and neon sign are examples of industrial applications, one of the most striking developments in modern society has been the explosion of applications of lasers. From delicate eye surgery to metal cutting, from the physics laboratory to research on fusion studies, gas lasers function according to the theoretical concepts and experimental techniques developed in this area. Areas that involve this branch of knowledge include medicine, electrical and mechanical engineering, environmental studies, defense applications, just to name a few. In nature, lightning and aurora lights are spectacular examples of electron interaction with gas molecules, while the everyday occurrence of red evening sky is a reminder of such interaction even when tranquil conditions prevail. The relatively innocuous electron–molecule interactions culminate in the awesome power of plasma, both in nature and the laboratory, in the destructive power of lightning, and the magnificent northern lights.

The study of gaseous electronics is over one hundred years old, beginning with the discovery of cathode rays in 1876, though one could arguably refer to the sparks observed with the Leyden jar circa 1750. In 1860, Maxwell's classical treatment of molecules as a group of particles and the velocity distribution within the group defined many concepts that would later be carried into the study of electron motion in gases. Rapid development of ingenious experimental techniques by Townsend in England and Ramsauer in Germany laid the solid foundation for studies of electrons in swarms and beams respectively. The advent of quantum mechanics gave a powerful tool for theoretical development of electron–molecule scattering, and the laser technology added impetus for renewed interest in this area.

For some time a need has been felt for a volume on gaseous electronics, considering the explosion of scientific literature published on all aspects of electron interaction with and without the application of an external electric field. At times this vast wealth of knowledge has appeared to be scattered in a seemingly hopeless disarray, discouraging even a modest attempt to classify and categorize the available information. The availability of online journals and modern software on personal computers for drawing graphs, digitizing for numerical integration and interpolation, and so on, combined with the long, cold Canadian, winters, has prodded the author to make such an attempt.

The present volume is intended to serve the following objectives:

1. To serve as a graduate and senior-level undergraduate textbook.
2. To provide experimental data with adequate but not overwhelming theoretical discussion. Excellent treatment of theoretical aspects have been presented in books by Massey and Burhop,[1] McDaniel,[2] Loeb,[3] MacDonald,[4] and Hasted.[5] Books by Meek and Craggs[6] and Roth[7] present different aspects of the discharge and plasma phenomena respectively. The present volume is meant to serve as complementary to this list.
3. To classify the data on cross sections, drift and diffusion, and ionization phenomena. The book concentrates largely on the critical evaluation of the available data in many gases, although the sheer volume of such data has precluded consideration of all gases.

4. To supply a resource material for established researchers and scholars.
5. To offer a source book for industrial and nonacademic users who seek data without needing to plough through the niceties of theoretical analyses and experimental sophistication.

Chapter 1 begins with an introduction to electron–neutral collision physics and, in view of the literature previously referred to, the treatment has been kept at a level that is easy to follow. The meaning of velocity space is explained as this is the central concept in the understanding of the energy distribution. The various cross sections are defined and the relation that exists between them is described. The quantum mechanical approach to scattering is introduced as this is the basis on which the Ramsauer–Townsend effect is understood. Though ion mobilities are not included as a separate chapter of presentation, the basic theory is included in view of its role in space charge build-up and secondary effects at the cathode. Since each subsequent chapter begins with a limited exposition of the theory necessary to understand the topic, the first chapter is made desirably concise.

Chapter 2 attempts to provide an overview of experimental techniques that are employed to measure collision cross sections. The methods for measuring other quantities such as swarm coefficients, drift velocities etc. are treated in later chapters. A large number of ingenious techniques, employing crossed beams, have been developed since the early beam experiments of Ramsauer and Brode (Figure 2.1). The principles involved in these measurements have been explained with selected reference to the measures adopted to improve accuracy and repeatability.

The techniques chosen for description are by no means exhaustive but have relevance to scattering cross sections discussed in Chapters 3 to 5. The bias in choosing which method to include has been the parameters that are relevant to discharge phenomena, and therefore the methods for measuring electronic excitation and ro-vibrational excitation have not been dealt with at great length. Recent advances in the measurement of angular differential cross sections, adopted by Cubric et al. (Figure 2.12), dispense with the need to rotate the relative position of the detector with respect to the collision region. Adoption of this technique to several gases should yield data that do not require interpolation at very low and very large angles, improving the accuracy. The ionization cross sections measured by Rapp and Englander-Golden in about fifteen gases have set a standard for accuracy and reliability. Their method is explained (Figure 2.13) and the more recent measurement method of Straub and colleagues (reference 71 of Chapter 2) has been described in a later chapter. Due to limitations of space it has not been possible to include methods for measuring attachment cross sections, though a brief explanation of the swarm technique (reference 116 of Chapter 2) is given.

Chapter 3 deals with scattering cross sections in rare gases with all aspects of measurements taken into account. Data available in the literature have been compiled and systematically categorized. A critical analysis is carried out and it is believed that this is the first time that such a comprehensive review has been made available. The gases considered are arranged in alphabetical order to avoid repetition—which becomes inevitable in the traditional method of dealing with gases with increasing atomic weight; helium first, neon next and so on. The cross sections compiled are mainly those measured during the past twenty-five years, though earlier publications are referred to as required. Results of new computations are included, as found necessary, for obtaining the momentum transfer and elastic scattering cross sections from differential cross sections measured as a function of angle of scattering.

All cross sections of each gas have been added to obtain the total cross sections in the energy range 0 to 1000 eV and compared with the measured total cross section. This kind of information, it is believed, has been provided for the first time over the entire energy range though excellent reviews (reference 10 of Chapter 3) are available that cover fewer energy values. To facilitate comparison, most of the curves of cross sections as a function of energy have been redrawn and grid lines have been retained for finding the approximate value in rapid mode. This method of providing cross section data, in addition to the tabular form, has been adopted throughout the volume.

Analytical representation of cross sections as a function of energy is required for the purposes of modeling, energy distribution computations, and simulation studies. Many such equations have been provided, though more work needs to be done to represent momentum transfer cross sections as a function of energy. As far as the author is aware there has been only a single equation available for argon (reference 107 of Chapter 3) and the rapid variation of cross section as a function of energy due to the Ramsauer–Townsend effect and the rather broader variation at higher energies due to shape resonance render the problem difficult.

Cross sections in the very low energy range are also represented analytically by the modified effective range theory (MERT). Each gas is discussed in the light of this theory and appropriate information is given. Again, it is thought, this is the first time that such a compilation has been made available in a single volume and the author expresses the opinion that this powerful technique has not been used adequately for molecular gases. Significant experimental investigations of low-energy inelastic collision cross sections in several gases have now become available (reference 41 of Chapter 4).

Chapter 4 continues with the presentation of cross sections of diatomic gases. The gases considered are carbon monoxide (CO), hydrogen (H_2), nitrogen (N_2), oxygen (O_2), and nitric oxide (NO). CO and NO are polar and electron attaching. O_2 is electron attaching without possessing a permanent dipole moment. The remaining gases (H_2 and N_2) are both nonpolar and nonelectron attaching. The long-range dipole interaction between the electron and molecule in polar gases presents difficulties for complete theoretical under-standing and experimental measurements are the main source, unlike the case with rare gases (Chapter 3) where theory can supplement experiment. The influence played by dipolar moment and electron attachment is highlighted and a broad interpretation of shape resonance as applicable to the gases considered is provided. The interaction potential is an integral part of the theory and a brief description of the potentials is included in Chapter 1. The similarities and differences in the cross section–energy behavior of isoelectronic molecules are dealt with.

The presentation of scattering cross section data is continued in Chapter 5, with attention focusing upon a variety of complex molecules. Polyatomic molecules such as SF_6 and CO_2 are nonpolar but electron attaching. On the other hand, there are a number of gases which are both attaching and polar. Extensive discussion is not presented for a few gases that have been analyzed thoroughly in recent years (reference 174 in Chapter 5). These gases include SF_6, CCl_2F_2, CF_4, and selected fluorocarbons. Gases of environmental concern such as NO_2, N_2O, SO_2, and O_3 are also considered, though in some cases the data appear in a later chapter. Chapter 5 concludes with a discussion of how the ionization cross section can be understood by using the most common parameters of a gas neutral: the ratio of the maximum of the ionization cross section to the ionization cross section at a given energy. Another attractive formulation is due to Hudson (reference 302 of Chapter 5) and involves the polarization of the molecule and its ionization potential. Considerable scope exists for original research in exploring this idea with necessary modifications to accommodate the specifics of a molecule under investigation.

While scattering cross sections are measured by using beam techniques, the focus now shifts to electron swarms. At the turn of the twentieth century the discovery of the electron and the advent of quantum mechanics gave birth to two schools of investigators. One school, led by Ramsauer and colleagues, adopted beam techniques; the other school, led by Townsend, adopted the swarm technique in which the electrons move through the gaseous medium under the influence of an applied external field. From the results of these investigations the details of electron–neutral interactions were deciphered. The method of measuring the drift velocity by employing grids (reference 6 of Chapter 6) and the availability of the oscilloscope facilitated the measurement of drift velocity of electrons. Development of the theory of diffusion by Huxley (reference 9 of Chapter 6) and the measurement of the diffusion coefficient by the use of the concentric and insulated collector led the method of approach. The experimental discovery of the lateral diffusion coefficient (reference 92 of Chapter 6) was followed by advancement of the theory to explain the observed results (references 93 and 94 of Chapter 6).

Chapter 6 summarizes the data on drift and diffusion of electrons in several gases as a function of reduced electric field E/N ($E =$ electric field, $N =$ gas number density). Early measurements of these parameters at low values of E/N (reference 118 of Chapter 6) have been extended to larger E/N values in various laboratories and a compilation of these is presented. Analytical expressions for a wide range of E/N are given, with an analysis of the range of applicability and the limits of accuracy. It is appropriate to comment that the best fitting equation given for the purpose of simulation etc. does not imply that there is theoretical background for that form of equation.

Swarm parameters have been measured as an end in themselves, and also to obtain low-energy momentum transfer cross sections where experimental difficulties render the measurements less accurate. The method of unfolding the swarm parameters to obtain the low-energy momentum transfer cross sections was pioneered in the early 1960s (reference 156 of Chapter 6) and extended to many gases. The results of these investigations have been blended into the data presented in Chapter 6.

Chapter 7 continues the presentation of these data to more complex molecules. Hydrocarbon gases, nitrogen compounds, and plasma industrial gases have been considered over a wide range of E/N.

Chapter 8 deals with the ionization process and presents the first ionization coefficient in nonattaching gases. Both the steady state method and time-resolved current methods have been employed and the data cover a wide range of E/N. At higher values of E/N the drift velocity and diffusion coefficients are subject to ionization effects and results obtained by simulation or theoretical computation are not excluded, though attention has been drawn to situations where experimental confirmation is desirable.

Chapter 9 extends the presentation of ionization coefficients to electron-attaching gases. Electron attachment is a process that depletes electrons from the ionization region. It may be a two-body process (electron and molecule) or a three-body process (electron and two molecules). Dissociative attachment involves the dissociation of the molecule and the attachment of the electron to one of the fragments. Dissociative attachment cross section is dominant at relatively low energies while at higher energies ion pair formation is more frequently encountered. The change of the familiar Townsend's semiempirical relation due to attachment is explained by several examples. The chapter provides attachment cross sections for several gases in addition to ionization and attachment coefficients. A point to note is that collision cross sections have been provided for some gases for which these data were not given earlier.

Chapter 10 shifts the focus to high-voltage phenomena in gaseous electronics, though in a compact form necessitated by limitation of space. For our present purpose high voltage is defined as that above 200 kV with no restriction on electrode geometry, gap length, or polarity of the voltage. This definition is, of course, purely arbitrary to serve the purpose of limiting the topics for inclusion. Only large air gaps of relevance to high-voltage power transmission and sulfurhexafluoride at elevated gas pressures have been considered. As an introduction to the chapter, methods of generating high voltages in the laboratory are described, though measurement aspects have had to be deleted. Standard volumes (reference 2 of Chapter 10) deal exhaustively with these methods. Switching impulse breakdown of large air gaps and volt–time characteristics of compressed gases have been briefly considered.

Chapter 11 concerns ionization and breakdown in crossed electric and magnetic fields. This area of research is still only moderately explored, relatively to the volume of literature available on other areas of gaseous electronics. The potential industrial uses of this type of discharge are at least as promising as those of other areas of research and the author is aware of just a single review paper (reference 2 of Chapter 11), published in 1980. The chapter begins by describing the motion of charge carriers in crossed fields in vacuum and extends the discussion to phenomena in the presence of gaseous neutrals. The effective reduced electric field concept is described, as is the influence of a crossed magnetic field on the ionization coefficients. Quantitative data on Townsend's first ionization coefficient in gases as a function of reduced electric and magnetic fields have been compiled for the first time, to the extent available. The effects of a crossed magnetic field on breakdown, time lags, and corona formation in nonuniform fields are described. Results obtained by computational methods are commented upon, with brief comments on the research to be completed.

The final chapter deals with high-frequency breakdown, included for completeness in view of the needs of beginners and students. RF discharges have assumed an important role in view of the explosive electronic industry, and discharge phenomena are described. A software package available (reference 11 of Chapter 12) has, in the author's opinion, served well to elucidate the complexities of this type of discharge and to provide visual images of the influence of various parameters on the discharge phenomena. Both microwave break-down and laser breakdown are dealt with, largely for the sake of completeness.

The present volume is the culmination of forty-seven years of the author's interaction with the study of gaseous electronics, beginning with his first entry to the Department of High Voltage Engineering at the Indian Institute of Science in 1958 as a graduate student. The topics chosen to be included have a personal bias, of course, though he has personally studied and researched in all the topics chosen, some with greater intensity than others. It is realized that topics such as ion mobilities, photo-ionization cross sections, and recent advances in lightning research have not been included due to limitations of space.

REFERENCES

1. Massey, H. S. W. and E. H. S. Burhop, *Electronic and Ionic Impact Phenomena*, Oxford University Press, Oxford, 1952.
2. McDaniel, E. W., *Collision Phenomena in Ionized Gases*, John Wiley & Sons, New York, 1964.
3. Loeb, L. B., *Basic Processes of Gaseous Electronics*, University of California Press, Berkeley, 1965.
4. MacDonald, A. D., *Microwave Breakdown in Gases*, John Wiley & Sons, New York, 1966.
5. Hasted, J. B., *Physics of Atomic Collisions*, Elsevier, New York, 1972.
6. Meek, J. M. and J. D. Craggs, *Electrical Breakdown of Gases*, John Wiley & Sons, New York, 1978.
7. Roth, J., *Industrial Plasma Engineering, vol 1., Principles*, Institute of Physics Publishing, Bristol, 1995.

About the Author

Gorur Govinda Raju was born in 1937. He obtained the B.Eng. degree in electrical engineering from the University of Bangalore, India, and the Ph.D. degree from the University of Liverpool, United Kingdom. He then worked in research laboratories of Associated Electrical Industries, United Kingdom. He joined the Department of High Voltage Engineering, Indian Institute of Science, Bangalore, and became its head from 1975 to 1980. He has held the Leverhulme Fellowship and Commonwealth Fellowship at the University of Sheffield, United Kingdom. He joined the University of Windsor, Ontario, Canada, in 1980 and became the Head of the Department of Electrical and Computer Engineering during 1989–1997 and 2000–2002. He is currently an Emeritus Professor at the University of Windsor. He has published over 130 research papers and two previous books. He is a Registered Professional Engineer and Fellow of the Institute of Engineers, India.

Acknowledgments

I am extremely grateful to many colleagues who were kind enough to provide me with reprints and reports: Dr. M. Fréchette, IREQ, Montreal; Dr. L. C. Pitchford, CPAT, Toulouse; Dr. J. W. McConkey, University of Windsor; Dr. A. E. D. Heylen, University of Leeds; Dr. Vishnu Lakdawala, Old Dominion University; Dr. L. G. Christophorou, NIST; Dr. J. K. Olthoff, NIST; Dr. de Urquiho, Universidad National Autónoma de Mexico; Dr. Hernández Ávila, Universidad Autónoma Metropolitana-Azcapotzalco, Mexico; and Dr. Kobayashi, Osaka City University. I also thank Dr. L. C. Pitchford, Dr. L. G. Christophorou, Dr. J. K. Olthoff, NIST, and Mr. A. Raju for permission to reproduce figures and data as indicated in the text. Dr. A. E. D. Heylen has been kind enough to read parts of the book and to make valuable suggestions. Dr. V. Agarwal has also made valuable suggestions at the planning stage of the volume, which is gratefully acknowledged. Dr. SriHari Gopalakrishna, Northwestern University, and Professor K. J. Rao of the Indian Institute of Science, have kindly read through Appendix 2. Professor P. H. Alexander of the University of Windsor has advised on parts of Chapter 12.

I am grateful to a number of graduate students who have studied with me several aspects of gaseous electronics, often asking probing questions which have helped me immensely. Drs. C. Raja Rao, S. Rajapandiyan, G. R. Gurumurthy, K. Dwarakanath, A. D. Mokashi, M. S. Dincer, Jane Liu, Nandini Gupta, and Mr. N. Weeratunga have all contributed to this book in their own way. It is a pleasure to acknowledge my association of many years with many colleagues: Drs. Vijendra Agarwal, Soli Bamji, Steve Boggs, Ed Cherney, Ravi Gorur, Reuben Hackam, Shesha Jayaram, Vishnu Lakdawala, and Tangli Sudarshan. The Indian Institute of Science, Bangalore, and the University of Windsor, Ontario, have provided all available facilities without reservation. The personal encouragement of Professor Neil Gold, University of Windsor, has contributed immensely to completion of the book. Thanks are due in no small measure to NSERC (Natural Science and Engineering Research Council) Canada, for supporting my research continuously for more than twenty-five years.

It is a pleasure to acknowledge Dr. Nagu Srinivas, DTE Energy, who provided facilities to complete parts of the book. Professor C. N. R. Rao, President of the Jawaharlal Nehru Center for Advanced Scientific Research, provided facilities for working on parts of the manuscript. Mrs. and Mr. N. Nagaraja Rao have been extraordinarily generous towards me as personal friends; I could depend upon their assistance under any circumstance. I have received generous hospitality and kindness from Mrs. and Dr. N. Rudraiah, Bangalore University, both academically and personally. Mrs. and Lt. Gen. Ragunath (retd.) have generously forgiven my intrusion into their home, often without notice.

I have made sincere efforts to obtain permission from various publishers to reproduce as indicated in the text. I thank all the publishers for giving such permissions. Any omission is entirely inadvertent and I fully apologize for it. Dr. S. Chowdhry has advised me whenever I had problems with software refusing to cooperate. Mr. F. Cicchello and D. Tersigni have spared time to keep my computers up to date. I thank S. Marchand for unhesitating secretarial assistance throughout. Many thanks are due to Helena Redshaw and Jill Jurgensen, who have patiently helped me in editing and improving the presentation of the volume. Mr. Andy Baxter of Keyword Group has kindly tolerated delays due to my international travels. A very special mention is due to Mr. Ian Guy for superb expertise in editing my draft, making proofreading an educational and less-arduous task.

While I am the beneficiary of receiving so much help I wish to state that any errors and omissions are entirely my own responsibility.

Finally, I would like to thank my wife Padmini and our son Anand who have generously forgiven my inevitable nonparticipation in social and family responsibilities, which results from an undertaking such as this. Their inexhaustible patience and the blessings of my parents have been a source of continuous strength all these years.

Windsor, Ontario

Contents

Chapter 6

Drift and Diffusion of Electrons—I 329

Contents

Chapter 7

Drift and Diffusion of Electrons—II. Complex Molecules ... 407

Chapter 8

Ionization Coefficients—I. Nonelectron-Attaching Gases ... 453

Chapter 9

Ionization and Attachment Coefficients—II. Electron-Attaching Gases.... 495

Chapter 12

1 Collision Fundamentals

Analysis of the motion of charge carriers in a gaseous medium is fundamental to a proper understanding of the various manifestations of the discharge phenomena, ranging from low electron density electrical coronas to very high density fusion plasmas. It is nearly 150 years ago (1860) that James Clerk Maxwell showed that the velocities of all particles in a gas are not the same. He derived a distribution of velocities and, from the distribution, derived a number of properties of the collection of the molecules as a single entity. The Maxwell distribution served as a cornerstone for developing the theory of energy distribution of electrons in a gas, the number density of electrons being much smaller than that of molecules. We simply state the Maxwell distribution for molecular velocity, noting that a reference to a book on the kinetic theory of gases supplies the proof.

1.1 COORDINATE SYSTEMS

Collisions between two particles may be analyzed using two alternative systems of coordinates.

1.1.1 LABORATORY COORDINATES

A particle with mass M_a is moving with velocity $\mathbf{V}_a$ and approaches a particle of mass M_b at rest in gravity-free space (Figure 1.1). If there is no collision, the moving particle continues its journey without being affected. If there is a collision, its velocity is changed to $\mathbf{V}'_a$, making an angle θ_a with its initial direction. The second particle acquires a velocity $\mathbf{V}'_b$ after collision. Conservation of momentum before and after collision dictates that the velocity vectors $\mathbf{V}_a$, $\mathbf{V}'_a$, and $\mathbf{V}'_b$ lie in the same plane. Further conservation of momentum yields

$$M_a \mathbf{V}_a = M_a \mathbf{V}'_a + M_b \mathbf{V}'_b \tag{1.1}$$

In component form this transforms to

$$M_a V_a = M_a V'_a \cos\theta_a + M_b V'_b \cos\theta_b \tag{1.2}$$

$$M_a V'_a \sin\theta_a - M_b V'_b \sin\theta_b = 0 \tag{1.3}$$

Conservation of energy leads to

$$M_a V_a^2 = M_a V'^2_a + M_b V'^2_b \tag{1.4}$$

If we suppose that the initial given quantities are M_a, M_b, and V_a, we are required to find four quantities, namely V'_a, V'_b, θ_a, and θ_b. We need a fourth equation, usually involving force, but some simple relations may be derived.[1]

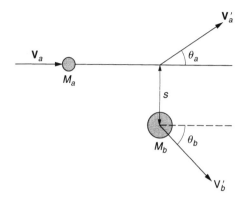

FIGURE 1.1 Elastic collision between a particle moving with a velocity $\mathbf{V}_a$ and a particle at rest in the laboratory coordinates system. The velocities after collision are $\mathbf{V}'_a$ and $\mathbf{V}'_b$ and $\mathbf{s}$ is the impact parameter. θ_a and θ_b are scattering angles.

TABLE 1.1
Range of Scattering Angle for Various Particle Types

Particle Type	Mass Description	Θ_m	Range of θ_a
Electron–atom	$M_b \gg M_a$	$\gg 1$	$0 \le |\theta_a| \le \pi$
Atom–atom	$M_b = M_a$	1	$0 \le |\theta_a| \le \pi/2$
Molecule–atom	$M_b \ll M_a$	$\ll 1$	$0 \le |\theta_a| \le \Theta_m$

From Johnson, R. E., *Introduction to Atomic and Molecular Collisions*, Plenum Press, New York, 1982.

Equations 1.2 to 1.4 may be combined to yield, after substituting $\Theta_m = M_b/M_a$,

$$\tan \theta_a = \frac{\Theta_m \sin 2\theta_b}{1 - \Theta_m \cos 2\theta_b} \tag{1.5}$$

Equation 1.5 imposes restrictions on the scattering angle, depending upon the relative masses of the particles as shown in Table 1.1. We note that Equation 1.5 may be used to determine M_b by measuring θ_a. Defining reduced mass M_{RM} as

$$M_{\mathrm{RM}} = \frac{M_a M_b}{M_a + M_b} \tag{1.6}$$

and applying the principles of conservation of both energy and momentum, the following relationships are obtained for the velocities after collision:[2]

$$V'_b = \frac{2V_a M_{\mathrm{RM}}}{M_a} \cos \theta_b \tag{1.7}$$

$$V'_a = V_a \left(1 - \frac{4M_{\mathrm{RM}}^2}{M_a M_b} \cos^2 \theta_b \right)^{1/2} \tag{1.8}$$

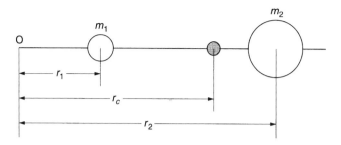

FIGURE 1.2 The center of mass is located at a distance of r_c from a chosen axis.

As is demonstrated above, the laboratory coordinate system does not yield the lowest number of unknowns for the analysis of the collision and the center of mass coordinate system is often preferred since it simplifies the analysis.[2]

1.1.2 CENTER OF MASS COORDINATES

In the previous discussion the second particle (B) was considered to be initially at rest. However, this is not a necessary condition and both particles may have an initial velocity. To deal with such situations, which are more common, the center of mass system is advantageous and results in a smaller number of unknowns. A brief recapitulation of the concept of center of mass is given first.

Consider any two masses (m_1 and m_2), not necessarily atomic particles, situated at a distance r apart, as shown in Figure 1.2. The distance of the center of mass from any fixed point (O) situated along the axis joining the centers (this is not essential) is given as

$$r_c = \frac{m_1 r_1 + m_2 r_2}{m_1 + m_2}$$

If the point lies between the masses, the center of mass divides the line joining the masses in the inverse ratio of masses. It is closer to the heavier mass.

Let $\mathbf{V}_a$ and $\mathbf{V}_b$ be the velocities of particles having masses M_a and M_b respectively before collision in the laboratory frame. The velocities of the same particles after collision are $\mathbf{V}'_a$ and $\mathbf{V}'_b$. The velocity of the center of mass is $\mathbf{W}_c$, which remains constant before and after collision. As seen from the center of mass, the relative velocities are in exactly opposite directions, both before and after collision, and their velocities are inversely proportional to their masses.

The velocities of the particles with respect to the center of mass are $\mathbf{W}_a$ and $\mathbf{W}_b$ before collision. After collision the velocities change to $\mathbf{W}'_a$ and $\mathbf{W}'_b$. The following relationships then hold true:

$$\left.\begin{aligned}
\mathbf{W}_a &= \mathbf{V}_a - \mathbf{W}_c \\
\mathbf{W}_b &= \mathbf{V}_b - \mathbf{W}_c \\
\mathbf{W}'_a &= \mathbf{V}'_a - \mathbf{W}_c \\
\mathbf{W}'_b &= \mathbf{V}'_b - \mathbf{W}_c
\end{aligned}\right\} \tag{1.9}$$

The momenta (**p**) of the particles in the laboratory frame are

$$\mathbf{p}_1 = M_a \mathbf{V}_a + M_b \mathbf{V}_b = M_a \mathbf{V}'_a + M_b \mathbf{V}'_b \tag{1.10}$$

Substituting Equation 1.9 into 1.10, one gets

$$\mathbf{p}_1 = (M_a + M_b)\mathbf{W}_c + M_a\mathbf{W}_a + M_b\mathbf{W}_b \tag{1.11}$$

The sum of the momenta ($\mathbf{p}_{sum}$) with respect to the center of mass of the particles is zero and therefore

$$\mathbf{p}_{sum} = M_a\mathbf{W}_a + M_b\mathbf{W}_b = 0 \tag{1.12}$$

Substituting Equation 1.12 into 1.11,

$$\mathbf{p}_1 = (M_a + M_b)\mathbf{W}_c = M_a\mathbf{V}_a + M_b\mathbf{V}_b \tag{1.13}$$

The total momentum is that of a particle having a mass $(M_a + M_b)$ situated at the center of mass.

The center of mass has the following properties in the analysis of the collision:

1. The velocity of the center of mass ($\mathbf{W}_c$) with respect to the laboratory frame remains the same before and after the collision (Figure 1.3).

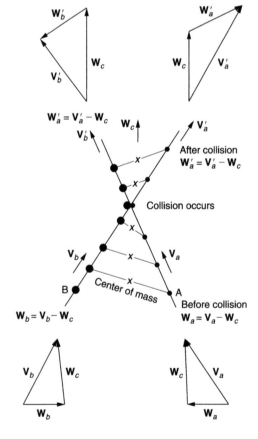

FIGURE 1.3 Particle collision and velocity relationships in the laboratory frame and center of mass frame. The velocity of the center of mass remains the same before and after collision. The relative velocities of the particles with reference to the center of mass are opposite in direction before collision and also after collision.

2. The velocities of the particles, before collision, relative to the center of mass ($\mathbf{W}_a$, $\mathbf{W}_b$) are exactly opposite in direction. Their magnitudes are inversely proportional to the masses (Equation 1.12). This is also true for their relative velocities ($\mathbf{W}'_a$, $\mathbf{W}'_b$) after collision.
3. The sum of the momenta of the particles with reference to the center of mass, before and after collision, is zero (Equation 1.12).
4. The total momentum of the particles is the same as though the total mass were centered at the center of mass (Equation 1.13).
5. In the center of mass system, reduced mass is defined by Equation 1.6. By using this fictitious mass we need to deal with only the relative velocities of the particles with respect to the center of mass and ignore the masses of the individual particles.

Figure 1.3 has been drawn to show both particles moving with velocities $\mathbf{V}_a$ and $\mathbf{V}_b$ respectively for the purpose of demonstrating the motion of the center of mass. Considerable simplification can be made by substituting $\mathbf{V}_b = 0$, that is, particle B is initially at rest. This leads immediately to the result, from Equation 1.13,

$$\mathbf{W}_c = \frac{M_a \mathbf{V}_a}{M_a + M_b} = \frac{M_{\mathrm{RM}} \mathbf{V}_a}{M_b} \tag{1.14}$$

The angle of scattering after collision is the same for both particles. The total linear momentum of the system is zero at all times. The magnitude of the initial and final velocity of each particle remains unaltered due to the collision as we are dealing with the relative velocity of the particles with respect to the center of mass. The relationships between selected quantities in both systems are shown in Table 1.2.

Conversion from the center of mass system to the laboratory system is accomplished by adding the velocity of the center of mass ($\mathbf{W}_c$) to the velocity of each particle in the CM system. The initial momentum in the CM system is $M_{\mathrm{RM}} \mathbf{W}_c$ and in the laboratory system it is $M_a \mathbf{V}_a$. One can show they are identical, assuming that $M_a \ll M_b$ and $\mathbf{V}_b = 0$. The situation corresponds to electron–neutral scattering.

1.2   MEANING OF VELOCITY SPACE

The number of molecules in a unit volume is so large that velocities of individual molecules cannot be determined. However, a statistical description of the distribution of velocities is possible and the function that describes this distribution is known as the velocity distribution function.

Consider a differential volume $dx\,dy\,dz$ in the Cartesian coordinate system, situated at the coordinates (x,y,z). The distance of the differential volume from the origin of the coordinates is a vector $\mathbf{r}$. In the differential volume there are particles that have a wide range of velocities, from very low to very high, in relative terms. Of these particles a certain number will have a velocity with components W_x, W_y and W_z. The velocity components are subject to the restriction

$$\left. \begin{array}{l} W_y \leq W'_y \leq W_y + dW_y \\ W_z \leq W'_z \leq W_z + dW_z \end{array} \right\} \tag{1.15}$$

TABLE 1.2
Relationships between Laboratory and Center of Mass Coordinates

Laboratory Quantities	CM Quantities
$M = M_a + M_b$	$M = \dfrac{M_a M_b}{M_a + M_b}$
$M_a \mathbf{V}_a = (M_a + M_b)\mathbf{W}_c$	$\mathbf{p}_{sum} = m_a \mathbf{W}_a + m_b \mathbf{W}_b = 0$
Total KE before collision $= \dfrac{1}{2}M_a V_a^2$	Total kinetic energy $=$ $\dfrac{1}{2}M_a W_a^2 + \dfrac{1}{2}M_b W_b^2$
Total KE after collision $=$ $\dfrac{1}{2}M_a V_a'^2 + \dfrac{1}{2}M_b V_b'^2$	

Velocities

$$\mathbf{W}_c = \frac{M_a \mathbf{V}_a}{M_a + M_b}$$

$$\mathbf{V}_a = \frac{(M_a + M_b)}{M_a}\mathbf{W}_c$$

$$\mathbf{W}_{rel} = \mathbf{W}_a - \mathbf{W}_b$$

$$\mathbf{W}'_{rel} = \mathbf{W}'_a - \mathbf{W}'_b$$

Johnson, R. E., *Introduction to Atomic and Molecular Collisions*, Plenum Press, New York, 1982.

We can say that, in a coordinate system with axes $\mathbf{W}_x$, $\mathbf{W}_y$, and $\mathbf{W}_z$, the velocity vector $\mathbf{W}$ joins the origin to the differential volume $dW_x\, dW_y\, dW_z$ (Figure 1.4). This coordinate system is referred to as velocity space.

Some ambiguity exists in the term "velocity space"[3] and it is useful to remember that measurable space is real (also referred to as configuration space) in the sense that it has physical dimensions. However, velocity space is a mathematical contrivance that identifies a group of particles, all of which have their velocity components according to the relationships in Equation 1.15 at a distance $\mathbf{W}$ from the origin. Though we have adopted Cartesian coordinates for the velocity space for the sake of simplifying the explanation, it is customary to use spherical coordinates. The relation between the velocity $\mathbf{W}$ and its components is

$$W^2 = W_x^2 + W_y^2 + W_z^2 \tag{1.16}$$

The relation between the radius vector $\mathbf{r}$ and its components is

$$r^2 = x^2 + y^2 + z^2 \tag{1.17}$$

Even in the early years of the theoretical development of molecular velocity distribution it was recognized that the number of molecules, dN, having velocities between $\mathbf{W}$ and $\mathbf{W} + d\mathbf{W}$

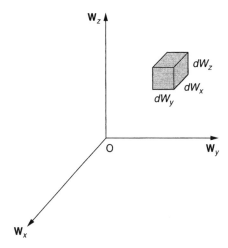

FIGURE 1.4 Definition of velocity space. Vectors ending in the differential velocity space represent the probability of distribution. It is conventional to write that $dW = dW_x dW_y dW_z$.

is proportional to $\mathbf{W}$. All points in the differential volume in the velocity space have the same velocity, subject to the restriction in Equation 1.15. The number of particles in the differential volume $dxdydz$ will encompass all possible velocities and will be a fraction of the total number of molecules N in the container. The differential volume in the velocity space is therefore a subset of differential volume in the measurable space, which is a subset of the set, the container.

The number of molecules having velocities satisfying the conditions of Equation 1.15 is given by

$$dn = f(\mathbf{W}_x, \mathbf{W}_y, \mathbf{W}_z)d\mathbf{W}_x d\mathbf{W}_y d\mathbf{W}_z \qquad (1.18)$$

The function f is called the velocity distribution function. Within the container of the gas it may have different values at different locations. It may also change with time. In the general case where these variables influence the distribution function, Equation 1.18 may be expressed as

$$dn = f(\mathbf{W}_x, \mathbf{W}_y, \mathbf{W}_z, x, y, z, t)d\mathbf{W}_x d\mathbf{W}_y d\mathbf{W}_z \qquad (1.19)$$

This equation can be concisely expressed as

$$dn = f(\mathbf{W}, \mathbf{r}, t)d\mathbf{W} \qquad (1.20)$$

In this format of the distribution function $\mathbf{W}$ is the velocity vector, $\mathbf{r}$ is the radius vector within the container, and $d\mathbf{W}$ is the differential volume in the velocity space.

If the distribution function is independent of $\mathbf{r}$ and t, Equation 1.20 simplifies to

$$\frac{dn}{d\mathbf{W}} = f(\mathbf{W}) \qquad (1.21)$$

If the distribution function is dependent on $\mathbf{r}$, the distribution becomes inhomogeneous. The physical conditions may be such that the non-homogeneity is removed or minimized due to collisions over a period of time, leading to a homogeneous distribution. The dependence of the distribution function on the velocity vector $\mathbf{W}$ may be of two kinds. If the distribution depends on the magnitude of the velocity vector only, it is known as isotropic. If the distribution depends on the magnitude and direction of the vector, then the distribution becomes anisotropic.

1.3 MAXWELL'S DISTRIBUTION FUNCTION

Although the assumptions made by Maxwell in the derivation of the velocity distribution function are well known, we shall restate them for the purpose of establishing familiarity with the fundamental concepts. The assumptions are:

1. A gas is composed of a large number of molecules, each of the same mass and size, moving in all possible directions.
2. The distance between any two molecules is always greater than their diameter.
3. The interaction between molecules is only through collisions between them.
4. The collision between two molecules is elastic, similar to the collision of billiard balls.

The distribution of velocities of molecules in a gas at a temperature T in equilibrium with the temperature of the container is given by the Maxwellian distribution

$$f(W) = \frac{4N}{\sqrt{\pi}} \left(\frac{m}{2kT}\right)^{3/2} W^2 \exp\left(-\frac{mW^2}{2kT}\right) \tag{1.22}$$

where m is the mass of the molecule, N the number of molecules, and k the Boltzmann constant. The total number of molecules in the assembly may be obtained by integrating Equation 1.22 over all possible velocities,

$$\int_0^\infty f(W)dW = N \tag{1.23}$$

which is the expected result.

By differentiating Equation 1.22 and equating it to zero the velocity at which the distribution function attains a maximum value may be determined. This velocity, known as the most probable velocity, is given by the expression

$$W_{\max} = \left(\frac{2kT}{m}\right)^{1/2} \tag{1.24}$$

The left-hand side of Equation 1.23, equal to the area under $f(W)$, is known as the zeroth moment of the distribution. The ratio $f(W)/N$, which lies between zero and one, is the fraction that is usually calculated.

The first moment divided by N is defined by the expression

$$\frac{1}{N}\int_0^\infty W f(W)\, dW = \left(\frac{8kT}{\pi m}\right)^{1/2} \tag{1.25}$$

and gives the arithmetic mean speed, also known as the mean thermal velocity, $\langle W \rangle$.

From the second moment the root mean square velocity, W_{rms}, may be calculated according to

$$W_{\mathrm{rms}} = \frac{1}{\pi}\left(\int_0^\infty W^2 f(W)dW\right) = \left(\frac{3kT}{m}\right)^{1/2} \tag{1.26}$$

Using simple algebra, we can rewrite this expression as

$$\frac{1}{2}mW^2_{\text{rms}} = \frac{3}{2}kT \tag{1.27}$$

The right-hand side of Equation 1.27 is known as the mean energy of the molecule.

Figure 1.5 shows the calculated Maxwellian velocity distribution at different temperatures. The distribution is narrower at lower temperatures, the peak of the distribution gets smaller, and the distribution extends to higher energies as the temperature increases. If the mean energy is non-Maxwellian, the zeroth moment, Equation 1.23, remains the same but the velocities W_{max}, $\langle W \rangle$, and W_{rms} assume different values.[4]

It is important to bear in mind that the molecules in neutral gases and liquids have velocities in random directions and the notation used for speed is commonly v. We have preferred to retain W as the symbol for random velocity. Under the action of an electric field E, positive ions move parallel and electrons anti-parallel to the direction of E, acquiring a drift velocity. Mobility (μ) is defined as

$$W = \mu E \tag{1.28}$$

It has the unit of V^{-1} m^2s^{-1} and is related to the collision frequency v (s^{-1}), considered in greater detail in Section 1.5.8, by the expression

$$\mu = \frac{e}{mv} \tag{1.29}$$

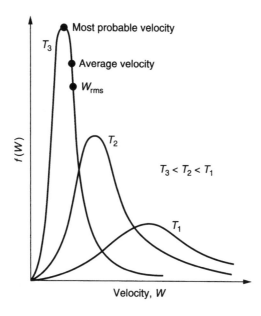

FIGURE 1.5 Velocity distribution of molecules in a gas at three different temperatures. As the temperature increases, the most probable velocity decreases and the distribution spreads to larger energies. The average velocity is the arithmetic mean and is also known as mean thermal speed.

In the literature on collision phenomena in gases the energy is the most frequently used parameter. The kinetic energy of a moving particle is related to its velocity according to

$$\varepsilon = \frac{1}{2}mW^2 \tag{1.30}$$

If we make this substitution in Equation 1.22, the energy distribution in terms of ε is given by

$$f(\varepsilon) = \frac{2N}{\sqrt{\pi}} \frac{\varepsilon^{1/2}}{(kT)^{3/2}} \exp\left(-\frac{\varepsilon}{kT}\right) \tag{1.31}$$

A slightly different form of energy distribution is given in Equation 1.122. The most probable energy is found by differentiating Equation 1.31 and equating the differential to zero. The most probable energy is

$$\langle \varepsilon \rangle = \frac{1}{2}kT \tag{1.32}$$

The mean energy of the molecules is given by the first moment of Equation 1.31,

$$\langle \varepsilon \rangle = \frac{1}{N} \int_0^\infty \varepsilon f(\varepsilon) d\varepsilon \tag{1.33}$$

A direct substitution of Equation 1.31 into 1.33 gives

$$\langle \varepsilon \rangle = \frac{3}{2}kT \tag{1.34}$$

in agreement with Equation 1.27. Equations 1.27 and 1.34 are examples of the equipartition energy principle which defines the way the energy is divided for each degree of freedom when a large number of particles are enclosed in a container and, due to collisions, attain a steady-state energy distribution. The average energy is proportional to the absolute temperature and equal to $kT/2$ for each degree of freedom. In the situation under study the molecule has three degrees of freedom, along the x, y, z directions, and the mean energy is $(3/2)kT$.

The colliding particles may be in thermodynamic equilibrium or kinetic equilibrium.[4] Thermodynamic equilibrium is described as the condition in which the colliding particles are at the same temperature as the container and energy flow within the gas is negligible. The gas acts as a black body, with energy inflow being equal to outflow. Thermodynamic equilibrium is a necessary and sufficient condition for the application of the mathematical techniques of classical thermodynamics to a gas. A gas in kinetic equilibrium can have different species of particles at different temperatures. Each species has attained a steady temperature and energy inflow to the gas is possible. The Maxwell distribution is applicable to both classes of equilibria.

1.4 MEAN FREE PATH

The average distance traveled between collisions is known as the mean free path. From elementary kinetic theory we know that the mean free path is $1/NQ$, where N is the number of molecules per m^3 and Q is the effective collision cross section (m^2). The free paths of particles between individual collisions are not equal, some being shorter than others. Using a simple probability calculation it is possible to derive the number of free paths which deviate from the mean value, in other words, to derive a distribution of free paths.

Let us assume that N molecules are to be considered and let n be the number that travel a distance x without collision. The number of molecules dn that undergo collision between a short interval of distance dx is

$$dn = -pn\,dx \tag{1.35}$$

where p is a constant. The negative sign indicates that the number undergoing collision is effectively removed from the ensemble. Integration of Equation 1.35 gives

$$n = A\exp(-px) \tag{1.36}$$

where A is the integration constant. At $x = 0$, $n = N$ and hence $A = N$. Equation 1.36 then transforms to

$$n = N\exp(-px) \tag{1.37}$$

The constant p may be expressed in terms of the mean free path. Let dN be the number of molecules having a free path between x and $x + dx$. The mean free path λ is then given by

$$\lambda = \int_0^N \frac{x}{N}\,dN \tag{1.38}$$

Now

$$dN = dn = pn\,dx = pN\exp(-px)\,dx \tag{1.39}$$

Substituting Equation 1.39 into 1.38, one gets

$$\lambda = \int_0^\infty \frac{pNx\exp(-px)}{N}\,dx = \frac{1}{p} \tag{1.40}$$

The distribution of free paths is given by substituting Equation 1.40 into 1.37,

$$n = N\exp\frac{-x}{\lambda} \tag{1.41}$$

The number of particles having a free path greater than λ decreases exponentially. The mean free path is a useful parameter to calculate the energy gained by the electron in an electric field, which is $eE\lambda$ (joules).

1.5 PARTICLE COLLISIONS

Collisions between particles are a means of exchanging energy till the volume of gas in the container attains a steady mean energy. If the particles are atoms or molecules the collision is visualized as that between hard billiard balls; the kinetic energy is conserved and the trajectory of each particle will usually be different before and after collision. The internal energy of each particle remains the same before and after collision. Such collisions are called elastic collisions. However, a collision between a charge carrier and a neutral atom may also result in a change of the internal energy of one or both particles and such collisions are called inelastic collisions. The change of the internal energy occurs in the form of dissociation of a molecule, excitation to various levels of the atom, ionization, recombination, etc. Two such processes may occur sequentially with negligible time interval. An example is the dissociative attachment during which an electron collides with a molecule, dissociation occurs, and the electron attaches to one of the dissociation products.

Inelastic collisions are associated with loss of energy, the energy inflow being maintained from an external agency such as a power source or laser beam. Another type of collision that occurs between excited atoms or molecules and electrons is known as a superelastic collision. Successful theoretical analyses of discharges and plasmas depend upon detailed knowledge of the internal energy exchange mechanisms and the onset energies for each inelastic collision. In order to have a general framework into which such a wide range of collision phenomena can be built, retaining the uniqueness of each process, some fundamental parameters have been defined. We shall consider these in the following sections.

Collisions between two charged species belong to the category known as Coulomb interaction. Physical contact is not visualized as in the hard sphere model. The interaction is mainly by the Coulomb force between two electrical charges, which varies as the inverse square of the distance between the two charges. A charged particle can have Coulomb interaction with a number of charged species simultaneously, in contrast with binary collisions between atoms or between an electron and an atom. A collision between two electrons or two ions also occurs through Coulombic interaction and is usually called "scattering." Rutherford derived a simple formula for the scattering cross section between two electrons.

Dissociative excitation and predissociative excitation also belong to the inelastic collision category. In the former the electron dissociates the molecule and the excess energy goes towards exciting the fragment atom. In predissociative excitation the electron excites the molecule and the molecule returns to the ground state, dissociating into fragments.

1.5.1 ELASTIC COLLISIONS

Atoms have a strong electrostatic repulsion, due to valence electrons, that decreases very sharply with increasing separation as a result of the inverse square law. However, this short-range force gives place to weak long-range attraction that extends to larger (in atomic dimensions) distances. This fact permits collision between atoms to be treated as a collision between hard spheres. The elastic collision cross section in the hard sphere model is defined in a simple way by reference to Figure 1.6. Two particles, each having a radius r, collide with each other if their centers come within a distance of $2r$. The cross section for elastic collision, Q_{el}, is defined as $4\pi r^2$ (m^2). Since the radius of a molecule is $\sim 10^{-10}$ m, Q_{el} is approximately equal to 10^{-19} m^2.

In the older literature the collision cross sections are usually given in terms of πa_0^2 (sometimes a_0^2), where a_0 is the Bohr radius ($a_0 = 5.292 \times 10^{-11}$ m), or in terms of NQ_{el}, where N is the number of molecules per m^3 of the gas. The number N (m^{-3}) at pressure p (Pa) and temperature T(K) is given by the formula

$$N = \frac{7.244 \times 10^{22} p}{T} \tag{1.42}$$

At standard temperature and pressure (273 K, 101.325×10^3 Pa) the number of molecules in a gas is approximately 2.65×10^{25} m^{-3}. The quantity NQ has the physical meaning of the number of collisions and units of m^{-1}.

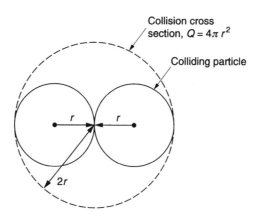

FIGURE 1.6 Schematic diagram of collision between two particles each of radius r. The collision cross section is $4\pi r^2$. The particles are assumed to be hard spheres.

The collision cross section is an alternative way of quantifying the average number of collisions per meter length of travel distance or, what amounts to the same thing, the mean distance traveled between collisions. The latter quantity is known as the mean free path (λ) and its relation to collision cross section may be established by considering a container having two species of particles. The smaller particles have radius r_a, and move with velocity W_a and the larger particles, molecules having radius r_b, are assumed to be stationary. The collision cross section for each encounter is Q (Figure 1.6) and the volume swept by the smaller particle in a time interval t along its zig-zag path is QW_at. The number of collisions in this interval is NQW_at. All velocities are possible for the smaller particle, and as the collision cross section is a function of velocity we have to substitute the mean velocity ($\langle W \rangle$) and average cross section $\langle Q \rangle$ in place of W_a and Q respectively, to include all velocities. The averaging is carried out according to

$$\langle Q W \rangle = \frac{1}{N} \int_{-\infty}^{\infty} Q(W)\, W f(W)\, dW \qquad (1.43)$$

The mean free path is obtained by dividing the average distance traveled by the number of collisions in the same time interval. Therefore

$$\langle \lambda \rangle = \frac{\text{distance traveled in time } t}{\text{number of collisions in time } t} = \frac{\langle W \rangle t}{N \langle QW \rangle t} = \frac{1}{NQ} \qquad (1.44)$$

The number of collisions per second (ν) is obviously $N \langle QW \rangle$ and it follows that

$$\nu = N \langle Q W \rangle \qquad (1.45)$$

The mean free path is the reciprocal of the number of collisions per meter (P), which allows the use of the relationship $P = NQ$.

The distribution of free paths is calculated in a relatively simple way. Consider a thin slab of electrons that has not suffered scattering, at a distance x from the origin. The slab moves through a distance dx and, due to scattering, the number of particles in the layer decreases. The decrease is given by

$$dn = -n(x)NQ\,dx \tag{1.46}$$

The negative sign denotes the decrease in the number. Integrating Equation 1.46, the number of particles at x is obtained as

$$n(x) = n_0 \exp(-NQx) \tag{1.47}$$

where n_0 is the initial number of particles at $x = 0$. Substituting Equation 1.44 into 1.47, the distribution of free paths is given as

$$n(x) = n_0 \exp\left(-\frac{x}{\langle\lambda\rangle}\right) \tag{1.48}$$

The exponential distribution shows that 36.8% of the initial number of particles arrive at the distance of $\langle\lambda\rangle$. The same result was obtained for atom–atom collisions in Equation 1.41.

Equation 1.44 assumes that the second particle is initially at rest, $W_b = 0$. This is not true in an electron–gas or in a gas 1–gas 2 ensemble because the particles possess thermal velocity. Here the cross section for collision as defined in Figure 1.6 should be multiplied by a factor

$$\Lambda = \sqrt{\frac{M_a + M_b}{M_b}} \tag{1.49}$$

The mean free path defined by Equation 1.44 now becomes

$$\langle\lambda\rangle = \frac{1}{\pi N(r_a + r_b)^2 \sqrt{1 + \dfrac{M_a}{M_b}}} \tag{1.50}$$

The reasoning may be extended to an assembly composed of several species such as electrons $(M_1, r_1, N_1, \lambda_1)$, ions $(M_2, r_2, N_2, \lambda_2)$, atoms of gas 1 $(M_3, r_3, N_3, \lambda_3)$, etc. The mean free path for particle 1 is

$$\langle\lambda_1\rangle = \frac{1}{\pi \sum_{j=1}^{j=n} N_j(r_1 + r_j)^2 \sqrt{\dfrac{M_1 + M_j}{M_j}}} \tag{1.51}$$

For an atom in a monospecies gas, $r_a = r_b$, $M_a = M_b$, and Equation 1.50 gives

$$\langle\lambda_a\rangle = \frac{1}{4\sqrt{2}\,\pi r_a^2 N} \tag{1.52}$$

For an electron in a gas, $M_a \ll M_b$, $r_a \ll r_b$, Equation 1.50 gives

$$\langle\lambda_e\rangle = \frac{1}{\pi r_b^2 N} \tag{1.53}$$

TABLE 1.3
Mean Free Paths of Common Gases at 15°C and 101 kPa

Gas	H_2	O_2	N_2	CO_2	H_2O
λ (nm)	117.7	67.9	62.8	41.9	41.8
Mol. Wt.	2.016	32.00	28.02	44.00	18.00

From Roth, J. R., *Industrial Plasma Engineering*, Institute of Physics, Bristol, 1995. With permission.

TABLE 1.4
Energy Gain during Successive Collisions in SF₆ at Various Gas Number Densities

Pressure (bar)	Voltage (kV)	E (MV/m)	λ ($\times 10^{-6}$ m)	ε (eV)
1	265	6.36	3.953	25.14
2	340	8.16	1.976	16.12
3	405	9.72	1.317	12.81
4	470	11.28	0.988	11.15
5	530	12.72	0.790	10.05
6	580	13.92	0.659	9.17
7	625	15.00	0.565	8.47
8	650	15.60	0.494	7.71

The relation between voltage and electric field depends on the electrode geometry.
From Espel, P., et al., *J. Appl. Phys.: Appl. Phys.*, 34, 593, 2001. With permission.

Expressing this equation in terms of the mean free path of the atom,

$$\langle \lambda_e \rangle = 4\sqrt{2}\langle \lambda_a \rangle = 5.66\langle \lambda_a \rangle \tag{1.54}$$

The electron has a longer free path than that of the atom. Typical values of the mean free path in several gases are shown in Table 1.3.

We shall now consider the energy gained by an electron as it moves through the gas under the influence of an electric field. If we denote the mean free path of the electron by λ, the energy gained between successive collisions, that is during the mean free path, is

$$\varepsilon = \frac{1}{2}mW^2 = eE\lambda \quad \text{(J)} \tag{1.55}$$

Knowing the collision cross section (Q_M) and using Equation 1.44, one can calculate the energy gained during a mean path. As the gas number density increases, the energy gained will, of course, be smaller since the mean free path will be shorter. Espel et al.[5] have calculated the energy gain as shown in Table 1.4 for SF_6, assuming a constant collision cross section of 10^{-20} m².

1.5.2 Energy Transfer in Elastic Collisions

We consider an elastic collision between an electron having a mass m and a molecule having a mass M that is considered to be initially at rest. The velocities of the electron before and after impact are W_a and W'_a. The velocity of the molecule after collision is W'_b (Figure 1.7).

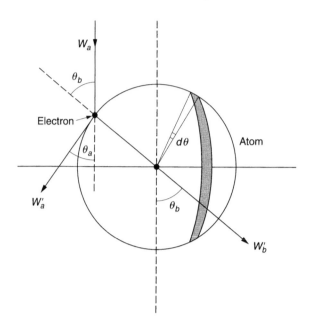

FIGURE 1.7 Elastic collision between an electron and a molecule. The electron is deflected by an angle θ_a. W_a is the initial velocity of the electron and the atom is initially at rest. The velocities of the two particles after collision are W'_a and W'_b, respectively.

The electron is scattered through an angle θ_a and the molecule is deflected along the line joining the centers.

The fractional energy loss during a collision is

$$\Delta E = \frac{\Delta W_{KE}}{W_{KE}} = \frac{W_a^2 - W_a'^2}{W_a^2} \tag{1.56}$$

Conservation of energy and momentum yield the following equations:

$$M_a W_a^2 = M_a W_a'^2 + M_b W_b'^2 \tag{1.57}$$

$$M_a W_a = M_a W_a' \cos\theta_a + M_b W_b' \cos\theta_b \tag{1.58}$$

$$M_a W_a \sin\theta_a = M_b W_b' \sin\theta_b \tag{1.59}$$

Solving for W'_b, one obtains

$$W_b' = \frac{2 M_a W_a \cos\theta_b}{M_a + M_b} \tag{1.60}$$

Substituting Equation 1.60 into 1.56, the fractional loss of energy is obtained as

$$\frac{\Delta W_{KE}}{W_{KE}} = \frac{4 M_a M_b \cos^2\theta_b}{(M_a + M_b)^2} \tag{1.61}$$

To find the mean fractional energy loss per collision we should consider collisions at all angles $0 \le \theta \le \pi/2$. The probability of collision (P) taking place between θ and $\theta + d\theta$ is the ratio of the shaded area in Figure 1.7 to the whole area, the latter being the collision cross section.

$$P(\theta)d\theta = \frac{2\pi(r_a + r_b)\sin\theta\cos\theta(r_a + r_b)d\theta}{\pi(r_a + r_b)^2}$$

$$= \sin 2\theta\,d\theta, \quad \left(0 \le \theta \le \frac{\pi}{2}\right) \tag{1.62}$$

$$P(\theta)d\theta = 0, \qquad \left(\frac{\pi}{2} < \theta < \pi\right) \tag{1.63}$$

where r_a and r_b are the radii of the electron and the atom respectively. The factor $\cos\theta$ arises because only a fraction of the area is presented to the incoming particles.[5] Considering all angles, the mean fractional loss of energy is obtained as

$$\langle \Delta E \rangle = \frac{\int_0^{\pi/2} P(\theta)\Delta E\,d\theta}{\int_0^{\pi/2} P(\theta)\,d\theta} \tag{1.64}$$

Substituting Equations 1.61 and 1.62 into 1.64 and integrating, one obtains

$$\langle \Delta E \rangle = \frac{2M_a M_b}{(M_a + M_b)^2} \tag{1.65}$$

If the incoming particle is an electron, the loss of energy is small as $M_a \ll M_b$ and the fractional energy loss in an elastic collision is $2M_a/M_b$. A collision of an electron with a nitrogen molecule results in a fractional energy loss of 4×10^{-5}, which is very small. On the other hand, if the collision is between an ion and a molecule, $M_a \approx M_b$ and the fractional loss of energy is ½, which is very large. Therefore ions lose far more energy than electrons during elastic collisions.

1.5.3 Differential Scattering Cross Section

We consider a beam having particles of type A with density N_a that encounters a beam having particles of type B with density N_b. If scattering occurs it is by no means certain that equal numbers of particles are scattered into the solid angle inclined at the angle θ (Figure 1.8). If the elastic scattering cross section for particle A is Q_{el}, then the differential scattering cross section is defined as

$$Q_{el} = 2\pi \int_0^\pi Q_{diff}(\theta)\sin\theta\,d\theta \tag{1.66}$$

It may also be expressed as

$$Q_{el} = \int_\Omega Q_{diff}(\theta)\,d\Omega \tag{1.67}$$

For isotropic scattering, the number scattered into the elementary solid angle $d\Omega$ is independent of (θ, φ), which simplifies the integral in Equation 1.67,

$$Q_{el} = Q_{diff}\int_\Omega d\Omega = 4\pi Q_{diff} \tag{1.68}$$

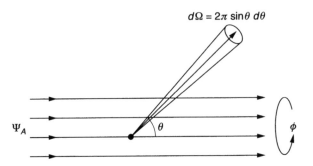

FIGURE 1.8 Scattering of particles from a beam into the solid angle $d\Omega$. ψ_A is the flux, defined as the number of particles crossing a unit area in one second.

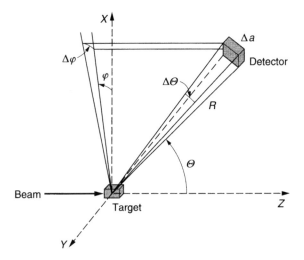

FIGURE 1.9 Schematic arrangement for the measurement of differential scattering cross section. The distance to the detector R is larger than the target volume. θ and φ are the angular positions of the detector. The detector spans a solid angle about θ and φ indicated by the angles $d\theta$ and $d\varphi$. Reproduced from Johnson, R. E., *Introduction to Atomic and Molecular Collisions*, Plenum Press, New York, 1982.

If two particles interact in more than one way, then there will be a differential cross section for each type of encounter. Let us suppose that an encounter results in elastic collision or one of several inelastic collisions; then the total scattering cross section is given as, by an extension of Equation 1.67,

$$Q_{\mathrm{T}} = \int_{\Omega} \left[Q_{\mathrm{diff}}(\theta) + \sum_{j=1}^{j=n} Q_j(\theta) \right] d\Omega \qquad (1.69)$$

where the subscript T means total cross section, and n types of inelastic collisions occur.

Figure 1.9 demonstrates the principle involved in the measurement of differential scattering cross section. A beam of electrons with a well-defined energy impinges on a target gas and some of the electrons are scattered at various angles. A detector with variable angular disposition with respect to the direction of the beam is employed to measure the number of scattered electrons. The relative signals at various angular dispositions of the detector are placed on an absolute scale by employing a reference gas, in which the differential scattering cross section is known, as target gas.

The distance between the target gas and the detector is assumed to be greater than the sides of the volume of the target gas. The angular positions of the detector are θ and φ. The detector spans a solid angle about θ and φ indicated by the angles $d\theta$ and $d\varphi$. The area seen by the detector (Δa) is

$$\Delta a = R d\theta \times R \sin \theta d\varphi \tag{1.70}$$

The element of solid angle subtended ($d\Omega$) is

$$d\Omega = \frac{\Delta a}{R^2} = \sin \theta \, d\theta \, d\varphi \tag{1.71}$$

By moving the detector to all angular positions without changing R, an area of $4\pi R^2$ and a solid angle of 4π may be covered.

As mentioned above, the definition of differential scattering holds well for inelastic collisions, namely electronic excitation and vibrational excitation collisions. The differential cross section is a function of both the scattering angle and the electron energy. Figure 1.10 shows the differential scattering cross section in helium for the first three levels of helium for electron impact energies of 30, 40, and 50 eV.[6]

The experimental results in Figure 1.10 are those of Cubric et al.[6] who adopted a novel technique for measuring the differential cross section of inelastically scattered

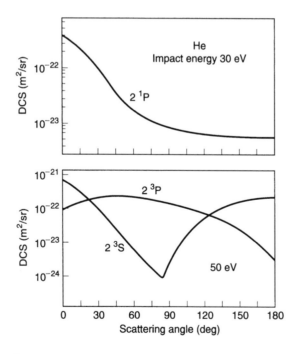

FIGURE 1.10 Differential cross sections as a function of scattering angle for excitation cross section of helium for the three excitation levels for electron energies of 30 and 50 eV. Good agreement is observed with the following results (not shown for clarity's sake): Asmis and Allan[8]; Trajmar et al.[7]; Fursa and Bray[9]; Bartchat et al.[10] Reproduced from Cubric, D. et al., *J. Phys. B, At. Mol. Opt Phys.*, 32, L-45, 1999. With permission.

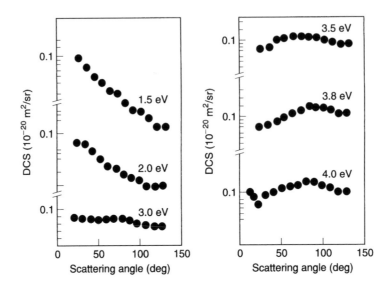

FIGURE 1.11 Differential scattering cross sections in carbon dioxide for the bending modes (010) at selected electron impact energies. Not shown are the results of: Antoni et al.[13]; Register et al.[14]; Kitajima et al.[11]; Takekawa and Itikawa[15] (theory); Takekawa and Itikawa.[16] Reproduced from Kitajima, M. et al., *J. Phys. B, At. Mol. Opt Phys.*, 34, 1929, 2001. With permission.

electrons. Note that those results cover the entire range of the angle 0 to 180° whereas previous results[7–10] are restricted to a range of approximately 30° to 150°. The excitation cross section at a given electron impact energy is the integral of the scattering cross section curve. Before the advent of Cubric's[6] technique it was necessary to interpolate at the low and high angle ends in order to calculate the inelastic scattering cross section.

A second example of differential scattering cross section for inelastic collisions is shown in Figure 1.11, which deals with vibrational excitation cross sections in carbon dioxide.[11] Carbon dioxide is a linear triatomic molecule that has three modes of vibration: a bending mode designated as (010) and (020); a symmetric stretching mode designated as 100; and an asymmetric stretching mode designated as 001. The molecule does not possess a permanent dipole moment in the ground state,[12] but a dipole is induced when either the bending or the asymmetric bending mode is excited. These excitations occur in the infrared wavelength region of the spectrum, 10^{12} to 10^{14} Hz. The symmetric stretching mode is referred to as Raman active. A discussion of these results is given in Chapter 5. The original publication of Kitajima et al.[11] includes other data.[13–16]

1.5.4 MOMENTUM TRANSFER CROSS SECTION

During collision between an electron and an atom there is transfer of momentum equal to

$$\Delta p = mW_e(1 - \cos\theta) \tag{1.72}$$

where m and W_e are the mass and velocity of the electron and θ is the angle of scattering. Strictly speaking, Equation 1.72 should be written in the CM system, but for $M_b \gg M_a$ the distinction between laboratory and CM system becomes negligible. The momentum transfer cross section is defined as

$$Q_M = \int_0^\pi \int_0^{2\pi} Q_{\text{diff}}(\theta)(1 - \cos\theta)\sin\theta \, d\theta \, d\phi \tag{1.73}$$

The momentum transfer cross section (some times referred to as the diffusion cross section) is a measure of loss of forward momentum of electrons drifting through a gas under the influence of an electric field. If the differential cross section is independent of θ, that is, scattering is isotropic, Q_M will be equal to integrated Q_{diff}. The diffusion coefficient is inversely related to the momentum transfer cross section and Q_M is alternatively referred to as the diffusion cross section, though this practice has been almost discontinued. The diffusion phenomenon is briefly discussed in the next section. If scattering is independent of the angle φ, then Equation 1.73 simplifies to

$$Q_M = 2\pi \int_0^\pi Q_{diff} \sin\theta(1 - \cos\theta)d\theta \tag{1.74}$$

It is important to realize that momentum transfer occurs during every collision, elastic or otherwise. However, the cross section one encounters in the literature is generally assumed to be that due to elastic collisions.

1.5.5 DIFFUSION COEFFICIENT

The diffusion phenomenon is the transport of particles from a higher density region to a lower density region. Higher densities result in higher transport rate, which is mathematically expressed as

$$\mathbf{J} = -D\nabla n \tag{1.75}$$

where $\mathbf{J}$ is the flux (number of particles crossing an area of one square meter per second), D is called the diffusion coefficient ($m^2 s^{-1}$), and ∇n is the concentration gradient. The negative sign indicates that diffusion occurs from a region of higher concentration to one of lower concentration. In a binary mixture of gases, diffusion of each species occurs from a higher concentration to a lower concentration region, irrespective of the partial pressure of the other constituent. If air which contains 21% oxygen is mixed with 100% nitrogen, oxygen in the air diffuses toward the pure nitrogen and pure nitrogen diffuses in the opposite direction.

Diffusion does not occur instantaneously as the molecules have finite speed. The time-dependent diffusion equation is expressed as

$$\frac{dn}{dt} = \nabla^2[D\,n(r,t)] \tag{1.76}$$

where the number density n is both time and space dependent. Let us suppose that the volume density of particles decays exponentially with time according to

$$n = n_0 \exp\left(-\frac{t}{\tau}\right) \tag{1.77}$$

where n_0 is the initial density and τ is called the diffusion time constant. Substituting Equation 1.76 into 1.77, one gets

$$\nabla^2 n + \frac{n}{D\tau} = 0 \tag{1.78}$$

The solutions of Equation 1.78 for a number of geometries are given by Hasted.[2]

The diffusion coefficient may be calculated approximately using the relationship[4]

$$D \approx v_c \lambda^2 \tag{1.79}$$

where ν_c is the frequency of collisions between the particles (s^{-1}) and λ is the mean free path. The collision frequency is equal to the mean thermal velocity $\langle W \rangle$ given by Equation 1.25, divided by λ. The diffusion coefficient now becomes

$$D \approx \langle W \rangle^2 \tau \tag{1.80}$$

where τ is the mean time between collisions.

Diffusion is process of electron loss from a high concentration region to a low concentration region. In the theoretical treatment of collision processes one often writes an equation for balancing the loss and gain of electrons in a given volume. The continuity equation for electrons in the presence of diffusion loss is written as

$$G - \frac{dn}{dt} - \nabla^2[Dn(r \cdot t)] = 0 \tag{1.81}$$

where G is a net electron production rate. Generally G will be equal to $n\nu_i$, which is the number of electrons generated per second by a single electron by ionization. Making this substitution, one gets the equation

$$\frac{dn}{dt} = -\nabla^2[Dn(r \cdot t)] + n\nu_i \tag{1.82}$$

The solution of this equation depends on the boundary conditions. In microwave breakdown the solution obtained is of the form

$$n = n_0 \exp\left[\left(\nu_i - \frac{D}{\Lambda^2}\right)t\right] \tag{1.83}$$

where Λ is called the characteristic diffusion length, which is determined by the boundary conditions. For a right circular cylinder of radius R and length L the diffusion length is[17]

$$\frac{1}{\Lambda^2} = \left(\frac{\pi}{L}\right)^2 + \left(\frac{2.405}{R}\right)^2 \tag{1.84}$$

In many types of discharge one would like to be able calculate the loss of electrons from the ionizing region; diffusion is one such phenomenon. The solution of the diffusion equation depends, of course, on the geometry of the electrodes and the boundary conditions. A spark channel is typical of a coaxial cylindrical geometry. The number of electrons between any two points separated by a small distance dx varies according to a Gaussian function[18]

$$n = \frac{n_0}{\sqrt{4\pi Dt}} \exp\left(-\frac{x^2}{4Dt}\right) dx \tag{1.85}$$

where n_0 is the initial number of electrons at $x = 0$ and $t = 0$, n is the number of electrons at time t and at a slab of thickness dx situated at distance x, and D is the diffusion coefficient. Equation 1.85 is known as the Einstein equation. It shows that, as time increases, the ratio n/n_0 decreases, rendering the distribution of electrons uniform. If the particles are diffusing uniformly in three dimensions x, y, z, one can substitute r in place of x, where

$$r^2 = \sqrt{x^2 + y^2 + z^2} \tag{1.86}$$

The electron density at a distance r from the origin then becomes

$$n = \frac{n_0}{(4\pi Dt)^{3/2}} \exp\left(\frac{-r^2}{4Dt}\right)$$ (1.87)

A rearrangement gives the radial distance for a given ratio of n/n_0 as

$$r^2 = -4Dt \ln\left[\frac{n}{n_0}(4Dt)^{3/2}\right]$$ (1.88)

The average displacement for cylindrical geometry simplifies to

$$\bar{r} = 2\sqrt{Dt}$$ (1.89)

For a spherical distribution of electrons the mean distance is expressed as

$$\bar{r} = \sqrt{3Dt}$$ (1.90)

This equation is often used to calculate the electric field at a distance $\bar{r}$ due to diffusing electrons (Chapter 8).

Under plasma conditions one encounters both positive charges and electrons and if their density is 10^{14} m^{-3} or more one cannot ignore the interaction between them. Electrons diffuse toward the walls and regions of lower density more readily than positive ions and charge separation will take place. The effect of the separated charge layers on the electric field is such that the electrons are retarded and positive ions are accelerated. A steady-state condition will be reached when charges of both polarities diffuse with the same velocity; this phenomenon is known as ambipolar diffusion. Further consideration will be given in Chapter 12.

1.5.6 Einstein Relationship

From the study of Brownian motion in liquids, Einstein derived the relationship

$$\frac{D}{\mu} = kT$$ (1.91)

where D and μ are defined by Equations 1.75 and 1.28 respectively. The term "characteristic energy," first used by Frost and Phelps,[19] signifies that the ratio D/μ is approximately a measure of the mean energy of the molecules. The exact ratio between the characteristic energy and the mean energy is a function of electron energy distribution and is 2/3 for a Maxwellian distribution.

1.5.7 Inelastic Collisions

Inelastic collisions cover a vast spectrum in the study of collision phenomena in gases, and Hasted[2] lists the various possibilities that include collision participants such as electrons, ions, and atoms. An inelastic collision is defined as a collision in which the internal energy of one or both particles changes. If the net change in internal energy is $\Delta\varepsilon_p$, the inelastic collision is defined as shown in Table 1.5.

If part of the kinetic energy of the incoming particle is converted to increase the potential energy of the target particle, the inelastic collision is known as endothermic. If the collision

TABLE 1.5
Classification of Collisions

$\Delta\varepsilon_p$	Classification
0 (no change in K.E.)	Elastic
> 0	Endothermic (inelastic)
< 0	Exothermic (inelastic)
0 (change in K.E.)	Resonant (inelastic)

Note: $\frac{1}{2}M_a^2 = \frac{1}{2}M_a'^2 + \frac{1}{2}M_b'^2 + \Delta\varepsilon_p$

results in a lowering of the potential energy of the target, the collision is known as exothermic. If there is an exchange of potential energy between the projectile and the target, but the net change of potential energy is zero, the collision is known as resonance. The masses of the projectile and target need not remain the same, though the total mass of the system before and after collision remains constant, viz.,

$$M_a \neq M_a'; \; M_b \neq M_b'; \; M_a + M_b = M_a' + M_b' \tag{1.92}$$

If ionization results from an inelastic collision between an electron and a neutral or excited atom the resulting three particles have the same total mass as the mass of the projectile and target before collision.

The fractional loss of energy during an inelastic collision may be obtained by the method adopted to calculate the fractional loss of energy during an elastic collision (Equation 1.65). In an inelastic collision the conservation of momentum will be in accordance with Equation 1.1, assuming that the masses of the projectile and target remain the same,

$$M_a\mathbf{W}_a = M_a\mathbf{W}_a' + M_b\mathbf{W}_b' \tag{1.1}$$

However, the conservation of energy before and after collision should take into account the increase in the potential energy of the molecule after collision. Accordingly,

$$\frac{1}{2}M_aW_a^2 = \frac{1}{2}M_aW_a'^2 + \frac{1}{2}M_bW_b'^2 + e\varepsilon_p \tag{1.93}$$

where ε_p is the increase in potential energy in eV. Substituting Equation 1.1 into 1.93 and rearranging gives

$$e\varepsilon_p = \frac{1}{2}\left[M_a(W_a^2 - W_a'^2) - \frac{M_a^2}{M_b}(W_a - W_b)^2\right] \tag{1.94}$$

For a constant energy of the electron, the energy transferred is a maximum when the differential of Equation 1.94 with respect to W_a' is equal to zero, viz.,

$$\frac{d(e\varepsilon_P)}{d(W_a')} = 0$$

or

$$W_a' = W_a\left(\frac{M_a}{M_a + M_b}\right) \tag{1.95}$$

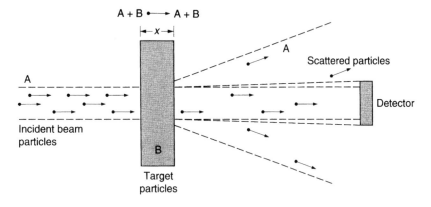

FIGURE 1.12 Schematic diagram of the principle of total cross section measurement. The detector measures the signal due to electrons that are not scattered.

The maximum energy transferred may be obtained by substituting Equation 1.95 into 1.94. Two special cases arise.

1. The incoming particle is an electron, $M_a \ll M_b$, and the atom is at rest before collision. The maximum potential energy transferred is, according to Equations 1.94 and 1.95,

$$W_{p\max} = \frac{1}{2e}M_a W_a^2 \qquad (1.96)$$

All of the kinetic energy of the electron is transferred, increasing the potential energy of the atom.

2. The incoming particle is an ion, $M_a \cong M_b$, and the maximum kinetic energy transferred is

$$W_{p\max} \cong \frac{1}{4e}M_a W_a^2 \qquad (1.97)$$

The velocity of an ion is usually below 1% of the velocity of the electron in a swarm and the energy transferred by an ion is therefore a small fraction in comparison with the energy transferred by the electron.

Figure 1.12 shows schematically the principle involved in the measurement of total cross section, defined as the sum of all individual cross sections. A beam of electrons with a well-defined energy interacts with the target gas. The signal due to the unscattered electrons is measured by the detector and compared with the signal received with no target molecules present. From the ratio of the signals the total cross section is calculated.

1.5.8 COLLISION FREQUENCY

Collision between particles is characterized by the parameter collision frequency (ν) which is defined as the average number of collisions per second. In a gas, with no charge carriers present, a simple relationship exists between the average speed of the molecule and the mean free path as

$$v = \frac{\langle W \rangle}{\langle \lambda \rangle} \qquad (1.98)$$

One has to distinguish between electron–neutral collisions and collisions between neutrals in a charge-free medium. Table 1.6 gives a collection of collision frequencies along with selected relevant parameters for collisions between neutrals.[29] The table demonstrates the dependence of the collision frequency for momentum transfer on the nature of the neutral gas.[4]

The electron–molecule collision frequency is an important parameter that determines the energy distribution and all the quantities that are determined by it, namely the drift velocity, diffusion coefficient, ionization and attachment coefficients. It is alternately defined by the equation

$$v_M = N Q_M(\varepsilon) W \qquad (1.99)$$

where W is the drift velocity of electrons and the momentum transfer cross section has been denoted as dependent on electron energy. Each collision process has a frequency associated with it and the total collision frequency is the sum for all processes. Since the elastic collision cross section is the largest, the elastic collision frequency is the largest of all collision frequencies. Before considering the energy dependence of the collision frequency, we should point out that the unit for v at any gas pressure is s^{-1} and it is in the range of 10^8 to $10^{10} s^{-1}$. The quantity v/N is therefore in the range of 10^{-14} to $10^{-12} s^{-1} m^3$ and at any gas number density this quantity should be multiplied by the appropriate number density. If the energy of the electron ε is expressed in electron volts, Equation 1.99 may be rewritten as

$$\frac{v_M}{N} = Q_M \sqrt{\frac{2e\varepsilon}{m}} \quad (m^3 s^{-1}) \qquad (1.100)$$

where e is the electronic charge.

TABLE 1.6
Mean Velocity, Diameter, Mean Free Path, and Collision Frequencies in Molecular Gases

Gas	Mol. Wt.	$\langle V \rangle \times 10^3$ (m/s)	Diameter (nm)	$\langle \lambda \rangle$ (μm)[a]	$v \times 10^9$ (s^{-1})
H_2	2.016	1.740	0.274	0.118	14.8
He	4.002	1.235	0.218	0.186	6.6
CH_4	16.03	0.618	0.414	0.052	12.0
NH_3	17.03	0.598	0.443	0.045	13.3
H_2O	18.02	0.582	0.460	0.042	13.9
Ne	20.18	0.550	0.259	0.132	4.2
N_2	28.02	0.467	0.375	0.063	7.4
C_2H_4	28.03	0.467	0.495	0.036	12.9
C_2H_6	30.05	0.451	0.530	0.032	14.3
O_2	32.00	0.437	0.361	0.068	6.4
HCl	36.46	0.409	0.446	0.044	9.2
A	39.94	0.391	0.364	0.067	5.9
CO_2	44.00	0.372	0.459	0.042	8.8
Kr	82.9	0.271	0.416	0.051	5.3
Xe	130.2	0.217	0.485	0.038	5.8

[a]15°C, 101.3 kPa.
From McDaniel, E. W., *Collision Phenomena in Ionized Gases*, John Wiley & Sons, New York, 1964. With permission.

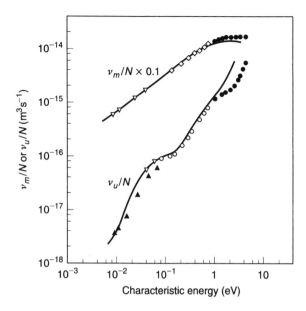

FIGURE 1.13 Collision frequency for elastic scattering and energy exchange in nitrogen at 300 K as functions of characteristic energy, calculated using Equations 1.103 and 1.104. Data for W are given in Table 6.25 and characteristic energy in Figure 6.46. The full line shows ν_M/N as a function of E/N.

Each process, including elastic and individual inelastic processes, has a collision frequency associated with it. The total collision frequency may be calculated using the total collision cross section Q_T. Figure 1.13 shows the calculated collision frequencies in nitrogen. Collision frequencies computed for momentum transfer and inelastic collisions in molecular nitrogen[20] are shown in Figure 1.14 and Figure 1.15. Figure 1.16 shows ν/N for several common gases.

The collision frequency given by Equation 1.100 closely reflects the elastic collision cross section, but there is no one-to-one correspondence, due to the square root term. In fact, much of the earlier literature used $\sqrt{\varepsilon}$ as the variable for the energy axis in representing the elastic collision cross sections, though for the reason that the velocity of the electron is proportional to $\sqrt{\varepsilon}$. It should be remembered that the collision frequencies shown are for mono-energetic electrons and it is assumed that collisional interaction with molecules is not of the swarm type. The collision frequency at energies lower than about 0.2 eV decreases as the energy of the electron is increased and this is attributed to the Ramsauer–Townsend cross section, as described later.

From the fundamental definition of collision frequency as the number of a particular type of collisions per second one can express the electron–molecule collision frequency in a swarm as

$$\nu(\varepsilon) = \frac{N \int_0^\infty Q(\varepsilon)\varepsilon f(\varepsilon)d\varepsilon}{\int_0^\infty f(\varepsilon)d\varepsilon} \tag{1.101}$$

Here N is the number of gas molecules, and the fraction of electrons having energy between ε and $\varepsilon + d\varepsilon$ is given by $f(\varepsilon)\ d\varepsilon$. Their total energy is $\varepsilon\ f(\varepsilon)\ d\varepsilon$ and the scattering cross section for these electrons is $Q\ \varepsilon f(\varepsilon)\ d\varepsilon$. The integral sign represents the addition of each subgroup as the energy is varied from 0 to ∞. The denominator is the total number of electrons and therefore Equation 1.101 gives the effective collision frequency as a function of energy for a given type of collision.

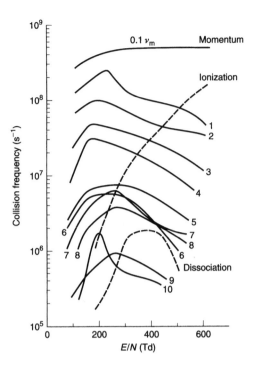

FIGURE 1.14 Collision frequency in nitrogen as a function of E/N. The reduced collision frequency is obtained by dividing the ordinate by N. The numbers show the vibrational states. From Liu, J., Ph.D thesis, University of Windsor, 1993.

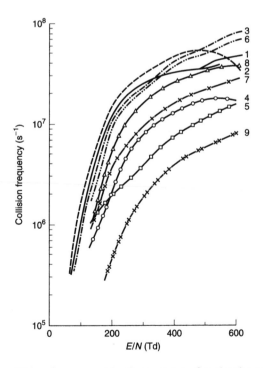

FIGURE 1.15 Excitation collision frequency of valence states of molecular nitrogen as a function of E/N. Curve numbers refer to the states: (1) $A\,^3\Sigma_u^+$; (2) $B\,^3\Pi_g$; (3) $W\,^3\Delta_u$, (4) $B'\,^3\Sigma_u^-$; (5) $a'\,^1\Sigma_u^-$; (6) $a\,^1\Pi_g$, (7) $W\,^1\Delta_u$; (8) $C\,^3\Pi_u$; (9) $a'\,^1\Sigma_g^+$. From Liu, J., Ph.D. thesis, University of Windsor, 1993.

In the situation of electrons colliding with gas molecules one encounters electrons having a wide range of energy. Equation 1.100 is not applicable and one has to define the average collision frequency that is dependent on energy; this energy is characteristic of the collision process and covers electrons of all energies. One represents such a characteristic energy[21] according to

$$\varepsilon_k = e \frac{D}{\mu} \qquad (1.102)$$

where e is the electronic charge, D the diffusion coefficient and μ the mobility. Engelhardt and Phelps[21] define the effective elastic collision frequency as

$$\frac{\nu_M}{N} = \frac{e}{m} \left[\frac{E}{N} \frac{1}{W} \right] \qquad (1.103)$$

where m is the mass of the electron and W the drift velocity. The parameters E and N are measured experimentally and the collision frequency may be calculated. The quantity ν_M is primarily sensitive to changes in the elastic collision cross section and is affected only slightly by the inelastic cross section. A constant collision frequency implies that $W \propto E/N$. However, energy loss also occurs due to inelastic collisions; an energy exchange frequency is defined according to

$$\frac{\nu_u}{N} = eW \left(\frac{E}{N} \right) \frac{1}{(\varepsilon_k - kT)} \qquad (1.104)$$

From Equation 1.104, ν_u is identical with the power input per electron divided by the excess energy above thermal energy. Figure 1.13 shows the representative trend of variation of ν_M and ν_u as a function of characteristic energy. Both quantities increase with energy, as expected. Below the first electronic excitation inelastic losses are due to rotational and vibrational excitation. Table 1.7 shows the influence of gas temperature on ν/N in several gases.

TABLE 1.7
Momentum Transfer Collision Rates for Various Gas Species at Low Background Neutral Gas Temperatures

	ν/N [10^{-14} m^{3}s^{-1}]		
Gas species	$T = 300$ K	$T = 500$ K	$T = 1000$ K
N_2	0.594	0.959	1.77
O_2	0.286	0.437	0.804
CO_2	10.05	9.68	7.48
H_2O	77.90	56.25	34.6
He	0.763	1.01	1.20
Ne	0.07	0.118	0.217
Ar	0.220	0.146	0.094
Kr	1.775	1.315	0.713
Xe	5.29	3.81	1.90
Dry air	0.529	0.845	1.550

N is gas density in m^{-3}.
From Roth, J. R., *Industrial Plasma Engineering*, Institute of Physics, Bristol, U.K., 1995. With permission.

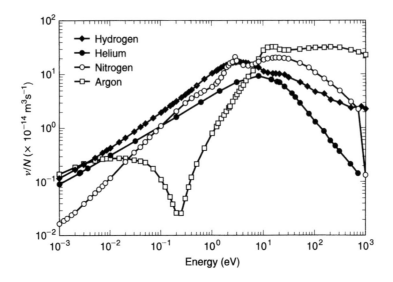

FIGURE 1.16 Calculated collision frequency according to formula 1.100. The data Q_M are obtained as follows: argon—Table 3.3 in this volume; helium—Table 3.13; hydrogen—Table 4.8; nitrogen—Table 4.14.

1.5.9 RATE COEFFICIENTS AND CONSTANTS

In studies on plasma processes, inelastic collision processes are often expressed in a different form known as rate coefficients. A reaction may involve only one species of particles, the rate of decay being dependent on its density. A real-life analogy is the number of starvation deaths per day in a famine-stricken country. The greater the density of population, the greater is the rate of deaths.

Time rate of decay of population of electrons due to diffusion and radiative decay are examples of this type. The differential equation that describes the decay is

$$\frac{dN_a}{dt} = -K_1 N_a \tag{1.105}$$

where K_1 is the rate coefficient of the first order, having the units of s^{-1}. The solution of this equation is

$$\ln\left(\frac{N_a}{N_{a0}}\right) = -K_1 t \tag{1.106}$$

where N_{a0} is the initial concentration of electrons. Confusion should not arise between the two definitions of K_1 because they refer to the same decay process and the units clarify which definition applies.

Reactions involving two species of particles (electron–atom or electron–ion) and resulting in a third species are also described by resorting to the rate coefficient. If the densities of both species are equal the decay occurs according to

$$\frac{dN_a}{dt} = -K_2 N_a^2 \tag{1.107}$$

where K_2, the reaction coefficient of the second order, has the units of $m^3 \, s^{-1}$. The solution is given by

$$\frac{1}{N_a} = \frac{1}{N_{a0}} + K_2 t \tag{1.108}$$

If the densities of both species are not the same, as in attachment of electrons to molecules, the decay occurs according to

$$\frac{dN_a}{dt} = \frac{dN_b}{dt} = -K_2 N_a N_b \tag{1.109}$$

If N_b is expressed in terms of gas pressure according to Equation 1.42, K has the units of torr s^{-1}. The negative sign indicates that there is a decrease in the population of N_a, as in electron–ion recombination. The solution is

$$\ln\left(\frac{N_a}{N_b}\right) = K_2(N_a - N_b)\,t + \ln\left(\frac{N_{a0}}{N_{b0}}\right) \tag{1.110}$$

In the case of electron–molecule attachment, the density of molecules is much greater than the density of electrons, $N_b \gg N_a$. Further, the initial density of molecules remains approximately independent of time because of the small number of attachments that occur, $N_{b0} \cong N_b$. Substituting these approximations into Equation 1.110, one gets

$$\ln\left(\frac{N_a}{N_{a0}}\right) = -K_2 N_b \, t \tag{1.111}$$

which is rewritten as

$$N_a = N_{a0} \exp(-K_2 N_b t) \tag{1.112}$$

The decay of electrons is therefore exponential with time.

It is useful to note here that the fraction of particles having a velocity between V and $V + dV$ at the instant of collision is given by the normalized velocity distribution $f(V)dV$. The cross section for each collision of the relevant type is $Q(V)$ and the rate coefficient is given by

$$K_2 = \int_0^\infty VQ(V)f(V)\,dV \tag{1.113}$$

with the integration carried out over all velocities. If the cross section is independent of the velocity, Equation 1.113 simplifies, because of Equation 1.25, to

$$K_2 = Q\langle V \rangle \tag{1.114}$$

The three-body rate coefficient, analogous to the two-body rate coefficient in which two species are involved, is defined according to the number of species involved in the reaction. Three cases arise:

Case A. All particles belong to the same species, $N_a = N_b = N_c$. The rate equation simplifies to

$$\frac{dN_a}{dt} = -K_3 N_a^3 \tag{1.115}$$

The units of K_3 for a three-body process are $m^6 s^{-1}$ and its value is some 10^{-20} times smaller than the two-body rate coefficient. The presence of a third body in such collisions serves the purpose of removing the excess energy that otherwise has to be dissipated as electromagnetic radiation.[2]

The solution of Equation 1.115 is

$$\frac{1}{N_a^2} - \frac{1}{N_{a0}^2} = 2K_3 t \tag{1.116}$$

where N_{a0} is the initial population.

Case B. Two species are identical and the third species is different. In this case the rate equation is

$$\frac{dN_a}{dt} = -K_3 N_a^2 N_b \tag{1.117}$$

and the solution is

$$(N_{b0} - N_{a0})\left(\frac{1}{N_a} - \frac{1}{N_{a0}}\right) + \ln\left(\frac{N_a N_{b0}}{N_b N_{a0}}\right) = (N_{b0} - N_{a0})^2 K_3 t \tag{1.118}$$

where the subscript zero refers to the initial population.

Electron attachment in oxygen is an example of this type of three-body process and the attachment coefficient in this case is dependent on the square of the gas pressure. The reaction is

$$O_2 + e + O_2 \rightarrow O_2^- + O_2$$

The measured attachment coefficients due to Chanin et al.[22] at low values of E/N ($1\,\text{Td} = 1 \times 10^{-21}\,\text{V}\,\text{m}^2$) are shown in Figure 1.17 for swarm experiments. These authors report a reaction rate for the three-body attachment of $\sim 2.8 \times 10^{-42}\,\text{m}^6\,\text{s}^{-1}$.

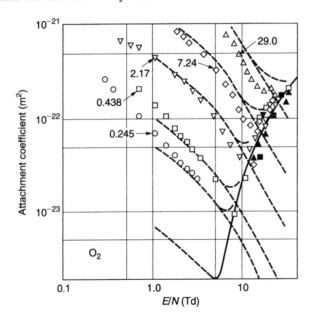

FIGURE 1.17 Attachment coefficients at low values of E/N, showing pressure dependence arising from the three-body process. For details see Figure 9.22.

Case C. All three species are different. The rate equation is

$$\frac{dN_a}{dt} = -K_3 N_a N_b N_c \tag{1.119}$$

For $N_a \ll N_b$, $N_a \ll N_c$, and if the changes in the populations of N_b and N_c due to the reaction are insignificant, the solution is an exponential decay, according to

$$N_a = N_{a0} \exp(-K_3 N_b N_c t) \tag{1.120}$$

The relation between the rate coefficient and cross section is defined as[23]

$$K(\langle \varepsilon \rangle) = \left(\frac{2}{m}\right)^{1/2} \int_0^\infty Q(\varepsilon) \varepsilon^{1/2} f(\varepsilon) d\varepsilon \quad \text{m}^3\text{s}^{-1} \tag{1.121}$$

where m is the mass of the electron, Q is the cross section for which process the rate is being calculated, $f(\varepsilon)$ is the energy distribution of electrons, and $\langle \varepsilon \rangle$ is the mean energy. If we assume a Maxwellian energy distribution, $f(\varepsilon)$ is given as

$$f(\varepsilon) = \frac{2}{\sqrt{\pi}} \left(\frac{3}{2\langle \varepsilon \rangle}\right)^{3/2} \varepsilon^{1/2} \exp\left(-\frac{3\varepsilon}{2\langle \varepsilon \rangle}\right) \tag{1.122}$$

To find an analytical expression for K from Equation 1.122, one has to find an analytical form for representing $Q(\varepsilon)$. One such expression is

$$Q(\varepsilon) = Q_0 \exp\left(-\frac{\varepsilon}{\gamma}\right) \tag{1.123}$$

where γ is a constant having the dimension of energy. It is usually necessary to represent cross sections for energy segments using an equation of type 1.123, in which case one represents the cross section as

$$Q(\varepsilon) = Q_0 \sum_{j=1}^n \exp\left(-\frac{\varepsilon}{\gamma_j}\right) \tag{1.124}$$

where the summation index j covers each energy segment. For a two-segment representation of attachment cross sections in SF_6 and $CFCl_3$,[23] substitution of Equations 1.124 and 1.122 into 1.121 yields

$$K(\langle \varepsilon \rangle) = \left[\left(\frac{1.233 \times 10^4}{\langle \varepsilon \rangle^{3/2}}\right)\right] \sum_{j=1}^2 a_j \Gamma_j^2 \left[\left(1 + \frac{\varepsilon}{\gamma_j}\right) \exp\left(-\frac{\varepsilon}{\Gamma_j}\right)\right]_{\varepsilon_{uj}}^{\varepsilon_{lj}} \tag{1.125}$$

where ε_{lj} and ε_{uj} represent the lower and upper ranges of each energy segment, a_j is the relative weighting of each term, and the symbol Γ is the gamma function.[24] Expressions similar to Equation 1.125 have been derived for swarm properties of a number of electron-attaching gases by Raju et al.[25-27] Evaluation of Equation 1.125 is a formidable task since it involves computation of incomplete gamma functions. For the sake of completeness we provide below the representation of attachment cross section in SF_6 for the formation of SF_6^- adopted by Chutjian[23]:

$$Q(\varepsilon) = 5.20 \times 10^{-18} \times \begin{cases} \exp\left(-\dfrac{\varepsilon}{44.4}\right) & m^2, \quad 0 \leq \varepsilon \leq 45\,meV \\[3mm] 0.868\exp\left(-\dfrac{\varepsilon}{51.6}\right) & m^2, \quad 45 \leq \varepsilon \leq 200\,meV \end{cases} \tag{1.126}$$

A representative value of attachment rate at a mean energy of $\langle 38.8 \rangle$ meV at 300 K is $2.28 \times 10^{-13}\,m^3s^{-1}$.

Finally, the rate coefficient, e.g., for the attachment process of Equation 1.113 is given in terms of the energy as

$$K_a\left(\frac{E}{N}\right) = \left(\frac{2}{m}\right)^{1/2} \int_0^\infty \varepsilon^{1/2} f(\varepsilon, E/N)\, Q_a(\varepsilon)\, d\varepsilon \tag{1.127}$$

where Q_a and K_a are the attachment cross section and rate respectively. The reduced attachment coefficient and the rate coefficients are related by the equation

$$K_a(E/N) = \frac{\eta}{N}(E/N) \times W(E/N) \tag{1.128}$$

where W is the drift velocity. Often the attachment coefficient is measured by a small additive of an attaching species in a nonattaching carrier gas, and Equation 1.128 then becomes

$$K_a(E/N) = \frac{\eta}{N_a}(E/N) \times W(E/N) \tag{1.1.29}$$

in which N_a is the number density of the attaching species.

1.5.10 ION MOBILITY

The mobility of a charged particle in an electric field has already been defined in Equation 1.28. An expression can be derived for the mobility of ions in a gas drifting under the influence of the electric field[28] on the basis of the kinetic theory of gases. A few general comments are in order before the derivation is given.

The electric field imparts energy to the ion, which gains energy between collisions with atoms and loses energy during collisions. The energy gained is a function of the parameter E/N and determines the mean energy of the ion. This energy is superimposed on the thermal energy, which is determined by the temperature of the gas. It is relatively easy to determine the ratio of the parameter E/N that imparts to the ion, an energy that is approximately equal to the thermal energy. The collision time, that is the interval between collisions during which the ion acquires accelaration is τ (s) and the velocity just before a collision is $eE\tau/m$ (m/s). Since τ is proportional to $1/N$ the energy gained from the electric field is a function of $(E/N).^2$ Retaining the terminology of McDaniel[29] "field energy" for the energy gained from the field, the field energy is

$$\text{Energy gain} = \frac{M}{m}eE\lambda \tag{1.130}$$

where M and m are the masses of the neutral and the ion respectively, and $eE\lambda$ is the energy gained by the ion in moving through a distance λ. For a singly charged ion

$M/m \approx 1$ and $\lambda = 1/NQ$, where N is the number of molecules per cubic meter (typically $\approx 50 \times 3.52 \times 10^{22}$ m^{-3}) and Q is the collision cross section between an ion and a molecule (typically $\approx 50 \times 10^{-20}$ m^2). The electric field is typically 1×10^4 V/m. Substituting these values in Equation 1.130, one finds that the energy gain is 0.01 eV. The thermal energy is 0.026 eV. Hence for values of $E/N \approx 5 \times 10^{-21}$ Vm^2 (≈ 5 Td) one can write that

$$\frac{M}{m} eE\lambda \ll kT \tag{1.131}$$

Under conditions that satisfy Equation 1.131 the drift velocity is small compared to the thermal velocity. The average time interval between two successive collisions, assuming that it is independent of the electric field, is given by

$$\langle \tau \rangle = \frac{\langle \lambda_i \rangle}{\langle W_i \rangle} \tag{1.132}$$

where $\langle \lambda_i \rangle$ and $\langle W_i \rangle$ are respectively the mean free path and mean thermal velocity, Equation 1.25, of the ions. During the time interval between collisions (τ) the ion is accelerated by the electric field, and moves a distance of

$$s = \frac{1}{2} \frac{eE}{m} \tau^2 \tag{1.133}$$

The drift velocity of the ion is therefore

$$W_i = \frac{1}{2} \frac{eE}{m} \tau = \frac{1}{2} \frac{eE}{m} \frac{\langle \lambda \rangle_i}{\langle W_i \rangle} \tag{1.134}$$

The ion mobility, defined as W_i/E, becomes

$$\mu_i = \frac{1}{2} \frac{e}{m} \frac{\langle \lambda_i \rangle}{\langle W_i \rangle} \tag{1.135}$$

In deriving Equation 1.135 it has been assumed that the mean free path of the ion is the same for all ions. This is not true, because the random velocity is superimposed on the drift velocity. We have seen that the free path decreases exponentially with the mean free path, i.e., there are many more free paths longer than the mean free path (Equation 1.48). To take this factor into account one rewrites Equation 1.133, employing 1.132, as

$$s = \frac{1}{2} \frac{eE}{m} \left(\frac{x}{W_i} \right) \tag{1.136}$$

where x is the distance traveled by the ion between collisions. The average distance traveled ($\langle s \rangle$) is given by

$$\langle s \rangle = \frac{1}{2} \frac{eE}{m} \frac{1}{W_i^2} \frac{\int_0^\infty \dfrac{x^2 e^{-x/\langle \lambda_i \rangle}}{\langle \lambda_i \rangle} dx}{\int_0^\infty \dfrac{e^{-x/\langle \lambda_i \rangle}}{\langle \lambda_i \rangle} dx} \tag{1.137}$$

Evaluating the integrals, one obtains the expression

$$\langle s \rangle = \frac{eE}{m\langle W_i \rangle^2} \langle \lambda_i \rangle^2 \tag{1.138}$$

Substituting Equation 1.132 into 1.138, one obtains

$$\langle s \rangle = \frac{eE}{m} \langle \tau \rangle^2 \tag{1.139}$$

The drift velocity is given by

$$W_i = \frac{\langle s \rangle}{\langle \tau \rangle} = \frac{e}{m} \frac{E}{\langle W_i \rangle} \langle \lambda_i \rangle \tag{1.140}$$

and the mobility by

$$\mu_i = \frac{W_i}{E} = \frac{e}{m} \frac{\langle \lambda_i \rangle}{\langle W_i \rangle} \tag{1.141}$$

Comparing Equation 1.141 with 1.135, one finds that the distribution of free paths leads to ion mobility that is twice as large. A further refinement is to take into account the initial velocity of the ion. Langevin derived a relationship more exact than Equation 1.141:

$$\mu_i = \frac{0.815\,e\langle \lambda \rangle}{M\,v} \sqrt{\frac{m+M}{m}} \tag{1.142}$$

where m and M are the masses of the ion and the gas neutral respectively, v is the r.m.s. velocity of agitation of the gas neutrals, and $\langle \lambda \rangle$, the mean free path of the ions, is an approximation to the mean free path of the neutrals ($\langle \lambda \rangle \simeq \langle \lambda_i \rangle$). A general form of Equation 1.142 is

$$\mu = \frac{A}{\sqrt{\rho(\varepsilon_r - 1)}} \left(1 + \frac{M}{m}\right)^{1/2} \tag{1.143}$$

where ρ is the gas density and ε_r is the dielectric constant of the gas. A is a function of the parameter γ defined by the equation[30]

$$\gamma^2 = \frac{8\pi p D_{12}^4}{(\varepsilon_r - 1)e^2} \tag{1.144}$$

where p is the gas pressure and D_{12} is the sum of the radii of the ion and the molecule. γ^2 is a dimensionless temperature having values in the range of 0.0 to 4.0. At thermal equilibrium

$$\frac{1}{2} m \langle v_i \rangle^2 = \frac{1}{2} M \langle v \rangle^2 = \frac{3}{2} kT \tag{1.145}$$

where v_i is the r.m.s. velocity of the ion, k is the Boltzmann constant, and T is the absolute temperature.

TABLE 1.8
Mobility of Singly Charged Ions at 101.3 kPa and 273 K in Units of 10^{-4} m^2 (Vs)$^{-1}$

Gas	μ^-	μ^+	Gas	μ^-	μ^+
Air (dry)	2.1	1.36	HCl	0.95	1.1
A	1.7	1.37	H$_2$S	0.56	0.62
A (very pure)	206.0	1.31	He	6.3	5.09
Cl$_2$	0.74	0.74	He (very pure)	6.3	5.09
CCl$_4$	0.31	0.30	N$_2$	1.84	1.27
C$_2$H$_2$	0.83	0.78	N$_2$ (very pure)	145.0	1.28
C$_2$H$_5$Cl	0.38	0.36	NH$_3$	0.66	0.56
C$_2$H$_5$OH	0.37	0.36	N$_2$O	0.90	0.82
CO	1.14	1.10	Ne		9.9
H$_2$	8.15	5.9	O$_2$	1.80	1.31
H$_2$ (very pure)	7900.0		SO$_2$	0.41	0.41

Kuffel, E. and W. S. Zaengel, *High Voltage Engineering*, Pergamon Press, Oxford, 1974.

Equation 1.142 may be applied to positive ions, negative ions, and electrons. In the case of electron mobilities one makes the approximation $m \ll M$, obtaining

$$\mu_e = 0.815 \frac{e}{m} \frac{\langle \lambda_e \rangle}{v} \tag{1.146}$$

where $\langle \lambda_e \rangle$ is the mean free path of the electrons.

The mobility of ions at standard temperature (273 K) and pressure (101 kPa) is of the order of 0.5 to 10×10^{-4} m^2 (Vs)$^{-1}$. Positive and negative ions in a gas have approximately the same mobilities in a given gas. The mobility varies inversely as the gas pressure, Equation 1.135, and as long as the temperature remains constant the mobility is inversely proportional to gas number density. The mobility is often expressed in a reduced form. If μ is the mobility in m^2 (Vs)$^{-1}$ at gas number density N (m^{-3}) and temperature T (K), the reduced mobility is given as

$$\mu_0 = \mu \frac{N}{2.69 \times 10^{25}} \frac{273}{T} \tag{1.147}$$

The ion mobility depends upon the ion species and the gas in which it moves, even though E/N remains the same. For example, He$^+$ ions in He have a reduced mobility of 10.2×10^{-4} m^2 (V s)$^{-1}$ at $E/N = 10$ Td, compared with a mobility of 30.6×10^{-4} m^2 (V s)$^{-1}$ for H$^+$ ions.[31]

Tables 1.8 and 1.9 list selected mobility values for ions in parent gas as a function of E/N. The ion attracts a molecule and binds itself to it, forming a cluster. A clustered ion is subjected to repeated collisions with the molecule and the kinetic energy transferred to the cluster is of the order of the thermal energy if the E/N ratio is low. If the thermal energy of the cluster becomes greater than the binding energy, the cluster will break up. The ions interact between themselves and two effects are discernible.[2]

The first is the space charge effect produced by widely separated ions. The physical dimensions of the containing vessel influence this phenomenon. The criterion for negligible space charge due to n number of ions is given as

$$n \ll \frac{E}{4\pi e L} \tag{1.148}$$

TABLE 1.9
Reduced Mobilities of Selected Ions ($10^{-4}\,m^2V^{-1}s^{-1}$) in Parent Gas as a Function of E/N (Td)

E/N	He^+/He	Ne^+/Ne	Ar^+/Ar	Kr^+/Kr	H^+/H_2	N_2^+/N_2
6	10.3	4.07			16.0	1.90
8	10.2	4.05	1.53		16.0	1.89
10	10.2	4.04	1.53		16.0	1.88
12	10.1	4.02	1.53		16.0	1.88
15	10.0	3.98	1.52		15.9	1.87
20	9.90	3.91	1.51		15.8	1.85
25	9.74	3.84	1.49		15.7	1.84
30	9.60	3.76	1.47		15.5	1.83
40	9.28	3.61	1.44	0.838	15.2	1.80
50	8.97	3.48	1.41	0.828	14.9	1.76
60	8.67	3.35	1.38	0.816	14.5	1.72
80	8.12	3.13	1.32	0.791	13.9	1.66
100	7.67	2.96	1.27	0.767	13.4	1.60
120	7.25	2.81	1.22	0.743	13.2	1.54
150	6.78	2.61	1.16	0.711	13.1	1.47
200	6.12	2.36	1.06	0.666	13.1	1.37
250	5.60	2.17	0.99	0.627	13.2	1.28
300	5.19	2.02	0.95	0.592	13.3	1.20
400	4.58	1.80	0.85	0.546	13.7	1.10
500	4.17	1.63	0.78	0.491		1.02
600	3.81	1.51	0.72	0.453		0.95
700	3.57	–				0.85
800		1.32	0.63	0.398		
1000		1.19	0.56	0.359		
1200		1.09	0.51	0.329		
1500		0.99	0.46	0.294		
2000			0.40	0.259		
3000				0.220		

Ellis, H. W. et al., *Atomic Data and Nuclear Data Tables*, 17, 177, 1976.

where L is the length of the vessel. The second effect is the random fluctuation of the number of ions. This will, in turn, change the velocity distribution, derived under the assumption that mutual ionic interaction is negligible. Space charge becomes significant at number densities of $\sim 10^{14}/m^3$ and velocity distribution is affected at number densities of $\sim 10^{17}/m^3$.

The restriction that the energy transferred from the field to the ions should be smaller than the thermal energy (Equation 1.131) is not so central to the development of the equation for ion mobility. Wannier[32,33] has extended the theory to higher electric fields. At lower fields the drift velocity is lower than the random thermal speed. The velocity distribution of ions is then almost Maxwellian and the electric field is treated as a perturbation upon this distribution. Wannier[32] applied the Boltzmann method of gaseous kinetics to the problem of positive ions moving through a gas under the influence of a static, uniform electric field. As in the case of low electric fields, the ion density is taken to be small. Velocity averages were computed in preference to calculating the entire velocity distribution which is a much more complex problem. Extracting the drift velocity average from the Boltzmann distribution function was accomplished by assuming that the mean free time between collisions of ions and molecules is constant. This is the case for the so called polarization force between ions and molecules which predominated over other forces at low temperatures. The method was extended to nonuniform ionic densities.[32]

Wannier's results[33] are briefly summarized here. If it is assumed that the mean free path is constant, then the drift velocity varies according to

$$W \simeq \sqrt{\frac{E}{N}} \quad \text{(high } E/N, \text{ constant } \lambda) \tag{1.149}$$

On the other hand, if it is assumed that the free time is constant, the drift velocity varies according to

$$W \simeq \frac{E}{N} \quad \text{(high } E/N, \text{ constant } \tau) \tag{1.150}$$

At low E/N, either assumption of constant λ or constant τ leads to the same result:

$$W \simeq \frac{E}{N} \quad \text{(low } E/N) \tag{1.151}$$

Further refinements of the theory of ionic mobility involve considerable mathematical expertise and are beyond the scope of this book.

1.6 POTENTIAL FUNCTIONS FOR PARTICLE INTERACTIONS

Consider a collision between a particle of mass M_R, the reduced mass of the system, and a fixed target. There is a force acting on the particles because they have electrical charges and the relative velocity (W_R) changes as a function of time in response to the force between the particles. This force, which is a center-of-mass quantity, depends on the separation between the particles, their orientations, and relative velocities. However, the force may be treated as velocity independent and it can be expressed in terms of an interaction potential, $V(R)$, such that $F = -\nabla V$. V is considered as a function of the separation distance only (R) and, obviously, it is zero at infinite distance.

The angle of scattering in the CM system is given in terms of the potential as[34]

$$\Theta(s) = -\frac{d}{dR_0} \left[\frac{1}{2\varepsilon} \int_{-\infty}^{\infty} V(R)dZ \right] \tag{1.152}$$

where ε = energy of the electron, R = distance of the electron from the target (assumed to be stationary), R_0 = distance of closest approach (Figure 1.18), s = impact parameter, and $R^2 = R_0^2 + Z^2$. For distant collisions one can use the approximation $R_0 \approx s$. Changing the integration variable, one gets

$$\Theta(s) \simeq -\frac{1}{\varepsilon} \int_0^{\infty} \left(\frac{dV}{dR} \right) \frac{R_0 dR}{R \left[1 - \left(\frac{R_0}{R} \right)^2 \right]^{0.5}} \tag{1.153}$$

Equation 1.153 is often referred to as the classical impulse approximation or the momentum approximation to the deflection function. For small-angle scattering the penetration is small

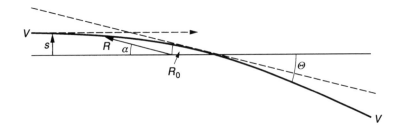

FIGURE 1.18 Reduced mass particle collision with velocity V and impact parameter s subject to a constant attractive force. The quantity Θ is the CM scattering angle, R the distance to the fixed center, α the angular position of the particle along the trajectory, and R_0 the distance of the closest approach. For small angle scattering $R_0 \simeq s$. The tangent at R_0, shown by the dashed line, indicates before and after symmetry of the collision. Adapted from Johnson, R. E., *Introduction to Atomic and Molecular Collisions*, Plenum Press, New York, 1982.

and one can substitute $R_0 \simeq s$ in Equation 1.153. To evaluate the integral in Equation 1.153 one needs to know the potential function $V(R)$. For example, the simple inverse power law for the potential is expressed as

$$V(R) = \frac{C_n}{R^n} \tag{1.154}$$

For the simplest case, the interaction between two charged particles, Coulomb potential applies with $n=1$ and $C_1 = (Q_A Q_B)/4\pi\varepsilon_0$ where Q_A and Q_B are the charges of the colliding particles. Substituting Equation 1.154 into 1.153 and carrying out the integration yields the deflection angle as[1]

$$\Theta(s) \cong a_n \left[\frac{V(R_0)}{\varepsilon} \right] \tag{1.155}$$

where

$$a_n = n \int_0^{\pi/2} \sin^n \alpha \, d\alpha = \left(\pi^{\frac{1}{2}} \right) \frac{\Gamma[(n+1)/2]}{\Gamma(n/2)} \tag{1.156}$$

Here, the symbol Γ denotes the gamma function. For $n=1$, $a_n=1$; for $n=2$, $a_n=\pi/2$ etc. The quantity $\Theta(s)\varepsilon$ in Equation 1.153 is related to the differential cross section and therefore the potential function $V(R)$ has a direct bearing in the calculation of the cross sections (see also Section 3.3.2). A discussion of potential functions may be found in McDaniel.[35]

Let D be the radius of the interacting particles and r the distance between them.

Case 1: Smooth elastic spheres. The particles are assumed to be two rigid impenetrable spheres and the potential varies according to

$$V(r) = \begin{cases} \infty & (r<D) \\ 0 & (r>D) \end{cases} \tag{1.157}$$

Case 2: Point centers of attraction or repulsion. Electrons interacting with atoms experience both attractive force (long range) due to the nucleus and repulsive force (short range) due to the orbiting electrons of the atom. The potential is defined by

$$V(r) = \begin{cases} \dfrac{-a}{r^n} & \text{(attractive)} \\ \dfrac{a}{r^n} & \text{(repulsive)} \end{cases} \tag{1.158}$$

where a is a positive constant. It is easily recognized that $n = 1$ gives the Coulomb potential. The interaction between molecules is approximated with values between $n = 9$ and $n = 15$. Maxwell suggested the value $n = 4$. The mathematical complexity of using this potential depends upon the chosen value of n.

Case 3: The square well. The potential extends up to a distance of D_0 outwards with an attractive well depth of V_0. The core with diameter D is impenetrable.

Case 4: The Lennard–Jones potential. This is a combination of repulsive and attractive force with exponents m and n for r, respectively, according to

$$V(r) = \frac{a}{r^m} - \frac{b}{r^n} \tag{1.159}$$

When a and b in Equation 1.159 are expressed as a function of D according to

$$V(r) = 4V_0 \left[\left(\frac{D}{r}\right)^{12} - \left(\frac{D}{r}\right)^6 \right] \tag{1.160}$$

this potential is known as the Lennard–Jones (6–12) potential. The exponent 12 in the first term (repulsive) is a mathematical expediency, while the second term (attractive) with an exponent of 6 is the induced dipole–induced dipole reaction. Figure 1.19 shows these potentials. An expression for the so-called Yukawa double potential, frequently used by theoreticians, is given in Equation 9.59, Section 9.3.5.4.

The attraction of a molecule to a charged ion is easily obtained by the following reasoning. As the ion approaches the molecule a dipole is induced in the latter. The polarization of the gas ($\mathbf{P}$), defined as the dipole moment per unit volume, is given by the expression.[12]

$$\mathbf{P} = \mathbf{E}\varepsilon_0(\varepsilon - 1) \tag{1.161}$$

where ε is the relative dielectric constant of the gas, usually slightly larger than unity, ε_0 the permittivity of free space, equal to 8.854×10^{-12} F/m, and $\mathbf{E}$ the electric field. The dipole moment per molecule ($\boldsymbol{\mu}$) is given by

$$\boldsymbol{\mu} = \frac{\mathbf{P}}{N} = \frac{\varepsilon_0(\varepsilon - 1)}{N} \mathbf{E} \tag{1.162}$$

where N is the number of molecules per unit volume.

In the presence of an ion having a charge q, the electric field experienced by the molecule is

$$E = \frac{q}{4\pi\varepsilon_0 r^2} \tag{1.163}$$

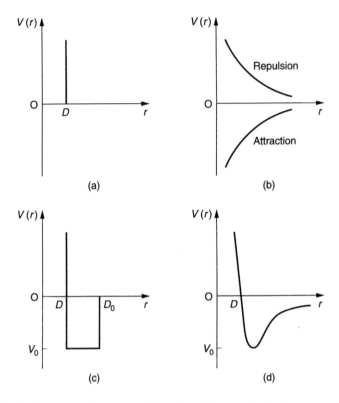

FIGURE 1.19 Spherically symmetric potential functions: (a) smooth elastic spheres; (b) point centers; (c) the square well; (d) the Lennard–Jones potential. Adapted from McDaniel, E. W., *Collision Processes in Ionized Gases*, John Wiley & Sons, New York, 1964.

where r is the distance between the ion and the molecule. Equation 1.163 is applicable as long as the distance is not very small. Substituting Equation 1.163 into 1.162, one gets

$$\mu = \frac{(\varepsilon - 1)}{4\pi N} \frac{q}{r^2} \tag{1.164}$$

The electric field along the line drawn from the ion to the dipole axis is given by

$$E = \frac{\mu \cos \theta}{2\pi \varepsilon_0 r^3} \tag{1.165}$$

where θ is the angle subtended. Since the dipole is induced by the charge q this angle is always zero, $\theta = 0$. The force between the ion and the dipole is therefore

$$F = \frac{\mu q}{2\pi \varepsilon_0 r^3} \tag{1.166}$$

Substituting Equation 1.164 into 1.166, one gets

$$F = \frac{(\varepsilon - 1)}{8\pi^2 N \varepsilon_0} \frac{q^2}{r^5} \tag{1.167}$$

The mutual potential energy between the ion and the molecule is given by

$$V(r) = -\int_r^\infty F\,dr \qquad (1.168)$$

Substituting Equation 1.167 into 1.168, one gets

$$V(r) = -\int_r^\infty \frac{(\varepsilon - 1)}{8\pi^2 N\varepsilon_0} \frac{q^2}{r^5}\, dr \qquad (1.169)$$

or

$$V(r) = -\frac{(\varepsilon - 1)}{32\pi^2 N\varepsilon_0} \frac{q^2}{r^4} \qquad (1.170)$$

The interaction potential varies according to r^{-n}, where $n = 4$. Polarization interaction is long range and is always attractive. This model of potential is usually referred to as the r^{-4} potential model with a point charge colliding with a polarizable sphere. The electron is therefore subject to two potentials, the static and the polarization. At high electron energies the static potential becomes the only significant interaction.

Classical theory of elastic scattering shows[36] that the differential scattering cross section in the CM system is given by the expression

$$Q_{\text{diff}}(\Theta) = v_0^{-4/n} \qquad (1.171)$$

where v_0 is the initial relative velocity of approach.

1.7 QUANTUM MECHANICAL APPROACH TO SCATTERING

Although we have used the results of the quantum mechanical formulation of the scattering phenomenon, it is appropriate to briefly look into the method, in its most basic form.[2] The methods were developed by Faxen and Holtsmark[37] in 1927 and by Massey, Mott, and others[38] in the early 1930s and applied to a number of gases, with increasing sophistication, during the past decades. In the classical approach to the impact phenomenon, the trajectory of each particle was determined in terms of initial velocity and impact parameter. In quantum mechanics, however, the projectile orbit is never defined exactly, and one speaks of the average values of a large number of projectiles interacting with the target.

A plane wave of constant amplitude travels along the $+Z$ direction toward a fixed scattering center situated at the origin of the coordinate system. The plane wave represents an infinitely wide beam of homogeneous particles with uniform particle density per unit area of cross section, in a direction perpendicular to its propagation. The particles are dissimilar to those forming the scattering center and each particle has the same energy, that is, the beam is mono-energetic. The scattering occurs in a spherically symmetric field, with the interaction potential energy varying only as the spatial coordinate.

The total elastic scattering cross section under these assumptions is defined as

$$Q_{\text{el}} = \int_0^\pi \int_0^{2\pi} Q_{\text{diff}}(\theta) \sin\theta\, d\theta\, d\phi = 2\pi \int_0^\pi Q_{\text{diff}}(\theta) \sin\theta\, d\theta \qquad (1.172)$$

Equation 1.172 is the same as Equation 1.66 and one wishes to calculate Q_{diff} as a function of both electron energy and angle of scattering. Integration then yields the elastic scattering cross section.

The plane wave representing the incident electrons is expressed by a wave function

$$A e^{j(k z - \omega t)}$$

in which A is the amplitude, k the wavenumber, defined as the number of waves per meter, ω the angular frequency of the wave $= 2\pi\nu$, ν is the frequency in Hz, and t the time. This function is familiar to electrical engineers. k is related to the de Broglie wavelength λ according to

$$k = \frac{2\pi}{\lambda} = \frac{M_{RM}\nu}{\hbar} \qquad (1.173)$$

where $\hbar$ is the Planck's constant and the reduced mass is defined according to

$$M_R = \frac{mM}{(m + M)} \qquad (1.174)$$

where m is the mass of the projectile (electron) and M the mass of the target particle (atom). The wave frequency and its energy are related according to

$$\omega = \frac{\varepsilon}{h} \qquad (1.175)$$

where ε is the energy in joules and h, the Planck's constant, is in joule seconds.

During elastic scattering the energy of the wave does not change and it is customary to omit the term $e^{-j\omega t}$ as it is present in every term for the wave. The time-independent wave function for the incoming plane wave is

$$\psi_{\text{inc}}(r,\theta) = A e^{jkz} \qquad (1.176)$$

Electrical engineers are familiar with the concept of an incoming wave getting reflected at a junction of changing impedance, a component of the incoming wave being transmitted.[39] The incoming wave is the sum of the reflected and transmitted waves. Applying the same concept to the electron wave, the total wave function consists of a component of the incoming plane wave and another component consisting of the scattered wave. The outgoing scattered wave is spherical and decreases in amplitude according to $1/r$. Since the number of particles scattered remains the same, the density of particles will decrease according to the inverse square of the distance. The wave function for the scattered wave has the form

$$\psi_{\text{scatt}} \approx \frac{A}{r} f(\theta) e^{jkr} \qquad (1.177)$$

where $f(\theta)$ is the scattering amplitude. It is defined by the relationship to $Q_{\text{diff}}(\theta)$ according to

$$Q_{\text{diff}}(\theta) = |f(\theta)|^2 \qquad (1.178)$$

The solution for the wave function has the asymptotic form, from Equations 1.176 and 1.177,

$$\psi = \psi_{\text{inc}} + \psi_{\text{scatt}} \approx A\left[e^{jkz} + \frac{e^{jkr}}{r} f(\theta) \right] \qquad (1.179)$$

The amplitude is usually set as unity, mainly for convenience, i.e., $A = 1$.

The time-independent wave equation given by Schroedinger is

$$\nabla^2 \psi + \frac{2M_R}{\hbar^2}[E - V(r)]\psi = 0 \tag{1.180}$$

Here, $V(r)$ is the potential energy of the electron as a function of the radial distance r, and E is the total energy. For completeness we state that the solution of Equation 1.180 should be everywhere single valued, finite, continuous, and vanish at infinity. This equation, with the restrictions mentioned, replaces the fundamental equations of motion of classical mechanics. The potential energy $V(r)$ is expressed in the same form as in classical theory because the laws of force are unchanged in both classical and quantum viewpoints. The solution of Equation 1.180 exists only for certain values of E, called *eigenvalues*, and the corresponding functions ψ, are known as the *eigenfunctions*. The eigenvalues are the discrete quantum numbers for the atomic states.

A simplified form of Equation 1.180 is

$$\nabla^2 \psi + [k^2 - U(r)]\psi = 0 \tag{1.181}$$

where we have made the substitutions

$$k = \sqrt{\frac{2M_R E}{\hbar^2}} \tag{1.182}$$

and

$$U(r) = \frac{2M_R V(r)}{\hbar^2} \tag{1.183}$$

In spherical coordinates Equation 1.181 becomes, in the symmetrical case of $\phi = 0$,

$$\frac{1}{r^2}\frac{\partial}{\partial r}\left(r^2\frac{\partial\psi}{\partial r}\right) + \frac{1}{r^2\sin\theta}\frac{\partial}{\partial\theta}\left(\sin\frac{\partial\psi}{\partial\theta}\right) + [k^2 - U(r)]\psi = 0 \tag{1.184}$$

To find the solution of Equation 1.184 one assumes that the solution depends only on r. The variables r and θ are separated to yield two equations, one of which contains only r and the other only θ. The wave equation is written as

$$\psi(r,\theta) = L(r)Y(\theta) \tag{1.185}$$

where $L(r)$ is called the radial function and $Y(\theta)$ the spherical harmonic. The former expresses the radial dependence and the latter the angular dependence of the wave function. Substituting Equation 1.185 into 1.184 and rearranging terms so that each side contains only one variable, one obtains

$$\frac{1}{L}\frac{d}{dr}\left(r^2\frac{dL}{dr}\right) + r^2[k^2 - U(r)] = -\frac{1}{Y}\left[\frac{1}{\sin\theta\,d\theta}\frac{d}{d\theta}\left(\sin\theta\frac{dy}{d\theta}\right)\right] \tag{1.186}$$

The left side of Equation 1.186 is a function of r only, and the right side of θ only. The variables have been separated. If both sides are to be equal for all values of r and θ, both must be equal to some constant, let us say $l(l+1)$. Then,

$$\frac{1}{L}\frac{d}{dr}\left(r^2\frac{dL}{dr}\right) + r^2\left[k^2 - U(r)\right] = l(l+1) \tag{1.187}$$

and

$$-\frac{1}{Y}\left[\frac{1}{\sin\theta}\frac{d}{d\theta}\left(\sin\theta\frac{dy}{d\theta}\right)\right] = l(l+1) \tag{1.188}$$

Rewriting these equations, one gets

$$\frac{1}{r^2}\frac{d}{dr}\left(r^2\frac{dL}{dr}\right) + \left[k^2 - U(r) - \frac{l(l+1)}{r^2}\right]L = 0 \tag{1.189}$$

and

$$\frac{1}{\sin\theta}\frac{d}{d\theta}\left(\sin\theta\frac{dY}{d\theta}\right) + l(l+1)Y = 0 \tag{1.190}$$

Equation 1.190 is known as Legendre's equation. It is a second-order equation and has two linearly independent solutions, each of which may be expressed as a power series in $\cos\theta$.[40] Both solutions become infinite (i.e., not acceptable) for $\theta = 0$ unless $l = 0$ or a positive integer. The solution of Equation 1.185 is written as[41]

$$\psi(r,\theta) = \sum_{l=0}^{\infty} A_l P_l(\cos\theta)L_l(r) \tag{1.191}$$

where the A_l are arbitrary constants and L_l are the solutions of Equation 1.189. The functions $P_l\cos\theta$ are known as Legendre polynomials, which are the series solutions of Legendre's equation (1.190). The solution of Legendre's equation is usually expressed as[42]

$$Y = a_0 F(\cos\theta) + a_1 G(\cos\theta) \tag{1.192}$$

where the function F contains only the even powers of $\cos\theta$ and G only the odd powers. Further, a_0 and a_1 are arbitrary constants. For simplicity's sake we can write P_l for both F and G remembering that when $l = 0, 2, 4, \ldots$ the associated constant is a_0; when $l = 1, 3, 5, \ldots$ the associated constant is a_1. The first few terms are given by the following expressions:

$$\left.\begin{array}{ll} l = 0; & P_l(\cos\theta) = P_0(\cos\theta) = 1 \\[2mm] l = 1; & P_l(\cos\theta) = P_1\cos(\theta) = \cos\theta \\[2mm] l = 2; & P_l(\cos\theta) = P_2(\cos\theta) = \frac{1}{2}\left(3\cos^2\theta - 1\right) \\[2mm] l = 3; & P_l(\cos\theta) = P_3(\cos\theta) = \frac{1}{2}\left(5\cos^3\theta - 3\cos\theta\right) \\[2mm] l = 4; & P_l(\cos\theta) = P_4(\cos\theta) = \frac{1}{8}(35\cos^4\theta - 30\cos^2\theta + 3) \end{array}\right\} \tag{1.193}$$

It is noted here that the constants a_0 and a_1 are customarily chosen, for standardizing purposes, so that each solution has a magnitude of 1 at $\cos\theta = 1$ or at $\theta = 0$. When $l = 0$, choose $a_0 = 1$; when $l = 1$, choose $a_1 = 1$; when $l = 2$, choose $a_0 = -1/2$; when $l = 3$, choose $a_1 = -3/2$; and so on.

We now determine $L_l(r)$ by expressing the function as

$$L_l(r) = \frac{G_l}{r} \tag{1.194}$$

This is in fact the standard way of expressing the solution of Equation 1.189. $G_l(r)$ is the solution of the equation

$$\frac{d^2 G_l(r)}{dr^2} + \left[k^2 - U(r) - \frac{l(l+1)}{r^2} \right] G_l(r) = 0 \tag{1.195}$$

At large values of r the potential energy $U(r)$ tends to zero, as does also the term containing $1/r^2$. One expects the asymptotic form of the solution to be

$$G_l(r) = \sin(kr + C_l) \tag{1.196}$$

where C_1 is a constant. The proof for this form of solution may be found in McDaniel.[29] The solution of Equation 1.189, which is finite at the origin, will then have the asymptotic form

$$L_l(r) \approx \frac{1}{kr} \sin\left(kr - \frac{l\pi}{2} + \eta_l \right) \tag{1.197}$$

where η_l is a constant for a given k (electron energy) and $U(r)$ (given atom). η_l is called the phase shift of the lth partial wave due to the action of the scattering potential. If $U(r) = 0$, then η_l should be zero; that is, no phase shift if there is no scattering. The term $(-l\pi/2)$ is introduced to account for this. Substituting Equation 1.197 into 1.191, one gets the expression for the total wave as

$$\psi(r,\theta) = \sum_{l=0}^{\infty} \frac{1}{kr} A_l P_l(\cos\theta) \sin\left(kr - \frac{l\pi}{2} + \eta_l \right) \tag{1.198}$$

It now remains to determine A_l so that the solution in the form 1.191 may be completed. The incident wave is (see Equation 1.176)

$$\psi_{\text{inc}} = e^{jkz} \tag{1.199}$$

where the amplitude A has been equated to 1 for the sake of convenience. The incident wave is expanded in partial waves:[43]

$$\psi_{\text{inc}} = e^{jkz} = e^{jkr\cos\theta} = \sum_{l=0}^{\infty} (2l+1))j^l P_l(\cos\theta) J_l(kr) \tag{1.200}$$

where $J_l(kr)$ is known as the spherical Bessel function. It is defined in terms of the ordinary Bessel function of half-integral order by the equation

$$J_l(kr) \equiv \left(\frac{\pi}{2kr} \right)^{1/2} J_{l+1/2}(kr) \tag{1.201}$$

The values of the first few spherical Bessel functions are[44]

$$J_0 = \frac{\sin k r}{k r} \tag{1.202}$$

$$J_1 = \frac{\sin(k r)}{(k r^2)} - \frac{\cos(k r)}{k r} \tag{1.203}$$

$$J_2 = \left[\frac{3}{(k r)^3} - \frac{1}{k r} \right] \sin(k r) - \frac{3}{(k r^2)} \cos(k r) \tag{1.204}$$

Figure 1.20 shows the first few spherical Bessel functions, which constitute the radial part of the expansion of the incident wave according to expression 1.200. It is seen that the first and the largest maximum of $J_l (kr)$ occurs near $kr \equiv r/\lambda \approx 1.5l$. The physical significance is that most of the scattered electrons whose angular momentum quantum number is l will be found somewhere within the shell having radii $l\lambda$ and $(l+1)\lambda$.

The spherical Bessel function has asymptotic values at both large and small values of r. For large values the asymptotic value is given by

$$J_l(k r) \approx \frac{1}{k r} \sin\left(k r - \frac{l\pi}{2} \right) \tag{1.205}$$

The asymptotic form of the incident wave for large values of r then becomes, by substitution of Equation 1.205 in 1.200,

$$\psi_{\text{inc}} = e^{jkz} \approx \sum_{l=0}^{\infty} (2l + 1))j^l P_l(\cos\theta) \left[\frac{1}{k r} \sin\left(k r - \frac{l\pi}{2} \right) \right] \tag{1.206}$$

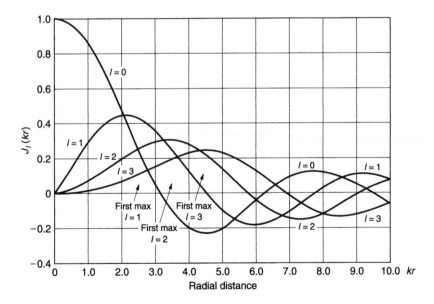

FIGURE 1.20 First four terms of the spherical expansion of the incident wave for various values of angular momentum of the electron l. $l=0$ is called the s-wave and is scattered with spherical symmetry about the scattering center; $l=1$ is called the p-wave, $l=2$ the d-wave, $l=3$ the f-wave, and so on, as in spectroscopy.

Subtracting Equation 1.206 from the asymptotic form of the total wave gives the scattered wave, ψ_{scatt}, according to

$$\psi_{\text{scatt}} = \psi(r,\theta) - \psi_{\text{inc}} \tag{1.207}$$

One then gets

$$\psi_{\text{scatt}} \approx \sum_{l=0}^{\infty} \frac{1}{kr} P_l(\cos\theta) \left\{ \left[A_l \sin\left(kr - \frac{l\pi}{2} + \eta_l\right) - (2l+1)j^l \sin\left(kr - \frac{l\pi}{2}\right) \right] \right\} \tag{1.208}$$

The scattered wave is outward bound. The terms in curly brackets in expression 1.208 are often expressed in exponential form, using the trigonometric identity

$$\sin(kr) = \frac{e^{jkr} - e^{-jkr}}{2j} \tag{1.209}$$

One then gets the two terms as, after rearrangement,

$$A_l \left[\frac{e^{j(kr - \frac{l\pi}{2} + \eta_l)} - e^{-j(kr - \frac{l\pi}{2} + \eta_l)}}{2j} \right] - (2l+1)j^l \left[\frac{e^{(kr - \frac{l\pi}{2})} - e^{-(kr - \frac{l\pi}{2})}}{2j} \right] \tag{1.210}$$

A further rearrangement of Equation 1.210 gives, in a straightforward manner,

$$\frac{1}{2j} \left\{ e^{j(kr - \frac{l\pi}{2})} \left[A_l e^{j\eta_l} - j^l(2l+1) \right] - e^{-j(kr - \frac{l\pi}{2})} \left[A_l e^{-j\eta_l} - j^l(2l+1) \right] \right\} \tag{1.211}$$

At this juncture we have to resort to the physical interpretation that terms with the negative exponentials (of the type e^{-x}) are incoming spherical waves, as opposed to terms with positive exponentials (of the type e^x) which are outward-bound waves. Since the scattered wave, does not have a component of the incident wave, the second term in expression 1.211 is zero, i.e.,

$$A_l e^{-j\eta_l} - j^l(2l+1) = 0 \tag{1.212}$$

We have thus determined the value of the constant A_l as

$$A_l = (2l+1)j^l e^{j\eta_l} \tag{1.213}$$

Substituting Equation 1.213 into 1.191, the total wave function is obtained as

$$\psi(r,\theta) = \sum_{l=0}^{\infty} (2l+1)j^l e^{j\eta_l} L_l(r) P_l(\cos\theta) \tag{1.214}$$

One can also derive an expression for the scattering cross section using Equations 1.177 and 1.214. The scattering cross section is then seen to be

$$\frac{e^{jkr}}{r} f(\theta) = \sum_{l=0}^{\infty} \frac{1}{kr} P_l(\cos\theta) \left[A_l \sin\left(kr - \frac{l\pi}{2} + \eta_l\right) - (2l+1)j^l \sin\left(kr - \frac{l\pi}{2}\right) \right] \tag{1.215}$$

This will be true only if the right side has the form

$$f(\theta) = \frac{1}{2jk} \sum_{l=0}^{\infty} (2l+1)\left(e^{2j\eta_l} - 1\right) P_l(\cos\theta) \tag{1.216}$$

Equation 1.216 shows that $f(\theta)$ is complex, having the form

$$f(\theta) = P + jQ \tag{1.217}$$

where

$$P = \frac{1}{2k} \sum_{l=0}^{\infty} (2l+1) \sin 2\eta_l P_l(\cos\theta) \tag{1.218}$$

$$Q = \frac{1}{2k} \sum_{l=0}^{\infty} (2l+1)(1 - \cos 2\eta_l) P_l(\cos\theta) \tag{1.219}$$

Note that Equations 1.218 and 1.219 are obtained from Equation 1.216 using the trigonometric identities

$$e^{jx} = \cos x + j\sin x \tag{1.220}$$

$$\cos 2x = \cos^2 x - \sin^2 x \tag{1.221}$$

The scattered cross section is defined as

$$Q_{\text{scatt}} = |f(\theta)|^2 \tag{1.222}$$

and one gets

$$Q_{\text{scatt}} = |f(\theta)|^2 = P^2 + Q^2 \tag{1.223}$$

Substituting Equations 1.218 and 1.219 into 1.223, one gets the scattering cross section as

$$Q_{\text{scatt}} = \frac{1}{k^2} \left| \sum_{l=0}^{\infty} (2l+1) e^{j\eta_l} \sin \eta_l P_l \cos\theta \right|^2 \tag{1.224}$$

The total elastic scattering cross section is given by integrating Equation 1.224 over the complete solid angle according to

$$Q_{\text{el}} = \int_0^{\pi} Q_{\text{scatt}} 2\pi \sin\theta \, d\theta \tag{1.225}$$

To carry out this integration we use the standard integration[45]

$$\int_0^{\pi} P_m(\cos\theta) P_n(\cos\theta) \sin\theta \, d\theta = \frac{2}{2m+1} \delta_{mn} \tag{1.226}$$

where δ_{mn} is the Kronecker's delta function, which equals unity if $m=n$, and zero if $m \neq n$. The result of integration is

$$Q_{el} = \frac{4\pi}{k^2} \sum_{l=0}^{\infty} (2l+1) \sin^2 \eta_l \qquad (1.227)$$

An important implication to note here is that the total scattering cross section given by Equation 1.227 is a function of the scattered amplitude in the forward direction $(\theta = 0)$, denoted by $f(0)$. To express Q_{el} in terms of $f(0)$ we first note that $P_l (\cos 0) = 1$[46] for all values of l,

$$f(0) = \frac{1}{2ik} \sum_{l=0}^{\infty} (2l+1)(e^{2j\eta_l} - 1) \qquad (1.228)$$

The complex conjugate of $f(0)$ is denoted by $f(0)^*$, and accordingly

$$f^*(0) = \frac{1}{2ik} \sum_{l=0}^{\infty} (2l+1)(e^{-2j\eta_l} - 1)$$

Therefore

$$f(0) - f^*(0) = \frac{1}{2jk} \sum_{l=0}^{\infty} (2l+1)(e^{2j\eta_l} - 1) - \frac{1}{2jk} \sum_{l=0}^{\infty} (2l+1)(e^{-2j\eta_l} - 1) \qquad (1.229)$$

Simplification of this expression yields[47]

$$f(0) - f^*(0) = -\frac{2}{jk} \sum_{l=0}^{\infty} (2l+1) \sin^2 \eta_l \qquad (1.230)$$

Substituting Equation 1.227 into 1.230, one easily obtains

$$f(0) - f^*(0) = \frac{jkQ_{el}}{2\pi} \qquad (1.231)$$

or

$$Q_{el} = \frac{2\pi}{jk}[f(0) - f^*(0)] = \frac{4\pi}{k} \mathscr{I}[f(0)] \qquad (1.232)$$

where $\mathscr{I}$ denotes the imaginary component. This result is known as the optical theorem.

The method of partial waves was first developed by Lord Raleigh in the treatment of reflection of sound waves by spherical particles.[48] The method was first applied in 1927 to the scattering of electrons by rare gas atoms by Faxén and Holtsmark.[49]

The phenomenon of Ramsauer–Townsend minimum in the scattering cross section at low electron impact energies is explained qualitatively by McDaniel.[29] The rare gas atoms have a complete outer shell and therefore the force they exert on an incoming electron is short range; it is much stronger at small radial distances. The situation may be approximated to a narrow, deep potential well. The radius of the well is short compared with the wavelength of the electron and only the s-wave ($l = 0$) will make any contribution to the scattering cross section. For certain impact energies the s-wave of the incident electron will be scattered with a phase shift of multiple integrals of π radians. This will result in zero contribution of the s-wave to the scattering cross section, according to Equation 1.227. Because of the low contribution of higher order waves to the total scattering cross section to the narrow potential well previously mentioned, the sum of all contributions to the scattering section will be very small.

The potential wells become narrower and deeper with increasing atomic number of the rare gas atom, which qualitatively explains the most pronounced Ramsauer–Townsend effect in xenon (see Figure 3.54). As we move toward lighter atoms, krypton and argon, the effect becomes less pronounced. Neon and helium do not show the Ramsauer–Townsend effect. Certain atoms and molecules do show the effect.

Holtsmark[50] first applied the partial wave technique to argon, obtaining very good agreement. Further literature may be found in Massey and Burhop[51] and Burke and Smith.[52]

REFERENCES

1. Johnson, R. E., *Introduction to Atomic and Molecular Collisions*, Plenum Press, New York, 1982.
2. (a) Hasted, J. B., *Physics of Atomic Collisions*, 2nd ed., Am. Elsevier Pub. Co., New York, 1972; (b) McDaniel, E. W. *Collision Phenomena in Ionized Gases*, John Wiley & Sons, New York, 1964.
3. Delcroix, J. L., *Introduction to the Theory of the Ionized Gases*, InterScience Publishers, New York, 1960.
4. Roth, J R., *Industrial Plasma Engineering*, Institute of Physics, Bristol, U.K., 1995.
5. Espel, P., A. Gibert, P. Domens, J. Paillol, and G. Riquel, *J. Phys. D.: Appl. Phys.*, 34, 593, 2001.
6. Cubric, D., D. J. L. Mercer, J. M. Channing, G. C. King, and F. H. Read, *J. Phys. B.: At. Mol. Opt. Phys.* 32, L-45, 1999.
7. Trajmar, S., D. F. Register, D. C. Cartwright, and G. Csanzk, *J. Phys. B.: At. Mol. Opt. Phys.*, 25, 4889, 1992.
8. Asmis K. R., and M. Allan, *J. Phys. B: At. Mol. Opt. Phys.*, 30, 1961, 1997.
9. Fursa D.V., and I. Bray, *J. Phys. B: At. Mol. Opt. Phys.*, 30, 757, 1997.
10. Bartchat, K., E. T. Hudson, M. P. Scott, P. G. Burke, and V. M. Burke, *J. Phys. B: At. Mol. Opt. Phys.*, 29, 2875, 1996.
11. Kitajima, M., S. Watanabe, H. Tanaka, M. Takekawa, M. Kimura, and Y. Itikawa, *J. Phys. B: At. Mol. Opt. Phys.*, 34, 1929, 2001.
12. Raju, Gorur G., *Dielectrics in Electric Fields*, Marcel Dekker, New York, 2003.
13. Antoni, Th., K. Jung, H. Erhardt, and E. S. Chang, *J. Phys. B: At. Mol. Phys.*, 19, 1377, 1986.
14. Register, D.F., H. Nishinura, and S. Trajmar, *J. Phys. B: At. Mol. Phys.*, 13 1651, 1980.
15. Takekawa M., and Y. Itikawa, *J. Phys. B: At. Mol. Opt. Phys.*, 31, 3245, 1998.
16. Takekawa, M., and Y. Itikawa, *J. Phys. B: At. Mol. Opt. Phys.*, 32, 4209, 1999.
17. Hirsch M.N., and H. J. Oskam (Ed.), *Gaseous Electronics*, vol. 1, Academic Press, New York, 1978.
18. Nasser, E., *Fundamentals of Gaseous Ionization and Plasma Electronics*, John Wiley & Sons, New York, 1971, p. 112.
19. Frost L.S., and A. V. Phelps, *Phys. Rev.*, 127, 1621, 1962.
20. Liu, J., Ph.D. thesis, University of Windsor, 1993.
21. Engelhardt, A.G., and A. V. Phelps, *Phys. Rev.*, 131, 2115, 1963.
22. Chanin, L.M., A. V. Phelps, and M. A. Biondi, *Phys. Rev. Lett.*, 2, 344, 1959.
23. Chutjian, A., *Phys. Rev. Lett.*, 46, 1511, 1981.

24. Abramowitz, M., and I. A. Stegun, *Handbook of Mathematical Functions*, Dover, New York, 1970.
25. Govinda Raju, G.R. and R. Hackam, *J. Appl. Phys.*, 53, 5557, 1982.
26. Govinda Raju, G.R. and R. Hackam, *J. Appl. Phys.*, 52, 3912, 1981.
27. Govinda Raju, G.R. and R. Hackam, *Proc. IEEE*, 69, 850, 1981.
28. Kuffel, E. and W. S. Zaengel, *High Voltage Engineering*, Pergamon Press, Oxford, 1974.
29. McDaniel, E.W. *Collision Phenomena in Ionized Gases*, John Wiley & Sons, New York, 1964, chap. 9.
30. McDaniel, E.W., *Collision Phenomena in Ionized Gases*, John Wiley & Sons, New York, 1964, p. 432.
31. Ellis, H.W., R. Y. Pai, E. W. McDaniel, E. A. Mason, and L. A. Viehland, *Atomic Data and Nuclear Data Tables*, 17, 177, 1976.
32. Wannier, G.H., *Phys. Rev.* 83, 281, 1941.
33. Wannier, G.H., *Phys. Rev.* 87, 795, 1952.
34. Johnson, R.E., *Introduction to Atomic and Molecular Collisions*, Plenum Press, New York, 1982. pp. 51 and 260.
35. McDaniel, E.W., *Collision Phenomena in Ionized Gases*, John Wiley & Sons, New York, 1964, pp. 25–30.
36. Yang, K. and T. Ree, *J. Chem. Phys.*, 35, 588, 1961.
37. Faxén, H. and J. Holtzmark, *Z. Physik*, 45, 307, 1927.
38. Massey, H.S.W. and C. B. O. Mohr, *Proc. Roy. Soc. (London)*, A 141, 434, 1933. Mott, N.F. and H. S. W. Massey, *The Theory of Atomic Scatterings*, 2nd ed., Oxford University Press, Oxford, 1952.
39. Greenwood, A., *Electrical Transients in Power Systems*, John Wiley & Sons, New York, 1991.
40. O'Neil, P.V., *Advanced Engineering Mathematics*, 2nd ed., Wadsworth Pub. Co., Belmont, CA, 1986, p. 328.
41. O'Neil, P.V., *Advanced Engineering Mathematics*, 2nd ed., Wadsworth Pub. Co., Belmont, CA, 1986, p. 332. The last term of the solution is expressed here as $(1/r) \, (G(r)/r)^n$ and the solution becomes the same as Equation 1.191 because of Equation 1.194 in the text.
42. O'Neil, P.V., *Advanced Engineering Mathematics*, 2nd ed., Wadsworth Pub. Co., Belmont, CA, 1986, p. 282. The method is to substitute $x = \cos\theta$ in Legendre's equation (1.190) and express the solution as

$$y = \sum_{n=0}^{\infty} a_n x^n$$

Differentiating twice and substituting in the original equation leads to a series solution, which is regrouped as shown in Equation 1.192.

43. Mott, N.F. and H. S. W. Massey, *Theory of Atomic Scatterings*, University of Oxford Press, London, 1952, pp. 20–22.
44. Morse, P.M., *Vibration and Sound*, McGraw-Hill Co., New York, 1936, pp. 244–245.
45. Gradshteyn, I.S. and I. M. Ryzhik, *Table of Integrals, Series and Products*, 4th ed., Academic Press, New York, 1965: Formula 1 of article 7.221. The standard form is

$$\left. \begin{aligned} \int_{-1}^{1} P_n(x) P_m(x)\, dx &= 0 \qquad [m \neq n] \\ &= \frac{2}{2n+1} \qquad [m = n] \end{aligned} \right\}$$

Substituting $x = \cos\theta$ and changing the limits appropriately, Equation 1.227 is obtained. The term $e^{2j\eta_l}$ vanishes because its magnitude is unity.

46. This is easily seen by substituting $\theta = 0$ or $\cos\theta = 1$ in each equation of expressions 1.193.
47. The trigonometric identity used here is

$$\left(e^{2jx} - e^{-2jx}\right) = -4\sin^2 x$$

48. Lord Raleigh, *The Theory of Sound Waves*, Dover Publications, New York, 1945.

49. Faxén, H. and J. Holtsmark, *Z. Phys.*, 45, 307, (1927).

50. Holtsmark, J., *Z. Phys.*, 55, 437, (1929).

51. Massey, H.S.W. and E. H. S. Burhop, *Electronic and Ionic Impact Phenomena*, Oxford University Press, Oxford, 1952.

52. Burke, P.G. and K. Smith, *Rev. Mod. Phys.*, 34, 458, (1962).

2 Experimental Methods

A knowledge of experimental methods is indispensable to understand, evaluate, and choose data for further study. This chapter covers the experimental methods employed for the measurement of collision cross sections. The methods for measuring other quantities such as swarm coefficients, drift velocities, plasma diagnostics, etc. will be treated later on in the appropriate chapters.

A number of ingenious techniques have been developed since the early years of collision cross section measurement (1920). The parameters employed in the experiments are controlled and measured with increasing sophistication. The techniques chosen for brief description, by no means exhaustive, have yielded a wealth of data and these methods are linked to the data provided and discussed in subsequent chapters. For each quantity measured, the principle of the method is explained first and is then followed by a brief description of more recent advances and a typical set of results, to stress the quantities measured and parameters controlled. The bias is towards data in the range of parameters that are relevant to discharge phenomena, rather than towards physics of collisions or different experimental techniques.

Collision cross sections are the fundamental data that are required to interpret theoretical results, calculate electron energy distribution functions, simulate discharges by the Monte Carlo technique, and so on. We will not attempt an exhaustive treatment of experimental methods, but will restrict ourselves to those techniques that are relevant to our purposes. Earlier experimental techniques and results are extensively discussed in books by McDaniel,[1] Hasted,[2] Gilardini,[3] and Huxley and Crompton.[4] A review of techniques and results obtained up to the year 1971 is also given by Bederson and Kieffer.[5] A few general comments are, however, in order.

2.1 TOTAL COLLISION CROSS SECTIONS

The total collision cross section is the sum of elastic and all inelastic cross sections. The first cross sections to be measured were of this type because of the relative simplicity of the concept. The methodology of these experiments belonged to the category known as the transmission method.

Total collision cross sections are measured using a number of different techniques that have been developed to improve accuracy and reduce discrepancies between the results obtained by different methods and theoretical analyses. The fundamental principle involved in these experiments is to measure the number of electrons that survive scattering or that get scattered as a result of collisions. Though some authors distinguish between these two techniques, the principle involved is not substantially different. A beam of electrons is generated by a suitable means such as photoelectric emission,[6] thermionic emission,[7,8] field emission,[2] and electron guns.[9] The electrons have low energy when emitted and are then accelerated through a set of grids to allow them to acquire the desired energy. Divergence of the beam is minimized by a set of electron lenses or the application of a magnetic field.

The electrons enter a collision chamber and are scattered as they pass through the chamber. The number of electrons scattered or the number that survive scattering is measured and related to the number entering the collision chamber to obtain the total scattering cross section. Sources of error in this type of experiment are:

1. The electron beam is not monoenergetic in the true sense and the energy spread may be of the order of 100 meV. This spread introduces greater errors at low electron energies.
2. Not all scattered electrons are collected by the collector.
3. Secondary electrons are not confined.
4. Since the experiment measures relative electron numbers or currents, obtaining the absolute cross section involves uncertainties of gas pressure and gas chosen as reference standard.

The electrons should be accelerated in vacuum in the grid region to attain their final energy, because the presence of gas molecules will result in collisions and the spread of energy of the beam will become so large that accurate measurements are not possible. Collisions also destroy the monoenergetic nature, making it impossible to assign a cross section to a definite energy. However, the collision region should necessarily have gas molecules and therefore operate at higher pressure. These conflicting requirements of vacuum in the grid region and higher pressure in the interaction region demand a differential pumping arrangement; Ramsauer and Kollath[6] and Brode[7] did not adopt such an arrangement. This is usually achieved by a baffle with a pinhole, each side being connected to a separate pumping system. The higher resistance of the pinhole to pumping facilitates the maintenance of a pressure differential.

In an attempt to reduce the errors, ingenious methods are developed and the different techniques are classified as:

1. Ramsauer technique
2. Modified Ramsauer technique
3. Linear transmission method
4. Crossed beams method
5. Time-of-flight methods

These methods are explained below with a brief discussion of their salient features.

2.1.1 RAMSAUER TECHNIQUE

The earliest measurements of the total cross sections are due to Ramsauer[10] and Brode,[11] who used the direct beam method in the range of 2 to 360 eV. Brode's experimental arrangement was similar to that adopted by Ramsauer, with minor modifications; Ramsauer used photoelectric emission for the initial electrons whereas Brode adopted thermionic emission. A schematic diagram is shown in Figure 2.1. Electrons are liberated from a hot thermionic filament and are bent into a circle by the application of a magnetic field perpendicular to the plane of the paper. A series of slits is arranged in a circular path and the electrons that reach the box B reach the ground terminal through a galvanometer. The current flowing through B and the filament current are measured, and the number of electrons that survive scattering is calculated from the absorption equation

$$I = I_0 \exp(-Q_T N L) \tag{2.1}$$

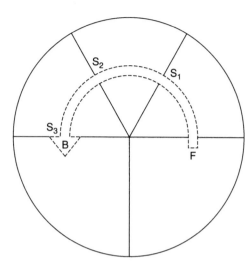

FIGURE 2.1 Schematic of Brode's experimental arrangement for measurement of total collision cross section. F is a thermionic filament that emits electrons which are bent by a magnetic field. They pass through slits S_1 to S_3 and arrive at the collector B. The path length of the electrons is 3.2 cm.

where I_0 is the current in the absence of scattering, Q_T is the total cross section, N the number of molecules per m³ in the chamber, and L the path length of the electrons. This equation is known as the Beer–Lambert relation. The difficulty of the transmission method is to determine I_0 without ambiguity so that Q_T may be determined accurately. Brode[11] measured the current at two different pressures, obtaining two current ratios, eliminating I_0, and then solving for Q_T.

Brode's setup[11] had a path length of 0.032 m and the experiments were carried out at a gas number density of $\sim 4 \times 10^{20}$ m⁻³ (~ 1 Pa). A ratio of $I/I_0 = 0.1$ yields a cross section of 2×10^{-19} m² per atom. Brode presented his results in terms of the quantity $N_0 Q$, which is the number of collisions per meter length of the path at $N_0 = 3.54 \times 10^{22}$ m⁻³; $N_0 Q = 7000$ m⁻¹ in this example. Early measurements showed that a minimum in the cross section, known as the Ramsauer–Townsend minimum, appears in some gases at approximately 1 eV energy. More recent results place the minimum at approximately 0.25 to 0.75 eV, depending on the gas. It is appropriate to remark here that the experimentally observed minimum is not satisfactorily explained by classical theory which predicts a monotonic decrease of the cross section with increasing energy. Quantum mechanical calculations in the early 1930s provided a reasonably satisfactory explanation, as is explained in Chapter 1.

Application of a magnetic field served the purpose of confining the secondary electrons and yielded a better resolution of the electron energy. The experiment of Brode[11] belongs to the category of absorption measurement; that is, the number of electrons surviving scattering is measured. Both elastic and inelastic scattering cross sections are obtained: the elastically scattered electrons do not reach the collector because they are deflected; the inelastically scattered electrons also do not reach the collector because their path curvature is reduced.[2]

The Ramsauer technique for measuring the total cross section was repeated by Golden and Bandel[8,12] using improved technology such as a bakable vacuum system to remove adsorbed gases, mercury-free pressure measurement, and other measures. Figure 2.2 shows schematically the improved version of Ramsauer's apparatus due to Golden and Bandel.[8] The chamber is of all-metal construction and employs ultrahigh vacuum techniques. The electrons are emitted by an indirectly heated cathode and pass through a grid, the potential of which is varied depending upon the desired beam conditions. The whole apparatus is

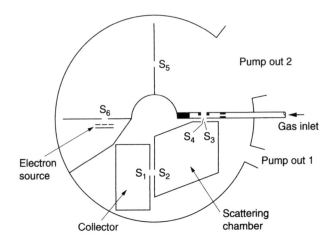

FIGURE 2.2 Schematic arrangement of apparatus used by Golden and Bandel[8] for measurement of scattering cross section in the 0.1–20 eV range. The pressure in the cathode region is lower than the pressure in the scattering chamber. Measurements were carried out in helium[8] and argon.[12]

placed in a magnetic field perpendicular to the plane of the page, similar to Ramsauer's arrangement. The electrons are selected for momentum by the three slits S_6, S_5, S_4 and enter a collision chamber, passing then to the collector. An improved technique is employed by having two regions in the chamber that could retain a differential gas density. While the filament region is pumped out, the collision region is maintained at the required pressure. For this purpose a pressure-dropping baffle with a small orifice is employed. The entire apparatus is free of mercury vapor, which was one of the major sources of contamination in the early work. Golden and Bandel[8] report a Ramsauer–Townsend minimum energy of 0.28 eV and a cross section of $1.5 \times 10^{-21}\,\mathrm{m}^2$ in argon, with a very sharp rise to the left of the minimum.

The review article of Brode[7] provides a collection of early data in a number of gases. Books by McDaniel[13] and Hasted[2] may be referred to for the results obtained up to the year 1972.

2.1.2 MODIFIED RAMSAUER'S TECHNIQUE

Dalba et al.[14] introduced an improvement to the Ramsauer technique in the collector part of the total cross section measurement setup. The interaction chamber is split into two electrically connected parts: the first chamber limits the gas region and the second chamber is a pumped region in which the pressure is held constant at a lower value than that of the gas region. If this pressure is held constant, there is no need to correct the path length of the electrons in the gas region for the presence of the adjoining region.

The idea behind splitting the interaction chamber into two parts by a diverter valve is explained in Figure 2.3. The traditional interaction chamber and the collector are shown in (a). Electrons in trajectory A undergo collision, with narrow angle scattering, but the scattering angle is such that the electron enters the collection chamber. The effect is that too small a number of scattering events is counted, resulting in a lower total cross section. Similarly, electrons along trajectory B undergo collision, closer to the exit aperture of the interaction chamber. After the scattering event they too enter the collector, again resulting in too few events being counted. The improvement effected is shown in (b). Electrons along both trajectories end up in the lower pressure segment of the now divided interaction chamber. The collector current determines the scattering cross section more accurately.

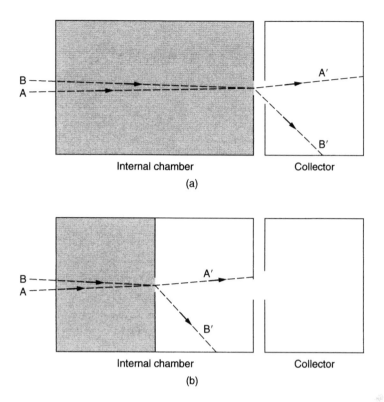

FIGURE 2.3 Improved Ramsauer technique for increasing the accuracy of measurement of total scattering cross section. (a) Standard setup with interaction chamber connected to the collector. Electrons along trajectories A and B are scattered, though counted as nonscattered electrons. (b) The interaction region is divided into two sections by a diverter valve. The hatched segment is at a higher pressure than the following segment. Electrons along trajectories A and B are scattered but they do not reach the collector, thus increasing the accuracy. Figure reproduced from Dalba, G. et al., *J. Phys. B*, 12, 3787, 1979. With permission of the Institute of Physics, U.K.

The latter method improves the angular resolution of the Ramsauer apparatus to $\sim 0.7°$ and the angular acceptance to 5×10^{-4} sr. Dalba et al.[14] used this apparatus to measure absolute total cross sections in helium over the electron energy range of 100 to 1400 eV, and subsequently extended their measurements to a number of gases. Since the parameters used in this study are typical of those employed in the Ramsauer technique, a brief summary of experimental conditions and accuracies is provided below.

1. *Magnetic field.* The bending magnetic field is provided by a pair of ironless Helmholtz coils. The mean radius of the coils is 688 mm and the mean distance between the coils is 635 mm. The nonuniformity is less than one part in 10^4.
2. *Pumping system.* A differential pumping system capable of attaining a vacuum better than ~ 1 Pa is employed. The difference of pressure between the beam segment and the interaction chamber is at least 1000. The pressures are measured, using a capacitance meter, to a stated accuracy of $\pm 1\%$.
3. *Current measurements.* Currents are in the range of 10^{-7} A, which is quite large in the context of electron molecule collision studies. The stated accuracy of current measurement is 0.3 to 0.6%.
4. *Energy measurement.* The entire apparatus from the anode to the collector is equipotential. The stated accuracy of energy measurement is $\pm 0.2\%$.

TABLE 2.1
Typical Accuracies of the Modified Ramsauer Technique

Parameter	Accuracy
Current measurements	±0.6%
Pressure measurements	
—Instrument calibration	±1.0%
—Thermal transpiration	±0.2%
Temperature measurement	±0.3%
Interaction length determination	
—Geometrical measurement	±0.2%
—Trajectories spread	±0.4%
—End effects	±1.5%
Energy measurements	±0.2%
Overall error (quadratic sum)	±2%

5. *Interaction length determination.* The mean geometrical length of the interaction segment, the hatched region of Figure 2.3(b), is 147.7 ± 0.3 mm, resulting in a stated uncertainty of ±0.2% in the cross section.
6. The overall stated accuracy for the absolute total cross section is ±2%. The stated accuracies of the modified Ramsauer technique adopted by Dalba et al.[14] are shown in Table 2.1.

2.1.3 LINEAR TRANSMISSION METHOD

The linear transmission experiments are of the attenuation type, in which electrons of known energy are transmitted through the scattering chamber and then detected by a Faraday cup.

Nickel et al.[15] describe a linear transmission technique for measuring total scattering cross sections in rare gases. Linear transmission devices are considered to be important because the accuracy of the results obtained depends on the geometry of the experimental setup and there is no need for an indirect normalization procedure. The technique, shown in Figure 2.4, does not employ a magnetic field, which is a distinct advantage in simplifying the number of parameters to control and determining the exact path length.

The setup is composed of four sections that are demountable and bakable. The first section has a cylindrical electron gun capable of generating a beam having an energy that is variable in the range of 4 to 300 eV. The gun chamber is differentially pumped and connected to the second section through an orifice. The purpose of this section is to remove the molecules that stream from the scattering chamber by a second pumping arrangement, the scattering chamber being at a higher pressure than the gun section. This section also has a Faraday cup to monitor the stability of the beam and orthogonal electrostatic deflectors to control the beam. The third section is the scattering chamber and is provided with a leak (gas admittance) valve and a capacitance manometer to measure the gas pressure without contaminating the gas.

The last section of the setup carries a Faraday cup, which is provided with a grid to remove the electrons inelastically scattered in the forward direction. The current flowing through this Faraday cup is measured and the scattering cross section obtained by Equation 2.1.

Studies of Jost et al.,[16] Kauppila et al.,[17] and Wagenaar and de Heer[18] also fall into the category of transmission method.

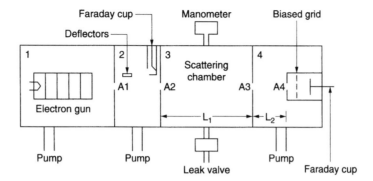

FIGURE 2.4 Linear transmission technique adopted by Nickel et al.[15] for the measurement of total scattering cross section in rare gases. The electron gun provides the beam and the beam enters the scattering chamber. The unscattered electrons are collected by the Faraday cup in the last section of the apparatus. Figure reproduced with permission of the Institute of Physics, U.K.

2.1.4 TIME-OF-FLIGHT METHOD

As stated earlier, the Ramsauer method and its variants are of the transmission type and the need to determine the current without scattering has been a source of uncertainty in cross section measurements. A method to determine I_0 in Equation 2.1 is to measure the electron current in the absence of scattering, that is, in vacuum. A second method is to measure the current due to unscattered electrons at different gas pressures and to treat the quantity I_0 as an unknown. The best estimate of I_0 is then used to plot the logarithm of ratios of currents as a function of N, according to Equation 2.1, and then to obtain the cross section from the slope of the plot.

The plot should be linear for the results to be acceptable. The y-intercept of such a plot should be zero according to Equation 2.1, but experimentally this is rarely accomplished, demonstrating that the approximate value of I_0 leads to an unsatisfactory situation. The intercept is the residual electron current in the absence of scattering and if this residual is not constant the scattering cross section derived from the slope of the plot will be in error. The reasons for the absence of a constant intercept are several. The gas molecules may diffuse backwards towards the source of electrons due to inadequate or complete absence of differential pumping of the cathode region. Space charge effects or charges accumulated on the solid surfaces may interfere with electron optics. Furthermore, the necessity of carrying out the cross section measurements at various pressures for the same electron energy renders the Ramsauer technique tedious. Setting the electron energy to a desired value at various gas pressures, in coordination with the transverse magnetic field, also introduces difficulties because the electron energy is not continuously variable. The discrepancies observed in the measured cross section are due to a combination of several of these effects.[19]

The time-of-flight method attempts to address some of these difficulties. In this method, an electron drifts through a field-free region. The electron energy is calculated from the measured flight time and drift distance and this can be of significant help at low energies and long paths. In the absence of collisions, the relationship between the energy of the electron and the time of flight is given by the simple equation

$$\varepsilon = \frac{1}{2} \frac{m}{e} v^2 = \frac{1}{2} \frac{m}{e} \frac{L^2}{t^2} \tag{2.2}$$

in which e/m is the charge-to-mass ratio of the electron, L the path length, and t the time. For a flight distance of $0.5\,m$ and an electron energy of $1\,eV$, the flight time is $\sim 1\,\mu s$.

Since microsecond flight times can be measured to an accuracy of a few nanoseconds, the time-of-flight method can very advantageously be adopted for determining the electron energy.

The method does not require a monoenergetic electron beam; flight times of individual electrons are measured, but these electrons may be from a distribution of energies. Thus a detected pulse shape yields energies of all electrons in the group and there is no loss of intensity due to picking out a small number of electrons having the desired energy. Further, the physical quantities measured are time and length, and not potential difference, which could become a significant source of error due to the presence of contact potentials. Contact potentials of the order of 10 mV are not uncommon and this aspect alone limits the lower energy to which the scattering experiments could be extended.

A principal advantage of the time-of-flight method is that electron energies are determined without the need for establishing energy calibration points. The disadvantage of the method is that the electron beam must be pulsed with a very small width in order to obtain the flight time information. The time-of-flight method has been applied to the measurement of vibrational cross sections[20] as well as total scattering cross sections.

The time-of-flight method—the apparatus has also been termed the electron time-of-flight spectrometer—essentially consists of generating short pulses of electrons, a field-free drift region, and an electron detector. The generating pulse and the subsequently detected pulse of electrons serve as time markers. A photocathode illuminated by a short burst of light[21] or a special electron gun[22] has been used for the purpose. A dc beam, swept across an aperture by an rf field or by a pulse, provides better control and more accurate timing. If the desired energy is low, a decelerating field is applied just before the electrons enter the drift space.[20]

Buckman and Lohmann[23] have adopted a new time-of-flight spectrometer to measure the absolute total cross section in the energy range 0.1 to 20 eV which includes the Ramsauer–Townsend minimum. The schematic diagram of the spectrometer is shown in Figure 2.5. It comprises, basically, an electron source in the form of a tungsten filament, a number of electron lenses, a beam pulsing system, a field-free flight tube, and an electron detector in the form of a channel electron multiplier. The principle of the method is to pulse the electron beam, thus providing a timing mark for the start of the time-of-flight measurements. An rf square pulse is applied to a pair of deflection plates and a second pulse is applied to a second pair of orthogonal plates approximately after an interval of half the width of the first pulse. The total scattering cross section is obtained by measuring the transmitted electron density with and without gas in the flight tube. The gas pressure employed is in the range of 0.03 to 0.3 Pa and the pressure ratio on the filament side is less than 0.5%.

The Ramsauer–Townsend minimum measured by Buckman and Lohmann[23] occurs at a higher energy and shows a higher cross section than those measured by Golden and Bandel.[12] Figure 2.6 shows this minimum and total cross sections in argon up to 20 eV energy. Selected data in the figure include: Ferch et al.,[24] Haddad and O'Malley,[25] McEachran and Stauffer,[26] Bell et al.,[27] Kauppila et al.,[28] Wagenaar and De Heer,[29] and Fon et al.[30]

The time-of-flight method developed by Kennerly and Bonham[19] for the measurement of absolute total cross section is a variation of the earlier method of Land and Raith,[20] and the basis of its capability is its utility for measuring the electron energy distributions of electrons ejected from gaseous or solid targets by electron impact. The experimental setup is shown in Figure 2.7.

Electrons are generated by an electron gun and the dc beam is allowed to pass through an aperture of 1 mm diameter. Pulses are generated by applying a rectangular voltage pulse of about 10 V and 30 ns duration to one plate of a pair of deflection plates. The dc beam is swept across the aperture by the voltage-pulse rise. In order to prevent the dc beam recrossing the aperture with the fall, the pulse is delayed by 15 ns and then applied to one plate of another

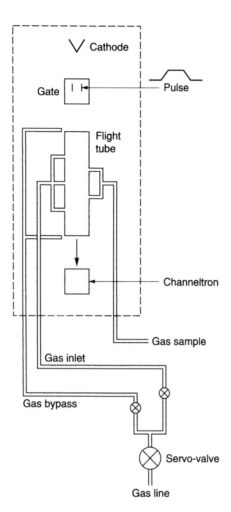

FIGURE 2.5 Time-of-flight spectrometer used by Buckman and Lohmann[23] for measurement of absolute total cross section for electron scattering of helium and argon. Electrons are emitted by a filament and are directed by a system of lenses to the flight tube which is free of electric field. The electrons which survive scattering arrive at the channeltron, which is a detector. The total cross section is measured by the relative intensities of electrons with and without gas in the flight tube. Figure reproduced with permission of the Institute of Physics, U.K.

pair, orthogonal to the first pair. The beam thus traces out a rectangle with one side centered on the aperture. The collected current ($\sim$1 μA) is passed to an external current measuring device. The duration of the electron pulse is controlled by varying the rise time of the voltage pulse in the range of 0.7 to 7 ns; this yields electron pulses having durations of 0.1 to 1 ns. The energy of the electrons is several keV and the repetition rate is several hundred kHz.

The electrons strike a solid target made of a platinum tube coated with either colloidal graphite or cesium iodide. The unscattered beam is trapped (Figure 2.7). Secondary electrons generated enter a flight tube and enter a pair of circular apertures at the detector end of the flight tube. The purpose of having two apertures is to provide better discrimination against small-angle scattering in the gas cell. Three orthogonal Helmholtz coils are employed to reduce the ambient magnetic field to 1×10^{-7} T, and the drift and detector regions are triple shielded to reduce the magnetic field to less than 1×10^{-7} T. The detector system consists of a time-to-amplitude converter (TAC) and a multichannel pulse height analyzer (MC PHA).

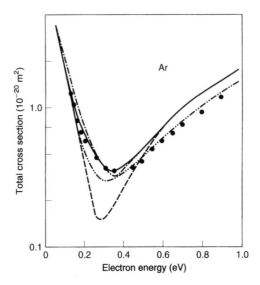

FIGURE 2.6 (a) Total cross section in argon at low energy (in units of $10^{-20}\,\mathrm{m}^2$). (•) Buckmann and Lohmann[23]; (——) swarm analysis of Haddad and O'Malley[25]; (- - -) McEachran and Stauffer[26]; (— - - —) Bell et al.[27] Not shown are: Ferch et al.[24]; Jost et al.[16]; Golden and Bandel[8]. Figures reproduced from Buckman, S. J. and B. Lohmann, *J. Phys. B*, 19, 2547, 1986. With permission of the Institute of Physics, U.K.

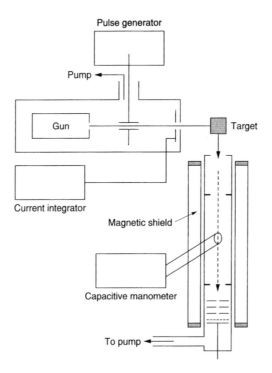

FIGURE 2.7 Schematic apparatus for the measurement of absolute total cross section in helium by the time-of-flight method. The electron gun generates a beam of electrons which is gated through an aperture to yield pulses of duration 0.1–1 ns. The pulses fall on a target, generating secondary electrons, and enter a flight tube. The detector pulse passes through a time–amplitude converter and a multi-channel pulse height analyzer (not shown).

A typical time-of-flight pulse has a duration of 1.5 μs and contains electrons in the energy range of 0.1 to 200 eV, the measurements being restricted to the range of 0.5 to 50 eV to retain accuracy.

2.1.5 PHOTOELECTRON SPECTROSCOPY

A new approach to the measurement of total scattering cross section, based on the idea of generating electrons by photoionization, is due to Kumar et al.[31] and Subramanian and Kumar.[32] Short ultraviolet (vuv) photons are allowed to impinge on a small region of the source gas, which in most scattering experiments is located close to the energy analyzer. Noble gases such as argon, krypton, and xenon are used as the source gases. When the gas is photoionized, electrons of two energies corresponding to the $^2P_{1/2}$ and $^2P_{3/2}$ states of the ion are produced. With photons having a single wavelength, and using two target gases, one at a time, two electron energies are accessible. With three wavelengths of photons (58.4, 73.6, and 74.4 nm) and three gases (Ar, Kr, Xe) as source gases, eighteen electron energies in the range of 0.7 eV to 10 eV are accessible. However, two of the energies are so close that they cannot be resolved and, for all practical purposes, seventeen energies are available.

Microwave discharges in helium produce resonant lines of 58.4 nm (He I) and in neon the discharge produces 73.6 and 74.4 nm radiation (Ne I). The resonant lines are monochromatic; the energy spread of these lines depends upon the pressure of the helium or neon gas used. The pressure of the emitting gas is kept constant throughout the experiment. The discharge is produced in a microwave cavity at 2450 MHz with an average microwave power of 100 W. The monochromatic light interacts with the target gas in the ionization region, generating photoelectrons at two discrete energies (Figure 2.8). The intensity of the photon beam is monitored using a beam splitter and a gate valve.

The photoelectrons are allowed to be scattered by the target gas and differential pumping allows different source and target gases to be employed. Current is measured first without the target gas, and then at various pressures of the target gas. The decrease in the current that occurs with increasing gas pressure of the target gas is a measure of the scattering cross sections. The advantage of the method is that different gases can be used for the source and

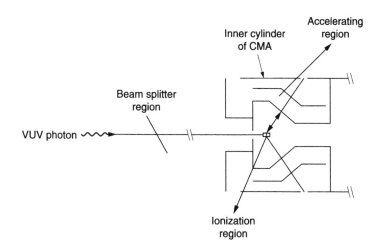

FIGURE 2.8 Photoelectron spectrometer for measurement of total scattering cross section in rare gases. The photon beam impinges on the target gas and causes photoionization. The released electrons are accelerated and a cylindrical mirror analyzer (CMA) with a channeltron is used to analyze the energy and count the scattered electrons. Reproduced from Kumar, V., et al., *J. Phys. B*, 20, 2899, 1987. With permission of the Institute of Physics, U.K.

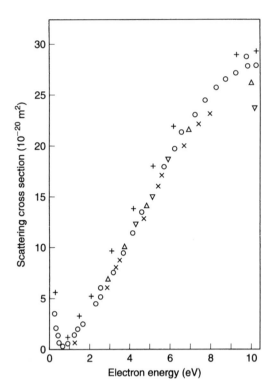

FIGURE 2.9 Total electron–xenon scattering cross sections as a function of incident electron energy from 0.7 to 10 eV. (o) Jost et al.[16]; (Δ) Dababneh et al.[33]; (□) Nickel et al.[9]; (×) Subramanian and Kumar[32]; (∇) Lam[34]; (+) Haberland et al.[35] Reproduced from Subramanian, K. P. and Vijay Kumar, *J. Phys. B*, 20, 5505, 1987. With permission of the Institute of Physics, U.K.

the target; the disadvantage is that the electron energy cannot be selected independently and exact comparison with other methods is rendered difficult. Figure 2.9 shows the cross section in xenon measured by this method and compares the results with selected previous measurements.[33–35] A more complete set of cross sections will be presented in Chapter 3.

2.2 DIFFERENTIAL CROSS SECTIONS

The differential cross section for elastic collisions is defined as

$$Q_{el} = 2\pi \int_0^\pi Q_{diff}(\theta) \sin\theta \, d\theta \tag{1.66}$$

where Q_{el} is the total elastic collision cross section. This equation gives the integral cross section for scattering in different directions as discussed in Chapter 1. The definition of Equation 1.66 is not limited to elastic collisions, and the integrated vibrational and momentum transfer cross sections, for example, are also expressed as

$$Q_v = 2\pi \int_0^\pi Q_{v,diff} \sin\theta \, d\theta \tag{2.3}$$

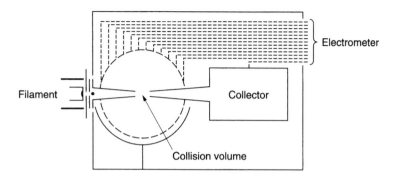

FIGURE 2.10 Experimental arrangement of Ramsauer and Kollath[36] for measurement of angular scattering. A thermionic filament emits electrons that enter a circular collision chamber and those electrons that survive collision enter the collector. The collecting electrodes, eleven in total, are arranged over a circular arc and the current to each gives the ratio of electrons scattered into that angle. The gas number density in the collision volume is $\sim 10^{19}\,\mathrm{m^{-3}}$ ($\sim 10^{-5}\,\mathrm{Pa}$) and the electron energies are limited to minimize scattering due to inelastic collisions.

and

$$Q_{\mathrm{M}} = 2\pi \int_0^\pi Q_{\mathrm{el,diff}} \sin\theta (1 - \cos\theta)\, d\theta \qquad (2.4)$$

where $Q_{\mathrm{v,diff}}$ and $Q_{\mathrm{el,diff}}$ are the differential cross sections at angle θ for vibrational and elastic collisions respectively.

2.2.1 BEAM SCATTERING TECHNIQUE

The earliest measurements of the angular distribution of scattered electrons are due to Ramsauer and Kollath,[36] whose setup is shown in Figure 2.10. A thermionic filament emits electrons which enter a circular collision chamber; those electrons that survive collision enter the collector. In this experiment the collecting electrodes, eleven in total, were arranged over a circular arc and the current to each gave the ratio of electrons scattered into that angle. The gas number density in the collision volume was $\sim 10^{19}\,\mathrm{m^{-3}}$ and the electron energies were limited to minimize scattering due to inelastic collisions. The apparatus had an overall length of about 20 cm, and pressures of the order of 0.1 Pa were employed in the scattering chamber.

Inelastic scattering becomes significant at low energies in molecular gases due to vibrational and rotational excitation. In rare gases, however, inelastic scattering occurs at considerably higher energies, in the range of ~ 8 to $20\,\mathrm{eV}$. The contribution of the elastically scattered electrons to the total scattering cross section may be filtered out by adopting energy selection methods that reject inelastically scattered electrons.[37] A factor to note is the presence of significant maxima and minima, which are attributed to the diffraction of the electron waves by the target atoms. The complexity of the angular distribution pattern increases, generally speaking, with the atomic number of the target particle.[1]

2.2.2 CROSSED BEAMS TECHNIQUE

The crossed beam technique for the measurement of angular scattering consists of a neutral beam of gas intersecting an mono-energetic beam of electrons having the desired energy. The scattered electrons at a particular scattering angle θ are detected and analyzed for energy. It is necessary that the interaction should take place in a cell where extraneous magnetic and electric fields do not exist. Use of multiple orthogonal Helmholtz coils and magnetic shielding of the interaction volume can reduce the ambient field to as low as 10^{-7} T.[38]

Experimental determination of differential cross sections yields relative values, which are converted into absolute values by adopting two different methods. The first is a phase shift analysis of an isolated resonance below the first inelastic threshold. An example is the 19.35 eV, ^{2}S resonance in helium. In recent years there has been some interest in adopting neon as a secondary standard.[39] The second method is to measure the ratios of elastic scattering signals in two gases, the target and the reference gas, under identical conditions. The latter method is known as the relative flow method and was developed at the Jet Propulsion Laboratory of the California Institute of Technology. It has the advantage of not requiring the determination of a number of parameters essential for the determination of the absolute cross section.

The principle of the relative flow method[40] is that two gases, one of which is the experimental gas and the other is a reference gas in which the cross sections are well known, are studied, employing exactly the same experimental conditions except for the gas flow rate. By adjusting the ratio of the gas flow rate, the ratio of the cross sections of the two gases can be determined. The crossed beam technique for measuring the cross sections has also been applied by Tanaka et al.,[41] who have measured the vibrational cross sections in CO_2 in the energy range 1.5 to 100 eV. The relative flow method described below is adopted.

A schematic representation of the crossed beam technique for the measurement of angular distribution of elastically scattered electrons is shown in Figure 2.11.[42] The apparatus essentially consists of three parts. Firstly, the input electron optical system: electrons are extracted from a filament placed in a small cylindrical chamber below the target gas inlet. A hemispherical energy analyzer is used to select the electron energy and the electrons are directed to the scattering chamber (not shown). Secondly, the target gas inlet system: the atomic target is created by effusion from a long cylinder, the pressure being 3 mPa. Thirdly, the analyzing system, consisting of a 90° analyzer, a focusing lens system, and a set of accelerating plates that increase the energy to 100 keV. Having reached this energy, the

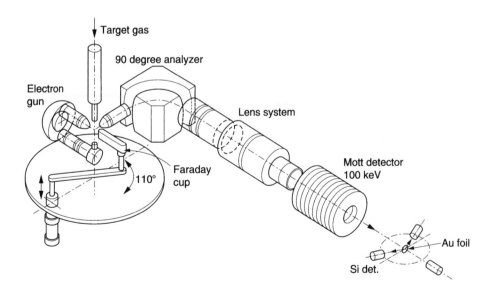

FIGURE 2.11 Experimental setup for the flow-through method of Qing et al.[42] for measurement of differential cross sections. Figure reproduced with permission of North Holland Publishing Co.

electrons are scattered on a gold foil. Two surface barrier electron detectors are employed to measure the polarization of the scattered electron beam impinging on the foil. The sum of the two count rates is related to the differential scattering cross section. The measured cross sections are relative with respect to the angle of scattering and the electron energy.

Nickel et al.[15] adopted the following procedure for determining the zero scattering angle for the beam. Helium gas is admitted by a capillary array and the electron energy is adjusted to 21 eV. The electrons that are inelastically scattered are detected on either side in the wing and the midpoint is taken as zero. The detector position is then shifted by a desired angle and both helium and the gas under study, in turn, are admitted and measurements taken. The ratio of the scattering cross section of the experimental gas is obtained as

$$\frac{Q_{d1}(\theta)}{Q_{d2}(\theta)} = \frac{N_{e1}}{N_{e2}} \frac{\delta_1^2}{\delta_2^2} \tag{2.5}$$

where N_{e1} and N_{e2} are the numbers of electrons scattered into the angle θ, and δ_1^2 and δ_2^2 are the molecular diameters respectively.

The technique has been applied successfully to measure differential cross sections of H_2 in the energy range of 3 to 75 eV and in the angular range of 20° to 135°, employing helium as the standard gas.[40] Measurements have also been made on carbon monoxide and nitrogen in the energy range of 20 to 100 eV and angular range of 20° to 120°,[15,43] and on carbon dioxide.[41] Selected earlier examples of differential cross section measurements are given in references 44–49.

Recent progress in the method of measuring the differential scattering cross section centers on extending the angular range from 0° to 180°, thus removing the need to extrapolate to these angles. Zubek et al.[50] adopt an ingenious technique of applying a localized static magnetic field to the interaction region of a conventional electron spectrometer. The magnetic field is produced by four coaxial solenoids, all of the same length and arranged as two pairs, namely an inner pair and an outer pair. The two members of each pair are separated by a short distance to form a slot through which the incident and scattered electrons can enter and leave the field. The current in the inner field flows in a direction opposite to that in the outer pair. The technique has been applied to measure the differential scattering cross section in Ar,[50] and to study electronic excitation[106] and inelastic scattering cross section of SF_6.[51]

The experimental technique of Cubric et al.[52] is a significant advance because the differential cross sections were measured over the complete range of 0° to 180°, overcoming the fundamental limitation of the angular measurements not covering very small and very large (<180°) angles. The technique, which is of the crossed beam type, is shown in Figure 2.12. There are two hemispherical deflection analyzers for energy selection and analysis of the incident and scattered waves respectively. The target region is surrounded by two pairs of solenoids. A gas beam effuses out of a hypodermic needle along the axis of the solenoids. The overall energy resolution is approximately 80 meV.

The two solenoids serve the purpose of creating a highly localized magnetic field. This is accomplished by controlling the current supplied to them. However, the overall magnetic dipole moment is zero, thereby not disturbing the operation of the selector and analyzer. Since the solenoids are axially symmetric, the axial component of the momentum of the electron in this field remains the same. This component is zero at the axis of the system and in the field-free region away from the axis for electrons initially directed toward the axis. Consequently, any such electrons change direction in the magnetic field but still pass through the field-free region. Similarly, electrons that originate from the interaction region move radially away from it once they reach the field-free region.

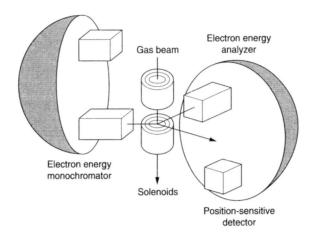

FIGURE 2.12 Schematic of the setup for measurement of the differential cross section for excitation of helium. Two pairs of solenoids are employed to separate the trajectories of inelastic electrons with zero or close to zero degree scattering. Changing the current through the solenoids changes the angle of scattering so that there is no need to move the analyzer to catch electrons scattered at various angles. Reproduced from Cubric, et al., *J. Phys. B*, 32, L45, 1999. With permission of the Institute of Physics, U.K.

For elastic scattering events the forward scattered electrons and the unscattered straight through-beam follow the same trajectory, thus preventing the forward scattered electrons at or near 0° from being measured. However, for the inelastically scattered electrons the trajectories are separated because of the energy difference. The backward scattered electrons also have a different trajectory from the trajectory of the incident beam, making it possible to cover the entire range of 0° to 180° for measurement. The angle of scattering depends on the current through the coils and, by varying this current, all angles are scanned without moving the analyzer.

2.3 IONIZATION CROSS SECTION

Measurement of ionization cross section is one of the most fundamental in the field of gaseous electronics. The cross sections are of vital importance in the development of discharges, plasma science and technology, developing theories of interaction between charged particles and neutral atoms, calculating the swarm parameters, simulating the discharge phenomena, and so on. In this section the methods used for measuring total ionization cross sections are described.

2.3.1 IONIZATION TUBE METHOD

The early measurements of ionization cross sections in several gases are due to Tate and Smith.[53] Their experimental arrangement is shown in Figure 2.13, which explains the principle of the method in a simple way. The tube is divided into two compartments, each connected to a separate pumping system. The filament emits electrons which pass through slits S_1 to S_4 and then to the ionization region, which consists of two parallel plates held at a constant potential difference. The potentials between S_1 and S_2 and between S_2 and S_3 are held at the same value, whereas a variable potential difference between S_3 and S_4 accelerates the electrons to the collision region. A longitudinal magnetic field from a solenoid is applied to ensure that the electrons are in a narrow beam. The electrons are collected by the trap and the ion current is given by measuring the current to P_1.

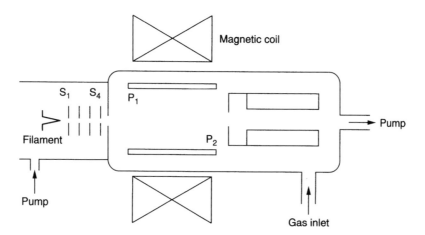

FIGURE 2.13 Schematic diagram for the measurement of ionization cross section. Filament F emits electrons which are accelerated into the collision chamber. A magnetic field from a solenoid is applied to ensure that the electrons are in a narrow beam. The cross section is obtained by measuring the positive ion current.

The ionization cross section is obtained as

$$Q_i = \frac{i_+}{i_- NL} \tag{2.6}$$

where i_+ and i_- are the ion and the electron current respectively, N is the gas number density, and L is the path length of electrons causing ionization. Tate and Smith[53] presented their results as the number of ionizing collisions (P_i) per centimeter path length at 133 Pa at 0°C, which may easily be converted to the cross section by dividing P_i by the number density.

The technique of Tate and Smith belongs to the category of beam-static gas measurements, though the gas is kept flowing through the tube. However, to convert the measured currents to absolute cross sections the pumps were cut off till steady state was achieved and gas pressure recorded. The results obtained by Tate and Smith[53] remained as a standard for nearly thirty years.

Several details need to be considered carefully for minimizing the errors in the determination of the cross sections.[54] The ion current and the electron curent must be completely collected. The collection should therefore be accomplished with increasing voltage during each experimental run and saturation obtained. It is observed that rare gases, due to their monatomic nature, require a lower collecting voltage to attain ion saturation, whereas molecular gases, due to dissociative ionization, have energetic ions that require much higher collection voltages. Furthermore, at higher electron energies ($\sim$100 eV) dissociative ionization results in even more energetic ions, requiring a further enhanced collecting field.

The path length can also be a source of error due to the magnetic field employed for collimating the beam. The electrons have a helical path in the magnetic field and trochoidal drift in the crossed electric and magnetic fields. The actual path of the electrons may therefore be greater than L and, for reasons of accuracy, this should not exceed a small fraction of 1% of L.

In determining the electron current, which is augmented due to the ionized electrons, one must ensure that the secondary electrons and the reflected electrons do not escape from the trap. The number of ionized electrons should be small compared with the number of electrons in the primary beam.

As the energetic electrons pass through the aperture, secondary effects set in, due to edge effects. The number of secondary electrons generated should therefore be minimum. Photoelectric action is also a source of unwanted secondary electrons and should be minimized. Asundi et al.,[55] Rapp and Englander-Golden,[56] and Schram et al.[57] have devoted considerable attention to these points and made a series of measurements using essentially the same type of apparatus. The relative currents measured are used to arrive at the absolute cross section by measuring the absolute cross section at a particular electron energy. This involves measuring the pressure in an independent manner and with the use of a McLeod gauge in combination with a cold trap to prevent contamination of mercury vapor which was often a source of uncertainty in the reported absolute measurements carried out before the year 1960. Availability of all-metal bakable capacitive manometers has rendered this difficulty only of historic interest.

It is generally agreed that the results of Rapp and Englander-Golden[56] constitute a benchmark set of data that has stood the test of time for nearly 40 years. These authors used a method that is similar in principle to that used by Tate and Smith.[53] An electron beam from a differentially pumped, indirectly heated cathode is collimated by apertures and confined by a strong magnetic field ($\sim$600 G) to pass through a chamber containing gas at $\sim$7 mPa and then into an electron collector. The entire system is bakable to a ultrahigh vacuum of 10^{-7} Pa. A uniform electric field maintained perpendicular to the electron beam removes the ions generated to an ion collector. The total ionization cross section is calculated from

$$Q_i = \frac{I_{\text{ion}}}{I_e N L} \tag{2.7}$$

where I_{ion} is the ion current, I_e is the electron current, N the target electron density and L the effective path length of electrons contributing to the ion collector. For a gas capable of multiple ionization, the total ionizing cross section is the weighted sum of individual cross sections,

$$Q_i = Q_{i1} + 2Q_{i2} + 3Q_{i3} + \cdots \tag{2.8}$$

where Q_{in} is the n-fold cross section.

Figure 2.14 shows the measurements of Rapp and Englander-Golden[56] in rare gases. Their measurements in several gases will be referred to extensively in the chapters to follow.

2.3.2 IONIZATION TUBE WITH *e/m* DISCRIMINATION

The ionization tube of Tate and Smith does not discriminate between the charges of the ion generated. If the energy of the electron exceeds the second ionization potential, there is a finite probability that some neutrals will be doubly ionized. To separate the cross sections for each ionization potential, charge-to-mass discrimination may be employed and Equation 2.8 will then be applicable to the total ionization cross section. Such experiments were carried out by Bleakney,[58] Tate and Smith,[59] and Harrison.[60] Many of the sources of error in the method of total ionization tube are also inherent in this method.

2.3.3 CROSSED BEAMS METHODS

The ionization tube method of Tate and Smith,[53] although simple in concept, is subject to a number of errors. The length of the electron trajectory within the ion collector region is uncertain due to the presence of the magnetic field that is employed to collimate the beam. Electron scattering and secondary effects may also cause errors. Measurement of target gas

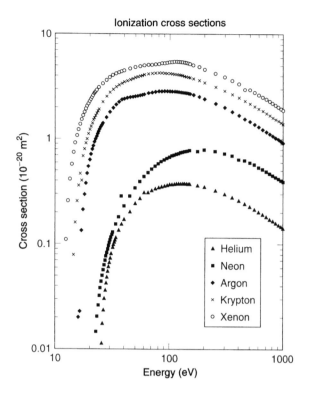

Ionization cross sections

FIGURE 2.14 Ionization cross sections in rare gases. From Rapp, D. and P. Englander-Golden, *J. Chem. Phys.*, 43, 1464, 1965.

density in the range of 0.01 to 1 Pa is difficult, and excluding contamination by mercury vapor, which has a low ionization potential, was a serious drawback in the experiments up to about 1965.

The crossed beam technique was originally developed by Fite and Brackmann[61] to measure the ionization cross section of the hydrogen atom by electron impact. Many atomic species such as H, N, O, and alkali metal atoms are unstable at room temperature and in this technique the atomic species is produced in adequate abundance and in an environment where the electron impact ionization can be measured. The technique has been extended to the study of electron collisions with rare gas atoms by Shah et al.[62] We shall describe the principle of the method before considering the several variations introduced since the first measurements of Fite and Brackmann.[61]

A primary beam of electrons is arranged to intersect a thermal beam of highly dissociated atoms which is chopped at a fixed frequency. The ions arising out of collision between electrons and the target particles get mixed with the unwanted ionization products of the background gas. The complexity is increased because the unwanted ions are of the same type that one wishes to detect. The number of ions from this source is much larger than the number from the intended target–electron collision. The undesired ions are the noise and the desired ions, though of the same kind, are the signals. Long observation periods are required to reduce the statistical fluctuations of the noise and improve the stability of the various parameters such as background pressure, beam magnitude, amplifier gains, etc.; these are extremely difficult to control. To overcome this problem, one or both the beams are chopped at a fixed frequency and the signal is analyzed with respect to specific frequency and phase shift. By comparing the signal to the signal obtained by using the undissociated molecular gas, the ionization cross section of the atom is determined.

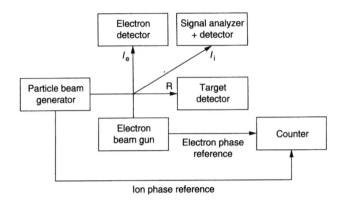

FIGURE 2.15 Block diagram of the crossed beam apparatus. A beam of electrons from a gun intersects a beam of target particles and the various measuring devices are connected as shown. I_e is the electron current after collision, I_i is the ion signal after collision, R is the number of particles per second arriving at the target detector.

Figure 2.15 shows schematically the principle involved.[54] A beam of electrons generated by a gun crosses the path of a beam of target particles. The ion signal is fed to a detector and an analyzer, followed by a phase-sensitive amplifier and counter. A reference pulse from the electron source and a reference pulse from the target beam generator also feed the same amplifier. The electron beam and the target beam are terminated, after collision, at an electron detector and a target detector respectively.

The ionization cross section is given by the equation

$$Q_i(\varepsilon) = \frac{I_i v_e v_t}{I_e RF(v_e^2 + v_t^2)^{1/2}} \tag{2.9}$$

where I_i = ion current, I_e = electron current after collision, v_e = electron velocity, v_t = target particle velocity, R = number of target particles per second arriving at the target detector, F = factor defining the measure of overlap of the two beams, and ε = relative energy of the target and electron, given by

$$\varepsilon_e + \left(\frac{m}{M}\right)\varepsilon_t \tag{2.10}$$

where m and M are the mass of the electron and the target particle respectively. The methods of evaluating R and F have been explained by Dolder et al.[63] The cross section may be total or partial, as defined by Equations 2.6 and 2.8 respectively. An absolute measurement of $Q_i(\varepsilon)$ involves measuring each of the quantities I_i, I_e, R, F, and v_t without normalizing to previous experiments or theory.

Crossed beam experiments generally fall into two categories: (1) the fast-neutral-beam method; (2) the pulsed electron beam method. The basic fast-neutral-beam method was first used by Cook and Peterson[64] to study atomic nitrogen, but it has been improved considerably for cross section studies in rare gases. Montague et al.[65] have adopted this technique to measure the single ionization of ground-state helium by electron impact. The method is capable of quantitative measurements of molecular fragment (dissociated atoms) cross sections as well as parent species, and of unstable atomic and molecular parents as well as stable atoms and molecules. Wetzel et al.[66] have paid particular attention to reducing the errors and improving repeatability.

The principle of the fast-neutral-beam method is to generate a fast beam of ions ($\sim$2 to 4 keV), mass filtered and then neutralized by symmetric charge transfer. The resulting beam

TABLE 2.2
Measured Threshold Energies (eV)

Gas	Spectroscopic[70]	Wetzel et al.[66]	Rapp et al.[56]	Stephan et al.[68]
He^+	24.59	24.56	24.57	24.33
Ne^+	21.56	21.50	21.86	22.22
Ar^+	15.76	15.82	16.06	15.30
Kr^+	14.00	13.99	14.25	13.59
Xe^+	12.13	12.03	12.19	
Ar^{2+}	43.39	41.8		
Kr^{2+}	38.36	38.2		
Xe^{2+}	33.34	32.7		
Kr^{3+}	75.31	69.2		
Xe^{3+}	65.46	62.9		

of fast neutral atoms retains the collimation of the ion beam. The flux of the neutral atoms is measured absolutely with a calibrated detector. The ionization cross section is measured by crossing the neutral beam with a well characterized electron beam. The resulting ion beam is then focused by an einzel lens, separated into different charge states with a hemispherical electrostatic analyzer, and counted with a channel electron multiplier. The einzel lens and the hemispherical analyzer are necessary, particularly in molecular gases, where 100% collection of ionization fragments is usually more difficult than in monatomic gases.

A few comments about the charge transfer method of generating an atomic beam are in order. Charge transfer collision, in which a fraction of the ions in a fast beam are converted to neutral atoms, is an inelastic process with very little momentum transfer and the neutral beam will retain almost the same velocity distribution as that in the ion beam. The charge transfer collisions are normally of the resonant type and in the lower velocity range the cross section is large, $\sim$1 to $3 \times 10^{-19} \, m^2$.

The resonance charge transfer cross section increases in the order of He, Ne, Ar, Kr, and Xe[67] for the same velocity. The large cross section discourages the production of metastables, which was one of the significant sources of error in the earlier stages of the development of the technique. The majority of collisions take place at large impact parameters (twice the atomic radius rather than the atomic radius in electron–atom collisions) and lead to very small scattering angles. The resulting neutral beam remains collimated, and has the velocity of the incident ion beam. Fast neutral atoms that are scattered by more than 1° are weeded out by the use of a very small aperture in the cell. The neutral beam is composed of nearly pure ground-state atoms, since all rare gas excited states are far from energy resonance with the charge transfer partner. Ionization measurements near threshold show that there are no metastables.

The threshold for single ionization is taken as the x-axis intercept of the signal vs. energy plot for the first ten points and a least-squares method is used for a linear fit. The energy scale is adjusted to make the intercept agree with the spectroscopic ionization potential. Table 2.2 shows these values, along with the values of Rapp and Englander-Golden,[56] and Stephan et al.[68]

2.3.4 PULSED CROSSED BEAM TECHNIQUE

The crossed beam technique for measuring of the ionization cross section of rare gas atoms has been successfully adopted by Shah et al.,[62] who introduced modifications to their apparatus for measuring ionization cross sections in atomic hydrogen.[69] The pulsed electron beam intersects a thermal neutral gas beam. The ions are collected as pulses and identified from the time of flight in accordance with their charge-to-mass ratio. These investigations are among the first to make use of the availability of the time-to-amplitude converter.

2.3.5 TIME-OF-FLIGHT METHOD

A variation of the time-of-flight method for the measurement of ionization cross sections has been developed by Straub et al.[71] and applied to a number of gases. The method is simple. The ionization chamber is filled with the experimental gas to a pressure of approximately 0.4 mPa. Pulses of electrons are directed through an interaction region located between two parallel electrodes, maintained at ground potential, and then collected by a Faraday cup. Following each electron pulse, an electric field is applied for a short period across the interaction region. The positive ions generated by ionization are driven toward the bottom plate. Some of the ions pass through an aperture in this plate and get collected by a position-sensitive detector that records both their arrival times and their positions. Using an automated computer system, partial ionization cross sections are measured from threshold to 1000 eV energy. These investigations, carried out during 1995 to 2002, form one of the most comprehensive and valuable sets from the point of view of applying the same technique to a number of gases. The time-of-flight mass spectrometer method has also been employed by Kobayashi et al.[72] for measurement of ionization cross section in rare gases.

The above descriptions reveal that ionization cross section measurements employ apparatus that has, generally speaking, three basic components: (1) electron beam production and control of the energy — the electron gun is the preferred choice and commercial brands are available; (2) interaction region — the crossed beams method is invariably adopted; (3) detection of the products and quantity of ionization. Several types of mass spectrometer have been employed: quadrupole mass spectrometer, time-of-flight mass spectrometer (see Section 4.1.6 for a brief description), cycloidal mass spectrometer, and double focusing section field spectrometer. Tian and Vidal[73] have given a brief description of the relative advantages and disadvantages of some of these methods.

One of the principal concerns is the complete collection of all the ions produced during the ionizing collisions. At higher impact energies or during dissociative ionization the ions acquire considerable kinetic energy, making complete collection difficult. A quadrupole mass spectrometer has the advantage of high mass resolution; however, the ions move in complicated trajectories under the radio frequency field. The transmission efficiency is therefore dependent on mass and energy. The mass-dependent transmission of ions can be calibrated[74]; however, the energy-dependent signal is almost impossible to correct since the energy of the product ions depends upon the incident electron energy, the dissociation energy, and the mass of the ions.

Table 2.3 provides selected references for these methods and gases studied.

The fast-neutral-beam technique of Wetzel et al.,[66] described previously, has been shown to accomplish complete collection for ionization of atoms or neutrals. However, this may not be the case for dissociative ionization.[73] The focusing time-of-flight spectrometer is more successful in ensuring that all ions generated are collected, since the signal strength as a function of collection voltage may be demonstrated to have attained saturation.

2.4 TOTAL EXCITATION CROSS SECTION

Measurements of the inelastic collision cross sections of gases have been carried out since the mid-1930s and a number of techniques have been developed to improve the accuracy of cross section and resolution of the energy level. Maier-Leibnitz[90] measured the inelastic collision cross sections in helium by a retarding potential technique, using an electron swarm as the

TABLE 2.3
Selected References for Ionization Cross Section Measurement Methods

Method[a]	Gases	Energy Range (eV)	Reference
A	CO_2	15–300	Crowe and McConkey[75]
	CH_4, C_2H_6, SiH_4, Si_2H_6	10–1000	Chatham et al.[76]
	H_2O, CO, CO_2, CH_4	15–510	Orient and Srivastava[74]
	He, Ne, Ar, Kr, Xe	Onset–1000	Krishnakumar et al.[77]
	N_2	20–1000	Krishnakumar and Srivastava[77]
B	NH_3 (ammonia)	15–1000	Rao and Srivastava[78]
	A, Kr, Xe	18–240	Syage[79]
	NH_3	10–270	Syage[80]
	Ar	Onset–1000 eV	Straub et al.[71]
	C_2H_2 (acetylene)	20–800	Zheng and Srivastava[81]
	N_2, O_2, CO, CH_4, C_2H_2	25–600	Tian and Vidal[73]
	CO	15–1000	Mangan et al.[82]
	Rare gases	Onset–1000 eV	Kobayashi et al.[72]
C	He, Ne, A, Kr, Xe	12–200	Wetzel et al.[66]
	N_2, CO, CO_2,	14–200	Freund et al.[83]
	SiF	Onset–200	Hayes et al.[84]
	SO_2	13–200	Basner et al.[85]
D	CD_4 (deuterated CH_4)	Onset–200	Tarnovsky et al.[86]
	CO_2	25–600	Adamczyk et al.[87]
E	CO_2	14.1–171.1	Märk and Hille[88]
	C_3H_8	Onset–950	Grill et al.[89]

[a] A = quadrupole mass spectrometer; B = time-of-flight mass spectrometer; C = fast-neutral-beam technique; D = cycloidal mass spectrometer; E = double focusing mass spectrometer.

source of electrons. His technique is usually referred to as the swarm method. Dorrestein[91] employed an electron beam to produce the metastable helium atoms and detected them by the secondary electrons. Woudenberg and Milatz[92] measured the metastables by the absorption of the 1083 nm line, i.e., the $2\,^3P$ level.[93] These experiments showed, for the metastables in helium, a sharp increase in the cross section and a peak in the cross section quite close to the threshold. These three techniques used different principles and laid the foundation for further improvements in accuracy and energy resolution. For example, Schulz and Fox[94] adopted the secondary electron emission method combined with a narrower electron energy distribution, using the method of retarding potentials for the latter purpose.

Energetic electrons colliding with gas neutrals excite the neutral and the excitation state may be short lived, generating a photon after deexcitation. The excited state may have a long life if it is a forbidden transition, and the atom is then called a metastable. Its detection usually involves some sort of secondary effect, either a released electron from a metal cathode or Penning ionization of a second gas that has a lower ionization potential. The experimental techniques employed till 1968 have been reviewed by Moiseiwitsch and Smith[93] in their classical paper.

The theories developed till 1968 made approximations for simplifying the atomic system. Many of these approximations were not satisfactory for electron energies below about 100 eV. This situation has been improved by theorists attempting to push the energy limit to lower values; for example, Kaur et al.[95,96] have succeeded in the calculation of differential and total cross sections in rare gases for electron energies in the range 15 to 100 eV. While the theoretician calculates transitions to particular levels, the experimentalist can only control the energy of the electron and infer the transitions. Further, the experiment gives the sum of several transitions, called cascading, which are of greater significance in the context of discharge modeling such as Boltzmann equation solutions and Monte Carlo simulations. The present state of knowledge on excitation cross sections is that theory has an edge over experimental data[97] and comparisons between the two are possible, though not over the entire range of electron energies. Reviews have been made by van der Burgt et al.,[98] Heddle and Gallagher,[99] and Buckman and Clark.[100]

Moiseiwitsch and Smith[93] classify the experiments on excitation cross section into three categories:

1. Intensities of the spectral lines excited by an electron beam are used as a measure of the excitation process. The measurements yield the photon excitation cross section and the probability of generating a photon of given energy per unit atom per electron particle. In this method, absolute calibration of the optical system should be carried out. The calibrating source, such as a tungsten strip lamp, is usually arranged at right angles to the electron beam or on a rotatable mirror if the two beams are opposite each other. It is essential to use an unpolarized light source, failing which a reduction factor should be applied, for the calculation of which theory is available.[93]

2. Cross sections for metastables are measured by providing a target at which the metastable is deactivated. The target may be a metallic cathode, the excitation energy being converted to the kinetic energy of the released electron and the work function of the metal. For example, the first metastable levels in helium are 19.82 and 20.61 eV above the ground state and the work function of most metallic cathodes falls in the range of 2.5 eV to 4.0 eV. The yield of secondary electrons provides information about the metastable. In this case the optical calibration difficulties are substituted by electron emission calibration, such as the efficiency of collection and precise knowledge of work function, which is a notoriously variable quantity.

   When two gases are mixed, the excitation potential of one of the gases may be higher than the ionization potential of the second and the ionization in the mixture will be larger than the sum of the ionizations of the individual constituents. This effect is known as the Penning effect, after the discoverer, and may be used in place of the metal target to deactivate the metastables.

   Another method of detecting metastables is to determine the selective absorption of light which transfers the atom from the metastable state to a higher level, or to record the radiation accompanied by the spontaneous decay of this higher excited state into the optically resolved state.[102] Application of an electric field or a magnetic field may transfer the metastable to a neighboring excited state from which a radiative transfer to the ground state or another excited state may occur. The resulting optical radiation is detected in the usual way.

3. A study of the energy loss spectrum of the scattered electrons provides valuable data on the excitation cross section. The method has many common features with the beam method of measuring scattering cross sections and ionization cross sections. A beam of electrons is passed through a gas. An electron energy analyzer is used to measure the energy spectrum of electrons scattered at a particular angle to the beam axis. Gas pressure is kept low to prevent multiple collisions.

A setting of approximately zero on the energy analyzer will give the elastic collision cross section. As the energy is increased a small background electron current is observed and, at the onset of the first excitation potential in atomic gases, a peak occurs. The peak is observed in both the elastic channel mode and the inelastic mode. For example, in helium, corresponding to the $2\,^3S$ transition, the elastic channel shows resonance at 19.3 eV and the

TABLE 2.4
Lifetime and Threshold Energy of Metastables in Rare Gases

Gas	Coupling Notation	Energy (eV)	Lifetime (s)
Argon	$3p^5$ $(^2P^o_{3/2})$ $4s[3/2]^0$ $[^3P_2]$	11.548	24.4
	$3p^5$ $(^2P^o_{3/2})$ $4s'[1/2]^0$ $[^3P_0]$	11.723	44.9
Helium	$2\ ^3S$	19.82	~10^4
	$2\ ^1S$	20.616	~0.02
Krypton	$4p^5$ $(^2P^o_{3/2})$ $5s[3/2]^0$ $[^3P_2]$	9.915	85.1
	$4p^5$ $(^2P^o_{3/2})$ $5s'[1/2]^0$ $[^3P_0]$	10.563	0.488
Neon	$2p^5$ $(^2P^o_{3/2})$ $3s[3/2]^0$ $[^3P_2]$	16.619	24.4
	$2p^5$ $(^2P^o_{3/2})$ $3p[1/2]$ $[^3P_0]$	18.716	430
Xenon	$5p^5$ $(^2P^o_{3/2})$ $6s[3/2]^0$ $[^3P_2]$	8.315	149.5
	$5p^5$ $(^2P^o_{3/2})$ $6s'[1/2]^0$ $[^3P_0]$	9.447	0.078

The atomic core is as shown.
From Fabrikant, I. I et al., *Phys. Reps.*, 159, 1, 1988.

inelastic channel shows resonance at 19.8 eV[101] (Table 2.4). The peak is attributed to resonance, first reported by Schulz and Fox,[94] and Schulz's findings are considered a major advance in the study of excitation cross sections.

2.4.1 RESONANCE NEAR EXCITATION ONSET

Studies of electron–neutral scattering and excitation cross sections have revealed sharp features that are attributed to resonances in the electron–atom interaction system. The resonance may be very close to an excitation level. The electron attaches to the neutral atom or molecule, forming a negative ion, and after a brief interval of time ($\sim 10^{-15}$ s) it autodetaches, resulting in the excitation of the neutral. The excited state may also be vibrational in nature. Schulz[103] and Buckman and Clark[100] have given reviews of negative ion resonances.

Figure 2.16 shows the energy dependence of inelastic collision cross sections in helium.[104] The solid line (lower curve) represents the differential cross section vs. electron energy for electrons scattered at 72° and having excited the $2\ ^3S$ state. The energy loss for this process is 19.8 eV. The top curve is the cross section for the total metastable production, due to Schulz and Fox,[94] who used the retarded potential difference (RPD) technique. Figure 2.17 shows the vibrational excitation cross sections of the nitrogen molecule in the energy range 1.8 to 3.0 eV.[104] The vibrational levels are excited by short-lifetime negative ions that undergo autodetachment.

The time-of-flight method for determining the metastable excitation cross section features an electron beam colliding with a gas beam consisting of neutral atoms or molecules.[105] The products of collision are electrons, ions, photons, and metastables. Ions and electrons are removed by the application of appropriate potentials and only the photons and metastables arrive at a suitable detector. From an analysis of the time-of-flight and time-to-amplitude conversion, the cross sections for metastable states are determined.

Zubek et al.[106] have studied resonance structures in the rare gases over the angular range of 100° to 180°. A special feature of this investigation is that a newly developed magnetic angle changing technique has been used, permitting the observation of resonance structures at and near to 180°.

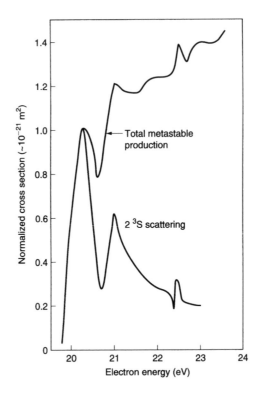

FIGURE 2.16 Schematic representation of energy dependence of inelastic collision cross sections in helium. The lower curve represents the differential cross section vs. electron energy for electrons scattered at a given angle (say 60°) and having excited the 2^3S state. The energy loss for this process is 19.8 eV. The top curve is the cross section for the total metastable production.

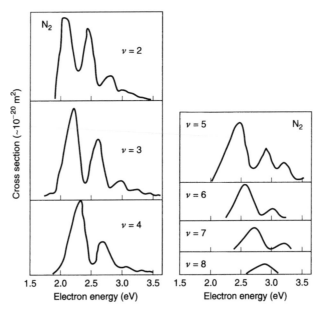

FIGURE 2.17 Schematic representation of vibrational excitation cross sections of the nitrogen molecule in the energy range 1.8 to 3.0 eV. The vibrational levels are excited by short-lifetime negative ions that undergo autodetachment.

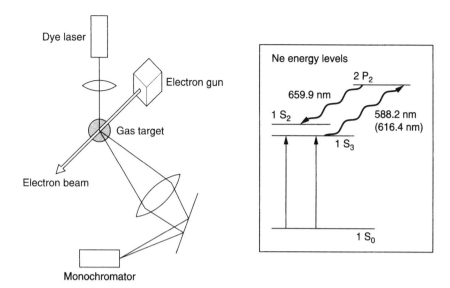

FIGURE 2.18 The laser fluorescence method of measuring the metastable cross section in neon. The electron gun provides the electrons that impinge on the target atoms and excite the atom to the metastable state. A tunable dye laser provides radiation which is absorbed by the metastable-excited atom, resulting in the atom being excited to a high level. An optically allowed transition from this higher state occurs and the emitted radiation is measured. The inset shows the energy level of the neon atom for these processes.

2.4.2 LASER FLUOROSCENCE METHOD

The principle of the method is that the metastable atom of a gas absorbs radiation of a definite wavelength and the transition of the atom to a higher state occurs. The attenuation of the radiation is measured to determine the metastable cross section. Alternatively, the atom, now in a higher state, falls into an optically allowed lower state, emitting a photon. Philips et al.[107] used this method to obtain the metastable cross section in neon.

Neon atoms in the ground state are excited to the $1S_3$ metastable state by electron collision; the threshold energy is 16.716 eV. The metastable absorbs radiation of wavelength 616.4 nm, attaining the optically allowed state of $2P_2$ as shown in the inset of Figure 2.18. Optical emission occurs from this level to the $1S_2$ level and the emitted radiation has a wavelength of 659.9 nm, which is recorded by a photomultiplier. As the tunable dye laser is tuned away from the 616.4 nm wavelength the intensity of the 659.9 nm wavelength varies, a measure of which gives the metastable cross section. The schematic of the method is shown in Figure 2.18. The method demands a complicated optical setup and suffers from low yield of the radiation as the atom falls to the $1S_3$ state.

2.4.3 SWARM METHOD FOR RO-VIBRATIONAL EXCITATION

Electrons having energy ε and scattered elastically by collision with neutrals suffer a fractional loss of energy $(2m/M)\varepsilon$ provided that the electron energy is less than the electronic excitation energy. While in rare gases and in atomic gases this holds true, in molecular gases the fractional loss of energy, especially at low values of E/N, has been recognized to be due to rotational and vibrational excitation that occur at low electron energies.

A review of the methods available for the determination of the cross sections for these inelastic processes is given by Phelps.[108] These methods are:

1. *Beam methods.* We have devoted sufficient attention to this technique for the measurement of various cross sections. The energy spread of the electron beam is of the order of 100 meV in the earlier experiments, close to 7.5 meV in more recent experiments.[109] The low energy resolution permits accurate measurements of both rotational and vibrational cross sections, as the threshold for the former process is of the order of tens of meV.

2. *Multiple scattering technique.* In this method electrons of the desired energy are injected into a region free of electric fields. The gas density is chosen such that the electrons undergo between 20 and 100 collisions as they move through the collision region. A small fraction of these collisions are inelastic, and electrons lose energy. The inelastically and elastically scattered electrons are separated by the application of retarding potentials at the collecting electrode. The attenuation of the elastically scattered component is then used to obtain the cross section for the inelastic process. The energy spread of the electrons is 200 meV or more and the technique is not applicable below 1.5 eV. The technique was used in the 1960s by Schulz and his colleagues[103] to measure vibrational cross sections in several nonpolar (H_2, N_2, O_2, CO_2) and polar (CO) gases.

3. *The swarm technique.* This method for the determination of rotational and vibrational cross sections consists of three procedures: (a) measuring the swarm parameters, that is the drift velocity of electrons (W_e) and the characteristic energy, defined as D/μ, where μ is the mobility of electrons, given by $W_e = \mu E$; (b) determining the electron energy distribution as a function of E/N; (c) calculating the swarm parameters and comparing them with the measured values in (a). Suitable adjustments are made to the assumed cross sections and the procedure is repeated till satisfactory agreement is obtained between measured and calculated parameters.

In the present context, the ratio E/N is the independent variable and in swarm experiments one independently varies E and N to make measurements of the required swarm parameter. The electron energy distribution is then obtained from the numerical solution of the Boltzmann equation by assuming an ε–Q curve, and the swarm parameters are calculated. A comparison between the measured and calculated swarm parameters, is used as a guide to change the assumed cross section, and the procedure is repeated iteratively. The same cross sections as a function of energy should yield swarm coefficients that agree with measured values over as wide a range of E/N as possible. The cross sections are independent of E/N and depend only on the electron energy.

The denominator in the parameter E/N is the number of molecules per unit volume at a given temperature and pressure. It is given by the equation

$$N = \frac{7.244 \times 10^{22} \times p}{T} \ \text{m}^{-3} \tag{2.11}$$

where p is the gas pressure in Pa and T the temperature in K. Equation 2.11 yields $N = 1.25 \times 10^{23}\,\text{m}^{-3}$, $3.54 \times 10^{22}\,\text{m}^{-3}$, and $3.22 \times 10^{22}\,\text{m}^{-3}$ at 77 K, 273 K, and 300 K respectively at 133 Pa (~ 1 torr) . The conversion factor to pascals (N/m^2) is obtained by 760 torr = 101.3 kPa. The parameter E/N has the dimension of V m^2 and has large negative exponentials ($\sim 10^{-22}$ to 10^{-19}). For convenient handling of numbers the unit 1 townsend (Td) = 10^{-21} V m^2 is used.

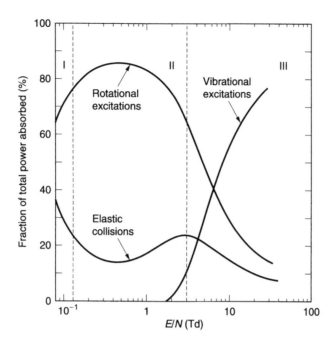

FIGURE 2.19 Fraction of total power absorbed as a function of E/N in molecular hydrogen. Region I, where elastic collision loss dominates, is shown enlarged for clarity. In region II rotational excitation and in region III vibrational excitation dominate. By an appropriate choice of E/N these cross sections may be derived as a function of electron energy. Reproduced from Morrison, M. A. et al., *Aust. J. Phys.*, 40, 239, 1987. With permission.

To decide upon the range of E/N suitable for calculating the rotational and vibrational cross sections we consider the loss of energy that occurs by several processes, as shown in Figure 2.19.[110] The fractional loss of energy of an electron in collision with a gas molecule depends upon the type of collision. The range of E/N may be divided into three regions, as shown. In region I, at very low values of E/N, there are very few electrons having energy greater than the rotational excitation threshold. In this range the momentum transfer cross section (Q_M) has the greatest influence on the swarm properties and hence this range is suitable for the determination of Q_M at low electron energies. In region II, the energy of individual electrons, and the mean energy of the swarm ($\langle\varepsilon\rangle$, eV), are not high enough to cause vibrational excitation. The fractional loss of power is almost entirely due to rotational excitations and its energy dependence may be determined in this range of E/N. In region III vibrational excitation sets in and increases rapidly, even as the rotational excitation and elastic collision losses decrease. This range is more appropriate for determining the energy dependence of the vibrational cross sections.

Note that lower values of E/N demand that higher values of N be used in experiments, to avoid the electrical field becoming unmanageably low from the practical point of view. As an example, at $E/N = 0.1$ Td and 10 m Pa gas pressure the electric field becomes ~ 3 V/m, which is very low from the experimental point of view. The gas pressure should be increased by a factor of at least 10^3 to achieve experimentally feasible values of E at low E/N. In the example quoted a gas pressure of ~ 10 kPa (0.1 atmospheres) is a convenient choice. England et al.[110] (1988) have used such high gas pressures to study drift velocities in mixtures of neon and hydrogen (1.2 to 2.9%) at $E/N = 0.12$ to 1.7 Td; the objective was to derive pure rotational cross sections in H_2.

Having measured the swarm parameters at the appropriate values of E/N, one calculates the electron energy distribution. The Boltzmann equation for the energy distribution is given by[1]

$$\frac{\partial F}{\partial t} + \mathbf{v} \bullet \nabla_r F_i + \mathbf{a} \bullet \nabla_v F_i = \left(\frac{\partial F}{\partial t}\right)_{coll} \tag{2.12}$$

Equation 2.12 is in notational form and is quite useless for practical application. Its meaning may briefly be described as follows. The energy distribution function for the ith particle is F_i which is a function of $(\mathbf{r}, \mathbf{v}, t)$, where $\mathbf{r}$ is the variable in the configuration space (x, y, z), $\mathbf{v}$ is the variable in the velocity space (v_x, v_y, v_z), and t is the time. $F_i (\mathbf{r}, \mathbf{v}, t) \, d^3r \, d^3v$ is the number of particles of species i in six-dimensional phase space at time t in the differential volume dx dy dz dv_x dv_y dv_z. The difficulty of finding the solution of this equation is mathematically insurmountable and one adopts the trick of separating the solution in the form

$$F(\mathbf{r}, \mathbf{v}, t) = n(\mathbf{r}, t) \cdot f(\mathbf{v}, t) \tag{2.13}$$

where $n(\mathbf{r}, t)$ is a spatially dependent function and $f(\mathbf{v}, t)$ is a spatially independent function. $f(\mathbf{v}, t)$ is expanded in spherical harmonics. The first two terms of the solution, denoted by $f_0(\varepsilon)$ and $f_1(\varepsilon)$, are generally known as the "two-term solution." In the steady state, of course, $f(\mathbf{v}, t) \equiv f(v)$.

Having determined the energy distribution, the next step in the process is to calculate the swarm coefficients using appropriate integrals (see Chapter 6, Section 6.3). Comparing the measured and calculated swarm properties, one revises the initially assumed $Q_M(\varepsilon)$ and the process is repeated. An iterative computation is carried out till there is a best fit with the final values of $Q_M(\varepsilon)$, W_e, and D_r/μ.

2.4.4 SEMIEMPIRICAL APPROACH

We have already described several experimental techniques to measure the total cross section in gases. The total excitation cross section may be obtained by summing all the individual excitation cross sections, that is, optical emission and metastable excitation, to every level. As already remarked, many of these excitation cross sections may be several orders of magnitude lower than the lowest ionization or vibrational excitation cross sections. From the discharge point of view such low cross sections do not make an appreciable difference in the analysis of energy distribution functions, Monte Carlo simulations, or energy analysis. An alternative way of arriving at the total cross section is to add the total cross section for individual processes; the resulting total cross section will be within about 5% of the true value. De Heer and Jansen,[111] and de Heer et al.[112] have adopted such an approach in helium and heavier rare gases respectively to arrive at individual cross sections. A detailed error analysis for both the experimental and theoretical values was adopted and these authors call their analysis semiempirical.

2.5 ATTACHMENT CROSS SECTION

Electron attachment is a process by which an electron attaches to a neutral atom or molecule. A number of different processes are possible. We restrict ourselves to dissociative attachment to a molecule, which is one of the most efficient ways of forming a negative ion. A low-energy electron collides with a diatomic or polyatomic molecule, causing dissociation, and attaches itself to one of the dissociation products. For certain molecules the cross section for

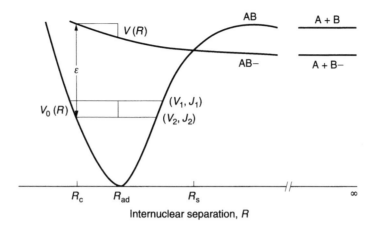

FIGURE 2.20 Mechanism of formation of negative ion by dissociation of a diatomic molecule.

dissociative attachment is strongly dependent on the rotational-vibrational cross section. If the molecule is excited this way initially, the cross section for negative ion formation by this process can increase dramatically, often by several orders of magnitude. Reviews of electron attachment to molecules at low energies are given by Domke[113] and Chutjian et al.[114]

The mechanism of formation of a temporary negative ion with an electron, near resonance energies, colliding with a diatomic molecule is

$$e_1 + AB_{v1} \rightarrow AB^- \begin{cases} \rightarrow A + B^- \\ \rightarrow e_2 + AB_{v2} \end{cases} \tag{2.14}$$

where e_1 and e_2 are the initial and final energies of the electron respectively, and v_1 and v_2 are the initial and final vibrational levels of the molecules respectively. The negative ion AB^- is in the resonance state and autodetaches after a lifetime of 10^{-14} s. A negative ion B^- is formed if the resonance state has a longer lifetime, whereas a molecule in the vibrational state AB_{v2} is formed otherwise. This mechanism is explained with reference to Figure 2.20.[114]

The potential energy of the neutral molecule is $V_0(R)$ and ε is the energy of the incident electron. The two nuclei are under the influence of the potential $V_0(R)$. The molecule is excited to the vibrational level (v_1, J_1) before attachment. After attachment the negative ion AB^- is formed. Its potential energy is shown by the curve designated as $V^-(R)$ and the two nuclei are under the influence of this potential. The two curves intersect at a internuclear separation of R_s and, for $R > R_s$, autodetachment is not permitted.

The probability of autodetachment is maximum when the energy difference between $V_0(R)$ and $V^-(R)$ is exactly equal to the energy of the incoming electron. Under these conditions the internuclear distance is R_c and is called the capture radius. If, however, autodetachment occurs when the internuclear distance has another value, such as R_{ad}, then the molecule acquires the vibrational level (v_2, J_2). If the lifetime of the resonance is longer, the nuclei move apart beyond a separation of R_s and autodetachment is not permitted. Dissociative attachment may occur, resulting in the formation of a stable negative ion. For some molecules and over a certain range of incident electron energies, it is possible to form more than one resonance state. Moreover, there can be more than one final electronic state of the molecule, following autodetachment of the electron in resonance. In either case, the total resonance widths can be viewed as the sum of several partial widths, each segment corresponding to an electronic state of the neutral molecule.[114]

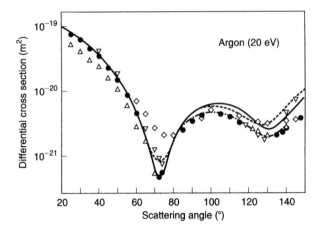

FIGURE 2.21 Differential cross sections for elastic scattering of 20 eV electrons in argon: (•) Panajotović et al.[121]; (Δ) Srivastava et al.[122]; (∇)Williams and Willis[123]; (◇) Mehr[124]; (——) McEachran and Stouffer[125]; (— —) Bartschat et al.[126]; (- - -) Nahar and Wadhera.[127]

Low-energy attachment in some gases such as SF_6 lead to the formation of the ion SF_6^- with the attachment cross section increasing as $\varepsilon^{-\frac{1}{2}}$, where ε is the electron energy. Attachment cross sections for low-energy electrons have been studied by a number of new techniques. We refer to a technique developed by Christophorou and colleagues[115–118] as it is more appropriate to the topics that follow. The method is known as the swarm technique and a brief description is as follows.

An electron source provides the initial electrons either at the cathode or at a well-defined plane close to the cathode. The electrons are not monoenergetic in this type of experiment, and their energy distribution is according to either Maxwell's distribution or some other function. The electron swarm, however, may be characterized by a mean energy $\langle \varepsilon \rangle$, and increasing values of E/N increase the mean energy.

The electron source in the experiments of Spyrou and Christophorou[117] is an alpha particle emitter, ^{252}Cf or ^{239}Pu, producing particles of energy 6.1 MeV/particle or 5.1 MeV/ particle. The decay of each ^{252}Cf alpha particle produces ∼2.3 × 10^5 electrons in argon and ∼1.7 × 10^5 electrons in nitrogen within a time of ∼5 ns at 133 kPa pressure and ∼0.2 ns at 3.2 Mpa.[116] The bursts of electrons obtained from the emitter are directed to drift towards the anode under the influence of a uniform electric field. The alpha particles are well-collimated so that they leave a well defined plane; this is essential for knowing the exact drift distance.

The pulse height distribution as a function of E/N is first recorded with a buffer gas (nitrogen or argon) in the drift tube and the number density of the target gas is added (1 part in 10^5 to 10^8) and the ratios of pulse height distributions are then recorded. This ratio is used to evaluate the reduced attachment coefficient η/N in units of m². The rate constant $K(\varepsilon)$, is related to the attachment coefficient according to

$$K(\varepsilon) = \frac{\eta}{N} W_e \qquad (2.15)$$

where W_e is the electron drift velocity. The drift velocity of the electrons is usually measured in a separate experiment with only the buffer gas present in the drift tube. Additional features are incorporated so that the pressure could be increased to ∼10 MPa and the temperature varied.

The attachment rate coefficient is related to the attachment cross section according to

$$k(\langle \varepsilon \rangle) = \left(\frac{2e}{m}\right)^{\frac{1}{2}} \int_0^\infty Q_a(\varepsilon)\, \varepsilon^{\frac{1}{2}} f(\varepsilon, E/N)\, d\varepsilon \qquad (2.16)$$

where m = mass of the electron, e = electronic charge, Q_a = attachment cross section, which is a function of electron energy, $f(\varepsilon)$ = electron energy distribution function, which is a function of E/N, and $\langle \varepsilon \rangle$ = mean energy. Calculation of $Q_a(\varepsilon)$ from Equation 2.16 requires a knowledge of the energy distribution function, making a special note that the rate constant is measured as a function of the mean energy of the electron swarm. A numerical solution of the Boltzmann equation is generally used, but the procedure depends upon the availability of elastic collision cross sections and vibrational excitation cross sections if the buffer gas is molecular.

Determining the electron energy distribution is, by itself, a formidable task, though its availability as a general tool has increased in recent years. For the unfolding procedure, one begins with a given E/N and determines the energy distribution function. The distribution function yields the mean energy also, and the measured rate constant for the calculated mean energy is used. For the deconvolution procedure to be successful, the same $Q_a(\varepsilon)$ should hold good for all values of $k(\langle \varepsilon \rangle)$ for which experiments have been carried out. An iterative procedure is required in which successive iterations between $Q_a(\varepsilon)$ and $k(\langle \varepsilon \rangle)$ are carried out. The computations are stopped when the attachment cross sections, evaluated in two successive iterations, attain a preset accuracy, usually $\sim 1\%$.

Three parameters are required for the computations: (1) the attachment rate at a selected mean energy $\langle \varepsilon_j \rangle$; (2) the energy distribution at the same mean energy $f[\varepsilon_j, (E/N)_j]$; and (3) the monoenergetic attachment rate $A(\varepsilon)$. The purpose of the iteration is to get the final curve Q_a–ε for which $A(\varepsilon)$ is the trial function. The three quantities are related according to

$$k(\langle \varepsilon_j \rangle) = \int_0^\infty A(\varepsilon)F[\varepsilon_j, (E/N)_j]\, d\varepsilon \qquad (2.17)$$

The iteration begins with a judicious choice of A–ε curve that makes $A(\varepsilon_{max})$ negligibly small at a high value of ε_{max}. Using Equation 2.17, $k_{calc}(\langle \varepsilon_j \rangle)$ is obtained and the ratio

$$\partial k \langle \varepsilon_j \rangle = \frac{k_{exp}\langle \varepsilon_j \rangle}{k_{calc}\langle \varepsilon_j \rangle} \qquad (2.18)$$

is computed. The procedure is repeated at other values of $\langle \varepsilon \rangle$ and these correction factors are then used to correct the previous $A(\varepsilon)$ so as to get the new $A(\varepsilon)$, according to

$$A_{n+1}(\varepsilon) = A_n(\varepsilon)\left[\frac{\sum\limits_j \partial k(\varepsilon_j)F[\langle \varepsilon_j \rangle, (E/N)_j]}{\sum\limits_j F[\langle \varepsilon_j \rangle, (E/N)_j]}\right] \qquad (2.19)$$

where A_{n+1} and A_n are two successive iterations. Equation 2.19 is now used in 2.17 and the computations are continued until the quantity

$$\frac{\sum\limits_j \left[k_{exp}(\langle \varepsilon_j \rangle) - k_{calc}(\langle \varepsilon_j \rangle)\right]^2}{\sum\limits_j \left[k_{exp}(\langle \varepsilon_j \rangle)\right]^2} \qquad (2.20)$$

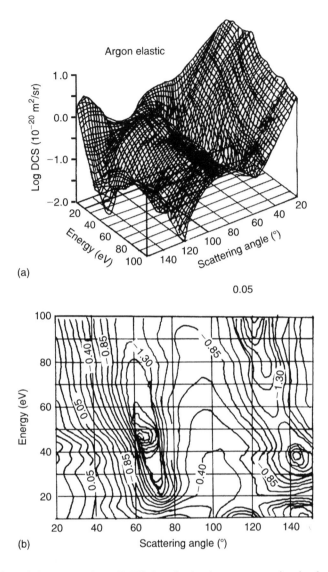

(a)

(b)

FIGURE 2.22 Differential cross sections (DCS) for elastic electron scattering in the angular range of 20° to 150° at incident energies of 10 to 100 eV: (a) three-dimensional representation of the DCS surface, shown on a logarithmic scale; (b) contours of constant DCS. Numbers correspond to the logarithm of the DCS values in units of 10^{-20} m^2 sr^{-1}. Figure reproduced from Panajotović, R. et al., *J. Phys. B*, 30, 5877, 1997. With permission of Institute of Physics, U.K.

attains a predetermined minimum value. The technique has been improved by Datskos et al.[119]

2.6   CONCLUDING REMARKS

The purpose of this chapter is to expose the reader to a number of different techniques that have been developed toward the measurement of cross sections for various elastic and inelastic processes. The number of parameters to be controlled and the accuracy of their measurement have improved steadily, and the complexity of the experiments has also increased. These aspects have been discussed by Filipelli et al.[120] for optical emission cross sections and by Buckman and Clark[100] for negative ion resonances.

Differential cross sections for elastic scattering have been a subject of considerable research in recent years. The cross sections usually show a minimum, and the smallest value that is measured is often used as a test of sensitivity of the experimental setup. They are also useful from theoretical points of view with regard to the shape of the curves, magnitude and dependence on energy. Calculated minima are very sensitive to atomic potentials with which the calculations are made, and the polarization of the scattered electrons reaches a maximum in the vicinity of the cross section minima. Theoretical electron polarization and total polarization are very sensitive to the scattering angle and the electron energy at the cross section minima.[121]

Panajotović et al.[121] have made extensive measurements of the differential elastic cross sections in argon in the energy range 10 to 100 eV, using a crossed beam technique. Figure 2.21 shows the differential cross sections measured for electrons of 20 eV energy. Comparisons are made with the data provided by Srivastava et al.,[122] Williams and Willis,[123] Mehr,[124] McEachran and Stauffer,[125] Bartschat et al.,[126] and Nahar and Wadhera.[127] Two minima are observed, designated as low energy and high energy minima, and compared with previous publications. The magnitudes of the minima and the scattering angles at which they occur are dependent on the electron energy. Figure 2.22 shows the three-dimensional plot and contour diagram for all energies investigated. The smallest values of the differential cross sections are found at $68.5° \pm 0.3°$, 41.30 ± 0.02 eV, and $143.5° \pm 0.3°$, 37.30 ± 0.02 eV.

A notable detail in the experimental procedure of Panajotović et al.[121] is the energy scale calibration. The standard procedure is to use the helium resonance structure of 2^2S at 19.34 eV or the resonances in argon at 11.08 and 11.27 eV. On the basis of the reasoning that calibration should be carried out at an energy level that is as close to the expected minima as possible and, if feasible, within the same gas, the energy scale was calibrated with the cadmium resonance that occurs at 29.07 eV with a width of 130 meV. A theoretical check of these measurements is provided by Sienkiewicz et al.[128]

Modern techniques for the measurement of ionization cross sections have been adopted by Rejoub et al.[129] in a large number of gases, employing a time-of-flight spectrometer and position-sensitive detector. Kobayashi et al.[130] have used a pulsed electron beam and pulsed ion extraction combined with a time-of-flight analysis of the charge to measure the ratios of multiple ionization cross sections. These developments will be referred to in the following chapters on cross section data.

REFERENCES

1. McDaniel, E. W., *Collision Phenomena in Ionized Gases*, John Wiley & Sons, New York, 1964.
2. Hasted, J. B., *Physics of Atomic Collisions*, 2nd ed., Am. Elsevier Pub. Co., New York, 1972.
3. Gilardini, A., *Low Energy Electrons in Gases*, John Wiley & Sons, New York, 1972.
4. Huxley, L. G. H. and R. W. Crompton, *The Diffusion and Drift of Electrons in Gases*, John Wiley & Sons, New York, 1974.
5. Bederson, B. and L. J. Kieffer, *Rev. Mod. Phys.*, 43, 601, 1971.
6. Ramsauer, C. and R. Kollath, *Ann. Phys. Lpz.*, 4, 91, 1931.
7. Brode, R. B., *Rev. Mod. Phys.*, 5, 257, 1933.
8. Golden, D. E. and H. W. Bandel, *Phys. Rev. A*, 138, 14, 1965.
9. Nickel, J. C., K. Imre, D. F. Register, and S. Trajmar, *J. Phys. B: At. Mol. Phys.*, 18, 125, 1985, Figure 1.
10. Ramsauer, C., *Ann. Phys., Lpz.*, (a) 64, 513, 1921; (b) 66, 546, 1921; (c) 72, 345, 1923.
11. Brode, R. B., *Phys. Rev.*, 25, 636, 1925; 39, 547, 1942.

12. Golden, D. E. and H. W. Bandel, *Phys. Rev. A*, 149, 58, 1966.
13. McDaniel, E. W., *Collision Phenomena in Ionized Gases*, John Wiley & Sons, New York, 1964, Chapter 4.
14. Dalba, G., P. Fornasini, I. Lazzizzera, G. Ranieri, and A. Zecca, *J. Phys. B: At. Mol. Phys.*, 12, 3787, 1979.
15. Nickel, J. C., K. Imre, D. F. Register, and S. Trajmar, *J. Phys. B: At. Mol. Phys.*, 18, 125, 1985; J. C. Nickel, C. Mott, I. Kanik, and D. C. McCollum, *J. Phys. B: At. Mol. Opt. Phys.*, 21, 1867, 1988. Covers gases CO and N_2.
16. Jost, K., P. G. F. Bisling, F. Eschen, M. Felsmann, and L. Walther, *Proc. 13th Int. Conf. on the Physics of Electronic Atomic* Collisions, Berlin, 1983, ed. J. Eichler et al., North Holland, Amsterdam, p. 91, cited by Nickel et al. (1985).
17. Kauppila, W. E., T. S. Stein, G. Jesion, M. S. Dababneh, and V. Pol, *Rev. Sci.*, 48, 822, 1977.
18. Wagenaar, R. W. and F. J. de Heer, *J. Phys. B: At. Mol. Phys.*, 13, 3855, 1980.
19. Kennerly, R. E. and R. A. Bonham, *Phys. Rev. A*, 17, 1844, 1978.
20. Land, J. E. and W. Raith, *Phys. Rev. A*, 9, 1592, 1974.
21. Baldwin, G. C. and S. I. Friedman, *Rev. Sci. Instr.*, 38, 519, 1967.
22. Nakai, M. Y., D. A. Labar, J. A. Harter, and R. D. Birkhoff, *Rev. Sci. Instr.*, 38, 820, 1967.
23. Buckman, S. J. and B. Lohmann, *J. Phys. B: At. Mol. Phys.*, 19, 2547, 1986.
24. Ferch, J., B. Granitza, C. Masche, and W. Raith, *J. Phys. B: At. Mol. Phys.*, 18, 967, 1985.
25. Haddad, G. N. and T. F. O'Malley, *Aust. J. Phys.*, 35, 35–9, 1982.
26. McEachran, R. P. and A. D. Stauffer, *J. Phys. B: At. Mol. Phys.*, 16, 4023, 1983.
27. Bell, K. L., N. S. Scott, and M. A. Lennon, *J. Phys. B: At. Mol. Phys.*, 17, 4757, 1984.
28. Kauppila, W. E., T. S. Stein, J. H. Smart, M. S. Dababneh, Y. K. Ho, J. P. Downing, and V. Pol, *Phys. Rev. A*, 24, 725, 1981.
29. Wagenaar, R. W. and F. J. de Heer, *J. Phys. B: At. Mol. Phys.*, 18, 2021, 1985.
30. Fon, W. C., K. A. Berrington, P. G. Burke, and A. Hibbert, *J. Phys. B: At. Mol. Phys.*, 16, 307, 1983.
31. Kumar, V., E. Krishnakumar, and K. P. Subramanian, *J. Phys. B: At. Mol. Phys.*, 20, 2899, 1987.
32. Subramanian, K. P. and Vijay Kumar, *J. Phys. B: At. Mol. Phys.*, 20, 5505, 1987.
33. Dababneh, M. S., W. E. Kauppila, J. P. Downing, F. Laperriere, V. Pol, J. H. Smart, and T. S. Stein, *Phys. Rev. A*, 2, 1872, 1980.
34. Lam, S. F., *J. Phys. B: At. Mol. Phys.*, 15, 119, 1982.
35. Haberland, R., L. Fritsche, and J. Noffke, *Phys. Rev. A*, 33, 2305, 1986.
36. Ramsauer, C. and R. Kollath, *Ann. Phys.*, 12, 529, 837, 1932.
37. Massey, H. S. W. and E. H. S. Burhop, *Electronic and Ionic Impact Phenomena*, Oxford University Press, Oxford, 1952, chapters I to IV.
38. Brunger, M. J., S. J. Buckman, D. S. Newman, and D. T. Alle, *J. Phys. B: At. Mol. Opt. Phys.*, 24, 1435, 1991.
39. Gulley, R. J., D. T. Alle, M. J. Brennan, M. J. Brunger, and S. J. Buckman, *J. Phys. B: At. Mol. Opt. Phys.*, 27, 2593, 1994.
40. Srivastava, S. K., A. Chutjian, and S. Trajmar, *J. Chem. Phys.*, 63, 2659, 1975.
41. Tanaka, H., T. Ishikawa, T. Masai, T. Sagara, and L. Boesten, *Phys. Rev.*, 57, 1798, 1998.
42. Qing, Z., M. J. M. Beerlage, and M. J. van der Wiel, *Physica*, 113C, 225, 1982.
43. Register, D. F., H. Nishimura, and S. Trajamar, *J. Phys. B: At. Mol. Opt. Phys.*, 13, 1651, 1980.
44. Tanaka, H., S. K. Srivatsava, and A. Chutjian, *J. Chem. Phys.*, 69, 5329, 1978.
45. McKoy, V., quoted by Nickel et al. (1988).
46. Jain, A., L. C. G. Freitas, M. T. Lee, and S. S. Tayal, *J. Phys. B: At. Mol. Phys.*, 17, L29, 1984.
47. Shyn, T. W. and G. R. Carignan, *Phys. Rev. A*, 22, 923, 1980.
48. Dubois, R. D. and M. E. Rudd, *J. Phys. B: At. Mol. Phys.*, 9, 2657, 1976.
49. Jansen, R. H. J., F. J. de Heer, H. J. Luyken, B. Van Wingergarden, and J. Blaaw, *J. Phys. B: At. Mol. Phys.*, 9, 185, 1976.
50. Zubek, M., N. Gulley, G. C. King, and F. H. Read, *J. Phys. B: At. Mol. Opt. Phys.*, 29, L239, 1996.
51. Cho, H., R. J. Gulley, and S. J. Buckman, *J. Phys. B: At. Mol. Opt. Phys.*, 33, L309, 2000.

52. Cubric, D., D. J. L. Mercer, J. M. Channing, G. C. King, and F. H. Read, *J. Phys. B: At. Mol. Opt. Phys.*, 32, L45, 1999.

53. Tate, J. T. and P. T. Smith, *Phys. Rev.*, 39, 270, 1932.

54. Kieffer, L. J. and H. Dunn, *Rev. Mod. Phys.*, 38, 1, 1966.

55. Asundi, R. K., J. D. Craggs, and M. V. Kurepa, *Proc. Phys. Soc. (Lond.)*, 82, 967, 1963.

56. Rapp, D. and P. Englander-Golden, *J. Chem. Phys.*, 43, 1464, 1965.

57. Schram, B. L., F. J. de Heer, M. J. van der Wiel, and J. Kistemaker, *Physica*, 31, 94, 1965.

58. Bleakney, W., *Phys. Rev.*, 34, 157, 1929.

59. Tate, J. T. and P. T. Smith, *Phys. Rev.*, 46, 773, 1934.

60. Harrison, H., quoted by Kieffer and Dunn[54], as their reference 62.

61. Fite, W. L. and R. T. Brackmann, *Phys. Rev.*, 112, 1141, 1958.

62. Shah, M. B., D. S. Elliot, P. McCallion, and H. B. Gilbody, *J. Phys. B: At. Mol. Opt. Phys.*, 21, 2751, 1988.

63. Dolder, K. T., M. F. A. Harrison, and P. C. Thonemann, *Proc. Roy. Soc. A*, 264, 367, 1961.

64. Cook, C. J. and J. R. Peterson, *Phys. Rev. Lett.*, 9, 164, 1962.

65. Montague, R. K., M. F. A. Harrison, and A. C. H. Smith, *J. Phys. B: At. Mol. Opt. Phys.*, 17, 3295, 1984.

66. Wetzel, R. C., F.A. Baiocchi, T. R. Hayes, and R. S. Freund, *Phys. Rev. A*, 35, 559, 1987.

67. Hasted, J. B., *Physics of Atomic Collisions*, 2nd ed., Am. Elsevier Pub. Co., New York, 1972, Figure 12.2.

68. Stephan, K., H. Helm, and T. D. Märk, *J. Chem. Phys.*, 73, 3763, 1980.

69. Shah, M. B., D. S. Elliot, and H. B. Gilbody, *J. Phys. B: At. Mol. Phys.*, 20, 3501, 1987.

70. Winters, R. E., J. H. Collins, and W. L. Courchene, *J. Chem. Phys.*, 45, 1931, 1966.

71. Straub, H. C., P. Renault, B. G. Lindsay, K. A. Smith, and R. F. Stebbings, *Phys. Rev. A*, 52, 1115, 1995.

72. Kobayashi, A., G. Fujiki, A. Okaji, and T. Masuoka, *J. Phys. B: At. Mol. Opt. Phys.*, 35, 2087, 2002. Rare gases are measured.

73. Tian, C. and C. R. Vidal, *J. Phys. B: At. Mol. Opt. Phys.*, 31, 895, 1998. Measured gases are CO, CH_4, and C_2H_4; *J. Phys. B: At. Mol. Opt. Phys.*, 31, 5369, 1998. Measured gases are N_2 and O_2.

74. Orient, O. J. and S. K. Srivastava, *J. Chem. Phys.* 78, 2949, 1983. Measured gases are H_2O, CO, CO_2, and CH_4.

75. Crowe, A. and J. W. McConkey, *J. Phys. B: At. Mol. Phys.*, 7, 349, 1974.

76. Chatham, H., D. Hills, R. Robertson, and A. Gallagher, *J. Chem. Phys.*, 81, 1770, 1984.

77. Krishnakumar, E. and S. K. Srivastava, *J. Phys. B: At. Mol. Opt. Phys.*, 21, 1055, 1988. Measured gases are He, Ne, Ar, Kr, Xe; *J. Phys. B: At. Mol. Phys.*, 23, 1893, 1990. Measured gas is N_2.

78. Rao, M. V. V. S. and S. K. Srivastava, *J. Phys. B: At. Mol. Opt. Phys.*, 25, 2175, 1992.

79. Syage, J. A., *Phys. Rev. A*, 46, 5666, 1992.

80. Syage, J. A., *J. Chem. Phys.*, 97, 6085, 1992.

81. Zheng, S.-H. and S. K. Srivastava, *J. Phys. B: At. Mol. Opt. Phys.*, 29, 3235, 1996.

82. Mangan, M. A., B. G. Líndsay, and R. F. Stebbings, *J. Phys. B: At. Mol. Opt. Phys.*, 33, 3225, 2000.

83. Freund, R. S., R. C. Wetzel, and R. J. Shul, *Phys. Rev. A*, 41, 5861, 1990.

84. Hayes, T. R., R. C. Wetzel, F. A. Biaocchi, and R. S. Freund, *J. Chem. Phys.*, 88, 823, 1995.

85. Basner, R., M. Schmidt, H. Deutsch, V. Tarnovsky, A. Levin, and K. Becker, *J. Chem. Phys.*, 103, 211, 1995.

86. Tarnovsky, V., A. Levin, H. Deutsch, and K. Becker, *J. Phys. B: At. Mol. Opt. Phys.*, 29, 139, 1996.

87. Adamczyk, B., A. J. H. Boreboom, and M. Lukasiewicz, *Int. J. Mass. Spectr. Ion Process.*, 9, 407, 1972.

88. Märk, T. D. and E. Hille, *J. Chem. Phys.*, 69, 2492, 1978.

89. Grill, V., G. Walder, D. Margreiter, T. Ruth, H. U. Poll, and R. D. Märk, *Z. Phys.*, 25, 217, 1993.

90. Maier-Leibnitz, H., *Z. Phys.*, 95, 499, 1936.

91. Dorrestein, R., *Physica*, 9, 447, 1942.
92. Woudenberg, J. P. M. and J. M. W. Milatz, *Physica*, 8, 871, 1941.
93. Moiseiwitsch, B. L. and J. L. Smith, *Rev. Mod. Phys.*, 40, 238, 1968.
94. Schulz, G. J. and R. E. Fox, *Phys. Rev.*, 106, 1179, 1957.
95. Kaur, S., R. Srivastava, R. P. McEachran, and A. D. Stauffer, *J. Phys. B: At. Mol. Phys.*, 31, 157, 1998.
96. Kaur, S., R. Srivastava, R. P. McEachran, and A. D. Stauffer, *J. Phys. B: At. Mol. Phys.*, 31, 4833, 1998.
97. Zecca, A., G. P. Karwasz, and R. S. Brusa, *La Rivista del Nuovo Cimento*, 19(3), 1, 1996.
98. Van der Bergt, P. J. M., W. B. Westerweld, and J. S. Risley, *J. Phys. Chem. Ref. Data*, 18, 1757, 1989.
99. Heddle, D. W. O. and J. W. Gallagher, *Rev. Mod. Phys.*, 61, 221, 1989.
100. Buckman, S. J. and C. W. Clark, *Rev. Mod. Phys.*, 66, 539, 1994.
101. Schulz, G. J., *Phys. Rev.*, 112, 150, 1958; *Phys. Rev. Lett.*, 13, 477, 1964. The second paper deals with excitation cross sections in helium, shown in Figure 2.16.
102. Fabrikant, I. I., O. B. Shpenik, A. V. Snegursky, and A. N. Zavilopulo, *Phys. Rep.*, 159, 1, 1988.
103. Schulz, G. J., *Rev. Mod. Phys.*, 45, 378, 1973.
104. Schulz, G. J., *Phys. Rev.*, 125, 229, 1962. This paper deals with vibrational excitation cross sections in nitrogen.
105. Mason, N. J. and W. R. Newell, *J. Phys. B: At. Mol. Opt. Phys.*, 20, 1357, 1987.
106. Zubek, M., B. Mielewska, J. Channing, G. C. King, and F. H. Read, *J. Phys. B: At. Mol. Opt. Phys.*, 32, 1351, 1999.
107. Philips, M. H., L. W. Anderson, C. L. Chun, and R. E. Miers, *Phys. Lett. A*, 82, 404, 1981.
108. Phelps, A. V., *Rev. Mod. Phys.*, 40, 399, 1968.
109. Randell, J., S. L. Hunt, G. Mroztek, J.-P. Ziesel, and D. Field, *J. Phys. B: At. Mol. Opt. Phys.*, 27, 2369, 1994.
110. England, J. P., M. T. Elford, and R. W. Crompton, *Aust. J. Phys.*, 41, 573, 1988; Morrison, M. A., R. W. Crompton, B. C. Saha, and Z. Lj. Petrović, *Aust. J. Phys.*, 40, 239, 1987.
111. De Heer, F. J. and R. H. J. Jansen, *J. Phys. B: At. Mol. Phys.*, 10, 3741, 1977.
112. De Heer, F. J., R. H. J. Jansen, and W. Van der Kaay, *J. Phys. B: At. Mol. Phys.*, 12, 979, 1979.
113. Domke, W., *Phys. Rep.*, 208, 97, 1991.
114. Chutjian, A., A. Garscadden, and J. M. Wadhera, *Phys. Rep.*, 264, 393, 1996.
115. Christophorou, L. G., *Atomic and Molecular Radiation Physics*, Wiley-Interscience, New York, 1971.
116. Hunter, S. R. and L. G. Christophorou, *J. Chem. Phys.*, 80, 6150, 1984.
117. Spyrou, S. M. and L. G. Christophorou, *J. Chem. Phys.*, 82, 2620, 1985.
118. Spyrou, S. M., S. R. Hunter, and L. G. Christophorou, *J. Chem. Phys.*, 83, 641, 1985.
119. Datskos, P. G., L. G. Christophorou, and J. G. Carter, *J. Chem. Phys.*, 98, 8607, 1993.
120. Filipelli, A. R., C. C. Lin, L. W. Anderson, and J. W. McConkey, in *Advances in Atomic, Molecular and Optical Physics*, eds. B. Bederson and H. Walther, vol. 33 (cross section data), Academic Press, New York, 1994.
121. Panajotović, R., D. Filipović, B.Marinković, V. Pejčev, M. Kurepa, and I. Vušković, *J. Phys. B: At. Mol. Phys.*, 30, 5877, 1997.
122. Srivastava, S. K., H. Tanaka, A. Chutjian, and S. Trajamar, *Phys. Rev. A*, 23, 2156, 1981.
123. Williams, J. F. and B. Willis, *J. Phys. B: At. Mol. Phys.*, 8, 1670, 1975.
124. Mehr, J., *Z. Phys.*, 198, 345, 1967.
125. McEachran, R. P. and A. D. Stauffer, *Phys. Lett. A*, 107, 397, 1985.
126. Bartschat, K., R. P. McEachran, and A. D. Stauffer, *J. Phys. B: At. Mol. Opt. Phys.*, 21, 2789, 1988.
127. Nahar, S. N. and J. M. Wadhera, *Phys. Rev.*, 35, 2051, 1987.
128. Sienkiewicz, J. E., V. Konopińska, S. Telega, and P. Syty, *J. Phys. B: At. Mol. Opt. Phys.*, 34, L409, 2001.
129. Rejoub, R., B. G. Lindsay, and R. F. Stebbings, *Phys. Rev. A*, 65, 2713, 2002.
130. Kobayashi, A., G. Fujiki, A. Okaji, and T. Masuoka, *J. Phys. B: At. Mol. Opt. Phys.*, 35, 2087, 2002.

3 Data on Cross Sections—I. Rare Gases

Electron–neutral scattering cross sections are important in a number of areas that deal with discharges, plasma chemistry, fusion research, the Boltzmann equation, Monte Carlo simulation, swarm experiments, etc.[1] Data on cross sections are one of the most fundamental requirements in the study of gaseous electronics, beginning with the measurements of Ramsauer[2] and Brode.[3] A large number of measurements were carried out in several gases and a summary of these early investigations is given by Massey and Burhop,[4] Bederson and Kieffer,[5] Brandsen and McDowell,[6] and in the books by McDaniel[7] and Hasted.[8] More recent compilations of cross section data are published by Trajmar and McKonkey[9] and Zecca et al.[10]

The total cross section is the sum of elastic and all inelastic scatterings. The latter are composed of excitation and ionization cross sections. There is momentum transfer during both elastic and inelastic scatterings and correspondingly the total momentum transfer is the sum of the momentum transfer for elastic and inelastic scattering. Similarly, differential scattering cross sections for elastic scatterings are different from those for excitation scatterings or vibrational excitation and rotation. Integration of the differential cross sections over the appropriate angles yields the corresponding integrated elastic or inelastic scattering cross section. In the following sections we present data for the cross sections in rare gases, predominantly relying on experimental values.

The cross sections compiled in this chapter are restricted mainly to the literature published since 1979, though reference has been made to earlier data if discussion is warranted. Experimental data have been preferred, though theoretical results are selected on the basis of necessity. Further, certain cross sections, such as elastic and momentum transfer, are measured at selected electron energies and recourse to theory is essential to fill the gaps. For each gas considered, a check on the total cross sections expressed as the sum of the individual cross sections has been provided and compared with the measured total cross sections. Electron–atom interaction is termed collision in the classical sense and scattering in the wave mechanical sense. Although we use the latter term, the former term is used occasionally and interchangeably.

3.1 ARGON

Argon gas has been studied from the discharge point of view for more than 100 years. Interest in this gas is due to several important fundamental and technical considerations. Its presence in high concentration plays a major role in the performance of the high pressure Ar–Kr–F_2 laser system and in direct nuclear-pumped lasing media using He–Ar mixtures.[11]

Argon gas has been subjected to intense study since the discovery of a minimum at low electron energies by Ramsauer[12,13] and Townsend and Bailey.[14,15] Easy availability of the gas and the interest aroused by the successful explanation of the effect by the application of the quantum mechanical method[16] added further interest. We summarize the data obtained during the past twenty-five years or so. Table 3.1 shows selected references for cross section measurements in argon. It is convenient to include other rare gases as well in this table, as several of these gases are usually measured by the same group, using the same apparatus.

An attempt to provide relative agreement between results of several methods and several groups is beset with the problem of choosing a reference cross section. Each group of authors

TABLE 3.1
Selected References of Scattering Cross Section Data in Rare Gases

Type	Gas	Energy Range	Method	Source
Q_T	Ar, He	0.1–100	Direct beam	Brode[3]
Q_T	—	20–400	Direct beam	Normand[21]
Q_{el}	—	16–40		Bullard and Massey[22]
Q_{diff}	—	0.6–12.5	Direct beam	Ramsauer et al.[23]
Q_{diff}	Ar, Kr	25–950	Beam–static gas	Webb[24]
Q_i	Ar, He, Ne, Kr, Xe	ε_i–100 eV	Beam–static gas	Asundi et al.[25]
Q_M	Ar	0–20	Swarm coefficients	Engelhardt et al.[26]
Q_i	Rare gases		Review	Kieffer et al.[27]
Q_{ex}	Ar	10–2000	Crossed beam	McConkey et al.[28]
Q_i	He, Ar	Onset–500	Beam–static gas	Fletcher et al.[29]
Q_i	Ar, He, Ne	ε_I–1000	Beam–static gas	Rapp et al.[30]
Q_T	He, Ar	0.1–20	Ramsauer	Golden and Bandel[31]
Q_i	Ar, He, Kr, Ne, Xe	0.5k–16k	Mass spectrometer	Schram et al.[32] (1966)
Q_i	Ar, He, Kr, Ne, Xe	0.6k–20k	Beam–static gas	Schram et al.[32] (1966)
Q_T	Ar	5–15	Transmission	Kauppila et al.[33]
Q_{el}, Q_T, $Q_{in,T}$, Q_{ex}	He	15–3000	Semiempirical	De Heer et al.[34]
Q_T	He, Ne	0.25–31	Beam transmission	Stein et al.[35]
Q_T	He	1–50	Time-of-flight	Kennerly et al.[36]
Q_{el}, Q_T, $Q_{in,T}$, Q_{ex}	Ar, Kr, Ne, Xe	20–3000	Semiempirical	de Heer et al.[37]
	He			Nesbet[38]
Q_{diff}	Ar, He, Ne	0.5–20 eV	Crossed beam	Williams[39]
Q_T	He	100–1400	Modified Ramsauer	Dalba et al.[40]
Q_T	Ar, He	500–5000	Beam–static gas	Nagy et al.[41]
Q_T	He, Ne	16–700	Transmission	Blaauw et al.[42]
Q_{el}, Q_{diff}, Q_m	He	5–200	Crossed beam	Register et al.[43]
Q_T	Ar, Ne, Kr	25–750	Transmission	Wagenaar et al.[44]
Q_{ex}	He	50–2000	Spectrometer	Van Zyl et al.[45]
Q_T	Ne	0–2.18	Swarm data	O'Malley et al.[46]
Q_{ex}	Ar, Kr, Xe	11.5–19.9	Swarm method	Specht et al.[47]
Q_i	Ar, He, Ne, Kr	Onset–200	Mass spectrometer	Stephan et al.[48]
Q_{ex}	Ar	16–100	Crossed beam	Chutjian et al.[11]
Q_T	Kr, Xe	1.9–750	Transmission	Dababneh et al.[49]
Q_M, Q_{diff}	Ar, Kr	3–100	Crossed beam	Srivastava et al.[50]
Q_T	Ar, He, Ne	15–800	Transmission, employing H	Kauppila et al.[51]
Q (set)	Ar, Xe	1–500	Compilation	Kücükarpaci et al.[52]
Q_M	Ar	0–4	Swarm analysis	Haddad et al.[53]
Q_{diff}	Ar, Kr	10–50	Crossed beam	Zhou Qing et al.[54]
Q_T	Ar	500–3000	Transmission, with H	Nogueira et al.[55]
Q_T	Ar, Kr, Xe	0.05–60	Transmission	Jost et al.[56]
Q_{diff}	Ar	3–150	Theory	Fon et al.[57]
Q_M, Q_{ex}	Xe	0–10000	Swarm method	Hayashi[58] (1983)
Q_i	Ar, Kr	Onset–150	Mass spectrometer	Mathur et al.[59] (1984)
Q_T	Ar, Ne	0–50	Theory	McEachran et al.[60]
Q_M	—	0–100	Swarm analysis	Yamabe et al.[61]
				Bell et al.[62]
Q_M	Ar	0.002–10	Compilation	Hunter et al.[63]
Q_d, Q_M,	Ne	1–100	Crossed beam	Register et al.[64]
Q_T	Ar, He, Ne, Xe	4–300	Linear transmission	Nickel et al.[65]

(continued)

TABLE 3.1
Continued

Type	Gas	Energy Range	Method	Source
Q_T	Ar	0.08–20.0	Time-of-flight, employing H	Ferch et al.[66]
Q_T	Ar, Kr, Xe	20–100	Transmission	Wagenaar et al.[67]
Q_T	Ar, He	0.12–20.0	Time-of-flight	Buckman et al.[68]
Q_T	Ar, Kr, Ne	700–6000	Transmission	Garcia et al.[69]
Q_{diff}	Ar, He, Ne, Kr, Xe	20–200	Linear transmission	Wagenaar et al.[70]
Q_T	Kr	0.175–20	TOF spectrometer	Buckman et al.[71]
Q_{diff}, Q_{el}, Q_M	Ar	300–1000	Crossed beam with H	Iga et al.[72]
Q_{Met}	Ar	12–142	Time-of-flight	Mason et al.[73]
Q_M, Q_{el}	Ar	3–300	Theory	Nahar et al.[74]
Q_i	Ar, He, Ne, Kr	0–200	Fast neutral beam	Wetzel et al.[75]
Q_T	Ar, Kr, Xe	0.7–10	PE spectroscopy	Subramanian et al.[76]
Q_T	He, Ne	0.7–10	PE spectroscopy	Kumar et al.[77]
Q_T	Ar, Ne	100–3000	Modified Ramsauer	Zecca et al.[78]
Q_{diff}	Ne	20–100	Relative flow	Nickel et al.[79]
Q_{el}	Ar	2.5–15	Swarm analysis	Nakamura[80]
Q_{i1}, Q_{i2}	He	26.6–10,000	Pulsed crossed beam	Shah et al.[81]
Q_i	All rare gases	ε_i–1000	Pulsed electron beam and mass spectrometer	Krishnakumar et al.[82]
Q_i	Ar, He, Kr, Ne, Xe	ε_i–10,000	Review	Lennon et al.[83]
Q_{diff}	Ar, Kr, Xe	0.05–2.0	Electron beam	Weyherter et al.[84]
Q_T and Q_{diff}	Ar	3–20	Time-of-flight	Furst et al.[85]
Q_T	Kr, Xe	81–4000	Modified Ramsauer	Zecca et al.[86]
Q_T	Kr	5–300	Linear attenuation	Kanik et al.[87]
Q_i	Ar^+, Ar^{2+}, Ar^{3+}	20–100	Time-of-flight MS	Bruce et al.[88]
Q_i	Ar^+, Ar^{2+}, Ar^{3+}	20–100	Time-of-flight MS	Ma et al.[89]
Q_{ex}	He	Onset–500	EI spectrometer	Cartwright et al.[90]
Q_{ex}	He	30–500	Theory	Trajmar et al.[91]
Q_M	Ar, He, Kr, Xe.	0.001–100	Swarm analysis	Pack et al.[92]
Q_i	Kr and Xe	0–470 eV	Time-of-flight	Syage[93]
Q_i	Ar^+ to Ar^{5+}	18–5300	Time of flight	McCallion et al.[94]
Q_{el}	He	1.5–50	Crossed beam	Brunger et al.[95]
Q_{diff}, Q_T	Ne	0.1–7.0	Crossed beam	Gulley et al.[96]
Q_{ex}	Ar	Onset–60	Optical absorption	Mityureva et al.[97]
Q_i	Ar	Onset–1000	Time-of-flight	Straub et al.[98]
Q_{diff}, Q_{el}, Q_M	Ar	1–10	Crossed beam	Gibson et al.[99]
Q_{ex}	Ar	11–1000	Photon counting	Tsurubuchi et al.[100]
Q_{diff}	Ar	10	Spectrometer	Zubek et al.[101]
Q (set)	—	0.001–1000	Compilation	Pitchford et al.[102]
Q_T, Q_{ex}, Q_i	Ar, Ne, Kr, Xe	10–10,000	Semiempirical	Brusa et al.[17]. For a correction see Zecca et al.[18].
Q_{diff}	Ar, Kr	20–235	EI spectrometer	Cvejanivić and Crowe[103]
Q_{diff}	Ar	10–100	Crossed beam	Panajotović et al.[104]
Q_i	Ne	140–1000	Electron beam	Sorokin et al.[105]
Q_{diff}	Ar, He, Kr, Ne, Xe	—	Spectrometer	Zubek et al.[106]
Q_m, Q_{ex}, Q_i	Ar	—	Semiempirical	Phelps et al.[107]
Q_i	Ar, Kr, Xe	140–1000	Electron beam	Sorokin et al.[108]
Q_{ex}	Ar	20–80	EI spectrometer	Filipovic et al.[109]
Q_{ex}	Ar	20–80	EI Spectrometer	Filipovic et al.[110]
Q_T	Ar, Ne, Kr, Xe	0.5 eV–10 keV	Semiempirical	Zecca et al.[18]
Q_{diff}	Ar	10.3–160.3	Theory	Sienkiewicz et al.[111]
Q_i	Ar, Ne, Kr, Xe	Onset–1000	Time-of-flight	Kobayashi et al.[112]

Q_T = total, Q_M = momentum transfer, Q_{el} = elastic, Q_{diff} = differential, Q_{ex} = excitation, $Q_{ex,T}$ = total excitation, Q_{Met} = metastable, Q_i = ionization, $Q_{in,T}$ = total inelastic, Q (set) = complete set. H = magnetic field, EI = electron impact, PE = photoelectron, MS = mass spectrometer. Energy in units of eV.

generally chooses its own values against which other measurements are compared. A compilation of these results, therefore, necessarily involves an arbitrary choice. We choose the analytical equations of Brusa et al.[17] and Zecca et al.[18] as reference, as they cover a broader range of electron energies. Further, the analytical equation provides the advantage of easier calculation at odd values of the electron energy (say 13.48 eV) without resorting to excessive interpolation.

A few remarks about the analytical equations for expressing the cross sections as a function of electron energy are in order. A number of equations are available in the literature for representing various cross sections. Such equations are useful for presenting data or a large number of gases, as otherwise one has to resort to extensive tables that are not the most efficient way of presenting data, particularly if a large energy range covering several decades is involved.

Further, the analytic equations are helpful to evaluate quickly the cross sections at a given energy, particularly in situations where repeated reference to closely spaced electron energies is required. Boltzmann equation analysis and Monte Carlo simulations are examples of such requirement. The functions for cross sections will not, of course, be in perfect agreement with measurements and the discrepancy between individual sets of measurements and the values obtained by the analytical function should be made available. The user can then determine the degree of accuracy required for a specific application and choose the data that satisfy this requirement.

The analytical functions for elastic scattering of electrons are useful for finding the zero energy cross section $Q(\varepsilon = 0)$ by extrapolation. This quantity is equated to $4\pi A^2$, where A has the dimension of length and is known as the scattering length. It appears in the expression for s-wave scattering shifts in the quantum mechanical description of scattering phenomena,[19] providing a link between theory and the analytic function. The relation of the analytical functions to the theory has been discussed by Inokuti et al.[20] and the types of functions useful for representing several different cross sections are also provided. The generally preferred analytical functions for the scattering cross sections are based on the Born–Bethe approximation, with fitting parameters for each gas. This procedure has been found to be more applicable in the high energy range, > 500 eV.

Zecca et al.,[113–115] March et al.,[116] and Brusa et al.[17] have, however, generated functions for both molecular and rare gases and the expressions are easy to use. One has to bear in mind that evaluation of the fitting parameters involves choosing data from one or few groups, thereby indirectly bestowing them the status of reference values because all other measurements are compared to the selected data. However, on the basis of stated accuracy of the measurements, the close agreement between the results of several groups renders the procedure justifiable. We adopt the procedure of Zecca et al.[10] in presenting data. However, the required number of decades is chosen in preference to their two decades, and a percentage difference graph is added for each cross section as calculated from the tabulated values in the original reference.

3.1.1 Total and Momentum Transfer Cross Sections in Ar ($20 \leq \varepsilon \leq 1000$ eV)

The significance of the lower energy considered in this section is that it is to the right side of the maximum of the ε–Q_T curve where analytical expressions of the type (3.1) apply.

De Heer and Jansen[34] calculated the total cross section of argon by considering the experimental and theoretical data available till then (1977) and their method is usually referred to as semiempirical in the literature. Their tabulated data were used by Zecca et al.[78] to evaluate a sixth-order double logarithmic dependence of the total cross section on energy (i.e., log Q_T–log ε); the resulting coefficients in order of ascending power of ε are: 0.673 24,

0.758 06, −0.253 55, −0.105 097, 0.047 59, −0.005 22, −0.000 06. A more recent function is
due to Brusa et al.[17] and Zecca et al.[18]

The analytical function for the total cross section given by Brusa et al.[17] is

$$Q_T = \frac{1}{A(B+\varepsilon)} + \frac{1}{C(D+\varepsilon)} + \frac{2}{\varepsilon}\sqrt{\frac{BD}{AC}}\frac{1}{|B-D|}\left|\ln\left(\frac{\frac{\varepsilon}{D}+1}{\frac{\varepsilon}{B}+1}\right)\right| \qquad (3.1)$$

in which A, B, C, and D are fitting parameters and ε is the electron energy in keV. The
parameters and the range of electron energies for which the analytic function (3.1) is
applicable are shown in Table 3.2. Data that are used to obtain the fitting parameters are
as follows: 10 to 300 eV range: Nickel et al.[65] for helium, neon, argon, xenon; Kanik et al.[87] in
xenon; in the 100 to 4000 eV range: Dalba et al.[40] for helium; Zecca et al.[78] for neon and
argon; Zecca et al.[86] for krypton and xenon. The criteria adopted for the choice of data were:

1. A wide coverage of energy range should be available to ensure better consistency.[17]
2. Data to come preferably from the same group measuring all the noble gases, to
   minimize systematic errors.

The number of experimental points ranged from 34 (neon) to 47 (krypton). In the
discussion of total cross sections in some rare gases we distinguish two ranges: (1) 20 to
1000 eV, (2) 0 to 20 eV. Since the same equation applies to several rare gases, a reference is
made to them. A comparison of selected results may be found in a recent reference.[199]

Figure 3.1 shows the total cross section in argon, calculated according to the function 3.1
and using the fitting parameters shown in Table 3.2. The curve is almost obscured by the

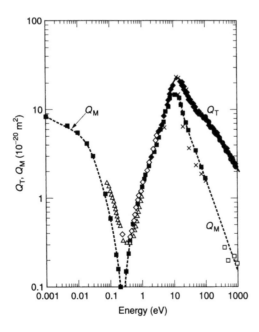

FIGURE 3.1 Total and momentum transfer cross sections in argon. Total cross sections: ▲, de Heer et al.[37];
●, Kauppila et al.[51]; ×, Jost et al.[56]; ○, Nickel et al.[65]; ◇, Ferch et al.,[66]; +, Wagenaar and de Heer[67]; △,
Buckman and Lohmann[68]; □, Zecca et al.[78]; ◆, Subramanian and Kumar[76]; —, empirical formula of
Brusa et al.[17] with the correction shown in Zecca et al.[18] Momentum transfer cross sections: ×,
Srivastava et al.[50]; ○, Iga et al.[72]; ■, Pack et al.[92] The broken line is drawn as a guide to the eyes.

TABLE 3.2
Parameters for the Total Cross Section and the Energy Range for the Analytical Function (Equation 3.1)

Gas	Energy Range (eV)	A (keV^{-1} × 10^{-20} m^2)	B (keV)	C (keV^{-1} × 10^{-20} m^2)	D (keV)
He	10–3000	37.2	0.0032	5.37	0.330
Ne	40–3000	4.120	0.106	2.07	1.031
Ar	14–3000	0.341	0.424	4.00	0
Kr	12–4000	0.261	0.986	3.59	0.0127
Xe	90–4000	0.101	20.16	0.407	0.219

Reproduced from Brusa et al., *Z. Phys. D*, 38, 279, 1996. With kind permission of Springer Science and Business Media.

density of experimental points lying on it, indicating the degree of accuracy that the function has achieved. The scattering cross sections are much larger in the heavier rare gases than in helium or neon, particularly at low electron energies. For example, at 10 eV electron impact energy the total cross section in argon is 20.06×10^{-20} m^2, which is approximately 400% of the cross section in helium at the same impact energy.

Momentum transfer cross sections measured by Srivastava et al.,[50] Iga et al.,[72] and Pack et al.[92] are also shown in Figure 3.1. This cross section controls the diffusion phenomenon and the ratio of diffusion coefficient to mobility of gases (D/μ). Srivastava et al.[50] employed an apparatus that consists of an electron scattering spectrometer, gas flow system, and a multichannel analyzer for detecting and storing the scattered electron signal. The experimental gas effused through a capillary, producing a well-defined beam of atoms at 90° to the electron beam. Differential cross sections covered the electron energy range of 3 to 100 eV and an angular range from 20° to 135°; normalization was with respect to helium. The differential elastic cross sections were integrated to derive momentum transfer and elastic scattering cross sections.

The momentum transfer cross sections cannot be measured below a certain minimum energy, due to experimental limitations. To extend the range of the electron energies toward the lower limit, Pack et al.[92] employed swarm experimental results to derive the ratio of the longitudinal diffusion coefficient (D_L) to mobility (μ) in the electron energy range from 0 to 100 eV. The principle of this method has been briefly touched upon in Chapter 2 and essentially consists of solving the Boltzmann equation and calculating the swarm coefficients. A comparison with experimental results is then used to apply corrections to the initial cross sections by a backward prolongation technique, first used by Frost et al.[145] The technique will be considered further in Chapter 6.

The Ramsauer–Townsend minimum in the momentum transfer cross section obtained by the swarm method is deeper at 0.25 eV with a cross section of 0.09×10^{-20} m^2, as shown in Table 3.3. In the range of 0.4 to 8 eV electron energy, the momentum transfer cross section agrees very well with the total cross sections of several investigators. For energies greater than 10 eV the momentum transfer cross section begins to get smaller than the total cross section; the differences increase with increasing energy.

Momentum transfer cross sections in the higher energy range from 300 to 1000 eV have been obtained by Iga et al.[72] in a crossed beam geometry with a small magnetic field (1.5 μT). Elastic differential cross sections are measured and, by integration, the integrated elastic and momentum transfer cross sections are derived in the usual manner. The values lie smoothly on the curve connecting the points of Pack et al. (see the broken line in Figure 3.1). There is

TABLE 3.3
Momentum Transfer Cross Section in Argon

Energy	Q_M	Energy	Q_M	Energy	Q_M
0.000	7.5	0.35	0.235	15	14.1
0.001	7.5	0.400	0.33	20	11.00
0.005	6.1	0.500	0.51	25	9.45
0.010	4.60	0.700	0.86	30	8.74
0.020	3.25	1.00	1.38	50	6.9
0.050	1.73	1.20	1.66	75	5.85
0.070	1.13	1.30	1.82	100	5.25
0.100	0.59	1.70	2.30	150	4.24
0.150	0.23	2.10	2.80	200	3.76
0.170	0.16	2.50	3.30	300	3.02
0.200	0.103	3.00	4.10	500	2.1
0.250	0.091	5.0	6.7	700	1.64
0.300	0.153	10.0	15	1000	1.21

Electron energy in units of eV and cross section in units of $10^{-20}\,\mathrm{m}^2$.
From Pitchford, L. C. et al., www.siglo.com. Courtesy of main author.

paucity of data in the 100 to 300 eV range of electron energies. Table 3.3 shows the momentum transfer cross sections.

3.1.2 TOTAL CROSS SECTIONS IN Ar ($0 < \varepsilon \leq 20$ eV)

Total scattering cross sections in this range of electron energy have been of particular interest from the earliest studies of Ramsauer and Townsend. The Ramsauer–Townsend minimum in rare gases occurs in this range of impact energies and the phenomenon of very low cross sections in the 0 to 3 eV range is of interest from the theoretical aspect of application of quantum mechanics to the argon atom.

The experimental technique of Golden and Bandel,[31] who used the Ramsauer beam technique, has been described in Section 2.1.1. Ferch et al.[66] used time-of-flight spectroscopy in the electron impact energy range of 0.08 to 20 eV. Milloy et al.[117] calculated the momentum transfer cross sections in argon, using the drift velocity, and D_T/μ up to 4 eV energy. Experiments carried out to determine the differential elastic cross sections also yield the momentum transfer cross section, by integration,

$$Q_M = 2\pi \int_0^\pi Q_{\mathrm{diff}}(1 - \cos\theta) \sin\theta d\theta \qquad (3.2)$$

These results are summarized in Figure 3.1. While the cross section minimum is observed to occur at approximately 0.25 to 0.285 eV, the magnitude of the minimum differs by a factor of four, the highest being observed by Ferch et al.[66] and the lowest by Pack et al.[92] (Table 3.4). The dominance of the p-wave contribution to the total cross section in the Ramsauer–Townsend region increases the forward scattering. It is possible that the results of Golden and Bandel[31] are too low in this energy region because of poor discrimination against forward scattering of electrons. In the time-of-flight method adopted by Buckman and Lohmann,[68] the energy resolution is estimated to be 20 meV whereas the Ramsauer–Townsend minimum has a width in excess of 150 meV. However, as the energy increases towards 1 eV the discrepancy between the results of Ferch et al.[66] and those of Buckman and

TABLE 3.4
Ramsauer–Townsend Minimum Cross Section in Argon

Authors	Method	Energy	Cross section
Golden and Bandel[31]	Total	0.28	0.15
Milloy et al.[117]	Momentum transfer	0.25	0.095
Ferch et al.[66]	Total	0.34	0.31
Buckman et al.[68]	Total	0.3	0.31
Pack et al.[92]	Momentum transfer	0.25	0.091

Energy in units of eV and cross section in units of $10^{-20}\,\mathrm{m}^2$.

Lohmann[68] increases, reaching a maximum of 12%. The reasons for this discrepancy are not clearly understood.[68]

The total cross section extrapolated to zero energy gives a cross section, $Q_T(0)$, that is usually expressed in terms of a length according to $Q_T(0) = 4\pi A^2$, where A is called the scattering length. At zero energy the total cross section contains only the s-wave scattering and equals the momentum transfer cross section as determined by swarm experiments. The scattering length determined by Pack et al.[92] is $89.2 \times 10^{-12}\,\mathrm{m}$. Earlier values are quoted by Zecca et al.[10]

3.1.3 ELASTIC AND DIFFERENTIAL CROSS SECTIONS IN Ar

The procedure followed for presenting the total cross sections is also adopted for presenting the elastic scattering cross sections. Brusa et al.[17] have demonstrated that Equation 3.1 is applicable for this purpose, with different values for the fitting parameters A, B, C, D for all the rare gases except helium, in the energy range 10 to 10,000 eV. The values of these coefficients are shown in Table 3.5. The elastic scattering cross section measurements usually have an uncertainty of 20 to 30%, arising from the fact that the measured differential cross sections are integrated over the scattering angle. The measurements are not carried out over the entire range of angles from 0 to 180°, and extrapolation is required at each end of the range. This extrapolation can introduce large errors. Further, for the determination of absolute differential cross sections, normalization is required, using other known experimental or theoretical cross sections.

The elastic scattering cross sections used for finding the fitting parameters were obtained indirectly by Brusa et al.[17] by subtracting the inelastic scattering cross sections from the total cross sections. This procedure yielded cross sections that were not dependent on direct measurements and the parameters evaluated are not biased towards an individual set of data.

Figure 3.2 shows selected elastic scattering cross sections published by several groups, along with the calculated values. For electron energies greater than 20 eV, the total cross section given by Zecca et al.[18] is also given again, to convey the contribution of inelastic scatterings to the total cross section. De Heer et al.[37] evaluated a set of cross sections in argon. The elastic scattering cross section is obtained by integrating the differential cross sections of several publications up to 1979, using the equation

$$Q_{el} = 2\pi \int_0^\pi Q_{diff}(\theta) \sin\theta \, d\theta \qquad (3.3)$$

TABLE 3.5
Fitting Parameters for Elastic Scattering Cross Sections in Rare Gases

Gas	No. of Points	Energy Range (eV)	A (keV^{-1} 10^{-20} m^2)	B (keV^{-1})	C (keV^{-1} 10^{-20} m^2)	D (keV)
Ar	52	10–10,000	0.632	0.593	3.65	0
Kr	52	10–10,000	4.115	0.0017	0.388	1.35
Ne	52	10–10,000	7.7076	0.0381	3.64	1.11
Xe	48	80–10,000	1.210	0.203	0.121	13.60

From Brusa, R. S. et al., *Z. Phys. D*, 38, 278, 1996. With kind permission of Springer Science and Business Media.

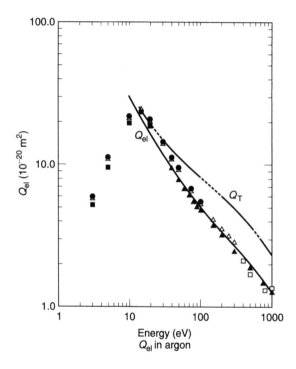

FIGURE 3.2 Integral elastic scattering cross sections in argon. The full line is the semiempirical formula of Zecca et al.[18]; (▲) semiempirical data of de Heer et al.[37]; (●) Srivastava et al.[50]; (□) Iga et al.[72]; (△) Nahar and Wadhera[74]; (■) Furst et al.[85]; (○) Gibson et al.[99] Total cross section according to Zecca et al.,[18] broken curve, is included to show the contribution of inelastic scattering cross sections. Data calculated by the present author (unpublished).

where Q_{diff} is the differential elastic cross section. The theoretical calculations of Nahar and Wadhera[74] in the energy range 3 to 300 eV are also shown. There is extremely good agreement between their values and the experimental results of Srivastava[50] over the entire range of electron energy. The measurements of Furst et al.[85] in the range 3 to 20 eV again show reasonable agreement. The higher energy range (400 to 1000 eV) is covered by Iga et al.[72] whose values lie smoothly on the semiempirical curve. A comparison of selected investigations of Q_M is given in a recent publication by the present author.[199]

Differential cross section data available up to 1977 have been reviewed by Bransden and McDowell.[118] Since then, accurate measurements have been published by: Srivastava et al.,[50]

3 to 100 eV, angular range 20° to 135°; Wagenaar et al.,[70] 20 to 200 eV, angular range 0° to 10°; Weyherter et al.,[84] 0.05 to 2 eV; Zubek et al.,[106] 10 eV; and Cvejanović and Crowe,[103] 20.4 to 110 eV, angular range 40° to 120°. The experimental setup of Wagenaar et al.[70] has the special feature of measuring the cross sections at low angles, with a resolution of 0.25°, at energies less than 100 eV. The experiments of Weyherter et al.[84] and Gibson et al.[99] have the special feature of extending the lower range of electron energy, covered by Zubek et al.,[106] to higher angles up to 180°, and the results of Cvejanović and Crowe[103] to continuously variable electron energy.

The differential scattering cross section is a function of both the angle of scattering and the electron energy. The direct determination of absolute cross sections is exceptionally difficult and the usual method adopted to achieve this goal is to measure the relative differential cross sections and then convert them to an absolute scale by normalization to cross sections of targets for which the absolute cross sections are reliably known.[85] Helium and neon targets are used for the purpose of normalizing the elastic scattering cross sections.

The normalization is accomplished by two different techniques: at sufficiently low energies where only elastic scattering occurs in rare gases, phase shifts of some of the lowest order partial waves may be obtained by fitting the angular distribution measurements. These extracted phase shifts, together with higher order phase shifts obtained by the Born approximation, are used to place the differential cross sections on an absolute scale. This method, known as the phase shift method,[50] has been successful in helium, making that gas a preferred choice for normalization. The second method is the relative flow technique that has been explained in Chapter 2. Briefly, the cross section in the target gas is measured and, under identical conditions, the target gas is replaced by helium and measurements are carried out at the same angles. The ratios of measured cross sections are related to the flow rates and the absolute cross section of the target gas is determined. This technique, known as the relative flow normalization technique, has also been employed in argon,[50] though the agreement between the two methods has been poor. The angular and energy dependence of the differential cross section shows that two steep minima occur in rare gases except in helium; the explanation for this experimental observation has been provided by Furst et al.[85] In helium the s-wave phase shift is dominant at all energies, though there is a negligibly small contribution from the p-wave phase shifts.[39] In other rare gases, with increasing energy the p- and d-waves make a larger contribution.

Figure 3.3 shows the angular variation of the cross sections at selected energies.[50,99,103] Srivastava et al.[50] normalized their measurements to helium, Gibson et al.[99] applied the phase shift analysis, and Cvejanović and Crowe[103] normalized their relative cross sections to the absolute cross sections of Srivastava et al.[50] The agreement between the measurements is good over the overlapping regions of energy.

At 1 eV and 3 eV there is only a broad hint of minimum in the experimental results, whereas at 10 eV the minimum is clearly seen. For a heavy rare gas atom such as argon, the validity of phase shift analysis at any energy is not generally successful. Due to the higher polarizability of the atom, higher order partial waves make a greater contribution to the scattering cross section at any energy. The effect of the higher order contribution is to cause a more pronounced forward peak in the differential scattering cross section. The relatively large d-wave contribution causes distinct minima in the differential cross sections at low energies. As the energy increases the d-wave contribution dominates, resulting in a shift of the minima with electron energy.[85] At 75 and 100 eV two minima can be seen (Figure 3.3), and theoretical calculations by Nahar and Wadhera[74] and others confirm the existence of two minima at higher energies.

A three-dimensional plot of the differential cross sections as a function of the scattering angle at various energies shows two global minima, one at a low angle and the other at a higher angle. This aspect of differential scattering has been referred to in Figure 2.22 of the

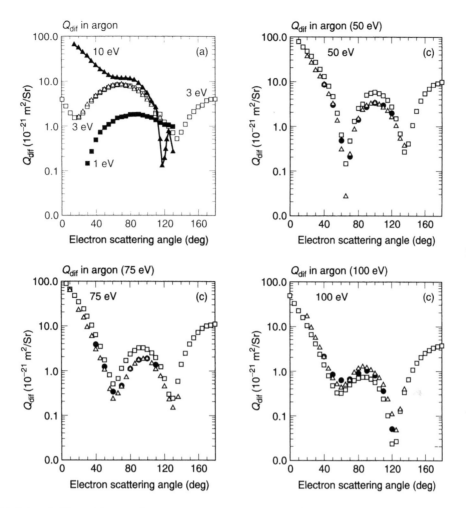

FIGURE 3.3 Differential scattering cross sections in argon as a function of scattering angle at various electron impact energies. Both experimental and theoretical results are shown. Due to experimental limitations, experimental results do not cover the entire range of the scattering angle. (a): (■) Gibson et al.,[99] 1 eV; (△) Srivastava et al.,[50] 3 eV; (□) Nahar and Wadhera,[74] 3 eV; (–▲–) Gibson et al.,[99] 10 eV. (b): (●) Cvejanović and Crowe,[103] 50 eV; (△) Srivastava et al.[50]; (□) Nahar and Wadhera.[74] (c): 75 eV. (d): 100 eV. Symbols for (c) and (d) same as for (b). Compiled by the author (unpublished).

previous chapter.[104] Theoretical verification of these results is provided by Sienkiewicz et al.,[111] Figure 3.4, and the observed critical minima, (39.3 eV, 68°) and (39.5 eV, 141°), agree with the measurements of Panajotović et al.[104] The critical minimum points serve as a check for the accuracy of measurements and calculations in argon.

3.1.4 Total Excitation Cross Sections in Ar

Excitation of atoms and molecules by electron impact results in the output of radiation and the generation of metastable states which return to ground state by indirect means, such as being excited back to the allowed level by scattering. The total excitation cross section is required for the purposes of discharge simulation, finding the solution of Boltzmann equation, and so on. Detailed discussion of excitation to every allowed level will not fall into the domain of our requirement; in argon alone there are over 75 lines. A review of electron

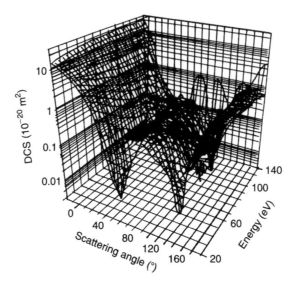

FIGURE 3.4 Differential cross sections in argon as a function of angle of scattering and electron impact energy, calculated by Sienkiewicz et al.[111] Two global minima are observed. For experimental results see Figure 2.22. Figure reproduced from Sienkiewicz, J. E. et al., *J. Phys. B*, 33, 2081, 2000. With permission of Institute of Physics, U.K.

impact excitation cross sections in atoms covering up to 1968 is given by Moiseiwitsch and Smith.[119] Electron excitation of metastable atoms has been reviewed by Fabrikant et al.[120] and electron impact optical excitation functions are reviewed by Heddle and Gallagher.[121]

Argon has 18 electrons, with the outer electron shell having the principle quantum number $n = 3$ in the $3s^2 p^6$ configuration. Within 14.3 eV of the ground state there are nearly thirty electronic states. Of the first four levels, level 1 (4s[3/2]$_2$, onset energy 11.548 eV) and level 3 (4s'[1/2]$_0$, onset energy 11.675 eV) are metastables; their lifetimes are 55.9 and 44.9 s respectively.[110] Level 2 (4s[3/2]$_1$, onset energy 11.631 eV) and level 4 (4s'[1/2]$_1$, onset energy 11.723 eV) radiate to ground level with emission wavelengths of 106.6 nm and 104.8 nm respectively (Table 3.5). A partial energy level diagram of argon is shown in Figure 3.5.

The cross sections for 23 states, due to Chutjian and Cartwright,[11] in the electron energy range of 16 to 100 eV, covering both the forward and backward hemispheres, have been recognized as benchmark measurements. The adopted values of electron energies are 16, 20, 30, 50, and 100 eV; the energy losses of the scattered electrons are measured. The scattering angle ranges from 5° to 138° and differential excitation cross sections are extrapolated to 0° and 180° to yield integral cross sections. The sums of their cross sections are shown in Figure 3.6. The total cross section reaches a peak in the region of 30 to 50 eV with a cross section of 0.9×10^{-20} m^2. Hayashi[122] has provided cross sections to an additional two states, making 25 in all, with the added advantage that the cross sections are presented in a range of electron energies, up to 1000 eV.

Optical measurements of excitation cross sections are classified as apparent excitation cross sections and cascade excitation cross sections. An electron beam traverses the gas, exciting some atoms to level i. As they fall to a lower level j, the resulting radiation is detected to measure the optical emission cross sections (Q_{ij}^{opt}). The sum of all emissions to lower states is termed the apparent excitation cross sections for the level i:

$$Q_i^{app}(\text{ex}) = \sum_{j<i} Q_{ij}^{app}(\text{ex}) \tag{3.4}$$

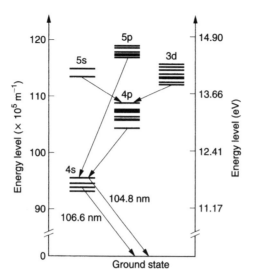

FIGURE 3.5 Partial energy level diagram of argon.

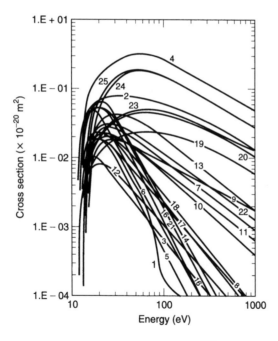

FIGURE 3.6 Excitation cross sections for 25 levels of argon.[122] Plotting is due to Saeed Ul-Haq. See reference [20] of Phelps and Petrović.[107] The numbers are levels as labeled in reference [199].

A level i may be populated both by direct electron impact excitation and by higher exciting levels cascading into it. The cascade cross section is the sum of all optical cross sections for the transition into the level from higher levels:

$$Q_i^{\text{cas}}(\text{ex}) = \sum_{k>i} Q_{ki}^{\text{opt}}(\text{ex}) \tag{3.5}$$

The direct electronic excitation cross section is obtained as the difference between the apparent cross section and the cascade contribution:[9]

$$Q_i^{dir}(ex) = Q_i^{app}(ex) - Q_i^{cas}(ex) \tag{3.6}$$

For example, McConkey and Donaldson[28] measured the apparent excitation cross sections by optical methods. Most of the cascade contribution in argon is in the infrared region and a Fourier transform spectrometer has been used by Chilton et al.[123] to determine the direct excitation cross section.

Cross sections for some or all of the first four levels, consisting of two metastable states and two resonance levels, have been measured by the following: McKonkey and Donaldson,[28] level 2 and level 4; Mason and Newell,[73] total metastables including level 1 and level 3; Mityureva and Smirnoff,[97] level 1 and level 3; Tsurubuchi et al.,[100] level 2 and level 4; Filipović et al.,[109] level 4; Filipović et al.,[110] levels 1, 2, and 3. The values of Tsurubuchi[100] for the lowest resonance states and the values of Mason and Newell[73] for the total metastable levels are shown separately in Figure 3.7. The more recent results of Chilton et al.[123] are not included as tabulated results are not given in the original publication. The total excitation scattering cross section obtained by adding these two cross sections in the range of 13 to 142 eV, merging smoothly with the cross sections of Tsurubuchi[100] for energies greater than 142 eV, are also shown. A comparison is shown in Table 3.6.

As the previous discussion reveals there are relatively very few data available on the total excitation cross section in the range from onset to 1000 eV. At onset energies in the range 12 to 30 eV there is considerable disagreement between various results. One of the more frequently used total excitation cross sections in argon is due to de Heer et al.,[37] who covered

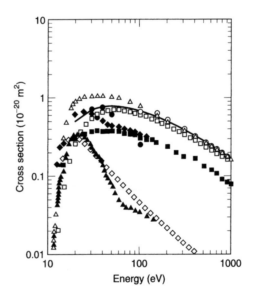

FIGURE 3.7 Selected excitation cross sections in argon as a function of electron energy. (○) de Heer et al.,[37] total excitation, semiempirical; (●) Chutjian and Cartwright,[11] sum of 23 excited states, crossed beam electron impact method; (▲) Mason and Newell,[73] total metastable excitation, time-of-flight method; (■) Tsurubuchi et al.,[100] sum of 106.6 nm and 104.8 nm radiations, measurement of total cascade cross section; (◆) calculated by the present author using Tsurubuchi's data, total excitation; (—) Brusa et al.,[17] semiempirical equation. (△) Hayashi,[122] sum of 25 states; (◇) Pitchford et al.,[102] (1996), total metastable, swarm calculation; (□) Pitchford et al.,[102] resonance levels, swarm calculations. Compiled by the author (unpublished).

TABLE 3.6

Comparison of Measured Excitation Cross Sections in Argon in Units of $10^{-23}\,\mathrm{m}^2$

Level 1 (4s[3/2]$_2$) [$^3P_2$], 11.55 eV

		Energy (eV)			
	Range (eV)	**20**	**40**	**60**	**80**
Chutjian et al.[11]	16–100	31.9			
Mityureva et al.[97]	Onset–60	300	145		
Schappe et al.[200]	Onset–100	40	7.6		
Filipović et al.[109]	20–80	53	18		

Level 2 (4s[3/2]$_1$) [$^3P_1$], 11.623 eV

		Energy (eV)			
	Range (eV)	**20**	**40**	**50**	**80**
McConkey et al.[28]	20–200	50	83	92	86
Chutjian et al.[11]	16–100	37		57.8	
Tsurubuchi et al.[100]	Onset–1000	153	135	120	110
Filipović et al.[110]	20–80	69	93	74	53

Level 3 (4s'[1/2]$_0$) [$^3P_0$], 11.72 eV

		Energy (eV)			
	Range (eV)	**20**	**40**	**50**	**80**
Chutjian et al.[11]	16–100	6.44			
Mityureva et al.[97]	Onset–60	75			
Schappe et al.[200]	Onset–100	13.5	3.4		
Filipović et al.[110]	20–80	12	6.3		

Level 4 (4s'[1/2]$_1$) [$^1P_1$], 11.83 eV

		Energy (eV)					
	Range (eV)	**16**	**20**	**30**	**40**	**50**	**60**
McConkey et.al.[28]	20–200		170	250	260	220	225
Chutjian et al.[11]	16–100	85.3	100	144		214	
Tsurubuchi et al.[100]	Onset–1000		145	194	237	257	269
Filipović et al.[109]	20–80	150	176	297	313	295	193
Mason & Newell[73a]	12–142	195	345	251	139	80	50

[a]Total metastable cross section only.

the range of 20 to 4000 eV. Since Tsurubuchi's results,[100] shown in Figure 3.9, cover only two optical levels, their cross sections are lower than those due to de Heer et al.[37]

Brusa et al.[17] have suggested a semiempirical expression from threshold to several keV energy,

$$Q_{\mathrm{ex}} = \frac{1}{F(G+\varepsilon)} \ln \frac{\varepsilon}{\varepsilon_{\mathrm{exc}}} \tag{3.7}$$

where ε is the incident electron energy, F and G the two adjustable parameters, and $\varepsilon_{\mathrm{ex}}$ is the excitation threshold energy (Table 3.7).

The sources for finding the fitting parameters are the following: He—Cartwright et al.[90]; Ne, Ar, Kr—de Heer et al.[37]; Xe—Hayashi.[58] The equation agrees very well with the semiempirical values of de Heer et al.[37] over the entire range of 20 to 1000 eV. In this range the total excitation cross sections adopted by Pitchford et al.[102] for solving the Boltzmann equation agree very well with the equation of Brusa et al.[17]

3.1.5 IONIZATION CROSS SECTIONS IN Ar

The technique and results of measurements of ionization cross sections in argon by Rapp and Englander-Golden[30] have been presented in Figures 2.13 and 2.14, respectively. These measurements, made in 1965, have stood the test of time, with a quoted uncertainty of 7%, and they may be employed as a basis to determine the differences between more recent literature. The methods adopted fall into the beam–static–gas category, though they employed the gas flow method to determine the pressure instead of a McLeod gauge. The total ionization cross sections are measured by collecting all the ions generated. Though the concept is rather simple, accurate measurements have proved to be difficult, due to the fact that uncertainties exist in the determination of various parameters involved.[27]

Partial ionization cross sections are defined as cross sections that describe the production by electron impact of an ion of specific charge n, according to $e + X \rightarrow X^n + (n + 1)e$. To determine the partial cross sections a mass spectrometer is required, and the earlier results of the 1930s suffered from errors due to the assumption of a constant ion collection efficiency for ions of differing charges. Stephan et al.[48] improved the mass spectrometer method, ensuring controlled extraction and collection of ions. Their results provided cross section data in all rare gases from the same laboratory using the mass spectrometer method. Relative cross sections were obtained, and from these absolute values were determined by normalizing to the cross section of Rapp and Englander-Golden[30] at a specific energy.

Electrons colliding with gas neutrals cause ionization if and only if their energy is equal to or higher than the first ionization potential (neglecting autoionization). Ionization cross sections may be measured by electron impact or photon impact; the ratio of the cross sections determined by both methods has been exploited to improve the accuracy of electron impact cross sections; this aspect will be referred to later on. As the electron energy increases, double or multiple ionization occurs and the total ionization cross sections (Q_i) may then be expressed in two different ways, depending upon the technique employed for measuring the cross section. In the first, the partial sum of the cross section at each ionizational potential is added according to

$$Q_{i,count} = Q_i^+ + Q_i^{2+} + Q_i^{3+} + \cdots \tag{3.8}$$

where Q_i^+ is the cross section for a singly charged ion, Q_i^{2+} is the cross section for doubly charged ion, and so on. The gross ionization cross section, on the other hand, is the charge-weighted sum of the partial ionization cross sections:

$$Q_{i,gross} = Q_i^+ + 2Q_i^{2+} + 3Q_i^{3+} + \cdots \tag{3.9}$$

Methods that employ measurement of total ion current fall into this category.

Rapp and Englander-Golden[30] measured the total ionization cross sections, but partial ionization cross sections could not be determined by their method. For energies below ~45 eV, however, only the first ionization cross section exists in argon and the partial and total cross sections are the same. For higher energies, the ratio of Q_{count}/Q_{gross} decreases from a value of 0.995 at 50 eV to 0.927 at 1000 eV.[37] Figures 3.8 and 3.9 show selected data of total and partial ionization cross sections in argon respectively.

Reviews of ionization cross sections are published by Kieffer and Dunn[27] and by Lennon et al.[83] Differences in the early results of Asundi and Kurepa[25] and Schram et al.[32] when compared with the results of Rapp and Englander-Golden[30] show that the path length and difficulties of pressure measurement are the sources of error. Contamination of mercury vapor with its low ionization potential was also a significant source of error.

Selected measured cross sections are shown in Figure 3.8. For the purpose of clarity of presentation the available data are presented as two sets of curves. In the first set (lower curve) the cross sections as determined by several authors are shown. In the second set (top curve) the cross sections are multiplied by a factor of ten. Fletcher and Cowling[29]

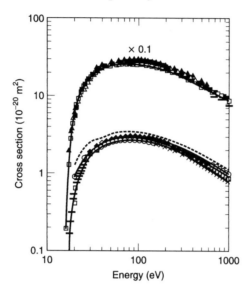

FIGURE 3.8 Absolute total ionization cross sections in argon. The data are presented by two curves for more clarity. The values shown by the top curve should be multiplied by 0.1. Experimental: (◆) Fletcher and Cowling[29]; (◇) Nagy et al.[41]; (◇) Stephan et al.[48]; (− −) Wetzel et al.[75]; (△) Krishnakumar and Srivastava[82]; (▲) Ma et al.[89]; (○) McCallion et al.[94]; (−■−) Syage[93]; (□) Straub et al.,[98]; (×) Sorokin et al.[108]; (−□−) Kobayashi et al.[124]; Semiempirical values: (●) de Heer et al.[37]; (— · —) Lennon et al.[83]; (——) Brusa et al.[17]; (−) Phelps and Petrović[107]; (●−●) this volume. Figure compiled by Raju (2002) (unpublished).

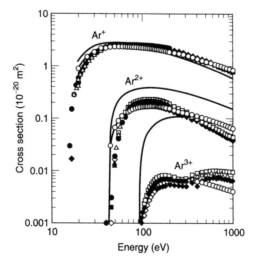

FIGURE 3.9 Partial ionization cross sections in argon. (◇) Stephan et al.[48]; (●) Wetzel et al.[75]; (□) Krishnakumar et al.[82]; (■) Ma et al.[89]; (△) McCallion et al.[94]; (◆) Straub et al.[98]; (—) Lennon et al.[83]; (−□−) Kobayashi et al.[124]; (○−○) Raju (2002), this volume. The results of Mathur and Badrinath (1984, 1986), and Syage[93] are not shown.

TABLE 3.7
Parameters for Total Excitation Cross Sections in Argon

Gas	No. of Points	Energy Range (eV)	ε_{ex} (eV)	F (keV^{-1} 10^{-20} m^2)	G (keV)
Ar	22	20–3000	11.5	25.19	23.6×10^{-3}
He	11	20–500	19.8	77.65	0
Ne	21	30–4000	16.619	85.97	31.7×10^{-3}
Kr	22	20–4000	9.915	22.0	23.3×10^{-3}
Xe	18	80–1000	8.315	18.27	0

From Brusa, R. S. et al., Z. *Phys. D*, 38, 279, 1996. With kind permission of Springer Science and Business Media.

have measured the absolute total cross sections in the range of 16 to 500 eV by using a pulsed electron gun. A special feature of their setup is that the ionization cross sections and ionization coefficients are measured simultaneously so that the purity of the gas remains the same in both measurements. The technique adopted for cross section measurements is similar to that of Rapp and Englander-Golden[30]: measuring the total ion current produced by a monoenergetic electron beam of well defined energy, passing through a layer of gas, in a beam–static–gas configuration.

Nagy et al.[41] have measured the absolute total cross sections in the relatively higher range of 0.5 to 5 keV by the beam–static–gas configuration method. Just three values (500, 700, and 1000 eV) are available in the range of our interest. A special feature of Nagy's setup is that the target gas is introduced into the source volume by two wide tubes and not through a collimated structure. Further, the scattering cell is divided into two interconnected parts, to ensure homogeneity of the target gas number density.

The measurements of Wetzel et al.[75] of the partial ionization cross sections in a crossed beam apparatus have a stated overall accuracy of 15%. Their experimental method is already given in Chapter 2 and covers the range of onset to 200 eV. The results of Stephan et al.[48] agree very well with those of Wetzel et al. and are therefore not shown. The partial cross sections given by Wetzel et al.[75] have been converted to gross cross sections by the present author and are shown in Figure 3.10. The double hump observed in the cross section of argon^{3+} is interesting; the first hump occurs at an energy of 180 eV and is due to direct multiple ionization. The second hump is much broader and has an onset energy of 250 eV; this hump is due to the Auger process.[82] In this process an inner shell electron is ionized by the impacting electron and an outer electron, in transferring into the vacancy, gives the transition energy to another outer electron. If this energy is high enough the outer electron is also ejected, thus leaving the atom ionized to one higher energy level.[27]

Krishnakumar and Srivastava[82] have measured the partial ionization cross sections in rare gases from threshold to 1000 eV using a pulsed electron beam and ion extraction technique. They used quadrupole mass spectrometer to collect charged particles and have demonstrated that failure to correct for ion transmission losses and to optimize ion extraction efficiency can lead to considerable errors in the measured cross sections.[89] They normalized their relative data with the results of Rapp and Englander-Golden[30] for energies below the production of doubly charged ions. This range of electron energy is often relevant to electron swarm simulation studies. The absolute cross sections of the multiply ionized species are determined from the relative flow technique. The total cross section is obtained by summation of the individual cross sections, ensuring proper weighting of the individual partial cross sections. A notable feature of the results obtained is that partial cross sections are obtained in all the rare gases, up to 1000 eV electron energy, in the same setup.

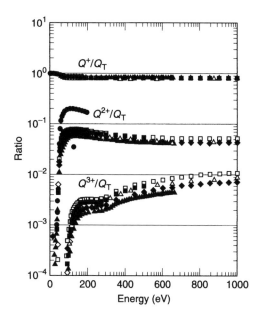

FIGURE 3.10 Ratios of partial cross sections to total ionization cross sections in argon. (◇) Stephan et al.[48]; (●) Wetzel et al.[75]; (□) Krishnakumar and Srivastava[82]; (■) Ma et al.[89]; (▲) Syage[93]; (△) McCallion et al.[94]; (◆) Straub et al.[98] Data compiled by the author (unpublished).

Ma et al.[89] have studied the absolute partial cross sections in argon, using an electron impact spectrometer that uses a pulsed electron source and time-of-flight detection of electrons and ions. The time-of-flight method is thought to overcome the serious transmission losses in electrostatic analyzers and mass spectrometers. The range of electron energies is from onset to 500 eV and absolute partial cross sections up to triply charged ions are measured. The absolute total cross section shown in Figure 3.10 is computed by the present author from their tabulated absolute partial cross sections. These measurements were repeated by Bruce and Bonham,[88] who obtained results greater by about 15%. McCallion et al.[94] have extended the time-of-flight technique to measure partial ionization cross sections, for five-fold charged ions, from onset energy up to 5300 eV. The absolute total cross section shown in Figure 3.8 is computed by the present author from their tabulated absolute partial cross sections.

The main sources of uncertainty in the measurements of electron ionization cross sections are[105]: (1) absolute measurement of the number of electrons striking the molecules; (2) the number of ions created; (3) the interaction path length; (4) the detector efficiency for differently charged ions; and (5) the target gas density at a pressure of the order of 10 mPa. The first four sources may be easily eliminated by adopting the measurement of cross sections by photon flux, and the measurements may be carried out at relatively higher pressure of the order of 100 Pa. The cross sections that have been determined in rare gases using photon beams are well known with an uncertainty of 1 to 3%, which is a considerable improvement over the electron impact ionization measurements.

Sorokin et al.[108] have used this idea to determine the absolute total electron impact ionization cross sections by measuring the total ion yield for both methods of ionization, in the range of electron energy of 140 to 4000 eV. The electron ionization cross sections are determined from the known photoionization cross sections and the ratios of ion yields. Kobayashi et al.[124] have measured the ratio of partial ionization cross section to the total ionization cross section in the energy range covering threshold to 1000 eV. For singly charged ions the authors claim good agreement with other reported data. Phelps and Petrović[125] have provided an analytical expression of high accuracy.

We shall now consider the semiempirical values and equations that are useful for modelers and for evaluating the total ionization cross sections at any desired electron impact energy. De Heer et al.[37] have provided total ionization cross sections in the range of 20 to 4000 eV, using the sum rules. The peak of the cross section is $2.95 \times 10^{-20}\,\mathrm{m}^2$, obtained at 90 eV electron impact energy. Their data are in very good agreement with those of Krishnakumar and Srivastava[82] above 100 eV. Lennon et al.[83] have proposed a semiempirical equation of the form

$$Q_i = \frac{1}{\varepsilon\,\varepsilon_i}\left[A\,\ln\!\left(\frac{\varepsilon}{\varepsilon_i}\right) + \sum_{j=1}^{N} B_j\!\left(1 - \frac{\varepsilon_i}{\varepsilon}\right)\right] \tag{3.10}$$

where A and B are the coefficients given in Table 3.8, ε the electron energy, ε_i the ionization energy, and N the number of terms in the summation term. The advantage of this equation is that it can also be used to calculate the partial ionization cross sections in addition to the first ionization cross section. The calculated values of the total ionization

---

TABLE 3.8
Constants for Using Equations 3.10 and 3.11

	Species		
Constant	Ar^+	Ar^{2+}	Ar^{3+}
A	2.532	2.086	1.170
B_1	−2.672	1.077	0.843
B_2	2.543	−2.172	−2.877
B_3	−0.769	0.809	1.958
B_4	0.008	—	—
B_5	0.006	—	—
$F(\varepsilon)$	$0.4045\,\varepsilon^{0.1844}$	0.0426	0.1333

A and B_j, taken from Lennon,[83] are in units of $10^{-23}\,\mathrm{eV^2 m^2}$. $f(\varepsilon)$ is evaluated by the author.

TABLE 3.9
Parameters of Brusa et al. to Calculate Ionization Cross Sections in Rare Gases

	Gas				
Parameter	Ar	He	Ne	Kr	Xe
n	52	50	50	52	54
Energy range (eV)	ε_i–5000	ε_i–5000	ε_i–5000	ε_i–5000	ε_i–5000
$L\ (10^{-20}\,\mathrm{m}^2)$	78.76	2000	7.92	33.76	1000
M	18.62	70.9	7.04	15.93	53.79
$N\ (10^{-20}\,\mathrm{m}^2)$	25.66	—	—	14.45	109.6
P (keV)	8.42×10^{-3}	1.49×10^{-3}	2.16×10^{-3}	12.6×10^{-3}	3.58×10^{-3}
ε_i (eV)	15.759	24.587	21.584	13.999	12.130

n is the number of points used to extract the parameters.
From Brusa, R. S. et al., *Z. Phys. D*, 38, 279, 1996. With kind permission of Springer Science and Business Media.

cross sections in argon, using the coefficients given in Table 3.9, are shown in Figure 3.9 along with selected measured data. While the shape of the total ionization cross section curve agrees well, the agreement between measured and calculated data is improved by modifying Equation 3.10 to

$$Q_i = \frac{f(\varepsilon)}{\varepsilon \, \varepsilon_i} \left[A \, \ln\left(\frac{\varepsilon}{\varepsilon_i}\right) + \sum_{j=1}^{N} B_j \left(1 - \frac{\varepsilon_i}{\varepsilon}\right) \right] \tag{3.11}$$

where $f(\varepsilon)$ is a correction function.

Brusa et al.[17] also provide a semiempirical equation for the total ionization cross section by using the data of Krishnakumar and Srivastava[82] and de Heer et al.[37]

$$Q_i = \left(\frac{L}{M+x} + \frac{N}{x}\right)\left(\frac{y-1}{x+1}\right)^{1.5} \times \left[1 + \frac{2}{3}\left(1 - \frac{1}{2x}\right)\ln\{2.7 + (x-1)^{0.5}\}\right] \tag{3.12}$$

where L, M, N, and P are the four free parameters (Table 3.9) determined by the best fitting procedure, $y = \varepsilon/\varepsilon_i$ and $x = \varepsilon/P$, ε_i is the ionization potential and ε the electron energy.

Total ionization cross sections in argon, calculated according to Equation 3.12, are included in Figure 3.8. Calculations for other rare gases will be discussed in the relevant subsection. The shape of the curve is reproduced remarkably well and a particular advantage of this formula is that the total ionization cross sections at the onset give relatively more accurate total ionization cross sections.

A more recent equation for the total ionization cross section is given by Phelps and Petrović[125] for modeling cold cathode discharges,

$$Q_i(\varepsilon) = 970\left[\frac{(\varepsilon - 15.8)}{(70 + \varepsilon)^2}\right] + 0.06(\varepsilon - 15.8)^2 \exp\left(\frac{-\varepsilon}{9}\right) \tag{3.13}$$

A special feature of their modeling is that fast-atom interactions, which are generated by symmetric charge transfer scatterings of Ar^+ with Ar, are considered for the first time in cold-cathode discharge. The values calculated according to Equation 3.13 are also shown in Figure 3.9.

The first and second ionization potentials of argon are 15.76 and 27.63, respectively.[75] The partial ionization cross sections shown in Figure 3.10 cover the period from 1980. Apart from the difficulties encountered in the measurement of total ionization cross sections, there are additional problems associated with partial ionization cross section measurements, such as variation in the ion collecting efficiency, which depends on the mass to charge ratio.

There are essentially three different types of data that have been reported: relative partial cross section functions, absolute partial cross section functions, and ratios of partial cross sections. The agreement between various measurements and with the semiempirical equation is poorer than in the case of total ionization cross sections.

The measurements of Stephan et al.,[48] using the mass spectrometer, belong to the category of determination of relative partial ionization cross sections. Absolute total cross sections were obtained by normalizing to the data of Rapp and Golden-Englander.[30] Stephan et al. observed a structure (also confirmed by Wetzel et al.[75] around 50 eV for Ar^+ in the shape of a very low second maximum, possibly due to autoionization according to the process $Ar + e \rightarrow Ar^+ + 2e$. In this process excitation occurs from an inner orbital (3s) electron to ^1D and ^1S states of highly excited neutral argon which autoionize into the adjacent Ar^+ continuum. Further, these authors did not find any peculiarity in the cross section for

TABLE 3.10
Authors and Energy Range within a Band of ± 5% of the Ionization Cross Sections Due to Rapp and Englander-Golden.[30]

Authors	Energy Range (eV)	Reference
De Heer et al. (1979)	20–500	37
Stephan et al. (1980)	20–180	48
Wetzel et al. (1987)	18, 19, 25–75	75
Krishnakumar and Srivastava (1988)	30–650	82
Ma et al. (1991)	20–500	89
McCallion et al. (1992)	18, 57–1000	94
Syage (1992)	18, 24–660	93
Straub et al. (1995)	25, 100–1000	98
Brusa et al. (1996)	25–100	17
Phelps and Petrović (1999)	18.5–750	107
Sorokin et al. (2000)	140–1000[a]	108
Kobayashi et al. (2002)	16–1000[b]	124
Raju	30–90, 300–600	This volume

[a]Multiply by 1.1.
[b]Multiply by 1.05.

Ar^{3+} because of their energy limitation to 180 eV, though the Auger process previously explained has been observed by other groups for electron energies greater than about 250 eV. A comparison of measured absolute total cross sections by selected authors has been published recently by the present author.[199]

There is considerably more disagreement in the partial ionization cross sections reported by several groups. Figure 3.10 shows the ratios of Ar^+/Ar(total), Ar^{2+}/Ar(total), and Ar^{3+}/Ar(total) as a function of electron energy; these ratios are chosen to avoid the need for choosing a reference. While there is very good agreement between various measurements for the ratio Ar^+/Ar(total), the remaining two ratios show considerable disagreement, sometimes by as much as a factor of two. The main reason stems from the differences in the partial ionization cross sections.

Table 3.10 shows the range of electron impact energies that fall within a band of ±5% of the cross sections measured by Rapp and Englander-Golden,[30] with appropriate references. The tabulated values and a figure are shown at the end of the section.

3.1.6 VERIFICATION OF THE SIGMA RULE FOR Ar

The sigma rule, also called the sum check, provides a method of judging systematic differences, at least partially, that may exist in several sets of data from different laboratories. The method is to add up the partial cross sections and compare the sum with the measured total cross sections. Such checks also help in choosing a set of cross sections for modeling and determine the gaps in energy where more accurate experimentation is desirable. Zecca et al.[10] have carried out this kind of analysis at selected electron energies. Table 3.11 and Figure 3.11 show the quality of the agreement over the energy range 12 to 1000 eV. Below the first excitation level the total scattering cross section is the same as the elastic scattering cross section.

At very low energies there are relatively few measurements of the total cross section; the data of Ferch et al.[66] cover the range 0.08 to 0.11 eV. The energy range of 0.12 to 10 eV

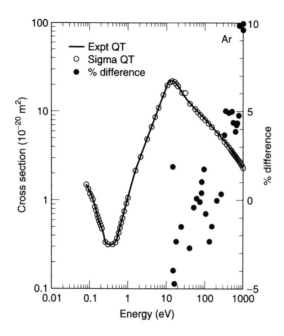

FIGURE 3.11 The sigma rule in argon. —, experimental total scattering cross section as shown in Table 3.11; ○, sum of partial cross sections; ●, percent difference between the two. Positive values of the ordinate mean that experimental values are higher.

is covered by Buckman and Lohmann.[68] The inelastic scattering cross section is composed of total excitation and ionization cross sections.

Mason and Newell[73] have measured the total metastable cross sections in the range of 12 to 140 eV and Tsurubuchi et al.[100] have measured the lowest resonance cross sections. The sum of these two cross sections is taken as the total excitation cross sections on the reasonable assumption that the total metastable cross sections above 140 eV make about 1% contribution (Figure 3.7).

The empirical formula of Zecca et al.[10] up to 1000 eV is remakably satisfactory. In the high energy range, >750 eV, the measured total cross section of Noguiera et al.[55] joins smoothly to the data of Wagenaar and de Heer.[67] It is noted that Noguiera et al.[55] have provided two sets of data: the measured ones, and those to which a theoretical correction is applied to account for forward low-angle scattering. The latter are about 6% higher. Since the uncorrected values satisfy the sigma criterion it is possible that the applied correction is on the high side. In other words, the influence of forward low-angle scattering on the total cross section is less than estimated. It is concluded that the sigma rule is satisfied in argon.

The elastic scattering cross sections show the greatest differences between various measurements. Very few measurements have been carried out above 20 eV since 1980. The measurements of Srivastava et al.[50] are rather too low and the more recent measurements of Gibson et al.[99] are limited to 10 eV electron energy. Therefore one has to look for theoretical calculations or experiments prior to 1980.

It is concluded that the sigma rule applies to argon quite well in the energy range studied. The cross sections shown in Table 3.11 should be regarded as a recommended set because, as pointed out by Phelps and Petrović,[125] the swarm parameters have been calculated and compared with measured data. The best agreement is obtained by the momentum transfer cross section of Pack et al.[92]

TABLE 3.11
Comparison of Sigma Rule and Measured Q_T for Argon (Columns 5 and 6)

Energy (eV)	Q_{el} $10^{-20}(m^2)$	Q_{ex} $10^{-20}(m^2)$	Q_i $10^{-20}(m^2)$	Sigma(Q_T) $10^{-20}(m^2)$	Q_T(measured) $10^{-20}(m^2)$	Difference (%)
0.08	1.50[F,85]			1.5	1.50[F,85]	
0.1	1.19[F,85]			1.19	1.19[F,85]	
0.5	0.374[B,86]			0.374	0.374[B,86]	
1	1.07[B,86]			1.07	1.07[B,86]	
5	8.81[B,86]			8.81	8.81[B,86]	
10	20.06[B,86]	T+M		20.06	20.06[B,86]	
15	23.22[F,89]	0.254	R,65	23.474	22.58(I)[B,86]	−3.959
16	22.30(I)[F,89]	0.262	0.02	22.963	21.93[B,86]	−4.710
18	19.52(I)[F,89]	0.565	0.294	20.433	19.97[B,86]	−2.318
20	18.60[F,89]	0.643	0.627	19.738	18.30[B,86]	−7.858
25	16.27[d,79]	0.676	1.302	16.45	16.21[N,85]	−1.481
30	13.94[d,79]	0.619	1.803	16.362	14.52[N,85]	−12.686
40	9.51[d,79]	0.511	2.393	12.414	12.09[N,85]	−2.680
50	7.74[d,79]	0.457	2.533	10.73	10.69[N,85]	−0.374
70	6.15[d,79]	0.406	2.771	9.327	9.320[N,85]	−0.075
90	5.00[d,79]	0.378	2.859	8.237	8.386[N,85]	1.777
100	4.86[d,79]	0.344	2.850	8.054	7.997[N,85]	−0.713
150	3.79(I)[d,79]	0.275	2.683	6.748	6.652[N,85]	−1.443
200	3.2[d,79]	0.237	2.393	5.83	5.831[N,85]	0.017
250	2.833(I)[d,79]	0.201	2.173	5.207	5.225[N,85]	0.344
300	2.466[d,79]	0.177	1.979	4.622	4.798[N,85]	3.668
350	2.290(I)[d,79]	0.175(I)	1.812	4.277	4.505[W,85]	5.061
400	2.115[d,79]	0.173	1.68	3.968	4.175[W,85]	4.958
450	2.0(I)[d,79]	0.154(I)	1.548	3.702	3.898[W,85]	5.028
500	1.885	0.134	1.46	3.479	3.662[W,85]	4.997
550	1.806(I)	0.128(I)	1.372	3.306	3.458[W,85]	4.396
600	1.728	0.121	1.302	3.151	3.279[W,85]	3.904
650	1.649(I)	0.116(I)	1.223	2.988	3.122[W,85]	4.292
700	1.571	0.111	1.161	2.843	2.974[W,85]	4.405
800	1.448(I)	0.091	1.064	2.603	2.89[N,82]	9.931
900	1.361	0.083	0.985	2.429	2.70[N,82]	10.037
1000	1.274	0.079	0.915	2.268	2.51[N,82]	9.641

I	Interpolated.
T+M	Tsurubuchi et al.[100] + Mason and Newell.[73]
R,65	Rapp and Englander-Golden.[30]
d,79	De Heer et al.[37]
F,85	Ferch et al.[66]
B,86	Buckman and Lohmann.[68]
W,85	Wagenaar and de Heer.[67]
N,82	Nogueira et al.[55]

3.2 HELIUM

Helium is one of the simplest atoms and therefore it is considered first, by those engaged in this field, for use as a standard for measuring atomic scattering data. It is preferred for testing theoretical approaches because of the simplicity of its atom. It is one of the gases in which the cross sections have been measured in a number of studies, as Table 3.1 shows. Electron–helium scattering phenomena have played a central role in the development of the quantum theory of scattering and of many body systems in general.

3.2.1 TOTAL AND MOMENTUM TRANSFER CROSS SECTIONS IN **He**

The total cross section, that is the sum of elastic and inelastic scatterings, was one of the first cross sections to be measured due to its practical importance and the relative simplicity of the transmission-type experiments. The measured absolute cross sections are generally known to be the most accurate available and serve as a benchmark for the accuracy of various theoretical treatments of scattering phenomena. The scattering cross sections in helium are used for calibration and normalization of experimental data.[36] Below 19.8 eV, which is the first excitation potential, the total scattering cross section is equal to the elastic scattering cross section. Figure 3.12 shows the various measurements and cross sections calculated using semiempirical expressions available in the literature.

Golden and Bandel[31] used the Ramsauer technique to measure the total cross section in the energy range of 0.3 to 28 eV, with a stated accuracy of 3%. Helium is a rare gas that does not show the Ramsauer–Townsend minimum. The cross section is observed to decrease from a maximum of 5.6×10^{-20} m^2 at 1.2 eV to 2.2×10^{-20} m^2 at about 28.0 eV. The decrease with increasing energy is sharp up to about 19.3 eV, which is about 0.5 eV lower than the first excitation potential. Resonance at 19.8 eV was first discovered experimentally by Schulz[125] and has been explained adequately in Chapter 2.

De Heer and Jansen[34] evaluated a set of total cross sections for scattering of electrons over a wide energy range of 0 to 3000 eV. The method adopted was to analyze the experiments and theories on total cross sections for electron scattering, ionization, and excitation, and on differential cross sections for elastic and inelastic scattering. The total scattering cross sections are stated to have an accuracy of 5% over a large part of the energy range.

The time-of-flight method for electron velocity analysis has been adopted, with electrons released by a photoelectric mechanism at the cathode, but these studies are restricted

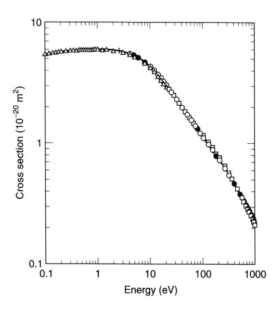

FIGURE 3.12 Absolute total cross sections in helium. (▲) de Heer and Jansen,[34] semiempirical; (+) Kennerly and Bonham[36]; (□) Dalba et al.[40]; (◇) Blaauw et al.[42]; (●) Kauppila et al.[51]; (○) Nickel et al.[65]; (△) Buckman and Lohmann[68]; (–) Brusa et al.,[17] with the correction applied according to Zecca et al.[18]; (– –) empirical formula.

to low-energy electrons. Kennerly and Bonham[36] adopted the time-of-flight method with a pulsed secondary electron emission source in the range of 0.5 to 50 eV. The method adopted had the advantage of a continuous distribution of electron energy and very good discrimination against small angle scattering. The stated accuracy is 3% and the results are about 10% higher than those due to Golden and Bandel.[31] For example, Kennerly and Bonham[36] obtain a total cross section of $2.2 \times 10^{-20}\,m^2$ at 28 eV, and the maximum cross section observed is $6.24 \times 10^{-20}\,m^2$ at 0.5 eV, which is the lowest energy in their study.

Low-angle scattering is a source of error in the total scattering cross section measurements because the scattered electrons are not differentiated from unscattered ones. Dalba et al.[40] introduced a modification of the Ramsauer technique by partitioning the scattering chamber into two interconnected parts. A series of investigations was conducted, of which helium was the forerunner. This method has been explained in Chapter 2. The energy range of electrons was 100 to 1400 eV with a stated uncertainty of 3.5%.

Blaauw et al.[42] used the transmission method with no magnetic field applied except for compensation of the Earth's magnetic field with Helmholtz coils. Linearization of the Ramsauer technique yields two advantages: (1) the angular resolution is determined exactly; (2) differential scattering measurements may also be carried out simultaneously. The Beer Lambert law for the measurement of total cross section is

$$Q_T = \left(\frac{1}{NL}\right) \ln\left(\frac{I}{I_0}\right) \tag{3.14}$$

where N is the gas number density in the absorption cell, L the effective path length of the electrons through the gas, and I/I_0 is the ratio of currents after and before scattering occurs. This is also the ratio of currents with and without the target gas in the absorption cell. Equation 3.14, however, applies to an ideal situation in which the beam is infinitely narrow and the solid angle of the detector is zero. Of course, this situation is not realized in practice and the expression for the total cross section is modified as

$$Q_T = \frac{1}{NL}\left[\mathrm{Ln}\left(\frac{I}{I_0}\right) + N \int_0^L dx \int_0^{\Delta\Omega(x)} \left(\frac{dQ}{d\Omega}\right) d\Omega \right] \tag{3.15}$$

where $\Delta\Omega\,(x)$ is the solid angle of the detector as seen by a scattering event taking place at position x on the beam axis. The ratio $(dQ/d\Omega)$ is the differential cross section due to elastic or elastic plus inelastic scattering cross sections, depending on the type of the detector. The range of electron energy covered is 15 to 750 eV with a stated accuracy of 4%.

Kauppila et al.[51] measured the total scattering cross sections in the intermediate energy range (30 to 600 eV) and their investigations have been referred to briefly in connection with the results of argon. These authors argue that the sum rule does not hold for electron scattering in helium, but it is noted that the sum rule frequently referred to in the literature has quite a different meaning because it involves the scattering length and the first Born amplitudes for elastic direct and exchange scattering.

Nickel et al.[65] measured the absolute total cross sections by the linear transmission method in the range of 4 to 300 eV electron impact energy to a stated accuracy of 2 to 4%. The design of the setup was such that the error due to low-angle forward scattering was negligible.

Buckman and Lohmann[68] used the time-of-flight measurements to obtain absolute total cross sections in the energy range 0.1 to 20 eV by gating the electron beam to provide a mark for the start of the flight time measurements. In the time-of-flight spectrometer employed, the absolute energy of the electrons as they pass through the target gas is determined from the time they take to traverse the field free drift region. The absolute total cross section as a function of energy in the region of 0.1 to 12 eV may be represented by the polynomial

$$Q_T = \left(2 \times 10^{-5}\, \varepsilon^5 - 0.0012\, \varepsilon^4 + 0.0232\, \varepsilon^3 - 0.1925\, \varepsilon^2 + 0.4217\, \varepsilon + 5.7716\right) \times 10^{-20}\, \mathrm{m}^2$$

(3.16)

A comparison of the results of various investigations is carried out in Figure 3.13. As in the case of argon, the reference values are the calculated values from Equation 3.1 for the energy range $12 \leq \varepsilon \leq 1000$ eV. For the lower energy range, $0.1 \leq \varepsilon \leq 12$ eV, the measurements of Buckman and Lohmann[68] are taken as reference. Very good agreement is observed between various investigations in the energy range from 0.1 to 700 eV. Table 3.12 gives the range of energies which fall within a band of ±4% of the two reference data.

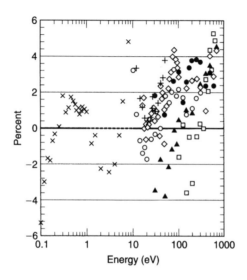

FIGURE 3.13 A comparison of total scattering cross sections in helium. The reference data are reported by Buckman and Lohmann[68] in the range $0.1 \leq \varepsilon \leq 12$ eV and by Zecca et al.[18] in the range $12 \leq \varepsilon \leq 1000$ eV. (▲) de Heer and Jansen[34] semiempirical; (+) Kennerly and Bonham[36]; (□) Dalba et al.[40]; (◇) Blaauw et al.[42]; (●) Kauppila et al.[51]; (○) Nickel et al.[65]; (×) Raju (2003), Equation 3.16. Positive values of the ordinate mean that the reference values are higher.

TABLE 3.12
Data That Lie within a Band of ± 4% of the Reference Data of Buckman and Lohmann, in the Range $0.1 \leq \varepsilon \leq 12$ eV, and Zecca et al.,[18] in the Range $12 \leq \varepsilon \leq 1000$ eV

Authors	Energy Range (eV)	Reference
Raju (2003)	$0.1 \leq \varepsilon \leq 10$	This volume
de Heer and Jansen (1977)	$30 \leq \varepsilon \leq 500$	34
Kennerly and Bonham (1978)	$12 \leq \varepsilon \leq 50$	36
Blaauw et al. (1980)	$17 \leq \varepsilon \leq 700$	42
Kauppila et al. (1981)	$30 \leq \varepsilon \leq 600$	51
Nickel et al. (1985)	$10 \leq \varepsilon \leq 300$	65

TABLE 3.13
Momentum Transfer Cross Section of Helium

Energy (eV)	Q_M $(10^{-20}\,m^2)$	Energy (eV)	Q_M $(10^{-20}\,m^2)$	Energy (eV)	Q_M $(10^{-20}\,m^2)$
0.000	4.95	8.000	5.50	100.0	0.22
0.0025	5.00	14.000	3.60	150.0	0.12
0.0036	5.10	18.00	2.90	200.0	0.07
0.010	5.27	20.00	2.69	250.0	0.05
0.032	5.52	25.00	2.00	300.0	0.036
0.200	6.20	35.00	1.26	500.0	0.016
0.600	6.66	40.00	1.00	700.0	0.009
1.400	6.98	50.00	0.70		
3.000	6.93	75.00	0.36		

From Pack, J. L. et al. *J. Appl. Phys.* 71, 5363, 1992. Reprinted with permission of AIP (USA).

The total cross section extrapolated to zero energy gives a cross section, $Q_T(0)$, that is usually expressed in terms of a length according to $Q_T(0) = 4\pi A^2$, where A is called the scattering length. At zero energy the total cross section contains only the s-wave scattering and equals the momentum transfer cross section as determined by swarm experiments. The scattering length determined by Pack et al.[92] is 62.8×10^{-12} m. Earlier values quoted by Zecca et al.[10] are in the range of 61.4×10^{-12} to 63.24×10^{-12} m.

Differential cross section measurements yield both the momentum transfer and the elastic scattering scattering cross sections through appropriate integrations, Equations 3.2 and 3.3 respectively. The differential cross sections cannot be measured at very low energies with sufficient accuracy and, in such cases, the swarm coefficients at low values of E/N are employed to obtain the momentum transfer cross sections.[126,127] Momentum transfer cross sections obtained in helium by several methods are shown in Table 3.13. Excellent agreement exists between various investigations, the data of Pack et al.[92] covering the widest energy range, 0 to 700 eV.

The momentum transfer cross sections of Crompton et al.,[128] 0.008 to 6.0 eV, are derived from swarm studies with a stated accuracy of ±2% in the range $0.01 \le \varepsilon \le 3.0$ eV. Other results are from: Milloy and Crompton,[127] from drift velocities, in the range of 4 to 12 eV, with a stated accuracy of ±3% in the range of $4 \le \varepsilon \le 7$ eV and ±5% in the range of $7 \le \varepsilon \le 12$ eV; Shyn,[129] from crossed beam differential cross section measurements in the range of $2 \le \varepsilon \le 400$ eV; Register and Trajmar,[43] from crossed beam studies in the range of $5 \le \varepsilon \le 200$ eV with a stated accuracy of 5 to 9%; Pack et al.[92] from the ratios of diffusion coefficient D to mobility μ in the range of $0 \le \varepsilon \le 700$ eV, and Ramanan and Freeman.[130] Very good agreement exists between these investigations. Table 3.13 shows Q_M in the energy range $0 \le \varepsilon \le 700$ eV.

3.2.2 ELASTIC AND DIFFERENTIAL CROSS SECTIONS IN He

Register and Trajmar[43] measured the angular distributions of electrons scattered elastically in a crossed beam geometry and below the first excitation potential; the angular distributions are placed on an absolute scale by use of a phase shift analysis. The wave function ψ describes the colliding electron at a large distance r from the scattering center, and is given by[43]

$$\psi = \exp[jkz] + \frac{f(\theta)}{r}\exp(jkr) \tag{3.17}$$

where the z-axis is the direction of the incoming plane wave, k the wave vector (electron momentum in atomic units), and $f(\theta)$ the scattering amplitude. In the early years of quantum mechanics Faxén and Holtsmark[131] showed that the spherical part of ψ may be split into partial waves according to their angular momentum (L) and that the scattering amplitude is given by

$$f(\theta) = \frac{1}{2\pi ik}\sum_{i=0}^{\infty}(2L+1)[\exp(2i\eta_L)-1]P_L\cos\theta \tag{3.18}$$

where η_L is the real phase shift and $P_L(\cos\theta)$ is the Lth Legendre polynomial. In theory the summation should be carried out for an infinite number of terms, but in practice the first few terms are the dominant terms and the remaining terms may be approximated by the effective range theory (ERT)[132] and Thompson's expression for the contribution of the partial waves beyond a certain cutoff number L. The phase shift according to the ERT is

$$\eta_L = \tan^{-1}\left(\frac{2\pi\alpha E_0}{(2L+3)(2L+1)(2L-1)}\right) \tag{3.19}$$

where α is the atomic polarizability in atomic units. Thompson's expression for the scattering amplitude is given by

$$f(\theta) = \frac{1}{2\pi ik}\sum_{i=0}^{L}(2L+1)[\exp(2i\eta_L)-1]P_L\cos\theta + C_L(\theta) \tag{3.20}$$

where

$$C_L(\theta) = \pi ak\left(\frac{1}{3}-\frac{1}{2}\sin\left(\frac{\theta}{2}\right)-\sum_{L=1}^{L}\frac{P_L(\cos\theta)}{(2L+3)(2L-1)}\right) \tag{3.21}$$

Note that the upper limit for the summation in Equation 3.20 is L and not infinity. In helium the number of partial waves considered by Register and Trajmar[43] is 2 to 5, depending on the energy of the electron; higher energies require more terms.

The differential cross section is defined in quantum mechanics as

$$\frac{dQ}{d\omega} = |f(\theta)|^2 \tag{3.22}$$

where $f(\theta)$ is the amplitude. The calculated differential cross sections are then used for placing the measured cross sections below the first excitation potentials on an absolute scale.

The total cross section is defined as[10]

$$Q_T = \frac{4\pi}{k^2}\sum_{0}^{\infty}(2L+1)\sin^2\eta_L \tag{3.23}$$

and the momentum transfer cross section as

$$Q_{\mathrm{M}} = \frac{4\pi}{k^2} \sum_{L=0}^{\infty} (L+1) \sin^2(\eta_L - \eta_{L+1}) \qquad (3.24)$$

At very low energies only the lowest partial waves contribute to the scattering amplitude. In the limit of zero energy, the differential scattering cross section is expected to be independent of the scattering angle and the momentum transfer cross section is then equal to the total scattering cross section. The total cross section, Equation 3.23, can reach very low values if the dominating phase shifts pass through multiples of $\pi/2$. As the higher order partial waves contribute to the scattering amplitude, the valley in the cross section can be considerably deep. This is the Ramsauer–Townsend effect observed experimentally. The scattering amplitude in helium is predominantly due to the s-wave and the Ramsauer–Townsend effect is not observed in helium. If a single partial wave contributes to the scattering amplitude, then the curve of differential cross section as a function of the scattering angle will be symmetrical on either side of $\pi/2$.[10]

Register and Trajmar[43] measured the differential cross section for elastically scattered electrons in a crossed beam geometry. Since the elastic scattering cross sections are the same as the total cross sections below $\sim$20 eV, their results are further commented upon in the section under elastic cross sections. Register's measurements[43] are complementary to the measurements of Srivastava et al.[50] in heavier noble gases. Register and Trajmar show that the sigma rule, that is, the sum of the individual cross sections equals the total cross section, holds valid in helium to an error within 3 to 7%.

Figure 3.14 shows the compilation of elastic scattering cross sections: de Heer and Jansen,[34] 15 to 1000 eV; Register and Trajmar,[43] 1 to 1000 eV; Shyn,[129] 2 to 400 eV. While the cross sections of these authors agree within a few percent below 40 eV, at higher energies they differ by as much as 25%. For example, at 80 eV the elastic scattering cross sections are 0.796, 0.712, and 0.94, all in units of 10^{-20} m^2 respectively. However, at higher energies, >100 eV, the differences get smaller. In spite of this, the average cross sections of these three investigations are shown by the full line in Figure 3.14 and used as reference for comparison

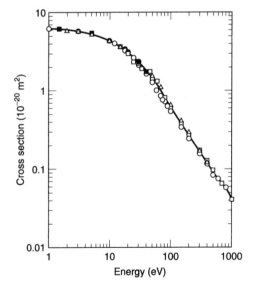

FIGURE 3.14 Elastic scattering cross sections in helium. □, de Heer and Jansen[34]; △, Shyn[129]; ○, Register and Trajmar[43]; —, average of these three data; ●, Brunger et al.[95]

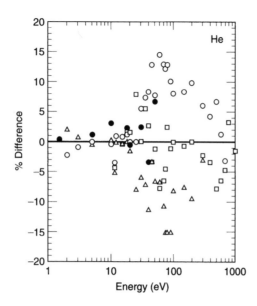

FIGURE 3.15 Comparison of elastic scattering cross sections in helium. Average values of de Heer and Jansen[34] Shyn,[129] and Register and Trajmar[43] are taken as reference. □, de Heer and Jansen[34]; △, Shyn[129]; ○, Register and Trajmar[43]; —, average of these three data; •, Brunger et al.[95]

of other data (Figure 3.15). Agreement in the low-energy region $1 \leq \varepsilon \leq 50$ eV and the high-energy region $200 \leq \varepsilon \leq 1000$ eV is very good, whereas in the intermediate region the disagreement is as high as ±15%. No analytical function could be found to fit the measured data; the same conclusion has been drawn by Brusa et al.[17]

The differential cross sections measured by Williams[39] are referred to in order to demonstrate the effect of electron energy on phase shifts. The angular range of 15° to 150° and the energy range of 0.5 to 50 eV were covered in these crossed beam experiments. A phase shift analysis has been carried out as shown in Figure 3.16, for the entire energy range. The phaseshift is predominantly due to the s-wave, p-wave, and d-waveshifts being a fraction at all energies. As the electron energy increases the s-waveshift decreases and the other two waveshifts increase, though not making a substantial contribution to the scattering amplitude (Figure 3.16). As noted previously, the gas does not exhibit the Ramsauer–Townsend minimum.

The differential cross section measurements of Shyn[129] are shown in Figure 3.17 to demonstrate the effect of electron energy on angular scattering. These data are selected because, at low energy (2 eV) and higher energies respectively, backward and forward scattering is clearly demonstrated. At 2 eV there is good symmetry about the zero angle (see Figure 3.17a) and a prominent backward scattering, which is present up to about 5 eV. At 10 eV there is pronounced forward scattering, and a minimum occurs at about 60°. As the energy increases there is increased forward scattering and decreased backward scattering. The minimum also moves toward a higher angle (see Figure 3.17b). At 100 eV and 400 eV there are no minima in the differential cross section.

3.2.3 Total Excitation Cross Sections in He

Excitation of atoms and molecules by electron impact results in radiation output and generation of metastable states which return to ground state by indirect means, such as being excited back to the allowed level by scattering. The total excitation cross section is required for the purposes of discharge simulation, finding the solution of Boltzmann equation,

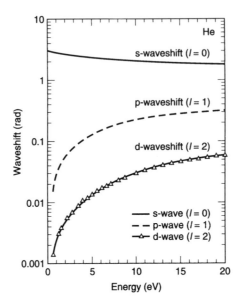

FIGURE 3.16 Phase shifts for electron scattering from helium as a function of energy. Drawn from the tabulated data of Williams, J. F., *J. Phys. B*, 12, 265, 1979. With permission of Institute of Physics, U.K.

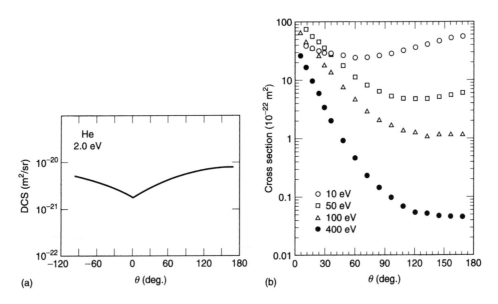

FIGURE 3.17 Differential cross sections for elastic scattering of electrons in helium. (a) At 2.0 eV (schematic) both backward and forward scattering is observed. (b) Differential cross sections for electron energies as shown.

and so on, and detailed discussion of excitation to every allowed level will not fall into the domain of our requirement; in argon alone there are over 75 lines. A review of electron impact excitation cross sections in atoms covering the years to 1968 is given by Moiseiwitsch and Smith.[119] Electron excitation of atoms to metastable states has been reviewed by Fabrikant et al.,[120] and electron impact optical excitation functions are reviewed by Heddle and Gallagher.[121]

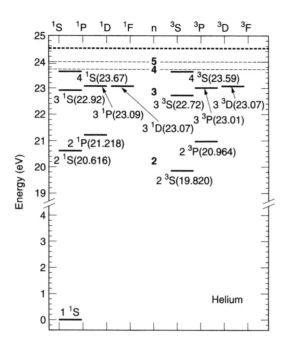

FIGURE 3.18 Energy levels of the helium atom.

The ground state configuration of the He atom is $1\,s^2$. The lowest optically allowed transition has a threshold of 21.21 eV from ground state to the $2\,^1P$ level (Figure 3.18). This transition corresponds to the 58.4 nm resonance line. The lower levels $2\,^3S$ (19.82 eV), and $2\,^1S$ (20.61 eV), and $2\,^3P$ (20.96 eV) are all metastables. The transition $4\,^1P \rightarrow 1\,^1S$ occurs at 23.74 eV with radiation of 52.2 nm. Figure 3.18 shows simplified energy level diagrams for the helium atom.[34]

Differential cross section measurements for excitation are usually carried out at selected electron energies and the cross sections given by different authors are usually for different transitions.[91] The situation is rendered more difficult because some authors do not present the integrated excitation cross sections. Further cascading effects occur; for example, the $2\,^3S$ level may be populated from the $3\,^3P$ level (threshold 23.01 eV, 388.9 nm) and from the $2\,^3P$ level (threshold 20.96 eV, 1083.0 nm). However, on the plus side, many of these cross sections make a relatively minor contribution to the total excitation cross section and the sum of cross sections for $2\,^3S$ (19.82 eV), $2\,^1S$ (20.61 eV), and $2\,^3P$ (20.96 eV) makes the largest contribution to the total excitation cross section. Trajmar et al.[133] have reviewed the differential and integral cross sections for various levels of excitation up to 500 eV covering the period up to the year 1992, followed by Röder et al.[134] (1996) for $2\,^3P$ excitation and Asmis and Allan[135] (1997). Cubric et al.[135] have measured the differential cross section for the first four levels ($2\,^3S$, $2\,^1S$, $2\,^3P$, $2\,^1P$).

Agreement between various measurements and theoretical calculations is poor, varying by as much as an order of magnitude. The values of Alkhazov[136] for seven states calculated using semiempirical equations are shown in Figure 3.19; the highest cross section obtains for the $2\,^1P$ (21.21 eV) state. Total excitation cross section for these levels is obtained by addition and is shown in Figure 3.20.

The semiempirical cross sections due to de Heer and Jansen[34] and cross sections calculated using the semiempirical formula of Brusa et al., (Equation 3.7) with the parameters shown in Table 3.8 are also plotted in Figure 3.20. Very good agreement exists, which is not surprising considering that the formula is meant to fit the values. The theoretical

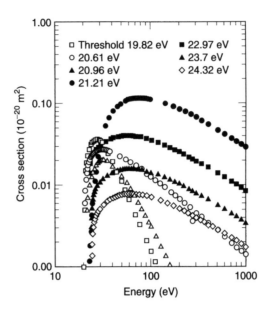

FIGURE 3.19 Excitation cross sections in helium reported by Alkhazov.[136] The threshold energies are: (□) 19.81 eV (2 ^{3}S); (○) 20.61 ev (2 ^{1}S); (△) 20.96 eV (2 ^{3}P); (•) 21.21 eV (2 ^{1}P); (■) 22.97 eV (3 ^{1}S); (▲) 23.7 eV (4 ^{3}P); (◇) 24.32 eV (7 ^{3}F).

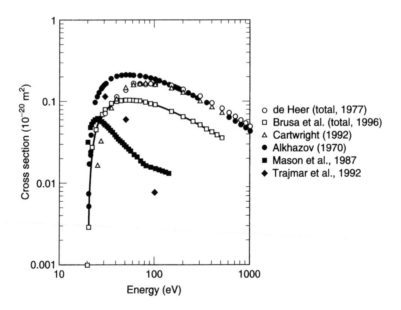

FIGURE 3.20 Total excitation cross section in helium. •, Alkhazov,[136] sum of seven levels; ▲, Alkhazov,[136] sum of first six levels; ○, de Heer and Jansen[34]; ■, Mason and Newell,[73] total metastables; △, Cartwright et al.,[90] theory; ◆, Trajmar et al.,[91] experimental; –□–, Brusa et al.,[17] empirical.

results of Cartwright et al.[90] also agree well with the formula of Brusa et al.[17] up to 40 eV; for higher energies the former are lower, possibly due to the fact that not all cross sections were considered in the summation provided by Cartwright et al.[90] The time-of-flight method of Mason and Newell[73] gives the total metastable cross section and it is lower, as expected, because the optical cross sections are excluded.

3.2.4 IONIZATION CROSS SECTIONS IN He

Ionization of helium by electron impact is a common phenomenon that occurs in celestial plasma. In laboratory plasmas it is particularly interesting because helium is the atomic byproduct of the deuterium–tritium fusion reaction in controlled thermonuclear devices. Further, there are no complicated effects due to inner shell ionization in comparison with gases such as argon. Autoionization through excitation of two electrons makes a small contribution to the total ionization cross section.[137]

Figure 3.21 shows the results of eleven groups of authors. The technique and results of measurements of ionization cross sections in argon by Rapp and Englander-Golden[30] have been presented in Figures 2.13 and 2.14. These measurements have stood the test of time, with a quoted uncertainty of 7%, and they may be employed as a basis to determine the differences between more recent literature. The method adopted falls into the beam–static-gas category, though the authors employed the gas flow method to determine the pressure instead of a McLeod gauge. The total ionization cross sections are measured by collecting all the ions generated. Though the concept is rather simple, accurate measurements have proved to be difficult, due to the fact that uncertainties exist in the determination of various parameters involved.[27]

One of the earliest measurements of total ionization cross sections was made by Smith.[138] For thirty years these were the only data available covering the range up to 100 eV. Asundi and Kurepa[25] measured the absolute cross sections using a setup similar to that of Tate and Smith[138] the currents due to positive ions and electrons are measured and the absolute cross sections are calculated according to

$$Q_i = \frac{i_1}{i_e} \frac{T}{273} \frac{133.28}{p \times L \times 3.53 \times 10^{22}} \, m^2 \tag{3.25}$$

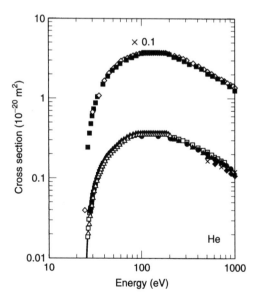

FIGURE 3.21 Ionization cross sections in helium. The data are divided into two groups for clarity. The cross sections of the upper curve should be multiplied by 0.1. Legends are for both curves. (▲) Asundi and Kurepa[25]; (◆) Schram et al.[32] (1966b); (●) de Heer and Jansen[34]; (×) Nagy et al.[41]; (○) Stephan et al.[48] (1980); (□) Montague et al.[137]; (△) Wetzel et al.[75]; (■) Shah et al.[81]; (◇) Krishnakumar and Srivastava[82] (– –) Lennon et al.[83]; (—) Brusa et al.[17]

where T is the ambient temperature, p is the gas pressure in the ionization tube in pascals, L the length of the ionization chamber in m, i_1 is the ion current, and i_e the electron current. Accurate measurement of pressure was a source of uncertainty in the determination of absolute total ionization cross sections before capacitance-type manometers became available.

Schram et al.[32] used both the ionization tube method and a mass spectrometer to measure the partial ionization cross sections and covered the energy range 100 to 16,000 eV. The high energy range of 500 to 5000 eV was covered by Nagy et al.,[41] who also used the mass spectrometer to measure partial ionization cross sections. In the overlapping narrow range of electron energies, these two sets of data (only three points) agree well.

Till about 1978 measurements of absolute total ionization cross sections in rare gases were carried out using the beam–static gas method developed by Tate and Smith. The fast atom beam method was adopted for the study of atomic nitrogen and hydrogen as early as 1962, and was extended to helium by Montague et al.[137] The sources of errors in the Tate and Smith apparatus have already been mentioned in Chapter 2. The crossed electron–fast atom beam method adopted by Montague et al.[137] is free from these errors because magnetic fields are not required to collimate the electron beam.

Further, the number density of target atoms is determined from their flux and velocity. The scattering volume and the beam density profiles are well known with increasing accuracy. The magnetic selection of the product ion ensures that only the single-process cross section is measured and not the total ionization cross section. However, the method requires an atomic beam that contains only the ground-state atoms. The target beam is produced by a charge exchange scattering of the parent ion beam in a gas cell; the atomic beam is susceptible for contamination by long-lifetime metastables and highly excited atoms. These are removed by ionization, employing high field ionizing bars. The data obtained cover a wide range of electron impact energy, 26 to 750 eV.

Wetzel et al.[75] also measured the absolute total ionization cross sections in all the rare gases using the fast atom beam technique, with a stated accuracy of $\pm 15\%$. The absolute cross section determination was made at a single energy near the maximum of the cross section for single ionization. From the ratios of ion currents to electron currents various parameters involved in calculating the cross section were determined. The energy dependence of the cross section and multiple to single ion ratios, both relative measurements, were then made and scaled to an absolute cross section.

Krishnakumar and Srivastava[82] extended the range of measurement of Wetzel et al.[75] in all the rare gases up to 1000 eV, using a pulsed electron beam and mass spectrometer. Normalization was with respect to the measurements of Rapp and Englander-Golden[30] below the second ionization potential, 54.4 eV. The authors assumed that second-level ionization cross sections are negligible, notwithstanding the higher impact energy employed. The upper energy range was extended to 10,000 eV by Shah et al.[81] who employed a pulsed electron beam and time-of-flight spectroscopy. Figure 3.21 shows the calculated absolute total cross section, by adding the weighted partial cross sections. The ratios of double ionization cross section to single ionization cross section are in the range of 10^{-3}, justifying the asumption of Krishnakumar and Srivastava.[82]

The semiempirical formulas of Lennon et al.[83] and the empirical equation of Brusa et al.[17] give good agreement with the measured total ionization cross sections. The former gives a much better fit, as is shown in Figure 3.22. The differences between various sets of data are also shown in this figure, with Rapp and Englander-Golden[30] taken as reference according to

$$\text{Percent difference} = \left(\frac{Q_{i(R-EG)} - Q_{test}}{Q_{i(R-EG)}} \right) \times 100 \qquad (3.26)$$

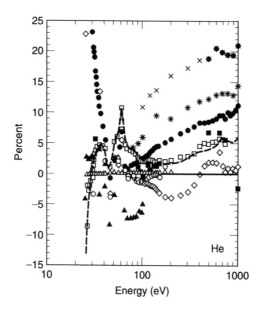

FIGURE 3.22 Comparison of ionization cross sections in helium. (▲) Asundi and Kurepa[25]; (×) Schram et al.[32] (1966a); (◆) Schram et al.[32] (1966b); (×) de Heer and Jansen[34]; (○) Stephan et al.[48] (1980); (□) Montague et al.[137]; (△) Wetzel et al.[75]; (■) Shah et al.[81]; (◇) Krishnakumar and Srivastava[82], (– –) Lennon et al.[83]; (●) Brusa et al.[17]; (—) Rapp and Englander-Golden,[30] as reference. Cross sections higher than those of Rapp and Englander-Golden fall on the negative side of the ordinate. Author's computations (unpublished).

TABLE 3.14
Authors and Energy Range within a Band of ±7% of the Cross Sections Due to Rapp and Englander-Golden.[30]

Authors	Energy Range (eV)	Reference
Asundi and Kurepa (1963)	25–100	25
de Heer and Jansen (1977)	30–100	34
Stephan et al. (1980)	25–180	48
Montague et al. (1984)	26.5–750	137
Wetzel et al. (1987)	26–200	75
Krishnakumar and Srivastava (1988)	26–1000	82
Shah et al. (1988)	26–1000	81
Lennon et al. (1988)	26–1000	83
Brusa et al. (1996)	40–300	17

---

where the subscripts (R-EG) and test mean cross sections due to Rapp and Englander-Golden[30] and cross sections tested, respectively. Higher values of Q_{test} fall on the negative side of the ordinate.

Table 3.14 shows the data that fall within ±7% of the results of Rapp and Englander-Golden.[30]

3.2.5 VERIFICATION OF THE SIGMA RULE FOR He

The measured total cross sections are compared with the sum of partial cross sections for various types of scattering, as shown in Table 3.15. The elastic scattering cross section is equal to the total cross section up to the first excitation level. The data of Buckman and Lohmann[68] for the total scattering cross section fit the equation (Figure 3.12)

$$Q_T = 2 \times 10^{-5}\varepsilon^5 - 0.0012\varepsilon^4 + 0.0232\varepsilon^3 - 0.1925\varepsilon^2 + 0.4217\varepsilon + 5.7716 \qquad (3.27)$$

TABLE 3.15
Verification of the Sigma Rule with Measured Total Cross Sections in Helium

Energy (eV)	Q_{el} $(10^{-20}$ m$^2)$	Q_{ex} $(10^{-20}$ m$^2)$	$Q_i^{R,65}$ $(10^{-20}$ m$^2)$	$\Sigma (Q_T)$	Q_T(measured) $(10^{-20}$ m$^2)$	Difference (%)
10.00	4.12[B,86]				4.12[B,86]	
20.00	2.995[Av]	0.0028[Br,96]		2.998	2.99[B,86]	1.232
22.00	2.800[Av]	0.0268[Br,96]		2.827	2.867[Bl,80]	1.410
24.00	2.556[Av]	0.0448[Br,96]		2.601	2.713[Bl,80]	4.142
26.00	2.471[Av]	0.0586[Br,96]	0.0175	2.548	2.576[Bl,80]	1.106
28.00	2.346[Av]	0.0692[Br,96]	0.0425	2.458	2.456[Bl,80]	−0.086
30.00	2.239[Av]	0.0775[Br,96]	0.0668	2.383	2.372[Bl,80]	−0.494
32.50	2.162[Av]	0.0842[Br,96]	0.0985	2.345	2.248[Bl,80]	−4.284
35.00	1.994[Av]	0.0910[Br,96]	0.1249	2.210	2.156[Bl,80]	−2.501
37.50	1.860[Av]	0.0946[Br,96]	0.1500	2.105	2.055[Bl,80]	−2.404
40.00	1.726[Av]	0.0983[Br,96]	0.1715	1.996	1.952[Bl,80]	−2.267
45.00	1.635[Av]	0.1020[Br,96]	0.2023	1.939	1.826[Bl,80]	−6.231
50.00	1.369[Av]	0.1036[Br,96]	0.2428	1.715	1.728[Bl,80]	0.707
55.00	1.279[Av]	0.1039[Br,96]	0.2709	1.654	1.627[Bl,80]	−1.660
60.00	1.189[Av]	0.1033[Br,96]	0.3079	1.600	1.529[Bl,80]	−4.671
65.00	1.082[Av]	0.1023[Br,96]	0.3079	1.492	1.448[Bl,80]	−3.077
70.00	0.975[Av]	0.1009[Br,96]	0.3211	1.397	1.378[Bl,80]	−1.405
75.00	0.877[Av]	0.0988[Br,96]	0.3342	1.310	1.310[Bl,80]	0.027
80.00	0.816[Av]	0.0976[Br,96]	0.3439	1.258	1.254[Bl,80]	−0.251
85.00	0.756[Av]	0.0958[Br,96]	0.3510	1.203	1.215[Bl,80]	1.020
90.00	0.695[Av]	0.0941[Br,96]	0.3571	1.146	1.176[Bl,80]	2.533
95.00	0.652[Av]	0.0923[Br,96]	0.3615	1.106	1.142[Bl,80]	3.159
100.0	0.610[Av]	0.0906[Br,96]	0.3659	1.066	1.112[Bl,80]	4.058
150.0	0.381[Av]	0.0900[Br,96]	0.3686	0.840	0.871[Bl,80]	3.588
200.0	0.274[Av]	0.085[Br+H]	0.3466	0.701	0.722[Bl,80]	3.023
250.0	0.220[Av]	0.080[Br+H]	0.3211	0.621	0.624[Bl,80]	0.616
300.0	0.155[Av]	0.080[Br+H]	0.2964	0.531	0.554[Bl,80]	4.144
350.0	0.142[Av]	0.080[Br+H]	0.2753	0.498	0.506[Bl,80]	1.715
400.0	0.120[Av]	0.08[Br+H]	0.2568	0.452	0.459[Bl,80]	1.602
500.0	0.090[Av]	0.075[Br+H]	0.2243	0.373	0.378[Bl,80]	1.297
600.0	0.077[Av]	0.072[H,77]	0.1997	0.349	0.333[Bl,80]	−4.613
700.0	0.063[Av]	0.068[H,77]	0.1803	0.308	0.291[Bl,80]	−5.878
800.0	0.060[Av]	0.061[H,77]	0.1645	0.264	0.256[D,80]	−3.347
900.0	0.050[Av]	0.056[H,77]	0.1504	0.231	0.227[D,80]	−1.687
1000.0	0.041[Av]	0.052[H,77]	0.1407	0.202	0.207[D,80]	2.505

Av	Average of de Heer and Jansen,[34] Register and Trajmar,[43] and Shyn.[129]
B,86	Buckman and Lohmann.[68]
Bl,80	Blaauw et al.[42]
Br,96	Brusa et al.[17]
D,80	Dalba et al.[40]
Br + H	Average of Brusa et al.[17] and de Heer and Jansen.[34]
R,65	Rapp and Englander-Golden.[30]

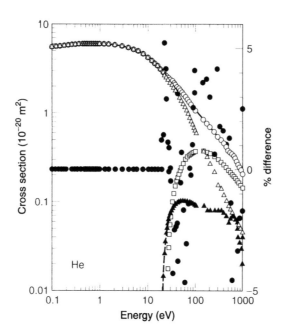

FIGURE 3.23 Verification of the sigma rule in helium. (△) Elastic scattering; (–▲–) excitation scattering; (–□–) ionization; (–○–) total cross section (summation), (—) total cross section (experimental); (●) percentage difference between the last two cross sections. Positive values of the ordinate mean that experimental cross sections are higher.

in the range of $0.1 \leq \varepsilon \leq 20\,\text{eV}$. A wide range of $20 \leq \varepsilon \leq 700\,\text{eV}$ is covered by the data of Blaauw et al.[42] Above $700\,\text{eV}$, the data of Dalba et al.[40] fit neatly to extend the range up to $1000\,\text{eV}$. These data are all within $\pm 4\%$ of the semiempirical equation of Zecca et al.[18] for Q_T (Figure 3.13). The elastic scattering cross sections for electron energies greater than $20\,\text{eV}$ are taken as the average of de Heer and Jansen,[34] Register and Trajamar,[43] and Shyn.[129] These cross sections agree very well between themselves.

Further, as pointed out earlier, elastic scattering cross sections over a wide range of energy are scarce because differential cross sections are usually measured at selected electron energies only. The excitation cross sections are obtained from the analytical formula of Brusa et al.[17] and for electron energies greater than $50\,\text{eV}$ they decrease rapidly. In the range of $150 \leq \varepsilon \leq 500\,\text{eV}$ the average of Brusa et al.[17] and, de Heer and Jansen[34] is taken. In the range of $500 \leq \varepsilon \leq 1000\,\text{eV}$, the excitation cross sections of de Heer and Jansen[34] are satisfactory. As the last column of Table 3.15 shows, the sigma criterion is satisfied in helium up to $1000\,\text{eV}$ within a maximum discrepancy of $\pm 4.6\%$ if the two isolated values at 650 and $700\,\text{eV}$ are ignored. As noted by Zecca et al.[10] the contribution of the elastic scattering cross section to the total cross section begins to drop off sharply for electron energies greater than $30\,\text{eV}$. For energies greater than $125\,\text{eV}$, the ionization cross sections are the largest contributors to the total cross section. Figure 3.23 shows the relative magnitudes of the various cross sections and the quality of agreement with the sigma rule.

3.3 KRYPTON

Krypton is a heavy rare gas that plays an important role in laser, plasma, and lighting technology. Mixtures of argon, krypton, and fluorine have been found to be high efficiency laser media, utilizing both direct electron beam pumping and electron beam stabilized discharge pumping. Krypton is also used in plasma diagnostics, where the electron and ion concentration may be determined from measurements of the intensities of emissions from excited rare gas atoms in the plasma. Krypton gas is also used in electrodeless discharges.[152]

3.3.1 TOTAL CROSS SECTIONS IN Kr

Though a large number of investigations have been carried out in krypton, low energy data (<20 eV) are few. The careful study of Buckman and Lohmann,[71] extending down to 0.175 eV, deserves special mention. They used a linear time-of-flight spectrometer in which a high-energy electron beam (~ 150 eV) is pulsed and retarded in energy by a series of electrostatic lenses; the beam then enters a long scattering cell. Those electrons that do not scatter, exit through an aperture. The electrons are accelerated and focused to a channel electron multiplier detector. The absolute energy of the electrons in the attenuation cell is determined by their time of flight. The total cross section is determined by the ratio of electron flux with and without the target gas in the scattering chamber. Since no magnetic field is employed, the physical length of the chamber is the effective scattering length for the application of the Beer–Lambert law of attenuation.

Absolute total cross sections in krypton have been measured by several groups; Figure 3.24 shows the results of selected authors. These are: Wagenaar and de Heer,[44] 22.5 to 750 eV, linear Ramsauer technique; Dababneh et al.,[49] 1.9 to 800 eV, transmission technique with magnetic field employed; Jost et al.,[56] 7.5 to 60 eV (for tabulated data see Wagenaar and de Heer[67]); Garcia et al.,[69] 700 to 6000 eV, linear transmission technique; Buckman and Lohmann,[71] 0.175 to 20.0 eV, time-of-flight electron spectrometer; Subramanian and

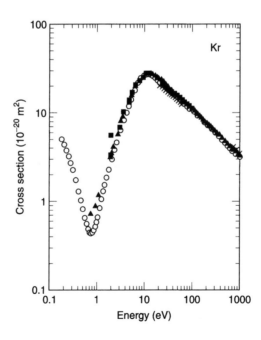

FIGURE 3.24 Absolute total scattering cross sections for electrons on krypton. (●) Wagenaar and de Heer[44]; (■) Dababneh et al.[49] (1980); (○) Dababneh et al.[49] (1982); (– –) Jost et al.[56]; (□) Wagenaar and de Heer[67]; (×) Garcia et al.[69]; (–○–) Buckman and Lohmann[71]; (▲) Subramanian and Kumar[76] (◇) Zecca et al.[86]; (◆) Kanik et al.[87]

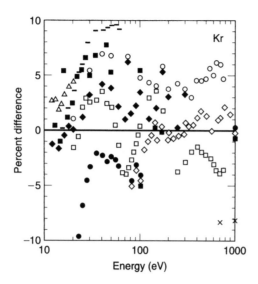

FIGURE 3.25 Percent differences in total cross sections in the energy range $12 Ès È 1000$ eV. (•) Wagenaar and de Heer[44]; (■) Dababneh et al.[49] (1980); (○) Dababneh et al.[49] (1982); (–) Jost et al.[56]; (□) Wagenaar and de Heer[67]; (×) Garcia et al.[69]; (–△–) Buckman and Lohmann[71]; (◇) Zecca et al.[86]; (◆) Kanik et al.[87]; (—) Zecca et al.,[18] reference values. Positive values of the ordinate mean that reference values are higher.

Kumar,[76] 0.73 to 9.14 eV, photoelectron spectroscopy; Zecca et al.[86] and Kanik et al.[87] 5 to 300 eV, linear attenuation technique.

Figure 3.25 shows the relative differences in the total cross sections between various sets of data for electron energies greater than 12 eV using the empirical formula of Zecca et al.[18] as reference. Using this formula has the added advantage that comparisons may be carried out at unusual values of electron energy employed in some measurements.

Table 3.16 shows selected total cross sections for electron energies up to 20 eV. An examination of this table shows that at low energies the agreement between various results is poorer than at higher energies. At 5 eV the result of Kanik et al.[87] is 18.8% higher than the cross section due to Buckman and Lohmann.[71] At higher energies the agreement improves: the discrepancy between the two data at 10 eV is 4.1% (Figure 3.25). The agreement with theoretical calculations[139,140] is also less than satisfactory. The cross sections of Subramanian and Kumar,[76] obtained using photoelectron spectroscopy, are intermediate between the results of Buckman and Lohmann[71] and Kanik et al.[87] The former authors do not consider that the forward scattered electrons are responsible for their low values of cross sections, particularly near the Ramsauer–Townsend minimum. Multiple scattering, that is, when a scattered electron while exiting the scattering cell undergoes a second scattering in the forward direction and is directed back to the cell, is a possibility. However, the cross sections will be dependent on the target gas pressure in that case, and this was not observed experimentally.

3.3.2 MODIFIED EFFECTIVE RANGE THEORY (MERT)

A brief description of the modified effective range theory is in order at this juncture. The concept of using phase shifts to derive scattering cross sections using the effective range theory had its roots in neutron–neutron and neutron–proton scattering studies.[141] Interpretation of scattering data generally proceeded according to the scheme that experimental cross sections were used to derive the phase shifts, which could also be

TABLE 3.16
Total Cross Sections in Krypton in the Low Energy Range

Kanik et al.[87]		Buckman et al.[71]		Dababneh et al.[49] (1980)	
Energy (eV)	Q_T $(10^{-20}\,m^2)$	Energy (eV)	Q_T $(10^{-20}\,m^2)$	Energy	Q_T $(10^{-20}\,m^2)$
5	16.73	0.175	5.03	1.9	3.14
6	20.22	0.20	4.38	2.9	6.96
8	25.98	0.225	3.77	3.6	10.23
10	28.16	0.25	3.22	4.7	14.06
12	28.42	0.275	2.67	5.6	21.64
14	27.09	0.30	2.31	9.7	26.4
16	25.54	0.35	1.75	11.6	28.6
18	24.24	0.40	1.30	12.9	27.3
20	23.32	0.45	1.04	14.5	26.6
		0.50	0.828	15.9	24.1
		0.55	0.649	17.5	24.3
Subramanian and Kumar[76]		0.60	0.570	19.5	23.0
		0.65	0.495		
Energy (eV)	Q_T $(10^{-20}\,m^2)$			**Sin Fai Lam[139] (Theory)**	
0.73	0.74	0.70	0.455		
0.91	0.90	0.72	0.450	Energy (eV)	Q_T $(10^{-20}\,m^2)$
1.09	1.19	0.74	0.441	0.01	27.05
2.00	3.54	0.76	0.451	0.05	14.01
2.18	4.17	0.80	0.458	0.1	8.49
2.66	5.79	0.90	0.530	0.2	3.76
2.85	6.36	1.00	0.672	0.5	0.74
3.23	8.33	1.50	1.84	1	1.06
3.41	9.14	2.00	3.02	2	4.30
4.59	13.10	2.50	4.68	5	19.07
4.77	13.84	3.00	6.36	10	31.48
5.28	16.12	3.50	8.24	14	28.01
5.46	17.31	4.00	10.10	20	22.83
6.55	20.17	5.00	14.08	25	19.83
7.22	22.20	6.00	18.03	30	17.63
7.78	23.29	7.00	21.67		
9.14	25.20	8.00	24.65		
		10.0	27.04		
		12.0	27.29		
		13.0	26.57		
		14.0	26.14		
		15.0	25.40		
		16.0	24.59		
		18.0	23.34		
		20.0	22.29		

The first excited state threshold is 9.915 eV and below this energy the total cross section is equal to the elastic scattering cross section.

obtained from theoretical nuclear potentials. In other words, the phase shifts were the meeting ground between theory and experiment. In view of large amount of computing involved to achieve this objective, semiempirical formulas were developed to shorten the labor of computation for any one form of the assumed nuclear potential. Experimental data are used to infer certain properties of the nuclear potential, and the semiempirical methods provide a relatively simple method to determine what properties of the nuclear potential can and cannot be inferred from the experimental data.

The phase shifts vary with energy of the projectile for any one of the assumed nuclear potentials. Commonly assumed potentials are square well, Gaussian well, exponential well, and Yukawa well. The nuclear potential energies are expressed as functions of the well depth and the intrinsic range beyond which the potential does not exist. The intrinsic range is never observed directly and it must be inferred from the experimental values by defining two parameters, called effective range r_0 and scattering length A, both in, units of length. The effective range also depends on the well depth, decreasing with increasing well depth. The relative values of the effective range and the intrinsic range of the nuclear potential depend on the well shape assumed.

At very low energies, scattering is predominantly s-wave. The essence of the effective range approach is the prediction of a simple functional form for the variation of the phase shifts with energy under very general assumptions about the nuclear potentials. This functional form involves some undetermined parameters. This shifts the meeting ground between theory and experiment from phase shifts to the variational parameters, reducing the computational work involved. Generally speaking, for an assumed potential energy three parameters need to be found rather than the large number of phase shifts. The scattering properties are revealed by the study of the parameters that are determined. A least-square-fit approach is adopted to determine the functional form of the phase shifts and to determine the errors with the phase shifts.

Denoting the phase shift as η and the wavenumber as k, the product $k\eta$ may be expressed as

$$k \cot \eta = -\frac{1}{A} + \frac{1}{2} r_0 k^2 \tag{3.28}$$

The following physical meaning can be ascribed to Equation 3.28. As stated previously, r_0 is the effective range, having the dimension of length, and the parameter A is the scattering length evaluated at zero energy, also having the dimension of length. The factor ½ makes the quantity fall somewhere near the edge of the potential well. Equation 3.28 has only two parameters, the scattering length and the effective range. The formula is known as the shape-independent approximation, since one can always adjust the two parameters to fit any shape of the well, the well depth, and the range of the well. The link with the Schrödinger equation is provided with the relationship that the square of the wavenumber is expressed as

$$k^2 = \frac{2m\varepsilon}{\hbar^2} \tag{3.29}$$

where ε is the energy and m is the reduced mass, given by

$$m = \frac{m_1 m_2}{m_1 + m_2} \tag{3.30}$$

where m_1 and m_2 are the masses of the colliding particles. The cross section is given by

$$Q = \frac{4\pi \sin^2 \eta}{k^2} \tag{3.31}$$

This equation is quite general and depends on three conditions for its validity [141]:

1. The state of the nuclear two-body system can be described by a wave function ψ.
2. The usual symmetries hold, that is, the center of mass motion and the angular motion can be factored out of ψ.
3. The wave function satisfies a Schrödinger-type equation.

The scattering cross section at zero energy is given by

$$Q_0 = 4\pi A^2 \tag{3.32}$$

For short-range potentials and angular momentum L, Equation 3.28 is modified to

$$k^{2L+1} \cot \eta(L) = -\frac{1}{a(L)} + \frac{1}{2} r_0(L) k^2 \tag{3.33}$$

It is useful to recall that Equation 3.28 is shape independent and gives nearly identical results for scattering for any chosen shape. The usefulness of the effective range theory lies in the fact that the scattering amplitude in a certain energy region is completely determined, to a certain accuracy, by a small number of parameters. A few experimental points anywhere in the region may be used to determine these parameters.

For neutron–proton scattering, the characteristic length is of the order of 10^{-15} m for a particular r_0, and one therefore expects the shape-independent approximation to be useful for energies up to perhaps 10 MeV. There are severe restrictions in adopting the effective range theory from the nuclear area to the atomic scattering domain. The effective range theory briefly described above does not hold good for electron–atom scattering because of the $1/r^4$ interaction at long distances. However, it has been possible to modify the effective range theory to scattering of particles from neutral particles that are polarizable.[142] These equations are called the modified effective range theory (MERT). The parameters of the theory are chosen to fit the experimental cross sections in the low energy range, up to a few eV, overlapping the Ramsauer–Kollath experiments. The result is an extrapolation of the results to zero energy. It then becomes possible to compare these extrapolated results with other estimates near zero energy and with drift velocity measurements in the meV energy region.

The momentum transfer and total cross sections are defined, in terms of the angular momentum, as[142]

$$Q_M = \frac{4\pi}{k^2} \sum_{L=0}^{\infty} (L+1) \sin^2(\eta_L - \eta_{L+1}) \tag{3.34}$$

$$Q_T = \left(\frac{4\pi}{k^2}\right) \sum_{L=0}^{\infty} (2L+1) \sin^2(\eta_L) \tag{3.35}$$

where

$$k^2 = \left(\frac{2m}{\hbar^2 a_0^2}\right) \varepsilon \tag{3.36}$$

and ε is the electron energy. It is pointed out here that the minimum in Q_T occurs when $\eta_0 \simeq 0$. The minimum in Q_M occurs when $(\eta_0 - \eta_1) \simeq 0$.[142]

The phase shifts for electron–atom scatterings in the MERT formulation are:

$$\tan \eta_0 = -Ak - \left(\frac{\pi}{3a_0}\right)\alpha k^2 - \frac{4}{3a_0}\alpha Ak^3 \ln(ka_0) \tag{3.37}$$

$$\tan \eta_1 = \left(\frac{\pi}{15a_0}\right)\alpha k^2 - A_1 k^3 \tag{3.38}$$

$$\tan \eta_L = \pi[(2L+3)(2L+1)(2L-1)a_0]^{-1} \times \alpha k^2; \quad (L>1) \tag{3.39}$$

where a_0 is the Bohr radius $(5.292 \times 10^{-11}$ m), and α is the electronic polarizability of the atom. The numerical values of α in rare gases are[143]: He—0.224, Ne—0.437, Ar—1.814, Kr—2.737, and Xe—4.452, in units of $10^{-40}\,F\,m^2$. Application of Equations 3.37 to 3.39 is limited to atoms that do not possess an electric quadrupole moment and for values of L that result in $k^2 \ll 1$. Recognizing that $\tan \eta$ on the left-hand side of Equations 3.37 to 3.39 may be replaced by $\sin \eta$, and substituting Equations 3.37 to 3.39 into 3.34 and 3.35, the low energy expansion equations for the momentum transfer and total cross section are

$$Q_M = 4\pi\left[A^2 + \left(\frac{4\pi}{5a_0}\right)\alpha A k + \left(\frac{8}{3a_0}\right)\alpha A^2 k^2 \ln(ka_0) + Bk^2 + \cdots\right] \tag{3.40}$$

and

$$Q_T = 4\pi\left[A^2 + \left(\frac{2\pi}{3a_0}\right)\alpha A k + \left(\frac{8}{3a_0}\right)\alpha A^2 k^2 \ln(ka_0) + Ck^2 + \cdots\right] \tag{3.41}$$

for sufficiently small $k \ll 1$. B and C are parameters to be determined from experimental data for the appropriate cross section. Both cross sections have the same zero energy limit and Q_M has a much steeper slope (see Figure 3.1). O'Malley[142] also gives the expansion for the differential cross section as

$$Q(\theta) = A^2 + \left(\frac{\pi}{a_0}\right)\alpha A k \sin\left(\frac{\theta}{2}\right) + \left(\frac{8}{3a_0}\right)\alpha A^2 k^2 \ln(ka_0) + \cdots \tag{3.42}$$

The differential cross sections are functions of both the scattered angle θ and the electron energy.

For the s-waveshift, Equation 3.37 is applicable $(L=0)$ and at sufficiently low energy one may make the approximation

$$\tan \eta_0 \simeq \eta_0 \simeq -Ak - \left(\frac{\pi}{3a_0}\right)\alpha k^2 \tag{3.43}$$

If A is positive, as it is in helium and neon, the phase shift will decrease monotonically. If, however, A is negative it will increase from a positive multiple of π at zero energy to some value, reaching a maximum. The phase shift then decreases, passing through zero as the two terms on the right-hand side of Equation 3.43 become equal (see Figure 3.26). Setting this condition, one has

$$A \simeq -\left(\frac{\pi}{3a_0}\right)\alpha k \tag{3.44}$$

which gives the scattering length at the Ramsauer minimum. To calculate A, we need the experimental value at which the minimum cross section is observed. For argon the

Ramsauer minimum energy is $0.30\,\text{eV}$,[68] and to show the calculation for A using Equation 3.44 we substitute the following values, remembering to use atomic units:

$$k^2 = \left(\frac{2m}{\hbar^2 a_0^2}\right)\varepsilon; \quad m = 1; \quad \hbar = 1; \quad \varepsilon = \frac{0.3}{27.2} = 0.011; \quad \alpha = 11.0\,a_0^3$$

$$k = \frac{\sqrt{2 \times 0.011}}{a_0} = \frac{0.1485}{a_0}$$

$$A \simeq -\left(\frac{\pi}{3a_0}\right)\alpha k = -\left(\frac{\pi}{3a_0}\right) \times 11.0\,a_0^3 \times \frac{0.1485}{a_0} = -1.71\,a_0$$

The equivalent equations for Equations 3.36 and 3.44 in SI units are

$$k^2 = 1.1 \times 10^{-19}\,e\left(\frac{2m}{\hbar^2 a_0^2}\right)\varepsilon \tag{3.45}$$

$$A = -9.0 \times 10^9 \left(\frac{\pi}{3a_0}\right)\alpha_e\,k \tag{3.46}$$

where e = electronic charge $(1.60 \times 10^{-19}\,\text{C})$, m = mass of electron $(9.11 \times 10^{-31}\,\text{kg})$, h = Planck's constant $(6.626 \times 10^{-34}\,\text{J s})$, a_0 = Bohr radius $(5.292 \times 10^{-11}\,\text{m})$, ε = electron energy (eV), and α_e = electronic polarizability, in units of $3 \times 10^{-9}\,a_0^3$. Table 3.17 summarizes the calculations for three rare gases.

The momentum transfer cross sections in some rare gases also show the Ramsauer–Townsend effect and the analysis for the minimum in the cross section as a function of energy may be carried out in a similar manner. The expansion for the momentum transfer cross section, Equation 3.34, involves the factor η_1 even for the s-wave, unlike the total scattering cross section for which this factor is present only in the p-wave. Equation 3.34 will have a minimum when $(\eta - \eta_1)$ is a minimum, approximately at zero degrees and multiples of π. Proceeding along the lines just described, O'Malley[142] (1963) shows that the Ramsauer minimum in the momentum transfer cross section occurs at energy (25/36)ths of the minimum for the total cross section. In argon the minimum in the momentum transfer cross section occurs at $0.25\,\text{eV}$,[92] which is lower than the minimum in the total scattering cross section.

TABLE 3.17
Calculation of Scattering Length from Measured Ramsauer–Townsend Minimum Energy

Gas	Ramsauer minimum (eV)	k, Equation 3.45 $(\text{m}^{-1}) \times (a_0)^{-1}$	α_e $(\text{F m}^2)\ (\times 10^{-9}\ a_0^3)$	A, Equation 3.46 $(\text{m})\ (\times a_0)$
Ar	0.3	0.148	1.224	−1.71
Kr	0.6	0.209	1.847	−3.63
Xe	0.65	0.218	3.00	−6.16

The analytical expressions for the waveshifts for the gases Ar, Kr, Ne, and Xe, neglecting higher phases, are:[142]

$$Ar : \frac{(\sin \eta_0)}{k} = 1.70 - 3.13\varepsilon^{1/2} + 0.92\varepsilon \ln \varepsilon + 1.23\varepsilon \tag{3.47}$$

$$Kr : \frac{(\sin \eta_0)}{k} = 3.7 - 4.74\varepsilon^{1/2} + 3.01\varepsilon \ln \varepsilon + 1.84\varepsilon \tag{3.48}$$

$$Ne : \frac{(\sin \eta_0)}{k} = -0.24 - 0.756\varepsilon^{1/2} - 0.031\varepsilon \ln \varepsilon + 0.317\varepsilon \tag{3.49}$$

$$Xe : \frac{(\sin \eta_0)}{k} = 6.5 - 7.68\varepsilon^{1/2} + 8.58\varepsilon \ln \varepsilon + 6.10\varepsilon \tag{3.50}$$

where ε is the energy in eV and is related to k by

$$\varepsilon = 13.605 (ka_0)^2 \tag{3.51}$$

It is to be remembered that Equations 3.37 to 3.39 are related to the Ramsauer minimum energy whereas Equations 3.47 to 3.50 are related to a range of electron energy that covers a few eV. Helium is not included in Equations 3.47 to 3.50 because the phase shifts for this gas at very low energies (up to ~2 eV) are predominantly due to the s-wave. Further, the s-waveshift decreases monotonically with increasing energy and does not pass through zero (see Figure 3.16). Figure 3.26 shows the theoretical calculations of phase shifts[185] in krypton, with the s-wave and p-waveshifts passing through zero, Equations 3.37 and 3.38.

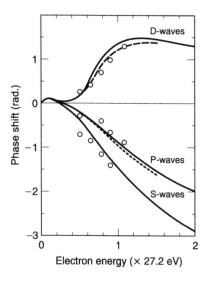

FIGURE 3.26 Phase shifts for elastic scattering of electrons from krypton. (—) McEachran and Stauffer[185]; (- - -) spin up and (— —) spin down results of Sin Fai Lam[139]; (○) Srivastava et al.[50] S-waves and p-waves pass through zero at low electron energies. Figure reproduced from McEachran and Stauffer, *Phys. Lett.*, 107A, 397, 1985. With permission of Institute of Physics, U.K.

Buckman and Lohmann[71] have analyzed their cross section results using the MERT approach. However, the expansion they use for the s-waveshift includes two higher terms of k^2, according to

$$\tan \eta_0 = -Ak - \left(\frac{\pi\alpha}{3a_0}\right)k^2 - Ak\left(\frac{4\alpha}{3a_0}\right)k^2 \ln(ka_0) + Dk^3 + Fk^4 \tag{3.52}$$

The expansions for the p- and higher waves remain the same as Equations 3.38 and 3.39. The momentum transfer and total cross sections are calculated from Equations 3.34 and 3.35.

The analysis of Buckman and Lohmann[71] recognizes the fact that the low-energy limit for the applicability of MERT has not been established either for the determination of scattering length or for comparing the relative values of Q_T and Q_M. In fact, in helium the energy dependence of the elastic scattering cross sections is not strong enough to enable a unique determination of s- and p-waveshifts using the MERT analysis. Values of constants obtained by fitting the cross sections in the range of 0.175 to 0.5 eV are: $A/a_0 = -3.19$, $A_1/a_0^3 = 12.12$, $D/a_0^3 = 184.75$, $F/a_0^4 = -300.8$. The results of these calculations will be referred to in the following section.

3.3.3 Momentum Transfer Cross Sections in Kr

Momentum transfer cross sections control the drift and diffusion phenomena in gases. Measured differential cross sections may be integrated to yield the momentum transfer cross sections according to[50]

$$Q_M = 2\pi \int Q_{\text{diff}}(\theta) \sin\theta (1 - \cos\theta) d\theta \tag{3.53}$$

Srivastava et al.[50] provided such data in the energy range 3 to 100 eV, using helium normalization. They used two different methods. First, experimentally, the relative flow technique was used to measure the elastic differential cross sections and these cross sections were integrated to derive the momentum transfer cross sections. The second method was to apply phase shift analysis. The former method is considered to be more reliable. The cross section reaches a peak at 9 eV and declines thereafter for higher energies. Sin Fai Lam[139] has made theoretical calculations of the phase shifts, taking into account direct and indirect relativistic effects, and the phase shifts have yielded the momentum transfer cross sections on both sides of the Ramsauer minimum (0.1 to 30 eV). Danjo[201] has measured the differential elastic cross sections in the energy range of 5 to 200 eV, using the crossed beam apparatus. Electron impact energy was calibrated using the helium resonance, and the absolute cross sections were determined by the relative flow method, using helium as the comparative gas.

Frost and Phelps[144] derived the momentim transfer cross sections in krypton from the drift velocity of electrons[145] and these results will be further referred to in a later chapter. Pack et al.[92] measured the ratio of longitudinal diffusion coefficient D_L to mobility μ and verified their results using a set of assembled momentum transfer cross sections as follows: $0.0 < \varepsilon \leq 0.3$ eV due to Hayashi[58]; $0.3 < \varepsilon < 1.2$ eV due to Koizumi et al.[188]; and $\varepsilon > 1.2$ eV, Hayashi.[58] Table 3.18 shows these cross sections.

Figure 3.27 shows selected cross sections Q_M. The momentum transfer cross section on either side of the Ramsauer minimum is steeper than the total cross section and therefore steeper than the elastic scattering cross sections, in agreement with the MERT analysis. The minima in Q_M and Q_T occur at electron energies of 0.5 eV and 0.74 eV respectively,

TABLE 3.18
Momentum Transfer Cross Sections in Krypton

Energy (eV)	Q_M $(10^{-20}\,\mathrm{m}^2)$	Energy (eV)	Q_M $(10^{-20}\,\mathrm{m}^2)$	Energy (eV)	Q_M $(10^{-20}\,\mathrm{m}^2)$
0.000	39.70	0.25	1.30	2.50	4.40
0.001	39.70	0.300	0.86	3.00	6.00
0.003	35.00	0.350	0.55	4.00	10.00
0.005	30.00	0.400	0.26	5.00	14.00
0.0085	27.00	0.500	0.10	6.00	16.00
0.010	26.20	0.540	0.11	7.00	17.00
0.020	21.40	0.600	0.15	8.00	16.50
0.040	15.32	0.700	0.27	10.00	15.50
0.060	11.68	0.800	0.42	12.00	13.50
0.080	9.13	1.00	0.80	20.00	6.00
0.100	7.23	1.20	1.30	50.00	1.55
0.16	4.00	1.60	2.00		
0.200	2.37	2.00	3.00		

From Pack, J. L. et al., *J. Appl. Phys.*, 71, 5363, 1992. With permission of AIP (USA).

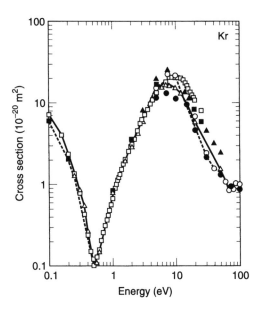

FIGURE 3.27 Momentum transfer cross sections in krypton. (–●–) Srivastava et al.[50]; (–■–) Sin Fai Lam[139]; (–▲–) McEachran and Stauffer[185]; (–○–) Danjo[199]; (□) Mitroy[146]; (–△–) Pack et al.[92]

in satisfactory agreement with the MERT analysis. The momentum transfer cross sections derived by Pack et al.[92] from characteristic energy ($\varepsilon_k = D_L/\mu$) measurements are shown in Table 3.18.

The momentum transfer cross sections derived by Frost and Phelps[144] from measured characteristic energy near the minimum (0.74 eV) are much higher than those obtained by Buckman and Lohmann[71]; the possible reason is the increased forward scattering of the electrons in the latter experiment. Moreover, the swarm-derived Q_M is affected more by the ratio of lateral diffusion coefficient to mobility (D_T/μ) than by drift velocity near the Ramsauer minimum. This factor may have also contributed to the higher value obtained by Frost and Phelps.[144] According to Koizumi et al.[188] the method of analysis of drift velocities for deriving the momentum transfer cross sections is particularly advantageous for thermal energy electrons (~ 0.025 eV at 300 K) but it is not sensitive to the detailed shape of the Ramsauer minimum, which occurs at energies that are an order of magnitude higher (~ 0.3 to 0.8 eV). The accuracy of MERT also decreases as the energy is increased beyond 0.2 to 0.5 eV. However, the ratios of D_r/μ are very sensitive to the rapid change in the cross section in the energy range of 0.1 to 1.0 eV and provide a better basis for comparison. This is the method followed by Koizumi et al.[188] and Pack et al.[92] Mitroy[146] has used the drift velocities in conjunction with more complicated expressions than Equations 3.38 and 3.39 for MERT analysis. Mimnagh et al.[147] theoretically computed the momentum transfer cross sections in the energy range 0.01 to 40 eV.

3.3.4 ELASTIC AND DIFFERENTIAL CROSS SECTIONS IN Kr

The total cross section is equal to the elastic scattering cross section below the threshold of the first excitation potential (9.915 eV). The accurate measurements of Buckman and Lohmann[71] yield the elastic scattering cross sections up to this energy level (Table 3.16). The elastic scattering cross sections in the range of 10 to 10,000 eV are given by the analytical expression of Zecca et al.[18]:

$$Q_T = \frac{1}{A(B + \varepsilon)} + \frac{1}{C(D + \varepsilon)} + \frac{2}{\varepsilon}\sqrt{\frac{BD}{AC}} \frac{1}{|B - D|} \left| \ln\left(\frac{\frac{\varepsilon}{D} + 1}{\frac{\varepsilon}{B} + 1}\right)\right| \tag{3.54}$$

where ε is the electron energy in keV and the constants have the values $A = 4.115$ (keV^{-1} 10^{-20} m^2), $B = 0.0017$ (keV), $C = 0.388$ (keV^{-1} 10^{-20} m^2), and $D = 1.35$ (keV). Figure 3.28 shows the calculated cross sections with selected results: de Heer et al.,[37] 30 to 3000 eV,

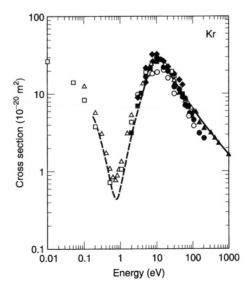

FIGURE 3.28 Elastic scattering cross sections in krypton. ($\blacktriangle$) de Heer et al.[37]; ($\blacksquare$) Dababneh et al.[49] (1980); ($\bigcirc$) Srivastava et al.[50]; ($\square$) Sin Fai Lam[139]; ($\triangle$) Fon et al.[57] (1984); ($\blacklozenge$) McEachran and Stauffer[185]; (- - -) Buckman et al.[71]; ($\bullet$) Danjo[199]; (—) Zecca et al.[18]

semiempirical; Dababneh et al.,[49] 1.9 to 10 eV, from total cross section measurements; Srivastava et al.,[50] 3 to 100 eV, from differential cross section measurements with helium used for normalization; Sin Fai Lam,[139] 0.01 to 30 eV, theory; Fon et al.,[57] (1984), 0.1 to 120 eV, theory; McEachran and Stauffer,[185] 3.0 to 50.0 eV, theory; Buckman and Lohmann,[71] 0.175 to 10 eV, from total scattering cross sections; and Cvejanović and Crowe.[103]

As stated previously, the elastic scattering cross sections are obtained by measuring the differential cross sections over a range of angles and integrating. The Ramsauer minimum observed by Buckman and Lohmann[71] at 0.76 eV is deeper than the theoretical values of Sin Fai Lam[139] and Fon et al.[57]

For the purpose of comparing the results of various investigations, both theoretical and experimental, two energy ranges, $\varepsilon \leq 10$ eV and $10 \leq \varepsilon \leq 1000$ eV, may be delineated. For the first range the measured elastic cross sections of Buckman and Lohmann[71] are taken as reference; for the second range, the empirical equation of Zecca et al.[18] is taken as reference. In the first range the differences between various measurements and theoretical calculations are as large as 50% in certain cases. For example, Buckman and Lohmann[71] obtain a cross section of 6.72×10^{-21} m^2 at 1 eV, which is compared with 1.06×10^{-20} m^2 reported by Sin Fai Lam[139] at the same energy. The same conclusion may be drawn when the calculated cross sections due to Fon et al.[57] (1984) are compared. The large differences between theory and measurements in the low energy range have not been resolved. Points denoting differences in this range have not been included in Figure 3.29.

With regard to the higher energy range, one can only look for measured or calculated cross sections that lie within a band of ±15% from the empirical equation. Over the range 10 to 120 eV the theoretical values of Fon et al.[57] show the best agreement. At higher values, data of de Heer et al.[37] show the least differences. Again, the disagreement between theory and measurements is less than satisfactory.

The differential cross sections calculated from theory[57] over a wide range of electron energies are shown in Figure 3.30 and compared with selected calculations and

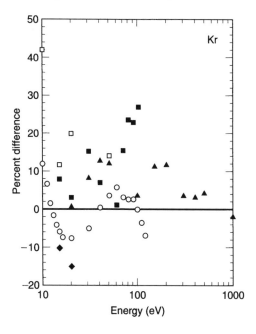

FIGURE 3.29 Comparison of elastic scattering cross sections in krypton as function of electron energy. (▲) de Heer et al.[37]; (□) Srivastava et al.[50]; (◆) McEachran and Stauffer[185]; (○) Fon et al.[57] (1984); (■) Danjo[201]; (—) Zecca et al.[18] Author's computations (unpublished).

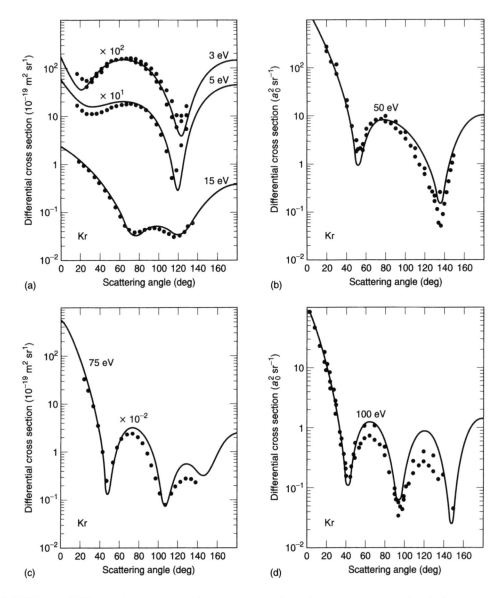

FIGURE 3.30 Differential cross sections in krypton at various electron impact energies. At low energies there are both forward and backward peaks in scattering. Above 50 eV a triple dip is observed due to dominance of f-wave. Symbols for (a) and (c): (•) Srivastava et al.[50]; (○) Heindörff et al.[149] For (b) and (d); (•) Williams and Crowe[148]; (○) Srivastava et al.[50]; (—) Fon et al.[57] for all. Figure reproduced from Fon, W. C. et al., *J. Phys. B*, 16, 307, 1983. With permission of Institute of Physics, U.K.

measurements. It is recalled that the integrated elastic scattering cross section of Fon et al.[57] (1984) is in good agreement with that of Zecca et al.[18] and also covers a wide range (0.1 to 120 eV). General features discerned from the angular distribution of cross sections are summarized by[57]

1. A sharp forward peak, as in Figure 3.30(a).
2. A peak in the backward direction.
3. The formation of double minima at energies below 70 eV, as in Figure 3.30(b).
4. A third minimum slowly taking shape above 70 eV, as in Figure 3.30(c).

TABLE 3.19
Comparison of Ramsauer–Townsend Minima in Krypton

Authors	Method	Minimum (eV)	Cross section (10^{-20} m^2)	Reference
Sin Fai Lam (1982)	Elastic	0.5	0.74	139
	Momentum transfer	0.5	0.10	139
Koizumi et al. (1986)	Momentum transfer	0.5	0.10	188
Buckman and Lohmann (1987)	Total	0.74	0.441	71
Mitroy (1990)	Total	0.5	0.1182	146

5. A simplification of differential cross section at very high energies ($\sim$200 eV) as observed by Williams and Crowe[148] and Cvejanović and Crowe.[103] This aspect is not shown in Figure 3.30.

The sharp forward peak, for example[149] at 75 eV, arises mainly from high partial waves and becomes more pronounced as the incident energy increases. The ground-state dipole static polarizability of krypton is higher than that of helium, neon, and argon,[143] and the contributions of higher partial waves to the cross section increase and the forward peaks become sharper, as a comparison between the 15 eV and 50 eV curves in Figure 3.30 shows.

The peak in the backward direction reflects the difficulty of the electron's penetrating the core. As the target atom becomes heavier the probability of the electron being reflected also increases. The multiple-dip structure of the differential cross section is entirely due to the dominance of a particular partial wave. The d-wave dominates the scattering amplitude for electrons having impact energy less than 60 eV. At low energies the d-wave dominance is compensated by the close interactions arising due to s- and p-waves. Hence, the double structure is not clearly established (Figure 3.36(a)). However, at energy greater than 70 eV, the f-wave takes over from the d-wave as the more dominant wave, resulting in a triple-dip structure in the differential cross section. Fon et al.[57] attribute the multiple-dip structure in the differential cross section to the size of the target atom. If this is true, other heavier atoms may also exhibit similar structure.

Finally, Table 3.19 shows a comparison of the Ramsauer–Townsend minimum in krypton observed by different methods.

3.3.5 TOTAL EXCITATION CROSS SECTIONS IN Kr

The ground state of krypton is $1^2\,2s^2p^6\,3s^2p^6d^{10}\,4s^2p^6$, with 36 electrons. Table 3.20 shows the energy corresponding to ninteen levels.[150] The first excitation potential is 9.915 eV to the 5s[1 ½], $J = 2$ level, which is a metastable state, and the ionization potential of krypton is 13.996 eV. In this energy interval the difference between the total and elastic scattering cross sections gives the total excitation cross section. The next higher level is the 5s[1 ½], $J = 1$ level with a threshold energy of 10.033 eV. This is a resonance level. Trajmar et al.[150] have measured the excitation cross sections of the nineteen levels shown in Table 3.20 by a crossed beam method combined with measurement of energy loss of inelastically scattered electrons. The total cross sections are obtained by adding the individual cross sections (Table 3.21). Figure 3.31 shows the relative magnitude of the first six levels: the resonance level has the highest excitation cross section with reasonable agreement with theoretical calculations.[151] Recent electron impact studies are due to Chilton et al.[152] using the optical method and Yuan et al.[153] using the energy loss spectrometer.

TABLE 3.20
Energy Levels for Excitation States in Krypton

Designation[a]	Designation[b]		J	Designation[c] [154]	Energy (eV)
Ground state	$4p^6\ ^1S$		0		0.0
1	$5s[1\,\tfrac{1}{2}]$		2	A	9.915
2	$5s[1\,\tfrac{1}{2}]$		1	B	10.033
3	$5s'[\tfrac{1}{2}]$		0	C	10.563
4	$5s'[\tfrac{1}{2}]$		1	D	10.644
5	$5p[\tfrac{1}{2}]$		1	E	11.304
6	$5p[2\,\tfrac{1}{2}]$	§[d]	3	F	11.443
	$5p[2\,\tfrac{1}{2}]$	§	2		11.445
7	$5p[1\,\tfrac{1}{2}]$		1		11.526
8	$5p[1\,\tfrac{1}{2}]$		2	G	11.546
9	$5p[\tfrac{1}{2}]$		0	H	11.666
10	$4d[\tfrac{1}{2}]$		0		11.966
11	$4d[\tfrac{1}{2}]$		1	I	12.037
12	$5p'[1\,\tfrac{1}{2}]$		1		12.101
13	$4d[1\,\tfrac{1}{2}]$	§	2		12.112
	$4d[1\,\tfrac{1}{2}]$	§	4		12.126
14	$5p'[0\,\tfrac{1}{2}]$	§	1	J	12.141
	$5p[1\,\tfrac{1}{2}]$	§	2		12.144
15	$4d[3\,\tfrac{1}{2}]$		3		12.179
16	$5p'[\tfrac{1}{2}]$	§	0	K	12.257
	$4d[2\,\tfrac{1}{2}]$	§	2		12.258
17	$4d[2\,\tfrac{1}{2}]$		3		12.284
18	$6s[1\,\tfrac{1}{2}]$	#	2	L	12.352
	$4d[1\,\tfrac{1}{2}]$	#	1		12.355
19	$6s[1\,\tfrac{1}{2}]$	#	1		12.386
	$6p[2\,\tfrac{1}{2}]$	§	2,3	M	12.785
	$4d'[1\,\tfrac{1}{2}]$	§	2		12.800
	$6p[\tfrac{1}{2}]$	#	0	N	12.856
	$4d'[2\,\tfrac{1}{2}]$	#	2		12.860
	$4d'[1\,\tfrac{1}{2}]$	§	1	P	13.005
	$6s'[\tfrac{1}{2}]$	§	1		13.037
	$5d[1\,\tfrac{1}{2}]$	#	1	Q	13.100
	$7s[1\,\tfrac{1}{2}]$	#	1		13.115
	$7s'[\tfrac{1}{2}]^e$	§	1	X	13.764
	$9d[1\,\tfrac{1}{2}]$	§	1		13.781
	$11s[1\,\tfrac{1}{2}]$	§	1		13.784

[a]Trajmar et al.[150]
[b]J-L coupling notation.
[c]Notation of Delâge and Carette.[154]
[d]§ and # mean that the levels are not distinguished.
[e]Ten levels in the energy range 13.285 to 13.730 eV are omitted for the sake of brevity.

The total excitation cross sections expressed as continuous functions of energy are obtained through analytical equations, because experimental measurements are made at specific electron impact energies. The analytical equation of Brusa et al.[17] for the energy range $20 \le \varepsilon \le 4000\,\text{eV}$ is

$$Q_{\text{ex}} = \frac{1}{F(G + \varepsilon_1)} \log\left(\frac{\varepsilon}{\varepsilon_{\text{ex}}}\right) \tag{3.55}$$

TABLE 3.21
Integral and Momentum Transfer Cross Sections for All Levels Combined

Energy (eV)	Q_{ex} $(10^{-23}\,m^2)$	Q_M $(10^{-23}\,m^2)$
15	485.1	478.8
20	1269.0	1083.0
30	847.8	400.4
50	743.8	173.0
100	150.9	79.3

From Trajmar et al., *Phys. Rev. A*, 23, 2167, 1981.

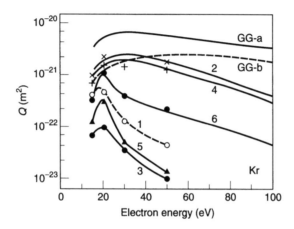

FIGURE 3.31 Measured excitation cross sections in krypton. See Table 3.20 for designations of numbers 1 to 6. The solid and dashed curves, marked as GG-a and b respectively, represent the calculations of Ganus and Green.[151]

where ε is the electron energy, ε_{ex} is the excitation energy (9.915 eV in krypton), and the constants are $F = 22.0$ (keV$^{-1}10^{-20}\,m^2$), $G = 23.3 \times 10^{-3}$ (keV).

Specht et al.[155] measured the Townsend's first ionization coefficients and, by the application of the Boltzmann equation, determined the total excitation cross section. The cross sections at low electron energies, $10 \leq \varepsilon \leq 30$ eV, are particularly useful in discharge simulation calculations. Mason and Newell[73] measured the total metastable cross section by the time-of-flight method. Danjo[201] measured the optical resonance cross sections for levels 2 and 4, using the electron spectrometer. Normalization was carried out by using the absolute differential elastic scattering cross section in the same gas.

Theoretical values have been provided by Kaur et al.[156] for six levels (5 to 8, 12, and 14) by two different methods (indicated as (a) and (b) in Figure 3.32), with considerable disagreement between the two. Guo et al.[157] have provided experimental integrated cross sections for 15 and 20 eV electron energy. In a follow-up paper[158] these authors have provided differential cross sections for the first 20 levels (Table 3.20), but integrated values are not available for comparison. Figure 3.32 shows selected excitation cross sections; the dashed curve in the low energy region (< 20 eV) is a visual best fit.

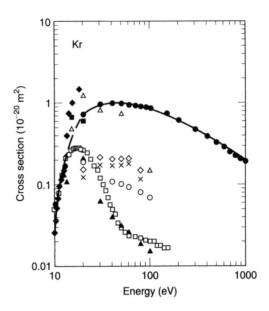

FIGURE 3.32 Excitation cross sections in krypton. (●) de Heer et al.,[37] semiempirical; (◆) Specht et al.,[155] total excitation, from ionization coefficients; (△) Trajmar et al.,[150] total of ninteen levels, experimental; (□) Mason and Newell,[73] total metastable, experimental; (▲) Kaur et al.,[156] total of 4p⁵5p states, theory (method 1); (○) Kaur et al.,[156] total of 4p⁵5p states, theory (method 2); (◇) Danjo,[199] level 2, experimental; (×) Danjo,[199] level 4, experimental; (—) Brusa et al.,[17] semiempirical; (■) Guo et al.,[157, 158] sum of 4p⁵5s level; experimental; (+) Chilton et al.[123] (2000), experimental. Dashed curve is visual best fit for total cross section in the 10 to 20 eV energy range.

3.3.6 IONIZATION CROSS SECTIONS IN Kr

Total and partial ionization cross sections in krypton have been measured by eleven groups of authors. Figure 3.33 shows selected results in the range of 14.5 to 1000 eV and Figure 3.34 shows the comparison between various measurements and the total cross sections of Rapp and Englander-Golden[30] in the same energy range. For the purpose of clarity; Figure 3.33 has two curves; values of the top curve should be divided by 10. Figure 3.34 is drawn in such a way that the positive values of the ordinate indicate that the reference values of Rapp and Englander-Golden[30] are higher.

Selected cross section measurements are due to Asundi and Kurepa[25] 15 to 100 eV, using a Tate and Smith apparatus employing a magnetic field of 0.03 to 0.04 T directed along the axis of the ionization chamber to constrain the electrons in a narrow beam. The results are higher than those of Rapp and Englander-Golden[30] by more than 30% over most of the whole range of electron energy and are therefore not shown in Figure 3.34. The probable reason is that a Macloed gauge was used for measuring target gas pressure; this method is known to present difficulty in the determination of absolute gas pressure. Further, the prevention of mercury vapor entering the ionization chamber is possibly not complete, though Asundi and Kurepa[25] used a cold trap. Corrections due to thermal transpiration were not applied.

Schram et al.[32] used the ionization tube method in the energy range 0.6 to 20 keV, followed by a second study in the 100 to 600 eV range.[32] Nagy et al.[41] covered the relatively high energy range of 500 to 5000 eV for the determination of absolute ionization cross sections. Absolute partial ionization cross sections up to Kr⁴⁺ were measured by Stephan et al.[48] in the energy range of 15 to 180 eV, using the mass spectrometer. The data shown in

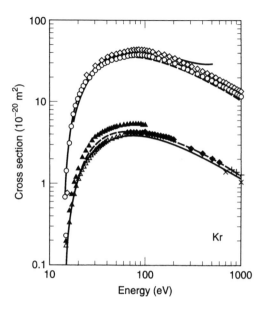

FIGURE 3.33 Total ionization cross sections in krypton. (▲) Asundi and Kurepa[25]; (–) Schram et al.[32] (1965); (♦) Schram et al.[32] (1966c); (×) Nagy et al.[41]; (○) Stephan et al.[48] (1980); (△) Wetzel et al.[75]; (—) Brusa et al.,[17] analytical equation; (– –) Krishnakumar and Srivastava,[82] analytical equation. Top curve, multiply the ordinates by 0.1. (◇) Krishnakumar and Srivastava,[82] experimental; (—) Syage[93]; (– –) Sorokin et al.[105]; (–○–) Kobayashi et al.[112]

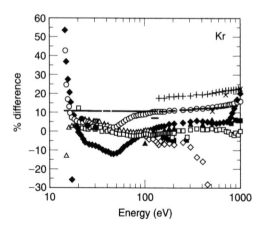

FIGURE 3.34 Comparison of total ionization cross sections in krypton. (■) Schram et al.[32] (1965); (▲) Schram et al.[32] (1966c); (×) Nagy et al.[41]; (●) Stephan et al.[48] (1980); (△) Wetzel et al.[75]; (□) Krishnakumar and Srivastava,[82] experimental; (◇) Syage[93]; (○) Brusa et al.,[17] analytical function; (+) Sorokin et al.[105]; (—) Kobayashi et al.[112]; (♦) Raju (2003, unpublished), analytical. The reference data for comparison are Rapp and Englander-Golden.[30] Positive values of the ordinate mean that the cross sections of Rapp and Englander-Golden[30] are higher.

Figures 3.33 and 3.34 appertain to the total cross sections calculated by the author, using the weighted sum of the partial cross sections.

The careful measurements of Wetzel et al.,[75] using the fast-neutral-beam method, cover the range of 15 to 200 eV at close intervals of the electron energy. Krishnakumar and

Srivastava[82] measured the partial ionization cross sections, using the mass spectrometer in the energy range 15 to 1000 eV. The weighted totals of the partial cross sections are shown in Figures 3.33 and 3.34. Further improved techniques, employing the crossed beams and time-of-flight spectrometer, were adopted by Syage[93] to measure the partial cross sections up to Kr^{6+} in the energy range 18 to 466 eV. More recent measurements, due to Sorokin et al.[108] compare the ion yields due to photoionization and electron impact ionization over the energy range of 140 to 4000 eV. The method avoids the irksome problem of measuring gas number densities below ~0.01 Pa, which is the typical gas pressure in ionization cross section measurements.

Kobayashi et al.[112] adopted the pulsed electron beam method combined with time-of-flight analysis of the charge of the ion. The observed ion yields were converted to partial cross sections by using the total cross sections of Rapp and Englander-Golden[30] in the low-energy range ($14.5 \leq \varepsilon \leq 140$ eV); in the range ($140 < \varepsilon \leq 1000$ eV) Kobayashi et al.[112] used the cross sections of Sorokin et al.[105] To merge the two sets of data smoothly the cross sections of the former were multiplied by a certain factor.[159] The resulting cross sections are uniformly lower than those of Rapp and Englander-Golden[30] by ~11% (Figure 3.40) up to ~700 eV and by ~15% at higher energies. Recalling that the cross sections of Kobayashi et al.[112] are counting cross sections, multiplying them by a factor of 1.1 yields excellent agreement with the cross sections of Rapp and Englander-Golden.[30]

With regard to the analytical functions for the ionization cross sections, Krishnakumar and Srivastava[82] suggest the function

$$Q_{p}(\varepsilon) = \frac{10^3}{\varepsilon_i \varepsilon} \left[A \ln\left(\frac{\varepsilon}{\varepsilon_i}\right) + \sum_{i=1}^{N} B_i \left(1 - \frac{\varepsilon_i}{\varepsilon}\right)^i \right] \tag{3.56}$$

where Q_p is the partial ionization cross section in 10^{-20} m^2, ε the electron energy in eV, ε_i the ionization potential in eV, A and B_i are constants, and N is the number of terms. The cross sections for each level may be computed using this function and the total ionization cross sections obtained by weighted addition. In view of the large computation involved, the cross section given by Equation 3.56 may be multiplied by a function $f(\varepsilon)$, given by

$$f(\varepsilon) = 9 \times 10^{-10} \varepsilon^3 - 10^{-6} \varepsilon^2 + 0.0002\varepsilon + 0.8449 \tag{3.57}$$

The cross sections calculated according to Equation 3.56 in conjunction with Equation 3.57 are also shown in Figure 3.34,[143] using the constants given by Krishnakumar and Srivastava.[82] These are: $A = 4.1675$, $B_1 = -4.0890$, $B_2 = 0.72968$, $B_3 = -2.14$, $B_4 = 2.2089$. The agreement with the cross sections given by Rapp and Englander-Golden[30] is within 10% over most of the energy range.

The analytical function of Brusa et al.[17] has already been referred to in connection with total ionization cross sections in argon and helium. While the shape of the cross section curve is reproduced very well, the percentage differences with Rapp and Englander-Golden[30] exceed 10% beyond 100 eV.

Figure 3.35 shows the absolute partial ionization cross sections obtained by mass spectrometry or time-of-flight methods. Partial ionization cross section measurements fall into three groups: (1) relative partial cross sections; (2) absolute partial cross sections; and (3) partial cross section ratios. The differences between various measurements increase as the charge of the resulting ion increases. Evidence for autoionization, leading to production of higher charged ions, is not as pronounced as in argon (Figure 3.10, Ar^{3+} curve).

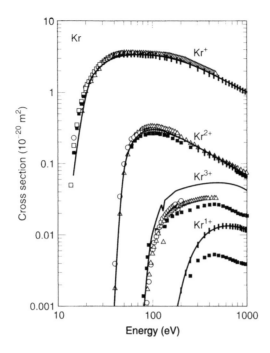

FIGURE 3.35 Partial absolute ionization cross sections in krypton. (○) Stephan et al.[48] (1980); (□) Wetzel et al.[75]; (–•–) Krishnakumar and Srivastava[82]; (△) Syage[93]; (■) Kobayashi et al.[112]

3.3.7 VERIFICATION OF THE SIGMA RULE FOR Kr

The total scattering cross sections measured by Buckman and Lohmann[71] in the energy range $0.175 \leq \varepsilon \leq 10\,\text{eV}$ are accurate to a stated accuracy of ±3%. The uncertainties are the greatest in the energy region 10 to 15 eV which covers the thresholds for excitation and ionization processes. In this narrow region several measured and theoretically calculated elastic scattering cross sections are greater than the total cross section, which cannot be true. The agreement in the 14 to 20 eV range is excellent considering the wide variations in the reported cross sections close to threshold regions. The measured total cross sections of Zecca et al.[86] in the high-energy range of 350 to 1000 eV are a few percent (<5%) higher than the cross sections given by their analytical formula[18] and the latter are shown in Table 3.22. Figure 3.36 shows the relative magnitudes of these cross sections over the electron energy range considered.

3.4 NEON

Neon has considerable practical applications such as neon signs, He–Ne lasers, and in plasma diagnostics. Absolute cross sections, particularly in the low energy region ($\leq 20\,\text{eV}$), are determined by measuring the cross sections and placing them on an absolute scale by using a reference gas, usually helium, below the first excitation level of 19.8 eV. The absolute total cross section in helium has been measured by several groups (see Figures 3.12 and 3.13) within a few percent. The theoretical cross sections using phase shift analysis also show very good agreement with the measured values. Helium has been the preferred choice as the reference gas.

The relative flow technique requires that the measurements be made in the target gas and the reference gas under identical scattering cross sections, that is, the mean free path of the

TABLE 3.22
Verification of the Sigma Rule with Measured Total Cross Sections in Krypton

Energy (eV)	Q_{el} $(10^{-20}\,m^2)$	Q_{ex} $(10^{-20}\,m^2)$	$Q_i^{R\text{-}EG,65}$ $(10^{-20}\,m^2)$	$\Sigma(Q_T)$	Q_T(measured) $(10^{-20}\,m^2)$	%Difference (%)
0.175					5.03[B,87]	
10.0	27.04[B,87]	0.06[S,81]		27.1	27.04[B,87]	−0.222
11.0	27.27[B,87]	0.08[M,87]		27.35	28.29[K,92] (I)	3.323
12.0	27.29[B,87]	0.13[M,87]		27.42	28.42[K,92]	3.519
14	26.26[B,96]	0.24[M,87]	0.078	26.58	27.09[K,92]	1.883
16	23.81[B,96]	0.28[M,87]	0.358	24.45	25.54[K,92]	4.268
18	21.83[B,96]	0.28[M,87]	0.799	22.91	24.24[K,92]	5.487
20	20.20[B,96]	0.74[B,96]	1.223	22.16	23.32[K,92]	4.974
25	17.13[B,96]	0.87[B,96]	2.090	20.09	21.04[K,92]	4.515
30	14.98[B,96]	0.94[B,96]	2.771	18.69	19.05[K,92]	1.890
40	12.14[B,96]	1.00[B,96]	3.492	16.63	16.53[K,92]	−0.605
50	10.34[B,96]	1.00[B,96]	3.835	15.18	14.91[K,92]	−1.811
60	9.09[B,96]	0.98[B,96]	4.090	14.16	13.87[K,92]	−2.091
70	8.17[B,96]	0.95[B,96]	4.213	13.33	13.00[K,92]	−2.538
80	7.46[B,96]	0.92[B,96]	4.257	12.64	12.09[K,92]	−4.549
90	6.89[B,96]	0.88[B,96]	4.231	12.00	11.32[K,92]	−6.007
100	6.43[B,96]	0.85[B,96]	4.196	11.48	10.59[K,92]	−8.404
125	5.56[B,96]	0.78[B,96]	4.037	10.38	9.69[K,92]	−7.121
150	4.96[B,96]	0.71[B,96]	3.826	9.50	8.73[K,92]	−8.820
200	4.17[B,96]	0.61[B,96]	3.457	8.24	7.44[K,92]	−10.753
250	3.66[B,96]	0.54[B,96]	3.131	7.33	6.82[K,92]	−7.478
300	3.30[B,96]	0.48[B,96]	2.867	6.65	6.00[K,92]	−10.833
350	3.03[B,96]	0.43[B,96]	2.656	6.12	5.63[K,92] (I)	−8.703
400	2.81[B,96]	0.40[B,96]	2.463	5.67	5.313[Z,00]	−6.723
450	2.63[B,96]	0.37[B,96]	2.296	5.30	4.977[Z,00]	−6.481
500	2.48[B,96]	0.34[B,96]	2.164	4.98	4.691[Z,00]	−6.163
550	2.35[B,96]	0.32[B,96]	2.041	4.71	4.442[Z,00]	−6.031
600	2.24[B,96]	0.30[B,96]	1.944	4.48	4.223[Z,00]	−6.082
650	2.14[B,96]	0.28[B,96]	1.847	4.27	4.028[Z,00]	−5.999
700	2.05[B,96]	0.27[B,96]	1.759	4.08	3.854[Z,00]	−5.878
750	1.97[B,96]	0.25[B,96]	1.680	3.9	3.695[Z,00]	−5.538
800	1.90[B,96]	0.24[B,96]	1.601	3.74	3.551[Z,00]	−5.312
850	1.84[B,96]	0.23[B,96]	1.539	3.61	3.420[Z,00]	−5.571
900	1.78[B,96]	0.22[B,96]	1.487	3.49	3.298[Z,00]	−5.816
950	1.72[B,96]	0.21[B,96]	1.434	3.36	3.186[Z,00]	−5.460
1000	1.67[B,96]	0.20[B,96]	1.390	3.26	3.082[Z,00]	−5.774

I means interpolated. B,96 means Brusa et al.[17] G,86 means Garcia et al.[69] K,92 means Kanik et al.[87] R-EG,65 means Rapp and Englander-Golden.[30] B,87 means Buckman and Lohmann.[71] S,81 means Specht et al.[155] Z,00 means Zecca et al.[18]

target gas and that of the reference gas should be equal. However, Gulley et al.[96] have observed that the cross section of the atomic beam of helium becomes narrower as the pressure increases. To overcome this disadvantage neon has been suggested as a secondary reference, to narrow down the differences between various measurements.

3.4.1 TOTAL CROSS SECTIONS IN Ne

Absolute total cross sections in neon have been reported by several groups and Figure 3.37 shows the results of selected authors. These are: de Heer et al.,[37] 20 to 3000 eV, semiempirical

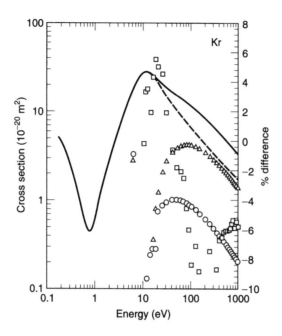

FIGURE 3.36 Verification of the sigma rule in krypton. (—) Measured absolute total cross sections; (–•–) total cross section from summation of individual cross sections; (– –) elastic scattering cross sections; (○) total excitation cross sections; (△) total ionization cross sections; (□) percent difference between measured and summed total cross sections. Positive values of the ordinate mean that measured values are higher. Sources for cross sections are shown in Table 3.22.

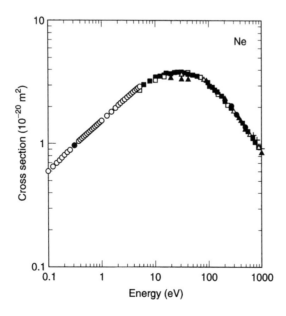

FIGURE 3.37 Total cross section of neon. (▲) de Heer et al.[37]; (◆) Wagenaar and de Heer[44]; (•) O'Malley and Crompton[142]; (△) Kauppila et al.[51]; (□) Register and Trajmar[43]; (■) Nickel et al.[65]; (+) Garcia et al.[69]; (◇) Zecca et al.[78]; (○) Gulley et al.[96]; (—) Zecca et al.,[18] analytical expression. Not shown are the results of Kumar et al.[77] Data compiled by the author (unpublished).

method; Wagenaar and de Heer,[67] 22.5 to 750 eV, linear Ramsauer technique; O'Malley and Crompton,[142] 0 to 2.18 eV, phase shifts calculated from drift velocity data; Kauppila et al.,[51] 20 to 700 eV, transmission method with magnetic field; Register and Trajmar,[43] 5 to 100 eV, averaged values of several authors; Nickel et al.[65] 4 to 300 eV, linear transmission method; Garcia et al.[69] 700 to 6000 eV, linear transmission method; Kumar et al.[77] 0.7 to 10 eV, photoelectron spectroscopy, not shown in the figure; Zecca et al.,[78] 100 to 3000 eV, modified Ramsauer technique; Gulley et al.,[96] 0.1 to 5.0 eV, time of flight attenuation measurements; and Zecca et al.,[18] 10 to 1000 eV, analytical expression.

The analytical formula, applicable in the range of 40 to 3000 eV, is

$$Q_T = \frac{1}{A(B+\varepsilon)} + \frac{1}{C(D+\varepsilon)} + \frac{2}{\varepsilon}\sqrt{\frac{BD}{AC}}\frac{1}{|B-D|}\left|\ln\left(\frac{\frac{\varepsilon}{D}+1}{\frac{\varepsilon}{B}+1}\right)\right| \tag{3.58}$$

where ε is the electron impact energy in keV, A, B, C, D are constants having the values: $A = 4.120$ (keV^{-1} 10^{-20} m^2), $B = 0.106$ (keV), $C = 2.07$ (keV^{-1} 10^{-20} m^2), $D = 1.031$ (keV). The continuous line of Zecca et al.[18] in Figure 3.37 is obscured by the experimental points, thus demonstrating the quality of the agreement.

Figure 3.38 shows the relative differences of Q_T as a function of electron energy. The references for comparison are as follows: Gulley et al.,[96] 0 to 5 eV; Nickel et al.,[65] 5 to 60 eV; Zecca et al.,[18] 60 to 1000 eV. Table 3.23 presents the total cross sections measured by the first two groups of authors. The rationale for selecting these results is as follows. We first note that the absolute cross section measurements for neon in the low energy region are fewer than in helium or argon. In the low energy region, $0.1 \le \varepsilon \le 5.0$ eV, the measurements of Gulley et al.[96] have a stated absolute error less than 4%. The only other experimental data available

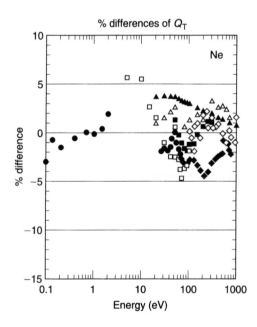

FIGURE 3.38 Percentage differences in total cross section between selected data. Reference values are: Gulley et al.,[96] 0 to 5 eV; Nickel et al.,[65] 5 to 60 eV; Zecca et al.,[18] 60 to 1000 eV. Positive values of the ordinate mean that reference values are higher. (▲) de Heer et al.[37]; (●), O'Malley and Crompton[142]; (◆) Wagenaar and de Heer[44]; (△) Kauppila et al.[51]; (□) Register and Trajmar[43]; (■) Nickel et al.[65]; (○) Zecca et al.[78] Data compiled by the author (unpublished).

TABLE 3.23
Total Scattering Cross Sections in Neon in the Energy Range $0.1 \leq \varepsilon \leq 300$ eV

Gulley et al.[96]		Gulley et al.[96]		Nickel et al.[65]	
Energy (eV)	$Q_T\ (10^{-20}\,m^2)$	Energy (eV)	$Q_T\ (10^{-20}\,m^2)$	Energy (eV)	$Q_T\ (10^{-20}\,m^2)$
0.100	0.595	0.760	1.415	4	2.565
0.120	0.658	0.780	1.425	5	2.843
0.140	0.690	0.800	1.446	6	2.984
0.160	0.734	0.820	1.457	8	3.26
0.180	0.766	0.840	1.471	10	3.443
0.200	0.808	0.860	1.490	12	3.555
0.220	0.845	0.880	1.502	14	3.625
0.240	0.878	0.900	1.506	16	3.668
0.250	0.884	0.920	1.511	18	3.706
0.360	1.042	0.940	1.520	20	3.727
0.380	1.057	0.960	1.544	25	3.766
0.400	1.084	0.980	1.559	30	3.78
0.420	1.105	1.000	1.569	40	3.709
0.440	1.133	1.250	1.708	50	3.613
0.460	1.153	1.500	1.827	60	3.509
0.480	1.177	1.750	1.944	70	3.398
0.500	1.200	2.000	2.060	80	3.283
0.520	1.227	2.250	2.160	90	3.172
0.540	1.238	2.500	2.260	100	3.041
0.560	1.262	2.750	2.339	125	2.803
0.580	1.276	3.000	2.425	150	2.58
0.600	1.291	3.250	2.488	200	2.25
0.620	1.311	3.500	2.569	250	2.008
0.640	1.326	3.750	2.617	300	1.827
0.660	1.341	4.000	2.691		
0.680	1.354	4.250	2.752		
0.700	1.371	4.500	2.812		
0.720	1.360	4.750	2.852		
0.740	1.413	5.000	2.902		

are those of O'Malley and Crompton,[142] which agree within an uncertainty of ±2%. McEachran and Stauffer[60] have theoretically calculated the elastic scattering and momentum transfer cross sections and these results will be discussed in the appropriate sections.

In the energy range $5 \leq \varepsilon \leq 60$ eV, the linear transmission measurements of Nickel et al.[65] have a stated accuracy of ± 3%. Although these authors provide absolute cross sections up to 300 eV, we have used the upper range of 60 eV because, at this energy, the third range of Zecca et al.[18] merges smoothly, within a discrepancy of 0.15%. For higher energies too, the cross sections of Zecca et al.[18] agree very well with the cross sections of Nickel et al.[65] and an analytical function is much more convenient for the purpose of comparison. Almost all available data over the entire energy range fall within a band of ±5% of the reference values, a much more satisfactory situation.

3.4.2 MOMENTUM TRANSFER CROSS SECTIONS IN Ne

Momentum transfer cross sections have been derived by Robertson[160] over the range of E/N (E = electric field, N = gas number density) from 15.18×10^{-24} to 20.03×10^{-22} Vm²

$(1.518 \times 10^{-2}$ to 2.003 Td) by using the drift velocities of electrons measured at 77 K and 293 K. The procedure adopted is to assume an initial set of values for the momentum transfer cross sections as a function of the electron energy and to calculate the electron energy distribution function. The electron energy is used in turn to calculate the drift velocities and the latter are compared with the measured drift velocities. The assumed cross sections are revised and the procedure repeated till satisfactory agreement with a specified accuracy ($\sim$0.2%) is obtained. If measured drift velocities are available at two different temperatures the accuracy of the derived cross sections is checked by calculating the drift velocities at the second temperature as well. Robertson[160] measured the drift velocities at 77 K and 293 K and derived the momentum transfer cross sections in the electron energy range 0.03 to 7 eV (Figure 3.39). He deduced a scattering length of 0.24 a_0.

O'Malley and Crompton[142] improved on the rather circuitous procedure of Robertson[160] by eliminating the unnecessary intermediate step of determining the numerical cross section. The momentum cross section is fitted to an improved version of MERT for the s-waveshift,

$$\eta_0 = -Ak\left[1 + \left(\frac{4\alpha}{3a_0}\right)k^2\ln(ka_0)\right] - \left(\frac{\pi\alpha}{3a_0}\right)k^2 + Dk^3 + Fk^4 \tag{3.59}$$

where α is the polarizability, and the wavenumber k is related to the electron energy ε (in eV) by

$$\varepsilon = 13.605\left(ka_0^2\right) \tag{3.60}$$

The higher order phase shifts are calculated from Equations 3.38 and 3.39. O'Malley and Crompton[142] used a different version, given by

$$\eta_1 = \frac{0.560k^2 - A_1k^3}{1 + B_1k^2} \tag{3.61}$$

$$\eta_2 = 0.080k^2 - A_2k^5 \tag{3.62}$$

Equations 3.61 and 3.62 give the phase shifts for p- and d-waves respectively. O'Malley and Crompton[142] give the following values for the constants in Equations 3.59, 3.61, and 3.62: $A = 0.2135\ a_0$, $A_1 = 1.846$, $A_2 = -0.037$, $B_1 = 3.29$, $D = 3086\ a_0^3$, $F = -2.656\ a_0^4$, $\alpha = 2.672\ a_0^3$. Substituting these values in Equations 3.59, 3.61, and 3.62 for an electron energy of 0.544 eV, one obtains, in a straightforward calculation, $\eta_0 = -0.1097$ rad., $\eta_1 = 0.0067$ rad., $\eta_2 = 0.0032$ rad. These values are then substituted into the expression for momentum transfer cross sections,

$$Q_M = \left(\frac{4\pi}{k^2}\right)\sum_{L=0}^{\infty}(L+1)\sin^2(\eta_L - \eta_{L+1}) \tag{3.63}$$

The contribution of the s-waveshift ($L = 0$) to Q_M is

$$Q_{M1} = \frac{4\pi}{k^2}\left[\sin^2(\eta_0 - \eta_1)\right] = 4.238\ a_0^2 \tag{3.64}$$

TABLE 3.24
Momentum Transfer Cross Sections in Neon at Low Impact Energies

Energy (eV)	Q_M ($10^{-20}\,m^2$)	Energy (eV)	Q_M ($10^{-20}\,m^2$)	Energy (eV)	Q_M ($10^{-20}\,m^2$)
0.000	0.142	0.425	1.230	1.200	1.695
0.025	0.386	0.450	1.257	1.225	1.703
0.050	0.502	0.475	1.282	1.250	1.710
0.075	0.593	0.500	1.307	1.300	1.725
0.100	0.670	0.544	1.347	1.350	1.738
0.125	0.738	0.550	1.352	1.400	1.751
0.136	0.765	0.600	1.393	1.450	1.763
0.150	0.799	0.650	1.430	1.500	1.774
0.175	0.854	0.700	1.465	1.600	1.795
0.200	0.904	0.750	1.497	1.700	1.813
0.225	0.951	0.800	1.526	1.800	1.830
0.250	0.994	0.850	1.553	1.900	1.845
0.275	1.035	0.900	1.578	1.965	1.854
0.300	1.072	0.950	1.601	2.000	1.858
0.325	1.108	1.000	1.622	5.000	2.073
0.350	1.141	1.050	1.643	10.000	2.419
0.375	1.173	1.100	1.661	20.000	2.962
0.400	1.202	1.150	1.679	50.000	2.798

From McEachran, R. P. and A. D. Stauffer, *Phys. Lett.*, 107A, 387, 1985. With permission of Institute of Physics, U.K.

The contribution of the p-waveshift ($L=1$) to Q_M is

$$Q_{M2} = \frac{4\pi}{k^2}\left[2\sin^2(\eta_1 - \eta_2)\right] = 7.697 \times 10^{-3}\,a_0^2 \qquad (3.65)$$

The contribution of higher terms is even smaller and may be neglected. The momentum transfer cross section is given by

$$Q_M = Q_{M1} + Q_{M2} = 1.36 \times 10^{-20}\,m^2$$

which is in excellent agreement with the results of McEachran and Stauffer[185] at the same energy (Table 3.24).

Figure 3.39 shows the momentum transfer cross sections, including the theoretical results of Fon and Berrington[161] for the range 5 to 200 eV. For the purpose of comparison the total cross section has also been plotted. In the low energy region (≤ 1.5 eV), as noted by Zecca et al.,[18] the momentum transfer cross section is larger than the total cross section. The momentum transfer cross sections of Gupta and Rees[162] in the energy range of 100 to 500 eV are considerably lower than the cross sections given by Zecca's formula[18] and their results will be discussed in the following section.

3.4.3 ELASTIC AND DIFFERENTIAL CROSS SECTIONS IN Ne

Integral elastic scattering cross section data in neon have been provided by the following authors: Gupta and Rees,[162] 1 to 500 eV; de Heer et al.,[37] 20 to 1000 eV; Fon and

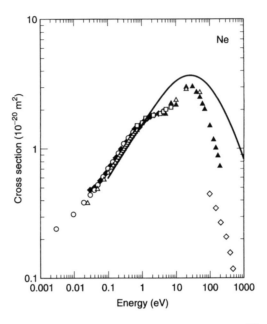

FIGURE 3.39 Momentum transfer cross sections in neon. (◆) Robertson[160]; (◇) Gupta and Rees[162]; (○) O'Malley and Crompton[142]; (▲) Fon and Berrington[161]; (△) McEachran and Stauffer[185]; (□) Gulley et al.[96] (—) Total cross sections are shown for comparison. The momentum transfer cross sections below ~1.5 eV are higher than the total cross sections.

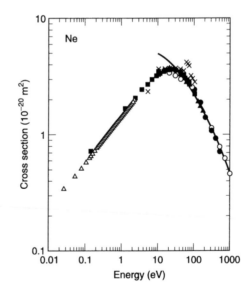

FIGURE 3.40 Integral elastic cross sections in neon. (•) Gupta and Rees[162]; (○) de Heer et al.[37]; (▲) Fon and Berrington[161]; (□) Register and Trajmar,[43] method A (see text); (×) Register and Trajmar,[43] method B (see text); (△) McEachran and Stauffer[185]; (■) Saha[163]; (—) Zecca et al.[18] Data compiled by the author (unpublished).

Berrington,[161] 4 to 200 eV; Register and Trajmar,[43] 5 to 100 eV; McEachran and Stauffer,[185] 0 to 50 eV; and Saha.[163] Figure 3.40 shows these cross sections.

Gupta and Rees[162] have measured the differential cross sections in the high energy range. As far as the author is aware, these are the only high energy (> 100 eV) experimental results available in the literature. The observed relative differential cross sections are converted to an absolute scale by measuring the differential cross sections in 5% and 10% mixtures of neon in helium and applying the absolute cross sections of helium to the mixtures. It is surprising that, whereas the absolute elastic cross sections of these authors are in reasonable agreement with the cross sections of other workers (Figure 3.40), the momentum transfer cross sections are considerably lower (Figure 3.39).

De Heer et al.[37] adopted the semiempirical approach to arrive at the elastic scattering cross sections. Following an elastic encounter of an electron with an atom, spin polarization of the electron occurs and theoretical calculations provide a tool for the sudy of this phenomenon in relatively light rare gases such as neon. Total spin polarization occurs at electron energies and electron scattering angles that are close to the electron energy where a cross section minimum attains its smallest value, the so-called critical energy and critical angle. The theoretical treatment of Fon and Berrington[161] takes into account the full static dipole polarizability. Register and Trajmar[43] measured the differential scattering cross sections and, using phase shift analysis, obtained the integral elastic cross sections. They also evaluated the integral elastic cross sections by the semiempirical approach. Both sets of these data are shown in Figure 3.40. These data must be supplemented with the total cross sections below ~ 16.62 eV (Figure 3.37), which is the first excitation potential.

Saha[163] has made *ab initio* calculations of momentum transfer and total elastic cross sections in the energy range 0.136 to 70 eV. A more recent investigation of differential cross sections is reported by Shi and Burrow[164] in the low energy range of 0.25 to 7.0 eV. The excellent agreement obtained with the results of O'Malley and Crompton[142] has been reasoned to suggest that neon may be used as a secondary standard in calibrating elastic scattering cross sections. This suggestion has been further examined by Gulley et al.[96] with a view to using neon in the relative flow method. The reason adduced is that at higher pressures the beam of helium atoms, which is the preferred gas, becomes narrower than with most other gases, even when the driving pressures are adjusted such that the mean free paths of the gases are identical.

Figure 3.41 shows a relative comparison of total elastic scattering cross sections. For the purpose of comparison the energy range is split into two regions, $0 \leq \varepsilon \leq 20$ eV and $21 \leq \varepsilon \leq 1000$ eV. The cross sections of McEachran and Stauffer[185] and of Zecca et al.[18] are used as references for the first and second range respectively. As noted by Gulley et al.,[96] up to 2 eV the cross sections of Gulley et al.[96] show the best agreement, the differences being

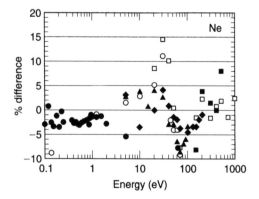

FIGURE 3.41 Comparison of selected integral elastic cross sections in neon. References are: McEachran and Stauffer,[185] 0 to 30 eV; Zecca et al.,[18] 30 to 1000 eV; (■) Gupta and Rees[162]; (□) de Heer et al.[37]; (◆) Fon and Berrington[161]; (▲) Register and Trajmar[43]; (○) Saha[163]; Gulley et al.[96] Positive values of the ordinate mean that the reference values are higher.

within 2.5%. The derived cross sections of O'Malley and Crompton[142] are also in excellent agreement.

Except for isolated energies, the differences lie within ±5% in the first range. Gulley et al.[96] conclude that the use of neon as secondary standard is acceptable up to 10 eV, if one sets an accuracy limit of ±5%, for both the integrated elastic and the angular dependent differential cross sections. Concerning the differential scattering cross sections as a function of electron energy, recent measurements of Gulley et al.[96] are in the low energy region and they are the most accurate data available so far. In the higher energy range the calculations of Fon and Berrington[161] provide the cross sections in the complete range of 0 to 180°, which is a distinct advantage for integration purposes

The essential features of differential scattering cross sections are demonstrated in Figure 3.42 using the results of these two groups. The general features of the elastic differential cross sections are:[161] (1) a sharp forward peak all energies except at the lowest; (2) the formation of a deep minimum at all energies of 20 eV or above.

Figure 3.42a compares differential scattering cross sections of selected authors at a low energy of 0.75 eV. A very good agreement is obtained between the results of Gulley et al.,[96] who observed both forward and backward scattering, Shi and Burrow,[164] O'Malley and Crompton[142] and McEachran and Stauffer.[185] That the different methods adopted, namely experimental, theoretical, and MERT analysis, result in such good agreement supports the view that neon may be used as a secondary standard for low energies.

The sharp forward peak is derived from the contribution of higher partial waves to the differential cross sections. As the incident energy increases, the higher partial waves become more dominant and the forward peak sharper. To demonstrate this point, the phase shifts evaluated as a function of energy by Williams[39] are given in Table 3.25. Note that the p-waveshift as a function of energy passes through zero at low impact energies, as predicted by the MERT analysis due to O'Malley and Crompton.[142] As already stated, at lower energies the s- and p-partial waves dominate and there is no forward peak. As the impact energy increases a single cross section minimum emerges (as in Figure 3.42b) and the valley becomes deeper, attaining the smallest cross section before rising again. The critical energy and the critical angle are 64 eV and 103.6°. Above this energy the valley becomes

TABLE 3.25
Selected Phase Shifts in Radians for the Elastic Scattering of Electrons from Neon Atoms

Energy (eV)	s-wave Phase Shift	p-wave Phase Shift	d-wave Phase Shift
0.58	−0.014	+0.002	0.002
1.22	−0.212	−0.002	0.004
3.00	−0.379	−0.038	0.016
4.00	−0.456	−0.066	0.021
6.00	−0.594	−0.119	0.038
8.00	−0.710	−0.171	0.057
10.00	−0.800	−0.220	0.076
11.00	−0.837	−0.243	0.086
12.00	−0.877	−0.265	0.099
14.00	−0.951	−0.309	0.125
16.00	−1.029	−0.348	0.148
18.00	−1.092	−0.388	0.172
20.00	−1.140	−0.423	0.196

From Williams, J. F., *J. Phys. B, At. Mol. Phys.*, 12, 265, 1979. With permission of the Institute of Physics, U.K.

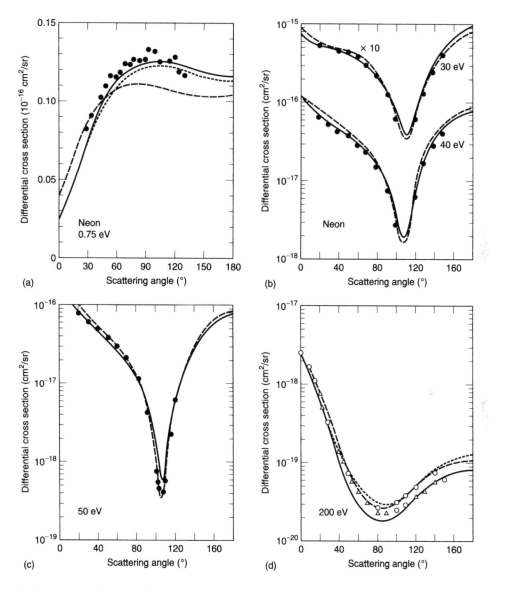

FIGURE 3.42 Experimental and theoretical differential scattering cross sections in neon. (a): (●) Gulley et al.[96]; (□) Shi and Burrow[164]; Williams[39]; (- - -) O'Malley and Crompton[142]; (– –) McEachran and Stauffer[185]; (—) Saha[163] (b), (c), (d): (●) Williams and Crowe[148]; (–) Fon and Berrington[161]; (– –) Thompson[165]; (○) Jansen et al. (quoted in Williams and Crowe[148]); (△) Gupta and Rees[162]; (–●–) McCarthy et al.[166] (a) reproduced from Gulley, R. J. et al., *J. Phys. B*, 27, 2593, 1994; (b), (c), and (d) reproduced from Fon, W. C. and K. A. Berrington, *J. Phys. B*, 14, 323, 1981. With permission of the Institute of Physics, U.K.

shallower, moving toward lower angles (compare Figure 3.42c and Figure 3.42d). Additional results shown in the figure are due to Thompson[165] and to McCarthy et al.[166]

3.4.4 Total Excitation Cross Sections in Ne

Neon has the configuration $1s^2\, 2s^2\, 2p^6$ or simply $2p^6$ with ten electrons. Like other rare gases, its outer shell is completely filled. The excitation levels of neon are shown in Table 3.26

TABLE 3.26
Excitation Levels and Corresponding Energies in Neon

Paschen Notation	Manifold	Level designation	J	Energy (eV)	L-S Coupling Notation	Feature
		$2p^6\ {}^1S_0$	0	0 (ground state)	$1S_0$	1
$1s_5$	3s	$3s[3/2]_2$	2	16.619	${}^3P_2$	2
$1s_4$	3s	$3s[3/2]_1$	1	16.671	${}^3P_1$	3
$1s_3$	3s	$3s'[1/2]_0$	0	16.716	${}^3P_0$	4
$1s_2$	3s	$3s'[1/2]_1$	1	16.848	${}^1P_1$	5
$2p_{10}$	3p	$3p[1/2]_1$	1	18.382	${}^3S_1$	6
$2p_9, 2p_8$	3p	$3p[5/2]_3, 3p[5/2]_2$	3, 2	18.556, 18.576	${}^3D_3, {}^3D_2$	7
$2p_7, 2p_6$	3p	$3p[3/2]_1, 3p[3/2]_2$	1, 2	18.613, 18.637	${}^3D_1, {}^1D_2$	8
$2p_3, 2p_5$	3p	$3p[1/2]_0, 3p'[3/2]_1,$	0,1,	18.712, 18.694,	${}^3P_0, {}^1P_1,$	9
$2p_4, 2p_2$	3p	$3p'[3/2]_2, 3p'[1/2]_1$	2, 1	18.704, 18.727	${}^3P_2, {}^3P_1$	9
$2p_1$	3p	$3p'[1/2]_0$	0	18.966	${}^1S_0$	10
$2s_5, 2s_4$	4s	$4s[3/2]_2, 4s[3/2]_1$	2, 1	19.664, 19.689	—	11
$2s_3, 2s_2$	4s	$4s'[1/2]_0, 4s'[1/2]_1$	0, 1	19.761, 19.780	—	12
$3d_6, 3d_5.$	3d	$3d[1/2]_0, 3d[1/2]_1,$	0,1,	20.025, 20.027,	—	13
$3d'_4, 3d_4,$	3d	$3d[7/2]_4, 3d[7/2]_3,$	4,3,	20.035, 20.035,	—	13
$3d_3, 3d_2,$	3d	$3d[3/2]_2, 3d[3/2]_1,$	2,1,	20.037, 20.041,	—	13
$3d_1, 3d'_1$	3d	$3d[5/2]_2, 3d[5/2]_3,$	2,3	20.049, 20.049	—	13
$3s'''_1, 3s'''_1$	3d	$3d'[5/2]_2, 3d'[5/2]_3,$	2,3,	20.137, 20.137,	—	14
$3s''_1, 3s'_1$	3d	$3d'[3/2]_2, 3d'[3/2]_1,$	2,1,	20.138, 20.140,	—	14
$3p_{10}$	4p	$4p[1/2]_1$	1	20.150	—	14
$3p_9, 3p_8$	4p	$4p[5/2]_3, 4p[5/2]_2,$	3, 2,	20.189, 20.197,	—	15
$3p_7, 3p_6$	4p	$4p[3/2]_1, 4p[3/2]_2,$	1, 2	20.211, 20.215,	—	15
$3p_3,$	4p	$4p[1/2]_0,$	0	20.260	—	16
$3p_5, 3p_4,$	4p	$4p'[3/2]_1, 4p'[3/2]_2,$	1, 2	20.291, 20.298,	—	16
$3p_2, 3p_1$	4p	$4p'[1/2]_1, 4p'[1/2]_0$	1, 0	20.298, 20.369	—	16

Configuration $2p^5 3s$ is shown as manifold 3s.
From Meneses, G. D. et al., *J. Phys. B*, 35, 3119, 2002. Energy data are taken from Register, D. F. et al., *Phys. Rev. A*, 29, 1785, 1984.

with corresponding energies.[167,168] Figure 3.43 shows a simplified diagram of the energy levels. Reviews of theoretical and experimental excitation cross sections can be found in Machado et al.,[169] Andersen et al.,[170] and Becker et al.[171] The latter authors' summary of experimental and theoretical work shows that cross section data in a continuous range of electron energy are relatively few. Following the benchmark measurements of Register et al.[167] for the lowest forty excited states, using the energy loss method, theoretical values in the range of 20 to 100 eV are provided by Meneses et al.[168]

The first excited configuration $2p^5 3s$ has four levels with $J = 2, 1, 0, 1$ (Table 3.26). Of these, features 2 (16.619 eV) and 4 (16.716 eV) are metastables whereas feature 5 (16.848 eV) is an optical-level resonance of wavelength 73.6 nm. The metastable level ${}^3P_0$ has a long lifetime of 430 s and absorbs radiation of 588.2 nm to attain the excitation level ${}^3P_1$. It now falls to ${}^1P_1^0$ level, emitting 659.9 nm radiation. The experimental method for recording this observation[172] and an energy level diagram are given in Chapter 2 (Figure 2.18). Both absolute and direct cross sections are given in the electron impact energies from threshold to 300 eV.

The next ten levels in the $2p^5 3p$ configuration (features 6 to 10, energy range 18.382 to 18.966 eV) fall within a range of 550 to 810 nm.[173] Chilton et al.[173] have measured both the direct and apparent excitation cross sections of the $2p^5 3p$ configuration (there are

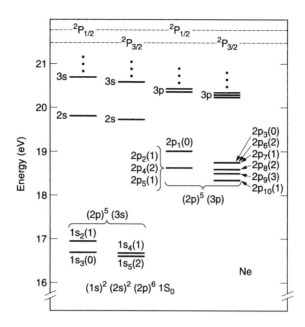

FIGURE 3.43 Energy levels associated with the ground, $(2p)^5 (3s)$, and $(2p)^5 (3p)$ configuration of the neon atom. A symbol like $1s$, $2s$, $2p$,..., without parentheses labels an energy level in Paschen notation, identical with column 1 of Table 3.26. The numeral inside the parentheses after $1s_2$,... $1s_5$, $2p_1$,... $2p_{10}$ refers to the value of the total angular momentum J, identical with column 4 of Table 3.26 (Phillips et al.[172]). For neon, Paschen notation is preferred because the L-S coupling notation is less than satisfactory.

ten levels) in the energy range 25 to 200 eV. Cascading from $2p^5 4s$ and $2p^5 3d$ into the $2p^5 3p$ configuration emits radiation in the 740 to 1720 nm range, part of which is in the infrared region.[173]

Sharpton et al.[174] measured the cross sections for some fifty states from the $2p^5 ns$, $2p^5 np$, and $2p^5 nd$ configurations by optical methods, from which they determined the apparent cross sections. We recall that the apparent cross section of a level is the sum of direct excitation cross sections and cascade excitations into the level. These authors find that the corrections due to cascading from the upper levels are important only for the np states and typically amount to 50% of the total population.

Configurations belonging to the $2p$ family (Paschen designation) and the $3p$ family both have ten levels, though the latter are more closely spaced than the corresponding $2p$ levels. All of the transitions out of the $2p^5 3p$ ($2p$ in Paschen notation) levels are in the visible region. Each ns family is composed of four energy levels (ns_2, ns_3, ns_4, and ns_5) and the allowed transitions to ground level are ns_2 and ns_4 levels. Each nd family is composed of twelve energy levels. These are designated in the Paschen notation by the symbols nd'_1, nd_1, nd_2, nd_3, nd_4, $nd_4{}'$, nd_5, nd_6, ns'_1, ns_1, ns_1, and ns''''_1. The d levels are very closely spaced.

As already mentioned, Register et al.[167] have measured differential and total cross sections for the forty lowest lying states of neon. The cross sections included those for the excitation of $1s_3$ (16.716 eV) and $1s_5$ (16.619 eV) states at incident energies of 25, 30, and 50 eV. The procedure followed was to integrate the measured relative differential cross sections. The absolute cross sections were obtained by comparing the cross sections to the differential cross sections for elastic scattering at selected energies. Teubner et al.[175] adopted the time-of-flight technique to measure the total cross sections for the production of

TABLE 3.27
Radiative Lifetime of 19 Excited States

State	Lifetime (ns)	State	Lifetime (ns)
$1s_4$	31.7	$2p_2$	18.8
$1s_2$	1.87	$2p_3$	17.6
$2s_4$	9.67	$2p_4$	19.1
$2s_2$	7.78	$2p_5$	19.9
$3d_5$	13.2	$2p_6$	19.3
$3d_2$	7.25	$2p_7$	19.9
$3s'_1$	12.3	$2p_8$	19.8
$3s_4$	19.5	$2p_9$	19.4
$3s_2$	23.1	$2p_{10}$	24.8
$2p_1$	14.4		

The states are designated in Paschen notation.
From Sharpton, F. A. et al., *Phys. Rev. A*, 2, 1305, 1970.

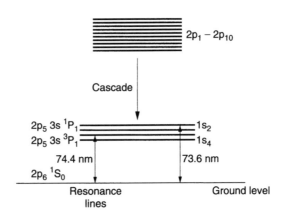

FIGURE 3.44 A simplified level diagram for the lower two optical emission lines in neon. On the right-hand side the Paschen notation is shown and on the left-hand side the electron configuration with terms and levels is shown.

metastables in the energy range of 17 to 500 eV. It has already been stated that the lifetimes of these states are relatively long, compared to radiative lifetimes. Table 3.27 shows the latter for selected levels. The relative cross sections were converted to absolute cross sections by using the known cross sections for the production of metastables in helium.

Time-of-flight method has also been adopted by Mason and Newell[73] in the energy range 17 to 147 eV; the relative cross sections were normalized using the cross sections of Teubner et al.[175] at 26 eV. The resonance lines corresponding to $1s_4$ and $1s_2$ levels are 74.4 and 73.6 nm respectively (Figure 3.44), in the extreme ultraviolet. Kanik et al.[176] provide absolute cross sections over the electron energy range of 20 to 400 eV.

Figure 3.45 shows selected excitation cross sections in neon, and Table 3.28 provides the details of the cross sections included in the figure. An examination of this table shows that the sum of absolute cross sections of Teubner et al.[175] with two metastable states, Kanik et al.[176] with the lowest two resonance states, and Sharpton et al.[174] with the ten optical states gives the total excitation cross sections with reasonable accuracy.

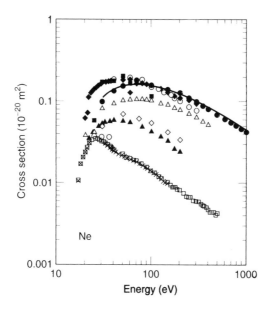

FIGURE 3.45 Selected excitation cross sections in neon. (▲) Sharpton et al.[174]; (●) de Heer et al.[37]; (■) Register et al.[167]; (○) Phillips et al.[172]; (□) Teubner et al.[175]; (×) Mason and Newell[73]; (—) Brusa et al.[17]; (△) Kanik et al.[176]; (◇) Chilton et al.[173]; (◆) Teubner et al.[175] + Kanik et al.[176] + Sharpton et al.,[174]; sum of three separate measurements.

TABLE 3.28
Summary of Data Included in Figure 3.45

Authors	Reference	Levels	Energy Range (eV)	Comments
Sharpton et al. (1970)	174	$2p_1 - 2p_{10}$	22–200	10 levels in $2p$ states
de Heer et al. (1979)	37	Total	30 – 1000	Semiempirical
Register et al. (1984)	167	$1s_5 - 3p_1$	25–100	Total of 40 levels
Phillips et al. (1985)	172	$1s_5 - 1s_2$	25–300	4 lowest levels
Teubner et.al. (1985)	175	$1s_3$ and $1s_5$	17–500	Metastables
Mason and Newell (1987)	73	$1s_3$ and $1s_5$	17–147	Metastables
Brusa et al. (1996)	17	Total	40 – 1000	Analytical
Kanik et al. (1996)	176	$1s_2$ and $1s_4$	20–400	Two resonance states
Chilton et al. (2000)	173	$2p_1 - 2p_{10}$	25 – 200	10 levels in $2p$ states

The levels shown are in Paschen notation.

Figure 3.45 also includes the curve for this shown as TRTFB + KAJ +SJLF. Instead of drawing on Sharpton et al.[174] for the ten optical states one could also use the cross section data of Chilton et al.[173] without affecting the accuracy, but the latter requires more interpolation as the energy intervals adopted are larger.

3.4.5 IONIZATION CROSS SECTIONS IN Ne

Ionization cross sections in neon have been provided by the following authors: Asundi and Kurepa,[25] Tate and Smith apparatus, 22 to 100 eV; Schram et al.,[32] ionization tube

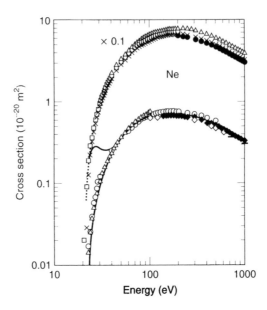

FIGURE 3.46 Absolute total ionization cross sections in neon. Ordinates of the top curve should be multiplied by 0.1. (△) Asundi and Kurepa[25]; (◇) Schram et al.[32] (1965); (-) Schram et al.[32] (1966a); (×) Schram et al.[32] (1966b); (○) Fletcher and Cowling[29]; (+) Nagy et al.[41]; (○) Stephan et al.[48] (× 0.1); (□) Wetzel et al.[75]; (– –) Lennon et al.[83]; (△) Krishnakumar and Srivastava[82] (× 0.1); (◆) Almeida et al.[177]; (●) Sorokin et al.[105], (× 0.1); (×) Kobayashi et al.[124] (× 0.1).

method, 0.6 to 20 keV (1965) and 100 to 600 eV (1966A); Schram et al.,[32] mass spectrometer method, 0.5 to 16 keV (1966b); Fletcher and Cowling,[29] ionization tube method, 22 to 500 eV; Nagy et al.[41] and Stephan et al.,[48] mass spectrometric method, 25 to 180 eV; Wetzel et al.,[75] crossed beam technique, 22 to 200 eV; Lennon et al.[83] and Krishnakumar and Srivastava,[82] 25 to 1000 eV, mass spectrometric method; Almeida et al.,[177] time-of-flight technique, 140 to 3000 eV; Brusa et al.,[17] threshold to 5000 eV, semiempirical formula, 21.6 to 1000 eV; Sorokin et al.,[105] ionization tube method, 140 to 4000 eV; and Kobayashi et al.,[124] time-of-flight method, 22 to 1000 eV.

Following the method adopted for other rare gases, the fourteen sets of data have been divided arbitrarily into two groups for better presentation in Figure 3.46. Ordinates of the top curve should be divided by 10 to obtain the cross section. The absolute total cross sections of Rapp and Englander-Golden[30] are not shown as they have been presented previously (Figure 2.14). Two semiempirical equations are available, due to Lennon et al.[83] and Brusa et al.[17] Lennon et al.[83] give the equation as

$$Q_i = \frac{1}{\varepsilon \, \varepsilon_i} \left\{ A \ln\left(\frac{\varepsilon}{\varepsilon_i}\right) + \sum_{j=1}^{N} B_j \left[\left(1 - \frac{\varepsilon_i}{\varepsilon}\right)^j \right] \right\} \tag{3.66}$$

where ε is the electron energy in eV, ε_i is the ionization potential in eV, N is the number of terms in the summation, $A = 2.192$, $B_1 = -0.447$, $B_2 = -7.006$, $B_3 = 5.927$. A and B_j are in units of 10^{-17} (eV)^2 m^2. Strictly speaking, Equation 3.66 is applicable for the first partial ionization cross sections, with similar equations for the higher partial cross sections. In view of the laborious calculations involved the higher partial cross sections are not computed.

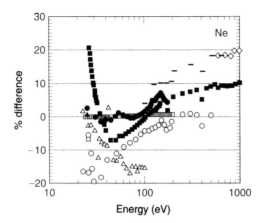

FIGURE 3.47 Relative differences between selected data, with the absolute total cross sections of Rapp and Englander-Golden[30] as reference. Positive values of the ordinate denote that the reference values are higher. (△) Asundi and Kurepa[25]; (◇) Schram et al.[32] (1965); (-) Schram et al.[32] (1966a); (○) Fletcher and Cowling[29]; •, Stephan et al.[48]; □, Wetzel et al.[75]; ■, Brusa et al.[17]

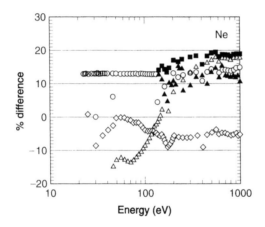

FIGURE 3.48 (continuation of Figure 3.47) Relative differences between selected data, with the absolute total cross sections of Rapp and Englander-Golden[30] as reference. Positive values of the ordinate denote that the reference values are higher. (△) Lennon et al.[83]; (◇) Krishnakumar and Srivastava[82]; (▲) Almeida et al.[177]; (■) Sorokin et al.[105]; (○) Kobayashi et al.[124]

Brusa et al.[17] give a different equation (3.12), which is reproduced below for convenience:

$$Q_i = \left(\frac{L}{M+x} + \frac{N}{x}\right)\left(\frac{y-1}{x+1}\right)^{1.5} \times \left[1 + \frac{2}{3}\left(1 - \frac{1}{2x}\right)\ln\{2.7 + (x-1)^{0.5}\}\right] \qquad (3.12)$$

where L, M, N, P are the four free parameters (Table 3.9) determined by the best fitting procedure, $y = \varepsilon/\varepsilon_i$ and $x = \varepsilon/P$, ε_i is the ionization potential and ε the electron energy.

Figure 3.46 summarizes these data. Figures 3.47 and 3.48 show the relative differences between various measurements and calculations, with Rapp and Englander-Golden[30] as reference. Table 3.29 shows investigations that fall within a ±10% band of values of Rapp and Englander-Golden,[30] and the appropriate energy range.

Figures 3.47 and 3.48 show that the differences between various sets of data are considerably larger in neon than in other rare gases considered so far. The differences are

TABLE 3.29

References for Ionization Cross Sections That Fall within a Band of ±10% of Values of Rapp and Englander-Golden.[30]

Authors	Energy Range (eV)	Maximum Deviation	Reference
Asundi and Kurepa (1963)	23–44	−8.41% at 40 eV	25
Fletcher and Cowling (1973)	60–500	−8.86% at 60 eV	29
Stephan et al. (1980)	25–180	+7.16% at 150 eV	48
Lennon et al. (1988)	45–200	+7.78% at 200 eV	83
Wetzel et al. (1987)	24–200	+0.78% at 200 eV	75
Krishnakumar and Srivastava (1988)	25–1000	−8.91% at 175 eV	82
Brusa et al. (1996)	29–1000	+9.94% at 1000 eV	17
Kobayashi et al. (2002)[a]	22–1000	+2.95% at 1000 eV	124

[a] Multiply by 1.12.

as great as 30% between the results of Krishnakumar and Srivastava[82] and Sorokin et al.[105] Asundi and Kurepa,[25] Rapp and Golden-Englander,[30] and Fletcher and Cowling[29] provide gross ionization cross sections whereas the results of Nagy et al.[41] are partial sums of cross sections (counting). Stephan et al.[48] presented absolute partial cross sections, from which gross cross sections are calculated by the present author. Wetzel et al.[75] measured the single ionization cross sections Q_i^+ and the ratios Q_i^{n+}/Q_i^T; the gross ionization cross sections were then calculated. These results show the best agreement, with less than 1% deviation over the whole range of electron impact energy investigated.

As mentioned previously, one of the major sources of uncertainty in the determination of absolute cross section is the target gas pressure. For example, the results of Rapp and Englander-Golden[30] and Schram et al.[32] show deviations as large as 20% (Figure 3.53) which do not appear in Table 3.30. Both these studies used similar techniques, the significant difference being the method of target gas density determination. Rapp and Englander-Golden[30] used an effusive flow apparatus calibrated against pressure measurements in molecular hydrogen obtained with a McLeod gauge, whereas Schram et al.[32] used a wall cooled McLoed gauge corrected for the mercury pumping effect. Asundi and Kurepa[25] did not make any correction to take this effect into account. Fletcher and Cowling[29] measured the gas density using an ionization gauge calibrated against a capacitance manometer. These results are in very good agreement with the reference data.

Almeida et al.[177] adopted the time-of-flight method. Sorokin et al.[105] measured the ratios of electron impact to photoionization cross section. The two sets of data are considerably lower than the cross sections of Rapp and Englander-Golden[30]. Recent time-of-flight measurements of Kobayashi et al.[125] are uniformly about 12 to 13% lower than those of Rapp and Englander-Golden[30]; multiplying the former by a factor of 1.12 gives excellent agreement.

Figure 3.49 shows the absolute partial ionization cross sections obtained by mass spectrometer or time-of-flight methods. The agreement between results of the cross sections for the first ionization level is better than for second and third ionizations, as in other rare gases. The relative shapes of the curves are, however, in better agreement. Apart from the problems associated with the determination of absolute total cross sections, partial ionization cross section measurements present additional difficulties such as the mass-to-charge dependence of ion collection (assumed to be constant). There is no evidence of an Auger process at higher energies, unlike in the case of argon (see Figure 3.10, Ar^{3+}).

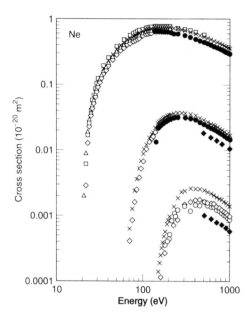

FIGURE 3.49 Partial ionization cross sections in neon. (◆) Schram et al.[32], (1966b); (□) Stephan et al.[48]; (△) Wetzel et al.[75]; (×) Krishnakumar and Srivastava[82]; (•) Almeida et al.[177]; (○) Kobayashi et al.[124] Data compiled by the author (unpublished).

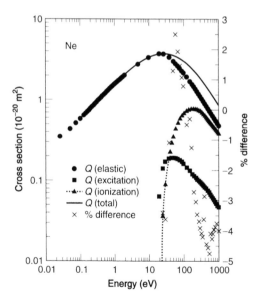

FIGURE 3.50 Verification of the sigma rule in neon. Percentage difference is with reference to measured total cross sections. Positive values mean that the total cross section obtained by adding individual cross sections is lower.

3.4.6 VERIFICATION OF THE SIGMA RULE FOR NE

Figure 3.50 and Table 3.30 show a comparison between the total cross sections, obtained by adding the elastic, excitation, and ionization cross sections, and the experimentally observed or theoretically calculated total cross sections. The elastic scattering cross sections

TABLE 3.30

Verification of the Sigma Rule in Neon. $Q_T = Q_{el} + Q_{ex} + Q_i$. For Comparison, Q_T (Measured) Are Used as Reference

Energy (eV)	Q_{el} $(10^{-20}\,m^2)$	Q_{ex} $(10^{-20}\,m^2)$	Q_i $(10^{-20}\,m^2)$	Q_T(sum) $(10^{-20}\,m^2)$	Q_T(measured) $(10^{-20}\,m^2)$	% Difference
20	3.7707[M,85]	0.063[a]	—	3.834	3.727[N,85]	−2.43
25	3.7146[M,85]*	0.141[a]	0.0380[b]	3.893	3.766[N,85]	−3.379
30	3.6585[M,85]*	0.170[a]	0.0888	3.917	3.780[N,85]	−3.627
40	3.3382[z,00]	0.183[a]	0.2278	3.749	3.709[N,85]	−1.070
50	3.0419[z,00]	0.187[a]	0.3378	3.567	3.613[N,85]	1.276
60	2.8008[z,00]	0.186[a]	0.4354	3.422	3.509[N,85]	2.485
70	2.6005[z,00]	0.182[a]	0.5137	3.296	3.361[z,00]	1.922
80	2.4309[z,00]	0.177[a]	0.5771	3.185	3.231[z,00]	1.451
90	2.2853[z,00]	0.167[a]	0.6281	3.080	3.113[z,00]	1.069
100	2.1588[z,00]	0.157[a]	0.6668	2.982	3.005[z,00]	0.759
125	1.9039[z,00]	0.144[a]	0.7354	2.783	2.770[z,00]	−0.490
150	1.7103[z,00]	0.130[a]	0.7424	2.583	2.574[z,00]	−0.323
175	1.5575[z,00]	0.120[a]	0.7617	2.439	2.409[z,00]	−1.257
200	1.4334[z,00]	0.113[a]	0.7811	2.328	2.266[z,00]	−2.728
225	1.3302[z,00]	0.108[a]	0.7688*	2.207	2.142[z,00]	−3.043
250	1.2429[z,00]	0.103[a]	0.7565	2.103	2.032[z,00]	−3.461
275	1.1680[z,00]	0.099[a]	0.7394*	2.006	1.935[z,00]	−3.678
300	1.1027[z,00]	0.095[a]	0.7222	1.920	1.848[z,00]	−3.910
325	1.0454[z,00]	0.092[a]	0.7042*	1.842	1.769[z,00]	−4.079
350	0.9945[z,00]	0.089[a]	0.6861	1.769	1.698[z,00]	−4.192
375	0.9490[z,00]	0.086[a]	0.6571*	1.692	1.633[z,00]	−3.625
400	0.9080[z,00]	0.082[a]	0.6281	1.618	1.573[z,00]	−2.866
450	0.8369[z,00]	0.0797[B,96]	0.6158	1.532	1.467[z,00]	−4.443
500	0.7774[z,00]	0.0745[B,96]	0.5867	1.439	1.376[z,00]	−4.554
550	0.7267[z,00]	0.0700[B,96]	0.5612	1.358	1.296[z,00]	−4.738
600	0.6828[z,00]	0.0660[B,96]	0.5278	1.277	1.226[z,00]	−4.097
650	0.6444[z,00]	0.0626[B,96]	0.5067	1.214	1.164[z,00]	−4.257
700	0.6106[z,00]	0.0595[B,96]	0.4838	1.154	1.108[z,00]	−4.102
750	0.5804[z,00]	0.0567[B,96]	0.4627	1.100	1.058[z,00]	−3.939
800	0.5533[z,00]	0.0542[B,96]	0.4442	1.052	1.012[z,00]	−3.873
850	0.5288[z,00]	0.0519[B,96]	0.4266	1.007	0.971[z,00]	−3.757
900	0.5066[z,00]	0.0498[B,96]	0.4134	0.970	0.933[z,00]	−3.975
950	0.4863[z,00]	0.0479[B,96]	0.3976	0.932	0.898[z,00]	−3.798
1000	0.4676[z,00]	0.0462[B,96]	0.3862	0.900	0.865[z,00]	−4.006

B,96 means Brusa et al.[17] M,85 means McEachran and Stauffer.[185] N,85 means Nickel et al.[65] z,00 means Zecca et al.[18]
[a]means sum of excitation cross sections due to Teubner et al.,[175] Kanik et al.,[176] and Sharpton et al.[174]
[b]means the entire column is due to Rapp and Englander-Golden.[30]
*means interpolated. Positive values of % difference mean that the added cross sections are lower.

reported by McEachran and Stauffer[185] are used up to 30 eV and, for higher energies, the cross sections of Zecca et al.[18] are used. At 30 eV the two cross sections merge smoothly with a discrepancy of ~1%. The excitation cross sections up to 400 eV are the sum of three investigations, as shown in Figure 3.50. The sigma rule is satisfied very well, with maximum and minimum discrepancies of +2% and −4.8%.

3.5 XENON

Development of rare gas halide high power lasers has focused attention to studies on scattering phenomena in rare gas atoms. ArF (193 nm), KrF (248 nm), XeCl (308 nm), and

XeF (351 nm) are examples of such devices that provide high output levels for industrial applications.

3.5.1 TOTAL CROSS SECTIONS IN Xe

Absolute total scattering cross sections have been measured by the following: Wagenaar,[178] 20 to 700 eV (not shown); Dababneh et al.,[49] transmission technique, 2.8 to 49.6 eV (1980); Wagenaar and de Heer,[44] 22.5 to 750 eV; Jost et al.,[56] linear Ramsauer technique; 7.5 to 60 eV; Dababneh et al.,[49] transmission technique, 20 to 750 eV (1982); Wagenaar and de Heer,[67] linear Ramsauer technique, 15 to 750 eV; Nickel et al.,[65] linear Ramsauer technique, 4 to 300 eV; Subramanian and Kumar,[76] photoelectron spectroscopy, 0.73 to 9.13 eV; Zecca et al.,[86] modified Ramsauer technique; 81 to 4000 eV; Ester and Kessler,[179] static-gas-target technique, 40 to 100 eV; and Szmytkowski et al.,[180] linear transmission technique, 0.5 to 220 eV.

Figure 3.51 shows the selected cross sections referred to above. For experimental data on xenon in the low energy range that includes the Ramsauer minimum, one has to refer to the 1933 data of Brode.[3] However, theoretical values are available as shown. The reference cross sections are reported by Nickel et al.[65] in the energy range 4 to 150 eV and by Zecca et al.[18] for higher energies. The agreement between several investigations is poorer than in other rare gases, as Figure 3.52 demonstrates. The best agreement exists between the results of Nickel et al.,[65] Wagenaar and de Heer,[67] Zecca et al.,[86] and Zecca et al.[18] Since the total cross section curve has a peculiar shape for energies below 90 eV, no analytical formula is available up to this energy.

Semiempirical cross sections at selected energies in the range of 20 to 4000 eV are given by de Heer et al.[37] An analytical expression for total elastic scattering cross section is given by Zecca et al.[18] and is applicable for the range of electron impact energy, 90 to 4000 eV. Their expression is reproduced below for the sake of convenience:

$$Q_T = \frac{1}{A(B+\varepsilon)} + \frac{1}{C(D+\varepsilon)} + \frac{2}{\varepsilon}\sqrt{\frac{BD}{AC}}\frac{1}{|B-D|}\left|\ln\left(\frac{\frac{\varepsilon}{D}+1}{\frac{\varepsilon}{B}+1}\right)\right| \tag{3.1}$$

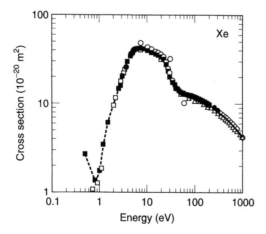

FIGURE 3.51 Absolute total scattering cross sections in xenon. Experimental: (●) Dababneh et al.[49] (1980); (○) Jost et al.[56]; (×) Wagenaar and de Heer[67]; (◆) Nickel et al.[65]; (□) Subramanian and Kumar[76]; (△) Zecca et al.[86]; (–■–) Szmytkowski et al.[180] Semiempirical: (○) de Heer et al.[37]; (—) Zecca et al.[18]

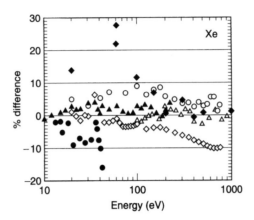

FIGURE 3.52 Comparison of absolute total cross sections in xenon. Reference values are Nickel et al.[65] for the energy range $4 \le \varepsilon \le 150\,\text{eV}$ and Zecca et al.[18] for the energy range $150 \le \varepsilon \le 1000\,\text{eV}$. Positive values of the ordinate mean that the reference values are higher. ($\blacklozenge$) de Heer et al.[37]; (O) Dababneh et al.[49] (1982); ($\bullet$) Jost et al.[56]; ($\diamond$) Wagenaar and de Heer[67]; ($\triangle$) Zecca et al.[86]; ($\blacktriangle$) Szmytkowski et al.[180]

where $A = 0.101$ (keV$^{-1} \times 10^{-20}\,\text{m}^2$), $B = 20.16$ (keV), $C = 0.407$ (keV$^{-1} \times 10^{-20}\,\text{m}^2$), and $D = 0.219$ (keV).

The Born approximation for the dependence of elastic scattering cross section on the incident energy ε is written as[181]

$$\frac{\varepsilon}{R} \frac{Q_{\text{el}}}{a_0^2} = \pi \left(A + B \frac{R}{\varepsilon} + \cdots \right) \tag{3.67}$$

where R is the Rydberg constant (13.595 eV) and a_0 is the Bohr radius. The total cross section for inelastic scattering Q_{inel} is given, according to the Born–Bethe approximation, by[182]

$$\frac{\varepsilon}{R} \frac{Q_{\text{inel}}}{a_0^2} = 4\pi \left[M_{\text{tot}}^2 \ln\left(4 C_{\text{tot}} \frac{\varepsilon}{R} \right) + \gamma_{\text{tot}} \frac{R}{\varepsilon} + \cdots \right] \tag{3.68}$$

where the constants M_{tot}^2, C_{tot}, and γ_{tot} have been given for all atoms from hydrogen to strontium by Inokuti et al.[183] One obtains the total cross section by adding the elastic and inelastic scattering cross sections. The total cross section in the energy range $80 \le \varepsilon \le 1000\,\text{eV}$ may then be written as[86]

$$Q_{\text{tot}} = A_1 + A_2 \left(\frac{R}{\varepsilon} \right)^{1/2} + A_3 \left(\frac{R}{\varepsilon} \right) \tag{3.69}$$

with the coefficients having values of $A_1 = -2.12$, $A_2 = 62.4$, $A_3 = -66.9$ for xenon. Figure 3.53 shows the total cross sections for all the rare gases in the high energy ($>100\,\text{eV}$) region, along with lines of slopes $(1/\varepsilon)^{1/2}$ and $(1/E)$. For all gases except helium, the total cross sections up to 1000 eV fall according to $(1/\varepsilon)^{1/2}$.

3.5.2 MOMENTUM TRANSFER CROSS SECTIONS IN Xe

Scattering of electrons on the heavier rare gas atoms is more susceptible to direct and indirect relativistic effects. The contribution of the direct effect increases with the charge of the

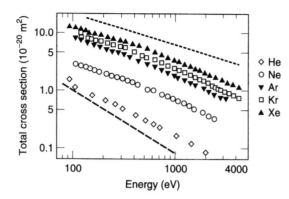

FIGURE 3.53 Total cross sections in the rare gases. The lower broken line has a slope of $1/\varepsilon$ and the upper broken line has a slope of $(1/\varepsilon)^{1/2}$. All rare gases except helium are parallel to the upper broken line for energies up to 1000 eV. ($\diamond$) He, Dalba et al.[40] (1979, 1981); ($\circ$) Ne, Zecca et al.[78]; (∇) Ar, Zecca et al.[78]; ($\square$) Kr, Zecca et al.[86]; ($\triangle$) Xe, Zecca et al.[86] Figure reproduced from Zecca, A. et al., *J. Phys. B*, 25, 2189, 1992. With permission of the Institute of the Physics, U.K.

nucleus of the target atom. Sin Fai Lam[139] has theoretically calculated the momentum transfer and elastic scattering cross sections in the energy range of 0.1 to 30 eV, obtaining the Ramsauer minimum at 0.5 eV. Partial waveshifts with $L = 0$ to 8 were calculated. The scattering of an electron by a heavy or moderately heavy atom is described by two scattering amplitudes,

$$f(\theta) = \frac{1}{2ik} \sum_{l=0}^{\infty} \left\{ (l+1)\left[\exp\left(i2\eta_l^+\right) - 1\right] + l\left[\exp\left(i2\eta_l^-\right) - 1\right] \right\} P_l(\cos\theta) \qquad (3.70)$$

$$g(\theta) = \frac{1}{2ik} \sum_{l=0}^{\infty} \left[\exp\left(i2\eta_l^-\right) - \exp\left(i2\eta_l^+\right)\right] P_l^1(\cos\theta) \qquad (3.71)$$

where $P_l(\cos\theta)$ and $P_l^1(\cos\theta)$ are the Legendre and associated Legendre functions, respectively. The plus and minus superscripts denote the spin up and spin down of the electron respectively.

For an unpolarized incident beam, the elastic differential cross section is given by

$$\frac{dQ_{\text{el}}}{d\Omega} = |f|^2 + |g|^2 \qquad (3.72)$$

Equations 3.70 to 3.72 may be treated as definitions, for our purpose, although Mott[184] has given a rigorous mathematical treatment. A comparison of Equation 3.70 with 3.20 and of Equation 3.72 with 3.22 shows the contribution of the spin down electron to the scattering amplitude. The spin polarization is given as[139]

$$P(\theta) = i\frac{(fg^* - f^*g)}{|f^2| + |g^2|} \qquad (3.73)$$

where the asterisk denotes the complex conjugate. The total elastic and momentum transfer cross sections are defined as

$$Q_T^{el} = \frac{4\pi}{k^2} \sum_l \left[(l+1) \sin^2 \eta_l^+ + l \sin^2 \eta_l^- \right] \tag{3.74}$$

$$Q_M^{el} = \frac{4\pi}{k^2} \sum_{l=0}^{l} \left[\frac{(l+1)(l+2)}{2l+3} \sin^2 \left(\eta_l^+ - \eta_{l+1}^+ \right) + \frac{l(l+1)}{2l+1} \sin^2 \left(\eta_l^+ - \eta_{l+1}^- \right) \right] \tag{3.75}$$

The momentum transfer cross sections (0.01 to 30 eV) derived by Sin Fai Lam[139] are shown in Figure 3.54. The Ramsauer minimum occurs at 0.6 eV with a cross section of $\sim 0.42 \times 10^{-20}$ m^2. A large peak, typical of heavier noble gases, is observed at 5 eV and is higher than the experimental elastic scattering cross sections.[49]

The swarm technique, applied to xenon,[58] yields the momentum transfer cross sections shown in Figure 3.54 and Table 3.31. These are the only data available up to the highest energy of 10,000 eV by the same author. The Ramsauer minimum is observed at 0.6 eV with a cross section of 0.80×10^{-20} m^2, almost twice the cross section stated by Sin Fai Lam.[139] The theoretical results of McEachran and Stauffer,[185] in the energy range 2.75 to 50 eV, agree well with the results of Sin Fai Lam,[139] although the methods are different. Register et al.[186] experimentally determined the momentum transfer cross sections in the 1 to 100 eV range, using the crossed beam technique, and normalization was achieved by empirically deriving the integral elastic scattering cross sections from the relationship $Q_{el} = Q_T - Q_i - Q_{ex}$. Good agreement is obtained with the values of Hayashi[58] up to 60 eV, and for higher energies Hayashi's values are higher.

Hunter et al.[187] have measured the drift velocities in xenon and derived the momentum transfer cross sections via the Boltzmann equation. The momentum transfer cross section shows the Ramsauer minimum at 0.64 eV, the minimum cross section being

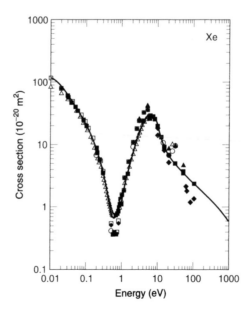

FIGURE 3.54 Momentum transfer cross sections in xenon. (○) Sin Fai Lam[139]; (−) Hayashi[58] (1983); (▲) McEachran and Stauffer[185]; (◆) Register et al.[186]; (●) Koizumi et al.[188]; (−△−) Hunter et al.[187]; (■) Pack et al.[92] The cross sections reported by Frost and Phelps[144] are not shown.

$0.753 \times 10^{-20}\,\mathrm{m}^2$, in closer agreement with the minimum Ramsauer cross section due to Hayashi.[58] Hunter et al.[187] have applied MERT theory to their cross sections in xenon and compared the four parameters, Equations 3.52 and 3.38, with other measurements, as shown in Table 3.32. The scattering length (parameter A/a_0) obtained by Hunter et al.[187] from drift velocities is in very good agreement with the theoretical value of Sin Fai Lam,[139] and that of

TABLE 3.31
Momentum Transfer Cross Sections in Xenon

Energy (eV)	Q_M (10^{-20} m²)	Energy (eV)	Q_M (10^{-20} m²)	Energy (eV)	Q_M (10^{-20} m²)
0	176	1.2	2.55	120	2.15
0.01	116	1.5	4.0	150	1.9
0.02	80	2	7.5	200	1.65
0.03	61.3	2.5	11.5	250	1.45
0.04	48.0	3	16	300	1.3
0.05	39.5	4	24.5	400	1.08
0.06	33.5	5	28	500	0.94
0.08	25.6	6	28	600	0.83
0.1	20.4	8	26	800	0.70
0.13	15.1	10	20	1000	0.58
0.16	12.0	12	13.5	1200	0.48
0.2	8.4	15	9.5	1500	0.37
0.25	5.35	20	7.0	2000	0.255
0.3	3.3	25	5.9	2500	0.19
0.4	1.6	30	5.1	3000	0.15
0.5	0.955	40	4.2	4000	0.103
0.6	0.80	50	3.6	5000	0.075
0.7	0.82	60	3.2	6000	0.057
0.8	1.05	80	2.7	8000	0.038
1	1.7	100	2.4	10000	0.027

From Hayashi, M. J., *J. Phys. D: Appl. Phys.*, 16, 581, 1983. With permission of Institute of Physics, U.K.

TABLE 3.32
Comparison of MERT Parameters in Xenon; Refer to Equations 3.38 and 3.52

Authors	A/a_0	D/a_3^0	F/a_4^0	A_1/a_3^0	Energy Range (eV)
Swarm method (Drift velocity, D/μ)					
Hunter et al.[187]	−6.09	490.2	−627.5	22.0	0.01–0.75
O'Malley[142 a]	−6.0				
Pack et al.[92]	−7.072				
Electron beam method					
O'Malley[142 b]	−6.50	388.0		23.2	
Jost et al.[56]	−5.83	490.0	−708.0	22.8	0.1–0.5
Weyherter et al.[84]	−6.527	517.0	−717.8	21.65	0.05–0.5
Theory					
Sin Fai Lam[139]	−6.04				
McEachran and Stauffer[185]	−5.232				

[a] Drift velocities from Pack et al.[145] were used.
[b] Beam studies from Ramsauer and Kollath[2,13] were used.
From Hunter, S. R. et al., *Phys. Rev. A*, 38, 5539, 1988.

O'Malley,[142] who used the drift velocities of Pack et al.[145] The values obtained from electron beam methods differ considerably.

The characteristic energies D/μ as a function of the parameter E/N were measured by Koizumi et al.,[188] using Townsend's method, and the results were used to derive the low energy (0.01 to 5.0 eV) momentum transfer cross sections. The Ramsauer minimum was observed at 0.6 eV, in agreement with other investigations, but the cross section was 0.4×10^{-20} m^2. Pack et al.[92] have measured the ratios of the longitudinal diffusion coefficient to mobility D_L/μ over five decades of the parameter E/N (0.001 to 100 Td) and derived the momentum transfer cross sections over the range of energy 0 to 100 eV (Table 3.31). The scattering length obtained by Pack et al.[92] is the largest reported in the literature and the Ramsauer minimum cross section is also the lowest reported, 0.38×10^{-20} m^2 at 0.6 to 0.7 eV. Particular attention was paid in this investigation to the purity of the gases investigated. Over the overlapping range of electron energy the cross sections are in agreement with those derived by Koizumi et al.[188] It is noted that the latter authors measured the ratios of radial diffusion coefficients to mobility, using an annular electrode for collecting electrons.

3.5.3 ELASTIC AND DIFFERENTIAL CROSS SECTIONS IN Xe

Below the first excitation threshold (8.32 eV) the elastic scattering cross sections are equal to the total scattering cross sections and the results shown in Figure 3.51 will not be repeated up to this energy. Elastic scattering cross sections have been measured by Klewer et al.[189] over the range 2 to 300 eV, Nishimura et al.[190] over the range 5 to 200 eV, and Ester and Kessler[179] over the range 15 to 100 eV. More recently, Gibson et al.[191] have made an experimental and theoretical study of elastic scattering in the low energy range of 0.67 to 50 eV.

Elastic differential cross sections are measured and the measured cross sections are placed on an absolute scale by the use of the relative flow technique in conjunction with standard cross sections for helium. Integral elastic and momentum transfer cross sections are obtained by the phase shift analysis technique. Figure 3.55 shows selected cross sections. The semiempirical elastic scattering cross sections of de Heer et al.[37] and the analytical expression of Zecca et al.,[18] which is applicable above 80 eV, are also included. Table 3.33 summarizes the Ramsauer minimum cross section and the corresponding electron energy for elastic and momentum transfer scatterings.

We conclude this section with a brief reference to the measured angular dependence of elastic differential cross sections in xenon. A general observation may be made that, for low energy electrons scattering from the heavy rare gas atoms in the region of the Ramsauer minimum, the contribution for the elastic scattering from the partial s-wave ($l=0$) is small. Higher order phase shifts, p-wave and d-wave phase shifts contribute in substantial magnitude.[191]

As in the cases of argon and krypton, the dominant feature of the differential cross sections is the diffraction pattern peaks and valleys. At low energy (~ 1 eV) two valleys are observed; with increasing energy (~ 20 eV) three shallow minima are observed. A further increase in energy (~ 60 eV) increases the minima to four, clearly demonstrating the dominance of the $l=4$ partial wave. As in krypton, some of the minima are not evident or become less pronounced at the higher energies, and obviously many partial waves contribute to the observed pattern.[148] Figure 3.56 demonstrates these features at selected electron energies.

3.5.4 TOTAL EXCITATION CROSS SECTIONS IN XE

Xenon has 54 electrons with the ground state configuration of $(1s^2\ 2s^2\ 2p^6 3s^2 3p^6 3d^{10} 4s^2 4p^6 4d^{10} 5s^2\ 5p^6)\ ^1S_0$, and excitation to the lowest electronic states occurs via the transition of

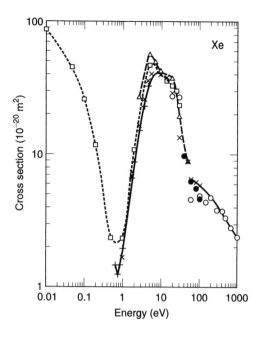

FIGURE 3.55 Total elastic scattering cross section in xenon. ○, de Heer et al.[37]; −□−, Sin Fai Lam[139]; (−△−) McEachran and Stauffer[185]; (×) Register et al.[186]; (●) Ester and Kessler[179]; (−+−) Gibson et al.[191]; (—) Zecca et al.[18] The analytical formula of Zecca et al.[18] shows poor agreement with experimental results in xenon, in contrast with the quality of agreement in other rare gases.

TABLE 3.33
Ramsauer Minimum Cross Section and Energy in Xenon

Authors	Momentum Transfer		Integral Elastic		Reference
	Energy (eV)	Q_M (10^{-20} m^2)	Energy (eV)	Q_{el} (10^{-20} m^2)	
Sin Fai Lam (1982)	0.5	0.42	0.5	2.32	139
Hayashi (1983)	0.6	0.80	—	—	58
Koizumi et al. (1986)	0.7	0.4	—	—	188
Subramanian and Kumar (1987)	—	—	0.73	1.10	76
Hunter et al. (1988)	0.64	0.7530	—	—	187
Pack et al. (1992)	0.7	0.38	—	—	92
Gibson et al. (1998)	0.67	0.41	0.75	1.24	191

the 5p electrons to 6s or some of the nearest higher free orbits. The large atomic number provides an opportunity to examine effects that are dependent on the size of the target atom, such as alignment and orientation of the atom after scattering. Angular distributions associated with the various excitations range from sharp forward peaking to nearly isotropic, indicating the contribution of long-range coulombic and short-range forces.[193]

The first excitation level is 6s[3/2]$_2$, with a threshold level of 8.315 eV. The twenty lowest levels and the threshold energy for each level are shown in Table 3.34. Levels 6s[3/2]$_2$ and 6s′ [1/2]$_0$ are metastable states of the atom. Figure 3.57 shows a simplified diagram of the first four levels. Integral and differential cross sections for selected levels at selected energies have been measured by the following: Williams et al.,[192] levels 1 to 18, at a fixed

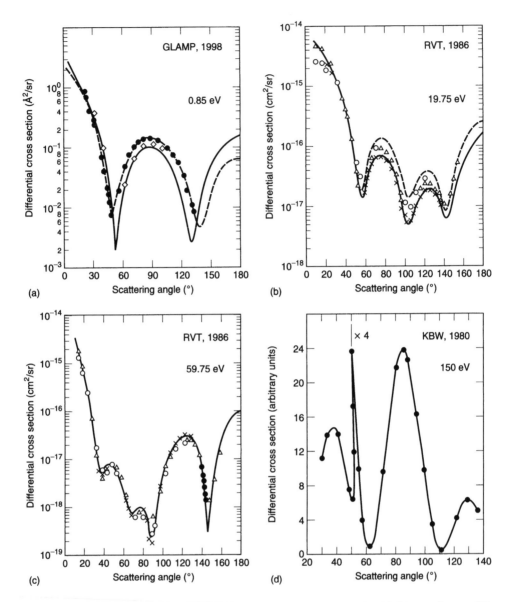

FIGURE 3.56 Elastic differential scattering cross sections in xenon. (a): (•) 0.85 eV, experiment, Gibson et al.[191]; (◇) experiment, Weyherter et al.[84]; (—) relativistic calculations, Gibson et al.[191]; (– – –) phase shift analysis, Gibson et al.[191] (b): (×) experiment, 19.75 eV, Register et al.[186]; (– – –) theory, 20 eV, McEachran and Stauffer[185]; (—) phase shift fit, 19.75 eV, Register et al.[186] (c): (×) experiment, 59.75 eV, Register et al.[186]; (○) experiment, 60 eV.[202] (d): (–•–) experiment, Klewer et al.[189] Sources: (a) Gibson, J. C. et al., *J. Phys. B*, 31, 3949, 1998; (b) and (c) Register, D. F. et al., *J. Phys. B*, 19, 1685, 1986; (d) Klewer, M. et al. *J. Phys. B*, 13, 571, 1980. Reproduced with permission of the Institute of Physics, U.K.

energy of 20 eV; Nishimura et al.,[190] levels 1 and 2, energy range 12 to 120 eV; Mason and Newell,[73] levels 1 and 3, 9 to 150 eV; Filipović et al.[193] levels 1 to 12, 15 to 80 eV; Suzuki et al.,[194] optically allowed transitions $5p^6(^1S_0)?5p^5(^2P_{1/2})6s$ and $5p^5(^2P_{3/2})6s$ at 100, 400, and 500 eV energy; Ester and Kessler,[179] levels 1 and 2, 15 to 100 eV; Khakoo et al.,[195] levels 1 to 4, 30 eV; Mityureva and Smirnoff,[197] levels 1 and 2, 10 to 40 eV; Khakoo et al.,[196]

TABLE 3.34
Designation of Lower Excitation Levels and Threshold Energy for Each Level

Level No.	Level Designation	Energy (eV)	Remarks
0	$5p^6\ {}^1S_0$	0.0	Ground level
1	$6s[3/2]_2$	8.315	Metastable
2	$6s[3/2]_1$	8.437	
3	$6s'[1/2]_0$	9.447	Metastable
4	$6s'[1/2]_1$	9.570	
5	$6p[1/2]_1$	9.580	
6	$6p[5/2]_2$	9.686	
7	$6p[5/2]_3$	9.721	
8	$6p[3/2]_1$	9.789	
9	$6p[3/2]_2$	9.821	
10	$5d[1/2]_0$	9.891	125.0 nm to ground
11	$5d[1/2]_1$	9.917	
12	$6p[1/2]_0$	9.934	
13	$5d[7/2]_4$	9.943	
14	$5d[3/2]_2$	9.959	
15	$5d[7/2]_3$	10.039	
16	$5d[5/2]_2$	10.159	
17	$5d[5/2]_3$	10.220	
18	$5d[3/2]_1$	10.401	119.2 nm to ground
19	$7s[3/2]_2$	10.562	
20	$7s[3/2]_1$	10.593	

From Khakoo, M. A. et al., *J. Phys. B: At. Mol. Opt. Phys.*, 28, 3477, 1986.

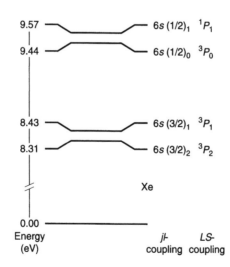

FIGURE 3.57 A simplified diagram of the first four excitation levels of xenon, with corresponding energy threshold.

levels 6 to 20, 10 to 30 eV; Khakoo et al.,[197] levels 1 to 5, 10 to 30 eV; and Fons and Lin,[198] threshold to 150 eV.

Total excitation cross sections have been measured or calculated by: de Heer et al.[37]; Specht et al.,[155] 11.5 to 13.75 eV; and Hayashi,[58] 8.32 to 10,000 eV. An analytical formula, valid over the electron impact energy range 80 to 10,000 eV and based on the total excitation cross sections due to Hayashi[58] is given by Brusa et al.[17] as

$$Q_{ex} = \frac{1}{F\varepsilon} \ln \frac{\varepsilon}{\varepsilon_{ex}} \tag{3.76}$$

where $F = 18.27$ keV^{-1} 10^{-20} m^2 and ε_{ex} = first excitation energy threshold, 8.315 eV for xenon.

Electrons having energy up to 30 eV show cross sections that are the highest for levels 10 to 14 combined, followed by levels 2 and 11. Kaur et al.[156] have calculated theoretically, by the relativistic distorted wave approximation, the cross sections for eight excited $5p^5 6p$ states in the electron energy range 20 to 100 eV. These levels are 5 to 9 in Table 3.34 and also the $6p'[3/2]_1$, $6p'[1/2]_1$, and $6p'[3/2]_2$ levels. The sums of these levels at each energy, obtained by simple addition and in conjunction with selected references, are shown in Figure 3.58.

Total metastable cross sections reported by Mason and Newell[73] show the relative contribution of metastables to the total excitation cross section. As in other rare gases, the metastable cross section reaches a peak at an electron energy above the first ionization potential. The peak in xenon occurs at 14.7 eV, only 6.4 eV above the excitation threshold. The shape follows a similar pattern to those of the other rare gas atoms, with a rapid rise to a distinct maximum followed by a subsequent fall to some almost consant value. Further, there is a sharp decrease of about 85% in the cross section as the electron energy is increased from 25 to 27 eV, which is not observed in other rare gases. The reason for this sudden decline is not clear. It is noted that the two metastable states, $6s[3/2]_2$ and $6s'[1/2]_0$, have widely differing lifetimes, 149.5 and 0.078 s respectively (Table 2.4). Ester and Kessler[179] have measured the excitation cross sections for the two lowest levels, one of which is an optically allowed transition, and the sudden decrease is not observed.

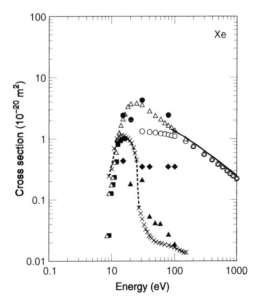

FIGURE 3.58 Excitation cross sections in xenon. ($\bigcirc$) Total, de Heer et al.[37]; ($\blacksquare$) total, Specht et al.[47]; ($\triangle$) Hayashi[58] (1983); ($-\times-$) Mason and Newell[73]; ($\bullet$) Filipović et al.[193]; ($\blacklozenge$) Ester and Kessler[179]; (—) Brusa et al.[17]; ($\blacktriangle$) Kaur et al.[156]

3.5.5   IONIZATION CROSS SECTIONS IN Xe

Total and partial ionization cross sections in xenon have been measured by several groups, as already discussed in connection with the other rare gases considered. Figure 3.59 shows selected data. A comparison with the cross sections due to Rapp and Englander-Golden,[30] adopted as reference, is shown in Figure 3.60. As in other gases, positive differences mean that the reference values are higher than the values under consideration.

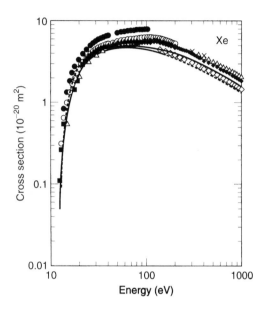

FIGURE 3.59 Absolute total ionization cross sections in xenon. Experimental: (•) Asundi and Kurepa[25]; (▲) Schram et al.[32] (1965); (×) Schram et al.[32] (1966a); (□) Nagy et al.[41]; (○) Wetzel et al.[75]; (△) Krishnakumar and Srivastava[82]; (–•–) Syage[93]; (⊙) Sorokin et al.[108]; (◆) Kobayashi et al.[112]. Semiempirical: (– –) Krishnakumar and Srivastava[82]; (—) Brusa et al.[17] The cross sections reported by de Heer et al.[37] are not shown.

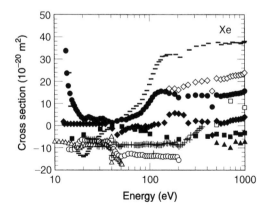

FIGURE 3.60 Comparison of ionization cross sections in xenon. Reference cross sections are due to Rapp and Englander-Golden.[30] Positive values mean that reference values are higher. (▲) Schram et al.[32] (1965); (■) de Heer et al.[37]; Nagy et al.[41]; (○) Wetzel et al.[75]; (–) semiempirical equation, Krishnakumar and Srivastava[82]; (△) experimental, Krishnakumar and Srivastava[82]; (+) Syage[93]; (•), Brusa et al.[17]; (◇) Sorokin et al.[108]; (◆) Kobayashi et al.[112]

The data come from several sources Asundi and Kurepa,[25] working in the energy range 15 to 100 eV, used a Tate and Smith apparatus employing a magnetic field of 0.03 to 0.04 T directed along the axis of the ionization tube to constrain the electrons in a narrow beam. The results are higher than others', the differences exceeding 30% over most of the energy range considered and therefore not shown in Figure 3.60. In 1965, Schram et al.[32] used the ionization tube method in the energy range (600 to 20,000 eV). This was followed in 1966 by a second study[32] covering the range of 100 to 600 eV. Nagy et al.[41] covered the higher range of 500 to 5000 eV and measured the partial ionization cross sections from which the absolute, weighted, gross ionization cross sections have been calculated. Stephan et al.[48] used the mass spectrometer in 1984 to measure partial ionization cross sections in the energy range, from threshold to 180 eV.

Absolute partial cross sections were measured in 1980 by Stephan et al.[48] in the energy range of 15 to 180 eV up to Xe^{6+}, using the mass spectrometer. Again the total cross sections, calculated by the weighted sum, have been shown. The careful measurements of Wetzel et al.,[75] using the crossed beam method, cover the range of 15 to 200 eV. Krishnakumar and Srivastava[82] extended the range of mass spectrometer measurements to the energy range of 15 to 1000 eV. Further improved techniques, employing the crossed beams and time-of-flight spectrometer, were adopted by Syage[93] to measure partial cross sections up to Xe^{6+} in the energy range of 18 to 466 eV. More recent measurements are due to Sorokin et al.,[108] who compared the ion yields due to photoionization and electron impact ionization in the range of 140 to 4000 eV. The method avoids the irksome problem of measuring gas number densities below $\sim$0.01 Pa, which is a typical gas pressure in ionization cross section measurements. Kobayashi et al.[112] covered the range of energy 12.5 to 1000 eV, adopting the pulsed electron beam method combined with the time-of-flight analysis of the charge of the ion.

Krishnakumar and Srivastava[82] have proposed a semiempirical equation similar to the equation of Lennon et al.[83]

$$Q_i^+ = \frac{10^3}{\varepsilon \varepsilon_i} \left\{ \left[A \ln \frac{\varepsilon}{\varepsilon_i} \right] + \sum_{j=1}^{N} B_j \left(1 - \frac{\varepsilon_i}{\varepsilon} \right)^j \right\} \tag{3.77}$$

where Q_i^+ is in units of 10^{-20} m^2, and the constants have the following values in units of $\times 10^5$ (eV m)^2: $A = 5.3180$, $B_1 = -5.5172$, $B_2 = 4.3084$, $B_3 = -10.138$, $B_4 = 4.3057$. Strictly speaking, Equation 3.77 yields the first ionization cross section and a similar equation with different constants and number of terms is given by these authors for partial cross sections. In view of the amount of calculation involved, only the first cross section is shown in Figure 3.59. This process lowers the cross sections by a maximum of 8.5% at 140 eV, with lower reductions on either side of this energy.

A better fit with the measured values of Rapp and Englander-Golden[30] from threshold energy to 5000 eV is given by the equation of Brusa et al.,[17] Equation 3.12, which is reproduced here for convenience,

$$Q_i = \left(\frac{L}{M+x} + \frac{N}{x} \right) \left(\frac{y-1}{x+1} \right)^{1.5} \times \left[1 + \frac{2}{3} \left(1 - \frac{1}{2x} \right) \ln\{2.7 + (x-1)^{0.5}\} \right] \tag{3.12}$$

where L, M, N, P are the four free parameters (Table 3.9) determined by the best fitting procedure, $y = \varepsilon/\varepsilon_i$ and $x = \varepsilon/P$, ε_i is the ionization potential and ε the electron energy.

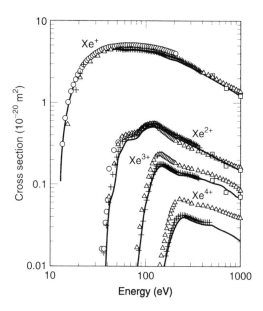

FIGURE 3.61 Partial ionization cross sections in xenon. (□) Nagy et al.[41]; (○) Wetzel et al.[75]; (△) Krishnakumar and Srivastava[82]; (+) Syage[93]; (–) Kobayashi et al.[112]

The constants have the values $L = 1000$ $(10^{-20} m^2)$, $M = 53.79$, $N = 109.6$ $(10^{-20} m^2)$, and $P = 3.58 \times 10^{-3}$ (keV).

A discussion of ionization cross sections has been given earlier in connection with krypton and will not be repeated here. The major difference to note with xenon is that the most recent results of gross cross sections reported by Kobayashi et al.[112] agree excellently with the cross sections due to Rapp and Englander-Golden,[30] falling within ±1% except for isolated deviations that do not exceed 5%.

As in other rare gases, the agreement between various studies becomes poorer as the charge of the ion produced increases. Figure 3.61 shows the partial ionization cross sections for multiple charged ion production. The cross sections for the singly charged (Xe^+), doubly charged (Xe^{2+}), and triply charged (Xe^{3+}) ions show, similarly to argon, double peaks. The first peak for Xe^+ occurs in the range of 40 to 50 eV, which shifts to ~115 eV for Xe^{2+} and 150 eV for Xe^{3+}. In addition, the cross section for Xe^{2+} shows a step at about 70 eV which is attributed to, on the basis of photoionization data, ejection of the 4d electron. Absence of the smooth nature of ionization cross sections above 300 eV is attributed to the onset of another inner shell ionization.

3.5.6 VERIFICATION OF THE SIGMA RULE FOR Xe

The measured total cross sections are compared with the sum of the partial cross sections in Table 3.35 and Figure 3.62. The measured total cross sections are reported by: Szmytkowski et al.,[180] 0.5 to 4.0 eV; Nickel et al.,[65] 4 to 150 eV; Zecca et al.,[18] 150 to 1000 eV; the latter results are based on the earlier measurements of Zecca et al.[86] The elastic scattering cross sections up to 9 eV are the same as the total cross sections. In the range of 10 to 100 eV, Figure 3.62 shows the elastic scattering cross sections reported by Register et al.[186] combined with those reported by Zecca et al.[18] Theoretical or experimental values of elastic scattering cross sections in the range of 10 to 80 eV in xenon are required.

TABLE 3.35
Verification of the Sigma Rule in Xenon

Energy (eV)	Q_{el}	$Q_{ex}^{H,83}$	$Q_i^{REG,65}$	$Q_T = \Sigma Q_i$	Q_T (Expt)	Difference (%)
9	$40.53^{R,86}$*	0.126	—	40.656	$41.38^{N.85}$*	1.75
10	$39.36^{R,86}$*	0.18	—	39.540	$40.15^{N.85}$	1.52
12	$38.96^{R,86}$*	0.84	—	39.80	$38.64^{N.85}$	−3.00
14	$37.68^{R,86}$*	1.70	0.5717	39.952	$37.27^{N.85}$	−7.20
16	$35.30^{R,86}$*	2.55	1.2314	39.081	$36.99^{N.85}$	−5.65
18	$32.26^{R,86}$*	3.35	1.8032	37.413	$36.63^{N.85}$	−2.14
20	$29.22^{R,86}$*	3.73	2.2782	35.228	$35.79^{N.85}$	1.57
25	$21.30^{R,86}$*	3.85	3.2326	28.383	$27.77^{N.85}$	−2.21
30	$13.39^{R,86}$*	3.57	3.8526	20.813	$20.80^{N.85}$	−0.06
40	$11.12^{R,86}$*	2.85	4.4772	18.447	$15.71^{N.85}$	−17.42
50	$8.84^{R,86}$*	2.40	4.8378	16.078	$13.95^{N.85}$	−15.25
60	$6.57^{R,86}$*	2.10	5.0313	13.701	$13.19^{N.85}$	−3.88
70	$6.39^{R,86}$*	1.85	5.1193	13.359	$12.67^{N.85}$	−5.44
80	$6.21^{R,86}$*	1.66	5.1808	13.051	$12.37^{N.85}$	−5.50
90	$6.02^{R,86}$*	1.52	5.2688	12.809	$12.13^{N.85}$	−5.60
100	$5.82^{R,86}$*	1.38	5.3832	12.583	$11.90^{N.85}$	−5.74
125	$5.563^{Z,00}$	1.19*	5.4095	12.222	$11.13^{N.85}$	−9.81
150	$5.286^{Z,00}$	1.00	5.1896	11.476	$10.63^{N.85}$	−7.95
200	$4.825^{Z,00}$	0.80	4.5827	10.208	$9.921^{N.85}$	−2.89
250	$4.455^{Z,00}$	0.684	4.2221	9.361	$9.207^{N.85}$	−1.68
300	$4.151^{Z,00}$	0.568	3.8966	8.616	$8.548^{N.85}$	−0.79
350	$3.896^{Z,00}$	0.516*	3.5976	8.011	$7.585^{Z,00}$	−5.61
400	$3.678^{Z,00}$	0.465	3.3513	7.495	$7.100^{Z,00}$	−5.56
450	$3.490^{Z,00}$	0.430*	2.9555	6.876	$6.681^{Z,00}$	−2.90
500	$3.325^{Z,00}$	0.395	2.9467	6.667	$6.317^{Z,00}$	−5.54
550	$3.180^{Z,00}$	0.370*	2.7619	6.312	$5.996^{Z,00}$	−5.27
600	$3.050^{Z,00}$	0.344	2.6212	6.015	$5.711^{Z,00}$	−5.33
650	$2.934^{Z,00}$	0.323*	2.4893	5.746	$5.456^{Z,00}$	−5.31
700	$2.829^{Z,00}$	0.302	2.3837	5.515	$5.227^{Z,00}$	−5.50
750	$2.733^{Z,00}$	0.289*	2.2782	5.302	$5.020^{Z,00}$	−5.62
800	$2.646^{Z,00}$	0.277	2.1814	5.104	$4.831^{Z,00}$	−5.67
850	$2.566^{Z,00}$	0.264*	2.0934	4.924	$4.658^{Z,00}$	−5.72
900	$2.492^{Z,00}$	0.252	2.0143	4.758	$4.499^{Z,00}$	−5.76
950	$2.424^{Z,00}$	0.242*	1.9439	4.609	$4.353^{Z,00}$	−5.88
1000	$2.361^{Z,00}$	0.231	1.8823	4.474	$4.218^{Z,00}$	−6.08

H,83 means Hayashi.[58] R,86 means Register et al.[186] Z,00 means Zecca et al.[18] N,85 means Nickel et al.[65] REG means Rapp and Englander-Golden.[30] Z,00 means Zecca et al.[18] Cross sections in units of 10^{-20} m^2. Positive sign of the difference means that experimental values are higher. * means interpolated value.

The excitation cross sections reported by Hayashi[58] cover the entire range from 9 to 1000 eV, and the ionization cross sections are due to Rapp and Englander-Golden.[30] The percentage differences between the total cross sections obtained by summing the partial cross sections and the measured cross sections are very good (< 5%), though at 40 to 50 eV range the differences are as large as ~19%. This large difference is attributed to the peak observed in the Xe$^+$ cross section and a step at ~70 eV, due to 3d electron ionization, resulting in high values for the summed cross sections. The reasons for this large increase of Q_i near the peak not showing up in the measured total cross sections need further investigation.

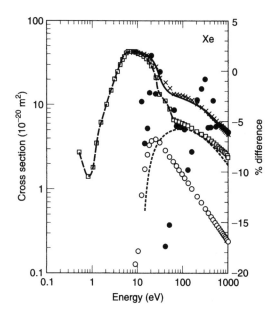

FIGURE 3.62 Verification of the sigma rule in xenon. (–□–) Q_{el}; (○) Q_{ex}; (– –) Q_i; (×) Q_T (sum) = $Q_{el} + Q_{ex} + Q_{iz}$; (—) Q_T (experimental); (•) percentage difference between Q_T (sum) and Q_T (experimental). Positive differences mean that experimental values are higher. For cross sections see Table 3.35.

TABLE 3.36
Page Numbers for Locating Cross Sections

Gas	Location (page number)				
	Q_T	Q_{el}	Q_M	Q_{ex}	Q_i
Argon	97, 116	101, 116	99	105, 116	112, 116
Helium	117, 130	122, 130	120	126, 130	127, 130
Krypton	133, 152	142, 152	141	148, 152	149–151, 152
Neon	155, 170	158, 170	157	165, 170	166–169, 170
Xenon	171, 184	177, 184	175	180, 184	181, 184

3.6 CONCLUDING REMARKS

A more recent review of cross sections with more accurate expressions for selected cross sections has been published by Raju.[199] A comprehensive analysis of the various cross sections in rare gases has been provided in this chapter, which has identified the gaps in the energy range where experimental or theoretical data are needed. In most situations, such as development of theory, discharge simulation, or calculations of swarm coefficients, tabulated data over a wide range of electron energy are desired. Data in this chapter are organized in such a way that the numerical values of cross sections may be found, using a graphical presentation, table, or analytical equation. Table 3.36 is provided to facilitate this objective.

REFERENCES

1. Meek, J. M. and J. D. Craggs, *Electrical Breakdown in Gases*, John Wiley & Sons, New York, 1979; Märk, T. D. and G. H. Dunn. (eds.), *Electron Impact Ionization*, Springer, Wien, 1985.
2. Ramsauer, C., *Ann. Phys. (Lpz.)*, 66, 546, 1921.
3. Brode, R. E., *Phys. Rev.*, 25, 636, 1925; *Rev. Mod. Phys.*, 5, 257, 1933.
4. Massey, H. S. W. and E. H. S. Burhop, *Electronic and Ionic Impact Phenomena*, Oxford University Press, Oxford, 1952, chapters 1 to 4.
5. Bederson, B. and L. J. Kieffer, *Rev. Mod. Phys.*, 43, 601, 1971.
6. Bransden, B. H. and M. R. C. McDowell, *Phys. Rep.*, 46, 249, 1978.
7. McDaniel, E. W., *Collision Phenomena in Ionized Gases*, John Wiley & Sons, New York, 1964.
8. Hasted, J. B., *Physics of Atomic Collisions*, 2nd ed., Am. Elsevier Pub. Co., New York, 1972.
9. Trajmar, S. and J. W. McKonkey, in *Advances in Atomic, Molecular, and Optical Physics*, Bederson and Walther (eds.), 33, 63, 1994. (See the chapter by Fillipelli et al. for excitation cross section.)
10. Zecca, A., G. P. Karwasz, and R. S. Brusa, *La Rivista del Nuovo Cimento*, 19(3), 1, 1996.
11. Chutjian, A. and D. C. Cartwright, *Phys. Rev. A*, 23, 2178, 1981.
12. Ramsauer, C. and R. Kollath, *Ann. Phys. Lpz.*, 3, 536, 1929.
13. Ramsauer, C. and R. Kollath, *Ann. Phys. Lpz.*, 12, 529, 1932.
14. Townsend, J. S. and V. A. Bailey, *Phil. Mag.*, 44, 1033, 1922.
15. Townsend, J. S. and V. A. Bailey, *Phil. Mag.*, 46, 661, 1923.
16. Holtzmark, J., *Z. Phys.*, 44, 437, 1929.
17. Brusa, R. S., G. P. Karwasz, and A. Zecca, *Z. Phys. D*, 38, 279, 1996.
18. Zecca, A., G. P. Karwasz, and R. S. Brusa, *J. Phys. B: At. Mol. Opt. Phys.*, 33, 843, 2000.
19. O'Malley, T. F., L. Spruch, and L. Rosenberg, *J. Math. Phys.*, 2, 491, 1961.
20. Inokuti, M., M. Kimura, M. A. Dillon, and I. Shimamura, in *Advances in Atomic, Molecular, and Optical Physics*, Bederson and Walther (eds.), 215, 63, 1994.
21. Normand, C. E., *Phys. Rev.*, 35, 1217, 1930.
22. Bullard, E. C. and H. S. W. Massey, *Proc. Roy. Soc. A.*, 130, 579, 1931.
23. Ramsauer, C. and R. Kollath, *Ann. Phys. Lpz.*, 12, 837–848, 1932.
24. Webb, G. M., *Phys. Rev.*, 47, 379–383, 1935.
25. Asundi, R. K. and M. V. Kurepa, *J. Electron. Control*, 15, 41–50, 1963.
26. Engelhardt, A. G. and A. V. Phelps, *Phys. Rev. A*, 133, 375–380, 1964.
27. Kieffer, L. J. and G. H. Dunn, *Rev. Mod. Phys.*, 38, 1, 1966.
28. McConkey, J. W. and F. G. Donaldson, *Can. J. Phys.*, 51, 914, 1973.
29. Fletcher, J. and I. R. Cowling, *J. Phys. B: At. Mol. Phys.*, 6, L258, 1973.
30. Rapp, D. and Paula Englander-Golden, *J. Chem. Phys.*, 13, 1464, 1965.
31. Golden, D. E. and H. W. Bandel, *Phys. Rev.*, 138, A14, 1965; 149, 58, 1966.
32. Schram, B. L., F. J. de Heer, F. J. van Der Wieland, and J. Kistemaker, *Physica*, 31, 94, 1965; Schram, B. L., H. R. Moustafa, H. R. Schutten, and F. J. de Heer, *Physica*, 32, 734, 1966a; Schram, B. L., J. H. Boereboom, and J. Kistemaker, *Physica*, 32, 185, 1966b; Schram, B. L., *Physica*, 32, 197, 1966c.
33. Kauppila, W. E., T. S. Stein, G. Jesion, M. S. Dababneh, and V. Pol, *Rev. Sci.*, 48, 822, 1977.

Helium

34. De Heer, F. J. and R. H. J. Jansen, *J. Phys. B: At. Mol. Phys.*, 10, 3741, 1977.
35. Stein, T. S., W. E. Kauppila, V. Pol, J. H. Smart, and G. Jesion, *Phys. Rev. A*, 17, 1600, 1978.
36. Kennerly, R. E. and R. A. Bonham, *Phys. Rev. A*, 17, 1844, 1978.
37. de Heer, F. J., R. H. J. Jansen, and W. van der Kaay, *J. Phys. B: At. Mol. Phys.*, 12, 979, 1979.
38. Nesbet, R. K., *Phys. Rev. A*, 20, 58–70, 1979.
39. Williams, J. F., *J. Phys. B: At. Mol. Phys.*, 12, 265, 1979.
40. Dalba, G., P. Fornasini, I. Lazzizzera, G. Ranieri, and A. Zecca, *J. Phys. B: At. Mol. Phys.*, 12, 3787, 1979; Dalba, G., P. Fornasini, R. Grisenti, I. Lazzizzera, G. Ranieri, and A. Zecca, *Rev. Sci. Instr.*, 52, 979, 1981.

41. Nagy, P., A. Skutzlart, and V. Schmidt, *J. Phys. B: At. Mol. Phys.*, 13, 1249, 1980.
42. Blaauw, H. J., R. W. Wagenaar, D. H. Barends, and F. J. de Heer, *J. Phys. B: At. Mol. Phys.*, 13, 359, 1980.
43. Register, D. F. and S. Trajmar, *Phys. Rev. A*, 21, 1134, 1980.
44. Wagenaar, R. W. and F. J. de Heer, *J. Phys. B: At. Mol. Phys.*, 13, 3855, 1980.
45. Van Zyl, B., G. H. Dunn, G. Chamberlain, and D. W. O. Huddle, *Phys. Rev. A*, 22, 1916, 1980.
46. O'Malley, T. F. and R. W. Crompton, *J. Phys. B: At. Mol. Phys.*, 13, 3451, 1980.
47. Specht, L. T., S. A. Lawton, and T. A. De Temple, *J. Appl. Phys.*, 51, 166, 1980.
48. Stephan, K., H. Helm, and T. D. Märk, *J. Chem. Phys.*, 73, 3763, 1980. Gases studied are He, Ne, Ar, Kr ; *J. Chem. Phys.*, 81, 3116, 1984. Gas studied is Xe.
49. Dababneh, M. S., W. E. Kauppila, J. P. Downing , F. Laperriere, V. Pol, J. H. Smart, and T. S. Stein, *Phys. Rev.*, 22, 1872, 1980; Dababneh, M. S., Y. F. Hsieh, W. E. Kauppila, V. Pol, and T. S. Stein, 26, 1252, 1982.
50. Srivastava, S. K., H. Tanaka, A. Chutjian, and S. Trajmar, *Phys. Rev. A*, 23, 2156, 1981.
51. Kauppila, W. E., T. S. Stein, J. H. Smart, M. S. Dababneh, Y. K. Ho, J. P. Downing, and V. Pol, *Phys. Rev. A*, 24, 725, 1981.
52. Kücükarpaci, H. N. and J. Lucas, *J. Phys. D: Appl. Phys.*, 14, 2001, 1981.
53. Haddad, G. N. and T. F. O'Malley, *Aust. J. Phys.*, 35, 35, 1982.
54. Zhou Qing, M. J. M. Beerlage, and M. J. van der Wiel, *Physica*, 113C, 225, 1982.
55. Nogueira, J. C., Ione Iga, and Lee Mu-Tao, *J. Phys. B: At. Mol. Phys.* 18, 2539, 1982.
56. Jost, K., P. G. F. Bisling, F. Eschen, M. Felsmann, and L. Walther, *Proc. 13th Int. Conf. on the Physics of Electronic Atomic Collisions*, Berlin, ed. J. Eichler et al., North Holland, Amsterdam, 1983, p. 91, cited by Wagenaar and de Heer.[67] Tabulated values above 7.5 eV are given in the latter reference. The lower range of 0.05 eV is obtained from Nickel et al.[65]
57. Fon, W. C., K. A. Berrington, P. G. Burke, A. Hibbert, *J. Phys. B: At. Mol. Phys.*, 16, 307, 1983; Fon, W. C., K. A. Berrington, P. G. Burke, and A. Hibbert, *J. Phys. B: At. Mol. Phys.*, 17, 3279, 1984.
58. Hayashi, M. J., *J. Phys. D: Appl. Phys.*, 15, 1411, 1982; *J. Phys. D: Appl. Phys.*, 16, 581, 1983. For tabulated values see Pack et al.[92]
59. Mathur, D. and C. Badrinathan, *Int. J. Mass. Spectr. Ion Proc.*, 57, 167, 1984; 68, 9, 1986.
60. McEachran, R. P. and A. D. Stauffer, *Phys. Lett. A*, 107, 397, 1985.
61. Yamabe, C., S. R. Buckman, and A. V. Phelps, *Phys. Rev.*, 27, 1345, 1983.
62. Bell, K. L., N. S. Scott, and M. A. Lennon, *J. Phys. B: At. Mol. Phys.*, 17, 4757, 1984.
63. Hunter, S. R. and L. G. Christophorou, *J. Chem. Phys.*, 80, 6150, 1984. Gases studied are Kr and Xe.
64. Register, D. F. and S. Trajmar, *Phys. Rev. A*, 29, 1785, 1984.
65. Nickel, J. C., K. Imre, D. F. Register, and S. Trajmar, *J. Phys. B: At. Mol. Phys.*, 18, 125, 1985.
66. Ferch, J., B. Granitza, C. Masche, and W. Raith, *J. Phys. B: At. Mol. Phys.*, 18, 967, 1985.
67. Wagenaar, R. W. and F. J. de Heer, *J. Phys. B: At. Mol. Phys.*, 18, 2021, 1985.
68. Buckman, S. J. and B. Lohmann, *J. Phys. B: At. Mol. Phys.* 19, 2547, 1986.
69. Garcia, G., F. Arqueros, and J. Campos, *J. Phys. B: At. Mol. Phys.*, 19, 3777, 1986.
70. Wagenaar, R. W., A. de Boer, T. van Tubergen, J. Los, and F. J. de Heer, *J. Phys. B: At. Mol. Phys.*, 19, 3121, 1986.
71. Buckman, S. J. and B. Lohmann, *J. Phys. B: At. Mol. Phys.*, 20, 5807, 1987.
72. Iga, I., L. Mu-Tao, J. C. Nogueira, and R. S. Barbieri, *J. Phys. B: At. Mol. Phys.*, 20, 1095, 1987.
73. Mason, N. J. and W. R. Newell, *J. Phys. B: At. Mol. Phys.*, 20, 1357, 1987.
74. Nahar, S. N. and J. M. Wadhera, *Phys. Rev.*, 35, 2051, 1987.
75. Wetzel, R. C., F.A. Baiocchi, T. R. Hayes, and R. S. Freund, *Phys. Rev. A*, 35, 559, 1987.
76. Subramanian, K. P. and V. Kumar, *J. Phys. B: At. Mol. Phys.*, 20, 5505, 1987.
77. Kumar, V., E. Krishnakumar, and K. P. Subramanian, *J. Phys. B: At. Mol. Phys.*, 20, 2899, 1987.
78. Zecca, A., S. Oss, G. Karwasz, R. Grisenti, and R. S. Brusa, *J. Phys. B: At. Mol. Phys.*, 20, 5157, 1987.
79. Nickel, J. C., C. Mott, I. Kanik, and D. C. McCollum, *J. Phys. B: At. Mol. Opt. Phys.*, 21, 1867, 1988. Covers gases CO and N_2.
80. Nakamura, Y. and M. Kurachi, *J. Phys. D: Appl. Phys.*, 21, 718, 1988.

81. Shah, M. B., D. S. Elliot, and H. B. Gilbody, *J. Phys. B: At. Mol. Phys.*, 20, 3501, 1987; 21, 2751, 1988.
82. Krishnakumar, E. and S. K. Srivastava, *J. Phys. B: At. Mol. Phys.*, 21, 1055, 1988.
83. Lennon, M. A., K. L. Bell, H. B. Gilbody, J. G. Hughes, A. E. Kingston, M. J. Murray, F. J. Smith, *J. Phys. Chem. Ref. Data*, 17, 1285, 1988.
84. Weyherter, M., B. Barzick, A. Mann, and F. Linder, *Z. Phys. D.*, 7, 333, 1988.
85. Furst, J. E., D. E. Golden, M. Mahgerefteh, J. Zhou, and D. Mueller, *Phys. Rev. A*, 40, 5592, 1989.
86. Zecca, A., G. Karwasz, R.S. Brusa, and R. Grisenti, *J. Phys. B. At. Mol. Opt. Phys.*, 24, 2737, 1991.
87. Kanik, I., J. C. Nickel, and S. Trajmar, *J. Phys. B: At. Mol. Opt. Phys.*, 25, 2189, 1992.
88. Bruce, M. R. and R. A. Bonham, *Z. Phys. D*, 24, 149, 1992.
89. Ma, C., C. R. Sporeleder, and R. A. Bonham, *Rev. Sci. Instr.*, 62, 909, 1991.
90. Cartwright, D. C., G. Csanak, S. Trajmar, and D.F. Register, *Phys. Rev. A*, 45 1602, 1992.
91. Trajmar, S., D. F. Register, D. C. Cartwright, and G. Csanak, *J. Phys. B: At. Mol. Opt. Phys.*, 25, 4889, 1992.
92. Pack, J. L., R. E. Voshall, and A. V. Phelps, *J. Appl. Phys.*, 71, 5363, 1992.
93. Syage, J. A., *Phys. Rev. A*, 46, 5666, 1992.
94. McCallion, P., M. B. Shah, and H. B. Gilbody, *J. Phys. B: At. Mol. Phys.*, 25, 1061, 1992.
95. Brunger, M. J., S. J. Buckman, L. J. Allen, I. E. McCarthy, and Nd K. Rathnavelu, *J. Phys. B: At. Mol. Opt. Phys.*, 25, 1823, 1992.
96. Gulley, R. J., D. T. Alle, M. J. Brennan, M. J. Brunger, and S. J. Buckman, *J. Phys. B: At. Mol. Opt. Phys.*, 27 2593, 1994.
97. Mityureva, A. A. and V. V. Smirnov, *J. Phys. B: At. Mol. Opt. Phys.*, 27, 1869, 1994.
98. Straub, H. C., P. Renault, B. G. Lindsay, K. A. Smith, and R. F. Stubbings, *Phys. Rev. A*, 52 1115, 1995.
99. Gibson, J. C., R. T. Gulley, J. P. Sullivan, S. J. Buckman, V. Chan, and P. D. Burrow, *J. Phys. B: At. Mol. Opt. Phys.*, 29, 3177, 1996.
100. Tsurubuchi, S., T. Miyazaki, and K. Motohashi, *J. Phys. B: At. Mol. Opt. Phys.*, 29, 1785, 1996.
101. Zubek, M., N. Gulley, G. C. King, and F. H. Read, *J. Phys. B: At. Mol. Opt. Phys.*, 29, L239, 1996.
102. Pitchford, L. C., J. P. Boeuf, and J. P. Morgan, www.siglo-kinema.com, 1996. See also Fiala, A., L. C. Pitchford, and J. P. Boeuf, *Phys. Rev. E*, 49, 5607, 1994.
103. Cvejanović, D. and A. Crowe, *J. Phys. B: At. Mol. Opt. Phys.*, 30, 2873, 1997.
104. Panajotović, R., D. Filipović, B.Marinković, V. Pejćev, M. Kurepa, and I. Vušković, *J. Phys. B: At. Mol. Phys.*, 30, 5877, 1997.
105. Sorokin, A. A., L. A. Shmaenok, S. B. Bobashev, B. Möbus, and G. Ulm, *Phys. Rev. A*, 58, 2900, 1998. Cross sections are measured in neon.
106. Zubek, M., B. Mielewska, J. Channing, G. C. King, and F. H. Read, *J. Phys. B: At. Mol. Opt. Phys.*, 32, 1351, 1999.
107. Phelps, A. V. and Z. Lj. Petrović, *Plasma Sources Sci. Tech.*, 8, R21–R44, 1999.
108. Sorokin, A. A., L. A. Schmaenok, S. B. Bobashev, B. Möbus, M. Richter, and G. Ulm, *Phys. Rev. A*, 61, 022723-1, 2000. Cross sections are measured in Ar, Kr, Xe.
109. Filipović, D. M., B. P. Marinković, V. Pejčev, and L. Vušković, *J. Phys. B: At. Mol. Opt. Phys.*, 33, 677, 2000.
110. Filipović, D. M., B. P. Marinković, V. Pejčev, and L. Vušković, *J. Phys. B: At. Mol. Opt. Phys.*, 33, 2081, 2000.
111. Sienkiewicz, J. E., V. Konopińska, S. Telega, and P. Syty, *J. Phys. B: At. Mol. Op. Phys.*, 34, L409, 2001.
112. Kobayashi, A., G. Fujiki, A. Okaji, and T. Masuoka, *J. Phys. B: At. Mol. Opt. Phys.*, 35, 2087, 2002.
113. Zecca, A., G. P. Karwasz, and R. S. Brusa, *Phys. Rev. A*, 45, 2777, 1992.
114. Zecca, A., G. P. Karwasz, and R. S. Brusa, *Phys. Rev. A*, 46, 3877, 1992.
115. Zecca, A., J. C. Nogueira, G. P. Karwasz, and R. S. Brusa, *J. Phys. B: At. Mol. Opt. Phys.*, 28, 477, 1994.

116. March, N. H., A. Zecca, and G. P. Karwasz, *Z. Phys. D*, 32, 93, 1994.
117. Milloy, H. B., R. W. Crompton, J. A. Rees, and A. G. Robertson, *Aust. J. Phys.*, 30, 61, 1977.
118. Bransden, B. H. and M. R. C. McDowell, *Phys. Rep.*, 46, 249, 1978.
119. Moiseiwitsch, B. L. and S. J. Smith, *Rev. Mod. Phys.*, 40, 238, 1968.
120. Fabrikant, I. I., O. B. Shpenik, A. V. Snegursky, and A. N. Zavilopulo, *Phys. Rep.*, 159, 1, 1988.
121. Heddle, D. W. O. and J. W. Gallagher, *Rev. Mod. Phys.*, 61, 221, 1989.
122. Hayashi, M., Gaseous Electronics Institute, Sakae, Nagoya, Japan. See also Kosaki, K. and M. Hayashi, *Denkigakkai Zenkokutaikai Yokoshu*, Prepr. Natl. Meet. Inst. Electr. Eng. Jpn., 1992 (in Japanese).
123. Chilton, J. E., J. B. Boffard, R. S. Schaffe, and C. C. Lin, *Phys. Rev. A*, 57, 267, 1998; Chilton, J. E. and C. C. Lin, *Phys. Rev. A*, 60, 3712, 1999; Chilton, J. E., M. D. Stewart, Jr., and C. C. Lin, *Phys, Rev. A*, 61, 032704, 2000
124. Kobayashi, A., G. Fujiki, A. Okaji, and T. Masuoka, *J. Phys. B: At. Mol. Opt. Phys.*, 35, 2087, 2002. The author thanks Dr. Kobayashi for sending tabulated results of cross sections.
125. Phelps, A. V. and Z. Lj. Petrović, *Plasma Sources Sci. Technol.*, 8, (1999) R21.

Krypton

126. Robertson, A. G., *J. Phys. B: At. Mol. Phys.*, 5, 648, 1972; Crompton, R. W., M. T. Elford, and A. G. Robertson, *Aust. J. Phys.*, 23, 667, 1970.
127. Milloy, H. W. and R. W. Crompton, *Phys. Rev. A*, 15, 1847, 1977.
128. Crompton, R. W., M. T. Elford, and A. G. Robertson, *Aust. J. Phys.*, 23, 667, 1970.
129. Shyn, T. W., *Phys. Rev. A*, 22, 916–922, 1980.
130. Ramanan, G. and G. R. Freeman, *J. Chem. Phys.*, 93, 3120, 1990.
131. Faxén, H. and J. Holtsmark, *Ann. Phys.* 45, 307, 1927.
132. Mott, N. F. and H. S. W. Massey, *Theory of Atomic Collisions*, Oxford University Press, London, 1971.
133. Trajmar, S., D. F. Register, D. C. Cartwright, and G. Csanak, *J. Phys. B: At. Mol. Opt. Phys.*, 25, 4889, 1992.
134. Röder, J., H. Ehrhardt, I. Bray, and D. Fursa, *J. Phys. B: At. Mol. Opt. Phys.*, 29, L421, 1996.
135. Asmis, K. R. and M. Allan, *J. Phys. B: At. Mol. Opt. Phys.*, 30, 1961, 1997. Also see Cubric, D., D. J. L. Mercer, J. M. Channing, G. C. King, and F. H. Read, *J. Phys. B: At. Mol. Opt. Phys.*, 32, L45, 1999.
136. Alkhazov, G. D., *Sov. Phys. Tech. Phys.*, 15, 66, 1970. Tabulated values are taken from Pitchford et al.[102]
137. Montague, R. G., M. F. A. Harrison, and A.C. H. Smith, *J. Phys. B: At. Mol. Opt. Phys.*, 17, 3295, 1984.
138. Smith, P. T., *Phys. Rev.*, 36, 1293, 1930; Also see Tate, J. T. and P. T. Smith, *Phys. Rev.* 39, 270, 1932.
139. Sin Fai Lam, L. T., *J. Phys. B: At. Mol. Phys.* 15, 119, 1982.
140. McEachran, R. P. and A. D. Stauffer, *J. Phys. B: At. Mol. Phys.*, 17, 2507, 1984.
141. Blatt, J. M. and J. D. Jackson, *Phys. Rev.*, 26, 18, 1949.
142. O'Malley, T. F., *Phys. Rev.*, 130, 1020, 1963; O'Malley, T. F. and R. W. Crompton, *J. Phys. B: At. Mol. Phys.* 13, 3451, 1980.
143. Raju, Gorur G., *Dielectrics in Electric Fields*, Marcel Dekker, New York, 2003, p. 40.
144. Frost, L. S. and A. V. Phelps, *Phys. Rev. A*, 136, 1538, 1964.
145. Pack, J. L., R. E. Voshall, and A. V. Phelps, *Phys. Rev.*, 127, 2084, 1962.
146. Mitroy, J., *Aust. J. Phys.*, 43, 19, 1990.
147. Mimnagh, D. J. R., R. P. McEachran, and A. D. Stauffer, *J. Phys. B: At. Mol. Opt. Phys.*, 26, 1727, 1993.
148. Williams, J. F. and A. Crowe, *J. Phys. B: At. Mol. Phys.*, 8, 2233, 1975.
149. Heindörff, T., J. Hofft, and P. Dabkiewicz, *J. Phys. B: At. Mol. Phys.*, 9, 89, 1976.
150. Trajmar, S., S. K. Srivastava, H. Tanaka, H. Nishimura, and D. C. Cartwright, *Phys. Rev. A*, 23, 2167, 1981.

151. Ganas, P. S. and A. E. S. Green, *Phys. Rev. A*, 4, 182, 1971.
152. Chilton, J. E., M. D. Stewart, Jr., and C. C. Lin, *Phys. Rev. A*, 61, 052718–1, 2000.
153. Yuan, Z., L. Zhu, X. Liu, Z. Zhong, W. Li, H. Cheng, and K. Xu, *Phys. Rev. A.*, 66, 062701–1, 2002.
154. Delâge, A. and J.-D. Carette, *J. Phys. B: At. Mol. Phys.*, 9, 2399, 1976.
155. Specht, L. T., S. A. Lawton, and T. A. de Temple, *J. Appl. Phys.*, 51, 166, 1981.
156. Kaur, S., R. Srivastava, R. P. McEachran, and A. D. Stauffer, *J. Phys. B: At. Mol. Opt. Phys.*, 31, 4833, 1998.
157. Guo, X., D. F. Mathews, G. Mikaelien, M. A. Khakoo, A. Crowe, I. Kanik, S. Trajmar, V. Zeman, K. Bartschat, and C. J. Fontes, *J. Phys. B: At. Mol. Opt. Phys.*, 33, 1895, 2000.
158. Guo, X., D. F. Mathews, G. Mikaelien, M. A. Khakoo, A. Crowe, I. Kanik, S. Trajmar, V. Zeman, K. Bartschat, and C. J. Fontes, *J. Phys. B: At. Mol. Opt. Phys.*, 33, 1921, 2000.
159. Kobayashi, A., Personal communication, Jan. 2003.

Neon

160. Robertson, A. G., *J. Phys. B: At. Mol. Phys.*, 5, 648, 1972.
161. Fon, W. C. and K. A. Berrington, *J. Phys. B: At. Mol. Phys.*, 14, 323, 1981.
162. Gupta, S. C. and J. A. Rees, *J. Phys. B: At. Mol. Phys.*, 8, 417, 1975.
163. Saha, H. P., *Phys. Rev.*, 39, 5048, 1989.
164. Shi, X. and P. D. Burrow, *J. Phys. B: At. Mol. Opt. Phys.*, 25, 4273, 1992.
165. Thompson, D. G., *J. Phys. B: At. Mol. Phys.*, 4, 468, 1971.
166. McCarthy, J. E., C. J. Noble, B. A. Phillips, and A. D. Turnbull, *Phys. Rev. A*, 15, 2173, 1977.
167. Register, D. F., S. Trajmar, G. Steffensen, and D. C. Cartwright, *Phys. Rev. A*, 29, 1793, 1984.
168. Meneses, G. D., *J. Phys. B: At. Mol. Opt. Phys.*, 35, 3119, 2002.
169. Machado, L. E., E. P. Leal, and G. Csanak, 1984, *Phys. Rev. A*, 29, 1811, 1984.
170. Andersen, N., J. W. Gallagher, and I. V. Hertel, *Phys. Rep.*, 165, 1, 1988.
171. Becker, K., A. Crowe, and J. W. McKonkey, *J. Phys. B: At. Mol. Phys.*, 25, 3885, 1992.
172. Phillips, M. H., L. W. Anderson, and C. C. Lin, *Phys. Rev. A*, 32, 2117, 1985.
173. Chilton, J. E., M. D. Stewart Jr., and C. C. Lin, *Phys. Rev. A*, 61, 052708–1, 2000.
174. Sharpton, F. A., R. M. St. John, C. C. Lin, and E. Fajen, *Phys. Rev. A*, 2, 1305, 1970.
175. Teubner, P. J. O., J. L. Riley, M. C. Tonkin, J. E. Furst, and S. J. Buckman, *J. Phys. B: At. Mol. Phys.*, 18, 1985, 1985.
176. Kanik, I., J. M. Ajello, and G. K. James, *J. Phys. B: At. Mol. Phys.* 29, 2355, 1996.
177. Almeida, D. P., A. C. Fontes, and C. F. L. Godinho, *J. Phys. B: At. Mol. Op. Phys.*, 28, 3335, 1995.
178. Wagenaar, R. W., FOM report No. 43948. Tabulated values quoted in de Heer et al.[37]

Xenon

179. Ester, T. and J. Kessler, *J. Phys. B: At. Mol. Opt. Phys.*, 27, 4295, 1994.
180. Szmytkowski, C., K. Maciag, and G. Karwasz, *Physica Scripta*, 54, 271, 1996.
181. Inokuti, M. and M. R. C. McDowell, *J. Phys. B: At. Mol. Phys.* 7, 2382, 1974.
182. Inokuti, M., *Rev. Mod. Phys.*, 43, 297, 1971.
183. Inokuti, M., R. P. Saxon, and J. L. Dehmer, *Int. J. Radiat. Phys. Chem.*, 7, 109, 1975; Inokuti, M., J. L. Dehmer, T. Baer, and J. D. Hansen, 23, 95, 1981.
184. Mott, N. F., *Proc. Roy. Soc. A*, 124, 425, 1929.
185. McEachran, R. P. and A. D. Stauffer, *Phys. Lett.* 107A, 397, 1985.
186. Register, D. F., L. Viscovic, and S. Trajmar, *J. Phys. B: At. Mol. Phys.*, 19, 1685, 1986.
187. Hunter, S. R., J. G. Carter, and L. G. Christophorou, *Phys. Rev. A*, 38, 5539, 1988. Gases studied are Kr and Xe.
188. Koizumi, T., E. Shirakawa, and I. Ogawa, *J. Phys. B: At. Mol. Phys.*, 19, 2331, 1986.
189. Klewer, M., M. J. M. Beerlage, and M. J. van der Wiel, *J. Phys. B: At. Mol. Phys.*, 13, 571, 1980.
190. Nishimura, H., T. Matsuda, and A. Danjo, *J. Phys. Soc. Jpn.*, 56, 70, 1987.

191. Gibson, J. C., D. R. Lunt, L. J. Allen, R. P. McEachran, L. A. Parcell, and S. J. Buckman, *J. Phys. B: At. Mol. Phys.*, 31, 3949, 1998.
192. Williams, W., S. Trajmar, and A. Kupperman, *J. Chem. Phys.*, 62, 3031, 1975.
193. D. Filipović, B. Marinković, V. Pejčev, and L. Vusković, *Phys. Rev. A*, 37, 356, 1988.
194. Suzuki, T. Y., Y. Sakai, B. S. Min, T. Takayanagi, K. Wakiya, H. Suzuki, T. Inaba, and H. Takuma, *Phys. Rev. A*, 43, 5867, 1991.
195. Khakoo, M. A., C. E. Beckmann, S. Trajmar, and G. Csanak, *J. Phys. B: At. Mol. Opt. Phys.*, 27, 3159, 1994.
196. Khakoo, M. A., S. Trajmar, L. R. LeClair, I. Kanik, G. Csanak, and C. J. Fontes, *J. Phys. B: At. Mol. Opt. Phys.*, 29, 3455, 1996.
197. Khakoo, M. A., S. Trajmar, L. R. LeClair, I. Kanik, G. Csanak, and C. J. Fontes, *J. Phys. B: At. Mol. Opt. Phys.*, 29, 3477, 1996.
198. Fons, J. T. and C. C. Lin, *Phys. Rev. A*, 58, 4603, 1998.
199. Raju, G., *IEEE Trans. Di. Elec. Insul.*, 11, 649, 2004.
200. Schappe, R. S., M. B. Schulman, L. W. Anderson, and C. C. Lin, *Phys. Rev. A*, 50, 444, 1994.
201. Danjo, A. *J. Phys. B: At. Mol. Opt. Phys.*, 21, 3759, 1988.
202. Holt Kamp, G., and Jost, K. Cited by Register et al.[186] as Private Communication, 1985.

4 Data on Cross Sections—II. Diatomic Gases

The presentation of scattering cross sections is continued in this chapter, with specific reference to selected diatomic gases. The gases onsidered are carbon monoxide (CO), hydrogen (H_2), nitrogen (N_2), oxygen (O_2), and nitrous oxide (NO). The molecular structure of these gases is relatively simple and, except for CO and NO, they are all nonpolar. The long-range dipole interaction between the electron and the molecule in polar gases presents difficulties for complete theoretical understanding[1] and experimental methods are the main source, unlike in rare gases (Chapter 3) where theory can supplement measurements. A brief presentation of interaction potentials is given in Chapter 1 and Johnson[2] may be referred to for further details. For details of polarization refer to McDaniel[3] and Raju[4]. The classical theory for elastic scattering, which is only of academic interest, may be found in Yang and Ree[5] according to which the differential scattering cross section in the CM system is given by the expression

$$Q_{\text{diff}}(\Theta) = v_0^{-4/n} \tag{4.1}$$

where v_0 is the initial relative velocity of approach, Θ is the angle of scattering (Figure 1.18) and n a constant.

4.1 CARBON MONOXIDE (CO)

Carbon monoxide plays an important role in our own atmosphere and the interstellar medium. Its role in a wide variety of electrical discharges is particularly interesting, both from the application and theoretical points of view. It has potential use in a number of plasma applications, such as electron device technology, plasma deposition, and high power lasers. CO is also considerably important in astrophysics, since it is the most abundant molecule, after hydrogen, in molecular clouds. In astrophysical plasmas rotational excitation of CO at energies in the range of a few hundred meV and subsequent radiation emission play an important role in the energy balance in a number of different environments such as molecular clouds and cloud edges.[41] The energy range in this type of application is 1 meV to tens of meV. A comprehensive review of experimental methods and electron scattering data published up to 1983 is given by Trajmar et al.[6] During the preparation of this volume a detailed review of electron–molecule scattering cross sections in diatomic molecules has been published by Brunger and Buckman[7].

The CO molecule is isoelectronic with N_2, having fourteen electrons, and its ground state is $1\sigma^2\ 2\sigma^2\ 3\sigma^2\ 4\sigma^2\ 5\sigma^2\ 1\pi^4$; X $^1\Sigma^+$.[30] It has a closed-shell ground electronic-state configuration. In many respects electron scattering from CO has similarities with that of N_2, leading to the observation that the transport properties in both gases are similar.[15] Like nitrogen, CO does not show a Ramsauer minimum. The difference is that CO is a heteronuclear, nonsymmetric molecule. It is weakly polar, with an internuclear separation of 0.112 nm,[104] and has a permanent dipole moment of 0.112 D ($1\,\text{D} = 3.33 \times 10^{-30}\,\text{C m}$) in the ground state (in the excited state the polarizability will be different). The weak dipole moment suggests an almost, but not quite, homonuclear molecule. It is also weakly electron attaching.

Table 4.1 gives references for selected cross section data.

TABLE 4.1

Selected References for Scattering Cross Section Studies in CO

Type	Energy Range	Method	Reference
Q_T	2–210	Ramsauer technique	Brode[8]
Q_T	1.1–45.0	Ramsauer technique	Brüche[9]
Q_T	0.5–400	Ramsauer technique	Normand[10]
Q_T	0.18–1.2	Ramsauer technique	Ramsauer and Kollath[11]
Q_i	Onset–100	Ionization tube	Asundi et al.[12]
Q_i	Onset–1000	Beam–static gas	Rapp et al.[13]
Q_i	Onset–1000	Beam–static gas	Rapp and Englander-Golden[14]
Q_M, Q_{inel}	10^{-3}–10.0	Swarm parameters	Hake and Phelps[15]
Q_M, Q_{el}	0.005–10.00	Theory	Chandra[16]
Q_T	1.3–5.6	Attenuation method	Szmytkowski and Zabek[17]
Q_m, Q_v	0–100	Swarm method	Land[18]
Q_T	1.0–500.0	Transmission method; parallel H	Kwan et al.[19]
Q_M		Swarm method	Haddad and Milloy[20]
Q_T	1.15–403	Crossed beams	Sueoka and Mori[21]
Q_T	0.1–5.0	Theory	Jain and Norcross[22]
Q_T	0.5–5.0 eV	Time-of-flight spectrometer	Buckman and Lohmann[23]
Q_i	15–510	Crossed beams	Orient and Srivastava[24]
Q_{ex}	10–60	Time-of-flight method	Mason and Newell[25]
Q_{diff}	20–100	Crossed beams; technique	Nickel et al.[26]
Q_T	380–5200	Transmission method	Garcia et al.[27]
Q_T	10–5000	Theory	Jain and Baluja[28]
Q_{ex}	20, 200	Crossed beams; UV	James et al.[29]
$Q_{el}, Q_M,$	0.001–10	Theory	Jain and Norcross[30]
Q_T	5–300	Linear transmission method	Kanik et al.[31]
Q_{dis}	13.5–198.5	Crossed beams	Cosby[32]
Q_T	80–4000	Modified Ramsauer method	Karwasz et al.[33]
Q_v	20–50	Crossed beams	Middleton et al.[34]
Q(set)	1–1000	Compilation	Kanik et al.[35]
Q_{ex}	6–18	Theory	Morgan and Tennyson[36]
Q_T	400–2600	Linear transmission method	Xing et al.[37]
Q_{diff}	1–30	Crossed beams	Gibson et al.[38]
Q_T	50–2000	Theory	Joshipura and Patel[39]
Q_{diff}	6.5–15.0	Time-of-flight method	LeClair and Trajmar[40]
Q_{inel}	0.002–0.16	Crossed beams with axial H	Randell et al.[41]
Q_{diff}, Q_v	1–30	Crossed beams	Gibson et al.[42]
Q_{ex}	3.7–15.0	Crossed beams spectrometer	Zobel et al.[43]
Q_{ex}	3.7–15.0	Crossed beams spectrometer	Zobel et al.[44]
Q_T	0.4–250	Transmission method	Szmytkowski et al.[45]
Q_{diff}	300–1300	Crossed beams; technique	Maji et al.[46]
Q_i	17.5–600	Time-of-flight spectrometer	Tian and Vidal[47]
Q_i	15–1000	Time-of-flight spectrometer	Mangan et al.[48]
Q_{ex}	Onset–15	Crossed beams	Poparić et al.[49]

Q_T = total; Q_{diff} = differential; Q_{dis} = dissociation; Q_{ex} = excitation; Q_i = ionization; Q_{inel} = inelastic; Q_M = momentum transfer; Q (set) = set; Q_v = vibrational. Energy in units of eV.

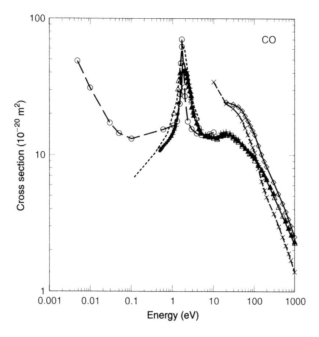

FIGURE 4.1 Absolute total cross sections in CO. Measurements: (—) Bruche[9]; (Δ) Kwan et al.[19]; (•) Buckman and Lohmann[23]; (■) Garcia et al.[27]; (▲) Kanik et al.[31]; (♦) Karwasz et al.[33]; (-+-) Szmytkowski et al.[45] Theory: (—○—) Chandra[16]; (— —) Jain and Norcross[30]; (-■-) Garcia et al.[27]; (-♦-) Jain and Baluja,[28] with anisotropic term; (-×-) Jain and Baluja[28]; without anisotropic term. Not shown are the results of Sueoka and Mori,[21] and Xing et al.[37] The increase towards thermal energy is due to the dipole moment.

4.1.1 TOTAL CROSS SECTIONS IN CO

Total cross sections have been measured by Kwan et al.[19] in the energy range 1 to 500 eV with a stated error of ±5%. This does not include the error due to incomplete discrimination against electrons that are scattered at small angles in the forward direction. This source of error, as already mentioned, results in a lower cross section than the true value. However, complete discrimination was accomplished in that study, covering inelastic scattering, electronic excitation, and ionization. Discrimination of rotational and vibrational excitation was incomplete, due to the small angle of scattering.

Figures 4.1 and 4.2 show the total cross section measured by several groups. Figure 4.2 is drawn expanded for the purpose of clarity. The striking feature of the energy–cross section curve is a sharp resonance peak due to $^2\Pi$ excitation centered near 1.9 eV, with a maximum Q_T value of $\sim 45 \times 10^{-20}\,\mathrm{m}^2$. The theoretical calculations of Chandra[16] are also shown; the latter cover the low energy range of 0.005 to 10 eV. Chandra[16] calculated a sharp resonance peak at 1.75 eV with a peak cross section of $69.27 \times 10^{-20}\,\mathrm{m}^2$. Szmytkowski and Zubek[17] observed experimentally that Q_T has a peak value of $35 \times 10^{-20}\,\mathrm{m}^2$ at an energy of 1.93 eV. On either side of this peak, structures were observed at 1.73 and 2.13 eV. No further confirmation of peaks has been reported at these two energies. According to Kwan et al.[19] the peak at resonance would have been lower and broader in energy if Chandra[16] had considered vibrational excitation in addition to rotational excitation. The broad weak feature around 30 to 40 eV is indicative of a resonance effect above the ionization potential; a similar effect is also observed with the N_2 molecule.

In an attempt to resolve the differences of over 30% in the peak cross section at 1.93 eV resonance, Buckman and Lohmann[23] used the time-of-flight measurement. The peak cross

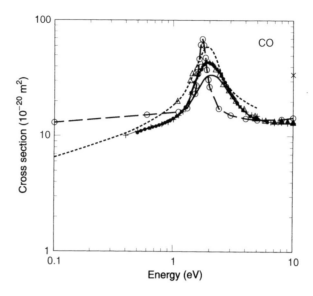

FIGURE 4.2 Expanded view of absolute total cross sections in CO. Symbols same as in Figure 4.1

TABLE 4.2
$^2\Pi$ Resonance Shape Parameters in CO

Authors	Method	Position (eV)	Width (eV)	Peak value $(10^{-20}\,m^2)$
Szmytkowski and Zubek[17]	Measurement	1.93	0.75	35.0
Tronc et al.[50]	Measurement	1.8	1.0	—
Kwan et al.[19]	Measurement	1.9	—	—
Buckman and Lohmann[23]	Measurement	1.90	—	43.50
Chandra[16]	Theory	1.75	—	29.45
Jain and Norcross[30]	Theory	1.85	0.95	49.5
Gibson et al.[38]	Measurement	1.91	—	35.8

---

section of $43.52 \times 10^{-20}\,m^2$ agrees well with the experimental results of Kwan et al.[19] but it is about 62% of the peak value obtained by Chandra[16] and about 30% lower than the calculations of Jain and Norcross.[30] Buckman and Lohmann[23] suggest that neglect of nuclear motion in the theory of Jain and Norcross[30] could be a source of the discrepancy between theory and experiment.

Table 4.2 shows the shape resonance parameters obtained from selected experiments and theories. Of course, the theoretical analyses are not expected to yield perfect agreement as they are dependent upon the assumptions made to simplify the computational work involved, whereas the experimental parameters are obtained within certain error limits by analysis of the integrated cross sections.

The total cross sections in the higher energy range ($380 \le \varepsilon \le 5300\,eV$) are measured by Garcia et al.[27] by the electron-beam attenuation method and are also shown in Figure 4.1. The high energy results join the low energy cross sections of Kwan et al.[19] smoothly. Often, analytical expressions are required for modeling discharges and in the high energy range ($>150\,eV$ in CO) the first choice is usually the Born–Bethe approximation. Garcia et. al.[27] have demonstrated the applicability in the following way.

The total cross section (Q_T) is the sum of elastic (Q_{el}) and inelastic (Q_{inel}) scattering cross sections. According to the Born approximation, Q_{el} is given as

$$Q_{el} = \pi a_0^2 \frac{R}{\varepsilon} \left[A + B\frac{R}{\varepsilon} + C\left(\frac{R}{\varepsilon}\right)^2 + \cdots \right] \tag{4.2}$$

According to the Born–Bethe approximation, Q_{inel} is given as

$$Q_{inel} = 4\pi a_0^2 \frac{R}{\varepsilon} \left\{ M_{tot}^2 \ln\left[4C_{tot}\frac{\varepsilon}{R}\right] + \gamma_{tot}\frac{R}{\varepsilon} + \cdots \right\} \tag{4.3}$$

In Equations 4.2 and 4.3, R is the Rydberg energy (13. 595 eV), ε the electron energy in eV, a_0 the Bohr radius, and A, B, M_{tot}, C_{tot}, γ_{tot} are constants defined in references[51] and [52] for electron–atom interactions. The method has been extended to homonuclear diatomic molecules by Liu.[53] The expression for CO given by Garcia et al.[27] for the total cross section is

$$Q_T = a_0^2 \frac{R}{\varepsilon} \left[-152 + 164.4\ln\left(\frac{\varepsilon}{R}\right) + 411\left(\frac{R}{\varepsilon}\right) \right] \tag{4.4}$$

As an example, substituting $\varepsilon = 381$ eV, $a_0 = 5.2917 \times 10^{-11}$ m, one obtains $Q_T = 4.103 \times 10^{-20}$ m^2; the measured cross section is 4.312×10^{-20} m^2, which is within 5%. Equation 4.4 is also plotted in Figure 4.1, showing the degree of agreement in the range $\varepsilon \geq 150$ eV.

Jain and Baluja[28] have theoretically calculated the total (elastic + inelastic) electron scattering cross sections in a wide energy range (10 to 5000 eV) for several diatomic and polyatomic molecules. They have also provided the constants for calculating the total cross sections using the Born and Born–Bethe equations, 4.2 and 4.3. The results show good agreement with the measured values above 100 eV impact energy, but the agreement for dipolar molecules like CO is poorer than that of the equation of García et al.[27] Equation 4.4. Jain and Baluja[28] have also provided a general formula similar to 4.4 above, which will be discussed in Chapter 5.

The total cross section measurements of Kanik et al.[31] are in excellent agreement with those of Kwan et al.[19] within ±2% over the entire electron energy range. Kanik et al.[31] attribute the broad maximum at 20 eV to the resonance observed in the vibrational channel by Chutjian et al.[54] The latter authors observed a large enhancement in the differential cross section around the same impact energy. Their calculations showed that this increase could not be attributed to a purely potential (nonresonant) scattering process.

The measurements of Karwasz et al.[33] are in very good agreement with those of Kwan et al.[19] and Kanik et al.[31] The computations of Jain and Baluja.[28] without an anisotropic term also show good agreement, while those with an anisotropic term are about 20% higher. Karwasz et al.[33] have also given an analytical expression for the total cross section, based on the Born approximation,

$$Q_T(\varepsilon) = \frac{1}{A + B\varepsilon} \tag{4.5}$$

where A and B are fitting parameters. These parametrs are not entirely empirical because of their relationships with the fundamental constants

$$A = \frac{\hbar^4}{16\pi m^2 a^4 V_0^2} \tag{4.6}$$

and

$$B = \frac{\hbar^2}{2\pi m\, a^2 V_0^2} \tag{4.7}$$

where $\hbar = h/2\pi$, $m = $ mass, $a = $ range of the molecular scattering potential, and V_0 is the scattering potential, given by

$$V(r) = \frac{V_0}{r}\exp\left(-\frac{r}{a}\right) \tag{4.8}$$

Karwasz et al.[33] have evaluated the values: $A = 0.067 \times 10^{20}\,\mathrm{m^{-2}}$, $B = 0.419 \times 10^{20}\,\mathrm{m^{-2}}$ keV^{-1}, $V_0 = 96.39$ (au) and $a = 0.140$ (au). Since we are using Equation 4.5 there is no need to use conversion from atomic to SI units. As an example, $\varepsilon = 150\,\mathrm{eV}$ yields a cross section of $7.70 \times 10^{-20}\,\mathrm{m^2}$, which is within $+2\%$ of the measured cross section. A slightly different version is given by Zecca et al.[55] Relatively recent measurements of Szmytkowski and Maciąg[45] are in good agreement with those of Kanik et al.[31] and Karwasz et al.[33] though the agreement with Kwan et al.[19] at lower energies ($<2\,\mathrm{eV}$) is less than satisfactory.

4.1.2 MOMENTUM TRANSVFER CROSS SECTIONS IN CO

Momentum transfer cross sections are obtained (1) by measurement of differential scattering cross sections, (2) by swarm analysis[15,18,20] or (3) by theoretical computations.[30] Table 4.3 gives the tabulated cross sections in the energy range 0 to 100 eV, due to Land,[18] and Figure 4.3 shows selected cross sections. The momentum transfer cross sections in CO have three characteristic features. There is a minimum at $\sim0.04\,\mathrm{eV}$, decreasing from the zero

TABLE 4.3
Momentum Transfer Cross Sections in CO

Energy (eV)	Q_m ($10^{-20}\,\mathrm{m^2}$)	Energy (eV)	Q_m ($10^{-20}\,\mathrm{m^2}$)	Energy (eV)	Q_m ($10^{-20}\,\mathrm{m^2}$)
0.0	60.0	0.30	12.1	4.0	13.8
0.0010	40.0	0.35	13.0	4.5	13.3
0.0020	25.0	0.40	13.8	5.0	12.9
0.0030	17.7	0.50	15.4	6.0	12.3
0.0050	12.3	0.70	16.5	7.0	11.8
0.0070	9.8	1.00	18.5	8.0	11.3
0.0085	8.6	1.20	28.0	10.0	10.6
0.0100	7.8	1.3	37.0	12.0	10.4
0.015	6.5	1.5	42.0	15.0	10.2
0.020	5.9	1.7	40.0	17.0	10.1
0.030	5.4	1.9	32.0	20.0	9.8
0.040	5.2	2.1	23.5	25.0	9.1
0.050	5.4	2.2	21.5	30.0	8.6
0.070	6.1	2.5	17.5	50.0	7.1
0.100	7.3	2.8	16.0	75.0	6.1
0.15	8.8	3.0	15.4	100.0	5.5
0.20	10.0	3.3	14.6		
0.25	11.2	3.6	14.2		

Reprinted from Land, J. E., *J. Appl. Phys.*, 49, 5716, 1978, © 1978, American Institute of Physics.

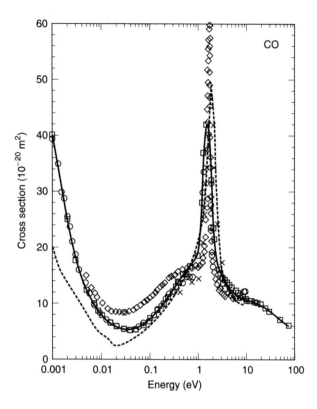

FIGURE 4.3 Momentum transfer cross sections in CO. (◆) Gibson et al.,[38] experimental; (○) Hake and Phelps[15] swarm analysis; (◇) Chandra,[16] theory; (—□—) Land[18] swarm analysis; (×) Haddad and Milloy,[20] swarm analysis; (— —) Jain and Norcross,[30] theory.

energy cross section of $60 \times 10^{-20}\,\mathrm{m}^2$. There is a CO^- attachment resonance peak near $\sim 1.9\,\mathrm{eV}$, and a negative slope for energies greater than $2\,\mathrm{eV}$. The cross sections obtained by the swarm method agree among themselves reasonably well, as shown in the figure, though the theoretically calculated cross sections show a much larger resonance capture cross section. The stated accuracy of Q_M due to Land[18] are; $\pm 10\%$ for energies of $3\,\mathrm{eV}$ and less, but as the energy increases the error possibly increases to as large as $\pm 30\%$.

4.1.3 Elastic Scattering Cross Sections in CO

Elastic scattering cross sections are given by Kanik et al.[35] in the electron impact energy range 1 to $1000\,\mathrm{eV}$ and are retained without the need to revise them (see Table 4.6 for values).

4.1.4 Rotational and Vibrational Cross Sections in CO

In addition to electronic excitation scattering (energy ε_{ex}), two other types of scattering at low energies contribute to electron energy loss in molecules. These are rotational excitation (energy ε_{rot}) and vibrational excitation (energy ε_v); the former has lower energy thresholds, denoted by quantum number $J = 0, 1, 2, \ldots$ etc. whereas the vibrational excitation levels are denoted by quantum numbers $v = 0, 1, 2, \ldots$. The energies decrease in the order $\varepsilon_{ex} > \varepsilon_v > \varepsilon_{rot}$. Both rotational and vibrational excitations are governed by selection rules. The scatterings are classified as vibrationally elastic or rotationally elastic if Δv, $\Delta J = 0$

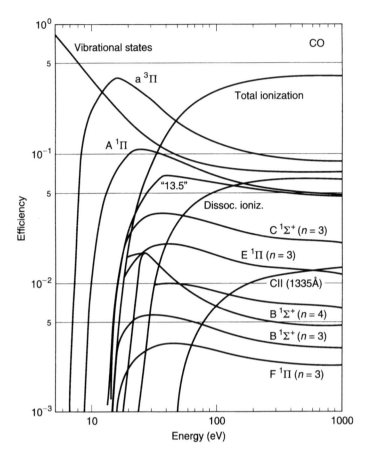

FIGURE 4.4 Relative efficiency (losses) of various inelastic scatterings in CO.[60] Vibrational losses are significant up to 20 eV. The symbols on the curves indicate the excitation processes and are explained in Section 4.1.4.

respectively; inelastic if Δv, $\Delta J \neq 0$.[56] The latter transitions are significant only at small angles. The relative efficiencies of various inelastic scatterings in CO are shown in Figure 4.4[60] and one sees that up to ∼20 eV energy vibrational energy losses are very significant.

Quantum mechanical calculations show that rotational excitation energy in CO is quite low, of the order of a few meV.[15] To put this energy in proper perspective, we note that the energies of a gas molecule are ∼7 and ∼26 meV at 77 K and 300 K respectively; the rotational excitation threshold is quite small compared to kT. Therefore one can expect to have neutrals excited rotationally, even at room temperature. Hake and Phelps[15] have derived the rotational excitation cross sections in CO from an analysis of the Boltzmann equation. Figure 4.5 shows these cross sections for excitation level $J = 4 \rightarrow 5$ and de-excitation level $J = 5 \rightarrow 4$. The computations apply to a gas temperature of 77 K and the cross sections contain a factor that takes into account the number of molecules that have the initial rotational level.

The energy regime for electron scattering measurements below 100 meV is technically difficult because of the problems associated with controlling the beam energy at these low values. Experimental measurement of rotational excitation cross sections in the very low energy range has become available only recently.[41] Electrons are produced by autoionization in argon, using radiation of 78.65 nm, close to the $^2P_{3/2}$ threshold; autoionization is the process in which two electrons in the same atom are excited to a higher level and one of the

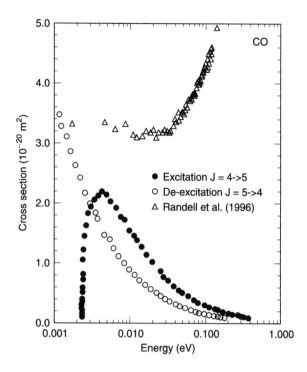

FIGURE 4.5 Rotational excitation cross section in CO at 77 K reported by Hake and Phelps.[15] For transitions: (●), $J = 4 \to 5$; (○), $J = 5 \to 4$. The measured cross sections are those of Randell et al.[41] at 293 K. Note that the measurements are the sum of elastic and rotationally inelastic scattering cross sections.

electrons returns to ground level, ionizing the atom in the process. The electrons are formed into a beam by an axial magnetic field, the energy spread being limited to 5 meV FWHM. These cross sections are also shown in Figure 4.5. A weak dip around 10 meV is noticeable. In contrast, in nitrogen and oxygen, both of which are nonpolar, there is no evidence of a minimum and the cross section decreases monotonically below a few tens of meV, as discussed in subsequent sections.

The application of MERT (modified effective range theory) has been explained in considerable detail in Chapter 3, and Randell et al.[41] have examined the applicability of this method for polar molecules. In view of the small polarizability of the CO molecule $(2.2 \times 10^{-40}\,\mathrm{F\,m^2})$ one gets a good agreement with the two-term MERT expressions for the phase shift:

$$\tan \eta_0 = -Ak\left[1 + \left(\frac{4\alpha_\mu}{3}\right)k^2 \ln k\right] - \left(\frac{\pi\alpha_\mu}{3}\right)k^2 + Dk^3 \tag{4.9}$$

$$\tan \eta_1 = \left(\frac{\pi\alpha_\mu}{15}\right)k^2 \tag{4.10}$$

$$\tan \eta_2 = \left(\frac{\pi\alpha_\mu}{105}\right)k^2 \tag{4.11}$$

where k is the wave vector (see Equation 3.36), α_μ is the polarizability due to the permanent dipole moment, A is the scattering length, and D is a fitting parameter. The range of energy used for fitting is 40 to 160 meV. The value of D is $44.1 \pm 3.9\ a_0^3$. The scattering length

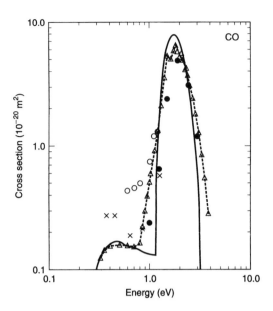

FIGURE 4.6 Vibrational excitation cross sections in CO. Experimental: ($\circ$) Schulz[57]; ($\times$) Sohn et al.[59]; ($\bullet$), Gibson et al.[38]; Swarm derivation: (——) Hake and Phelps[15]; ($-\triangle-$) Land.[18]

is 0.2×10^{-10} m and the total scattering cross section at zero electron energy ($= 4\pi A^2$) is 0.50×10^{-20} m^2.

Vibrational cross sections Q_v extend to a relatively higher energy range, up to $\sim$4 eV. The vibrational levels are separated by equal energy quanta and in CO the energy gap is 0.266 eV.[15] The transition $v = 0$, $v = 1$ is the only important one since transitions to the higher levels are relatively unpopulated. The vibrational cross sections derived from the swarm analysis by Hake and Phelps[15] are shown in Figure 4.6 and they are stated to be in very good agreement with the simple dipolar theory of vibrational excitation for electron energies below 1 eV.

Both theoretically calculated and derived vibrational cross sections are about 30 times higher than those in N$_2$ at the same energy. The differences are attributed to the effect of the dipole moment of the CO molecule, and the interactions that determine the low energy vibrational cross sections in the two gases differ considerably. The vibrational cross sections from 0.226 eV to 1.1 eV remain relatively flat, but for higher energies they increase steeply like a impulse function. The vibrational cross sections reported by Schulz.[57] on the other hand, increase relatively less rapidly.

Vibrational excitation of the CO molecule by electron impact can proceed by one of two processes.[18] At energies less than 1 eV, direct excitation takes place due to the permanent dipole moment of the molecule; this process is observed only at $v = 1$ excitation. At energies between 1 and 3 eV, resonant excitation takes place via the compound negative ion state, CO$^-$. This resonance is present in all cross sections, $v = 1, 2, 3, \ldots 10$. The derived cross sections reported by Land[18] are shown in Figure 4.6 and Table 4.4. Chutjian and Tanaka[58] have measured the vibrational cross sections for the $v = 0 \rightarrow 1$ excitation in the 3 to 100 eV range. The effect of the small dipole moment of the CO molecule (0.11 D) was found to be negligible in promoting the vibrational excitation in this energy range. Chutjian and Tanaka[58] suggest that the short range potentials due to polarization and quadrupole moment are more effective than the long range dipole potential. Further, there is a pronounced hump at 20 eV which is attributed to shape resonance. The crossed beam measurements of Sohn et al.[59] (1985) cover a narrow range (0.37 to 1.26 eV) to make meaningful comparison.

TABLE 4.4
Vibrational Excitation Cross Sections in CO

Land[18]						Chutjian and Tanaka[58]	
Energy	Q_v	Energy	Q_v	Energy	Q_v	Energy	Q_v
0.266	0.0	1.0	0.513	2.086	5.683	3	0.365
0.290	0.095	1.031	0.604	2.170	4.965	5	0.104
0.32	0.125	1.13	0.924	2.281	4.176	9	0.024
0.35	0.144	1.215	1.350	2.316	4.285	20	0.097
0.40	0.156	1.307	2.137	2.404	3.747	30	0.021
0.50	0.159	1.410	3.602	2.508	3.120	50	0.007(6)
0.60	0.157	1.514	5.377	2.688	2.455	75	0.005(4)
0.70	0.154	1.645	5.054	2.872	1.828	100	0.007(3)
0.80	0.165	1.740	5.934	3.072	1.290		
0.85	0.224	1.821	6.578	3.294	0.861		
0.90	0.300	1.902	5.843	3.528	0.555		
0.95	0.397	1.982	5.216	3.816	0.287		

Energy in units of eV; cross section in units of 10^{-20} m^2.

Measurements of Q_v by Gibson et al.[38] show that there is a peak at 1.91 eV with a cross section of 4.9×10^{-20} m^2 (Figure 4.6). At this energy, which is close enough to the $^2\Pi$ resonance, the cross section is the sum of elastic (nonresonant) and inelastic (resonant) components; the former is quite small, ~3% of the latter. There is very good agreement with the cross sections reported by Hake and Phelps[15] and Land.[18]

4.1.5 ELECTRONIC EXCITATION CROSS SECTIONS IN CO

Compared to the total cross section data available (Figure 4.3), studies of electronic excitation are relatively few. The total excitation cross sections derived from swarm studies or by summing cross sections for selected states, usually obtained by different investigators adopting different techniques, do not add up to a very satisfactory picture. Fortunately, relatively recent contributions of Zobel et al.[43,44] go a long way to alleviate this situation. A brief summary is given, in addition to references for the rather sparse earlier data.

Cross sections for absolute differential electronically excited states are measured either as a function of scattering angle at constant electron energies, or at constant scattering angles at selected energy values. The range used by Zobel et al.[43,44] in the crossed beam experiments is from threshold down to 3.7 eV. The potential energy curves and vibrational levels of the electronic states of CO up to the excitation energy of the 11.524 eV (E $^1\Pi$) state are shown in Figure 4.7. Table 4.5 lists the states and threshold energies ($v = 0$) for these states.

Sawada et al.[60] have given an empirical equation for the excitation cross section for a number of states; the equation is of the form

$$Q(\varepsilon) = \frac{4\pi a_0^2 R^2 f_0 c_0}{W^2} \left(\frac{W}{\varepsilon}\right)^a \left[1 - \left(\frac{W}{\varepsilon}\right)^b\right]^n \tag{4.12}$$

where a_0 is the Bohr radius, R the Rydberg constant ($= 13.59$ eV), and the fitting parameters $f_0 c_0$, a, b, and n have different values for different states. Figure 4.8 shows these cross sections. The curve for the state labelled "13.5 eV" has the following meaning. The electron

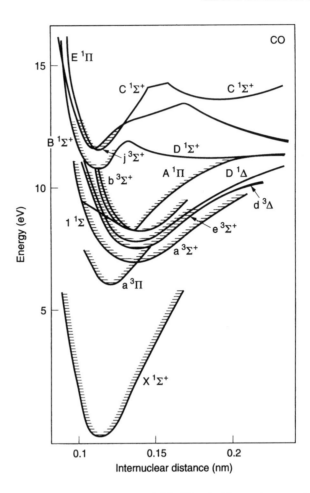

FIGURE 4.7 Potential energy diagrams of states of CO. Threshold energies are tabulated in Table 4.4. Figure reproduced from Zobel, J. et al. *J. Phys. B*, 29, 813, 1996. With permission of the Institute of Physics, U.K.

impact excitation spectra show, besides low lying levels and their vibrational structures, numerous peaks near the first ionization potential (14.01 eV), in particular a large broad peak near 13.5 eV. This state is not shown in the book by Herzberg[104] presumably because it is a composite of several different states and referred to by Sawada[60] as the "13.5-eV level." These cross sections were used by Land[18] in the swarm analysis.

CO is known to sustain several metastable states. The interest in identifying the states and the cross sections has increased since the Martian atmosphere obtained by the Mariner space probes has revealed the existence of a $^3\Pi$–X $^1\Sigma^+$ transitions. These transitions, known as the Cameron bands, are forbidden by the selection rules as they involve transitions from singlet to triplet states. Two channels exist for the metastable formation. The first one is a long-lived a $^3\Pi$ state with a lifetime of more than 1 ms, and the second one is a higher lying state, >9.5 eV.[61] A metastable state at about 10 eV has been detected, with a lifetime of $\sim100\,\mu s$, but controversy exists about the state. Mason and Newell[25] (1988) suggest that the 10 eV metastable is due to the transition X $^1\Sigma^+$–I $^1\Sigma^-$. The total metastable cross sections measured by these authors are shown in Figure 4.9, which also includes the total excitation cross sections due to several other sources. Before commenting on these, we will briefly refer to the major contributions of Zobel et al.[43,44]

TABLE 4.5
Electron Configuration and Important Properties of the Vibrational Level ($v = 0$) of the Electronic States of CO

Electronic State	Valence Orbitals				Rydberg Orbitals			Energy (eV)	$r_e(10^{-10}m)$
	1π	5σ	2π	6σ	3σ	$3p\pi$	$3p\pi$		
X $^1\Sigma^+$	4	2						0	1.128
a $^3\Pi$	4	1	1					6.006 (6.034)	1.206
a' $^3\Sigma^+$	3	2	1					6.863 (6.928)	1.352
d $^3\Delta$	3	2	1					7.513	1.369
e $^3\Sigma^-$	3	2	1					7.898	1.384
A $^1\Pi$	4	1	1					8.024 (8.065)	1.235
I $^1\Sigma^-$	3	2	1	—	—	—	—	8.137	1.391
D $^1\Delta$	3	2	1	—	—	—	—	8.241	1.399
b $^3\Sigma^+$	4	1	—	—	1	—	—	10.399 (10.387)	1.113
B $^1\Sigma^+$	4	1	—	—	1	—	—	10.777 (10.776)	1.119
D' $^1\Sigma^+$	3	2	1	—	—	—	—	10.995	
j $^3\Sigma^+$	4	1	—	—	—	1	—	11.269	1.144
C $^1\Sigma^+$	4	1	—	—	—	1	—	11.396 (11.393)	1.122
c $^3\Pi$	4	1	—	—	—	—	1	11.414	1.127
E $^1\Pi$	4	1	—	—	—	—	1	11.524	1.115

Core orbitals are $1\sigma^2 \, 2\sigma^2 \, 3\sigma^2 \, 4\sigma^2$. Energy figures in brackets are spectroscopic values from Herzberg.[104] r_e = internuclear distance at equilibrium. From Zobel et al.[44], *J. Phys. B*, 29, 813, 1996. With permission of the Institute of Physics U.K.

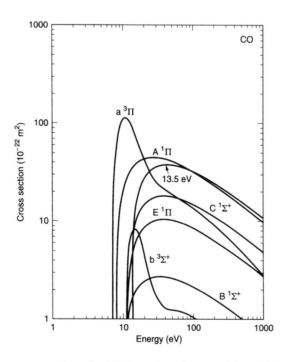

FIGURE 4.8 Excitation cross sections in CO for selected states. Threshold energies are shown in Table 4.4. Untabulated values are obtained by digitizing the original publication and then replotted. Figure reproduced from Sawada, T. et al., *J. Geophys. Res.*, 77, 4819, 19972. With permission.

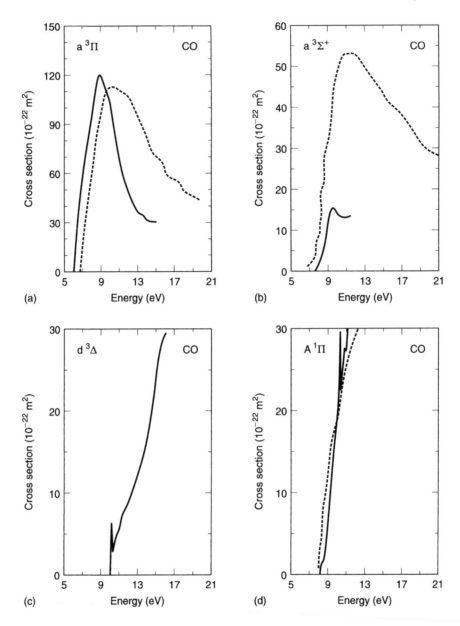

FIGURE 4.9 Excitation cross sections of CO near threshold energies, measured by Zobel et al.[44] full curves. The states are: (a) a $^3\Pi$, 6.006 eV threshold energy; (b) a' $^3\Sigma^+$, 6.863 eV threshold energy; (c) d $^3\Delta$, 7.513 eV threshold; (d) A $^1\Pi$, 8.024 eV threshold. Broken curves: Land.[18]

The orbital energy in the CO molecule increases in the order $1\sigma^2\,2\sigma^2\,3\sigma^2\,4\sigma^2\,1\pi^4\,5\sigma^2$ for the ground state, and the $1\sigma\,2\sigma$ orbitals retain their character during the excitation process. From the remaining ten valence electrons the lone pair orbitals 3σ and 4σ correspond to the 2s orbitals of the O and C atoms in the separated atom model. The 1p and 5s states have bonding character; the 2π and 6σ orbitals, which are not occupied in the ground state, are anti-bonding. The 2π orbital is populated in the lowest excitation state, a $^3\Pi$ state, and the energetic order of the bond is changed. In contrast to the ground state (see above), the 5σ orbital is more tightly bound than the 1π orbital in the excited state.[44]

With reference to Figure 4.7, A $^1\Pi$, b $^3\Sigma^+$ and B $^1\Sigma^+$ are optically allowed transitions whereas, in order of increasing energy, the states a, a', d, e, I, and D are metastables. The lifetimes of the first three are 5, 4, and 3.7 μs respectively. The lifetime of the I and D metastable is of the order of 80 ± 10 μs. Singlet to triplet transition, forbidden by selection rules, is not strictly obeyed in the molecule.[25] Further, weak bands are observed corresponding to the five forbidden metastable transitions. The lowest transition a $^3\Pi$ is also the more intense, followed by A $^1\Pi$ and B $^1\Sigma^+$.

The excitation cross section of the a $^3\Pi$ state occurs as a result of a 5σ electron jumping to a 2π orbital, resulting in the electron configuration $1\sigma^2\,2\sigma^2\,3\sigma^2\,4\sigma^2\,1\pi^4\,5\sigma^1\,2\pi^1$. Q_{ex} as a function of energy for this state shows a peak at ~9 eV (Figure 4.9a) with a maximum excitation cross section of $1.2 \times 10^{-20}\,\text{m}^2$. It is interesting to recall that the negative ion shape resonance CO^- at approximately ~1.9 eV (see Table 4.2) has the term $^2\Pi$ and the electronic configuration $1\pi^4\,5\sigma^2\,2\pi^1$ results, due to a 2π electron attaching to the ground state. The decay into the triplet state, a $^3\Pi$, after an interval of ~10^{-15} s, occurs by the ejection of a 5σ electron.[44]

The a' $^3\Sigma^+$ state results from the ejection of a deeper lying 1π electron to the 2π orbital, leading to the electron configuration $1\pi^3\,5\sigma^2\,2\pi^2$. There is also a peak just after 9 eV, but the magnitude of the peak is only about 12% of the peak of the a $^3\Pi$ state. A comparison with the excitation cross sections for the state a' $^3\Sigma^+$, used by Land[18] in swarm calculations, shows a large difference (Figure 4.9b), which is possibly due to the fact that Land[18] used theoretically derived cross sections.[62] The more recent theoretical calculations of Sun et. al.[63] also differ considerably from the experimental results shown.

The excitation cross sections for the d $^3\Delta$ state are quite low. The cross sections shown in Figure 4.9c are obtained by multiplying the measured cross sections by a factor of 9 to obtain agreement with the theoretically calculated cross sections.[44] The cross sections for the A $^1\Pi$ state rise smoothly from the threshold energy and are in reasonable agreement with those of Land.[18] The excitation cross sections shown in Figure 4.10 belong to the category of Rydberg states, loosely defined as electrons belonging to orbitals of high quantum numbers. The transitions have a common feature, namely that states with several principal quantum numbers decay into the state of the same lower energy. A comparison of the ordinates of Figures 4.9 and 4.10 shows that the excitation cross sections of the latter figure are generally lower (~10%) than those of the former, which are themselves rather low compared with momentum transfer cross sections (compare the ordinates of Figures 4.9 and 4.3).

Total excitation cross sections are shown in Figure 4.11. The curve labelled Sawada et al.[60] is obtained by adding the cross sections shown in Figure 4.8. Hake et al.[15] derived Q_{ex} in a narrow energy range of 7 to 10 eV. Hake's values agree well with the rising part of the Sawada curve. Saelee and Lucas[64] employed a much higher cross section for Monte Carlo simulation; they do not comment on their excitation cross sections.

The total metastable cross sections due to Mason and Newell[25] are quite low as they correspond to lifetimes of excited states, ~80 μs. The excitation cross sections recommended by Kanik et al.[35] are the sum of the four states (a $^3\Pi$, a' $^3\Sigma^+$, d $^3\Delta$, A $^1\Pi$), increased by 20%. They follow the Sawada curve in shape up to 1000 eV energy. The recommended excitation cross sections are the average of these two curves, as shown in Figure 4.11.

Dissociation occurs by the route of electronic excitation according to the reaction[32]

$$CO + e \rightarrow CO^* + e \qquad (4.13)$$

$$CO^* \rightarrow C(^3P) + O(^3P) + KE \qquad (4.14)$$

where KE means the kinetic energy of the atoms. The designation of states within brackets refers to the ground state of the atoms. The dissociation energy is 11.092 eV.[29] Since the

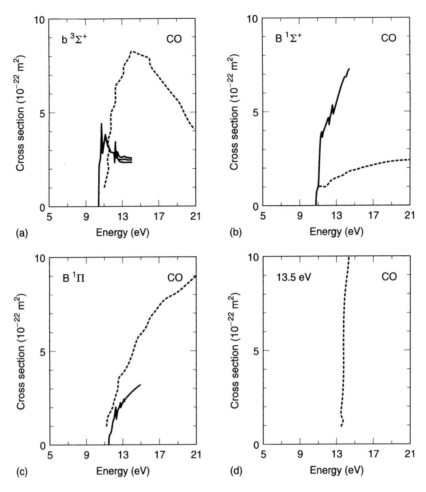

FIGURE 4.10 Excitation cross sections in CO near threshold energies, measured by Zobel et al.,[43,44] full curves. The states are: (a) b $^3\Sigma^+$, 10.399 threshold energy[44]; (b) B $^1\Sigma^+$, 10.777 eV threshold energy[43]; (c) E $^1\Pi$, 11.524 eV threshold energy.[43] Broken curves: Land.[18] (d) 13.5 eV loss.

excited state necessarily lies above the molecular dissociation limit, the excess energy is transferred to the atoms. Figure 4.11 also includes Cosby's dissociation cross sections.[32]

4.1.6 IONIZATION CROSS SECTIONS IN CO

Ionization cross sections have been measured by Tate and Smith,[65] Asundi et al.[12] and Rapp and Englander-Golden.[14] A set of ionization cross sections has also been recommended by Kanik et al.[35] and belongs to the category of total ionization cross section measurement. Partial ionization cross sections are measured by Vaughan.[66] Hille and Märk.[67] Orient and Srivastava.[24] Freund, et al.[143] Tian and Vidal,[47] and Mangan et al.[48]

To measure the cross sections for different ion species a mass spectrometer is required. The errors associated with this technique are discussed by Mangan et al.[48] These authors have used a time-of-flight mass spectrometer for the determination of absolute partial cross sections for the production of CO^+, C^+, O^+, and CO^{2+}. Basically, the technique involves a vacuum chamber which is filled with CO at a gas pressure of $\sim 0.7\,mPa$ (5 µtorr). The electron gun produces 20 ns long pulses each containing approximately 2500 electrons at a repetition

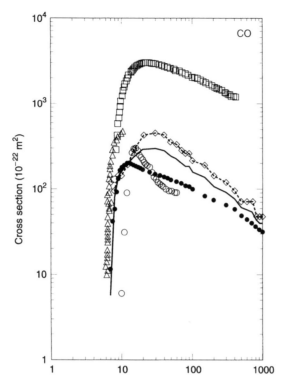

FIGURE 4.11 Total excitation cross sections in CO. (△) Hake and Phelps,[15] swarm method; (•) Sawada et al.,[60] compilation; (□) Saelee and Lucas,[64] Monte Carlo method; (○) Mason and Newell,[25] experimental—the cross sections shown should be multiplied by the factor 0.01; (—◆—) Kanik et al.,[35] compilation; (——) recommended in this volume.

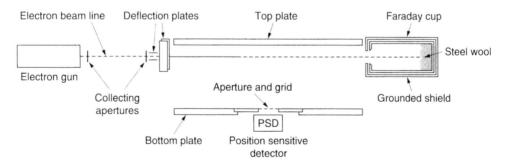

FIGURE 4.12 Time-of-flight spectrometer for the measurement of partial ionization cross sections in CO. PSD = position-sensitive detector which records the arrival times and positions of ions. Figure reproduced from Mangan, M. A. et al., *J. Phys. B*, 33, 3225, 2000. With permission of the Institute of Physics, U.K.

rate of the order of 2.5 kHz. These pulses are directed through an interaction region that is located between two plates maintained at ground potential, and are collected in a Faraday cup (Figure 4.12). The electron pulse is followed by a second voltage pulse of 3 kV magnitude to the top plate, after an approximate interval of 200 ns, to drive any positive ions formed by electron impact toward the bottom plate. The field generated is 48 kV/m. Some ions pass

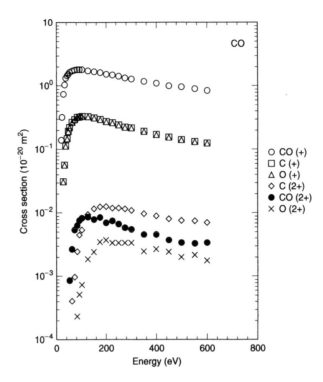

FIGURE 4.13 Ionization cross sections measured by Tian and Vidal.[47] The ions produced are shown as: (○) CO⁺; (□) C⁺; (△) O⁺; (◇) C²⁺; (●) CO²⁺; (×) O²⁺.

through an aperture and a grid, and fall on a position-sensitive detector (PSD) after being accelerated to 5.4 keV. The ion arrival times are used to identify their charge-to-mass ratios and the ion arrival positions are used to determine the effectiveness of product ion collection. The measured absolute cross sections have a stated uncertainty of ±5%.

Orient and Srivastava[24] used a quadrupole mass spectrometer to measure the ionization cross sections of the parent and fragment ions. Their total cross sections agree within 6% with those of Rapp and Englander-Golden.[14] Kanik et al.[35] have recommended the latter cross sections because they cover a wider range of electron energies. We just examine whether these cross sections need to be revised in the light of measurements that have been carried out since 1993 by Tian and Vidal.[47] and Mangan et al.[48] It is noted here that the measurements of Freund et al.[143] who used the fast neutral beam technique together with energy analyzers, are limited to the production of the parent ion, CO⁺. Figure 4.13 shows the cross sections for generation of fragment ions[47] and Figure 4.14 shows the total cross sections of several investigations. The earlier data of Vaughan[66] and Defrance and Gomet[68] are not included.

A concise discussion of several techniques available, with their advantages and disadvantages, is provided by Tian and Vidal[47] who have employed a time-of-flight spectrometer for the measurement of ionization cross sections of the parent and fragmented ion. The stated uncertainty in the cross sections is 10% for cross sections higher than 1×10^{-21} m² and 15% for lower cross sections. The total cross section is calculated according to

$$Q_i = CO^+ + C^+ + O^+ + 2(CO^{2+} + C^{2+} + O^+) \tag{4.15}$$

The partial cross sections decrease in the order CO⁺, C⁺, O⁺, C²⁺, CO²⁺, and O²⁺.

The cross sections reported by Orient and Srivastava,[24] Tian and Vidal,[47] and Mangan et al.[48] agree amongst themselves within 10%. A point to note is that they all have a positive

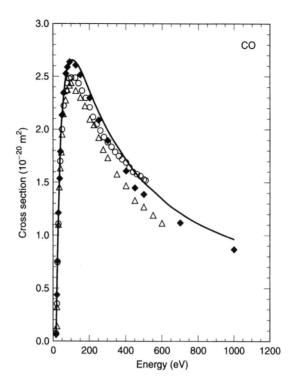

FIGURE 4.14 Absolute total ionization cross sections in CO. (———) Rapp and Englander-Golden[14]; (○) Orient and Srivastava[24]; (△) Tian and Vidal[47]; (◆), Mangan et al.[48]

difference, that is lower in magnitude, when compared with the cross sections of Rapp and Englander-Golden.[14] Figure 4.15 clarifies this point.

4.1.7 VERIFICATION OF THE SIGMA RULE FOR CO

The recommended set of cross sections is summarized in Table 4.6. The elastic scattering cross sections in the electron impact energy range 1 to 30 eV are due to Gibson et al.[38] whose careful measurements provide much needed data in this energy range. For higher energies, data on Q_{el} is patchy. The recommended cross sections of Kanik et al.[35] for higher energies do not merge smoothly with the measured values of Gibson et al.[38] for example, at 30 eV electron energy Kanik et al.[35] recommend $Q_{el} = 7.8 \times 10^{-20}$ m^2 which is about 24% lower than the measured value of 10.3×10^{-20} m^2. Nickel et al.[26] have measured the differential scattering cross sections up to 100 eV electron energy, and the integrated values are given by Brunger and Buckman.[7] For electron energies greater than 100 eV, the recommended cross sections of Kanik et al.[35] are accepted. Obviously there is a need for more experimental data in this energy range.

The vibrational cross sections are also due to Gibson et al.[38] These values agree well with those of Land.[18] The total cross sections in the last but one column of Table 4.6 are reported by Szmytkowski et al.[45] in the 1 to 200 eV range and by Karwasz et al.[33] for higher energies. We conclude that the sigma rule applies within an uncertainty of ±20%.

4.2 MOLECULAR HYDROGEN (H$_2$)

Molecular hydrogen is one of the most frequently studied gases from the cross sections point of view, since it has one of the simplest molecular structures. It is also a preferred choice of

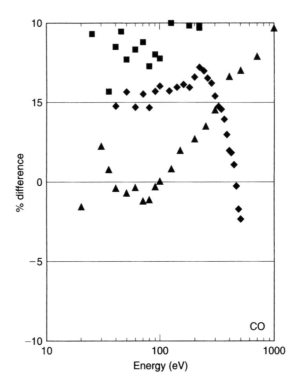

FIGURE 4.15 Comparison of absolute total ionization cross sections in CO. Reference cross sections (0% line) are due to Rapp and Englander-Golden1.[14] (♦) Orient et al.[24]; (■) Tian and Vidal[47]; (▲) Mangan et al.[48] Positive values of the ordinates mean that reference values are higher.

gas by theoreticians who are interested in the complexities that occur as one moves from electron–atom to electron–molecule interaction. In the former case one can assume a spherically symmetric potential whereas in the latter case one has to deal with the non-spherically symmetric nature of the potential.[72] Development of theory therefore begins with this molecule and experimental data are needed to test and improve the theoretical concepts. The gas is present in interstellar space and plasma modeling usually begins with this gas. Table 4.7 shows selected references for cross sections in H_2.

4.2.1 TOTAL CROSS SECTIONS IN H_2

Measured and theoretical absolute total cross sections are shown in Figure 4.16. A low energy feature is a peak around 3.5 eV, which is explained as follows. As stated previously (Section 2.4.1), resonance occurs near the onset of an excitation process and the electron attaches to the molecule; forming a short-lived negative ion. After an interval of $\sim 10^{-15}$ s the molecule autoionizes, with an excited molecule as a product, reaction 4.17, or dissociates with one of the atoms as a negative ion, reaction 4.18. The reactions are

$$(AB) + e \rightarrow (AB)^* + e \qquad \text{(Direct excitation)} \qquad (4.16)$$

$$(AB) + e \rightarrow (AB)^{*-} \rightarrow (AB)^* + e \qquad \text{(Negative ion resonance)} \qquad (4.17)$$

$$AB + e \rightarrow AB^{*-} \rightarrow A + B^- \qquad \text{(Dissociative attachment)} \qquad (4.18)$$

The potential curves for reactions 4.17 and 4.18 are shown in Figure 2.20.

TABLE 4.6
Recommended Cross Sections in CO

Energy	Q_{el}	Q_v	Q_{ex}	Q_i	Q_Σ	Q_T(expt)	% Diff.
1	15.40[G]	0.24[G]	—	—	15.64	14.00[S]	−11.71
2	34.87[G]	4.60[G]	—	—	39.47	43.40[S]	9.05
3	22.70[G]	1.20[G]	—	—	23.90	23.60[S]	−1.27
4	18.35[G]	—	—	—	18.35	16.40[S]	−11.89
5	14.00[G]	—	—	—	14.00	15.10[S]	7.28
6	12.90[G]	—	—	—	12.90	13.70[S]	5.84
8	12.00[G]	—	0.80[R]	—	12.80	13.20[S]	3.03
10	11.40[G]	—	1.62[R]	—	13.02	13.10[S]	0.61
12	11.30[G]	—	1.94[R]	—	13.24	13.30[S]	0.45
15	11.20[G]	—	2.39[R]	0.05[REG]	13.64	14.20[S]	3.94
20	11.00[G]	1.1[G]	2.91[R]	0.43[REG]	15.44	14.70[S]	5.03
30	10.30[G]	0.035[G]	3.00[R]	1.24[REG]	14.54	13.70[S]	−6.13
40	7.60[N]	—	2.84[R]	1.78[REG]	12.22	12.90[S]	5.27
50	6.15[N]	—	2.43[R]	2.12[REG]	10.70	11.80[S]	9.32
60	5.35[N]	—	2.26[R]	2.34[REG]	9.95	11.10[S]	10.36
70	4.71[N]	—	2.00[R]	2.50[REG]	9.21	10.70[S]	13.93
80	4.00[N]	—	1.85[R]	2.58[REG]	8.43	10.20[S]	17.35
90	4.05[N]	—	1.85[R]	2.63[REG]	8.53	9.75[S]	12.51
100	3.25[N]	—	1.85[R]	2.65[REG]	7.75	9.47[S]	18.16
200	2.80[K]	—	1.08[R]	2.36[REG]	6.24	6.48[S]	3.70
300	2.21[K]	—	0.92[R]	1.99[REG]	5.12	4.98[KBGZ]	−2.81
400	1.90[K]	—	0.77[R]	1.72[REG]	4.39	4.14[KBGZ]	−6.04
500	1.65[K]	—	0.60[R]	1.49[REG]	3.74	3.56[KBGZ]	−5.06
600	1.50[K]	—	0.57[R]	1.35[REG]	3.42	3.16[KBGZ]	−8.23
700	1.38[K]	—	0.55[R]	1.21[REG]	3.14	2.80[KBGZ]	−12.14
800	1.28[K]	—	0.44[R]	1.11[REG]	2.83	2.52[KBGZ]	−12.30
900	1.20[K]	—	0.41[R]	1.03[REG]	2.64	2.29[KBGZ]	−15.28
1000	1.10[K]	—	0.39[R]	1.01[REG]	2.50	2.10[KBGZ]	−18.97

G = Gibson et al.[38]; K = Kanik et al.[35]; KBGZ = Karwasz et al.[33]; N = Nickel et al.[26]; R = recommended by the author of the present volume; REG = Rapp and Englander-Golden[14]; S = Szmytkowski et al.[45] Energy in units of eV; Q in units of 10^{-20} m^2 positive value of percentage difference means that the measured Q_T is higher than Q_Σ.

In hydrogen the ion species H$^-$ was observed, resulting from dissociative attachment, by Schulz and Asundi[103] at 3.7 eV, and experimental evidence for the existence of $^2\Sigma_u^+$ shape resonance was obtained by Ehrhardt et al.[73] Ehrhardt et al. who observed that simultaneous rotational excitation and one quantum of vibrational excitation occur at 4.42 eV (see note in reference 73). At this energy excitation occurs predominantly through the lowest ($^2\Sigma_u^+$) compound state of H$_2^-$. They also observed that the shape of the differential scattering cross sections (Q_{dif}) as a function of scattering angle (θ) changed with electron energy for elastic scattering ($v = 0$). This is attributed to the fact that the phase shifts of the partial waves change with energy and the interference of these partial waves causes a change in the shape. In contrast, the differential scattering cross section for inelastic scattering is due only to a single partial wave that results from autoionization of the negative ion state of H$_2$, and therefore the shape of the Q_{dif}–θ curve remains the same for all impact energies.

The motivation for measuring the total cross sections at low electron energies ($< \sim 5$ eV) is that there may be resonances below 100 meV, presumably related to rotational excitation. There is no theoretical evidence for such resonances, though swarm studies at low values of E/N (E = electric field, N = gas number density) have suggested their possible existence.[77] The early measurements of Golden et al.[72] with a stated accuracy of 3%, and Ferch et al.,[77]

TABLE 4.7
Selected References for Cross Section Studies in H_2

Type	Energy Range	Method	Reference
Q_M	0.01–20	Swarm method	Engelhardt and Phelp[69]
Q_i	Onset–1000	Ionization tube method	Rapp and Englander—Golden[70]
Q_i	Onset–1000	Ionization tube method	Rapp et al.[71]
Q_T	0.25–15.0	Ramsauer method	Golden et al.[72]
Q_{el}, Q_v	0.5–10	Crossed beams	Ehrhardt et al.[73]
Q_v	1.0–6.0	Trapped electron	Burrow and Schulz[74]
Q_v, Q_{rot}	0–1.5 eV	Swarm method	Crompton et al.[75]
Q_{diff}	2–20	Crossed beams	Srivastava et al.[76]
Q_T	0.02–2	TOF spectrometer	Ferch et al.[77]
Q_T	0.2–100	Linear transmission method	Dalba et al.[78]
Q_T	25–750	Linear transmission method	van Wingerden et al.[79]
Q_{el}, Q_{ex}	2.0–200.0	Crossed beams	Shyn and Sharp[80]
Q_T	2–500	Transmission method	Hoffman et al.[81]
Q_T	6–400	Transmission method	Deuring et al.[82]
Q_{el}	1–19	Crossed beams	Furst et al.[83]
Q_T	1–50	TOF spectrometer	Jones et al.[84]
Q_{el}	25–500	Crossed beams	Nishimura et al.[85]
Q_{ex}	20–60	Crossed beams	Khakoo and Trajmar[86]
Q_v, Q_{rot}	0.0–10.0	Theory	Morrison et al.[87]
Q_M, Q_v	0–25	Swarm method	England et al.[88]
Q_T	0.21–9.14	PE spectrometer	Subramanian and Kumar[89]
Q_{diff}	1.5 eV	Crossed beams	Brunger et al.[90]
Q_i	100–3000	Time-of-flight method	Kossmann et al.[91]
Q_T, Q_{el}	10–5000	Theory	Jain and Baluja[28]
Q_T	4–300	Linear transmission	Nickel et al.[92]
Q_v	0–10	Theory	Morrison and Trail[93]
Q_{ex}	9.2–20.2 eV	Energy loss method	Khakoo and Segura[94]
Q_i	Onset–1000	TOF spectrometer	Krishnakumar and Srivastava[95]
Q_T	0.10 –0.175	Beam attenuation; axial H	Randell et al.[96]
Q_i	400–2000	Time-of-flight method	Jacobsen et al.[97]
Q_i	Onset–1000	Time-of-flight method	Straub et al.[98]
Q_T	0.4–250	Linear transmission	Szmytkowski et al.[45]
All	0.001–1000	Compilation	Zecca et al.[61]
Q_i	Onset–1000	Compilation	Stebbings and Lindsay[99]
Q_{diss}	7.5–14.5	Theory	Trevisan and Tennyson[100]
Q_v	0.01–0.6	Theory	White et al.[101]
Q_{ex}	17.5,20,30	Energy loss spectroscopy	Wrkich et al.[102]

Q_T = total; Q_{diff} = differential; Q_{diss} = dissociation; Q_{ex} = excitation; Q_i = ionization; Q_{inel} = inelastic; Q_M = momentum transfer; Q_{rot} = rotational; Q (set) = set; Q_v = vibrational;   DCS = differential cross section; TOF = time-of-flight;   PE = photoelectron. Energy in units of eV.

with a stated accuracy of 2.5%, do not show any structure at low energies. There is a peak between 2 and 4 eV, as observed in all measurements of Q_T, and this is attributed to shape resonance of the $^2\Sigma_u^+$ state of H_2^-, which occurs around 3 eV.[103] Golden et al.[72] have given a useful simplified formula for the MERT expression

$$Q_T = (A + B\varepsilon^{1/2} + C\varepsilon \ln \varepsilon + D\varepsilon) \times 10^{-20} \text{ m}^2 \qquad (4.19)$$

where $A = 5.53$, $B = 8.71$ (eV)$^{-1/2}$, $C = 0.931$ (eV)$^{-1}$, $D = 1.05$ (eV)$^{-1}$. The scattering length calculated by extrapolating to zero energy is 0.0663×10^{-9} m.[72]

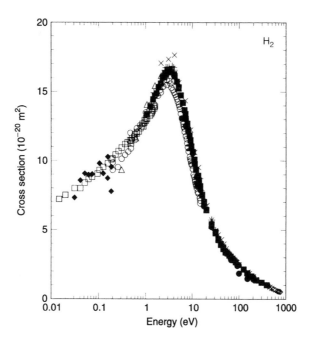

FIGURE 4.16 Absolute total scattering cross sections in H_2. (O) Golden et al.[72]; (□) Ferch et al.[77]; (△) Dalba et al.[78]; (♦) van Wingerden et al.[79]; (×) Hoffman et al.[81]; (●) Deuring et al.[82]; (■) Jones[84]; (▲) Nickel et al.[92]; (♦) Randell et al.[96]; (+) Szmytkowski et ai.[45]; (— —) Jain and Baluja.[28]

The energy range of electrons was extended to 100 eV by Dalba et al.[78] who used a linear transmission method. The stated accuracy of the measurements is 3.5% ($\varepsilon < 1$ eV) and 2.0% ($\varepsilon > 2$ eV). In the absolute total cross sections, structure was seen in the energy range of 11 to 15 eV and was attributed to H_2^- resonance; the parent state is $c^3\Pi_u$. The spectroscopic energy for this state is 11.87 eV, $v = 0$,[104] and the resonant state is $(1s\sigma_g)(2p\pi_u)^2$ $^2\Sigma_u^+$. Fluctuations in the total cross section are small and not seen unless plotted on a large scale. This structure was not observed in previous studies.

The results of Hoffman et al.[81] are in good agreement with those of van Wingerden et al.[79] and Dalba et al.[78] over the overlapping energy ranges. They are, however, consistently higher than the measurements of Golden et al.[72] by more than 10%. The measurements of Deuring et al.[82] which have a stated accuracy of 5%, agree reasonably well with other measurements. Jones[84] adopted the time-of-flight method and covered the energy range of 1 to 50 eV, with a stated accuracy of 2.8%. The measurements carried out using the linear transmission method are extended by Nickel et. al.[92] to H_2 and N_2 in the energy range of 4 to 300 eV, with a stated accuracy of 2 to 3%. Agreement with the previous measurements over the overlapping energies is very good, usually within 5%, though the results of Deuring et al.[82] are about 2 to 9% lower.

The measurements of Randell et al.[96] using the beam techniques, cover the lower energy range (10 to 175 meV); a notable feature is that the earlier determination of Q_T in this range was possible only by the swarm method. The energy resolution of Randell's[96] technique was 7.5 meV, low enough to identify rotational excitation loss of the H_2 molecule; the threshold energy for this process is 45 meV. Since the energy of the gas molecule at 300 K is 26 meV back scattering is the dominant mode. The cross section for back scattering (Q_B) at zero energy is related to Q_M according to $Q_B = \frac{1}{2} Q_M$. At other energies the derived phase shifts are used to calculate the momentum transfer cross sections and hence Q_T.

The total cross sections calculated by Randell et al.[96] are about 10% higher than the only other measurements in the low energy range, those of Ferch et al.[77] by the TOF technique. Combining these measurements, as noted by Zecca et al.[61] a set of reliable cross sections is available for Q_T.

4.2.2 ELASTIC SCATTERING CROSS SECTIONS IN H_2

Elastic scattering cross sections have been measured by: Srivastava et al.[76] Shyn and Sharp.[80] 2 to 200 eV; Furst et al.[83] 1 to 19 eV; Brunger et al.[90] 1.5 eV; and Brunger et al.[105] 1 to 5 eV. Below the first electronic excitation the total cross sections are the same as the elastic scattering cross sections, if we neglect the low energy inelastic scatterings. The latter are at least an order of magnitude lower; as an example, the total elastic cross section at 3 eV impact energy is $\sim 16 \times 10^{-20}$ m^2, compared with the total inelastic cross section of $\sim 7 \times 10^{-21}$ m^2 at the same energy. We include the total cross sections of Subramanian and Kumar[89] below the first excitation energy. These cross sections are shown in Figure 4.17 and there is good agreement between them. Differential elastic scattering cross sections have also been measured by Ehrhardt et al.[73] but integrated values are not available in either tabulated or graphical form.

Brunger et al.[90] conclude that there is good agreement between the various investigations, though Zecca et al.,[61] on the basis of essentially the same data, conclude that the agreement is less than satisfactory. Improving upon the semiempirical formula of Liu,[53] an empirical formula for total elastic scattering cross section is suggested as

$$Q_{el} = -0.252\varepsilon^2 + 2.125^*\varepsilon + 11.338 \text{ m}^2; \quad 0.2 \leq \varepsilon \leq 10 \tag{4.20}$$

$$Q_{el} = \frac{\pi \times 6.798 \times 10^{-20}}{\varepsilon}\left(4.2106 - \frac{27.19}{\varepsilon}\right)^2 \text{ m}^2; \quad 30 \leq \varepsilon < 1000 \tag{4.21}$$

where ε is the electron energy in eV. Figure 4.17 shows the cross sections. Note the linear scale of the ordinate for Figure 4.16 which explains the apparent change of shape of the Q_T curve.

4.2.3 MOMENTUM TRANSFER CROSS SECTIONS IN H_2

Momentum transfer cross sections are measured by: Srivastava et al.,[76] 3 to 75 eV; Shyn and Sharp,[80] 2 to 200 eV; and Brunger et al.[105] and 1 to 5 eV. Swarm-derived cross sections are reported by Engelhardt and Phelps[69] and England et al.[88] Theoretical calculations are due to Nesbet et al.[106] 0.14 to 14 eV. Figure 4.18 shows the cross sections. Reasonably good agreement exists between various investigations in the range 0.001 to 20 eV, except for the cross sections of Srivastava et al.[76] which differ by as much as 25%. Part of this discrepancy is possibly due to the fact that the stated uncertainty is ±18%.

It is significant that all three approaches, namely the beam methods, swarm technique, and theory, give cross sections that agree within acceptable uncertainty limits. Above 20 eV, the measurements of Shyn and Sharp[79] up to 200 eV have not been repeated, to the knowledge of the author of the present volume. The only other values available in the higher energy range are reported by Pitchford et al.[107] using their BOLSIG software. These two sets of data differ considerably (Figure 4.18) and the recommended cross sections are shown in the figure and in Table 4.8.

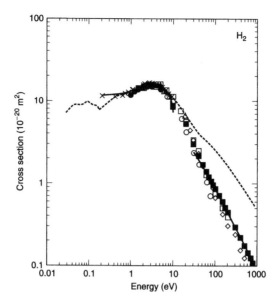

FIGURE 4.17 Elastic scattering cross sections in H_2. (O) Srivastava et al.[76a]; (□) Shyn and Sharp,[80]; (△) Furst et al.[83]; (◇) Nishimura et al.[85]; (×) Subramanian and Kumar[89]; (●) Brunger et al.[105]; (■) Jain et al.[28]; (——) formulas 4.20 and 4.21. Total cross sections (— —) have also been plotted for the sake of comparison: 0.03 to 0.18 eV, Randell et al.[96]; 0.4 to 250 eV, Szmytkowski et al.[45]; 250 to 750 eV, van Wingerden et al.[79]; 750 to 1000 eV, Jain and Baluja,[28] with cross sections multiplied by 1.45 for smooth joining. Note the linear scale of the ordinate in Figure 4.16, which explains the apparent change of shape of the Q_T curve.

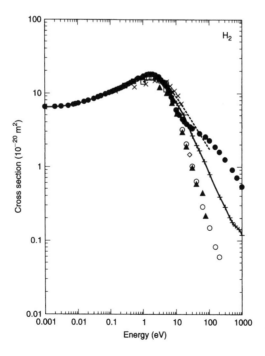

FIGURE 4.18 Momentum transfer cross sections in H_2. Swarm method: (— —) Engelhardt et al.[69]; (◇) England et al.[88] Experimental: (△) Srivastava et al.[76a]; (O) Shyn and Sharp[80]; (□) Brunger et al.[105] Theoretical: (×) Nesbet et al.[106] Compilation: (●) Morgan et al.,[107] Bolsig; (—+—) recommended.

TABLE 4.8
Recommended Momentum Transfer Cross Sections of Hydrogen

Energy	Q_M	Energy	Q_M	Energy	Q_M	Energy	Q_M
0.00	6.35	0.15	11.40	2.2	17.70	20	3.80
0.001	6.40	0.17	11.60	2.5	17.20	30	2.70
0.002	6.50	0.20	12.05	2.8	16.90	40	2.10
0.003	6.60	0.25	12.50	3	16.30	50	1.70
0.005	6.80	0.30	13.00	3.3	15.60	70	1.20
0.007	7.10	0.35	13.45	3.6	15.05	80	1.00
0.0085	7.20	0.4	13.90	4.0	14.75	100	0.80
0.01	7.30	0.5	14.75	4.5	13.90	200	0.40
0.015	7.65	0.7	16.30	5.0	13.10	300	0.29
0.02	8.05	1.0	17.40	6.0	11.50	400	0.21
0.03	8.50	1.2	17.80	7	8.90	500	0.18
0.04	8.96	1.3	18.05	8	7.85	600	0.16
0.05	9.28	1.5	18.25	10	6.80	800	0.15
0.07	9.85	1.7	18.15	12	6.20	1000	0.12
0.10	10.50	1.9	18.10	15	5.50		
0.12	10.85	2.1	17.90	17	5.20		

Energy in units of eV; cross sections in units of $10^{-20}\,\mathrm{m}^2$.

4.2.4 Ro-Vibrational Cross Sections in H_2

Rotational and vibrational excitations are inelastic scatterings that occur in molecules at low incident energies. Threshold energies for rotational excitation are lower than those for vibrational excitation till large quantum numbers J are attained. In this energy range, $v = 0$, $j = 0, 1, 2,\ldots$ and rotational excitation occurs by itself according to the selection rules $\Delta j = 0$, ± 2. At relatively higher energies vibrational excitation sets in according to the selection rules $\Delta v = 0, \pm 1$. At these energies both rotational and vibrational excitation may occur and the term "ro-vibrational" has been coined to denote the combined excitation. During vibrational excitation $v_{0\rightarrow 1}$ there may not be rotational excitation, that is, $\Delta j = 0$. Such a scattering is termed rotationally elastic. Similarly, if the vibrational quantum number $v = 0$ and remains the same even after a scattering, then the scattering is known as vibrationally elastic. In this section we consider all of these, that is, rotational excitation occurring singly, vibrational excitation occurring singly, and ro-vibrational excitation. Table 4.9 lists the threshold energies for some of the lowest rotational and vibrational states.[88] All of the states shown have a maximum cross section in the energy range ~3.5 to 4.0 eV.

The population of the molecules having a given initial rotational state depends upon the temperature of the gas. In normal hydrogen, at 293 K, the relative population of molecules having the rotational state $J = 1$ is 66.2% as opposed to 13.2% in the $j = 0$ state. At 77 K, the same populations are 75% and 24.8% respectively. The threshold for excitation $j = 1 \rightarrow j = 3$ is 73 meV (second row, Table 4.9). The latter excitation (that is, 73 meV threshold) therefore requires higher energy.[75] The energies for $v = 0 \rightarrow 2$, $v = 0 \rightarrow 3$, and $v = 0 \rightarrow 4$ are 1.03, 1.55, and 2.07 eV respectively. The threshold cross sections fall off rapidly from $v = 0 \rightarrow 1$ onwards so that we need to consider only this cross section.

Vibrational excitation cross sections in H_2 are studied by Engelhardt and Phelps[69] by the swarm method; Ehrhardt et al.,[73] crossed beams method; Crompton et al.,[75] swarm method; Burrow and Schulz,[74] trapped electron method; England et al.,[88] swarm method; Morrison et al.[87] theory; Brunger et al.,[90,105] crossed beams method. Figure 4.19 shows the cross sections for the states shown in Table 4.9.[88] The cross sections reach a maximum at

TABLE 4.9
Threshold Energy and Terminology for Low Energy Inelastic Scatterings

Transition	Energy	Q (maximum)	Terminology
$j = 0 \to 2$	0.046	1.758	Rotational
$j = 1 \to 3$	0.073	1.050	Rotational
$j = 2 \to 4$	0.101	0.802	Rotational
$j = 3 \to 5$	0.128	0.828	Rotational
$v = 0 \to 1,\ \Delta j = 0$	0.516	0.201	Vibrational, rotationally elastic
$v = 0 \to 1,\ \Delta j = 2$	0.558	0.267	Vibrational, rotationally elastic
$v = 1 \to 2$	1.03	7.42×10^{-3}	Vibrational
$v = 2 \to 3$	1.55	6.2×10^{-4}	Vibrational

Energy in units of eV; cross section in $10^{-20}\,\mathrm{m}^2$. The cross sections in the last two rows are at threshold and not maximum.

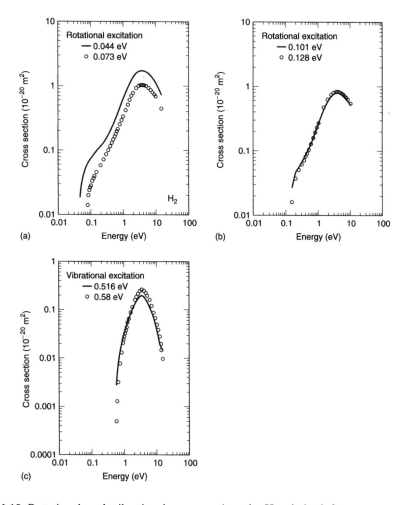

FIGURE 4.19 Rotational and vibrational cross sections in H_2, derived from swarm parameters. Rotational excitation cross sections with threshold energies: (a) (——) 0.044 eV, (○) 0.073 eV; (b) (——) 0.101 eV, (○) 0.126 eV. Vibrational excitation: (c) (——) 0.516 eV, (○) 0.56 eV.[88]

approximately 3.5 to 4.0 eV electron energy after a monotonic increase. The relatively large cross sections in (a) and (b) are explained on the basis of the quadrupole moment of the molecule.[108]

The loss of energy of an electron during elastic scattering, provided the energy ε is below the electronic excitation threshold, is $(2m/M)\varepsilon$ per scattering (Equation 1.67). While this holds good in the case of rare gases and other atomic gases, the loss of energy in molecular gases is considerably greater than the above fraction and the mechanisms of rotational and vibrational cross sections come into play. If the molecule is heteronuclear, the dipole moment of the molecule will have an electric field component along the axis joining the nuclei. The electron will then experience a long-range interaction potential, falling off as r^{-2}.[4] In other words, the cross section for scattering with such a molecule will be large. An example will make this clear. The maximum of Q_v is $\sim 5 \times 10^{-21}\,\mathrm{m}^2$ in H_2, $\sim 1.2 \times 10^{-20}\,\mathrm{m}^2$ in N_2, and $\sim 1.7 \times 10^{-21}\,\mathrm{m}^2$ in O_2. All these molecules are nonpolar. In contrast, CO, though isoelectronic with N_2, has a vibrational cross section as large as $\sim 1.0 \times 10^{-19}\,\mathrm{m}^2$,[109] as shown in Figure 4.6, due to the dipole moment.

However, in homonuclear molecules there is no permanent dipole moment and the energy loss at low electron energies, below that of the vibrational threshold, is attributed to the rotational excitation of the molecule. Diatomic molecules generally possess an electric quadrupole moment, so that their interaction with electrons also has a long-range tail, falling off as r^{-3}. Though the cross sections are smaller than those of polar molecules, the large cross section for rotational excitation cannot be neglected.

We now compare the ro-vibrational excitation cross sections obtained by the different methods. The cross sections in the low energy range 0 to 10 eV are shown in Figure 4.20.

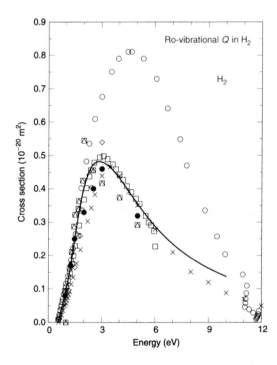

FIGURE 4.20 Vibrational cross sections ($Q_{v=0\to1}$) in H_2, determined by several methods. Swarm method: (○) Engelhardt and Phelps[69]; (◇) Crompton et al.[75]; (×) England et al.[88] Beam method: (△) Schulz and Asundi[103]; (□) Ehrhardt et al.[73]; (●) Brunger et al.[105] The cross sections of England et al. have been obtained by adding the rotationally elastic and inelastic $Q_{v=0\to1}$ cross sections. The curve is due to Morrison et al.[87]

The highlight of the several data is that there is reasonably good agreement between the beam experiments[73,103,105] and theoretical[87] vibrational cross sections. The swarm-derived cross sections above 2 eV[69,75] are considerably higher. Surprisingly, the vibrational cross sections deduced from drift velocity data in mixtures of H_2 and Ne at 294 K[88] show acceptable agreement (~10%) with theoretical values, and therefore with beam measurements, but this result does not remove the differences in Q_V obtained from swarm and beam methods in the parent H_2 gas.

Many publications have devoted attention to this disagreement (see, for example, references 75, 87, 88, 90, 93, and 101). White et al.[101] suggest that Monte Carlo analysis should be carried out to determine whether or not the discrepancy may be resolved. The author of the present volume also suggests that a bimodal electron energy distribution should be investigated, to prove or disprove that such distributions may account for the discrepancy. It is noted that bimodal distributions have been invoked to obtain better agreement between calculated and measured power in radio frequency discharges.[110] The basis for this suggestion is that the swarm at low values of E/N is likely to have a relatively larger fraction of back-scattered electrons than forward-scattered ones and it is not inconceivable that there will be two different energy distributions. A preliminary study carried out recently[227] shows that this suggestion is worthy of more detailed study.

In addition, White et al.[101] make the following suggestions: "Does the conventional kinetic theory ... suffer from certain basic flaws? Must the fermionic nature of the electrons and electron–electron interaction be incorporated into the transport analysis? Is it meaningful to compare quantities derived from transport analysis of swarm experiments with theoretical and beam measured cross section? Are the heretofore unaccounted for processes operative within the drift chamber?" Clearly this problem awaits a fresh approach for its solution.

4.2.5 ELECTRONIC EXCITATION CROSS SECTIONS IN H_2

The excitation cross sections of the H_2 molecule play an important role in the physics and chemistry of planetary atmospheres, interstellar space, plasma discharges, etc. Further excitation to certain levels leads to dissociation, producing atomic hydrogen which is reactive.[94] In view of the very large number of excited states (Herzberg[104] lists 38 states), the first task is to determine which of these make an appreciable contribution to the total cross section. Selected excited states of the H_2 molecule and the onset potentials are shown in Table 4.10. The cross sections for these states are shown in Figure 4.21.

TABLE 4.10
Excitation Levels of H_2 Molecule from the Ground State $X^1\Sigma_g^+$

State	Onset (eV)	Q_{ex} (max.) $(10^{-20}\,m^2)$	Remarks
$b\ ^3\Sigma_u^+$	6.9^{KS}	0.83	Unstable lower state of continuous spectrum. Leads to dissociation
$B\ ^1\Sigma_u^+$	11.183^W, 11.36^H	0.55	Optically allowed transition (Lyman bands)
$c\ ^3\Pi_u$	11.789^W, 11.87^H	0.20	Optically forbidden transition to triplet state
$a\ ^3\Sigma_g^+$	11.793^W, 11.89^H	0.25	Optically forbidden transition to triplet state
$C\ ^1\Pi_u$	12.295^W	4.25	Optically allowed transition (Werner bands)
$B'\ ^1\Sigma_u^+$	14.8^{MLM}	0.05	Optically allowed transition^Z

Superscripts: H, spectroscopic values from Herzberg[104]; KS, Khakoo and Segura[94]; W, Wrkich et al.[102]; MLM, Mu-Tao et al.[111]; Z, Zecca et al.[61] See the notes appended to reference 111.

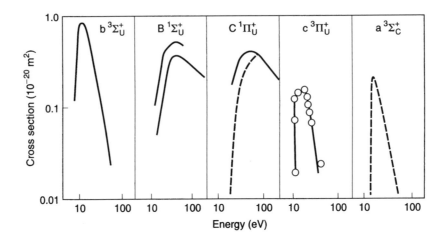

FIGURE 4.21 Electronic excitations in H_2.[61] b state: (•) Khakoo et al.[112] B state: (———) Fliplet and McKoy[115]; (— —) Shemansky et al.[116] C state: (— —) Shemansky et al.[116]; (———) Mu-Tao et al.[111] c state: (— —) Mu-Tao et al.[111] a state: (•) Khakoo and Trajmar.[86] Reprinted from Zecca, A. et al., *Rivista Del Nuov. Cim.* 10, 1, 1996. With permission of Societa Italiana di Fisica.

The b $^3\Sigma_u^+$ state has a threshold energy of 6.9 eV[94] and results in a continuum. Dissociation of the molecule occurs according to[111]

$$H_2 + e \rightarrow H_2^* \rightarrow H(1s) + H(2s) \tag{4.22}$$

The hydrogen atom in the 2s state is in a metastable state and its energy has a distribution that depends upon the impact energy of the electron. Since a negative ion does not result as the final product, the transition is known as *repulsive*, in contrast with the *bound* state.

The H_2^* excited state in the interaction (Equation 4.22) is obtained by the impact of the electrons with ground state molecules (X $^1\Sigma_g^+$) resulting in the b $^3\Sigma_u^+$ state (Figure 4.22). Higher triplets a $^3\Sigma_g^+$, c $^3\Pi_u$, and B' $^1\Sigma_u^+$ also lead to dissociation, though through different mechanisms. In the case of a $^3\Sigma_g^+$ and c $^3\Pi_u$, radiative transition to b $^3\Sigma_u^+$ occurs first,[94] followed by dissociation, whereas in the case of B' $^1\Sigma_u^+$ there is interaction 4.22 without an intermediate process.[61] The dissociative cross section of H_2 is shown in Figure 4.23. The sum of the cross sections shown in Figure 4.21 approximates the total excitation cross section. The references for these states are: Khakoo et al.[112]; Nishimura and Danjo[113]; Hall and Andrič[114]; Fliplet and McKoy,[115] theory; Shemansky et al.[116] optical measurements; Rescigno et al.[117] and Ajello and Shemansky,[118] optical measurements. Figure 4.22 shows selected excitation levels to help understand the transitions.[119] Figure 4.23 summarizes the excitation and dissociation cross sections in H_2, including the experimental data of Vroom and de Heer[120] and Corrigan.[121]

4.2.6 IONIZATION CROSS SECTIONS IN H_2

Electron impact ionization of the H_2 molecule occurs via the reactions[95]

$$H_2 + e \rightarrow H_2^+ + 2e \tag{4.23}$$

$$H_2 + e \rightarrow H + H^+ + 2e \tag{4.24}$$

$$H_2 + e \rightarrow H^+ + H^+ + 3e \tag{4.25}$$

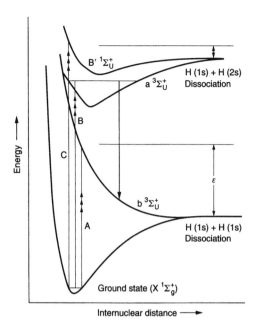

FIGURE 4.22 Potential energy curve of the H_2 molecule, illustrating the excitation and dissociation mechanism of the molecule according to electron–molecule interaction (Equation 4.22). The singlet states B and C are not shown. (A) is the direct excitation of the repulsive b $^3\Sigma_u{}^+$ state. (B) is the excitation to the a state which cascades into the b state, after which dissociation occurs. (C) is the higher optically allowed state (2) and dissociates directly.

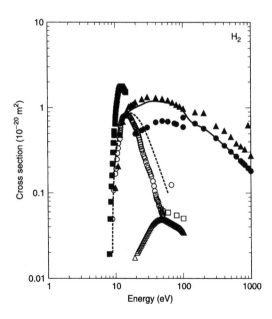

FIGURE 4.23 Excitation and dissociation cross sections in H_2 in units of 10^{-20} m². Experimental: (□) Vroom and de Heer[120]; (○) Khakoo et al.,[94] b $^3\Sigma_u{}^+$ continuum; (— —) Corrigan,[121] dissociation cross section; (▲) Phelps and Pitchford,[109b] sum of all states. Theory: (△) Mu-Tao et al.;[111] B′ $^1\Sigma_u{}^+$ state only; (◆) Trevisan and Tennyson.[100] (●) Excitation cross sections of van Wingerden et al.[79] for the Lyman and Werner bands are shown for comparison.

Impact ionization according to the process 4.23 is non-dissociative whereas processes 4.24 and 4.25 are dissociative. Rapp and Englander-Golden[70] have measured the total ionization cross sections and calculated the percentage of total cross sections due to the formation of H^+ ions (dissociative ionization) with initial kinetic energy greater than 0.25 eV.[71] These measurements were carried out up to an electron impact energy of 1000 eV, in the same ionization tube, with different potentials applied to the ion collector. If single ionization cross sections are desired, the cross sections for multiple ionization, interaction 4.25, should also be deducted, because the total cross section is defined as

$$Q_T = \sum n Q^{n+} \tag{4.26}$$

The dissociative ionization of Rapp et al.[71] was less accurate than the total ionization cross section because the ion collection efficiency was not known with greater certainty. The ion collection efficiency depends on (1) the retarding (positive) electric field to direct the ions to the collector, (2) the magnetic field employed to collimate the electron beam, and (3) the kinetic energy and angular distribution of the ions themselves. Van Zyl and Stephan[122] have recalculated the experimental data of Rapp et al.[71] and obtained H^+ cross sections that are considerably higher, with corresponding lowering of the H_2^+ cross sections.

Edwards et al.[123a] measured the absolute single (H_2^+) and double (H_2^{2+}) ionization cross sections in the high energy range 408 to 1906 eV. This was followed by absolute measurements of these cross sections by Kossmann et al.,[91] who used time-of-flight spectrometry. It is noted that the H_2^+ ($^2\Sigma_g^+$ state) ion has relatively low energy, very close to thermal energy, whereas H^+ ions have appreciable kinetic energy, 9.4 eV each.[91] This situation demands two separate techniques. Krishnakumar and Srivastava[95] have measured the cross sections for non-dissociative (H_2^+) and dissociative (H^+) ionizations using a pulsed electron beam and time-of-flight mass spectrometer. Jacobsen et al.[97] have measured the non-dissociative single ionization of molecular hydrogen by the time-of-flight method over a wide range of electron energy, 16.1 to 2101.2 eV.

Figure 4.24 shows the non-dissociative (H_2^+) and dissociative (H^+) ionization cross sections obtained from the studies mentioned above. The H^+ cross sections of Van Zyl and Stephan[122] and those of Krishnakumar and Srivastava[95] agree very well. These are considerably higher ($\sim$50%) than those obtained by Rapp and Englander-Golden,[70] for the reasons already mentioned. They are higher by a factor of 4 when compared with the data of Adamczyk et al.[123] The H_2^+ cross sections of Krishnakumar and Englander-Golden[70] are also higher than those of Kossman et al.[91]

Figure 4.24 also shows the total cross sections obtained by Rapp and Englander-Golden[70] and the theoretical computations of Rudd[123b]; they agree very well amongst themselves and, as expected, lie above the H_2^+ cross sections by $\sim$7 to 10%.

4.2.7 SIGMA RULE FOR H$_2$

The recommended cross sections to satisfy the sigma rule are tabulated in table 4.11. The more recent elastic scattering cross sections of Brunger et al.[105] over the range 0.5 to 5 eV are included. As noted by Zecca et al.[61] elastic scattering cross sections obtained in the same study over the whole range are not available. However, Jain and Baluja[28] have theoretically calculated the elastic and total cross sections, which agree very well with several experimental data (see Figures 4.16 and 4.17). The ro-vibrational cross sections are due to England et al.[88]

The neutral dissociation cross section of Khakoo and Segura[94] shows a peak of 0.83×10^{-20} m^2 at 15.2 eV and this value shows excellent agreement for calculating Q_Σ.

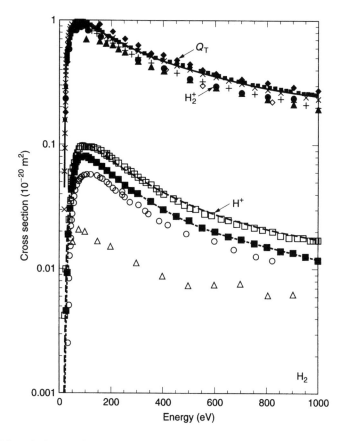

FIGURE 4.24 Dissociative and non-dissociative ionization cross sections in H_2. Dissociative: ($\triangle$) Adamczyk et al.[123]; ($\circ$) Rapp et al.[70]; ($\square$) Krishnakumar and Srivastava[95]; (— —) Van Zyl and Stephan[122]; (— ■ —) Straub et al.[98]; Non-dissociative: ($\diamond$) Edwards et al.[123a]; ($\diamond$) Kossmann et al.[91]; (■) Krishnakumar and Srivastava.[95] Total cross sections: ($\times$) Rapp and Englander-Golden[70]; (——) Rudd[123b]; ($\blacklozenge$) Krishnakumar and Srivastava[95]; (+) Straub et al.[98]; calculated by adding the two cross sections ($H_2^+ + H^+$).

Their measurements are for the b $^3\Sigma_u^+$ continuum; contributions from higher states increase their cross sections and the recommended cross sections in this volume are shown in Figure 4.23. The excitation cross sections are the sum of the Lyman and Werner bands as tabulated by van Wingerden et al.[79] The ionization cross sections are due to Krishnakumar and Srivastava[95] up to 60 eV electron energy, and for higher energies the more recent data of Straub et al.[98] (see also Figure 4.24) give very good agreement. As the last column of Table 4.11 shows, there is good agreement between the measured total cross sections and Q_Σ over the entire range except at 3 to 5 eV. The reason for this discrepancy has been discussed adequately in Section 4.2.1. It is noted that the sigma rule is verified on the basis of measured and theoretical calculations of individual cross sections, but is not based on derivation from swarm parameters which arguably may lead to a different conclusion.

4.3 MOLECULAR NITROGEN

As already stated, nitrogen is isoelectronic with CO and there are many similarities between the Q–ε characteristics of the two gases. From a theoretical point of view, molecular nitrogen has become almost a standard for the study of low energy scattering, particularly in the

TABLE 4.11
Verification of the Sigma Rule in H_2 and Recommended Cross Sections

Energy	Q_{el}	Q_{rot}	Q_v	Q_{dis}	Q_{ex}	Q_i	$Q\Sigma$	Q_T	% diff.
0.5	11.5[B]	0.61[E]	—	—	—	—	12.11	11.5[SM]	−5.30
1	11.68[B]	1.51[E]	0.057	—	—	—	13.25	13.5[SM]	1.87
2	14.56[B]	3.41[E]	0.260	—	—	—	18.23	17.1[H]	−6.61
3	15.52[B]	4.29[E]	0.441				20.25	17.4[H]	16.39
4	14.71[B]	4.44[E]	0.440	—	—	—	19.59	17.65[H]	10.99
5	13.9[B]	4.27[E]	0.350	—	—	—	18.52	15.6[H]	18.72
10	8.55[J]	2.93[E]	0.087	[0.24[C]]	0.12[PP]	—	11.69	10.8[SM]	−8.22
15	6.94[J]	—	0.025	[0.84[C]]	0.95[PP]	—	7.92	8.28[SM]	4.35
20	5.33[J]	—	—	[0.80[C]]	1.18[PP]	0.18[KS]	6.69	6.66[SM]	−0.45
30	3.00[J]	—	—	[0.45[C]]	1.35[PP]	0.46[KS]	4.81	4.88[SM]	1.43
40	2.179[J]	—	—	[0.25[C]]	1.33[PP]	0.81[KS]	4.32	4.17[SM]	−3.57
50	1.705[J]	—	—	[0.15[C]]	1.29[PP]	0.88[KS]	3.88	3.63[SM]	−6.75
60	1.397[J]	—	—	[0.09[C]]	1.23[PP]	1.02[KS]	3.65	3.36[SM]	−8.54
70	1.194[J]	—	—	[0.04[C]]	1.17[PP]	1.00[S]	3.36	3.15[SM]	−6.79
80	1.060[J]	—	—	[0.008[C]]	1.10[PP]	0.99[S]	3.15	2.93[SM]	−7.51
90	0.958[J]	—	—	[0.005[C]]	1.04[PP]	0.97[S]	2.97	2.80[SM]	−6.00
100	0.874[J]	—	—	[0.002[C]]	0.99[PP]	0.95[S]	2.81	2.61[SM]	−7.82
125	0.713[J]	—	—	—	0.73[W]	0.88[S]	2.32	2.22[SM]	−4.55
150	0.599[J]	—	—	—	0.66[W]	0.82[S]	2.08	2.04[SM]	−1.81
175	0.514[J]	—	—	—	0.62[W]	0.76[S]	1.89	1.91[SM]	1.05
200	0.450[J]	—	—	—	0.58[W]	0.70[S]	1.73	1.71[SM]	−0.98
300	0.296[J]	—	—	—	0.45[W]	0.55[S]	1.29	1.27[H]	−1.89
400	0.220[J]	—	—	—	0.36[W]	0.45[S]	1.03	1.04[H]	0.57
500	0.174[J]	—	—	—	0.31[W]	0.38[S]	0.86	0.87[H]	0.60
600	0.144[J]	—	—	—	0.27[W]	0.32[S]	0.73	0.736[L]	0.43
700	0.123[J]	—	—	—	0.24[W]	0.29[S]	0.65	0.648[L]	−0.90
800	0.107[J]	—	—	—	0.22[W]	0.26[S]	0.59	0.580[L]	−0.93
1000	0.086[J]	—	—	—	0.18[W]	0.22[S]	0.40	0.481[L]	14.26

Q_{el} = elastic; Q_{rot} = rotational; Q_v = vibrational; Q_{dis} = dissociative; Q_{ex} = excitation; Q_i = ionization; $Q\Sigma$ = sum of all cross sections; Q_T = total (measured). Energy in units of eV, Q in units of $10^{-20}\,m^2$. Superscripts: B, Brunger et al.[105]; C, Corrigan[121] [], not used in summation; E, England et al.[88]; H, Hoffman et al.[81]; J, Jain and Baluja[28]; KS, Krishnakumar and Srivastava[95]; L, Liu[53]; PP, Phelps and Pitchford[106]; R, recommended; S, Straub et al.[98]; SM, Szmytkowski et al.[45]; W, van Wingerden et al.[79] Negative signs in the last column denote that the measured total cross sections are lower than $Q\Sigma$.

region of shape resonance. The nitrogen molecule is reasonably aspherical and has an electronic structure that is computationally manageable.[146] The ground electronic state of the molecule which has 14 electrons, is $(1\sigma_g)^2 (1\sigma_u)^2 (2\sigma_g)^2 (2\sigma_u)^2 (1\pi_g)^4 (2\sigma_g)^2$. Extensive data on cross sections are available; Table 4.12 shows selected references.

4.3.1 TOTAL CROSS SECTIONS IN N_2

The most striking feature of the total cross section is that resonances in the ε–Q_T curve are displayed in the 1.5 to 5 eV region, as shown in Figure 4.25. These resonances were first discovered by Haas[125] and were studied in detail by Schulz,[127] as the peaks also appear in the inelastic channel (see Figure 2.31). Figure 4.25 shows the data of Kennerly,[138] who used the time-of-flight technique with an energy resolution as low as 2 meV, producing what is considered to be the best data available in this energy range. The vibrational excitation cross sections are quite large in the low energy region. As the energy for exciting the vibrational

TABLE 4.12
Selected References for Cross Section Data in Molecular Nitrogen

Type	Energy range	Method	Reference
Q_m	10^{-3}–20	Swarm analysis	Englehardt et al.[126]
Q_v	1.0–4.0	Trapped electron method	Schulz[127]
Q_i	Onset–1000	Ionization tube method	Rapp et al.[70,71]
Q_T	0.2–5.0	Modified Ramsauer method	Golden[128]
Q_T	0.3–1.6	TOF spectrometer	Baldwin[129]
Q_{ex}	0–2000	Theory	Chung and Lin[130]
Q_{diff}, Q_{el}	5–75	Crossed beams	Srivastava et al.[131]
Q_T	15–750	Linear transmission method	Blaauw et al.[132]
Q_{ex}	10–50	Energy loss method	Cartwright et al.[133]
Q_T	0.5–4.0	Transmission method	Mathur and Hasted[134]
Q_{diff}, Q_m	5–90	Crossed beams technique	Shyn et al.[135]
Q_T	5–750	Linear transmission method	Blaauw et al.[136]
Q_T	100–1600	Ramsauer method	Dalba et al.[137]
Q_T	0.50–50	TOF method	Kennerly[138]
Q_T	2–500	Transmission method	Hoffman et al.[81]
Q_M, Q_T	0.003–10^4	Swarm analysis	Phelps and Pitchford[109b]
All	0–1000	Compilation	Itikawa[139]
Q_T	100–2000	Theory	Liu[53]
Q_T	600–5000	Linear transmission method	Garcia et al.[140]
Q_v	2.1, 2.4, 3.0	Crossed beams	Brunger et al.[141]
Q_{ex}	15–50	Energy loss method	Brunger and Teubner[142]
Q_i	0–200	Fast neutral beam	Freund et al.[143]
Q_i	Onset–1000	TOF spectrometer	Krishnakumar and Srivastava[144]
Q_m	10^{-4}–1	Electron mobility	Ramanan and Freeman[145]
Q_{el}, Q_{diff}	1.5–5.0	Crossed beams	Brennan et al.[146]
Q_T	10–5000	Theory	Jain and Baluja[28]
Q_T	4–300	Linear transmission method	Nickel et al.[147]
Q_{dis}	18.5–148.5	Crossed beams	Cosby[148]
Q_T	80–4000	Modified Ramsauer method	Karwasz et al.[149]
Q_M, Q_T	0.01–0.175	Beam attenuation; axial H	Randell et al.[96]
Q_i	Onset–1000	Theory	van Zyl et al.[122]
Q_i	Onset–1000	Time-of-flight method	Straub et al.[98]
Q_T	0.4–250	Beam attenuation method	Szmytkowski et al.[45]
All	0.001–1000	Compilatiom	Zecca et al.[61]
Q_{ex}	6–30	TOF measurements	LeClair and Trajmar[150]
Q_{ex}	15–50	Crossed beam DCS	Campbell et al.[151]
Q_i	Onset–1000	Compilation	Stebbings and Lindsay[99]
Q_M, Q_T	0–100	Review	Brunger and Buckman[7]

Subscripts and abbreviations have the same meaning as Table 4.7. Energy in units of eV.

levels from $v = 1$ to $v = 8$ increases, the resonance peak decreases and in the elastic channel only five peaks are discernible.

Golden,[128] Mathur and Hasted,[134] and Brennan et al.[146] also observed these peaks, though, for the sake of clarity, their data are not shown. The resonances are due to vibrational excitation resulting from the formation of short-lived N_2^- negative ion ($^2\Pi_g$ state). The lifetime of the state is about the same as the vibration period, $\sim 10^{-14}$ s, and the peaks result from interference between the outgoing and reflected wave functions.[138]

A comparison with respect to the energy of the peaks and the cross sections at the peak is provided in Table 4.13. The agreement between cross sections is better than the agreement with the energy where the peak is observed. It should be mentioned that Golden[128]

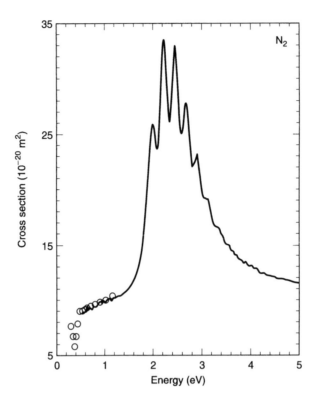

FIGURE 4.25 Total cross sections in N_2, showing $^2\Pi_g$ resonances, replotted from tabulated data. (○) Baldwin[129]; (———) Kennerly.[138] For clarity, data of Golden,[128] Mathur and Hasted,[134] and Brennan et al.[146] are not plotted.

TABLE 4.13
Comparison of Observed Resonance Peaks and Energies in N_2

Energy (eV)				Peak Cross Section (10^{-20} m^2)		
Schulz[a,127]	Golden[128]	Mathur and Hasted[b,134]	Kennerly[138]	Golden[128]	Mathur and Hasted[134]	Kennerly[138]
1.73	1.79	1.92	1.980	24.5	27.9 (27.8)[c]	25.95
2.03	2.06	2.15	2.210	32.3	32.3 (32.1)	33.58
2.31	2.32	2.40	2.440	30.8	30.9 (30.8)	33.03
2.56	2.57	2.62	2.665	22.0	27.2 (27.1)	23.28
2.80	2.82	2.87	2.900	22.0	23.8 (23.7)	23.28
3.05	3.06	3.11	3.12	18.6	21.1 (21.0)	—

[a] These values are given by Golden.[128]
[b] These values are obtained by digitizing the original publication. Estimated accuracy ±1%.
[c] Numbers in parentheses show digitized values and therefore indicate the accuracy of the digitizing process. Average spacing between levels = 0.26 eV.

also reported some additional peaks very close to 1 eV, though these have not been confirmed in other studies.

Szmytkowski et al.[45] who measured total cross sections in the energy range 0.4 to 250 eV, also observed the $^2\Pi_g$ resonance peaks. However, the energy interval adopted by them was larger, 0.5 eV, and is therefore not shown in Table 4.13.

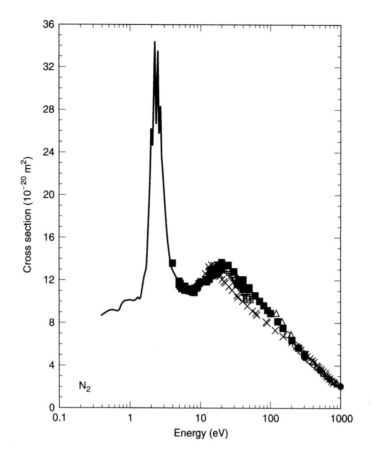

FIGURE 4.26 Total scattering cross sections in N_2. (O) Blaauw et al.[132]; (×) Blaauw et al.[136]; (△) Dalba et al.[137]; (□) Kennerly[138]; (+) Garcia et al.[140]; (■) Nickel et al.[147]; (●) Karwasz et al.[149]; (——) Szmytkowski et al.[45] It is noted that Szmytkowski et al.,[45] who measured total cross sections in the energy range 0.4 to 250 eV, have also observed the $^2\Pi_g$ resonance peaks. However, the energy interval adopted by them is larger, 0.5 eV, and is therefore not shown in Table 4.13.

We now consider the total cross sections at energies greater than 5.0 eV. Figure 4.26 shows these data and one observes that there is less agreement than in other gases. However, Nickel et al.[147] and Szmytkowski et al.[45] obtain cross sections that show excellent agreement over the entire range. There is a broad peak at ~22 eV which is attributed to increased vibrational excitations[152] and possibly to several excitation processes that occur in this energy vicinity. Some theoretical justification has also been provided.[153]

4.3.2 Momentum Transfer Cross Sections in N_2

There has been a considerable number of determinations of momentum transfer cross section and a selected list is given here to cover the range of 0 to 1000 eV: Erhardt and Willman,[154] 0.003 to 30 eV; Baldwin,[129] 0.30 to 1.56 eV; Ramanan and Freeman,[145] 10^{-4} to 1 eV; Srivastava et al.,[131] 5 to 75 eV; Shyn and Carignan,[155] 1.5 to 400 eV; Herrman et al.,[158] 90 to 1000 eV; and Phelps and Pitchford[109] 10 to 10^4 eV. Above 30 eV the more accurate cross sections given by Phelps and Pitchford[109] are considerably lower than those measured by Herrmann et al.[158] (not shown) in Fig. 4.27, the difference being as large as 100% at 400 eV. However, they agree very well with those of Shyn and Corignan[155] in the range from 90 to

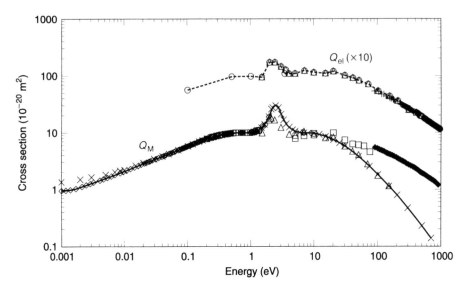

FIGURE 4.27 Momentum transfer cross sections in molecular nitrogen. Bottom curve: (○) Baldwin[129]; (□) Srivastava et al.[131]; (△) Shyn and Carignan[155]; (×) Phelps and Pitchford[109b]; (◇) Ramanan and Freeman[145]; (——) recommended cross sections. Top curve: (+) Herrmann et al.[158]; (△) Shyn and Carignan[155]; (—○—) recommended elastic scattering cross sections (see Table 4.19); ordinates have been multiplied by a factor of 10 for the sake of clarity.

1000 eV. Figure 4.27 shows the cross sections; recommended values are given in Table 4.14. The momentum transfer cross section shows a structure at 2.5 eV (compare with Figure 4.25). The scattering length A is obtained as 6.2×10^{-11} m,[145] and other reported values are 1.12×10^{-11} m and 0.60×10^{-11} m.[61] The zero energy momentum transfer cross section is given by $4\pi A^2$ and has the following values: Phelps and Pitchford,[109] 1.1; Morrison et al.[156] 1.4; Ramanan and Freeman,[145] 4.97; Randell et al.[96] 3.98—all in units of 10^{-20} m².

4.3.3 Elastic Scattering Cross Sections in N_2

As in the case of momentum transfer cross sections, in Figure 4.27 we show selected elastic scattering cross sections to cover the range of 0 to 1000 eV. These are: Shyn et al.,[135] 5 to 90 eV; Wedde and Strand,[157] 40 to 1000 eV; Srivastava et al.,[131] 5 to 75 eV; Herrmann et al.,[158] 90 to 1000 eV; DuBois and Rudd,[159] 20 to 800 eV; Shyn and Carignan,[155] 1.5 to 400 eV; and Jain and Baluja,[28] 10 to 5000 eV.

The ratio Q_{el}/Q_m is expected to remain the same if the scattering is isotropic, but Figure 4.27 shows that it is far from being isotropic in N_2 for energies greater than ~10 eV. Differential cross section measurements of Shyn and Carignan[155] show evidence of this aspect. The effects of anisotropic scattering on the transport properties of the electron swarm have been studied by Phelps and Pitchford,[109] and they find that the effects of anisotropy are more discernible in the transport properties at higher E/N, which is qualitatively consistent with Figure 4.27.

4.3.4 Ro-Vibrational Excitation in N_2

Rotational excitation by electron scattering occurs by the interaction of the electron with the weak electric field resulting from the quadrupole moment of the N_2 molecule, $1.04ea_0^2 = 4.67 \times 10^{-40}$ C m². The rotational quantum number changes in steps, $\Delta J = \pm 2, \pm 4$, etc.

TABLE 4.14
Momentum Transfer Cross Sections in N_2

Energy (eV)	Q_M (10^{-20} m^2)	Energy (eV)	Q_M (10^{-20} m^2)
0.0006	0.90	1.5	11.87
0.0008	0.90	1.7	13.47
0.0009	0.91	1.9	16.41
0.0010	0.91	2.1	16.85
0.0012	0.96	2.2	18.02
0.0014	0.97	2.5	17.92
0.0017	1.02	2.8	21.00
0.002	1.05	3.0	17.20
0.0025	1.21	3.30	15.30
0.003	1.23	3.60	13.96
0.004	1.37	4.0	12.42
0.0041	1.44	4.5	11.19
0.005	1.50	5.0	10.86
0.006	1.80	6.0	10.36
0.007	1.70	7.0	10.00
0.008	1.82	8.0	10.20
0.01	2.0	10.0	9.9
0.02	2.08	12.0	9.5
0.03	3.48	15.0	8.7
0.04	3.82	17.0	8.26
0.05	4.23	20	7.6
0.06	4.76	25	6.70
0.08	5.25	30	5.9
0.100	5.93	50	3.8
0.2	7.82	75	2.56
0.3	9.04	100	1.80
0.4	9.52	150	1.13
0.50	9.84	200	0.8
0.60	9.93	300	0.48
0.70	10.07	500	0.23
1.0	9.96	700	0.14
1.2	10.34	1000	0.01
1.3	10.92		

Cross sections are based on Ramanan and Freeman[145] for the range $0 < \varepsilon \leq 1$ eV; Morgan et al.[107] for the range $1 \leq \varepsilon \leq 1000$ eV. It is important to realize that the total Q_M is higher than that for elastic scattering only, particularly at high energies.

The rotational excitation of the N_2 molecule occurs at all energies; however, a threshold energy may be calculated from the rotational constant of 2.477×10^{-4} eV; the energy spacing between $J = 0$ and $J = 1$ levels is obtained as 1.486 meV.[156]

Phelps and Pitchford[109] have discussed the mechanism of rotational and vibrational excitation in homonuclear (H_2, N_2, O_2) and heteronuclear (CO, CO_2) molecules. Even at liquid nitrogen temperatures there are ~20 rotational states which must be considered. This reduces the rate of energy exchange between electrons and molecules in N_2 compared with the rate in H_2.[109] Rotational excitation, because of its lower energy thresholds, can be studied by high frequency relaxation methods or by using an electron beam with energy discrimination in the meV range. The rotational excitation cross sections in the vicinity of 2 eV are very large, as measured by Jung et al.[160] Table 4.15 shows the magnitude of the total cross sections near 2 eV.

TABLE 4.15
Total Cross Sections Q_T (ΔJ, Δv) for Electron Impact by N_2 (J, $v = 0$) in Units of $10^{-20}\,m^2$

Scattering Energy (eV)	2.47	2.2	2.25	2.47	2.54
Vibrational transition	$v = 0 \to 0$	$v = 0 \to 1$			
$\Delta J = 0$	19.1	2.33	2.58	1.19	1.89
$\Delta J = +2$	3.97	0.77	0.85	0.39	0.62
$\Delta J = -2$	3.04	0.58	0.64	0.30	0.47
$\Delta J = +4$	2.82	0.77	0.85	0.39	0.62
$\Delta J = -4$	1.65	0.44	0.49	0.22	0.35
$\Sigma\ \Delta J = 0, \pm 2, \pm 4$	30.58	4.89	5.41	2.49	3.95

From Jung, K. et al., *J. Phys. B*, 15, 1982, 1982. With permission of the Institute of Physics, U.K.

In view of the very low so-called threshold for rotational excitation (1.486 meV), experimental determinations of pure rotational cross sections, are not available and one has to resort to theoretical calculations. Unlike vibrational excitation cross sections, which show sharp spikes for some molecules, the rotational excitation cross sections rise smoothly from the threshold. However, in the case of the N_2 molecule theoretical calculations show a sharp spike at an energy near 80 meV as well as at higher energies. To demonstrate this, Figure 4.28 has been drawn to combine the rotational cross sections of Engelhardt et al.[126] with the theoretical cross sections of Morrison et al.[156]

An exhaustive study of rotational excitation of diatomic molecules at intermediate energies (10 to 200 eV) has been published by Gote and Ehrhardt,[160] but unfortunately integral cross sections are not given. Vibrational excitation has already been discussed in Section 4.3.1. In addition, Allan[162] has measured excitation functions for vibrational levels up to $v = 17$ in the energy range 0.25 to 5.0 eV. Sohn et al.[163] have measured absolute differential scattering cross sections for elastic scattering and $v = 1$ excitation from 0.1 to 1.5 eV. More recent data are reported by Brennan et al.[146] whose results in the $^2\Pi_g$ resonance region are shown in Table 4.16.

4.3.5 ELECTRONIC EXCITATION CROSS SECTIONS IN N_2

Anyone who has seen the aurora at night cannot but marvel at the emission processes in atmospheric nitrogen. At altitudes lower than about 200 km molecular nitrogen is the major constituent and the major part of the radiated energy falls into the molecular nitrogen bands. This is only one aspect of the excitation processes occurring in N_2; technological interest lies in molecular nitrogen lasers and gas discharge afterglows. Six different electronic systems are known to be of laser interest[133] and integral excitation cross sections are also of great interest. The classical measurements of Cartwright et al.[133a,b] have served as an indispensable source of data in this respect.

Figure 4.29 shows the potential energy diagram of selected levels. The singlet states are designated with lower case letters a, b, c, . . . , and triplet states by capital letters, A, B, C, . . . , unlike in other gases, mainly for reasons of historic continuity. The (a' $^1\Sigma_u^-$, a^1 π_g, w $^1\Delta_u$, and a'' Σ_g^+) states are the lowest levels of the singlet states, and the lowest level of the triplet state is A $^3\Sigma_u^+$. The onset energy of selected levels is shown in Table 4.17. States with designations A $^3\Sigma_u^+$, a $^1\Pi_g$, C $^3\pi_u$ (spin forbidden), and E $^3\Sigma_g^+$ (spin and symmetry forbidden) are the metastables.

Trajmar et al.[164] renormalized the cross sections of Cartwright et al.[133a] Integral cross sections are given by Cartwright et al.[133b] Mason and Newell[165] have measured the cross

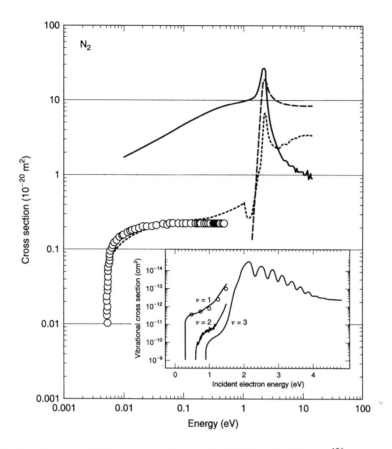

FIGURE 4.28 Rotational excitation cross sections in N_2. (○) Engelhardt et al.,[126] rotational excitation $J = 4 \to 6$; (- - - , short broken line) Morrison et al.,[156] $J = 0 \to 2$; (— —, long broken line) Morrison et al.,[156] $J = 0 \to 4$; (——— full line) Morrison et al.,[156] elastic scattering cross section included for comparison. Untabulated values are digitized from the original publication and replotted. The inset shows, for comparison, the vibrational cross sections measured by Allan.[162] Open circles in the inset are from Engelhardt et al.[126]

TABLE 4.16
Total Elastic, Total Vibrational, and Grand Total Cross Sections for N_2 Molecule, in Units of 10^{-20} m^2

Process	Energy (eV)			
	1.5	2.1	3.0	5.0
$0 \to 0$	12.7	18.4	18.7	11.3
$0 \to 1$	0.089	1.97	1.37	0.080
$0 \to 2$	—	1.32	0.90	—
$0 \to 3$	—	0.34	0.26	—
Grand total	12.8	22.0	21.2	11.4
Kennerly[138]	11.24	24.13	19.73	11.60

From Brennan, M.J. et al., *J. Phys.* B, 25, 2669, 1992. With permission of the Institute of Physics, U.K.

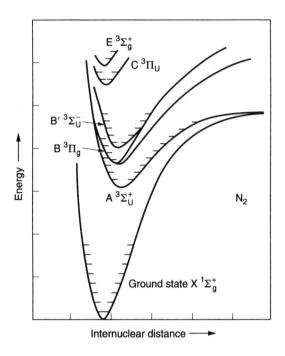

FIGURE 4.29 Schematic of triplet energy states of the N_2 molecule.

TABLE 4.17
Onset Energies of Selected States of N_2 Molecule

State	Onset (eV)	State	Onset (eV)	State	Onset (eV)	State	Onset (eV)
A $^3\Sigma_u^+$	6.17	a $^1\Pi_g$	8.55	b $^1\Pi_u$	12.58	F $^3\Pi_u$	12.98
B $^3\Pi_g$	7.35	w $^1\Delta_u$	8.89	D $^3\Sigma_u^+$	12.81	o $^1\Pi_u$	13.10
W $^3\Delta_u$	7.36	C $^3\Pi_u$	11.03	G $^3\Pi_u$	12.84	b' $^1\Sigma_u^+$	13.22
B' $^3\Sigma_u^-$	8.16	E $^3\Sigma_g^+$	11.88	c $^1\Pi_u$	12.91		
a' $^1\Sigma_u^-$	8.40	a'' $^1\Sigma_g^+$	12.25	c' $^1\Sigma_u^+$	12.94		

Reproduced from Leclair, L.R. and Trajmar, S.,[150] *J. Phys.* B, 29, 5543. With permission of Institute of Physics U.K.

sections for the metastable state a $1\Pi g$ (8.55 eV), which has a lifetime of $115 \pm 20\,\mu s$ in the energy range 9 to 141 eV. The excitation cross section rises rapidly near the threshold, as for many states, and reaches a maximum at 17 eV with $Q_{ex} = 3.5 \times 10^{-21}\,m2$. Brunger et al.[166] have measured the metastable excitation cross section in the threshold region for the state E $^3\Sigma g^+$, appropriately subtracting the contributions from the lower states A $^3\Sigma u^+$ and a $1\Pi g$. They find that near the onset potential (11.88 to 12.0 eV) negative ion resonances occur. The states E $^3\Sigma g^+$ and a'' $^1\Sigma g^+$ are the lowest Rydberg states.

Zubek and King[124] and Zubek[167] have measured the differential and integral cross sections for the C $^3\Pi_u$, E $^3\Sigma_g^+$, and a'' $^1\Sigma_g^+$ states at two values of energy, 17.5 eV and 20 eV. The C $^3\Pi_u$ state belongs to a group of excited states that have relatively higher cross section. It has a two-step decay, C $^3\Pi_u \to$ B $^3\Pi_g \to$ A $^3\Sigma_u^+$. The first step of this decay is a source of light emission in the visible region and the second step leads further into the metastable state A $^3\Sigma_u^+$.[168a] The A state is relatively long lived, with vibrationally dependent lifetimes ranging

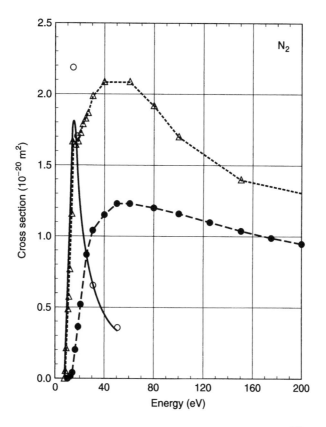

FIGURE 4.30 Total excitation cross sections in N_2. (———) Cartwright et al.[133]; (— △ —) Phelps and Pitchford[109b]; (○), Campbell et al.[151]; (— • —) Cosby[148]; dissociation cross sections.

from 16.0 μs in $v = 0$ to 7.6 μs in $v = 7$.[148] Poparić et al.[168a,b] have studied the excitation cross sections of the C $^3\Pi_u$ and E $^3\Sigma_g^+$ states.

This brief review shows that there are only three investigations that cover the first ten states: Cartwright et al.,[133b] Brunger and Teubner[142] and Campbell et al.[151] Figure 4.30 shows the cross sections of the first and last mentioned authors; Brunger and Teubner,[142] do not give integral excitation cross sections. A comparison with previous investigations for each of these states will not be presented as our focus is on the total excitation cross sections. The excitation cross sections used by Phelps and Pitchford[109] in the swarm calculations, with the correction for the C $^3\Pi_u$ state (multiplied by 0.67) as suggested in the website of Phelps and Pitchford[109] are also shown. The cross sections of Cartwright et al.[133b] and Campbell et al.[151] are in excellent agreement with each other.

Metastable states play an important role in low pressure discharges by their action on the cathode, which releases secondary electrons. The formative time of the discharge depends on the population of the secondary electrons released and therefore on the lifetimes of the metastable states. Table 4.18 lists selected transition lifetimes.

Transitions shown in the second column of Table 4.18 result in production of metastables and transitions shown in the third column are optical. The lifetimes of metastables are: A $^3\Sigma_u^+$ ($v = 0$), 1.9 s; a' $^1\Sigma_u^-$, 0.5 s; E $^3\Sigma_g^+$, 1.9×10^{-4} s; w $^1\Delta_u$ ($v = 0$–4) $\times$ (1 to 5) $\times 10^{-4}$ s; and a $^1\Pi_g$, 1.15×10^{-4} s. The transitions to the ground state a $^1\Pi_g \to X\ ^1\Sigma_g^+$ are known as Lyman–Birge–Hopfield (LBH) bands and extend from approximately 130 to 200 nm (VUV region), comprising emission from $v' = 0$ to 6 of the a $^1\Pi_g$ state.[169]

TABLE 4.18
Lifetimes, in Units of Seconds, of Electronic Triplet States of N_2

	Transition			
v'	$B\,^3\Pi_g \rightarrow A\,^3\Sigma_u^+ + W\,^3\Delta_u$	$C\,^3\Pi_u \rightarrow B\,^3\Pi_g$	$W\,^3\Delta_u \rightarrow B\,^3\Pi_g$	$B'\,^3\Sigma_u^- \rightarrow B\,^3\Pi_g$
0	13.4 (−6)^a	3.67 (−8)	31.6^b	4.60 (−5)
1	11.0 (−6)	3.65 (−8)	4.52 (−3)	3.61 (−5)
2	9.31 (−6)	3.69 (−8)	1.22 (−3)	3.01 (−5)
3	8.15 (−6)	3.77 (−8)	6.07 (−4)	2.59 (−5)
4	7.30 (−6)	3.94 (−8)	3.80 (−4)	2.30 (−5)
5	6.65 (−6)		2.68 (−4)	2.08 (−5)
6	6.16 (−6)		2.03 (−4)	1.90 (−5)
7	5.77 (−6)		1.62 (−4)	1.77 (−5)
8	5.46 (−6)		1.34 (−4)	1.65 (−5)
9	5.23 (−6)		1.15 (−4)	1.56 (−5)
10	5.05 (−6)		1.00 (−4)	1.48 (−5)
11	4.92 (−6)		8.88 (−5)	1.42 (−5)
12	4.82(−6)		7.96 (−5)	1.36 (−5)

Reproduced from Itikawa, Y. et al.,[139] *J. Phys. Chem. Ref. Data*, 15, 985, 1986.
^{a}a(−b) means a × 10^{-b}.
^{b}Decays also to ground state (X $^1\Sigma_g^+$); the lifetime is 4 s.

Turning our attention to dissociation cross sections, we find that electron impact dissociation occurs according to the reaction

$$N_2 + e \rightarrow N_2^* + e \rightarrow N + N + e + KE \tag{4.27}$$

where KE stands for kinetic energy of the atoms. This energy can be as high as 4 eV, though the distribution peaks at 1.0 eV with a smaller peak centered at ∼2 eV.[148] The first step of this reaction is the formation of the excited molecule, and the three possible states for the dissociated atoms are ^4S, ^2D, ^2P.[148] Cosby[148] has also measured the dissociation cross section using the crossed beams technique in the energy range 18.5 to 148 eV. The dissociation cross section is also shown in Figure 4.30.

4.3.6 Ionization Cross Sections in N_2

Ionization cross sections in the energy range considered consist of the following components:

1. Single ionization of the parent molecule (N_2^+).
2. Single ionization of the dissociated nitrogen atom (N^+).
3. Double ionization of the parent molecule (N_2^{2+}).
4. Double ionization of the dissociated nitrogen atom (N^{2+}).

The cross sections for N^+ and N_2^{2+} cannot be separated by a mass spectrometer for the reason that both species have almost identical charge to mass ratio. The total cross section is given by

$$Q_i(\text{total}) = Q_i(N_2^+) + Q_i(N^+ + N_2^{2+}) + Q_i(2N^{2+}) \tag{4.28}$$

Strictly speaking, one should add $2N_2^{2+}$ in place of N_2^{2+}, but the latter cannot be distinguished from N^+, as stated above. Moreover, the cross section for N_2^{2+} production is an order of magnitude lower than that for N^+ and therefore the error introduced is negligible.[144]

Total and dissociative ionization cross sections have been measured by Rapp and Englander-Golden[70] and Rapp et al.[71] respectively. The difference between the two cross sections gives the cross section for N_2^+. To be more accurate, the dissociation cross section should be increased by between 5 and 9% because Rapp et al.[70] considered only those ions that had a kinetic energy greater than 0.25 eV. This work was followed by that of Crowe and McConkey[170] who used a quadrupole mass spectrometer for measuring the dissociative cross sections N^+ and N^{2+}. They suggest a downward correction of 7.1% to the data of Rapp et al.[70] to account for McLeod gauge errors.

Märk[171] used a mass spectrometer to detect the ions N_2^+ and N_2^{2+} and obtain the ionization cross section for $(N_2^+ + 2N_2^{2+})$. For the purpose of comparison the relationship

$$Q_i(N_2^+) = Q_i(\text{total}) - Q_i(\text{diss}) \qquad (4.29)$$

was applied to the values of Rapp and Englander-Golden.[71] The agreement was not very good and Armentrout et al.[172] pointed out that the disagreement was due to the argon cross sections used by Märk[171] for normalization. Krishnakumar and Srivastava[144] have also used a quadrupole mass spectrometer to measure cross sections for ion production N_2^+, $N^+ + N_2^{2+}$, and N^{2+}. These authors give semiempirical formulas for ionization cross sections, which are useful for modeling purposes, as

$$Q_i(\varepsilon) = \frac{1}{\varepsilon_i \varepsilon} \left[A \ln\left(\frac{\varepsilon}{\varepsilon_i}\right) + \sum_{i=1}^{N} a_i \left(1 - \frac{\varepsilon_i}{\varepsilon}\right)^i \right] \qquad (4.30)$$

where ε_i is the ionization potential, ε the electron energy, A and a_i are coefficients obtained by fitting experimental data, and N is the number of terms.

Freund et al.[143] have extended their fast neutral beam method to measure the ionization cross sections of the parent molecule, N_2^+. Their results are lower than those of Rapp and Englander-Golden[70] by 5 to 20%. Van Zyl and Stephan,[122] in a re-evaluation of the data of Rapp and Englander-Golden[70] suggest a higher fraction for the ratio $Q_i(N^+)/Q_i(\text{total})$. Straub et al.[98] have used their time-of-flight mass spectrometer for measuring the ionization cross sections of the parent ion and dissociated atomic ions. These results are shown in Figure 4.31, with the recommended cross section. Table 4.19 shows the sigma rule check, taking into account some of the recent measurements and recommended cross sections. In the region where there is no structure, that is $\varepsilon > 10$ eV, the sigma check, is excellent, as is shown in the last column. Figure 4.32 shows the consolidated cross sections with the purpose of showing the relative contributions.

4.3.7 SIGMA RULE FOR N_2

The sigma rule is verified in Table 4.19 and consolidated cross sections are shown in Figure 4.32. The agreement is exceedingly good except in the range of 1 to 10 eV energy where the resonances dominate the total cross section (Figure 4.26). The ionization cross sections are reported by Straub et al.[98] and they are in good agreement with the results of Rapp and Englander-Golden.[70] As noted by Stebbings and Lindsay,[99] it is indeed remarkable that the cross sections of the two studies mentioned agree so well in several gases, though the measurements were made at an interval of some thirty years. The total cross sections of Szmytkowski et al.[45] in the energy range from 0.5 to 250 eV, combined with the total cross

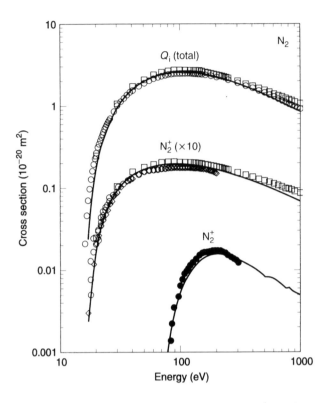

FIGURE 4.31 Ionization cross sections in N_2. The cross sections for N_2^+ are shown reduced for clarity purposes; the ordinates of this curve should be multiplied by ten to obtain the cross sections. The symbols for all the curves are: (○) Rapp and Englander-Golden[70]; (●) Crowe and McConkey[170]; (◇) Freund et al.[143]; (□) Krishnakumar and Srivastava[144]; (——) Straub et al.[98]

sections of Karwasz et al.[149] in the energy range from 300 to 1000 eV, give the best agreement, as the last column of Table 4.19 demonstrates.

A further discussion of the use cross sections for calculating the swarm parameters according to Boltzmann and Monte Carlo methods in the same gas, as published by Liu and Raju[173] is given in Chapter 6.

4.4 MOLECULAR OXYGEN (O_2)

Oxygen supports life and is the second most abundant gas in the Earth's atmosphere. Electron–molecule interaction is of fundamental interest in several areas of research. Reviews of scattering cross sections in the gas has been published by Trajmar et al.,[164] Itikawa et al.,[174] Kanik et al.[35] and Brunger and Buckman.[7] In view of this, only a brief discussion will be presented, supplementing data of previous years as required to cover the electron energy range 0 to 1000 eV. Table 4.20 provides selected references for cross section data.

4.4.1 TOTAL SCATTERING CROSS SECTIONS IN O_2

Figure 4.33 shows selected data of Q_T. The general features of the total cross sections in the 0 to 1000 eV range are as follows.

The lowest energy range covered is 25 to 200 meV. This is reported by Randell et al.[96] who have measured the backward scattering cross section (Q_B) of photo-ionized electrons.

TABLE 4.19
Sigma Check and Recommended Cross Sections in N_2

Energy (eV)	Q_{el} $(10^{-20}m^2)$	Q_{rot} $(10^{-20}m^2)$	Q_v $(10^{-20}m^2)$	Q_{ex} $(10^{-20}m^2)$	Q_{dis} $(10^{-20}m^2)$	Q_i $(10^{-20}m^2)$	Q_Σ $(10^{-20}m^2)$	Q_T (Expt) $(10^{-20}m^2)$	% Diff.
0.1	4.44^M	0.236^M	—	—		—	8.405	4.4^M	−6.27
0.5	8.1^S	0.305^M	—	—		—	8.405	9.13^{SZ}	7.94
1	8.1^S	0.41^M	0.009^Z	—		—	8.519	10.2^{SZ}	16.48
2	17.1^{SC}	10.54^M	3.5^Z	—		—	31.14	25.4^{SZ}	−22.60
3	14.8^{SC}	3.78^M	2.53^Z	—		—	21.11	20.3^{SZ}	−3.99
5	11.2^{SC}	4.1^M	0.06^Z	—		—	15.36	11.5^{SZ}	−33.57
10	11.7^{SC}	4.319^M	0.01^Z	—		—	16.029	12.5^{SZ}	−28.23
20	12.1^{SC}	0.001^M	0.02^Z	1.11^{C,T}	[0.52]^{CO}	0.218ST	13.449	13.8^{SZ}	2.54
30	10.0^{SC}			0.54^{C,T}	[1.04]^{CO}	1.033ST	11.573	13.1^{SZ}	11.66
40	9.4^{SC}			0.37^{C,T}	[1.15]^{CO}	1.648ST	11.42	12.4^{SZ}	7.90
50	8.5^{SC}			0.28^{C,T}	[1.23]^{CO}	2.040ST	10.820	11.2^{SZ}	−3.39
60	7.9^{SC}			0.27^R	[1.23]^{CO}	2.296ST	10.466	10.7^{SZ}	0.22
70	7.3^{SC}			0.25^R	[1.22]^{CO}	2.434ST	9.984	10^{SZ}	−0.16
80	6.73^{SC}			0.25^R	[1.20]^{CO}	2.542ST	9.525	9.45^{SZ}	−0.80
90	6.17^{SC}			0.21^R	[1.18]^{CO}	2.610ST	9.027	9.21^{SZ}	1.98
100	5.6^{SC}			0.19^R	[1.16]^{CO}	2.637ST	8.487	8.85^{SZ}	4.10
110	5.41^{SC}			0.18^R	[1.14]^{CO}	2.632ST	8.293	8.46^{SZ}	1.98
120	5.22^{SC}			0.16^R	[1.11]^{CO}	2.615ST	8.085	8.09^{SZ}	0.07
140	4.84^{SC}			0.14^R	[1.05]^{CO}	2.567ST	7.657	7.67^{SZ}	0.17
160	4.46^{SC}			0.10^R	[1.02]^{CO}	2.476ST	7.186	7.22^{SZ}	0.47
180	4.08^{SC}			—	[0.98]^{CO}	2.389ST	6.718	6.81^{SZ}	1.35
200	3.7^{SC}			—	[0.95]^{CO}	2.286ST	6.236	6.41^{SZ}	2.71
250	3.15^{SC}			—	—	2.057ST	5.457	5.67^{SZ}	3.76
300	2.6^{SC}			—	—	1.876ST	4.726	4.85^K	2.55
350	2.45^{SC}			—	—	1.738ST	4.438	4.53^K	2.03
400	2.3^{SC}			—	—	1.613ST	4.163	4.215^K	1.24
450	2.0^H			—	—	1.494ST	3.744	3.902^K	4.06
500	1.9^H			—	—	1.395ST	3.545	3.58^K	0.97
600	1.63^H			—	—	1.228ST	3.108	3.185^K	2.42
700	1.44^H			—	—	1.090ST	2.78	2.79^K	0.37
800	1.32^H			—	—	0.989ST	2.559	2.55^K	−0.35
900	1.21^H			—	—	0.922ST	2.382	2.31^K	−3.13
1000	1.1^H			—	—	0.856ST	2.210	2.08^K	−6.25

Q_{el} = elastic; Q_{rot} = rotational; Q_v = vibrational; Q_{dis} = dissociative; Q_{ex} = excitation; Q_i = ionization; Q_Σ = sum of all cross sections; Q_T = total (measured). Energy in units of eV, Q in units of 10^{-20} m^2. [] means not used in summation. Negative sign in the last column denotes that the measured total cross sections are lower than Q_Σ. Superscripts: C, T = Cartwright et al.[133] renormalized by Trajmar et al.[164]; CO = Cosby[148]; H = Herrmann et al.[158]; K = Karwasz et al.[149]; M = Morrison et al.[156]; R = recommended; S = Sohn et al[163]; SC = Shyn and Carignan[155]; ST = Straub et al.[98]; SZ = Szmytkowski et al.[45]; Z = Zecca et al.[61]

Noting that the total cross section is not given by the authors, unlike in H_2, one looks for Q_T in the compilations of Itikawa et al.[174] and Brunger and Buckman.[7] Both publications give the lower energy limit of 0.1 eV, and one has to look to theoretical analyses for lower energies, where the situation is less clear. The interest below 100 meV is due to the fact that the momentum transfer cross sections derived by Randell et al.[96] show a peak at 91.5 meV due to the doublet resonance, associated with electron attachment into the fourth vibrationally excited state of O_2^-. The thermal energy attachment to molecules will be commented upon in Chapter 7.

The total cross section increases from $\sim 4 \times 10^{-20}$ m^2 at 0.1 eV to a broad maximum of $\sim 11 \times 10^{-20}$ m^2 at ~ 11 eV; at higher energies Q_T decreases very slowly. There is no minimum

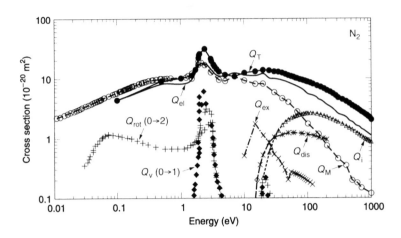

FIGURE 4.32 Cross sections in N_2 to satisfy sigma check (see Table 4.19). The rotational and excitation cross sections are reproduced from Itikawa et al.[174] The dissociation cross section is not included in the sigma check because it is composed of excitation and ionization components.

due to the Ramsauer–Townsend effect, though Fisk[175] in the early theoretical calculations of Q_T reported a shallow minimum at 0.3 eV. Zecca et al.[178] have measured the absolute total cross sections in the range 0.23 to 100 eV; at 0.23 eV, their cross section is $4.27 \times 10^{-20}\,m^2$, in very good agreement with the theoretical calculations. These authors also reported oscillatory structure attributed to vibrational excitation ($v' = 4$ to 13) of the ground state of O_2 via the $^2\Pi_g\ O_2^-$ state.

The initial rise of Q_T with energy is attributed to the rise in the elastic scattering cross section which also shows a peak around $\sim$11 eV.[189] The contribution of the inelastic processes in the energy region 1 to 12 eV is small, $\sim$1 to 7% of the total.[181]

The cross section maximum at 11 eV is partly due to the contribution of the negative ion resonance scattering state $^4\Sigma_u^-$ of O_2^-.[176] The slow decrease in the cross section beyond $\sim$12 eV is possibly due to the fact that there is onset of ionization; the ionization potentials of atomic and molecular oxygen are 13.61 and 12.07 eV respectively. There is good agreement between various measurements in the ranges of 0 to 1 eV and 80 to 1000 eV. In the intermediate range Zecca et al.[178] and Dalba et al.[177] get higher cross sections, whereas Subramanian and Kumar[180] get lower values. The cross sections recommended by Kanik et al.[35] fall between these results. The measurements of Kanik et al.[31] and the most recent data of Szmytkowski et al.[45] agree very well with their recommended values.

In the energy range where there is a large scatter between various measurements one looks for guidance in theoretical calculations. The number of theoretical studies that have been carried out in O_2 is comparatively small. One of the earliest (1936) is due to Fisk[175] who represented the diatomic molecules (H_2, N_2, O_2) by a simplified potential field and obtained the exact solution of Schroedinger's equation in spherical coordinates. Internuclear distances were taken from spectroscopic studies. Fisk's calculations are also shown in Figure 4.33. Agreement is very good in the energy range 3 to 9 eV.

4.4.2 MOMENTUM TRANSFER CROSS SECTIONS IN O_2

Momentum transfer cross sections are given by Hake and Phelps,[15] Shyn and Sharp,[184] Iga et al.,[185] Itikawa et al.,[174] Randell et al.[96] and Sullivan et al.[188] These and the recommended values are shown in Table 4.21 and Figure 4.34. The only experimental data in the energy range 300 to 1000 eV are due to Iga et al.[185] and the recommended cross sections in this range are in very good agreement with their data.

TABLE 4.20
Data on Scattering Cross Sections in Molecular Oxygen (O_2)

Type	Energy Range (eV)	Method	Reference
Grand total ionization cross sections			
Q_T	100–1600	Transmission method	Dalba et al.[177]
Q_T	0.2–100	Transmission method	Zecca et al.[178]
Q_T	5.2–500	Transmission method	Dababneh et al.[179]
Q_T	0.15–9.14	Photoelectron spectroscopy	Subramanian and Kumar[180]
Q_T	5–300	Linear transmission method	Kanik et al.[31]
Q_M, Q_T	0.01–0.175	Beam transmission method	Randell et al.[96]
Q_T	0.4–250	Linear transmission method	Szmytkowski et al.[45]
Momentum transfer and elastic cross sections			
Q_{el}, Q_{ex}	4–45	Electron impact spectrometer	Trajmar et al.[181]
Q_{el}	40–1000	Theory	Wedde and Strand[182]
Q_{ex}, Q_{el}	20–500	Electron impact spectrometer	Wakiya (1978 a,b)[183]
Q_{el}, Q_m	2–200	Crossed beams	Shyn and Sharp[184]
Q_{el}, Q_m	300–1000	Crossed beams	Iga et al.[185]
Q_{el}, Q_{ex}	0–15	Theory	Noble et al.[186]
Q_{el}	0–14	Theory	Higgins et al.[187]
Q_{el}, Q_m	1–30	Crossed beams	Sullivan et al.[188]
Q_{diff}, Q_{el}	5–20	Crossed beams	Wöste et al.[189]
Ro-vibrational excitation cross sections			
Q_{rot}, Q_v	5–15	Crossed beams	Shyn and Sweeney[190]
Q_v	4–15	Crossed beams	Noble et al.[191]
Q_v	5–20	Electron monochromator	Brunger et al.[192]
Excitation and dissociation cross sections			
Q_{el}, Q_{ex}	4–45	Electron impact spectrometer	Trajmar et al.[131]
Q_{ex}, Q_{el}	20–500	Electron impact spectrometer	Wakiya[183]
Q_{dis}	13.5–198.5	Crossed beams	Cosby[193]
Q_{ex}	5–20	Crossed beams	Middleton et al.[194]
Q_v	0.9–18	Electrostatic spectrometer	Allan[195]
Q_{ex}	9–20	Energy loss method	Green et al.[196]
Ionization cross sections			
Q_i	12.5–1000	Ionization tube	Rapp and Englander-Golden[70] Schram et al.[197]
Q_i	Onset–170	Mass spectrometer	Märk[198]
Q_i	Onset–1000	Mass spectrometer	Krishnakumar et al.[199]
Q_i	Onset–1000	Theory	Van Zyl and Stephan[122]
Q_i	Onset–1000	Theory	Hwang et al.[200]
Q_i	Onset–1000	Time-of-flight spectrometer	Straub et al.[98]
Q_i	Onset–1000	Time-of-flight spectrometer	Stebbings and Lindsay[99]
Q_i	20–2000	Theory	Joshipura et al.[201]
Review and compilation			
All	1–1000	Compilation	Itikawa et al.[174]
All	1–1000	Compilation	Kanik et al.[35]
All	0–100	Compilation	Brunger and Buckman[7]

4.4.3 Elastic Scattering Cross Sections in O_2

Elastic scattering cross sections are given by Trajmar et al.,[181] Shyn and Sharp,[184] Wakiya,[183] Iga et al.,[185] Itikawa et al.,[174] Kanik et al.,[35] Sullivan et al.,[188] Zecca et al.[61] and Brunger and

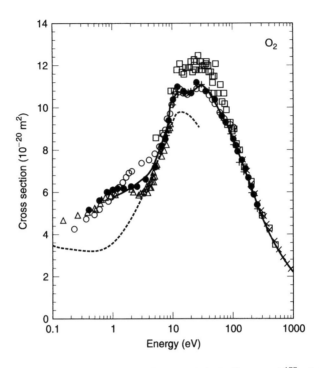

FIGURE 4.33 Total scattering cross sections in O_2. ($\times$) Dalba et al.[177]; ($\circ$) Zecca et al.[178]; ($\square$) Dababneh et al.[179]; ($\triangle$) Subramanian and Kumar[180]; ($+$) Kanik et al.[31]; (— —) Kanik et al.,[35] recommended; ($\bullet$) Szmytkowski et al.[45]; (— —) theoretical values, Fisk.[175]

TABLE 4.21
Recommended Momentum Transfer Cross Sections in Molecular Oxygen

Energy (eV)	Q_M $(10^{-20}\ m^2)$	Energy (eV)	Q_M $(10^{-20}\ m^2)$	Energy (eV)	Q_M $(10^{-20}\ m^2)$
0.1	2.5	10	7.65	100	1.55
0.25	3.9	15	6.9	150	1.22
0.4	5	20	6.1	200	0.92
0.6	5.65	25	5.3	300	0.55
0.8	6	30	4.8	400	0.4
1	6.3	40	4.02	500	0.35
2	6.5	50	3.5	600	0.25
3	6.1	60	2.85	700	0.15
4	6	70	2.4	800	0.08
5	6.28	80	2.1	900	0.076
7	6.95	90	1.75	1000	0.070

Buckman.[7] A detailed discussion has also been given by the latter authors, and we supplement this information to extend the electron energy range from 100 to 1000 eV. The only measurements in the energy range 500 to 1000 eV are due to Iga et al.[185] whose values are in very good agreement with the recommended value of Kanik et al.[35] On the theoretical side, the computations of Jain and Baluja[28] cover a wide energy range whereas the data of Higgins et al.[187] are the more recent. Figure 4.35 shows selected results and the recommended cross sections.

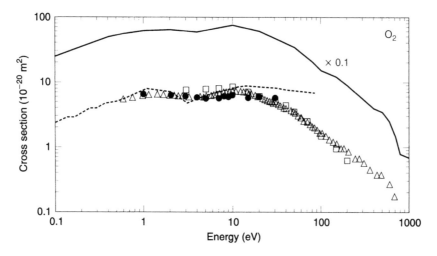

FIGURE 4.34 Momentum transfer cross sections in O_2. (– –) Hake and Phelps[15]; (□) Shyn and Sharp[184]; (△) Itikawa et al.[174]; (●) Sullivan et al.[188]; (——) recommended values; multiplied by 10 for easy readability.

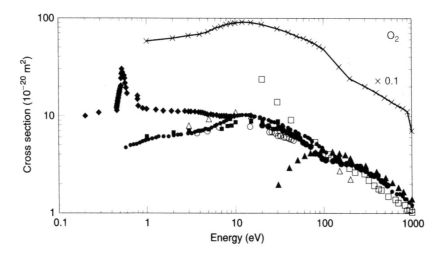

FIGURE 4.35 Elastic scattering cross sections in O_2. (○) Trajmar et al.[181]; (□) Wedde and Strand[182]; (●, big) Wakiya[183]; (△) Shyn and Sharp[184]; (◇) Iga et al.[185]; (●, small) Itikawa et al.[174]; (×—×—) Kanik et al.,[35] recommended; (■) Sullivan et al.[188] Theory: (▲) Jain and Baluja[28]; (◇) Noble and Burke.[186] The recommended cross sections are multiplied by 10 for easy readability.

4.4.4 Ro-Vibrational Excitation Cross Sections in O_2

The ground state of the O_2 molecule is X $^3\Sigma_g^-$ with an equilibrium internuclear distance of 2.3 a_0 (0.12 nm), and the energy for the vibrational excitation $v = 0 \to v = 1$ is 0.193 eV. In the energy range 0.193 to 0.936 eV only ro-vibrational excitation takes place and the electronic excitation is due to the states a $^1\Delta_g$ and b $^1\Sigma_g^+$. The threshold energies for these states are 0.977 and 1.627 eV respectively.[181] Hake and Phelps[15] arbitrarily assumed that the vibrational excitation cross sections are in the shape of narrow spikes delayed in energy relative to the vibrational excitation thresholds. Recent integral ro-vibrational cross sections are reported by Shyn and Sweeney,[190] Noble et al.,[191] and Brunger et al.[192] The contributions of the four resonances with $^2\Pi_g$, $^2\Pi_u$, $^4\Sigma_u^-$, and $^2\Sigma_u^-$ symmetries to the ro-vibrational excitation cross sections are discovered by the Australian group.[192] The first two resonances make a major contribution to the X $^3\Sigma_g^- \to$ a $^1\Delta_g$ and X $^3\Sigma_g^- \to$ b $^1\Sigma_g^+$ transitions respectively.[186] Table 4.22 and Figure 4.36 show the integral cross sections for vibrational levels X $^3\Sigma_g^-$ ($v=0$) $\to$ X $^3\Sigma_g^-$ ($v=1, 2, 3, 4$).

TABLE 4.22
Integral Ro-Vibrational Cross Sections in O_2, in Units of $10^{-22}\,m^2$

	Energy (eV)			
Transition	5	7	10	15
$0 \to 1$	9.5	30.5	31.2	5.7
$0 \to 2$	3.4	11.4	16.5	1.5
$0 \to 3$	—	4.5	7.5	0.65
$0 \to 4$	—	3.5	4.0	0.31

	Energy (eV)									
	5	6	7	8	9	9.5	10	11	15	20
$0 \to 1$	5.21	11.55	20.24	25.62	40.26	46.36	53.63	40.56	11.57	1.75
$0 \to 2$	—	2.26	6.04	9.60	16.32	20.11	25.52	17.47	3.89	—
$0 \to 3$	—	—	2.64	4.07	8.11	10.12	11.45	9.02	2.15	—
$0 \to 4$	—	—	1.15	2.28	4.73	5.82	6.53	5.30	1.20	—

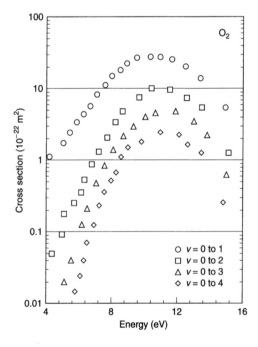

FIGURE 4.36 Recommended integral vibrational cross sections in units of $10^{-22}\,m^2$ for the levels shown.

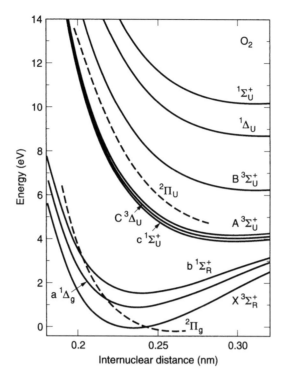

FIGURE 4.37 Schematic diagram of potential energy and internuclear separation of the O_2 molecule.

TABLE 4.23
Threshold Energy for Excitation Processes in O_2

State	Energy (eV)	State	Energy (eV)	State	Energy (eV)
$X\,^3\Sigma_g^-$	0	$A\,^3\Sigma_u^+$ (F)	4.340	$e\,(^1\Delta_{2u})$	9.346
$a\,^1\Delta_g$ (F)	0.977	$B\,^3\Sigma_u^-$	6.120	$\beta^3\Sigma_u^+$	9.355
$b\,^1\Sigma_g^+$ (F)	1.627	$^1\Pi_g$	8.141	$\alpha^1\Sigma_u$	9.455
$c\,^1\Sigma_u^-$ (F)	4.050	$d\,(^1\Pi_g)$	8.595		
$C\,^3\Delta_u$ (F)	4.262	$e'\,(^1\Delta_{2u})$	9.318		

F means an optically forbidden state.
From Itakawa, Y. et al., *J. Phys. Chem. Ref. Data*, 18, 23, 1989. With permission. © 1989, American Institute of Physics.

4.4.5 ELECTRONIC EXCITATION CROSS SECTIONS IN O_2

The potential energy diagram for the lower states is shown in Figure 4.37. Table 4.23 provides the threshold energies for selected states. Transitions from the ground state to the two lowest states, a $^1\Delta_g$ and b $^1\Sigma_g^+$, are known to give rise to the infrared ($\sim 10^{14}$ Hz, 1.3 μm) and red bands in the atmospheric spectra respectively.[104]

The two lowest states, a $^1\Delta_g$ and b $^1\Sigma_g^+$, have potential energy curves similar to the ground state of O_2.[193] They cannot be excited to the dissociative region of the molecule. They have received considerable attention since their discovery in certain auroras and in the light of the night sky. Electrons produced in the aurora are thought to excite the O_2 molecules to these two levels. The a $^1\Delta_g$ level radiates to ground level with a radiative lifetime of 2700 s.[194] The b $^1\Sigma_g^+$ level also decays by radiating to the ground level, with its characteristic wavelength of 762 nm and lifetime of 12 s.

Metastable atoms excited to these two levels are also formed during the dissociation process. In an electrical discharge the transition probabilities for these two states are 0.006 and 0.085 s^{-1}.[202] The mechanism of generation of the b $^1\Sigma_g^+$ metastable state is

$$e + O_2(X\ ^3\Sigma_g^-) \rightarrow O_2(B\ ^3\Sigma_u^-) + e \rightarrow O(^1D) + O(^3P) + e \qquad (4.31)$$

and

$$O(^1D) + O_2(X\ ^3\Sigma_g^-) \rightarrow O_2(b\ ^1\Sigma_g^+) + O(^3P) \qquad (4.32)$$

In reactions 4.31 and 4.32 the letters D and P designate the configuration of the atom generated (see Section 4.4.6).

The close lying c $^1\Sigma_u^-$, C $^3\Delta_u$, and A $^3\Sigma_u^+$ states are also metastables and are generally known as "6 eV states".[193] The A $^3\Sigma_u^+$ state gives rise to the strongest ultraviolet emissions in the night sky, known as Herzberg bands (~285 nm). The B $^3\Sigma_u^-$ state gives rise to Schumann–Runge bands and plays an important role in the production of metastable atoms (^1D, energy 1.97 eV) in the upper atmosphere. The potential curve for the state B $^3\Sigma_u^-$ has a shallow minimum (Figure 4.35) at an internuclear distance considerably larger than the equilibrium distance of the X $^3\Sigma_g^-$ ground state. As a result the X $^3\Sigma_g^- \rightarrow$ B $^3\Sigma_u^-$ absorption is composed mainly of a continuum over a few eV with a set of very weak bands at the low frequency end, the Schumann–Runge continuum.[203] The bands are in the ultraviolet region of the wavelength and the convergence limit lies at 176 nm, corresponding to 7.047 eV.[204]

Excitation to the states in the range 9.7 to 12.1 eV is mainly due to the optically allowed transitions from the ground state[183] (1978a). To show the relative magnitudes of the excitation cross sections, the measurements of Wakiya[183] (1978 a,b) are plotted on the same graph in the range of 20 to 500 eV. Over this range the Schuman–Runge bands dominate the total cross sections and, unless one is interested in calculating the contribution of each state to the scattering phenomena, the total cross section is equal to the cross sections for the optically allowed states, within the experimental error.

The excitation cross sections for selected states may be found in the following additional studies: a $^1\Delta_g$ and b $^1\Sigma_g^+$ states—Trajmar et al.,[181] Noble and Burke,[186] Shyn and Sweeney,[190] Middleton et al.[194]; A $^3\Sigma_u^+$ and c $^1\Sigma_u^-$ states—Teillet-Billy et al.,[205] Campbell et al.,[206] Noble and Burke.[207] Figures 4.38 and 4.39 show a comparison of selected studies.

4.4.6 Dissociation Cross Sections in O_2

The ground state of the oxygen atom is $1s^2\ 2s^2\ 2p^4\ [^3P_2]$ and the dissociation energy is 5.116 eV.[174] Upon dissociation, three states of the atom are possible: the ground state with the configuration ^3P and two metastable states, $^1D_2$ and $^1S_0$, with energies of 1.97 and 4.45 eV respectively. The lifetimes of these two states are long, 148 s and 0.8 s respectively. The next excited state is also a metastable state with the configuration $^5S_2{}^0$ and an energy of 9.14 eV. The lifetime of this state is 185 µs.[208] Excitation to the "6 eV states" causes dissociation into products O (^3P) + O (^3P); excitation into the B $^3\Sigma_u^-$ state yields the products $O(^1D) + O(^3P)$.

All the electronic excitation states above the dissociation energy result in dissociation. The dissociation cross section is the sum of states B $^3\Sigma_u^-$ and higher, and is almost equal to

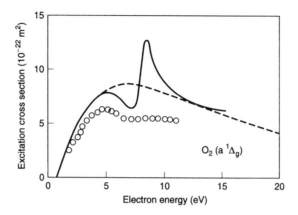

FIGURE 4.38 Integral cross sections for electron impact excitation of the a $^1\Delta_g$ state of O_2. (— —) Noble and Burke[207]; (——) Noble and Burke[186]; (○) theory.

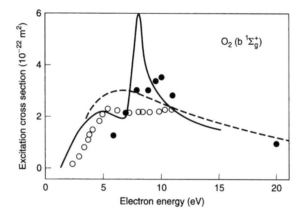

FIGURE 4.39 Integral cross sections for electron impact excitation of the b $^1\Sigma_g^+$ state of O_2. Symbols as in Figure 4.38 with the addition (●) Middleton et al.[194]

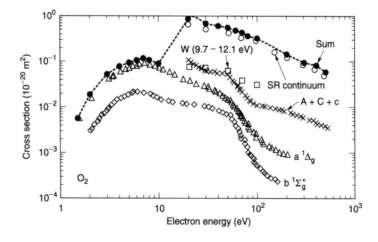

FIGURE 4.40 Excitation cross sections in O_2, showing the relative magnitudes of the selected states. SR, Schumann–Runge; W, optical emissions in the energy range shown; A + C + c is the abbreviation for (A $^3\Sigma_u^+$ + C $^3\Delta_u$ + c $^3\Sigma_u^-$) states; sum is the total of all states.

the SR-continuum shown in Figure 4.40.[174] The measured dissociation cross sections are given by Cosby.[193] An excitation of any of these states results in the dissociation of the molecule and leads to the formation of ozone (O_3). The formation of O_3 through dissociation proceeds according to the following reactions[209]:

(a) Electronic dissociation of O_2:

$$e + O_2 \rightarrow O_2^*(A\,^3\Sigma_u^+) \rightarrow O(^3P) + O(^3P) + e \tag{4.33}$$

$$e + O_2 \rightarrow O_2^*(B\,^3\Sigma_u^-) \rightarrow O(^1D) + O(^3P) + e \tag{4.34}$$

These reactions occur within nanoseconds.

(b) Formation of vibrationally excited ozone:

$$O + O_2 + M \rightarrow O_3^* + M \tag{4.35}$$

where M is a third body.

(c) Quenching of O_3^*

$$O_3^* + O \rightarrow O_3 + O \tag{4.36}$$

$$O_3^* + O_2 \rightarrow O_3 + O_2 \tag{4.37}$$

The reactions are completed within 10 μs. The energies taken by the various processes as a function of the ratio E/N (E = electric field, N = gas number density) are shown in Figure 4.41. Anyone working in a high-voltage laboratory cannot fail to notice the strong odour of O_3 and, since the ratio $E/N = 100$ Td (1 Td = 10^{-21} V m^2) at atmospheric pressure (1 bar), the 8.4 eV and 6.1 eV excitations are dominant.

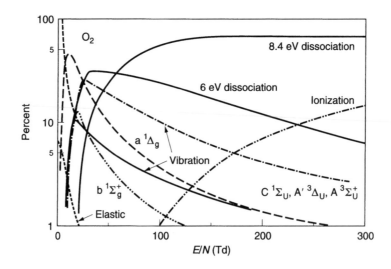

FIGURE 4.41 Energy deposition in the dissociation processes in O_2 that lead to the formation of O_3. At atmospheric pressure the ratio E/N has a value of ~100 Td and the dominant dissociation processes may be easily identified. Figure reproduced from Eliasson, B., and Kogelschatz, U., *J. Phys. B*, 19, 1241, 1986. With permission of the Institute of Physics, U.K.

4.4.7 IONIZATION CROSS SECTIONS IN O_2

The product ions generated due to ionization in O_2 are O_2^+, O^+, O_2^{2+}, and O^{2+}. The grand total ionization cross section is given by

$$Q_{i,T} = Q_i(O_2^+) + Q_i(O^+) + 2Q_i(O_2^{2+}) + 2Q_i(O^{2+}) \qquad (4.38)$$

Of the products shown on the right-hand side of reaction 4.38, the mass spectrometer method does not distinguish between ions O^+ and O_2^{2+} because both particles have the same charge-to-mass ratio. The total cross section is, therefore, taken as

$$Q_{i,T} = Q_i(O_2^+) + Q_i(O^+ + O_2^{2+}) + Q_i(O^{2+}) \qquad (4.39)$$

Strictly speaking, one should put $2\,O_2^{2+}$ in the middle term of reaction 4.39, but the ratio $Q_i(O_2^{2+})/Q_i(O^+)$ is negligibly small. The ionization potentials of the oxygen atom and molecule are 13.60 and 12.07 respectively.

Compared with other diatomic gases considered so far, the number of cross section measurements that have been made is relatively few. Rapp and Englander-Golden[70] have measured the total cross sections and Rapp et al.[71] have measured the cross sections for partial dissociative ionization. Krishnakumar et al.[199] and Straub et al.[98] have covered the wide range of onset to 1000 eV by using quadrupole and time-of-flight spectrometer respectively. In view of the relatively small number of cross section measurements we include the theoretical results of Hwang et al.[200] and Joshipura et al.[201] The agreement between various measurements shown in Figures 4.42 and 4.43 is very good for the total and partial cross sections.

Vibrationally excited molecules can have a different ionization cross section from that of ground state molecules. For example, Evans et al.[210] using a shock-heated molecular beam and a mass spectrometer, observed a five-fold increase of the ionization cross section in

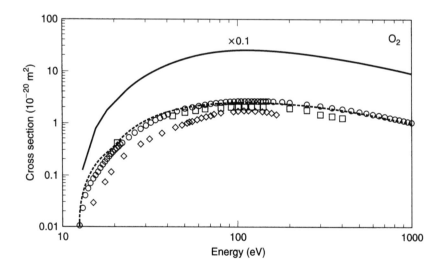

FIGURE 4.42 Ionization cross sections in O_2. (○) Rapp and Englander-Golden[70]; (◇) Märk[198]; (□) Evans et al.[210]; (— —) Hwang et al.,[200] theory; (——) Straub et al.[98]; recommended. The recommended cross sections are shown multiplied by the factor 10 for easy readability. The agreement between measurements is very good. The lower values of Evans et al.[210] are due to vibrationally excited O_2 at 5300 K.

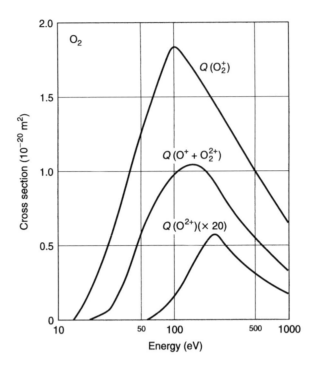

FIGURE 4.43 Averaged partial cross sections in O_2. References: Rapp et al.[71]; Märk[198]; Krishnakumar and Srivastava[199]; Straub et al.[98] The data of Rapp et al.[71] are for the production of ions with kinetic energies greater than 0.25 eV.

vibrationally excited CO. However, using the same techniques, Evans et al.[210] measured the ionization cross sections in vibrationally excited oxygen, which corresponds to vibrational temperatures in the range 1800 to 7000 K. While the dissociative ionization cross section increased by a small amount, the total cross sections agreed with those of Rapp and Englander-Golden[70] within experimental errors.

4.4.8 SIGMA RULE IN O_2

Kanik et. al.[35] have recommended a set of cross sections for O_2 and, since their publication, the total cross sections have been measured by Szmytkowski et al.,[45] ionization cross sections measured by Straub et al.[98] and a number of measurements of elastic scattering and excitation cross sections have been made by several authors (see Table 4.20). Table 4.24 updates the recommendations of Kanik et al.[35] in the light of these publications. It is noted that the sigma rule holds true extremely well for O_2, better than in H_2 and N_2, without adjustment for any of the cross sections. In the energy range above 300 eV, if the dissociation cross section of Chung and Lin[203] is added to the sum, the agreement is exemplarily good. Relative magnitudes of various cross sections are shown in Figure 4.43.

4.5 NITRIC OXIDE (NO)

Nitric oxide is a minor but significant gas in the atmosphere. It is classified as an open shell molecule and possesses a dipole moment of 0.157 D (1 debye = 3.33×10^{-30} C m). There have been relatively few cross section studies in the gas. A recent extensive review[7] has become available and we restrict ourselves to presenting data from the user point of view.

TABLE 4.24

Sigma Check and Recommended Cross Sections in O_2

Energy eV	Q_{el} $(10^{-20} m^2)$	$Q_v + Q_{rot}$ $(10^{-20} m^2)$	Q_{ex} $(10^{-20} m^2)$	Q_{dis} $(10^{-20} m^2)$	Q_i $(10^{-20} m^2)$	Σ $(10^{-20} m^2)$	Q_T (Expt) $(10^{-20} m^2)$	% diff.
0.15	4.7[S]	—	—	—	—	4.70	4.69[S]	−0.21
0.23	4.6[S]	—	—	—	—	4.60	4.27[Z]	−7.73
0.49	4.5[S]	0.1[H]	—	—	—	4.60	4.95[Z]	7.07
1	6.1[SU]	0.12[H]	—	—	—	6.22	6.12[SZ]	−1.63
2	6.7[SU]	0.04[H]	0.019	—	—	6.76	6.27[SZ]	−7.80
3	6.9[SU]	0.09[H]	0.051	—	—	7.046	6.46[SZ]	−9.00
5	7.1[SU]	0.13[N]	0.092	—	—	7.326	7.18[SZ]	−1.97
10	8.6[SU]	0.97[N]	0.088	(0.31)[CL]	—	9.66	10.4[SZ]	7.14
15	8.8[SU]	0.19[N]	0.350	(0.31)[C]	0.132[ST]	10.0	10.8[SZ]	7.43
20	8.7[SU]	0.02[N]	0.875	(0.55)[C]	0.354[ST]	9.95	10.7[SZ]	7.01
30	8.8[SU]	—	0.664	(0.61)[C]	1.132[ST]	10.52	11.2[SZ]	6.05
40	7.1[SU]	—	0.590	(0.60)[C]	1.687[ST]	9.38	10.0[SZ]	6.23
50	6.5[K]	—	0.543	(0.52)[C]	2.046[ST]	9.09	9.8[SZ]	7.26
60	6.1[K]	—	0.466	(0.44)[C]	2.287[ST]	8.85	9.5[SZ]	6.82
70	5.7[K]	—	0.419	(0.38)[C]	2.441[ST]	8.56	8.9[SZ]	3.82
80	5.4[K]	—	0.350	(0.36)[C]	2.536[ST]	8.29	8.55[SZ]	3.09
90	5.0[K]	—	0.325	(0.34)[C]	2.593[ST]	7.92	8.3[SZ]	4.60
100	4.8[K]	—	0.314	(0.32)[C]	2.621[ST]	7.74	8.1[SZ]	4.51
150	3.98[K]	—	0.227	(0.29)[C]	2.562[ST]	6.77	7.33[SZ]	7.65
200	3.15[K]	—	0.139	(0.29)[C]	2.394[ST]	5.68	5.85[SZ]	2.86
300	2.4[K]	—	0.092	(0.16)[CL]	2.053[ST]	4.54	4.85[SZ]	6.29
400	2.0[K]	—	0.080	—	1.781[ST]	3.86	4.00[SZ]	3.48
500	1.72[K]	—	0.058	(0.11)[CL]	1.57[ST]	3.35	3.5[SZ]	4.35
600	1.53[K]	—	—	—	1.405[ST]	2.94	3.1[SZ]	5.32
700	1.37[K]	—	—	—	1.272[ST]	2.644	2.8[SZ]	5.64
800	1.27[K]	—	—	—	1.164[ST]	2.434	2.55[SZ]	4.55
900	1.18[K]	—	—	—	1.073[ST]	2.254	2.35[SZ]	4.13
1000	1.12[K]	—	—	(0.93)[CL]	0.996[ST]	2.10	2.2[SZ]	4.73

Positive values in the last column denote that the experimental values are higher. Total dissociation cross sections are due to Cosby[193] and are not included in the summation. C = Cosby[193]; H = Hake and Phelps[15]; K = Kanik et al.[35]; N = Noble et al.[191]; S = Subramanian and Kumar[180]; ST = Straub et al.[98]; SU = Sullivan et al.[188]; SZ = Szmytkowski et al.[45]; W = Wakiya[183] (1978 a,b); Z = Zecca et al.[178] Q_{ex} from Figure 4.40.

4.5.1 Total Scattering Cross Sections (NO)

Grand total cross sections have been measured by: Zecca et al.,[211] 0.037 to 9.5 eV; Dalba et al.,[212] 121 to 1600 eV; Szmytkowski et al.[213] 0.5 to 160 eV; Alle et al.,[214] 0.2 to 5.0 eV; and Szmytkowski et al.,[215] 0.4 to 250 eV. The total cross sections are shown in Figures 4.44 and 4.45 for the low and higher energy ranges. A series of peaks is observed, superimposed on a monotonically increasing backbone; the peaks are attributed to the vibrational excitation caused by electron attachment, resulting in the NO^- ion.

The resonance phenomenon observed, due to short-lived negative ions, has been discussed in connection with the nitrogen and oxygen molecules. An electron can excite the vibrations of the molecule by providing an impulse; the mechanism is known as direct excitation. On the other hand, an electron attaches to a neutral molecule and the life of the negative ion may be so short ($\sim 10^{-14}$ s) that the electron is released by auto-detachment and the parent molecule will now have vibrationally excited levels. This mechanism of vibrational excitation is termed by Spence and Schulz[216] as arising due to a "compound state." The latter

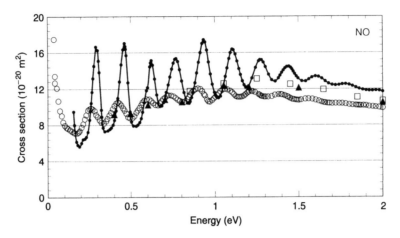

FIGURE 4.44 Grand total scattering cross sections in NO. (○) Zecca et al.[211]; (□) Szmytkowski and Maciąg[213]; (▲) Szmytkowski et al.[45]; (—●—) Alle et al.[214]

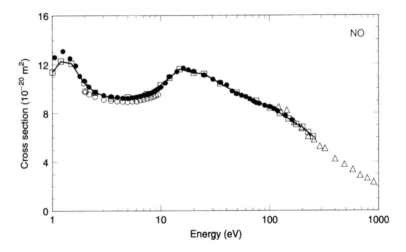

FIGURE 4.45 Grand total scattering cross sections in NO in the high energy range. (○) Zecca et al.[211]; (△) Dalba et al.[212]; (●) Szmytkowski and Maciąg[213]; (—□—) Szmytkowski et al.[215]

mechanism is favored in diatomic molecules such as N_2, CO, O_2, and H_2. The "direct" process is larger in polar molecules. Both O_2 and NO form stable negative ions, and therefore NO is likely to undergo vibrational excitation by both mechanisms.

If the lifetime of the ion is long when compared to the vibrational period of the nuclei ($\sim 10^{-12}$ s), the cross sections exhibit sharp structure corresponding to the vibrational levels of the compound state, as observed in Figures 4.44 and 4.35 (NO and O_2 respectively). If the lifetime is short, the resonant process appears as a broad hump with no structure, as in H_2 (Figure 4.16 in the 3 to 4 eV range). In the intermediate range, where the lifetime of the ion is comparable to the nuclear vibrational period, an oscillatory structure is observed, as in N_2 (Figure 4.26, 2 to 3 eV), CO (Figure 4.3, 1.7 eV), and NO (Figure 4.44, 1.5 eV). This oscillatory nature is attributed to the interference effect called the "Boomerang model".[217]

The mechanisms of formation of the negative ion in NO are as follows[217]:

$$1: \quad e + NO\,(^2\Pi, v = 0) \rightarrow NO^-\,(^3\Sigma^-, v') \rightarrow NO\,(^2\Pi, v) + e \qquad (4.40)$$

The ground state of the NO molecule is $^2\Pi_{1/2,\,3/2}$ and the ground state of the NO$^-$ ion is the same as that of O_2, $^3\Sigma_g^-$. The negative ion is stable and autoionization does not occur. The lowest vibrational level of NO$^-$ is 24 meV below the $v=0$ state of the NO molecule (see Figure 2, Appendix 2). The first step of the impact reaction (4.40) means that the negative ion is excited to the vibrational level v'. In the second step the ion decays to the ground state of NO, with excitation level v. It is shown below that $v=2$, 3, or 4, with energy not exceeding 0.77 eV.

$$2: \qquad e + NO(^2\Pi, v= 0) \rightarrow NO^-(^1\Delta, v') \rightarrow NO(^2\Pi, v) + e \qquad (4.41)$$

In this process the second step yields the ion in the $^1\Delta$ state.

$$3: \qquad e + NO(^2\Pi, v= 0) \rightarrow NO^-(^1\Sigma^+, v) \rightarrow NO(^2\Pi, v) \qquad (4.42)$$

In this process the second step yields the ion in the $^1\Sigma^+$ state. The energy for both states of the ion in reactions 4.41 and 4.42 is below 2 eV.

The vibrationally excited molecule has a relatively long lifetime $(10^{-10}$ s$)$[214] and the resonance peaks are sufficiently dominant that they appear in both the elastic scattering[216,217] and the total cross section measurements[211,214] as sharp peaks. The ion NO$^-$ is isoelectronic with the O_2 molecule (16 electrons), and a reference to Figure 4.37 shows that in O_2 the states $X\,^3\Sigma^-$ (0 ev), a $^1\Delta$ (0.98 eV), and b $^1\Sigma^+$ (1.63 eV) are the three lowest energy levels. The first three peaks that occur in NO at energies 0.293, 0.46, and 0.624 eV are attributed to the ground state of the NO$^-$ ion, for vibrational levels $v=2$, 3, 4 respectively. The fourth peak, at 0.768 eV, is suggested as a mix of the states $^3\Sigma^-$ $(v=4)$ and $^1\Delta$ $(v'=0)$.[214] The energy distance between states, Δv, for the first six states is 170 meV. Beyond 2 eV impact energy the measured grand total scattering cross section shows a broad peak at about 16 eV, decreasing thereafter. The results of the studies shown in Figure 4.45 show broad agreement among themselves.

4.5.2 MOMENTUM TRANSFER AND ELASTIC SCATTERING IN NO

Differential cross sections have been measured by Mojarrabi et al.[218] and a detailed discussion of the results is given by Brunger and Buckman.[7] Table 4.25 gives the momentum transfer and elastic scattering cross sections and also the ro-vibrational cross sections for $v=0 \rightarrow v=1$, 2. Theoretical calculations are due to Mu-Tao et al.[111] who calculated the differential cross sections at selected energy values up to 500 eV. The momentum transfer and elastic scattering cross sections calculated from their figures are shown in Table 4.26. While the elastic scattering cross section at 5 eV agrees very well with the recent value of Brunger et al.[219] the cross sections at 10 and 20 eV are 31% and 47% higher respectively. The higher values are attributed by Mu-Tao et al.[111] to their neglect of polarization effects. Above 40 eV there are no experimental values available for comparison. The broad peak at 20 eV, noticed for the first time in the present volume, agrees well with Q_T (see Figure 4.45).

4.5.3 ELECTRONIC EXCITATION CROSS SECTIONS IN NO

The lowest excitation level of NO is 5.48 eV (A $^2\Sigma^+$ state) and dissociation of the ground state occurs at 5.296 eV,[104] yielding N ($^4S°$) and O (^3P) atoms.[7] The spectroscopy of NO is

TABLE 4.25
**Momentum Transfer, Elastic Scattering, and Ro-Vibrational Cross
Sections in NO**

Energy	Q_M	Q_{el}	$Q_{rot, v, 0 \to 1}$	$Q_{rot, v, 0 \to 2}$	Q_T
1.5	8.415	10.473	—	—	10.473
3.0	7.044	9.604	—	—	9.604
5.0	6.296	9.239	—	—	9.239
7.5	5.797	9.123	0.028	—	9.123
10.0	5.539	9.241	0.074	0.014	9.329
15.0	5.116	9.714	0.270	0.073	10.057
20.0	4.232	9.707	0.097	0.022	9.826
30.0	3.547	9.314	0.022	—	9.336
40.0	2.546	8.214	0.014	—	8.228

Electron energy in units of eV, cross sections in units of $10^{-20}\,m^2$. Q_{el} and Q_T have been corrected according to Brunger et al.[219] Q_M has been corrected in the same ratio by the present author on the reasoning that the ratio $1/(1-\cos\theta)$ is presumably not in error.

TABLE 4.26
Q_M **and** Q_{el} **in NO**

Energy	5.0	10.0	20.0	50	100	200	500
Q_M	5.21	8.64	10.75	6.24	2.68	1.21	0.27
Q_{el}	9.45	12.14	14.33	7.66	4.33	3.22	1.04

Theoretical values of Mu-Tao et al.,[111] digitized and numerically integrated by the present author.[221] Energy in units of eV, cross sections in units of $10^{-22}\,m^2$.

extremely complicated, with some 30 excited electronic states.[219] Figure 4.46 identifies selected excitation levels, and Figure 4.47 shows the integral excitation cross sections for these states along with the total excitation cross section. The integral cross sections for six valence states and sixteen Rydberg states have recently been evaluated by Brunger et al.[220] and the total cross section, summed for all the states, is given in Table 4.27.

4.5.4   IONIZATION CROSS SECTIONS IN NO

Rapp and Englander-Golden[70] have measured the absolute total ionization cross sections, and Rapp et al.[71] have measured the fraction of dissociative to total ionization cross sections for production of ions of kinetic energy greater than 0.25 eV. This work was followed by that of Kim et al.[222] The products of ionization of NO are NO^+ (parent ion), N^+ (dissociative), O^+ (dissociative), and NO^{2+} (double ionization of parent molecule). Iga et al.[223] have measured the partial ionization cross sections, using the quadrupole mass spectrometer, and Lindsay et al.[224] have used a time-of-flight mass spectrometer with a position-sensitive detector. Theoretical calculations using the binary encounter Bethe (BEB) model are due to Kim et al.[225] who cover electron impact energy up to 1000 eV.

These results are shown in Figure 4.48, which displays the total ionization cross sections. The absolute total cross sections of Rapp and Englander-Golden[70] are higher than those of Lindsay et al.[224] by as much as 50% for energies greater than 30 eV. The theoretical results lie

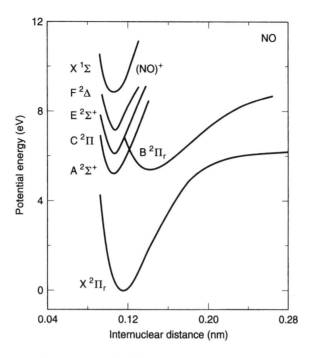

FIGURE 4.46 Lower excitation states of NO. The threshold energies are as follows, with spectroseopic values shown in brackets: A $^2\Sigma^+ = 5.484$ eV (5.449 eV); B $^2\Pi_r = 5.769$ eV (5.691 eV); C $^2\Pi = 6.499$ eV; E $^2\Sigma^+ = (7.514$ eV); F $^2\Delta = 7.722$ eV; X $^1\Sigma = (9.4$ eV). The energy for states is due to Brunger et al.[219]; Spectroscopic values from Herzberg.[104]

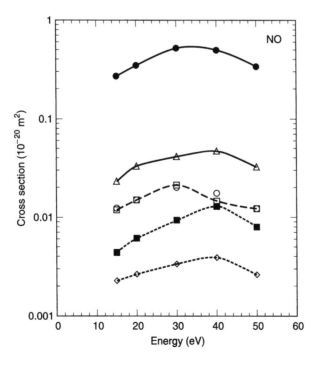

FIGURE 4.47 Integral excitation cross sections in NO. (—○—) A state; (—□—) B state; (—△—) C state; (—◇—) E state; (—■—) F state; (—●—) sum of all states.

TABLE 4.27
Integral Electronic Excitation Cross Sections of NO

Energy	15	20	30	40	50
Q_{ex}	0.270	0.346	0.521	0.497	0.344

Energy in units of eV, cross sections in units of 10^{-20} m^2.

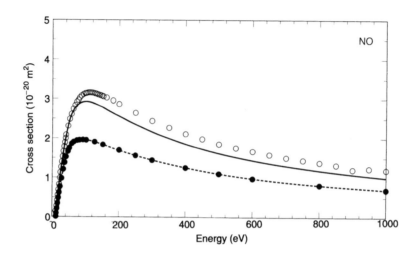

FIGURE 4.48 Absolute total ionization cross sections in NO. (○) Rapp and Englander-Golden[70]; (—●—) Lindsay et al.[224]; (——) Kim et al.[225]

between these two sets of data, but closer to those of Rapp and Englander-Golden.[70] The recommended cross sections are those of the latter authors as the sigma rule holds true very well up to 50 eV impact energy (Table 4.27). Partial cross sections are shown in Figure 4.49. The agreement between the various sets is less than satisfactory.

4.5.5 SIGMA RULE VERIFICATION FOR NO

Brunger et al.[7] have compiled a set of cross sections in the energy range 1.5 to 50 eV, and the range has been extended to 500 eV in the present volume. As far as the author of the present volume knows, there has been no experimental or theoretical determination of the integral elastic scattering and excitation cross sections beyond this energy. Table 4.28 shows that the sigma rule applies well up to 500 eV; these are the recommended cross sections.

4.6 CLOSING REMARKS

The diatomic gases considered in this chapter show certain similarities in the cross sections, particularly for isoelectronic molecules such as CO and N$_2$. The elastic differential cross sections in these gases at selected energies and the Q_{diff}–energy curve[61] show similar behavior (Figure 4.50). Further, the sharp and broad peaks observed in total cross sections may be related to the vibrational cross sections, as explained in Section 4.5.1. However, the total cross section in O$_2$[226] and NO show different structures in spite of the fact that both are attaching gases and form stable negative ions.

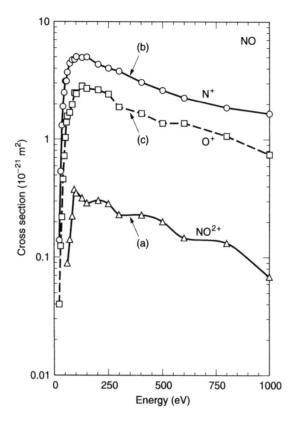

FIGURE 4.49 Partial ionization cross sections in NO. Ion species: (a) NO^{2+}; (b) N^+; (c) O^+. ($\triangle$) Kim et al.[222]; ($\square$) Iga et al.[223]; ($\bigcirc$) Lindsay et al.[224]

TABLE 4.28
Sigma Rule Verification in NO, 1.5 to 500 eV

Energy	Q_{el}	Q_v	Q_{ex}	Q_i^{REG}	Q_Σ	Q_T(Expt)	% Diff.
1.5	10.473	—	—	—	10.473	12.3[SM]	14.85
3.0	9.604	—	—	—	9.604	9.45 [SM]	−1.56
5.0	9.239	—	—	—	9.239	9.22 [SM]	−0.22
7.5	9.095	0.028			9.123	9.52 [SM]	4.17
10.0	9.241	0.088	-	0.018	9.347	10.1 [SM]	7.45
15.0	9.714	0.343	0.270	0.418	10.745	11.50 [SM]	6.56
20.0	9.707	0.119	0.346	0.813	10.985	11.40 [SM]	3.60
30.0	9.314	0.022	0.521	1.522	11.379	10.80 [SM]	−5.36
40.0	8.214	0.014	0.497	2.086	10.811	10.3 [SM]	−4.96
50.0	6.444	—	0.344	2.482	9.270	9.6 [SM]	3.44
100	433[L]		0.3[R]	3.14	7.78	8.48 [SM]	8.25
200	3.22 [L]		0.3 [R]	2.86	6.38	6.81[SMK]	6.31
500	1.03 [L]		0.3 [R]	1.86	3.19	3.55[D]	10.14

Energy in units of eV, cross sections in units of 10^{-20} m². The % difference is calculated using the formula $[(Q_T - Q_\Sigma)/Q_T] \times 100$. Superscript: D = Dalba et al.[212]; L = Lee et al.[111]; R = recommended; REG = Rapp and Englander-Golden[70]; SM = Szmytkowski and Mąciag[213]; SMK = Szmytkowski et al.[45]

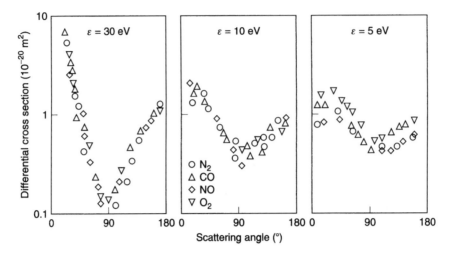

FIGURE 4.50 Comparison of elastic differential cross sections for N_2, CO, O_2, and NO at 5, 10, and 30 eV. References for data are 155, 183, 218, and 226.

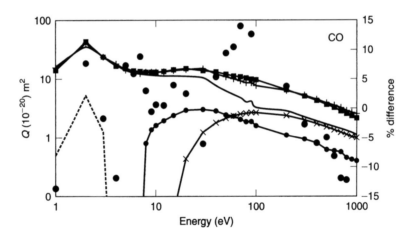

FIGURE 4.51 Relative cross sections as a function of electron energy in carbon monoxide. (———) elastic; (– – –) vibrational; (—•—) excitational; (—×—) ionization; (—+—) total sigma; (—■—) total experimental; (•) % difference (See Table 4.6).

The elastic scattering and excitation cross sections over certain energy ranges are incomplete. Momentum transfer cross sections in O_2 are in need of some clarification and the theory for elastic scattering needs to be improved. The high electron attachment at low energies may partly explain the differences. The sigma rule applies almost perfectly for O_2, but in H_2 and N_2 the agreement is not as good. The low energy momentum transfer cross sections obtained from beam studies in H_2 do not agree with those obtained from swarm derivation. In NO, experimental cross sections for excitation and elastic scattering are not available at all. The ionization cross sections in this gas also show as much as 50% difference at electron energies greater than 30 eV. The sigma rule holds reasonably well in CO.

Relative magnitudes of cross sections in CO and H_2 are shown in Figures 4.51 and 4.52 respectively as they were not shown earlier in the respective sections.

With this, we now consider industrial gases in Chapter 5.

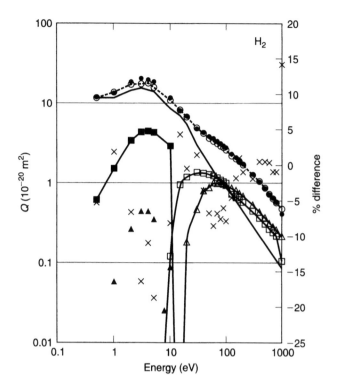

FIGURE 4.52 Consolidated cross sections on H_2, showing the relative magnitudes of various components. Tabulated values are shown in Table 4.11 (———) elastic; (—■—) rotational; (▲) vibrational; (—□—) excitation; (—△—) ionization; (●) total from sigma; (—○—) total, experimental; (×) % difference between the last two cross sections.

REFERENCES

1. Gibson, J. C., L. A. Morgan, R. C. Gulley, M. J. Brunger, C. T. Bundschu, and S. J. Buckman, *J. Phys. B: At. Mol. Opt. Phys.*, 29, 3197, 1996. See the introduction section.
2. Johnson, R. E., *Introduction to Atomic and Molecular Collisions*, Plenum Press, New York, 1982. See p. 50 and p. 261.
3. McDaniel, E. W., *Collision Phenomena in Ionized Gases*, John Wiley & Sons, New York, 1964, pp. 25–30.
4. Raju, G., *Dielectrics in Electric Fields*, Marcel Dekker, New York, 2003.
5. Yang, K., and T. Ree, *J. Chem. Phys.*, 35, 588, 1961.
6. Trajmar, S., D. F. Register, and A. Chutjian, *Phys. Rep.*, 97, 219, 1983. A number of gases are dealt with in this review.
7. Brunger, M. J., and S. J. Buckman, *Phys. Rep.*, 357, 215, 2002.
8. Brode, R. B., *Phys. Rev.*, 25, 636, 1925.

Carbon Monoxide

9. Brüche, E., *Ann. Phys.* (Leipzig) 83, 1065, 1927.
10. Normand, C. E., *Phys. Rev.*, 35, 1217, 1930.
11. Ramsauer, C., and R. Kollath, *Ann. Phys.*, (Leipzig) 4, 91, 1930.
12. Asundi, R. K., J. D. Craggs, and M. V. Kurepa, *Proc. Phys. Soc.*, 82, 967, 1963.
13. Rapp, D., P. Englander-Golden, and D. Briglia, *J. Chem. Phys.*, 42, 4081, 1965.
14. Rapp, D., and P. Englander-Golden, *J. Chem. Phys.*, 43, 1464, 1965.

15. Hake, R. D., Jr., and A. V. Phelps, *Phys. Rev.*, 158, 70, 1967.

16. Chandra, N., *Phys. Rev.*, A16, 80, 1977.

17. Szmytkowski, C., and M. Zubek, *Chem. Phys. Lett.*, 57, 105, 1978.

18. Land, J. E., *J. Appl. Phys.*, 49, 5716, 1978.

19. Kwan, Ch. K., Y. F. Hsieh, W. E. Kauppila, S. J. Smith, T. S. Stein, M. N. Uddin, and M. S. Dababneh, *Phys. Rev. A*, 27, 1328, 1983.

20. Haddad, G. N., and H. B. Milloy, *Aust. J. Phys.*, 36, 473, 1984.

21. Sueoka, O., and S. Mori, *J. Phys. Soc. Jpn*, 53, 2491, 1984.

22. Jain, A., and D. W. Norcross, in *Abstracts, Proceedings of the XIV International Conference on the Physics of Electronic Atomic Collisions*, Palo Alto, CA, 1985, ed. M. J. Coggiola, D. L. Huestis, and R. P. Saxon, p. 214.

23. Buckman, S. J., and B. Lohmann, *Phys. Rev.*, A34, 1561, 1986.

24. Orient, O. J., and S. K. Srivastava, *J. Phys. B: At. Mol. Opt. Phys.*, 20, 3923, 1987.

25. Mason, N. J., and W. R. Newell, (a) *J. Phys. B: At. Mol. Phys.*, 19, L203, 1986; this reference gives excitation cross sections of the c $^3\Pi_u$ ($v=0$) and B $^1\Sigma_u^+$ plus; states of H_2 over the electron impact energy range 11 to 14.5 eV. (b) *J. Phys. B: At. Mol. Phys.*, 19, L587, 1986; this reference gives total excitation cross sections of the c $^3\Pi_u$ state of H2 over the energy from threshold to 60 eV. (c) *J. Phys. B: At. Mol. Opt. Phys.*, 21, 1293, 1988: this reference gives electron impact excitation cross sections in CO.

26. Nickel, J. C., C. Mott, I. Kanik, and D. C. McCollum, *J. Phys. B: At. Mol. Opt. Phys.*, 21, 1867, 1988.

27. García, G. C., C. Aragon, and J. Campos, *Phys. Rev. A*, 42, 4400, 1990.

28. Jain, A., and K. L. Baluja, *Phys. Rev. A*, 45, 202, 1992.

29. James, G. K., J. A. Ajello, I. Kanik, B. Franklin, and D. E. Shemansky, *J. Phys. B: At. Mol. Opt. Phys.*, 25, 1481, 1992.

30. Jain, A., and D. W. Norcross, *Phys. Rev. A*, 45, 1644, 1992.

31. Kanik, I., J. C. Nickel, and S. Trajmar, *J. Phys. B: At. Mol. Opt. Phys.*, 25, 2189, 1992.

32. Cosby, P. C., *J. Chem. Phys.*, 98, 7804, 1993.

33. Karwasz, G., R. S. Brusa, A. Gasparoli, and A. Zecca, *Chem. Phys. Lett.*, 211 529, 1993.

34. Middleton, A. G., M. J. Brunger, and P. J. O. Teubner, *J. Phys. B: At. Mol. Opt. Phys.*, 25, 3541, 1992.

35. Kanik, I., S. Trajmar, and J. C. Nickel, *J. Geophys. Res.*, 98, 7447, 1993.

36. Morgan, L. A., and J. Tennyson, *J. Phys. B: At. Mol. Opt. Phys.*, 26, 2429, 1993.

37. Xing, S. L., Q. C. Shi, X. J. Chen, K. Z. Xu, B. X. Yang, S. L. Wu, and R.F. Feng, *Phys. Rev. A*, 51, 414, 1995.

38. Gibson, J. C., L. A. Morgan, R. J. Gulley, M. J. Brunger, C. T. Bundschu, and S. J. Buckman, *J. Phys. B: At. Mol. Opt. Phys.*, 29, 3197, 1996. Stated energy resolution is 50 meV.

39. Joshipura, K. N., and P. M. Patel, *J. Phys. B: At. Mol. Opt. Phys.*, 29, 3925, 1996.

40. Leclair, L. R., and S. Trajmar, *J. Phys. B: At. Mol. Opt. Phys.*, 29, 5543, 1996.

41. Randell, J., R. J. Gulley, S. L. Lunt, J.-P. Ziesel, and D. Field, *J. Phys. B: At. Mol. Opt. Phys.*, 29, 2049, 1996.

42. Gibson, J. C., L. A. Morgan, R. J. Gulley, M. J. Brunger, C. T. Bundschu, and S. J. Buckman, *J. Phys. B: At. Mol. Opt. Phys.*, 29, 3197, 1996. See Figures 4 and 5. Theoretical differential cross sections from 0° to 180° and 1 to 30 eV are also presented.

43. Zobel, J., U. Mayer, K. Jung, H. Ehrhardt, H. Pritchard, C. Winstead, and V. McKoy, *J. Phys. B: At. Mol. Opt. Phys.*, 28, 839, 1995.

44. Zobel, J., U. Mayer, K. Jung, and H. Ehrhardt, *J. Phys. B: At. Mol. Opt. Phys.*, 29, 813, 1996.

45. Szmytkowski, C., K. Maciag, and G. Karwasz, *Phys. Scripta*, 54, 271, 1996.

46. Maji, S., G. Basavaraju, S. M. Bharati, K. G. Bhushan, and S. P. Khare, *J. Phys. B: At. Mol. Opt. Phys.*, 31, 4975, 1998.

47. Tian, C., and C. R. Vidal, *J. Phys. B: At. Mol. Opt. Phys.*, 31, 895, 1998.

48. Mangan, M. A., B. G. Lindsay, and R. F. Stebbings, *J. Phys. B: At. Mol. Opt. Phys.*, 33, 3225, 2000.

49. Poparić, G., M. Vićić, and D. S. Belić, *J. Phys. B: At. Mol. Opt. Phys.*, 34, 381, 2001.

50. Tronc, M., R. Azria, and Y. LeCoat, *J. Phys. B.*, 13, 2327, 1980.

51. Inokuti, M., and M. R. C. McDowell, *J. Phys. B*, 7, 2382, 1974.
52. Inokuti, M., *Rev. Mod. Phys.* 43, 297, 1971.
53. Liu, J. W., *Phys. Rev.*, 35, 591, 1987.
54. Chutjian, A., D. G. Truhlar, W. Williams, and S. Trajmar, *Phys. Rev. Lett.*, 29, 1580, 1972.
55. Zecca, A., G. P. Karwasz and R. S. Brusa, *Phys. Rev. A*, 45, 2777, 1992.
56. Jung, K., Th. Antoni, R. Müller, K.-H. Kochem, and H. Ehrhardt, *J. Phys. B: At. Mol. Phys.*, 15, 3535, 1982.
57. Schulz, G. J., *Phys. Rev.* (a) 116, 1141, 1959; (b) 135, 988, 1964.
58. Chutjian, A., and H. Tanaka, *J. Phys. B: At. Mol. Phys.*, 13, 1901, 1980.
59. Sohn, W., K.-H. Kochem, K. Jung, H. Ehrhardt, and E. S. Chang, *J. Phys. B: At. Mol. Phys.*, 18, 2049, 1985.
60. Sawada, T., D. L. Sellin, and A. E. S. Green, *J. Geophys. Res.* 77, 4819, 1972.
61. Zecca, A., G. P. Karwasz, and R. S. Brusa, *Riv. Nu. Ci.*, 19, 1, 1996.
62. Chung, S., and C. C. Lin, *Phys. Rev. A*, 9, 1954, 1974. The theoretically calculated values are still higher than those shown in Figure 4.8 and Land[48] reduced them to ~35%.
63. Sun, Q., C. Winstead, and V. McKoy, *Phys. Rev. A*, 46, 6987, 1992.
64. Saelee, H. T., and J. Lucas, *J. Phys. D: Appl. Phys.*, 10, 343, 1977.
65. Tate, J. T., and P. T. Smith, *Phys. Rev.*, 39, 270, 1998.
66. Vaughan, A. L., *Phys. Rev.*, 1, 1687, 1931.
67. Hille, E., and T. D. Märk, *J. Chem. Phys.*, 69, 2492, 1978.
68. Defrance, A. M. M., and J. C. Gomet, *Methodes Physiques d'Analyse* (GAMS), July–Sept. 1965, p. 205.

Molecular Hydrogen

69. Engelhardt, A. G., and A. V. Phelps, *Phys. Rev.*, 131, 2115, 1963.
70. Rapp, D., and P. Englander-Golden, *J. Chem. Phys.* 43, 1464, 1965. The energy range covered is from threshold to 1000 eV in closely spaced intervals. Tabulated values are given in units of $\pi a_0^2 = 0.88 \times 10^{-20}$ m^2.
71. Rapp, D., P. Englander-Golden, and D. D. Briglia, *J. Chem. Phys.*, 42, 4081, 1965. Dissociative cross sections are measured in H_2, expressed as a fraction of the total ionization cross sections for ions of energy greater than 0.25 eV.
72. Golden, D. E., H. W. Bandel, and J. A. Salerno, *Phys. Rev.*, 146, 146, 1966.
73. Ehrhardt, H., L. Langhans, F. Linder, and H. S. Taylor, *Phys. Rev.*, 173, 222, 1968. The energy of 4.42 eV stated at the beginning of section 4.2.1 for H_2^- is due to Read F. H. and D. Andrick, *J. Phys. B: At. Mol. Phys.*, 4, 911, 1971.
74. Burrow, D. P., and G. J. Schulz, *Phys. Rev.*, 187, 1969, 1970.
75. Crompton, R. W., D. K. Gibson, and A. G. Robertson, *Phys. Rev. A*, 2, 1386, 1970.
76. Srivastava, S. K., A. Chutjian, and S. Trajmar, (a) *J. Chem. Phys.*, 63, 2659, 1975; this reference gives elastic and momentum transfer cross sections. (b) Srivastava, S. K. and S. Jensen, *J. Phys. B: At. Mol. Phys.*, 10, 3341, 1977; this reference gives excitation cross sections.
77. Ferch, J., W. Raith, and K. Schröder, *J. Phys. B: At. Mol. Phys.*, 13, 1481, 1980.
78. Dalba, G., P. Fornasini, I. Lazzizzera, G. Ranieri, and A. Zecca, *J. Phys. B: At. Mol. Phys.*, 13, 2839, 1980.
79. van Wingerden, B., R. W. Wagenaar, and F. J. de Heer, *J. Phys. B: At. Mol. Phys.* 13, 3481, 1980.
80. Shyn, T. W., and W. E. Sharp, *Phys. Rev. A*, 24, 1734, 1981.
81. Hoffman, K. R., M. S. Dababneh, Y. F. Hsieh, W. E. Kaippila, V. Pol, J. H. Smart, and T. S. Stein, *Phys. Rev. A*, 25, 1393, 1982.
82. Deuring, A., K. Floeder, D. Fromme, W. Raith, A. Schwab, G. Sinapius, P. W. Zitzewitz, and J. Krug, *J. Phys. B: At. Mol. Phys.*, 16, 1633, 1983.
83. Furst, J., M. Mahgerefteh, and D. Golden, *Phys. Rev. A*, 30, 2256, 1984.
84. Jones, R. K., *Phys. Rev. A*, 31, 2898, 1985.
85. Nishimura, H., A. Danjo, and H. Sugahara, *J. Phys. Soc. Jpn*, 54, 1757, 1985.
86. Khakoo, M. A., and S. Trajmar, *Phys. Rev. A*, 34, 146, 1986.

87. Morrison, M. A., R. W. Crompton, B. C. Saha, and Z. Lj. Petrovic, *Aust. J. Phys.*, 40, 239, 1987.

88. England, J. P., M. T. Elford, and R. W. Crompton, *Aust. J. Phys.*, 41, 573, 1988.

89. Subramanian, K. P., and V. Kumar, *J. Phys. B: At. Mol. Opt. Phys.*, 22, 2387, 1989.

90. Brunger, M. J., S. J. Buckman, and D. S. Newman, *Aust. J. Phys.*, 43, 665, 1990.

91. Kossmann, H., O. Schwarzkopf, and V. Schmidt, *J. Phys. B: At. Mol. Opt. Phys.*, 23, 301, 1990.

92. Nickel, J. C., I. Kanik, S. Trajmar, and K. Imre, *J. Phys. B: At. Mol. Phys.*, 25, 2427, 1992.

93. Morrison, M. A., and W. K. Trail, *Phys. Rev. A*, 48, 2874, 1993.

94. Khakoo, M. A., and J. Segura, *J. Phys. B: At. Mol. Opt. Phys.*, 27, 2355, 1994.

95. Krishnakumar, E., and S. K. Srivastava, *J. Phys. B: At. Mol. Opt. Phys.*, 27, L251, 1994.

96. Randell, J., S. L. Hunt, G. Mrotzek, J.-P. Ziesel, and D. Field, *J. Phys. B: At. Mol. Opt. Phys.*, 27, 2369, 1994.

97. Jacobsen, F. M., N. P. Frandsen, H. Knudsen, and U. Mikkelsen, *J. Phys. B: At. Mol. Opt. Phys.*, 28, 4675, 1995.

98. Straub, H. C., P. Renault, B. G. Lindsay, K. A. Smith, and R. F. Stebbings, *Phys. Rev. A*, 54, 2146, 1996.

99. Stebbings, R. F., and B. G. Lindsay, *J. Chem. Phys.*, 114, 4741, 2001.

100. Trevisan, C. S., and J. Tennyson, *J. Phys. B: At. Mol. Opt. Phys.*, 34, 2935, 2001.

101. White, R. D., M. A. Morrison, and B. A. Mason, *J. Phys. B: At. Mol. Opt. Phys.*, 35, 605, 2002.

102. Wrkich, J., D. Mathews, I. Kanik, S. Trajmar, and M. A. Khakoo, *J. Phys. B: At. Mol. Opt. Phys.* 35, 4695, 2002.

103. Schulz, G. J., and R. K. Asundi, *Phys. Rev.*, 158, 25, 1967. Also see *Phys. Rev. A*, 135, A 988, 1964. In this paper Schulz gives data for all three gases, CO, H_2, and N_2.

104. Herzberg, G., *Molecular Spectra and Molecular Structure*, Van Nostrand, Princeton, NJ, 1950.

105. Brunger, M. J., S. J. Buckman, D. S. Newman, and D. T. Alee, *J. Phys. B: At. Mol. Opt. Phys.*, 24, 1435, 1991.

106. Nesbet, R. K., C. J. Noble, and L. A. Morgan, *Phys. Rev. A.*, 34, 2798, 1986.

107. Morgan, W. L., J. P. Boeuf, and L. C. Pitchford, www. siglo-kinema.com, 1999.

108. Gerjuoy, E., and S. Stein, *Phys. Rev.*, 97, 1671, 1955.

109. (a)Phelps, A. V., *Rev. Mod. Phys.*, 40, 399, 1968. This paper reviews the ro-vibrational cross sections available up to 1968. (b) Phelps, A. V., and L. C. Pitchford, *Phys. Rev. A*, 31, 2932, 1985. In this paper a set of cross sections for N_2 derived and a detailed discussion of transport coefficients available up to 1985 is provided. Also see ftp://jila.colorado.edu/collision_data/electronneutral/electron.txt and reference 107.

110. Gupta, N., and G. R. Govinda Raju, *8th International Symposium on Gaseous Dielectrics*, Virginia Beach, VA, June 2–5, 1998; *J. Phys. D: Appl. Phys.*, 33, 2949, 2000.

111. Mu-Tao, L. R., R. Lucchese, and V. McKoy, *Phys. Rev. A*, 26, 3240, 1982. M. Hayashi (*J. Phys. Colloque c7*, 45, 1979) gives the cross sections derived from Monte Carlo simulation, as shown in Table 4.29.

112. Khakoo, M. A., S. Trajmar, R. McAdams, and T. W. Shyn, *Phys. Rev. A*, 35, 2832, 1987.

113. Nishimura, H., and A. Danjo, *J. Phys. Soc. Jpn*, 54, 3031, 1986.

114. Hall, R. I., and L. Andric, *J. Phys. B: At. Mol. Phys.*, 17, 3815, 1984.

115. Fliplet, A. W., and V. McKoy, *Phys. Rev. A*, 21, 1863, 1980.

116. Shemansky, D. E., J. M. Ajello, and D. T. Hall, *Astrophys. J.*, 296, 765, 1985.

117. Rescigno, T. N., C. W. McCurdy, V. McKoy, and C. F. Bender, *Phys. Rev. A*, 13, 216, 1976.

118. Ajello, J. M., and D. E. Shemansky, *Astrophys. J.*, 407, 820, 1993.

119. Chung, S., C. C. Lin, and T. P. Lee, *Phys. Rev. A*, 12, 1340, 1975.

120. Vroom, D. A., and F. J. de Heer, *J. Chem. Phys.*, 50, 580, 1969.

121. Corrigan, S. J. B., *J. Chem. Phys.*, 43, 4381, 1965.

122. Van Zyl, B., and T. M. Stephan, *Phys. Rev. A*, 40, 3164, 1994.

123. Adamczyk, B., A. J. H. Boerboom, B. L. Schram, and J. Kistemaker, *J. Chem. Phys.*, 44, 4640, 1966.

123a. Edwards, A. K., R. M. Wood, A. S. Beard, and R. L. Ezell, *Phys. Rev., A*, 37, 1664, 1991.

123b. Rudd, M. E., *Phys. Rev. A*, 44, 1644, 1991.

TABLE 4.29
Cross Sections Derived from Monte Carlo Simulation[111]

Level	Threshold (eV)	Peak energy (eV)	X-section at peak (10-20 m2)
b $^3\Sigma_u^+$	8.8	16	0.28
B $^1\Sigma_u^+$	11.37	40	0.48
c $^3\pi_u$	11.87	15	0.56
a $^3\Sigma_g^=$	11.89	15	0.09
C $^1\pi_u$	12.40	35	0.24
E $^1\Sigma_g^+$	12.40	50	0.076
e $^3\Sigma_u^+$	13.36	16.5	0.068
B' $^1\Sigma_u^+$	13.70	40	0.038
d $^3\pi_u$	13.97	18.5	0.034
D $^1\pi_u$	14.12	50	0.012
Ionization	15.425	70	0.972
r_{02}	0.0439	4	1.84
r_{13}	0.0727	4	1.07
v_{01}	0.516	3	0.50
v_{02}	1.003	4.3	0.040

Molecular Nitrogen

124. Zubek, M., and G. C. King, *J. Phys. B: At. Mol. Opt. Phys.*, 27, 2613, 1994.
125. Haas, R., *Z. Phys.*, 148, 177, 1957.
126. Engelhardt, A. G., A. V. Phelps, and C. G. Risk, *Phys. Rev.* 135, A1566, 1964.
127. Schulz, G. J., *Phys. Rev.*, 135, A988, 1964. Figure 11 of this reference gives the vibrational cross sections in H_2 and compares them with the cross sections derived by Engelhardt and Phelps[68] from transport coefficients. For energies greater than 2 eV the latter are considerably higher.
128. Golden, D. E., *Phys. Rev. Lett.*, 17, 847, 1966.
129. Baldwin, G. C., *Phys. Rev. A*, 9, 1225, 1972.
130. Chung, S., and C. C. Lin, *Phys. Rev. A*, 6, 988, 1972.
131. Srivastava, S. K., A. Chutjian, and S. Trajmar, *J. Chem. Phys.*, 64, 1340, 1976.
132. Blaauw, H. J., F. J. de Heer, R. W. Wagenaar, and D. H. Barends, *J. Phys. B: At. Mol. Phys.*, 10, L299, 1977.
133. Cartwright, D. C., S. Trajmar, A. Chutjian, and W. Williams, *Phys. Rev. A*, 16, 1013, 1977a. *Phys. Rev. A*, 16, 1041, 1977b.
134. Mathur, D., and J. B. Hasted, *J. Phys. B: At. Mol. Phys.*, 10, L265, 1977.
135. Shyn, T. W., R. S. Stolarski, and G. R. Corignan, *Phys. Rev. A*, 6, 1009, 1972.
136. Blaauw, H. J., R. W. Wagenaar, D. H. Barends, and F. J. de Heer, *J. Phys. B: At. Mol. Phys.*, 13, 359, 1980.
137. Dalba, G., P. Foransini, R. Grisenti, G. Ranieri, and A. Zecca, *J. Phys. B: At. Mol. Phys.* 13, 4695, 1980.
138. Kennerly, R. E., *Phys. Rev. A*, 21, 1876, 1980.
139. Itikawa, Y., M. Hayashi, A. Ichimura, K. Onda, K. Sakimoto, K. Takayanagi, M. Nakamura, H. Nishimura, and T. Takayanagi, *J. Phys. Chem. Ref. Data*, 15, 985, 1986.
140. García, G., A. Pérez, and J. Campos, *Phys. Rev. A*, 38, 654, 1988.
141. Brunger, M. J., P. J. O. Teubner, A. M. Weigold, and S. J. Buckman, *J. Phys. B: At. Mol. Opt. Phys.*, 22, 1443, 1989.
142. Brunger, M. J., and P. J. O. Teubner, *Phys. Rev. A*, 41, 1413, 1990.
143. Freund, R. S., R. C. Wetzel, and R. J. Shul, *Phys. Rev. A*, 41, 5861, 1990.
144. Krishnakumar, E., and S. K. Srivastava, *J. Phys. B: At. Mol. Phys.*, 23, 1893, 1990.

145. Ramanan, G., and G. R. Freeman, *J. Chem. Phys.*, 93, 3120, 1990.
146. Brennan, M. J., D. T. Alle, P. Euripides, S. J. Buckman, and M. J. Brunger, *J. Phys. B: At. Mol. Opt. Phys.*, 25, 2669, 1992.
147. Nickel, J. C., I. Kanik, S. Trajmar, and K. Imre, *J. Phys. B: At. Mol. Opt. Phys.*, 25, 2427, 1992.
148. Cosby, P. C., *J. Chem. Phys.*, 98, 9544, 1993.
149. Karwasz, G., R. S. Brusa, A. Gasparoli, and A. Zecca, *Chem. Phys. Lett.*, 211, 529, 1993.
150. Leclair, L. R., and S. Trajmar, *J. Phys. B: At. Mol. Opt. Phys.*, 29, 5543, 1996.
151. Campbell, L., M. J. Brunger, A. M. Nolan, L. J. Kelly, A. B. Wedding, J. Harrison, P. J. O. Teubner, D. C. Cartwright, and B. Mclaughlin, *J. Phys. B: At. Mol. Opt. Phys.*, 34, 1185, 2001.
152. Pavlovic, Z., M. J. W. Boness, A. Herzenberg, and G. J. Schulz, *Phys. Rev. A*, 6, 676, 1972.
153. Dill, D., and J. L. Dehmer, *Phys. Rev. A*, 16, 1423, 1977.
154. Ehrhardt, H., and K. Willman, *Z. Phys.*, 204, 462, 1967.
155. Shyn, W. T., and G. R. Carignan, *Phys. Rev. A*, 22, 923, 1980.
156. Morrison, M. A., B. C. Saha, and T. L. Gibson, *Phys. Rev.*, 36, 3682, 1987.
157. Wedde, T., and T. G. Strand, *J. Phys. B: At. Mol. Phys.*, 7, 1091, 1974.
158. Herrmann, D., K. Jost, and J. Kessler, *J. Chem. Phys.*, 64, 1, 1976.
159. DuBois, R. D., and M. E. Rudd, *J. Phys. B: At. Mol. Phys.*, 9, 2657, 1976.
160. Jung, K., Th. Antoni, R. Müller, K.-H. Kochem, and E. Ehrhardt, *J. Phys. B: At. Mol. Phys.*, 15, 1982, 1982.
161. Gote, M., and H. Ehrhardt, *J. Phys. B: At. Mol. Opt. Phys.*, 28, 3957, 1995.
162. Allan, M., *J. Phys. B: At. Mol. Phys.*, 18, 4511, 1985. The energy for 17 vibrational transitions from $0 \rightarrow 1$ to $0 \rightarrow 17$ is given.
163. Sohn, W., K.-H. Kochem, K.-M. Scheueriein, K. Jung, and H. Ehrhardt, *J. Phys. B: At. Mol. Phys.*, 19, 4017, 1986. In addition to vibrational excitation cross sections, both momentum transfer and total cross sections are given in the energy range 0.1 to 1.5 eV.
164. Trajmar, S., D. F. Register, and A. Chutjian, *Phys. Rep.*, 97, 219, 1983.
165. Mason, N. J., and W. R. Newell, *J. Phys. B: At. Mol. Phys.*, 20, 3913, 1987. Tabulated values and comparison; with previous works for this state in the energy range 9 to 141 eV are given.
166. Brunger, M. J., P. J. O. Teubner, and S. J. Buckman, *Phys. Rev. A*, 37, 3570, 1988. Tabulated values are given in the energy range 11.868 to 12.688 eV.
167. Zubek, M., *J. Phys. B: At. Mol. Opt. Phys.*, 27, 573, 1994.
168. Poparić, G., M. Vićić, and D. S. Belić, *Chem. Phys.*, 240, 289, 1999a; *Chem. Phys.*, 240, 1999b 283.
169. Marinelli, W. J., W. J. Kessler, B. D. Green, and W. A. M. Blumberg, *J. Chem. Phys.*, 91, 701, 1989.
170. Crowe, A., and J. W. McConkey, *J. Phys. B: At. Mol. Phys.*, 6, 2108, 1973.
171. Märk, T. D., *J. Chem. Phys.*, 63, 3731, 1975.
172. Armentrout, P. B., S. M. Tarr, A. Sori, and R. S. Freund, *J. Chem. Phys.*, 75, 2786, 1981.
173. Liu, J., and G. R. Govinda Raju, *J. Frank. Inst.*, 329, 181, 1992.

Molecular Oxygen

174. Itikawa, Y., A. Ichimura, K. Onda, K. Sakimoto, K. Takayanagi, Y. Hatano, M. Hayashi, H. Nishimura, and S. Tsurubuchi, *J. Phys. Chem. Ref. Data*, 18, 23, 1989.
175. Fisk, J. B., *Phys. Rev.*, 49, 167, 1936.
176. Wong, S. F., M. J. W. Boness, and G. J. Schulz, *Phys. Rev. Lett.*, 31, 969, 1973.
177. Dalba, G., P. Fornasini, R. Grisenti, G. Ranieri, and A. Zecca, *J. Phys. B: At. Mol. Phys.*, 13, 4695, 1980.
178. Zecca, A., R. S. Brusa, R. Grisenti, S. Oss, and C. Szmytkowski, *J. Phys. B: At. Mol. Phys.*, 19, 3353, 1986.
179. Dababneh, M. S., Y.-F. Hsieh, W. E. Kauppila, C. K. Kwan, S. J. Smith, T. S. Stein, and M. N. Uddin, *Phys. Rev. A*, 38, 1207, 1988.
180. Subramanian, K. P., and V. J. Kumar, *J. Phys. B: At. Mol. Phys.*, 23, 745, 1990.

181. Trajmar, S., D. C. Cartwright, and W. Williams, *Phys. Rev. A*, 41, 1482, 1971.
182. Wedde, T., and G. Strand, *J. Phys. B: At. Mol. Phys.*, 7, 1091, 1974.
183. Wakiya, K., *J. Phys. B: At. Mol. Phys.*, 11, 3913, 1978 a; *J. Phys. B: At. Mol. Phys.*, 11, 3931, 1978b.
184. Shyn, T. W., and W. E. Sharp, *Phys. Rev. A*, 26, 1982, 1982.
185. Iga, I., L. Mu-Tao, J. C. Nogueira, and R. S. Barbieri, *J. Phys. B: At. Mol. Phys.*, 20, 1095, 1987.
186. Noble, C. J., and P. G. Burke, *Phys. Rev. Lett.*, 68, 2011, 1992.
187. Higgins, K., C. J. Noble, and P. G. Burke, *J. Phys. B: At. Mpol. Opt. Phys.*, 27, 3203, 1994. For integrated values of cross section, see Brunger and Buckman 7.
188. Sullivan, J. P., J. C. Gibson, R. J. Gulley, and S. J. Buckman, *J. Phys. B: At. Mol. Opt. Phys.*, 28, 4319, 1995.
189. G. Wöste, C. J. Noble, K. Higgins, P. G. Burke, M. J. Brunger, P. J. O. Teubner, and A. G. Middleton, *J. Phys. B: At. Mol. Opt. Phys.*, 28, 4141, 1995.
190. Shyn, T. W., and C. J. Sweeney, *Phys. Rev. A*, 47, 1006, 1993.
191. Noble, C. J., K. Higgins, G. Woeste, P. Duddy, P. G. Burke, P. J. O. Teubner, A. G. Middleton, and M. J. Brunger, *Phys. Rev. Lett.*, 76, 1996, 3534.
192. Brunger, M. J., A. G. Middleton, and P. J. O. Teubner, *Phys. Rev. A*, 57, 208, 1998.
193. Cosby, P. C., *J. Chem. Phys.*, 98, 9560, 1993.
194. Middleton, A. G., M. J. Brunger, P. J. O. Teubner, M. W. B. Anderson, C. J. Noble, G. Wöste, K. Blum, P. G. Burke, and C. Fullerton, *J. Phys. B: At. Mol. Opt. Phys.*, 27, 4057, 1994.
195. Allan, M., *J. Phys. B: At. Mol. Opt. Phys.*, 28, 2163, 1995.
196. Green, M. A., P. J. O. Teubner, M. J. Brunger, D. C. Cartwright, and L. Campbell, *J. Phys. B: At. Mol. Opt. Phys.*, 34, L157, 2001.
197. Schram, B. L., H. R. Moustafa, J. Schutten, and F. J. de Heer, *Physica*, 32, 734, 1966.
198. Märk, T. D., *J. Chem. Phys.*, 63, 3731, 1975.
199. Krishnakumar, E., et al., *Int. J. Mass Spectr. Ion. Proc.*, 113, 1, 1992.
200. Hwang, M., Y. K. Kim, and M. E. Rudd, *J. Chem. Phys.*, 104, 2956, 1996. For tabulated values see *http://physics.nist. gov*/PhysRefData/Ionization
201. Joshipura, K. N., B. K. Antony, and M. Vinodkumar, *J. Phys. B: At. Mol. Opt. Phys.*, 35, 4211, 2002.
202. Lawton, S. A., and A. V. Phelps, *J. Chem. Phys.*, 69, 1055, 1978.
203. Chung, S., and C. C. Lin, *Phys. Rev. A*, 21, 1075, 1980.
204. Herzberg, G., *Molecular Spectra and Molecular Structure*, Van Nostrand Co., New York, 1950, p. 447. The products are $^3P + {}^3P$ and $^3P + {}^1D$.
205. Teillet-Billy, D., L. Malegat, and J. P. Gauyacq, *J. Phys. B: At. Mol. Opt. Phys.*, 20, 3201, 1987.
206. Campbell, L., M. A. Green, M. J. Brunger, P. J. O. Teubner, and D. C. Cartwright, *Phys. Rev. A*, 61, 022706, 2000.
207. Noble, C. J., and P. C. Burke, *J. Phys. B: At. Mol. Phys.*, 19, L35, 1986.
208. Mason, J. N., and W. R. Newell, *J. Phys. B: At. Mol. Phys.*, 23, 4641, 1990.
209. Eliasson, B., and U. Kogelschatz, *J. Phys. B: At. Mol. Phys.*, 19, 1241, 1986.
210. Evans, B., S. Ono, R. M. Hobson, A. W. Yau, S. Teii, and J. S. Chang, *Proceedings of the 13th International Symposium on shock tubes and waves*, SUNY Press, Albany, 1982, pp. 535–542.

Nitric Oxide

211. Zecca, A., I. Lazzizzera, M. Krauss, and C. E. Kuyatt, *J. Chem. Phys.*, 61, 4560, 1974.
212. Dalba, G., P. Fornasini, R. Grisenti, G. Ranieri, and A. Zecca, *J. Phys. B: At. Mol. Phys.*, 13, 4695, 1980.
213. Szmytkowski, C., and K. Maciag, *J. Phys. B: At. Mol. Opt. Phys.*, 24, 4273, 1991.
214. Alle, D. T., M. J. Brennan, and S. J. Buckman, *J. Phys. B: At. Mol. Opt. Phys.*, 29, L277, 1996.
215. Szmytkowski, C., K. Maciag, and G. Karwasz, *Phys. Scripta*, 54, 271, 1996.

216. Spence, D., and G. J. Schulz, *Phys. Rev. A*, 3 1968, 1971.
217. Tronc, M., A. Huetz, M. Landau, F. Pichou, and J. Reinhardt, *J. Phys. B: At. Mol. Phys.*, 8, 1160, 1975.
218. Mojarrabi, B., R. J. Gulley, A. G. Middleton, D. C. Cartwright, P. J. O. Teubner, S. J. Buckman, and M. J. Brunger, *J. Phys. B: At. Mol. Opt. Phys.*, 28, 487, 1995.
219. Brunger, M. J., L. Campbell, D. C. Cartwright, A. G. Middleton, B. Mojarrabi, and P. J. O. Teubner, *J. Phys. B: At. Mol. Opt. Phys.*, 33, 783, 2000.
220. Brunger, M. J., L. Campbell, D. C. Cartwright, A. G. Middleton, B. Mojarrabi, and P. J. O. Teubner, *J. Phys. B: At. Mol. Opt. Phys.*, 33, 809, 2000.
221. Gorur G. Raju, unpublished, 2003.
222. Kim, Y. B., K. Stephan, and E. Märk, *J. Chem. Phys.*, 74, 6771, 1981.
223. Iga, I., M. V. V. S. Rao, and S. K. Srivastava, *J. Geophys. Res.*, 101, 9261, 1996.
224. Lindsay, B. G., M. A. Mangan, H. C. Straub, and R.F. Stebbings, *J. Chem. Phys.*, 112, 9404, 2000.
225. Kim, Y.-K., W. Hwang, N. M. Weinberger, M. A. Ali, and M. E. Rudd, *J. Chem. Phys.*, 106, 1026, 1997.
226. Tanaka, H., S. K. Srivastava, and A. Chutjian, *J. Chem. Phys.*, 69, 5329, 1978.
227. Weeratunga, N., M.A.Sc. thesis, University of Windsor, 2005.

5 Data on Cross Sections—III. Industrial Gases

This chapter continues the presentation of scattering cross sections in major industrial gases. This term is applied rather loosely since complex molecules such as CO_2, NO_2, SF_6, etc. are considered, along with atomic gases such as mercury vapor and the common molecule of water, H_2O. A brief discussion of quantum mechanical interpretation of scattering has been provided in Chapter 1 as an introduction. Excellent treatment of this aspect may be found in Morse,[1] Massey and Burhop,[2] and McDaniel,[3] following the early treatment of Faxén and Holtsmark,[4] Holtsmark,[5] Rayleigh,[6] Massey and Mohr,[7] Mott and Massey,[8] Massey,[9] and Burke and Smith.[10]

5.1 CARBON DIOXIDE (CO_2)

Scattering in carbon dioxide (CO_2) has been studied by a number of groups since the early measurements of Brüche,[11] Ramsauer and Kollath,[12,13] and Kollath,[14] all using the Ramsauer technique. The practical importance of this gas in planetary atmospheres and for ecological purposes, combined with its technological importance in lasers and fusion plasmas, has increased researchers' attention since about 1960. CO_2 is a linear polyatomic molecule which acts as a convenient test choice for scattering theories[15,16] where one wishes to study the influence of polarity of several atoms in the molecule. It is nonpolar in the ground state, with an equilibrium distance of 0.116 nm between the carbon and oxygen atoms.[17] A dipole moment is induced in the excited states; in the bending mode the molecule acquires a dipole moment of 5.78×10^{-31} C m (0.17 debye)[70]; in the asymmetric stretching mode it acquires 0.44 D.[69]

Table 5.1 shows selected references for scattering cross sections in CO_2. In keeping with our objective, we limit ourselves to electron impact energies below 1000 eV and refer to the literature before 1980 only when essential. Theoretical treatments are due to Morrison et al.[15] and Morrison and Lane[16] for rotational excitation and elastic scattering respectively.

5.1.1 TOTAL SCATTERING CROSS SECTIONS IN CO_2

The total scattering cross sections will be discussed by dividing the energy range into two: 0 to 10 eV and 10 to 1000 eV. Figures 5.1 and 5.2 show the total scattering cross sections covering these two ranges. Note the linear scale of Figure 5.1. The main feature of the low energy range is that there is a prominent resonance at 3.8 eV.[19] Hoffman et al.[20] measure this resonance at 3.85 eV.

An oscillatory structure with several narrowly spaced peaks is observed near the resonance.[44] The resonance demonstrates the formation of a $^2\Pi_u$ compound state in the linear configuration of the nucleus. A brief description of vibrational modes is in order at this point.

CO_2 is a linear molecule with $^1\Sigma_g$ in the ground state. It has three modes of vibration: a bending mode designated as $(0n0)$ with Π_u symmetry, a symmetric stretching $(n00)$ with Σ_g symmetry, and an asymmetric stretching $(00n)$ with Σ_u symmetry.[48] In the ground state the

TABLE 5.1
Selected References for Scattering Cross Section Data in CO_2

Type	Energy Range	Method	Authors
		Total scattering cross section	
Q_T	1.5–8	Linear transmission method	Szmytkowski and Zubek[18]
Q_T	0.07–4.5	Time-of-flight spectrometer	Ferch et al.[19]
Q_T	2.0–50	Beam transmission method	Hoffman et al.[20]
Q_T	100–500	Beam transmission method	Kwan et al.[21]
Q_T, Q_M	500, 800, 1000	Crossed beam method	Iga et al.[22]
Q_T	0.5–80	Linear transmission method	Szmytkowski et al.[23]
Q_T	72.2–2916	Modified Ramsauer method	Szmytkowski et al.[24]
Q_T	0.5–10	Time-of-flight method	Ferch et al.[25]
Q_T	10–5000	Theory	Jain and Baluja[26]
Q_T	0.7–600	Time-of-flight method	Sueoka and Hamada[27]
Q_T	400–5000	Beam attenuation	Garcia and Manero[28]
		Elastic and momentum transfer cross sections	
Q_{el}, Q_M	0.02–100	Swarm method	Lowke et al.[29]
Q_{el}	3–90 eV	Crossed beams method	Shyn et al.[30]
Q_{el}, Q_M	4, 10, 20, 50	Crossed beams method	Register et al.[31a]
Q_{el}, Q_v	0–6	Crossed beams method	Kochem et al.[32]
Q_{el}, Q_v	0.1–5.0	TOF spectrometer	Buckman et al.[33]
Q_{diff}	20–100	Crossed beams method	Kanik et al.[34]
Q_{el}, Q_v	4 eV	Crossed beams method	Johnstone et al.[35]
Q_{el}, Q_M	0–100	Theory	Gianturco and Stoecklin[36]
Q_{el}	3–60	Theory	Takekawa and Itikawa[47]
Q_{el}	1.5–100	Crossed beams method	Tanaka et al.[37]
Q_{el}, Q_M	1.0–50.0	Crossed beams method	Gibson et al.[38]
Q_{el}		Theory	Gianturco and Stoecklin[39]
Q_{el}	0.1–100	Theory	Lee et al.[40]
		Ro-vibrational cross sections	
Q_v	0.003–0.008	Swarm method	Pack et al.[41]
Q_v	0–3.0	Swarm method	Hake and Phelps[42]
Q_v	Near threshold	Electron beam with H	Statamovic and Schulz[43]
Q_v	Near threshold	Retarded potential method	Sanchez and Schulz[44]
Q_v	3–5	Electron spectrometer	Boness and Schulz[45]
Q_v, Q_{rot}	2, 3.8	Crossed beams method	Antoni et al.[46]
Q_{el}	3–60	Theory	Takekawa and Ibikawa[47]
Q_v	1.5–30	Crossed beams method	Kitajima et al.[48]
		Excitation cross section	
Q_{ex}	3–20	Crossed beam spectrometer	Cvejanović et al.[49]
Q_{ex}	20–200 eV	Crossed beams method	Green et al.[50]
		Ionization cross section	
Q_i	Onset-1000	Ionization tube	Rapp and Englander-Golden[51]
Q_i	25–600	Mass spectrometer	Adamczyk et al.[52]
Q_i	Onset-300	Quadrupole mass spectrometer	Crowe and McConkey[53]
Q_i	14–200	Fast-neutral-beam method	Freund et al.[78]
Q_i	Onset-1000	TOF mass spectrometer	Straub et al.[54]
		Reviews, comments, etc.	
All	0–1000 eV	—	Trajmar et al.[55]
Q_T	0.3–100	—	Kimura et al.[56]
Q_i	Onset-1000	—	Stebbings and Lindsay[57]
All	0–1000	—	Karwasz et al.[58]

Energy in units of eV.

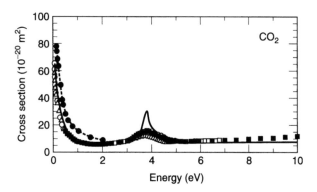

FIGURE 5.1 Selected total scattering cross sections in CO_2 in the low energy range. (—) Morrison et al.,[15] theory; (□) Szmytkowski and Zubek[18]; (△) Ferch et al.[18]; (○) Buckman et al.,[33] 310 K data; (-●-) Buckman et al.,[33] 573 K data; (■) Szmytkowski et al.[23] The higher temperature cross sections are larger since there is a larger percentage of vibrationally excited (bending mode) molecules.

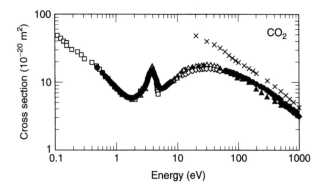

FIGURE 5.2 Selected total scattering cross sections in CO_2. (○) Szmytkowski and Zubek[18]; (□) Ferch et al.[19]; (△) Hoffman et al.[20]; (◇) Szmytkowski et al.[23]; (●) Szmytkowski et al.[24]; (×) Jain and Baluja,[26] theory; (▲) Sueoka and Hamada[27]; (◆) Garcia and Manero.[28]

molecule is nonpolar but a dipole moment is induced in the asymmetric or bending mode. A fourth mode, simply called "mixed mode," is designated $(1n0)$.[35] The ground vibrational mode is designated as (000), the lowest bending mode as (01^10), and the fundamental asymmetric stretch mode as (100). Each mode is a series, with n taking successive values. The direction of the dipole moment is time dependent and the resonance for these modes occurs in the infrared region of the electromagnetic spectrum.[59] The symmetric stretching mode has resonance in the Raman effect region. Which of these modes is active at a given energy is a question which we shall defer till the section on ro-vibrational cross sections.

The low energy electron scattering cross sections of Szmytkowski and Zubek,[18] Ferch et al.,[19] Buckman et al.,[33] and Szmytkowski et al.[23] are in good agreement with each other, though the peak cross sections at resonance do not agree very well. These are: Morrison and Lane,[16] 30.87; Szmytkowski and Zubek,[18] 15.45; Ferch et al.,[19] 13.18; and Buckman et al.,[33] 15.38; all in units of 10^{-20} m^2. Taking swarm measurements, Haddad and Elford[60] observed that the drift velocities measured in CO_2 at 573 K were incompatible with those at room and lower temperatures. This observation was attributed to the increased momentum transfer cross section for scattering by vibrationally excited molecules. Experimental confirmation of this reasoning was provided by Buckman et al.[33]

The measured total scattering cross sections at 573 K are also shown in Figure 5.1. At this temperature there are about 19.2% of vibrationally excited molecules, of which approximately 15.2% are in the first bending (010) mode. The increase does not show up for electron energies greater than 2 eV, as is evident from the fact that the 310 K and 573 K curves merge at this energy.

In the high energy range, the agreement between various measurements is again quite satisfactory, except for the theoretical computations of Jain and Baluja.[26] The data of Sueoka and Hamada[27] are about 15% lower than those of Szmytkowski et al.,[24] but this difference is within the combined uncertainties of the two experiments. The total cross section is obtained here by adding the elastic and inelastic scattering cross sections; the latter are calculated by the Born–Bethe approximation which is increasingly inaccurate toward the lower energy region. It is also known that the inelastic cross sections calculated according to the B–B approximation are asymptotic to the high energy cross sections, as is evident from Figure 5.2.

The elastic scattering cross section is calculated using the Born approximation (Equation 4.2) and the inelastic scattering cross sections are calculated using the Born–Bethe approximation (Equation 4.3). The constants in these equations have not been worked out from the wave functions of CO_2. Joshipura and Patel[61] have adopted the optical theorem to the forward scattering amplitude (Equation 1.223) and the independent atom model to obtain the total cross section as

$$Q_{T(BB)} = \frac{R}{\varepsilon} a_0^2 \left[675.3 + 106.3 \ln \frac{\varepsilon}{R} - 480.7 \frac{R}{\varepsilon} + \cdots \right] \tag{5.1}$$

where a_0 is the Bohr radius ($= 5.2917 \times 10^{-11}$ m), R is the Rydberg energy (13.595 eV), and ε is the electron energy in eV. As an example of application of Equation 5.1 we substitute $\varepsilon = 1600$ eV and obtain a cross section of 2.254×10^{-20} m^2; the measured cross section of Garcia and Manero[28] is 1.604×10^{-20} m^2. The differences at lower energies are even larger and these authors suggest a correction factor

$$f = \left[1 - 0.239 \exp\left(-\frac{1}{416} \frac{\varepsilon}{R} \right) \right] Q_{T(BB)} \tag{5.2}$$

where $Q_{T(BB)}$ is obtained from Equation 5.1. Application of this correction factor gives the cross section at 1600 eV as 1.848×10^{-20} m^2, a difference of ~15% from measurements.

5.1.2 ELASTIC AND MOMENTUM TRANSFER CROSS SECTIONS IN CO_2

Differential cross sections have been measured by Shyn et al.[30] and selected references since then are shown in Table 5.1. The energy range covered is patchy and in many publications integral values are not given. The theoretical results show poor agreement among themselves and with various experiments.

The dominant feature of the elastic scattering cross section is the $^2\Pi_u$ resonance that occurs at 3.8 eV, also showing up in the total scattering cross section (Figures 5.1 and 5.2). The increasing cross section with decreasing energy below the foot of the resonance also shows up in both elastic and total scattering. The dominant partial wave of the resonance has not been established with certainty, but the increase in scattering cross sections below 2 eV and toward thermal energy is attributed to s-wave scattering.[40] In quantum mechanics this

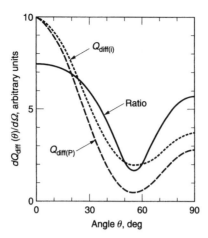

FIGURE 5.3 Schematic diagram of differential cross sections for rotational excitation of molecules. $Q_{\text{diff(i)}}$ values are the integrated rotational differential cross sections, $Q_{\text{diff(p)}}$ is the peak of the differential cross sections, and the ratio refers to peak/integral. The cross sections are shown in arbitrary units. The integral cross section is shallower than the peak cross section.

phenomenon is usually referred to as a virtual state, notwithstanding the fact that the electron wave is itself a measure of probability!

The physical meaning ascribed to the virtual state is that the electron is nearly attached to the CO_2 molecule and has an s-wave function $(l = 0)$. If the energy of the electron is the same as the thermal energy $(3/2\ kT)$ the electron will be attached. We prefer to use the term "thermal attachment" for this process. The scattering of such electrons is therefore expected to be isotropic, though experimental measurements of Kochem et al.[32] show significant forward scattering at energies as low as 155 meV.

The rotational and vibrational excitations produce a change in the shape of the angular dependence of the differential cross sections.[62,63] Figures 5.3 and 5.4 briefly explain

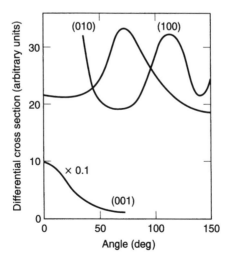

FIGURE 5.4 Differential cross sections for vibrational modes in CO_2. The differential cross section for the symmetric stretching mode (100) is not symmetric about the 90° angle and contains higher partial waves. The bending mode (010) is approximately symmetric about the 90° angle. The asymmetric stretching mode (001) has a magnitude of 10% of the other two modes.

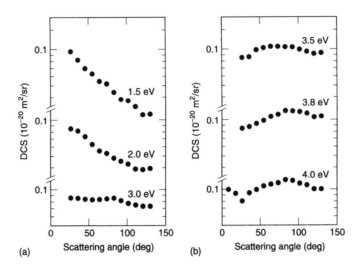

FIGURE 5.5 Absolute differential scattering cross sections for vibrational excitation in CO_2: (a) bending mode (010); (b) bending mode (020). Figure reproduced from Kitajima, M. et al., *J. Phys. B*, 34, 1929, 2001. With permission of the Institute of Physics (IOP), U.K.

the angular dependence of the differential cross sections for rotational and vibrational excitations respectively. Figure 5.3 shows the integrated differential cross section $Q_{diff(i)}$ for several rotational levels, the peak cross section $Q_{diff(p)}$, and the ratio of the two curves. The situation shown represents the rotational excitation of the N_2 molecule with 2 meV between successive levels[62] and one can see that $Q_{diff(p)}$ has a much deeper minimum than the other two curves. Read and Andrick[62] suggest plotting the integrated intensities for presenting differential cross sections.

Figure 5.4 shows schematically the differential cross sections in CO_2 for the different modes of vibrational excitation. For the symmetric stretching mode (100) the differential cross section is not symmetric about 90° and contains a number of higher partial waves. This mode also exhibits strong forward peaking. On the other hand, the differential cross section for the excitation of the bending mode (010) shows approximate symmetry about 90°. The asymmetric stretching mode (001) has a Q_{diff} that is about 10% of those due to the above two modes. Measured vibrational differential cross sections due to Kitajima et al.[48] in the 1.5 to 30 eV range, shown in Figure 5.5, demonstrate these features in a more or less pronounced way.

Returning to integral elastic scattering cross sections, these data have been compiled by Karwasz et al.[58]; Table 5.4 also provides tabulated values. Figure 5.6 shows the relative agreement between various studies in the energy range of 0.1 to 60 eV. Momentum transfer cross sections in the energy range 0 to 1000 eV are shown in Figure 5.7 and Table 5.2. The cross sections in Table 5.2 are composed of results from three ranges: 0 to 1.5 eV,[64] 1.5 to 100 eV,[37] and 100 to 1000 eV.[64] The values merge smoothly within ~8% at 1.5 eV, and within less than 1% at 100 eV.

5.1.3 RO-VIBRATIONAL EXCITATION CROSS SECTIONS IN CO_2

As stated earlier, an oscillatory structure with several narrowly spaced peaks is observed near the resonance.[44] To understand this phenomenon better we have to briefly recall the vibrational excitation via the shape resonance phenomenon in diatomic gases.

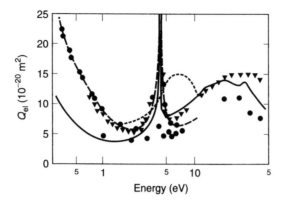

FIGURE 5.6 Integral elastic cross sections in CO_2. (●) Gibson et al.,[38] ANU data; (◆) Gibson et al.,[38] Flinders data; (–●●–) Morrison et al.[15]; (- - -) Takekawa and Itikawa[47]; (—) Lee et al.[40] Also shown for comparison are the total scattering cross sections of (▼) Szmytkowski et al.[23] Figure reproduced from Gibson, J. C. et al., *J. Phys. B*, 32, 213, 1999. With permission of IOP, U.K.

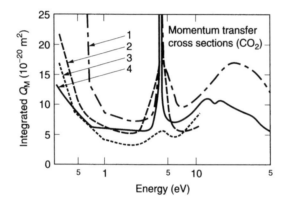

FIGURE 5.7 Momentum transfer cross sections in CO_2. Experimental: (●) Gibson et al.[38] Theory: (1) Morrison et al.[15]; (2) Nakamura.[31b] Swarm derived: (3) Lowke et al.[29]; (4) Lee et al.[40] Figure reproduced from Gibson, J. C. et al., *J. Phys. B*, 32, 213, 1999. With permission of IOP, U.K.

An electron attaching to the molecule forms a negative ion and the lifetime of the ion, in comparison with the vibration time of the nuclei, decides the nature of the resonance peak. If the lifetime is shorter than a typical vibration time, which may be easily calculated from the vibration frequency,[65] the resonance appears as a broad featureless hump in the total or elastic scattering cross section. A further feature of these short-lived compound states (negative ions) is that the peaks to successive vibrational levels get attenuated. Such resonances represent the short lifetime of the compound states. Beyond the highest experimentally discernible vibrational excitation level, the resonance phenomenon merges with the direct process of scattering. The 3 to 4 eV shape resonance in H_2 is an example of this kind of resonance (Figure 4.16).

When the lifetime of the negative ion becomes comparable to the vibrational times ($\sim 10^{-14}$ s), the vibrational cross sections develop an oscillatory structure. The location of the peaks and valleys forming the oscillatory structure shift towards higher energies as the vibrational energy of the final state is increased. This behavior is exhibited by N_2 around

TABLE 5.2
Momentum Transfer Cross Sections in CO_2

Energy	Q_M	Energy	Q_M	Energy	Q_M
0.000	600.0	1.000	5.55	8.0	8.07
0.001	540.0	1.2	5.02	9.0	9.24
0.002	387.0	1.3	4.90	10.0	9.94
0.010	170.0	1.5	4.83	15.0	11.19
0.020	119.0	2.0	4.53	20.0	10.17
0.080	58.0	3.0	5.96	30.0	7.51
0.100	52.0	3.8	7.69	60.0	4.15
0.150	40.0	4.0	7.22	100.0	2.65
0.200	31.0	5.0	5.66	200.0	1.08
0.300	20.30	6.0	6.69	300.0	0.66
0.400	14.30	6.5	6.56	500.0	0.36
0.500	10.90	7.0	6.56	1000.0	0.14

Electron energy in units of eV and cross sections in units of $10^{-20}\,m^2$.
Source: 0 to 1.5 eV and 100 to 1000 eV, Pitchford, L., personal communication, 2003; 1.5 to 100 eV, Tanaka, H. et al., *Phys. Rev. A*, 57, 1798, 1998.

2.3 eV (Figures 2.17 and 4.25), and in CO around 1.7 eV (Figure 4.6). The oscillatory structure is attributed to interference effects called the "boomerang effect".[66]

As the ion lifetime increases further, it becomes much larger than the vibration time of the compound state. Under these conditions the energy dependence of the vibrational cross sections exhibits very narrow, isolated spikes. These spikes remain at constant energy for all final vibrational states and these energies correspond to the vibrational levels. Scattering of electrons in the region of the 0 to 1 eV compound state in O_2 exhibits this kind of behavior (Figure 4.36). N_2 also exhibits a similar structure (Figure 4.44) in the region of 0.1 to 2.0 eV.

Table 5.3 shows the inelastic processes in CO_2. Scattering cross section measurements show a series of narrow structures (Figure 5.8), which is interpreted as increasing levels of the vibrational state. The spacing between successive peaks is approximately 130 meV. This structure extends from 3.0 to 5.0 eV and is attributed to the symmetric stretch of the vibrational mode of the resonance. Auto-ionization occurs within limits of 0.13 to 0.26 eV.[45]

Čadež et al.[67] suggest that the oscillations are due to the boomerang effect, which is the interference effect between the incoming and outgoing waves. This wave function changes with the negative ion energy and produces the sequence of oscillations when overlapped with the wave function of the ground state vibrational levels. With decreasing energy, below 1 eV, the cross section increases rapidly and this is attributed to the nearly bound $l = 0$ state in the CO_2 potential.

As mentioned earlier, differential scattering measurements are carried out only at selected energies because a large number of readings should be taken for scanning the angles. Recent measurements of Kitajima et al.[48] in the energy range 1.5 to 30 eV are an exception. Moreover, the excitation cross section for the bending mode (010) has a peak close to the elastic scattering peak. Thus one should look for theoretical studies of data on inelastic collisions over a wide range of energies. Takekawa and Itikawa[69] have studied theoretically the symmetric and asymmetric modes, extending to the bending mode.[70] Figure 5.9 shows these cross sections up to 100 eV.

TABLE 5.3
Selected Inelastic Processes and Threshold Energies in CO_2

Energy Loss	Threshold (eV)	Process	Remarks
0.083	0.083	$000 \rightarrow 010$	I bending mode
0.167	0.167	$000 \rightarrow 020 + 100$	Bending and symmetrical stretching
0.291	0.291	$000 \rightarrow 001$	I asymmetrical stretching
0.252	2.5	$000 \rightarrow 0n0 + n00$	Bending and symmetrical stretching
0.339	1.5	$000 \rightarrow 0n0 + n00$	Bending and symmetrical stretching
0.422	2.5	$000 \rightarrow 0n0 + n00$	Bending and symmetrical stretching
0.505	2.5	$000 \rightarrow 0n0 + n00$	Bending and symmetrical stretching
2.5	2.5	$000 \rightarrow 0n0 + n00$	Bending and symmetrical stretching
3.85	3.85	$e + CO_2 \rightarrow CO + O^-$	Dissociative attachment
7.0	7.0	$^3\Sigma_u$ state	Electronic excitation
10.5	10.5		Electronic excitation
13.3	13.3	$e + CO_2 \rightarrow CO_2^+ + 2e$	Ionization
20.9	20.9	$e + CO_2 \rightarrow CO^+ + O + 2e$	Dissociative ionization[68]
22.6	22.6	$e + CO_2 \rightarrow CO + O^+ + 2e$	Dissociative ionization[68]
24.6	24.6	$e + CO_2 \rightarrow C^+ + O_2 + 2e$	Dissociative ionization[68]

Main source: Lowke, J. J. et al., *J. Appl. Phys.*, 44, 4664, 1973. Last three rows: Crowe, A. and J. W. McConkey, *J. Phys. B: At. Mol. Phys.*, 7, 349, 1974. A more detailed energy list is available in Boness, M. J. W. and G. J. Schulz, *Phys. Rev. A*, 9, 1969, 1974.

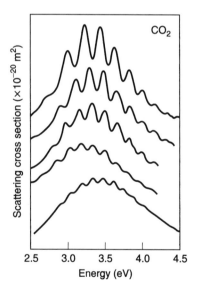

FIGURE 5.8 Schematic diagram of oscillations in vibrational excitation scattering cross sections in the resonance region. The approximate energy range is as shown. Energy levels for individual peaks in excitation mode are given by Čadež et al.[67]

5.1.4 ELECTRONIC EXCITATION CROSS SECTIONS IN CO_2

The excitation processes in CO_2 are quite complex and a presentation of the processes and assignment of states is beyond the scope of our concern. In view of the bending mode, the

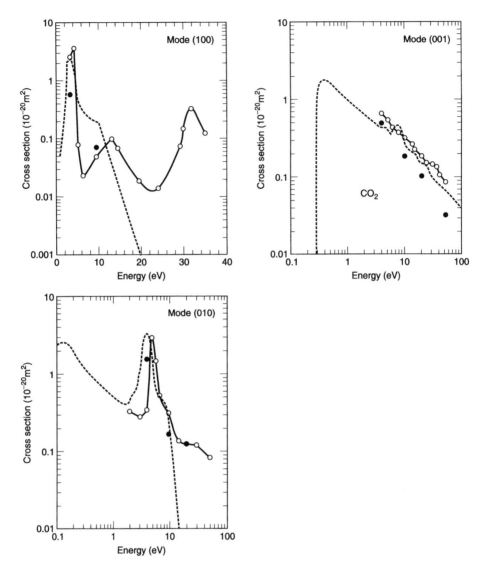

FIGURE 5.9 Selected vibrational excitation cross sections in CO_2. (—) Takekawa and Itikawa[69] for (100) and (001) modes; Takekawa and Itikawa[70] for (010) mode; (— —) Nakamura[31b]; (●) Register et al.[31a] Notice the similarities of the shape of cross sections in Figure 5.4. However, the asymmetric stretching mode (001) is not as small, relative to the other modes, as the theory predicts.

potential energy diagram changes with the angle and is not presented. We restrict ourselves to summarizing the integral cross sections from the discharge point of view, preceded by only a few general comments.

The electronic configuration of the ground state of CO_2, which has 22 electrons (16 valence electrons), is

$$
\left(1\sigma_g\right)^2\left(1\sigma_u\right)^2\left(2\sigma_g\right)^2\left(3\sigma_g\right)^2\left(2\sigma_u\right)^2\left(4\sigma_g\right)^2\left(3\sigma_u\right)^2\left(1\pi_u\right)^4\left(1\pi_g\right)^4
$$
$$
\left(2\pi_u^*\right)^0\left(5\sigma_g^*\right)^0\left(4\sigma_u^*\right)^0\left(1\delta_g^*\right)^0\left(2\pi_g^*\right)^0\cdots
\tag{5.3}
$$

Here the numbers appearing as superscripts are the number of electrons in the molecular orbit and the numbers within the brackets are the principal quantum numbers. The outermost $1\pi_g$ orbital has four electrons and the next higher orbital, $2\pi_u$, is vacant. Attachment of an electron to form the compound state CO_2^- results in an electron in this orbital. Most of the valence states are expected to be bent, whereas the Rydberg states are linear. The reader is referred to Herzberg[71] for an explanation of molecular structure.

The lowest excited state occurs when electrons from the occupied $1\pi_g$ orbital are excited into the first vacant orbital $2\pi_u$.[49] The theoretically computed threshold energy for the triplet state $^3\Sigma_u^+$ is 8.1 eV.[72] However, for the lowest energy triplets and $^1\Sigma_u^+$ singlet states, energy assignments of between 7 and 8.5 eV appear more accurate.[73] If we set the threshold energy for the lowest $^3\Sigma_u^+$ triplet state at 7.0 eV, then the threshold energies for the other states are: $^3\Delta_u$ 7.5 eV, $^{1,3}\Sigma_u^-$ 7.9 eV, and $^1\Delta_u$ 8.0 eV.[49] The singlet $^1\Delta_u$ and $^1\Pi_g$ transitions are located at 8.5 and 9.3 eV respectively.

Higher states correspond to alternative excitation of one or more of the $1\pi_u$, $3\sigma_u$, or $4\sigma_g$ electrons into the valence $2\pi_u$ orbital. The next higher orbital is $5\sigma_g$. Transitions can occur from any one of $4\sigma_g$, $3\sigma_u$, $1\pi_u$, $1\pi_g$ states to any one of $2\pi_u$, $5\sigma_g$, $4\sigma_u$, $1\delta_g$, $2\pi_g$ states, giving a total of 60 transitions! Each state is a member of one of four series. Each series converges to one of the four lowest electronic states of CO_2^+, designated as X $^2\Pi_g$, A $^2\Pi_u$, B $^2\Sigma_u^+$, C $^2\Sigma_g^+$.[72] Transitions from $1\pi_g$ orbital to $2\pi_u$ and $4\sigma_u$ lead to the $^1\Sigma_u^+$ (11.046 eV) and $^{3,1}\Pi_u$ (11.40 eV) states respectively. Transitions from $1\pi_g$ orbital to $1\delta_g$ and $2\pi_g$ are optically forbidden since they have the same symmetry and lead respectively to $^{1,3}\Delta_g$ (12.49 eV), $^{1,3}\Sigma_g^+$ (12.49 eV), and $^{1,3}\Pi_g$ (12.46 eV), $^{1,3}\Phi_g$ singlet–triplet splittings.

Carbon dioxide is one of the most widely studied molecules by electron energy loss spectroscopy. A large number of studies have revealed the following general structure, which may be divided, broadly, into three regions. The first region, in the 7 to 10 eV range, has broad, partly resolved structures. The second region consists of a series of vibrational excitations on a sharply rising backbone, occurring between 10 and 11 eV. The vibrational excitations are referred to as "Rathenau" progressions in spectroscopy literature. They are optically allowed transitions. Finally, the third region stretches from 11 eV to the ionization limit and has a nearly constant level.[49]

Dissociation of CO_2 occurs by several mechanisms, and LeClair and McKonkey[74] list eleven processes covering the threshold energies of 5.45 to 16.54 eV. We consider two mechanisms. The first is dissociative excitation, in which the electron, having sufficient energy, dissociates the molecule and one of the fragments is excited, according to

$$e + CO_2(X^1\Sigma_g^+) \rightarrow CO^*(a^3\pi) + O + e \tag{5.4}$$

Here the oxygen atom is in the ground state. The minimum energy for this process is 11.46 eV (5.45 eV for dissociation and 6.01 eV for excitation of CO). Ajello[76] has given the cross section for this process in terms of relative values. If one adopts the absolute cross section of 2.4×10^{-20} m^2 at 80 eV,[74] one gets the curve shown in Figure 5.10, labeled Ajello (see caption).

The second mechanism for dissociation is that the CO_2 molecule is first excited to the 1 $^1\Sigma_u$ state and then dissociates into fragments according to

$$e + CO_2(X^1\Sigma_g^+) \rightarrow CO(^1\Sigma^+) + O(^1S_0) \tag{5.5}$$

The oxygen atom is a metastable and the threshold energy for the process is 11.0 eV.[74] The cross section has been measured by LeClair and McConkey[74] and their results are also shown in Figure 5.10.

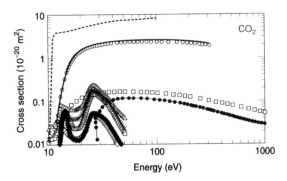

FIGURE 5.10 Excitation and dissociation cross sections in CO_2. (–•–) Strickland and Green,[75] $^1\Sigma_u^+$; (○) Ajello[76]; (– –) Lowke et al.[29]; (□) Leclair and McConkey[74]; (△) Lee et al.[40] X $^1\Sigma_g^+ \to {}^3\Sigma_u^+$; (◇) ibid., X $^1\Sigma_g^+ \to {}^1\Delta_u$; (×) ibid., X $^1\Sigma_g^+ \to {}^3\Delta_u$; (•) ibid., X $^1\Sigma_g^+ \to {}^1\Sigma_u^-$; (■) ibid., X $^1\Sigma_g^+ \to {}^3\Sigma_u^-$; (—) recommended. Untabulated values are digitized from the original publications and replotted.

The more recent theoretical calculations of Lee et al.[40] for the lowest five states in the energy range 10 to 50 eV and those of Strickland and Green[75] for the 1 $^1\Sigma_u^+$ state up to 1000 eV are also shown. The cross section for reaction 5.4 makes the largest contribution and the recommended cross sections are 10% higher than those due to Ajello[76] to account for contributions from other states. In the light of this, the excitation cross sections used by Lowke et al.[29] appear to be not as large as first thought.

5.1.5 Ionization Cross Sections in CO_2

Ionization cross sections have been measured by Rapp and Englander-Golden,[51] Orient and Srivastava,[77] Freund et al.,[78] and Straub et al.[54] The products of ionization are CO_2^+, CO^+, O^+, and C^+, with appearance potentials of 13.769, 20.9, 22.6, and 24.6 eV respectively.[53] Theoretical data are due to Hwang et al.[79] who used the BEB method to calculate the ionization cross sections. Figure 5.11 shows selected data; there is very good agreement between Rapp and Englander-Golden,[51] Straub et al.,[54] and Hwang et al.[79] Cross sections for CO_2^+ and CO_2^{2+} have been measured by Adamczyk et al.,[80] Märk and Hille,[81] and Krishnakumar.[82] The agreement between various investigations for production of CO_2^+ and other ions is less than satisfactory (Figure 5.12) which may partly explain the discrepancies that show up in the total cross section.

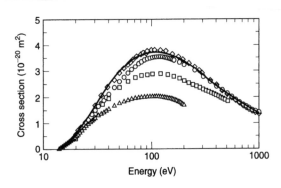

FIGURE 5.11 Total ionization cross sections in CO_2. (○) Rapp and Englander-Golden[51]; (□) Orient and Srivastava.[77]; (△) Freund et al.[78]; (◇) Straub et al.[54]; (—) Hwang et al.,[79] theory.

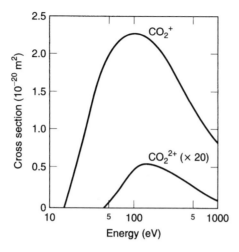

FIGURE 5.12 Partial ionization cross sections in CO_2.

TABLE 5.4
Sigma Check in CO_2

Energy	Q_{el}	Q_v	Q_{diss}	Q_i	Q_Σ	Q_T(expt)
0.155	25	1.3	—	—	26.3	35.3
1.05	5.8	1.4	—	—	7.2	7.7
2.0	4.62	1.69	—	—	6.3	5.84
4.0	11.0	3.90	—	—	14.9	15.4
10.0	10.61	0.59	—	—	11.2	12.0
20.0	14.59	0.26	1.4	0.56	18.46	18.5
50.0	11.7	0.05	2.3	2.66	16.7	17.0
60.0	11.0	—	2.3	3.24	16.5	16.2
90.0	7.5	—	2.5	3.23	13.2	13.4
100	6.8	—	2.5	3.81	13.1	12.5
500	2.98	—	—	2.14	5.12	5.47
800	2.31	—	—	1.57	3.88	3.88
1000	1.92	—	—	1.4	3.32	3.33

Reproduced from Karwasz, G. P. et al., *Riv. Nuovo Cimento*, 24, 1, 2001. With permission of Societa Italiana di Fisica. For sources of entries see original publication.
Energy in eV, and cross section in $10^{-20}\,m^2$.

5.1.6 SIGMA CHECK FOR CO_2

Karwasz et al.[58] have checked the sigma rule. Their data are shown in Table 5.4 without need for further revision.

5.2 HYDROCARBON GASES C_xH_y

Hydrocarbon gases have the general formula C_xH_y and have many applications in industry; moreover, a knowledge of the electron–molecule interactions leads to a better understanding of the constituents of both terrestrial and planetary atmospheres. The simplest hydrocarbon is CH_4 (methane), and its applications in gas radiation counters and diffuse discharges have been explored by Zecca et al.[83]

In view of the relative simplicity of the molecular structure and its almost spherical symmetry, it is a favored molecule for testing theoretical approaches against experimental results. In view of the voluminous literature that exists we consider this section from the user point of view, that is to provide data and references to earlier literature. This approach also has the advantage of examining the influence of molecular parameters such as number of carbon atoms, molecular weight, dipole moment if any, etc. We proceed in order of increasing carbon atoms in the molecule, realizing that every kind of cross section may not be available for each molecule. A recent review of hydrocarbons has been published by Karwasz et al.[84]

5.2.1 TOTAL SCATTERING CROSS SECTIONS IN $C_X H_Y$

Table 5.5 shows molecules in this series in the order of increasing number of carbon atoms. Table 5.6 provides selected references for total cross sections.

TABLE 5.5
Hydrocarbon Molecules

Molecule	Formula	Mol. Wt	Density (kg/m³)	Polarizability (10^{-40} F m²)
Methane	CH_4	16.04	466.0	2.88
Acetylene (or ethyne)	C_2H_2	26.04	620.8	3.71
Ethene (or ethylene)	C_2H_4	28.05	567.8	4.73
Ethane	C_2H_6	30.07	572.0	4.98
Propene (or propylene)	C_3H_6	42.08	519.3	6.97
Cyclopropane	C_3H_6	42.08	519.3	6.97
Propane	C_3H_8	44.11	585.3	7.0
1-butene	C_4H_8	56.12	595.1	8.87
n-butane	C_4H_{10}	58.12	601.2	9.13

Gases ending with the letters "ane" belong to the classification of alkanes and have the general formula C_nH_{2n+2}. Gases ending with letters "ene" belong to the classification of alkenes and have the general formula C_nH_{2n}. If the carbon backbone is a ring the prefix "cyclo" is added.

TABLE 5.6
Selected References for Total Scattering Cross Sections in Hydrocarbon Gases in the 0 to1000 eV Range

Gas	Formula	Authors and Energy Range (eV)
Methane	CH_4	Jones,[87] 1.3 to 50; Ferch et al.,[88] 5 to 400; Floeder et al.,[89] 5 to 400eV; Dababneh et al.,[90] 1.3 to 50; Zecca et al.,[83] 0.9 to 4000; Lohmann and Buckman,[91] 0.1 to 20; Nishimura and Sakae,[92] 1 to 1000 eV
Hydrocarbons	C_nH_{2n}; C_nH_{2n+2}	Floeder et al.,[89] 5 to 400 eV
Acetylene	C_2H_2	Sueoka and Mori,[93] 1 to 400 eV

5.2.1.1 Methane (CH$_4$)

The dominant feature of total cross section in CH$_4$ is the deep Ramsauer–Townsend minimum that occurs at ~0.38 eV. This is also the energy at which the vibrational modes of CH$_4$ set in. A broad structureless maximum exists around ~8 eV and is attributed to a number of resonances in the region.[83] At energies greater than a few hundreds of eV, Auger transitions and inner shell ionizations are possible from energy considerations, but these are not evident in the total scattering cross section, possibly because the processes are two weak in magnitude.[83] There is reasonable agreement on Q_T between various studies in the range of 1 to 1000 eV. Momentum transfer and elastic scattering cross sections are measured by Sohn et al.[85] for scattering energies between 0.2 and 5.0 eV, and by Sakae et al.,[86] 75 to 700 eV, crossed beams method.

5.2.1.2 Other Hydrocarbons

Figure 5.13 shows the total cross sections in selected hydrocarbons that belong to the series C$_n$H$_{2n}$ and C$_n$H$_{2n+2}$. As mentioned above, there is good agreement between various studies in CH$_4$. The total cross sections in the C$_n$H$_{2n+2}$ (Figure 5.13b) and C$_n$H$_{2n}$ groups (Figure 5.13c) show that the total cross sections increase with the increase of number of carbon atoms. Hydrocarbons with an equal number of carbon atoms (n) are close to each other, and well separated from those having $n+1$ carbon atoms. Differences in the configuration with the

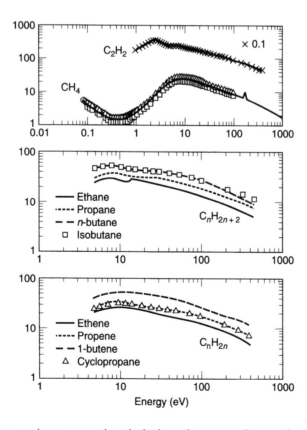

FIGURE 5.13 Total scattering cross sections in hydrocarbon gases. Sources for data are as follows. Methane: Ferch et al.[88]; Lohmann and Buckman[91]; and Zecca et al.[83] Acetylene: Sueoka and Mori.[93] C$_n$H$_{2n+2}$ and C$_n$H$_{2n+2}$: Floeder et al.,[89] 5 to 400 eV. Ethane has a low-lying shape resonance below 5 eV (not shown).[131] See Appendix 3 for details on CH$_4$.

same molecular structure are negligible. For example, propene and cyclopropane have the same molecular formula (C_3H_6) but different arrangements of carbon atoms. The total cross sections for the two gases are nearly the same (Figure 5.13c). Similar comments apply to *n*-butane and isobutane (Figure 5.13b).

Acetylene (C_2H_2) is a triple bond hydrocarbon with a 14 electron system. At low energies this system, along with N_2 and CO, is expected to exhibit a shape resonance phenomenon at about 2 eV due to $^2\Pi$ (or $^2\Pi_g$) symmetry of the molecule.[26] This is evident from Figure 5.13a where, due to the logarithmic scale, the peak appears to be rather small. It has a magnitude of $35.8 \times 10^{-20} \, m^2$.[93]

5.2.2 INELASTIC SCATTERING CROSS SECTIONS IN C_xH_y

5.2.2.1 Vibrational Excitation Cross Sections

Methane has four vibrational levels, scissoring (0.162 eV), twisting (0.190 eV), symmetrical stretch (0.362 eV), and asymmetrical stretch (0.374 eV).[94] They cannot be individually resolved and it is customary to combine them into two vibration bands, sometimes referred to as hybrid stretching and hybrid bending, having energy thresholds of 0.162 eV and 0.374 eV respectively.[95] Other recent references are: Boesten and Tanaka,[96] 1.5 to 100 eV; Mapstone and Newell[97]; and Bundschu et al.,[98] 0.6 to 5.4 eV.

5.2.2.2 Excitation Cross Sections

Excitation occurs at ~9 eV and all excited singlet states are unstable. The cross section for excitation into a singlet state is included in the dissociation cross section data.[89] There are no data on triplet states and, if they are unstable too, then the excitation cross sections are equal to dissociation cross sections and the total inelastic scattering is equal to the sum of ionization and dissociation cross sections. Dissociation cross sections have been measured by Nakano et al.[99] and are shown in Table 5.7.

The ionization potential of CH_4 is 12.51 eV and the fragments of ionization are CH_4^+, CH_3^+, CH_2^+, CH^+, C^+, (H_2^+), (H^+). A mass spectrum is shown in Figure 5.14. Ionization cross sections have been measured by: Rapp and Englander-Golden[51] in CH_4 and C_2H_4, onset to 1000 eV; Chatham et al.[100] in CH_4 and C_2H_6, onset to 300 eV; Orient and Srivastava,[77] CH_4, 10 to 510 eV; Durié et al.,[101] CH_4; Tarnovsky et al.[102] CH_4, 10 to 200 eV; Straub et al.,[103] CH^4, 15 to 1000 eV; Tian and Vidal[104] CH_4, 17.5 to 600 eV; and Zheng and Srivastava,[105] C_2H_2. Tarnovsky et al.[102] measured the cross sections in CD_x ($x = 1$ to 4); the ionization cross sections are insensitive to isotope effects and therefore a direct comparison is possible. Theoretical results are due to Hwang et al.[79] Figure 5.15 and Table 5.8 show the ionization cross sections in selected hydrocarbons, including the data of Nishimura and

TABLE 5.7
Dissociation Cross Sections in CH_4

Energy	10	15	20	30	50	80	100
Q_{diss}	0.30	1.35	2.0	1.65	1.2	0.85	0.6

Energy in units of eV, cross sections in units of $10^{-10} \, m^2$.
Source: Nakano, T. et al., *Jpn J. Appl. Phys.*, 29, 2908, 1991.

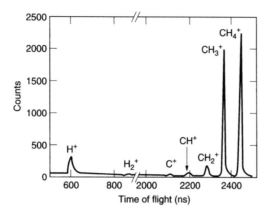

FIGURE 5.14 Mass spectrum of products of ionization of CH_4 for an electron impact energy of 200 eV. Figure reproduced from Tian, C. and C. Vidal, *J. Phys. B*, 31, 895, 1998. With permission of IOP, U.K.

Tawara[106] and Kim et al.[107a] The ionization cross sections increase with the total number of electrons in the molecule and also with molecular dipole polarizability, α_e.

In hydrocarbon gases the asymptotic Bethe formula for Q_i at sufficiently high energies (> 600 eV) is given by[106]

$$Q_i = \frac{4\pi a_0^2 R}{\varepsilon} M_c^2 \ln \frac{4\varepsilon C}{R} \tag{5.6}$$

where ε is the electron energy, a_0 is the first Bohr radius of atomic hydrogen, R is the Rydberg energy, M_c^2 is the square of the dipole matrix elements, and C is a constant. The values of M_c^2 are: CH_4, 4.8; C_2H_4, 9.3; C_2H_6, 12.0; C_3H_6 (cyclopropane), 16.0; propene, 17.0; C_3H_8, 20.0. The authors have also calculated the ratio of ionization cross section to the number of electrons (Q_i/z) and the ratio of ionization cross section to electronic polarizability (Q_i/α_e). For each electron energy each of the two ratios is found to be the same for all molecules of the family. This leads them to conclude that the incoming electron recognizes not only the number of electrons of the target molecule but also the electronic structure (polarizability).

Bart et al.[107b] have measured the ionization cross sections in a number of chlorocarbons and expressed the maximum in the ionization cross section as a ratio of molecular volume polarizability (α_e, units $F m^2$) to ionization threshold energy (ε_i, units eV), using the expression

$$Q_{i(max)} = c \left(\frac{\alpha_e}{\varepsilon_i} \right)^{\frac{1}{2}} \tag{5.7}$$

where c is a constant.

5.3 MERCURY VAPOR

Mercury vapor has important applications in the lighting industry with emission extending to ultraviolet wavelengths. It has been used in arc rectifiers for conversion of high voltage ac to dc and it is potentially a dissociation laser medium. It also has possible applications in space technology for ion thrusters.[108] It has 80 electrons and its ground state configuration is

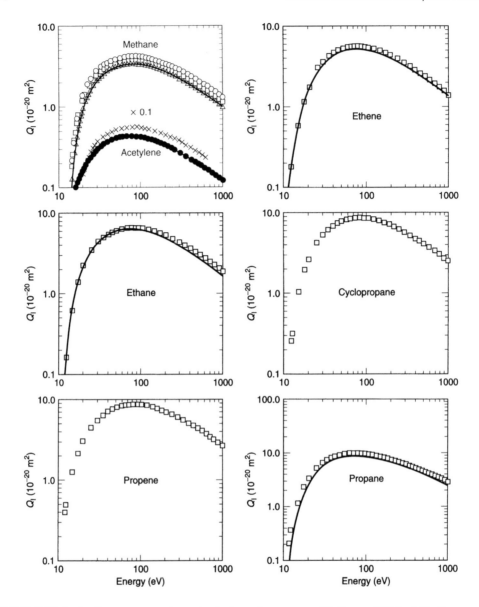

FIGURE 5.15 Ionization cross sections in hydrocarbon gases. Methane (CH_4): ($\triangle$) Rapp and Englander-Golden[51]; ($\square$) Nishimura and Tawara[106]; (—) Hwang et al.[79] Acetylene: ($\times$) Tian and Vidal[104b]; ($\bullet$) Kim et al.[107a]; the cross sections for this gas have been multiplied by a factor of 0.1 for better presentation. Remaining gases: (—) Kim et al.[107]; ($\square$) Nishimura and Tawara.[106]

$1s^2\ 2s^2\ p^6\ 3s^2\ p^6\ 3d^{10}\ 4s^2\ p^6\ d^{10}\ 4f^{14}\ 5s^2\ p^6\ 5d^{10}\ 6s^2$, combining to give the ground term $^1S_0$ (because of the closed subshell $6s^2$). An attaching electron enters, or excitation occurs to the p level.

Total cross sections have been measured by Jost and Ohnemus[109] in the range 0.1 to 500 eV and are shown in Figure 5.16a. In view of the large atom size, the cross section is very high. As an example, the maximum cross section observed is $\sim260 \times 10^{-20}\,m^2$, which is an order of magnitude larger than that in argon. The striking feature of the total cross section is that there is a peak at 0.63 eV which is attributed to $(6s^2\ 6p_{1/2})\ ^2P_{1/2}$ resonance. A brief explanation of this phenomenon is as follows.[110]

TABLE 5.8
Ionization Cross Sections in Hydrocarbon Gases

Energy	CH_4 Methane	C_2H_4 Ethene	C_2H_6 Ethane	C_3H_6 Cyclopropane	C_3H_6 Propene	C_3H_4 Propane
10				0.034	0.093	
12		0.097	0.074			0.206
12.5				0.312	0.498	
15	0.209	0.581	0.618	1.04	1.26	1.14
17.5	0.693	1.15	1.39	1.97	2.15	2.30
20	1.22	1.71	2.24	2.61	3.07	3.31
25	2.01	3.01	3.48	4.27	4.54	5.21
30	2.56	3.52	4.45	5.36	5.54	6.47
35	2.96	4.17	4.94	6.13	6.42	7.37
40	3.23	4.52	5.41	6.71	7.18	8.00
45	3.49	4.82	5.84	7.42	7.54	8.54
50	3.60	5.11	6.04	7.84	8.00	9.22
60	3.86	5.48	6.67	8.27	8.42	9.79
70	3.93	5.74	6.93	8.48	8.82	10.09
80	3.98	5.76	6.86	8.83	9.04	10.20
90	3.98	5.79	6.84	8.87	9.17	10.24
100	3.92	5.70	6.89	8.27	9.02	10.23
125	3.75	5.58	6.53	8.30	8.62	9.90
150	3.55	5.20	6.32	8.10	8.14	9.36
175	3.32	4.80	5.98	7.29	7.83	8.84
200	3.17	4.58	5.68	7.25	7.34	8.35
250	2.86	3.92	5.01	6.59	6.78	7.80
300	2.55	3.56	4.60	5.88	5.99	6.84
350	2.36	3.18	4.18	5.34	5.48	6.25
400	2.17	2.87	3.86	4.88	4.95	5.78
450	1.99	2.64	3.47	4.56	4.66	5.26
500	1.85	2.45	3.33	4.33	4.50	4.93
600	1.62	2.19	3.03	3.77	3.96	4.33
700	1.44	1.96	2.71	3.51	3.61	3.99
800	1.33	1.75	2.38	3.12	3.18	3.67
900	1.22	1.63	2.25	2.77	2.98	3.27
1000	1.13	1.52	2.03	2.58	2.79	3.05

Energy in units of eV, cross sections in units of $10^{-20}\,m^2$.
Source: Nishimura, H. and H. Tawara, *J. Phys. B: At. Mol. Opt Phys.*, 27, 2063, 1994. With permission of IOP, England.

In atoms that have an electron affinity, introduction of an extra electron into the lowest unoccupied orbit results in a bound state of the electron, the product of scattering being a negative ion. For atoms in certain columns of the periodic table, the rare gases most evidently, no bound state of the negative ion is known to exist (Chapter 3). Certain elements in the column IIA (Mg) and column IIB (Zn, Cd, Hg) also form negative ions which are unstable; the electron is said to exist in the repulsive state. The negative ion has a greater energy than the neutral atom and the ion is embedded in the continuum of states of the electron and the atom. Because of its occupation of the continuum states the lifetime is very short and at the energy of formation of the ion a sharp peak in the cross section or resonance is observed. For the nature of the resonance (sharp or broad), see the explanation in Chapter 4.

The differences in the shape resonances between rare gas atoms and group II elements are instructive. In the former, the shells are closed and the extra electron in the lowest unfilled orbit hovers around the outer periphery of the atom. It is not under the influence of the polarization potential of the atom; its wave packet is therefore not localized and the associated increase in the scattering cross section is broad. However, in the case of the Hg atom, only a subshell is filled ($6s^2$), which prevents the formation of a stable negative ion. The extra electron should occupy the lowest unfilled orbital (6p); notice that it has the same principal quantum number as the ground state atom. It therefore experiences a considerable polarization potential of the atom; the resonance is sharper in peak. If we recall the discussion of states of atoms (Appendix 3), the extra electron will have a total angular momentum of $J = \frac{1}{2}$ and, combining with $6s^2$ electrons, a state of $^2P_{1/2}$ results for the ion. The width of resonance in Hg is $\sim 0.4\,\text{eV}$.[110]

The absolute total cross sections, reported by Holtkamp et al.[111] and Peitzmann and Kessler,[112] are also shown in Figure 5.16a. The former authors measured the differential cross sections and obtained the total cross section by adding the inelastic scattering cross sections. The latter authors measured the differential cross sections for excitation and obtained the total cross sections. There is good agreement in the high energy range, as shown.

The momentum transfer cross sections shown in Figure 5.16b are reported by Rockwood[113] and England and Elford.[114] These were obtained from swarm data. Holtkamp et al.[111] have measured the differential cross sections in the 25 to 300 eV range, but integrated values are not given. McEachran and Elford[115] have recently calculated Q_m, the results agreeing with those of England and Elford[114] for the energy range 0.2 to 3.0 eV. The significantly higher values for lower energies are attributed to the presence of dimers, Hg_2.

The lower electronic excitation thresholds of the Hg atom are: $6\,^3P_0$, 4.67 eV; $6\,^3P_1$, 4.89 eV; $6\,^3P_2$, 5.46 eV; and $6\,^1P_1$, 6.7 eV; $6\,^3P_1$ and $6\,^1P_1$ are resonance states whereas the state $6\,^3P_2$ is a metastable. The resonance line $6\,^3P_1{}^0$ at 4.487 eV is the most characteristic line of 253.7 nm in the ultraviolet,[116] used in ultraviolet lamps. The largest contribution to the total excitation cross section is due to the $6\,^1S_0 \rightarrow 6\,^1P_1$ transition and the remaining cross sections add up to $\sim 10\%$.[109]

Peitzmann and Kessler[112] have measured the cross sections to this state at energies of 15, 60, 100 eV and obtain cross sections of 3.5, 3.5, $3.02 \times 10^{-20}\,\text{m}^2$ respectively. The excitation cross sections used by Liu and Raju[117] in the swarm analysis are shown in Figure 5.16c. Srivastava et al.[118] have calculated the differential and integral cross sections for several states, including $7\,^1P_1$, and the cross section for the $6\,^1S_0 \rightarrow 6\,^1P_1$ transition is considerably larger.

The ionization potential of mercury vapor is 10.437 eV. Data on ionization cross sections are not plentiful. The cross sections used by Liu and Raju[117] are higher than those quoted by Holtkamp et al.[111] and Pietzmann and Kessler[112] by a factor of almost two at the peak, but identical to those used by Rockwood.[113] The ionization cross section for production of Hg^+ is obscured by metastable excitations and a dominant presence of auto-ionization above the ionization threshold energies.[116] Obviously there is need for more experimental cross sections to clarify these points.

The sigma rule has been verified in a recent publication of Karwasz et al.[84]

5.4 NITROUS OXIDE (N_2O)

Nitrous oxide (N_2O) is a gas that has been found to be important in the upper atmosphere in the destruction of ozone. It has also been used as a laser medium and in medical applications. N_2O lasers have been used as a secondary frequency standard in areas of spectroscopy within the 10 µm region where the frequency range of the CO_2 laser is inadequate.[119] It is a linear asymmetrical molecule with the structure N–N–O, having 22 electrons; it is isoelectronic with CO_2. Recall that the isoelectronic systems N_2 and CO, with 14 electrons, have similar

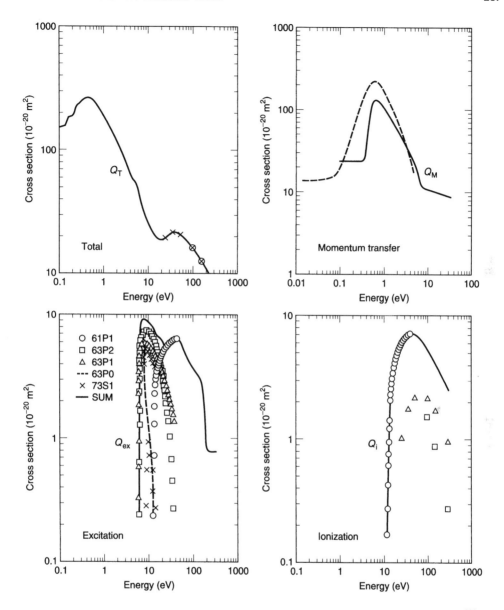

FIGURE 5.16 Absolute cross sections in mercury vapor. Total: (—) Jost and Ohnemus[109]; (×) Holtkamp et al.[111]; (o) Peitzmann and Kessler.[112] Momentum transfer: (—) Rockwood[113]; (— —) England and Elford.[114] Excitation: individual states are shown by points, (— —) sum of Liu and Raju[117]; (+) Srivastava et al.[118]; (—) recommended. Ionization cross sections: (△) Holtkamp et al.[111]; (■) Pietzmann and Kessler[112]; (o) Liu and Raju[117]; (—) recommended.

transport properties,[42] a similar resonance phenomenon and vibrational excitation,[120] and similar behavior in rotational excitation and elastic differential cross sections.[121] Since CO_2 has already been discussed at length, we give only the main features.

N_2O is a polar molecule with a permanent dipole moment of $0.167\,D$ and a polarizability of $3.37 \times 10^{-40}\,F\,m^2$. Kwan et al.[122] quote a dipole moment of $0.161\,D$ and Sarpal et al.[127] theoretically calculate a moment of $0.873\,D$. The bond lengths are $0.113\,nm$ and $0.119\,nm$ for the N–N and N–O bonds respectively, with a bond angle of $134°$.[124] In comparison, the CO_2 molecule is nonpolar, has a polarizability of $3.24 \times 10^{-40}\,F\,m^2$, and a bond length of $0.116\,nm$.

Total scattering cross sections have been measured by: Kwan et al.,[122] 1 to 500 eV; Szmytkowski et al.,[123] 0.4 to 40 eV; Szmytkowski et al.,[123] 40 to 100 eV; and Shilin et al.[124] For energies below 1 eV one has to go as far back as Ramsauer and Kollath,[125] 0.1 to 2 eV (1930), followed in 1984 by Kwan et al.,[122] 0.2 to 7 eV. These cross sections are shown in Figure 5.17.

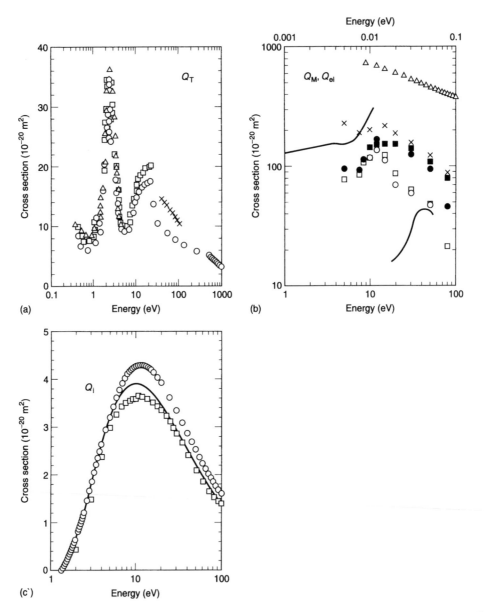

FIGURE 5.17 Scattering cross sections in N_2O. (a) Total scattering cross sections: (○) Kwan et al.[122]; (□) Szmytkowski et al.[123a]; (×) Szmytkowski et al.[123b]; (△) Strakeljahn et al.[138]; (◇) Shilin et al.[124] (b) Momentum transfer cross sections: (△) Singh.[132] Note that the abscissae are shown at the top. The results of Singh[132] and Pack et al.[41] are identical and therefore the latter are not shown separately; (○) Marinković et al.[135]; (□) Johnstone and Newell[119] Elastic scattering cross sections: (●) Marinković et al.[135]; (■) Johnstone and Newell[119]; (×) Michelin et al.[130] (c) Ionization cross sections: (○) Rapp and Englander-Golden[51]; (□) Iga et al.[139]; (—) Kim et al.[140] theory.

The total cross section demonstrates all the three essential features that we have individually observed in the gases discussed so far. There is a sharp resonance around 2.2 eV, attributed to the $^2\Pi$ state of the N_2O^- negative ion.[126] There is a broad maximum at $\sim$20 eV, followed by a rapid decline up to 1000 eV. At the low energy end, below 1 eV, the cross section increases with decreasing energy, the zero energy cross section being large, $5 \times 10^{-18}\,m^2$. This large increase is attributed to thermal resonance capture of electrons. Recall that the polarization of the molecule is a long-range force and attractive to both the particles (Chapter 1).

Theoretical calculations[127] have shown a second resonance at $\sim$8 eV due to the $^2\Sigma$ state of N_2O^-, though the resonance is too weak to be detected in total or elastic scattering cross sections. Vibrational excitation and dissociative attachments show a peak at this energy.[84] In this state the ion has a linear configuration but its energy is higher. The lower energy $^2\Pi$ state has a bent configuration. Recall that the isoelectronic molecule CO_2 has resonances at $\sim$3 eV and 8 eV due to the $^2\Pi_u$ and $^2\Sigma$ states respectively. The somewhat lower energy of the resonance (2.3 eV) in N_2O is attributed to the fact that the molecule is not symmetric, unlike CO_2, and therefore there may be more couplings of the partial waves in N_2O.[127]

At energies greater than 100 eV there is reasonably good agreement between the various studies. At 400 eV a minor fluctuation in the inelastic channel is attributed to the resonance due to inner shell excitation.[128]

Differential cross sections for elastic scattering are measured by Marinković et al.,[129] Johnstone and Newell,[119] Kitajima et al.,[126] and theoretically calculated by Michelin et al.[130] Low energy momentum transfer cross sections are due to Pack et al.,[131] and Singh.[132] Selected data are shown in Figure 5.17(b). Note that the energy range for the theoretical calculation of momentum transfer cross sections is shown at the top. Pack et al.[131] have evaluated the momentum transfer cross sections in the same low energy range as Singh,[132] and the results are in excellent agreement with each other; therefore the former are not shown in the figure. There is reasonable agreement between various studies on the elastic scattering cross section as well. There is paucity of experimental data in the energy range 0.1 to 10 eV for both Q_M and Q_{el}.

Vibrational excitation cross sections have been measured by Azria et al.[133] and Andrick and Hall.[134] In N_2O the states $^2\Pi$ and $^2\Sigma^+$ are situated close together at 2.2 and 2.3 eV respectively. The former leads to electron attachment with the negative ion having essentially the same geometry as the neutral, whereas the latter leads to vibrational excitation with a bent geometry.[127] The process here is dissociative attachment; the electron attaches to the oxygen atom. To study vibrational excitation one must have an energy resolution much better than 100 meV; Azria et al.[133] achieved a resolution of 22 meV. The excited modes are n 00, n 10, n 01, and n 02 series with the quantum number n. The energy quantum for vibrational excitation is 159 meV. The peak cross section for vibrational excitation occurs at 2.3 eV, $Q_v = 9.6 \times 10^{-20}\,m^2$ at the peak.[84]

The major differences in vibrational excitation between CO_2 and N_2O are:

1. The resonance occurs in CO_2 at a higher energy (3.8 eV) than that in N_2O (2.8 eV).
2. The energy dependence of the resonance in CO_2 is a series of sharp peaks and valleys (Figure 5.8) whereas the energy dependence in NO_2 is a smoother structure.
3. The vibrational quantum number n reaches a value up to 20 in CO_2 whereas in N_2O it is limited to a upper value of 7[133] Andrick and Hall[134] detected up to $n = 12$.

The energy loss spectrum of N_2O shows that, in the order of increasing energy, the states are $^3\Sigma^+$, $^3\Delta$, $^1\Delta$, $^3\Pi$, $^1\Pi$, and $^1\Sigma^+$.[135] The low-lying states in the energy range 5 to 9 eV overlap one another. Hall et al.[136] assign energies of 5.6 eV to the triplet states $^3\Sigma^+$ and $^3\Delta$, 6.6 eV to the singlet state $^1\Delta$. Electronic excitation to $C\,^1\Pi$ and $D\,^1\Sigma^+$ states has a threshold

TABLE 5.9
Integral Cross Sections for Excitation to C $^1\Pi$ (8.5 eV threshold) and D $^1\Sigma^+$ (9.6 eV Threshold) States

Energy	15	20	30	50	80
C $^1\Pi$	14.5	24	29.5	65	15.5
D $^1\Sigma^+$	250	500	1025	1500	475

Energy in units of eV, cross sections in units of $10^{-23}\,\mathrm{m}^2$. Correction factors of 0.5 and 2.5 have been applied to obtain absolute cross sections.
Source: Marinković, B. et al., *J. Phys. B: At. Mol. Opt. Phys.*, 19, 2365, 1986; *J. Phys. B: At. Mol. Phys.*, 32, 1949, 1999.

energy of 8.5 and 9.6 eV respectively. Differential scattering cross sections have been measured by Marinković et al.[137] and their integral cross sections are shown in Table 5.9. Data of Strakeljahn et al.,[138] Iga et al.,[139] and Kim et al.[140] are included in Figure 5.17.

Dissociation products of N_2O are $(N_2 + O)$ and $(NO + N)$ in various electronic states The cross sections have been measured by Allcock and McConkey,[141] Mason and Newell,[142] Leclair et al.,[143] and LeClair amd McConkey.[144] The lowest threshold is 7.9 eV, with the resulting fragments of N_2 excited to the A $^3\Sigma^+$ state and oxygen to the ^{3}P state.

The ionization potential of N_2O is 12.89 eV. Total absolute ionization cross sections have been measured by Rapp and Englander-Golden[51] and by Iga et al.[139] and theoretically calculated by the BEB (binary encounter Bethe) model by Kim et al.[140] Figure 5.17c shows the cross sections; there is reasonable agreement between these investigations.

The sigma rule has been verified by Karwasz et al.[84]

5.5   OZONE (O_3)

Ozone is a minor but important constituent of the Earth's atmosphere, but its presence in the stratosphere is a deciding factor in filtering the ultraviolet radiation from the Sun. It is an inherently unstable and explosive gas. It is extremely reactive and will rapidly decompose when in contact with many metallic and non-metallic surfaces.[159] It is a polar molecule with a dipole moment of 0.53 D and is formed during electrical discharges.[145] Electron scattering and ultraviolet light absorption are processes that occur in the atmosphere, and depletion of the ozone layer has become a major environmental threat. The mechanism of ozone production has been presented in Chapter 4. Experimental data are fragmentary, and theoretical results provide supplementary data.

Total scattering cross sections are measured by Gulley et al.,[146] 9 to 10 eV, and by de Pablos et al.,[311] 350 to 5000 eV; these measurements are supplemented by the theoretical calculations of Joshipura et al.[147] in the range 20 to 2000 eV. The high energy cross sections are in excellent agreement with each other, as is shown in Figure 5.18. In the low energy region, between a few meV and 100 meV Gulley et al.[146] find a sharp rise in the total cross section below ~30 eV, with a weak shoulder present between 40 and 60 meV. The low energy rise in cross section (it is $130 \times 10^{-20}\,\mathrm{m}^2$ at 9 meV) may be due to dissociative attachment, rotationally inelastic scattering, or elastic scattering. Above ~1 eV, there is a broad shape resonance at ~4 eV but no other discernible features for higher energies. Momentum transfer and elastic scattering cross sections are measured by Shyn and Sweeney[148] for scattering energy in the range 3.0 to 20.0 eV. There is a paucity of experimental data in the energy range of 10 to 350 eV.

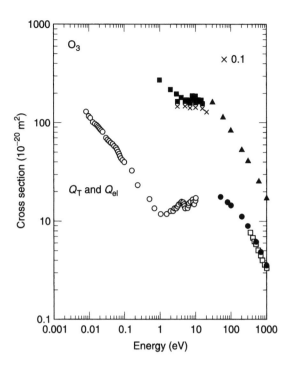

FIGURE 5.18 Total and elastic scattering cross sections in ozone. Q_T: (o) Gulley et al.[146]; (□) de Pablos et al.[311]; (●) Joshipura et al.[147] Q_{el}: ordinates are multiplied by a factor of 10 for clarity; (×) Shyn and Sweeney[148]; (■) Sarpal et al.[149]; (▲) de Pablos et al.[311]

Theoretical results of differential cross sections are derived by Sarpal et al.,[149] 1 to 15 eV, and Bettega et al.,[150] 6 to 30 eV. The latter authors observe structures at 8.5 eV and 13 eV, in agreement with the results of Allan et al.[156] However, the theoretical cross sections (for elastic scattering) are higher than the measured cross sections (Figure 5.18).

The lowest vibrational energy of O_3 is ~89 meV.[146] Vibrational excitation cross sections are measured by Davies et al.[312] in the range of 3 to 7 eV. Vibrational peaks are observed at 100, 260, and 380 meV. The peaks are therefore, composed of many vibrational levels. At ~4 eV evidence exists for the formation of low lying resonance. Allan et al.[151] have resolved the structure into six resonances in the 0.4 to 7.5 eV range, not including the elastic scattering peak due to ground state thermal attachment, forming O_3^-. These resonances are attributed to the following processes:

1. The thermal attachment resonance has a long tail, and formation of O_3^- is observed at 0.4 eV.
2. The dissociation energy is 1.05 eV. At 1.3 eV, formation of O_2^- occurs, with the formation of low kinetic energy O^- continuing to take place. The electron affinity of O_2 is 0.44 eV and the threshold for O_2^- production is at 0.61 eV.
3. At 3.2 eV formation of high kinetic energy O^- occurs.
4. At 4.2 eV there is vibrational excitation.
5. At 6.6 eV there is more vibrational excitation; both these resonances are attributed to the temporary attachment of the electron and excitation of the vibrational levels.
6. Resonance at 7.5 eV is also attributed to O^-.

Allan et al.[151b] have resolved the three modes of vibration: the stretching modes (001 + 100) and the bending mode (010). All three fundamental modes are infrared active.

As stated earlier, the rise in cross sections at low energies is due to the increase in vibrational cross sections. Allan et al.[151b] also find a sharp increase in the cross section between 3 and 6 eV (see above), peaking at 4.2 eV. At this energy the stretching modes dominate over the bending mode.

The leading molecular orbital representation of ozone in its ground electronic state is that of Walker et al.,[152]

$$core + (5a_1)^2(3b_2)^2(1b_1)^2(6a_1)^2(4b_2)^2(1a_2)^2(2b_1)^0(7a_1)^0$$

The electron affinity of ozone is 2.11 eV and the extra electron enters the $2b_1$ orbital. The dissociative electron attachment occurs through the processes.[152]

$$O_3 + e \rightarrow \left[O_3^-\right] \rightarrow \begin{cases} O^- + O_2 & \text{(a)} \\ O_2^- + O & \text{(b)} \\ O^- + 2O & \text{(c)} \\ O_3^* + e & \text{(d)} \end{cases} \quad\quad (5.8)$$

Dissociative electron attachment processes (a) to (c) produce stable negative ions whereas (d) is known as auto-detachment in which an excited molecule results. O^- and O_2^- have appearance potentials of 0 and 0.42 eV respectively. Further references to dissociative attachment in O_3 are provided by Walker et al.[152] and Skalny et al.[153]

Ozone has low-lying excited states near the dissociation limit of 1.05 eV.[151b] The triplet states $^3B_2$, $^3A_2$, $^3B_1$ and the singlet state $^1A_2$ lie close to this energy limit.[154] The energies calculated for these states are 3.27 eV, 5.58 eV, 6.5 eV and 6.37 eV respectively.[155] The energy loss spectrum of O_3 shows a very large elastic cross section peak at thermal energy, followed by smaller peaks called Chappius band and Hartley band, centered around ~2 eV and ~5 eV respectively.[148] Allan et al.[156] and Sweeney and Shyn[157] have reported on the electronic excitation levels and cross sections. The latter authors measured cross sections of Hartley bands (5 eV energy loss) as 69, 88, 95, and 60, all in units of 10^{-22} m², at 7, 10, 15, and 20 eV respectively.

The ionization potential of O_3 is 12.80 eV and the fragments of ionization are O^+, O_2^+, and O_3^+. Partial ionization cross sections are measured by Siegel[158] and Newson et al.[159] and theoretically calculated by Kim et al.[140] The latter authors suggest a normalization factor of 1.558 for the data of Newson et al.[159] Figure 5.19 shows both reported and renormalized values. There are no significant disagreements.

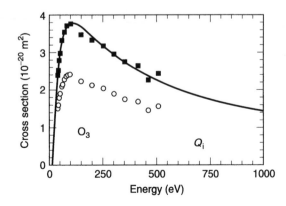

FIGURE 5.19 Ionization cross section in ozone. (o) Newson et al.[159] reported; (■) renormalized; (—) Kim et al.[107a]

5.6   SILANE (SiH$_4$)

Silane (SiH$_4$) is an industrially important gas used extensively in the semiconductor industry for etching processes. It is the silicon analog of methane (CH$_4$) with the silicon replacing the carbon atom. At room temperature it is a gas (boiling point 161 K) and it decomposes into silicon and hydrogen above 420°C, which is a property useful for chemical vapor deposition. Just as methane is a member of the group called alkane (C$_n$H$_{2n+2}$), silane is a member of a group having the general formula Si$_n$ H$_{2n+2}$. Silanes are less stable than their hydrocarbon analogs because the C–C bond is slightly more energetic than the Si–Si bond. The nomenclature for silanes is dependent on the number of silicon atoms: disilane for two, trisilane for three, tetrasilane for four silicon atoms, etc., though the compound with a single silicon atom is called just silane rather than monosilane. The silane molecule has 18 electrons, is nearly spherical, is nonpolar, the molecular diameter is 0.355 nm, its bond strength is 3.979 eV, the bond length is 0.148 nm, the bond angle is 109.5° and it has a rather large electronic polarizability of 6.05×10^{-40} F m^2.[161]

Total scattering cross sections (Figure 5.20) are measured by Wan et al.,[160] 0.5 to 12 eV; Zecca et al.,[161] 75 to 4000 eV; Sueoka et al.,[162] 1.0 to 400 eV; and Szmytkowski et al.[163] The theoretical results of Jain and Thompson,[164] Jain and Baluja,[26] and Jiang et al.[165] cover the desired range. Ohmori et al.,[166] Mathieson et al.,[167] and Kurachi and Nakamura[168] provide a set of cross sections from swarm studies. Momentum transfer and elastic scattering cross sections have been measured by Tanaka et al.[169] The highlights of the total scattering cross section are that there is a Ramsauer–Townsend minimum at 0.3 eV and a shape resonance in the vicinity of 1.8 to 2.2 eV. Jain and Thompson[164] place the Ramsauer–Townsend minimum at 2.8×10^{-20} m^2 for Q_T and 0.4×10^{-20} m^2 for Q_M. The swarm data of Kurachi and Nakamura[168] give a minimum value of 1.1×10^{-20} m^2 for Q_M. For higher energy, there is a monotonic decrease up to 1000 eV energy. Table 5.10 shows the momentum transfer cross sections derived by Kurachi and Nakamura[168] by the swarm method.

The silane molecule has two modes of vibration, the stretching mode (v_1, v_3) and the bending mode (v_2, v_4), with onset energies of 0.271 eV and 0.113 eV respectively.[168] Ohmori et al.[166] have determined the cross sections for these two modes, designated as $Q_{1,3}$ and $Q_{2,4}$. Kurachi and Nakamura[168] have also determined the same cross sections and, for the sake of clarity, we show only their data in Figure 5.20c. Each vibrational cross section has two distinctive peaks, one occurring at an energy just above the threshold. Scattering cross sections for rotationally elastic transitions in the energy range 0.001 to 20 eV have been theoretically calculated by Jain and Thompson.[164] The agreement, at best, is only marginal.

The ionization potential of silane is 11.65 eV. Ionization cross sections have been measured by Chatham et al.[170] Krishnakumar and Srivastava,[171] and Basner et al.[172] The products of ionization are SiH$_3^+$, SiH$_2^+$, SiH$^+$, Si$^+$, H$^+$, H$_2^+$; note that SiH$_4^+$ is not observed as it is very unstable. Theoretical cross sections by the binary encounter Bethe (BEB) method are given by Ali et al.[173] Figure 5.20c and Table 5.11 show selected data. The theoretical results agree well with the measurements of Basner et al.[172] up to 100 eV. The measurements of Krishnakumar and Srivastava[171] are lower by ~25% at the peak of the cross section; Karwasz et al.[84] attribute this to the normalization process.

5.7   SULFUR HEXAFLUORIDE (SF$_6$)

Sulfur hexafluoride (SF$_6$) is an important industrial gas that has applications in diverse fields. In the electrical power industry it has become the gas of choice for extinguishing arcs in high voltage circuit breakers during fault conditions; it is used as a high dielectric strength gas in laser-triggered spark gap closing switches; as a fluorine donor in, for example, rare gas–halide excimer lasers; a gas for plasma etching of silicon and GaAs-based

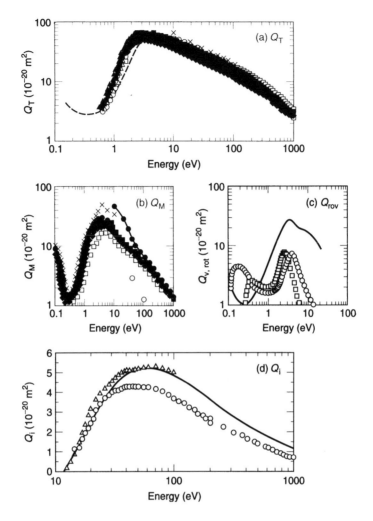

FIGURE 5.20 Scattering cross sections in silane (SiH_4). (a) Total scattering. Measurement: (o) Wan et al.[160]; (▲) Sueoka et al.[162]; (□) Zecca et al.[161]; (△) Szmytkowski et al.[163] Theory: (— —) Jain and Thompson[164]; (—) Jain and Baluja[26]; (–×–) Jiang et al.[165] (b) Momentum transfer cross sections: (—□—) Ohmori et al.[166]; (–◇–) Mathieson et al.[167]; (–×–) Kurachi and Nakamura[168]; (o) Tanaka et al.[169]; (— —) Jain and Thompson[164]; (—●—) elastic scattering cross section, Jain and Baluja,[26] plotted for comparison. (c) Ro-vibrational cross sections: (o) Kurachi and Nakamura,[168] bending mode; (□) Kurachi and Nakamura,[168] stretching mode; (—) Jain and Thompson,[164] rotationally summed. (d) Ionization cross sections: (o) Krishnakumar and Srivastava[171]; (△) Basner et al.[172]; (—) Ali et al.[173]; theory.

semiconductors; a blanket gas for magnesium casting; a reactive gas in aluminum recycling, to reduce porosity; a gas for thermal and acoustic insulation; and many other wide-ranging applications such as retinal detachment surgery, airplane tires, AWACS radar domes, and x-ray equipments.[174b]

SF_6 is a stable molecule, resisting chemical and photolytic degradation, and adding to environmental concerns. It absorbs infrared radiation efficiently. There have been a number of review articles, more recently by Christophorou and van Brunt,[174a] followed by Christophorou and Olthoff,[174b] this being the most comprehensive with nearly 350 citations, and Karwasz et al.[175c] We base our summary on the exhaustive publication of Christophorou and Olthoff,[174b] including their recommended cross sections. (Since the

TABLE 5.10
Momentum Transfer Cross Sections Derived by Kurachi and Nakamura[168] by the Swarm Method

Energy	Q_M	Energy	Q_M	Energy	Q_M
0.01	31.22	0.5	15.68	30	99.75
0.02	24.63	0.7	4.24	40	5.95
0.03	21.38	1	7.74 (6.867)	50	5.35
0.04	18.48	2	32.55 (52.03)	60	5.04
0.05	15.68	3	19.48	70	4.69
0.07	11.75	4	16.72 (55.98)	80	4.35
0.1	7.61 (2.823)	5	15.48 (46.46)	90	4.05
0.2	0.96 (0.921)	7	14.09	100	3.85
0.3	0.71	10	12.10 (20.95)		
0.4	1.45 (2.521)	20	8.82 (9.04)		

The cross sections are obtained by digitizing the original publication and interpolating at specific energies. The numbers in brackets are theoretical cross sections of Jain and Thompson.[164] Energy in units of eV, and cross sections in $10^{-20}\,m^2$.

TABLE 5.11
Ionization Cross Sections in SiH$_4$

Energy (eV)	$Q_i\,(10^{-20}\,m^2)$	Energy (eV)	$Q_i\,(10^{-20}\,m^2)$
15	1.10	200	2.66
20	2.18	250	2.40
30	3.89	300	2.10
40	4.27	400	1.74
50	4.25	500	1.47
60	4.24	600	1.31
70	4.18	700	1.13
80	4.06	800	0.96
90	3.85	900	0.83
100	3.66	1000	0.75
150	3.18		

Source: Krishnakumar, E. and S. K. Srivastava, *Contrib. Plasma Phys.*, 35, 395, 1995.

present volume was completed Christophorou and Olthoff[174c] have published a book on electron interaction in plasma processing gases.)

In the SF$_6$ molecule the sulfur atom is situated at the center, with the six fluorine atoms at the corners of a regular octahedron. Sulfur has 16 electrons, with a valency of six, and the atomic configuration is $1s^2 2s^2 p^6 3s^2 p^4$. Fluorine has nine electrons, with a valency of one and configuration $1s^2 2s^2 3p^5$. The molecule of SF$_6$ thus has 70 electrons, with 48 valence electrons in ten valence orbitals. It has a high molecular weight of 146.0 and tends to hang around the floor of the experimental cells. The distance between the S and F atoms is 2.95 a_0 ($= 0.156$ nm).[176] Its ground state configuration is given by Dehmer et al.[177,178] The ground state designation shown in Equation 5.9 is due to Herzberg.[71]

$$[(S)1s^2\,2p^6(F)1s^2]\,(4a_{1g})^2(3t_{1u})^6(2e_g)^4(5a_{1g})^2(4t_{1u})^6(1t_{2g})^6(3e_g)^4 \times$$
$$[(1t_{2u})^6(5t_{1u})^6](1t_{1g})^6 \tilde{X}^1A_{1g} \tag{5.9}$$

TABLE 5.12
Selected Data on the SF_6 Molecule

Physical Quantity	Value	Reference
Molecular weight	146	
Polarizability	$7.27 \times 10^{-40} \, F \, m^2$	CRC Handbook[265]
SF_5–F bond dissociation energy	$3.9 \pm 0.15 \, eV$	Hildenbrand[180]
S–F bond length of SF_6	$0.1557 \, nm$	Ischenko et al.[181]
F–S–F bond angle of SF_6	$90°$	Chase et al.[182]

Source: Christophorou, L. G. and J. K. Olthoff, *J. Phys. Chem. Ref. Data*, 29, 267, 2000.

The four lowest lying empty states are: $(6 \, a_{1g})^0$, $(6 \, t1_u)^0$, $(2 \, t_{2g})^0$, $(4 \, e_g)^{0,176}$ and the next four states are $(7 \, a_{1g})^0$, $(8 \, a_{1g})^0$, $(7 \, t_{1u})^0$, $(2 \, t_{2g})^{0.179}$ The energies corresponding to these states are: a_{1g}, 2.56 eV; t_{1u}, 7.05 eV; t_{2g}, 11.87 eV; e_g, 27.0 eV;[184] see also Table 6 of Christophorou and Olthoff.[174b] Table 5.12 lists selected properties of the molecule.

5.7.1 TOTAL SCATTERING CROSS SECTION IN SF_6

Rohr[183] measured the absolute differential cross section in the energy range from 0.3 eV to 10 eV and obtained integral cross sections. Total scattering cross sections have been measured by Kennerly et al.,[184] 0.5 to 100 eV; Ferch et al.,[185] 36 meV to 1.0 eV; Romanyuk et al.,[186] 0.1 to 22 eV; Dababneh et al.,[187] 0.98 to 500 eV; Zecca et al.,[188] 75 to 4000 eV; Wan et al.[189] 0.1 to 20 eV and Kasperski et al.,[180] 0.6 to 250 eV. More recent calculations are due to Jiang et al.[165] 10 to 1000 eV. These cross sections are shown in Figure 5.21 along with the cross sections recommended by Christophorou and Olthoff.[174b] The striking features of the total cross sections of SF_6 are as follows:

1. At low energies the total scattering cross sections are mainly due to the attachment processes forming both parent and fragment ions. Dissociation is preceded by the formation of $(SF_6^-)^*$, which has a lifetime of $> 1 \, \mu s$.[191] At near-zero energy the total cross section is $5.2 \times 10^{-18} \, m^2$,[192] which is 50 to 100 times greater than Q_T of other gases. At energies of 0.04 to 0.06 eV the attachment accounts for 60% of the total cross section.[185] As the electron energy decreases toward thermal energy the cross section increases rapidly, attaining a value of $\sim 5 \times 10^{-18} \, m^2$.[185]

    The total cross section is the sum of all contributing partial cross sections. Near zero energy, attachment of the electron to the neutral molecule, forming SF_6^-, is the largest contributor. The resonance at zero energy is measured as broader than it actually is, mainly because of instrumental effects.[174b] Christophorou and Olthoff[174b] recommend an energy of 1.06 eV for the electron affinity of the SF_6^- molecule.

    At 0.01 eV the attachment cross section is about $3.2 \times 10^{-18} \, m^2$.[192] At 0.35 eV there is a resonance in the attachment cross section as a result of the dissociative attachment, leading to $SF_5^- + F$.[185] However, this resonance does not show up in the total scattering cross section measurements since the peak of the resonance is about 7% of the total. It is noted that the total scattering cross section below 0.5 eV has been measured by only one group, namely Ferch et al.[185] who used the TOF spectrometer, and that the base cross section on which the resonance is superimposed lies above $21 \times 10^{-20} \, m^2$.

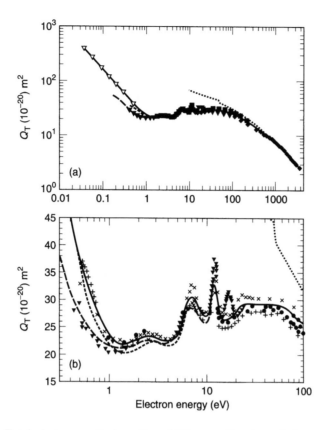

FIGURE 5.21 (a) Total electron scattering, Q_T, of SF_6 as a function of electron energy. (a) (---) Rohr[183]; (+) Kennerly et al.[184]; (∇) Ferch et al.[185]; ($\blacktriangledown$) Romanyuk et al.[186]; ($\bullet$) Dababneh et al.[187]; (o) Zecca et al.[188]; (- - -) Wan et al.[189]; ($\times$) Kasperski et al.[190]; (. . .) Jiang et al.[165]; (—) recommended by Christophorou and Olthoff.[174b] (b) Expanded view. Figure reproduced from Christophorou, L. G. and J. K. Olthoff, *J. Phys. Chem. Ref. Data*, 29, 267, 2000. With permission.

At very low energies the total cross section for electron scattering is almost equal to the scattering cross section of Rydberg atoms impacting on SF_6 molecules. As stated earlier, Rydberg atoms are those in which the electrons are excited to very high principal quantum numbers and the electron in the excited atom is in the "outhouse"; the electron is weakly bound to the nucleus and the core if the molecule is an ion. The scattering behavior of such an atom is, to a first approximation, given by the sum of the electron and the ion core scattering cross sections. The weakly bound electron in the Rydberg atom is considered "quasi-free" and has the same cross section as a free electron that has the same orbital velocity. The total cross section for scattering of Rydberg electrons is measured as $\sim 4.5 \times 10^{-18}\, m^2$,[193] which agrees very well with the total scattering cross section of Ferch et al.[185] at 36 meV.

2. As observed by Kennerly et al.[184] a resonant peak occurs at 2.56 eV and is assigned to the a_{1g} orbital, though an energy value as low as 2.3 eV is reported.[187] This resonance is possibly due to the parent ions SF_6^- and SF_5^-,[186] or, alternatively, due to the fragment negative ion F^-.[194]

3. Resonances at 7.05 eV and 11.87 eV are attributed to the t_{1g} and t_{2g} orbitals respectively. These resonances are due to total electron scattering. The second resonance was also observed by Trajmar et al.[195] in the elastic and vibrationally

inelastic channels at 12 eV. The 11.87 eV resonance has a peak cross section as large as 33.5×10^{-20} m^2.[187] Romanyuk et al.[186] also report a resonance peak at 16.8 eV, but this has not been confirmed by any other group.

4. A broad maximum occurs in the 25 to 55 eV range and is assigned to the e_g orbital, with several possible resonances occurring in this range. Theoretical calculations that take into account multichannel scattering predict a peak at 25 eV,[196] an effect attributed to the sulfur atom. In the energy range above 30 eV inelastic forward scattering yields a higher transmitted proportion and therefore a lower scattering cross section. In this energy range the measured cross sections of Kennerly et al.[184] are therefore a lower bound.

5. In the energy range 80 to 1000 eV the total scattering cross section decreases monotonically from 27.4×10^{-20} m^2 at 80 eV to 8.27×10^{-20} m^2 at 1000 eV.[184,187,188,190] The cross sections of Zecca et al.[188] are generally higher than those of Dababneh et al.[187] by 2 to 6% in the overlapping range from 75 to 700 eV. They are 3 to 5% higher than those of Kennerly et al.[184] in the overlapping range from 75 to 100 eV.

These results are shown in Figure 5.21.

5.7.2 MOMENTUM TRANSFER CROSS SECTIONS IN SF$_6$

Momentum transfer cross sections have been measured by Srivastava et al.[197] 5 to 75 eV; Sakae et al.,[86] 75 to 700 eV; Johnstone and Newell,[198] 5.0 to 75.0 eV; Cho et al.,[199] 2.7 to 75.0 eV. The swarm method has been used to compile momentum transfer cross sections, and the data of Phelps and Van Brunt[200] and of Liu and Raju[201] are cited as typical examples. The measurements of Srivastava et al.[197] were relative to helium cross sections, and when more accurate helium results became available it was realized that renormalization was required.[185] Discussions of these results are given by Cho et al.[199] and Christophorou and Olthoff.[174b] Figure 5.22 shows selected cross sections and Table 5.13 gives the values of Q_M suggested by Christophorou and Olthoff[174b] for the energy range from 2.75 to 700 eV.

The energy range from 0 to 2.75 eV is covered by the data of Phelps and Van Brunt.[200] The cross section for electron attachment at energies below 0.05 eV is at least 60% of the theoretical limit for spherical s-wave scattering. If the elastic scattering section is equated to the largest theoretically possible attachment cross section for s-wave electron scattering below 0.1 eV, the total cross section is the sum of the elastic scattering and inelastic scattering cross sections. Since each of these two is equal to the attachment cross section it is evident that the effective Q_M is simply twice the attachment cross section.

The values of momentum transfer cross section in the energy range below 2.75 eV should be treated with caution. An order of magnitude difference exists between swarm derived cross sections and beam data in the 100 to 200 meV range.[202] As stated above, the values of Phelps and van Brunt[200] should be multiplied by 1.45 to allow for smooth merging with the suggested cross section of Christophorou and Olthoff.[174b]

The effect of increasing Q_M on the transport coefficients has not been examined explicitly in SF$_6$, though an observation may be made that the transport coefficients are relatively insensitive to Q_M in the low energy range.

Chutjian[192] gives the attachment cross section in SF$_6$ at near-zero energy as

$$Q_{att} = 5.20 \times 10^{-18} \times \begin{cases} \exp\left(-\dfrac{\varepsilon}{44.4}\right) \text{ m}^2, & 0 \le \varepsilon \le 45 \text{ meV} \\[2ex] 0.868 \exp\left(-\dfrac{\varepsilon}{51.6}\right) \text{ m}^2, & 45 \le \varepsilon \le 200 \text{ meV} \end{cases} \tag{5.10}$$

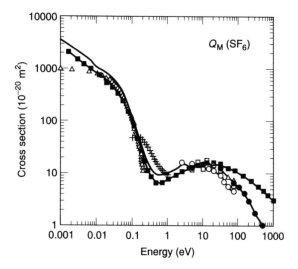

FIGURE 5.22 Momentum transfer cross sections in SF$_6$. (—○—) Srivastava et al.[197] renormalized by Trajmar et al.[195]; (–△–) derived from the attachment cross section of Chutjian[192]; (—■—) Phelps and van Brunt[200]; (●) Sakae et al.[86]; (□) Johnstone and Newell[198]; (◇) Cho et al.[199]; (—) Christophorou and Olthoff,[174b] supplemented by Raju (unpublished) in the range from 0 to 2.75 eV. (+) Elastic collision cross section of Randell et al.,[202] digitized from Figure 7 of Christophorou and Olthoff,[174b] plotted for comparison in the 0.1 to 1 eV range.

TABLE 5.13
Suggested Momentum Transfer Cross Sections in SF$_6$

Energy	Q_M	Energy	Q_M	Energy	Q_M
0.001	3453.8	6.0	15.1	40	10.3
0.002	2542.9	7.0	15.5	45	9.37
0.004	1817.1	8.0	14.8	50	8.65
0.007	1371.7	9.0	14.4	60	7.69
0.01	1054.0	10	15.1	70	7.06
0.02	782.4	11	16.7	75	6.74
0.04	472.5	12	17.6	80	6.46
0.07	223.2	13	17.1	90	6.03
0.1	127.9	14	15.8	100	5.70
0.2	26.0	15	14.9	125	4.92
0.4	10.3	16	14.5	150	4.16
0.7	9.6	17	14.7	200	2.98
1	10.3	18	15.0	250	2.23
2	13.3	19	15.4	300	1.76
2.75	16.0 (15.7)	20	15.7	350	1.47
3.0	15.4	22	15.7	400	1.28
3.5	14.5	25	15.0	450	1.13
4.0	14.0	27	14.3	500	1.02
4.5	13.9	30	13.2	600	0.82
5.0	14.1	35	11.5	700	0.66

Energy in units of eV, cross sections in units of 10^{-20} m^2. The cross sections in the energy range from 0.001 eV to 2.75 eV are due to Phelps and van Brunt,[200] multiplied by a factor of 1.45 to allow for smooth merging at 2.75 eV. The cross sections in the energy range from 2.75 eV to 700 eV are due to Christophorou and Olthoff.[174b] The value in brackets at 2.75 eV is due to Phelps and van Brunt.[200]

The relative agreement of Q_M, calculated on the basis of Equation 5.10, with selected data is also shown in Figure 5.22.

5.7.3 ELASTIC AND DIFFERENTIAL SCATTERING CROSS SECTIONS IN SF$_6$

The differential cross section has been measured by a number of authors, including Srivastava et al.,[197] whose original measurements were renormalized by Trajmar et al.[195]; Rohr[183]; Sakae et al.[86]; Johnstone and Newell[198]; Randell et al.[202] and Cho et al.[199] Rohr[183] had concluded that the cross section below 1 eV was due to direct scattering, but it is now established that attachment is the dominant process that contributes to the increase of cross section as zero energy is approached.

The measurements of Cho et al.[199] are significant in that they employed a magnetic angle-changing technique which extended the angular measurement up to 180°. This technique permits recording of elastic scattering events at 180°. The differential cross sections measured at two electron energies, 5 eV and 75 eV, and comparisons with selected measurements are shown in Figure 5.23. Theoretical results are provided by Benedict and Gyemant,[203] Dehmer et al.,[177] Gianturco et al.,[196] and Jiang et al.[204] A discussion of these results may be found in Christophorou and Olthoff.[174b] Figure 5.24 shows selected elastic scattering cross sections with the values suggested by these authors. Tabulated values are shown in Table 5.14.

5.7.4 VIBRATIONAL EXCITATION CROSS SECTIONS IN SF$_6$

The molecule does not possess either dipole or quadrupole moment; hence rotational excitation is not appreciable. In the energy region up to 3 eV vibrationally elastic and vibrational excitation are the significant processes that occur. SF$_6$ is a seven-atom molecule

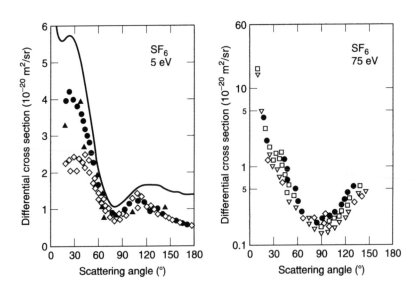

FIGURE 5.23 Differential cross section in SF$_6$ as a function of scattering angle for electrons of 5 eV energy. ($\diamond$) Srivastava et al.[197]; ($\blacktriangle$) Rohr[183]; ($\square$) Johnstone and Newell[198]; ($\bullet$) Cho et al.[199]; (—) Gianturco et al.[196] 75 eV energy: ($\diamond$) Srivastava et al.[197]; ($\triangledown$) Sakae et al.[86]; ($\square$) Johnstone and Newell[198]; ($\bullet$) Cho et al.[199] Figure reproduced from Cho, H. et al., *J. Phys. B*, 33, L309, 2000. With permission of IOP, U.K.

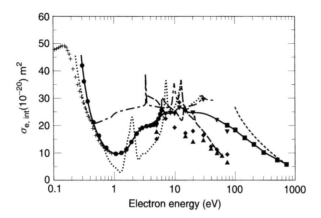

FIGURE 5.24 Elastic scattering cross section in SF_6 as a function of electron energy. Experimental: ($\bullet$) Rohr[183]; ($\blacktriangle$) Trajmar et al.[195]; ($\blacksquare$) Sakae et. al.[86]; ($\blacklozenge$) Johnstone and Newell[198]; (+) Randell et al.[202]; ($\blacktriangledown$) Cho et al.[188] Theoretical: (— —) Benedict and Gyemant[203]; ($\bullet\,\bullet\,\bullet$ — $\bullet\,\bullet\,\bullet$) Dehmer et al.[177]; ($\bullet$– $\bullet$–) Gianturco et al.[196]; (- - -) Jiang et al.[187] Suggested: (—) Christophorou and Olthoff.[174b] Figure reproduced from Christophorou, L. G. and J. K. Olthoff, *J. Phys. Chem. Ref. Data*, 29, 267, 2000. With permission.

TABLE 5.14
Elastic Integral Cross Sections of SF_6, Suggested by Christophorou and Olthoff[174]

Energy	Q_{el}	Energy	Q_{el}	Energy	Q_{el}	Energy	Q_{el}
0.30	45.6	3.5	19.9	17	24.9	90	19.1
0.35	33.0	4.0	20.1	18	24.8	100	18.4
0.40	26.2	4.5	20.6	19	24.8	125	16.9
0.45	21.8	5.0	21.3	20	24.7	150	15.5
0.50	18.7	6.0	23.6	22	24.7	200	13.3
0.60	14.8	7.0	24.2	25	24.7	250	11.8
0.70	12.5	8.0	24.6	30	24.4	300	10.6
0.80	11.1	9.0	24.5	35	24.0	350	9.71
0.90	10.2	10.0	24.8	40	23.5	400	8.95
1.0	9.72	11.0	26.1	45	22.8	450	8.31
1.2	9.73	12.0	26.6	50	22.2	500	7.74
1.5	10.9	13.0	26.5	60	21.3	600	6.76
2.0	14.8	14.0	26.1	70	20.5	700	5.94
2.5	17.8	15.0	25.6	75	20.2		
3.0	19.3	16	25.2	80	19.8		

Energy in units of eV, Q_{el} in units of $10^{-20}\,m^2$.
Source: Christophorou, L. G. and J. K. Olthoff, *J. Phys. Chem. Ref. Data*, 29, 267, 2000. With permission.

with the sulfur atom in the center and the six fluorine atoms symmetrically situated in a octahedral geometry. It belongs to the point group O_h.[205] It has six modes of vibration, v_1 to v_6: v_1 is the strong Raman-active totally symmetric breathing mode and v_3 the strong infrared mode associated with vertical displacement of the apical F atoms in the octahedral structure.[202] These two are the only important modes of vibrational excitation (Figure 5.25). Figure 5.26 shows the 3D plot of energy loss spectra, depicting the onset of v_1, v_3 and $2v_1$, $v_1 + v_3$ modes.

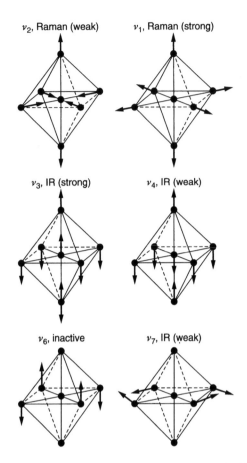

FIGURE 5.25 Normal vibrational modes of SF$_6$. ν_1 and ν_3 are the only significant excited modes. From Randell, J. et al., *J. Phys. B*, 25, 2899, 1992. With permission of IOP, U.K.

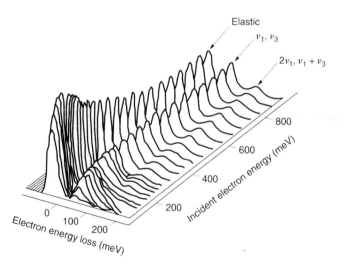

FIGURE 5.26 Energy loss spectra of SF$_6$ for vibrational modes and elastic scattering. $\nu_1 + \nu_3$ means overlapping excitation modes. Energy corresponding to the $2\nu_1$ mode is 190 eV. The elastic scattering intensity is approximately equal to that of vibrational modes. Figure reproduced from Randell, J. et al., *J. Phys. B*, 25, 2899, 1992. With permission of IOP, U.K.

TABLE 5.15
Vibrational Excitation Cross Section of SF_6, Deduced from Total Scattering Cross Section[174]

Energy	Q_V	Energy	Q_V	Energy	Q_V
0.09	1.9	0.50	21.6	4.5	2.0
0.10	7.0	0.60	19.7	5.0	2.4
0.12	21.3	0.70	18.1	6.0	4.4
0.15	30.6	0.80	16.4	7.0	6.5
0.17	30.6	0.90	15.1	8.0	4.6
0.20	34.9	1.0	13.9	9.0	3.1
0.22	35.5	1.2	12.6	10.0	2.5
0.25	33.6	1.5	10.8	11.0	3.5
0.28	30.9	2.0	8.0	12.0	6.3
0.30	29.4	2.5	5.6	13.0	2.4
0.35	26.8	3.0	3.8	14.0	0.5
0.40	25.4	3.5	2.8		
0.45	23.5	4.0	2.3		

Energy in units of eV, and cross section in units of $10^{-20}\,m^2$.
Source: Christophorou L. G. and J. K. Olthoff, *J. Phys. Chem. Ref. Data*, 29, 267, 2000. With permission.

The measurements of Randell et al.[202] are significant in that the energy range was low, 50 meV to 1 eV, with a resolution of 11.5 meV, making it possible to study the excitation of close-lying vibrational modes independently. From Figure 5.26 one sees that the inelastic scattering energy loss at $\sim$400 meV is approximately equal to that due to the elastic scattering. The vibrational energies for the ν_1 and ν_3 modes are 95 and 117 meV,[206] respectively, and the energy loss spectrum reveals that these modes are excited by different mechanisms.

The direct mechanism for exciting the vibration mode is that the free electron wave imparts an impulse to the nucleus of the target molecule. In the case of SF_6 the impulse arises from the long-range polarization force since the molecule does not possess either a dipole moment or a quadrupole moment.[202] The indirect mechanism is that the free electron gets attached to the neutral molecule, forming a negative ion. In the low energy range there is competition between channels for vibrational excitation, SF_6^- and SF_5^- attachment processes. From symmetry considerations, Randell et al.[202] suggest that the ν_1 mode is due to the indirect mechanism whereas the ν_3 mode is due to the direct mechanism.

Table 5.15 shows the vibrational cross sections deduced by Christophorou and Olthoff[174b] by deducting all other cross sections from the total scattering cross section.

5.7.5 ELECTRONIC EXCITATION CROSS SECTIONS IN SF_6

Electronic excitation of the SF_6 molecule is generally believed to lead to dissociation; in other words, the excited states are antibonding. Optical emission from the dissociated fluorine atom has been studied by Blanks et al.,[207] and the cross sections are of the order of $10^{-23}\,m^2$ for electron impact energy in the range from 50 to 400 eV. The light output from an electrical discharge in the gas is very low till the final phase of the spark. Corona exists from sharp points, as is evidenced from current flow, without the associated light output.

Spectroscopic study of the molecule has shown that, unlike most polyatomic molecules, the molecule does not exhibit sharp band systems.[208] Absorption of light occurs at 21 nm wavelength and the gas is transparent to 110 nm radiation. At still smaller wavelengths very

TABLE 5.16
Threshold Energy of SF$_6$ for Dissociation into Neutral Fragments

Reaction	Threshold Energy (eV)
$SF_6 + e \rightarrow SF_5 + F + e$	9.6
$SF_6 + e \rightarrow SF_4 + F_2 + e$	11.3
$SF_6 + e \rightarrow SF_4 + 2F + e$	12.1
$SF_6 + e \rightarrow SF_3 + F + F_2 + e$	15.2
$SF_6 + e \rightarrow SF_3 + 3F + e$	16.0
$SF_6 + e \rightarrow SF_2 + 2F_2 + e$	17.0
$SF_6 + e \rightarrow SF_2 + 2F + F_2 + e$	17.8
$SF_6 + e \rightarrow SF_2 + 4F + e$	18.6
$SF_6 + e \rightarrow SF + F + 2F_2 + e$	21.1
$SF_6 + e \rightarrow SF + 3F + F_2 + e$	21.9
$SF_6 + e \rightarrow SF + 5F + e$	22.7

Source: Itjo, M. et al., *Contrib. Plasma Phys.*, 35, 405, 1995.

broad absorption maxima occur at 105.4, 93.6, 87.2, and 83.0 nm wavelengths. The energy corresponding to these maxima are 11.76 eV, 13.25 eV, 14.22 eV, and 14.94 eV respectively. The various dissociation energies of the molecule, given by Ito et al.,[209] are shown in Table 5.16.

In addition to the normal neutral fragments, excited and metastable state neutral fragments have been observed as dissociation products. These are F^R, F_2^R, S^R (superscript R indicates Rydberg state), and other metastables; the energy range for these reactions is ∼18 to 45 eV.[210] Tabulated cross sections for dissociation as a function of electron energy in the range from 15 to 200 eV are given by Christophorou and Olthoff.[174b] The highest dissociation cross section is 3.7×10^{-20} m^2 at 25 eV.

Electronic excitation occurs under both optical and nonoptical conditions. In energy loss measurements optical conditions are associated with high energy and low angle of scattering, whereas nonoptical conditions are associated with low energy and high angle of scattering. The lowest level electronic transition is $1\,t_{1g} \rightarrow 6\,a_{1g}$ which is an orbital excitation that occurs at 9.6 eV and 9.9 eV for triplet and singlet components. Electronic excitation has been measured by Trajmar and Chutjian[179] in the energy range from 7 to 20 eV, and the cross section at 20 eV has an approximate value of 1.0×10^{-20} m^2. Energy loss measurements yield an excitation threshold of 9.8 eV for the triplet state of the orbital transition $1\,t_{1g} \rightarrow 6\,a_{1g}$.

5.7.6   IONIZATION CROSS SECTIONS IN SF$_6$

The electron impact ionization cross section has been measured by Asundi and Craggs,[211] Rapp and Englander-Golden,[51] Stanski and Adamczyk,[212] Margreiter et al.,[213,214] Rao and Srivastava,[215] and Rejoub et al.[216] Theoretical calculations are provided by Hwang et al.,[217] Tarnovsky et al.,[218] Kim and Rudd,[219] and Deutsch et al.[220] These data, along with those recommended by Christophorou and Olthoff,[174] are shown in Figure 5.27. Table 5.17 shows the ionization cross sections recommended by Christophorou and Olthoff[174] and Rejoub et al.,[216] the latter being published subsequently.

The ionization fragments and appearance potentials are SF_5^+ (15.7 eV), SF_4^+ (18.0 eV), SF_3^+ (19.0 eV), SF_2^+ (26.0 eV), SF^+ (31.0 eV), S^+ (37.0 eV), F^+ (35.8 eV), SF_4^{++} (40.6 eV) and SF_2^{++} (40.0 eV). SF_6^+ ion is not observed since it is highly unstable. Statistics of the relative abundance of ions show that SF_5^+ is the most abundant ion fragment and that the cross

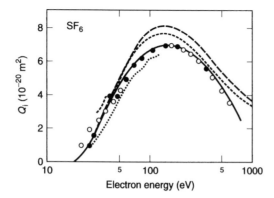

FIGURE 5.27 Ionization cross section in SF_6 as a function of electron energy. Measurements: (- - -) Asundi and Craggs[211]; (●) Rapp and Englander-Golden[51]; (o) Stanski et al.[212]; (- - -) Margreiter et al.[213]; (. . . .) Margreiter et al.[214]; (■) Rao and Srivastava.[215] Theory: (- — - —) Hwang et al.[217]; (–•–•–) Tarnovsky et al.[218]; (— ••• — ••• —) Kim and Rudd.[219] (—) Recommended by Christophorou and Olthoff.[174b] Figure reproduced from Christophorou, L. G. and J. K. Olthoff, *J. Phys. Chem. Ref. Data*, 29, 267, 2000. With permission.

TABLE 5.17
Total Ionization Cross Sections in SF_6

Energy (eV)	Q_i ($\times 10^{-20}$ m^2) Rejoub et al.[216]	Christophorou and Oltho[174b]	Energy (eV)	Q_i ($\times 10^{-20}$ m^2) Rejoub et al.[216]	Christophorou and Oltho[174b]
16.5		0.020	80	5.761	5.95
17.0		0.035	90	6.010	6.28
17.5		0.055	100	6.378	6.53
18	0.060	0.084	120	6.522	—
19.0		0.155	140	6.701	—
20	0.190	0.240	150		6.97
22.5	0.485	—	160	6.689	—
25	0.942	1.04	200	6.601	6.83
27.5	1.339	—	250	6.151	6.48
30	1.760	1.93	300	5.853	6.04
32.5	2.316	—	350		5.60
35	2.696	2.87	400	5.069	5.16
40	3.334	3.47	500	4.617	4.36
45	3.705	3.79	600	4.176	3.65
50	4.234	4.35	800	3.424	
60	5.012	5.09	1000	3.875	
70	5.482	5.65			

section for a doubly charged ion is small.[212,213] Figure 5.28 shows the relative cross sections from the most recent study of Rejoub et al.[216]

5.8 WATER VAPOR (H_2O)

Water is a gift to planet Earth and sustains life. Medical applications such as nuclear magnetic resonance imaging for noninvasive diagnosis of internal organs has become

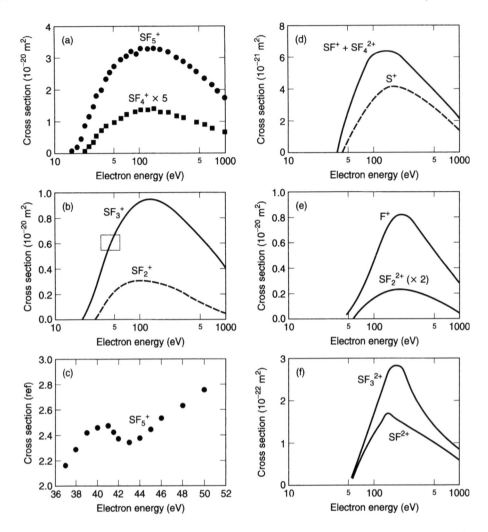

FIGURE 5.28 Partial impact ionization cross sections in SF_6. Structure is observed in the SF_3^+ cross section in the vicinity of 44 eV as shown in (c) with an enlarged abscissa. Figure reproduced from Rejoub, R. et al., *J. Phys. B*, 34, 1289, 2001. With permission of IOP, U.K.

possible because of the water content of human tissue. For example, human heart tissue has 80% water and that in the white matter of the brain is even higher at 84%.[221] Water is one of the few substances that exists in solid, liquid, and gaseous phases at temperatures experienced on the Earth. Yet the molecular structure is relatively simple with an inverted V shape, the OH bonds making an angle of ~103° with each other. It is a very good dielectric with high resistivity in pure conditions and is used as a coolant in a number of specialized applications such as electromagnets.[222] It has a dipole moment of 1.85 D and a high dielectric constant of ~81.[59] Microwave cooking has become possible because of the water content of food. Figure 5.29 provides a summary of data of various cross sections.

Total scattering cross sections have been measured by Sueoka et al.,[223] 1.0 to 400 eV; Zecca et al.,[224] 81 to 3000 eV; Nishimura and Yano,[225] 7 to 500 eV; Sağlam and Aktekin,[226] 25 to 300 eV; Sağlam and Aktekin,[227] 4 to 20 eV. The total cross section decreases monotonically up to 1000 eV energy. In the range of energy studied by Zecca et al.[224] the cross sections of diatomic molecules such as O_2 and N_2 are greater than that of H_2O. However, for low energy electrons with up to 35 eV energy, the cross section of H_2O is higher

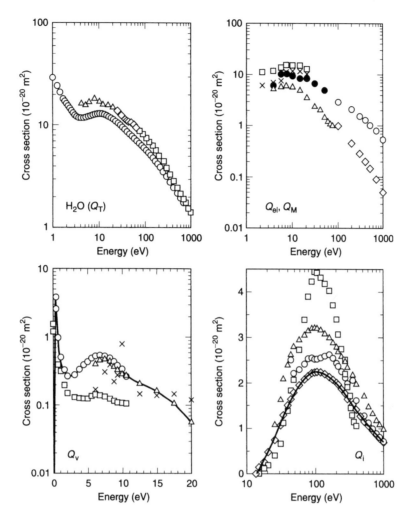

FIGURE 5.29 Scattering cross sections in the H_2O molecule. (a) Total cross section: (○) Sueoka et al.[223]; (□) Zecca et al.[224]; (◇) Sağlam and Aktekin[226]; (△) Sağlam and Aktekin.[227] (b) Elastic scattering and momentum transfer cross section: (○) Katase et al.,[229] Q_{el}; (□) Shyn and Cho,[230] Q_T. Q_M: (△) Danjo and Nishimura[228]; (◇) Katase et al.[229]; (×) Shyn and Cho[230]; (●) Johnstone and Newell.[231] (c) Vibrational excitation: (—●—) Seng and Linder,[236] 100 + 001 mode; (—□—) 010 mode; (—△—) El-Zein et al.,[239] 100 + 001 mode; (×) El-Zein et al.,[239] 010 mode. (d) Ionization cross section: (○) Bolorizadeh et al.[242]; (□) Orient and Srivastava[243]; (△) Rao et al.[245]; (◇) Straub et al.[246]; (- - -) Hwang et al.,[217] theory; (—) recommended.

than that of O_2. This may possibly be due to the fact that the contribution of the long-range potential due to the large dipole moment of H_2O is not as significant at higher electron energies.

Elastic scattering cross sections have been measured by Danjo and Nishimura,[228] 4 to 200 eV; Katase et al.[229] in the high energy range from 100 to 1000 eV; Shyn and Cho,[239] 2.2 to 20 eV; and Johnstone and Newell.[231] The differential cross section for elastic scattering shows strong backward scattering, which is consistent with theory that predicts such behavior for energy greater than 10 eV.

Momentum transfer cross sections have been derived by Shimamura,[232] and theoretical calculations have been published by Gianturco and Thompson,[233] Jain and Thompson,[234] and Rescigno and Lengsfield.[235]

The rotational excitation cross section has been measured by Seng and Linder,[236] Jung et al.,[237] (see also Appendix 3), 2.14 to 6.0 eV, Cvejanović et al.,[238] and El-Zein et al.[239] The vibrational modes of the H_2O molecule are (100), symmetric stretch; (001), asymmetric stretch; and (010), bending mode.[238] The vibrational energy quanta for the modes are 453 meV, 466 meV, and 198 meV respectively. In view of the small energy difference between the symmetric stretch (100) and the asymmetric stretch, they are difficult to resolve separately and the cross section is measured for the (100 + 001) modes together. Cvejanović et al.[238] have extended the study of vibrational levels up to 3 eV, that is, for progressions (n 0 1) and (n 1 1).

As stated earlier in connection with diatomic molecules, vibrational excitation occurs by a direct method due to an impulse imparted by the electron to the target molecule, or by resonance due to negative ion formation. In the H_2O molecule both mechanisms are superposed, that is, two resonances are superposed on a background of direct excitation.

Attachment of an electron to the molecule yields H_2O^- and the strong dipole moment of the molecule (1.85 D) results in bonding. The critical dipole moment, that is the minimum value for a stationary dipole to bind an electron, is 1.625 D,[236] which is just below that of the dipole moment of the molecule. The dipole field introduces new features in the scattering behavior and in H_2O one observes a sharp rise in the scattering cross section at $\sim$0.6 eV which is 0.15 eV above the (100 + 001) mode. This type of behavior is referred to as threshold structure[183] or dipole dominated resonance[240] in the literature. It is observed in several other polar molecules such as hydrogen fluoride (HF, dipole moment = 1.82 D) and hydrogen chloride (HCl, dipole moment = 1.11 D).

In the low energy region, 0 to 10 eV, one observes a sharp threshold resonance and a broad resonance region in the range of 6 to 8 eV. The second resonance is similar to that observed in H_2 at 3 eV and in N_2O at 2.3 eV. Figure 5.29c shows the vibration cross section in the 0 to 20 eV range.

Ionization cross sections have been measured by Mark and Egger,[241] Bolorizadeh and Rudd,[242] Orient and Srivastava,[243] Djurić et al.,[244] Rao et al.,[245] and Straub et al.[246] Theoretical calculations are given by Hwang et al.,[217] using the binary encounter Bethe (BEB) model. The products of ionization are H_2O^+, OH^+, H^+, O^+, O_2^+, and H^{2+}, depending on the electron energy. There are considerable discrepancies in the partial ionization cross sections and the agreement on the total ionization cross section is less than satisfactory. The theoretical calculations of Hwang et al.[217] agree excellently with those of Straub et al.[246] Figure 5.29d shows selected data along with the recommended cross sections; the latter are identical to those of Straub et al.[246]

5.9 PLASMA PROCESSING GASES

The gases considered in this section are important from the industrial point of view, and a series of excellent reviews has been published by Christophorou and colleagues at the National Institute of Standards and Technology (NIST). This allows us to provide consolidated cross section data with only a very brief description. The gases considered by these authors are: tetrafluoromethane (CF_4),[247,248] methyl fluoride (CH_3F),[249] dichlorodifluoromethane (CCl_2F_2),[250] hexafluoroethane (C_2F_6),[251] and perfluoropropane (C_3F_8).[252] Four of these gases are treated below.

5.9.1 Tetrafluoromethane (CF_4)

Tetrafluoromethane (CF_4) is a widely used feed gas in material processing operations that employ plasma. It belongs to the general category of gases called halides, incorporating halogen (F, Cl, Br, I) atoms. CF_4 has a tetrahedral structure with 42 electrons. It does not

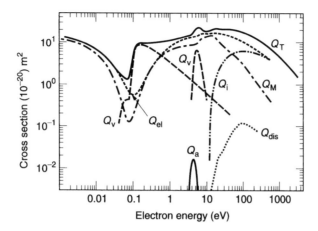

FIGURE 5.30 Scattering cross sections in CF_4 recommended by Christophorou et al.[247] A Ramsauer–Townsend minimum is seen in the total scattering cross section. Q_v^- means vibrational excitation due to negative ion resonance. For revised cross sections see Christophorou and Olthoff.[248] Figure reproduced from Christophorou, L. G. et al., *J. Phys. Chem. Ref. Data*, 25, 1341, 1996. With permission.

have a dipole or quadrupole moment and its electronic polarizability is between 3.04 and 3.26×10^{-40} F m². In common with other molecules containing halogens, it is moderately electronegative.

Figure 5.30 depicts the consolidated cross sections in CF_4. The total scattering cross section shows a Ramsauer–Townsend minimum at 0.16 eV and is consistent with the behavior of CH_4 in this respect (Figure 5.13). There are two broad resonances: one in the 6 to 8 eV range and the other at ~20 eV. Beyond this energy there is a monotonic decrease of Q_T up to 3000 eV. The momentum transfer and elastic scattering cross sections generally follow the trend of Q_T, the Ramsauer–Townsend minimum in Q_M being deeper than those in Q_T or Q_{el}.

The low energy inelastic scattering losses below 12.5 eV are due to vibrational excitation. The CF_4 molecule does not have either dipole moment or quadrupole moment; hence the rotational excitation is relatively negligible. It has four modes of vibration, designated as v_1 (stretch mode, energy quantum 0.112 eV), v_2 (symmetric bend, energy quantum 0.054 eV), v_3 (asymmetric stretch, energy quantum 0.157 eV), and v_4 (asymmetric bend, energy quantum 0.078 eV). Mostly the stretching modes ($v_1 + v_3$) and the bending modes ($v_2 + v_4$) are resolved together. The v_3 mode energy is almost equal to the Ramsauer–Townsend minimum and occurs as a result of direct excitation. The bending modes are attributed to the negative ion resonance, indicated as Q_v^- in Figure 5.30.

The excitation potential is 12.51 eV. In the excited state the molecule is highly unstable and dissociates or predissociates. The energy of dissociation into neutral fragments is equal to the lowest excitation energy, 12.5 eV. There is no light emission from the molecule. The ionization threshold potential is 16.20 eV and the parent ion CF_4^+ is not observed, due to dissociation. Fragments of ionization are CF_3^+, CF_2^+, CF^+, C^+, F^+, CF_2^{2+}, and CF_3^{2+}. The total ionization cross section increases monotonically, reaching a maximum at 120 eV. Data on ionization cross section by Sieglaff et al.,[253] threshold to 1000 eV, dissociative ionization cross section by Torres et al.,[254] up to 100 eV, and calculations by Torres et al.[255] have been published since the 1888 review of Christopher and Olthoff.[248]

A brief summary of several reaction processes for plasma modeling is given by Kurihara et al.[310] and is reproduced in Table 5.18. Table 5.19 gives the ionization cross sections of Sieglaff et al.[253] as they were published after the review of Christophorou and Olthoff.[248]

TABLE 5.18

Reaction Processes with Threshold Energy for Modeling of Discharge in CF_4

Type of Collision	Symbol	Reaction	Threshold (eV)
Momentum transfer	Q_M	$CF_4 + e \rightarrow CF_4 + e$	0
Vibrational excitation	Q_{v1}	$CF_4 + e \rightarrow CF_4\,(v_1) + e$	0.108
Vibrational excitation	Q_{v3}	$CF_4 + e \rightarrow CF_4\,(v_3) + e$	0.168
Vibrational excitation	Q_{v3}	$CF_4 + e \rightarrow CF_4\,(v_4) + e$	0.077
Electronic excitation	Q_{ex}	$CF_4 + e \rightarrow CF_4^j + e$	7.54
Dissociative attachment	Q_a	$CF_4 + e \rightarrow CF_3 + F^-$	6.4
Dissociative ionization	Q_{i1}	$CF_4 + e \rightarrow CF_3^+ + F + 2e$	16.0
Dissociative ionization	Q_{i2}	$CF_4 + e \rightarrow CF_2^+ + 2F + 2e$	21.0
Dissociative ionization	Q_{i3}	$CF_4 + e \rightarrow CF^+ + 3F + 2e$	26.0
Dissociative ionization	Q_{i4}	$CF_4 + e \rightarrow C^+ + 4F + 2e$	34.0
Dissociative ionization	Q_{i5}	$CF_4 + e \rightarrow F^+ + CF_3 + 2e$	34.0
Dissociative ionization	Q_{i6}	$CF_4 + e \rightarrow CF_3^{2+} + F + 3e$	41.0
Neutral dissociation	Q_{d1}	$CF_4 + e \rightarrow CF_3 + F + e$	12.0
Neutral dissociation	Q_{d2}	$CF_4 + e \rightarrow CF_2 + 2F + e$	17.0
Neutral dissociation	Q_{d3}	$CF_4 + e \rightarrow CF + 3F + e$	18.0

Source: M. Kurihara, et al., *J. Phys. D: Appl. Phys.*, 33, 2146, 2000. With permission of IOP, England.

TABLE 5.19

Total Ionization Cross Sections in CF_4[253]

ε	Q_i	ε	Q_i	ε	Q_i
18	0.041	60	3.23	250	4.55
20	0.16	70	3.92	300	4.25
23	0.40	80	4.20	400	3.72
27	0.84	90	4.43	500	3.29
30	1.26	100	4.66	600	2.96
35	1.79	120	4.84	800	2.45
40	2.19	140	4.87	1000	2.12
45	2.60	160	4.89		
50	3.02	200	4.77		

Electron energy in units of eV, cross section in units of $10^{-20}\,m^2$.
Source: Sieglaff, D. R. et al., *J. Phys. B: At. Mol. Opt. Phys.*, 34, 799, 2001.

5.9.2 PERFLUOROETHANE (C_2F_6)

Perfluoroethane is a moderately attaching gas with 66 electrons having D_{3d} symmetry. It is nonpolar, possessing an electronic polarizability lying between 5.1 and $7.23 \times 10^{-40}\,F\,m^2$. Cross sections, compiled on the basis of Karwasz et al.[257] are shown in Figure 5.31. Total cross sections are reported by Sanabia et al.,[257] Christophorou and Olthoff,[251] and Szmytkowski et al.[258] Elastic scattering cross sections are taken from Takagi et al.[259] and Christophorou and Olthoff.[251] Momentum transfer: Christiphorou and Olthoff.[251] Ionization: Nishimura et al.[256] and Christophorou et al.[251] Total dissociation: Winters and Inokuti.[262]

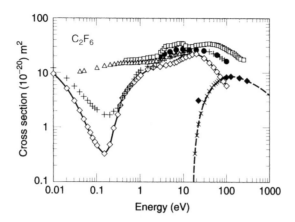

FIGURE 5.31 Scattering cross sections in C_2F_6. Total: (o) Sanabia et al.[257]; ($\triangle$) Christophorou and Olthoff[251]; ($\square$) Szmytkowski et al.[258] Elastic: ($\bullet$) Takagi et al.[259]; (+) Christophorou and Olthoff.[251] Momentum transfer: (—•—) Christiphorou and Olthoff.[251] Ionization: (— —) Nishimura et al.[256]; ($\times$) Christophorou and Olthoff.[251] Total dissociation: ($\blacklozenge$) Winters and Inokuti.[262]

Total scattering cross section measurements are reported by Sanabia et al.[257] and Szmytkowski et al.[258] Below 1 eV, towards zero, Q_T shows an increasing trend, though the peak has not actually been observed. Since $C_2F_6^-$ has not been observed to exist, this increase of Q_T toward lower energy is attributed to scattering showing Ramsauer–Townsend effects. Above 1.0 eV, two structures are observed at 5.0 and 9.0 eV; the origin of the first of these is uncertain. It is possibly due to dissociative attachment forming F^- and CF_3^- ions,[257] or to vibrational excitation.[259] The second resonance is attributed to the peak in the vibrational excitation cross section.[257] Christophorou and Olthoff[248] suggest values of total scattering cross section in the energy range from 0.04 to 20 eV. Szmytkowski et al.[258] observe a broad enhancement at impact energies ranging from 20 to 60 eV which is attributed to a general increase of both elastic scattering and ionization cross sections in this energy range.

The momentum transfer cross section suggested by Christophorou and Olthoff[251] has a deep Ramsauer minimum cross section of $0.32 \times 10^{-20}\,m^2$ at 0.15 eV. The elastic scattering minimum is not as deep as is expected from scattering theory. Elastic scattering and momentum transfer cross sections are obtained by Takagi et al.[259] in the energy range 2 to 100 eV. The vibrational excitation has complicated combinations. For example, ν_1 designates 3 C–F elongation + C–C contraction with an allowed frequency of 152.3 meV, ν_2 designates 3 FCC angle increase + C–C contraction with allowed frequency of 100.1 meV, ν_3 designates CF_3–CF_3 stretch with allowed frequency of 43.2 meV energy.[259] Hayashi and Niwa[260] have derived from swarm measurements the excitation cross section for several vibrational modes. Pirgov and Stefanov[261] have evaluated the vibrationally inelastic and momentum transfer cross sections from swarm data, using the Boltzmann code, in the energy range 82.5 meV to 8.0 eV.

Figure 5.31 does not show the attachment cross sections as they will be considered separately in Chapter 7. The ionization potential of the C_2F_6 molecule is 15.9 eV[251] and the total ionization cross section is entirely due to total dissociative ionization. Nishimura et al.[256] have measured the total ionization cross section in the energy range from 16 to 3000 eV since the review of Christophorou and Olthoff.[251] The total dissociation cross section of Winters and Inokuti,[262] which is the sum of the cross sections for dissociative ionization and neutral fragment production, with limited energy overlap, is also shown for comparison.

5.9.3 PERFLUOROPROPANE (C_3F_8)

The perfluoropropane (C_3F_8) molecule has 90 electrons and is electron attaching. Christophorou and Olthoff[252] state that the molecule is nonpolar whereas Tanaka et al.[263] quote a small dipole moment of 0.097 D. The calculated electronic polarizability has values of 7.20, 8.19, and 1.05×10^{-39} F m^2, depending on the model employed. The photon absorption cross section peaks at 119 nm (10.42 eV) in the nonvisible region of the spectrum. The cross sections compiled by Karwasz et al.[84] are shown in Figure 5.32.

The total scattering cross section has three maxima at about 3.9, 6.6, and 9.0 eV, possibly due to dissociative negative ion formation. In contrast with CF_4 and C_2F_6, a parent negative ion with a lifetime of 0.1 ns exists but there is no evidence of increasing scattering cross section toward zero energy. The competition between dissociative attachment and attachment to parent molecule increases in favor of the former as the temperature increases. CF_4 and C_2F_6 both show a Ramsauer–Townsend minimum and it is not unlikely that the same effect will be observed if measurements are extended towards energies lower than 25 meV, which is the lower limit of measurement by Sanabia et al.[257] The region between 0.2 and 2 eV shows enhanced scattering, possibly due to direct vibrational excitation. The recommended Q_T of Christophorou and Olthoff[252] is the same as that measured by Sanabia et al.[257] in the range of 25 meV to 32 eV. The range has been extended by Tanaka et al.[263] to 600 eV; these authors also observed a broad peak in the 20 to 40 eV range, with some small structure superposed.

From measurements of differential cross sections, Tanaka et al.[263] have obtained the elastic scattering cross section in the energy range from 1.5 to 100 eV. The cross section shape as a function of energy is broadly similar to that of CF_4 and C_2F_6, though the magnitude of Q_{el} in C_3F_8 is larger because of its increased molecular size. The momentum transfer cross section in the low energy region is due to Pirgov and Stefanov,[261] who observed a steep Ramsauer–Townsend minimum at 70 meV.

The molecule has D_{3d} symmetry,[264] in common with molecules such as C_2H_6 and C_2Cl_6. The C–F distance is 0.133 nm, the C–C distance 0.1546 nm, C–C–C angle 115.9°, and F–C–F angles 107° and 109°. The four major components of the vibrational excitation, ν_1 to ν_4, have energy quanta of 169.9, 161.0, 143.2, 96.7 meV respectively. Modes ν_6 and ν_7 have energy

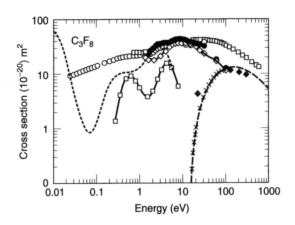

FIGURE 5.32 Scattering cross sections in C_3F_8. Total: (o) Sanabia et al.[257]; ($\square$) Tanaka et al.[263] Momentum transfer: (—•—) Christophorou et al.[252]; ($\blacktriangle$) Tanaka et al.[263] Elastic: (—• •—) Pirgov and Stefanov[261]; (+) Christophorou et al.[252]; (•) Tanaka et al.[263] Vibrational: (—$\square$—) Pirgov and Stefanov.[261] Ionization: (×) Christophorou et al.[248]; (— —) Nishimura et al.[256] Dissociation: ($\blacktriangle$) Winters and Inokuti.[262]

quanta of 82.4 and 47 meV respectively.[263] Pirgov and Stefanov[261] have evaluated the vibrationally inelastic and momentum transfer cross sections from swarm data, using Boltzmann code, in the energy range from 2 meV to 8.0 eV.

The ionization cross section has been measured by Nishimura et al.[256] since the review of Christophorou and Olthoff,[252] and hence only these two data are shown. The total dissociation cross section (ions + neutrals) has been measured by Winters and Inokuti,[262] as shown in Figure 5.32.

A comparison of the three gases considered, namely CF_4, C_2F_6, and C_3F_8 shows that the cross section increases as the molecular mass increases. Further, the Ramsauer–Townsend minimum becomes deeper with increasing molecular mass. The momentum transfer cross section shows a maximum around 3.5 to 4.5 eV for all three gases. The resonance peaks become larger but move toward lower energy as the molecular mass increases.

5.9.4 Dichlorodifluoromethane (CCl_2F_2)

Dichlorodifluoromethane (CCl_2F_2) is a plasma processing gas that was also used in the past as a refrigerant. It is a molecule with 58 electrons and is moderately polar having a dipole moment of 0.51 D.[265] Novak and Frechette,[266] Christophorou et al.,[250] and Karwasz et al.[84] have reviewed the available literature. Figure 5.33 shows the consolidated cross sections.

The total scattering cross section is built up of four studies: Randell et al.,[267] Jones et al.,[268] whose values are corrected by up to 7% for their stated systematic error due to scattering on the exit apertures, Underwood-Lemons et al.,[269] normalized by multiplying by with a factor of 1.33, and Zecca et al.,[270] Four resonances are seen in the 1 to 10 eV range, corresponding to negative ion formation.

The low energy, magnetically confined transmission experiment of Randell et al.[267] shows pure rotational excitation, ro-vibrational excitation, and a weak Ramsauer–Townsend effect.

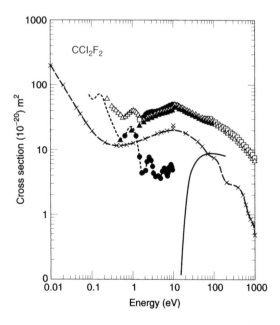

FIGURE 5.33 Scattering cross sections in CCl_2F_2. Total: (○) Jones[268]; (□) Zecca et al.[270]; (△) Christophorou et al.[250] Momentum transfer: (—×—) Hayashi and Niwu,[260] quoted in Christophorou et al.[250] Elastic: (▲) Mann and Linder.[271] Vibrational: (—●—) Mann and Linder[271]; (— — —) Randell et al.[267] Ionization: (—) Christophorou et al.[250]

TABLE 5.20
Negative Ion Production in Electron CCl_2F_2 Scattering[271]

Ion	Neutral Fragment	ε_A (eV)	ε_{max} (eV)	Rel. Int.
Cl^-	CF_2Cl	0.0	0.55*	50
Cl_2^-	CF_2	0.3	0.65*	0.7
FCl^-	$CFCl$	2.3	2.85*	1
Cl^-	$CF_2 + Cl$	2.1	3.0	1
F^-	$CFCl_2$	1.5	3.1	6.5
Cl_2^-	CF_2	0.3	3.1	0.3
$CFCl_2^-$	F	2.7	3.55*	2
Cl^-	$CFCl + F$	3.8	3.9*	1

ε_A is appearance potential and ε_{max} is the maximum position in the resonance capture. Asterisks indicate processes for peak positions in Q_T in the study of Jones.[268]
Source: Mann, A. and Linder, F., *J. Phys. B: At. Mol. Opt. Phys.*, 25, 1633, 1992. With permission of IOP, England.

The cross section decreases monotonically from ~ 10 meV to 450 eV except for a hump at ~ 160 meV, attributed to the onset of ro-vibrational excitation. Below 160 meV there is pure rotational excitation and it is stronger than ro-vibrational excitation. Since electron attachment does not occur below 100 meV the rise toward zero energy is attributed to the Ramsauer–Townsend effect which overlaps the rotational excitation.

The resonances observed are attributed to the capture of electrons in the lowest valence molecular orbitals, resulting in the formation of negative ions and neutral fragments. The resonances usually occur at a slightly higher energy than the appearance potential of the respective negative ion. Table 5.20 shows the appearance potentials and the positions of maxima with relative intensity for the dissociative attachments. In the high energy region (75 to 1000 eV) the total cross section reported by Zecca et al.[270] decreases monotonically. Between 75 and 1000 eV Q_T decreases by a factor of four.

Due to its low symmetry, the molecule possesses nine different modes of vibration. These are shown in Table 5.21. The total vibrational excitation cross section ($\nu_1 + \nu_2 + \nu_4 + \nu_6 + \nu_8$) is shown in Figure 5.33.[271]

Molecules that belong to the group called chlorofluoromethanes are $CClF_3$, CCl_2F_2, CCl_3F, and CCl_4 (carbon tetrachloride) and their cross sections are usually measured in the same apparatus. They show similar trends, though differing in detail, and are generally

TABLE 5.21
Vibration Modes and Energy Quanta of CCl_2F_2[271]

Mode	Nuclear Motion	Energy (meV)
ν_1	CF_2 symmetric stretch	136.1
ν_2	CF_2 bending	82.7
ν_3	CCl_2 symmetric stretch	56.6
ν_4	CCl_2 bending	32.5
ν_5	Torsion	39.9
ν_6	CF_2 asymmetric stretch	144.7
ν_7	CF_2-plane rocking	55.3
ν_8	CCl_2 asymmetric stretch	114.4
ν_9	CCl_2-plane rocking	53.9

reported as a group.[84,261,270] Appendix 5 shows selected cross sections for gases considered in this section.

5.10 OTHER GASES

Fragmentary data have been published on a number of gases. The following selected references serve as a guide for further study.

5.10.1 AMMONIA (NH_3)

Total scattering: Zecca et al.[272] Elastic scattering: Alle et al.[273] Ionization Tarnovsky et al.[274]

5.10.2 DISILANE (Si_2H_6)

Total scattering: Szmytkowski et al.,[275] Integral elastic scattering and momentum transfer: Dillon et al.[276] Ionization: Krishnakumar et al.[277]; Krishnakumar and Srivastava.[278]

5.10.3 GERMANE (GeH_4)

Total scattering: Możejko et al.[279] Elastic scattering: Dillon et al.[280] Momentum transfer: Soejima and Nakamura.[281] Ionization, theoretical: Ali et al.[282]

5.10.4 HEXAFLUOROPROPENE (C_3F_6)

Total scattering: Szmytkowski et al.[283] Ionization: Hart et al.[284] Attachment rate: Hunter et al.[285]

5.10.5 HYDROGEN SULFIDE (H_2S)

Total scattering: Szmytkowski and Maciąg[286]; Zecca et al.[287]; Jain and Baluja.[288] Ionization: Rao and Srivastava.[289]

5.10.6 NITROGEN DIOXIDE (NO_2)

Total scattering: Zecca et al.[290] Elastic scattering, theory: Curik et al.[291] Ionization, theoretical: Kim et al.[140]; Lindsay et al.[292]

5.10.7 SULFUR DIOXIDE (SO_2)

Total scattering: Zecca et al.[290] Elastic scattering: Gulley and Buckman[293]; Raj and Tomar.[294] Ionization: Basner et al.[295]

5.10.8 TETRACHLOROGERMANE ($GeCl_4$)

Total scattering: Szmytkowski et al.[163] Dissociative attachment: Guillot et al.[296]

5.10.9 URANIUM FLUORIDE (UF_6)

Differential and integral cross sections, and elastic cross sections: Cartwright et al.[297] Ionization cross sections: Compton[298]; Margreiter et al.[299]

5.11 CONCLUDING REMARKS

A reasonable attempt has been made in Chapters 3, 4, and 5 to present data on cross sections in gases, and a few general remarks will now be made. The total scattering cross section sums up the electron–neutral interactions in their various manifestations. A very succinct summary has been provided by Underwood-Lemons et al.[269] In the 0 to 10 eV energy range the scattering phenomenon is related to the physical size of the molecule; the cross section is typically of the order of 10 to $60 \times 10^{-20}\,\mathrm{m}^2$. The total cross section of helium (Figure 3.16) is almost without any structure and the incoming electron wave is scattered by a spherical target. As the atomic number increases, the radius of the target increases and the interaction assumes different patterns, as is revealed by the shape of the scattering cross section–energy curve.

As an electron approaches the neutral, if the energy of the electron is close to that of the valence electron there will be resonance between the incident and target electrons. The scattering cross section now shows resonance with a peak (Figure 3.57) superposed on a background of relatively simple variation. If the scattered wave has components that cancel out due to different phases, then one observes a minimum in the scattering cross section, usually around 1 eV. The wavelength of the electron is of the same order as the dimension of the molecule. This effect is the Ramsauer–Townsend minimum, for which there is no classical explanation. Relatively symmetrical molecules such as CH_4 also show the Ramsauer–Townsend effect (Figure 5.13).

The electron may temporarily get attached to the neutral, forming a negative ion; this capture process is called shape resonance. The scattering cross section increases over a narrow energy region because the electron gets detached after a short time of the order of $10^{-14}\,\mathrm{s}$. There are other mechanisms, such as vibrational excitation of the molecule, that show one or more peaks in the scattering cross section. This is very dramatically exhibited by NO (Figure 4.44). The negative ion is visualized as the electron being trapped in an unoccupied orbital of the molecule. If the potential energy of the molecule and the negative ion cross near zero energy, then the cross section rises sharply toward zero energy; the quantum mechanical explanation is that the eigenvalue of the vacant orbit is the same as the electron energy. This mechanism is characterized as s-wave scattering and is exhibited by SF_6 (Figure 5.21), resulting in the formation of SF_6^-.

In the complex patterns of resonance shapes and energies one looks for similarities in similar scattering processes and relates them to the atomic or molecular properties. Since scattering is related to the dimension of the target, one looks to relate the electronic polarizability, dipole moment, and quadrupole moment with cross sections. Jain and Baluja[26] discuss this aspect for several molecules. For the 14-electron systems (C_2H_2, CO, N_2) a common shape–resonance phenomenon is observed as a result of $^2\Pi$ (or $^2\Pi_g$) symmetry. At higher energies too, some resemblance may be discerned. The larger molecules have larger cross sections, as evidenced by hydrocarbons and fluorocarbons.

In an attempt to obtain an analytical expression for the total cross section, Jain and Baluja[26] suggest an expression of the type

$$\frac{\varepsilon}{R}\frac{Q_M}{4\pi a_0^2} = a' \ln \frac{\varepsilon}{R} + b'\frac{R}{\varepsilon} + c' \tag{5.11}$$

where a_0 is the Bohr radius (0.0529 nm), R the Rydberg energy (13.595 eV) and, a′, b′ and c′ are constants that have different values for different gases. The expression is valid only for high energies, and obviously it is not expected to hold for energy regions where resonance is observed. Brusa et al.[300] have provided reasonably successful expressions for total, elastic, excitation, and ionization cross sections. However, the momentum transfer cross section continues to present difficulties. In fact, as far the author of the present volume is aware, there is only one expression, given by Phelps and Petrović,[301] for momentum transfer cross section in argon. This equation is not easy to use and is not applicable in general form to other gases.

Several attempts have been made to express the ionization cross section as a function of electron energy with constants that assume different values depending upon the gas. The semi-theoretical expressions of Hwang et al.[217] have been very successful for a number of gases but the data to be supplied demand considerable knowledge of quantum mechanics. Another attractive formulation is due to Hudson et al.,[302] in which the maximum in the ionization cross section is obtained by the relatively simple formula

$$Q_{i,\,max} = c' \left(\frac{\alpha_e}{\varepsilon_i} \right)^{\frac{1}{2}} \tag{5.12}$$

in which $Q_{i,max}$ is the maximum of the ionization cross section, α_e the electronic polarizability, ε_i the ionization potential, and c' a constant. The only atomic quantity required is the electronic polarizability. Hudson et al.[302] have measured the ionization cross section in a number of chlorocarbons with the number of carbon atoms increasing from 1 to 5. Equation 5.12 is shown to apply to a number of gases, the measured $Q_{i,max}$ showing a linear relationship with the term having a square root. Since an analytical equation should be able to give the ionization cross section as a function of energy, the maximum Q_i calculated according to Equation 5.12 should be combined with the formulation of the type explored by Raju and Hackam[303] for general application.

Attempts to find scaling laws for the ionization cross section has also been moderately successful. Rost and Pattard[304] point out that the plot of $Q_i(\varepsilon)/Q_{max}$ as a function of $(\varepsilon/\varepsilon_{max})$ gives an identical shape for several atoms. The idea has been improved by Szhuińska et al.,[305] who suggested that plots of $Q_i(\varepsilon)/Q_{max}$ vs. $(\varepsilon/\varepsilon_i)$ for the atoms in the same column of the periodic table will be similar in shape; that is, the plot will be parameter free. Further, $Q_{i,max}$ itself is given by

$$Q_{i,\,max} = \alpha \exp(-\beta \varepsilon_i) \tag{5.13}$$

where α and β are constants that are the same for gases that lie in the same column of the periodic table. Using the most recent values of ionization cross sections in rare gases obtained by Rejoub et al.,[306] who used the time-of-flight spectrometer, the curves plotted are shown in Figure 5.34. They demonstrate the degree of validity of the idea of Szhuińska et al.[305] All of the rare gases show the same shape, confirming the parameter-free aspect of the choice of variables for the plot. Further, all gases except neon fall on the same curve, whereas the curve of neon is shifted to the right. It is important to realize that the curves are calculated using single ion production cross sections, though for all gases except xenon using the total cross section does not make a significant difference. This is demonstrated by plotting the ratio $Q_i(\varepsilon):Q_T$, and one obtains a curve with an entirely different shape (Figure 5.33).

Figure 5.34 shows the applicability of Equation 5.13 for a number of atomic gases, as compiled by Szhuińska et al.[302]

The collision cross sections for several attaching gases for electrons having very low energies, down to a few µeV, has been reviewed by Dunning.[307] The cross section for electron attachment quoted for SF_6 is $1.9 \times 10^{-16}\,m^2$ for 10 µeV energy and decreases linearly with energy up to 0.01 eV. Compare this value with the cross section given in Table 5.13. Beyond 0.1 eV energy the decrease is steeper. In CCl_4 the cross section is even larger, $2 \times 10^{-15}\,m^2$ for an electron energy of 1 µeV. A list of references for collision cross sections for a number of

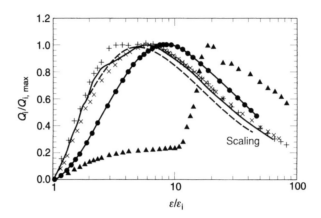

FIGURE 5.34 Scaling of ionization cross section in rare gases. The cross section for a singly charged ion is due to Rejoub et al.[306] All gases show the same shape and all gases except neon fall on the same curve. The total cross section of xenon when parameterized in the same way shows a different shape. Symbols are: (—) argon; (- - -) helium; (—●—) neon; ($\times$) krypton; (+) xenon (Xe^+ only); (▲) xenon (total up to Xe^{6+}).

polyatomic molecules is published by Huo and Kim.[308] Itikawa[309] has published a review of vibrational cross sections in several polyatomic molecules. These publications were not referenced earlier in the text.

REFERENCES

1. Morse, P. M., *Vibration and Sound*, McGraw-Hill, New York, 1936, pp. 244–245.
2. Massey, H. S. W. and E. H. S. Burhop, *Electronic and Ionic Impact Phenomena*, Oxford University Press, Oxford, 1952.
3. McDaniel, E. W., *Scattering Phenomena in Gases*, John Wiley and Sons, New York, 1964.
4. Faxén, H. and J. Holtzmark, *Z. Phys.*, 45, 307, 1927.
5. Holtsmark, J., *Z. Phys.*, 55, 437, 1929.
6. Lord, Rayleigh, *The Theory of Sound Waves*, Dover Publications, New York, 1945.
7. Massey, H. S. W. and C. B. O. Mohr, *Proc. Roy. Soc. London A* 141, 434, 1933.
8. Mott, N. F. and H. S. W. Massey, *Theory of Atomic Scatterings*, University of Oxford Press, London, 1952, pp. 20–22.
9. Massey, H. S. W., *Atomic and Molecular Collisions*, Taylor and Francis, London, 1979.
10. Burke, P. G. and K. Smith, *Rev. Mod. Phys.*, 34, 458, 1962.

CARBON DIOXIDE

11. Brüche, E., *Ann. Phys. Lpz.*, 83, 1065, 1927.
12. Ramsauer, C. and R. K. Kollath, *Ann. Phys. Lpz.*, 83, 1129, 1927.
13. Ramsauer, C. and R. Kollath, *Ann. Phys. Lpz.*, 4, 91, 1930.
14. Kollath, R. K., *Ann. Phys. Lpz.*, 15, 485, 1932.
15. Morrison, M. A., N. F. Lane, and L. A. Collins, *Phys. Rev.*, A 15, 2186, 1977.
16. Morrison, M. A. and N. F. Lane, *Phys. Rev. A*, 16, 975, 1977. A semiempirical polarization potential was adopted and the resonance peak was adjusted at 3.8 eV. There is good agreement of calculated cross sections with the measured cross sections of Ferch et al.[19]
17. Buenker, R. J., M. Honigmann, and H. Libermann, *J. Chem. Phys.*, 113, 1046, 2000.
18. Szmytkowski, C. and M. Zubek, *Chem. Phys. Lett.*, 57, 105, 1978.
19. Ferch, J., C. Masche, and W. Raith, *J. Phys. B: At. Mol. Phys.*, 14, L97–L00, 1981. In TOF measurements the energy resolution depends upon the energy of the beam, resolution increasing

with decreasing energy. Ferch et al. state a resolution better than 10 meV at 90 meV and 250 meV at 3.8 eV, at resonance.

20. Hoffman, K. R., M. S. Dababneh, Y.-F. Hsieh, W. E. Kauppila, V. Pol, J. H. Smart, and T. S. Stein, *Phys. Rev. A*, 25, 1393, 1982. This paper also gives tabulated scattering crossing sections in H_2 and N_2.

21. Kwan, Ch. K., Y.-F. Hsieh, W. E. Kauppila, S. J. Smith, T. S. Stein, M. N. Uddin, and M. S. Dababneh, *Phys. Rev. A*, 27, 1328, 1983.

22. Iga, I., J. C. Nogueira, and L. Mu-Tao, *J. Phys. B: At. Mol. Phys.*, 17, L185, 1984.

23. Szmytkowski, C., A. Zecca, G. Karwasz, S. Oss, K. Marciąg, B. Marinković, R. S. Brusa, and R. Grisenti, *J. Phys. B: At. Mol. Phys.*, 20, 5817, 1987.

24. Szmytkowski, C., A. Zecca, G. Karwasz, S. Oss, K. Marciąg, B. Marinković, R. S. Brusa, and R. Grisenti, *J. Phys. B: At. Mol. Phys.*, 20, 5817, 1987. Two different setups were used to obtain the wide range of energy. Very useful tabulated values over the entire range are provided.

25. Ferch, J., C. Masche, W. Raith, and L. Wiemann, *Phys. Rev. A*, 40, 5407, 1989.

26. Jain, A. and K. L. Baluja, *Phys. Rev. A*, 45, 202, 1992. This paper is a theoretical analysis and the same technique is applied to several diatomic and polyatomic gases. Elastic and total inelastic scattering cross sections are calculated separately.

27. Sueoka, O. and A. Hamada, *J. Phys. Soc. Jpn.*, 62, 2669, 1993.

28. Garcia, G. and F. Manero, *Phys. Rev. A*, 53, 250, 1996.

29. Lowke, J. J., A. V. Phelps, and B. W. Irwin, *J. Appl. Phys.*, 44, 4664, 1973.

30. Shyn, T. W., W. E. Sharp, and G. R. Carignan, *Phys. Rev. A*, 17, 1855, 1978. Tabulated values of elastic and momentum transfer cross sections in the energy range 3 to 90 eV are presented.

31. (a) Register, D. F., H. Nishimura, and S. Trajmar, *J. Phys. B: At. Mol. Phys.*, 13, 1651, 1980. Tabulated vibrational excitation cross sections for $v = 0 \rightarrow 1$ to 10 are provided for the four electron energies stated. (b) Nakamura, Y., *Aust. J. Phys.*, 48, 357, 1995.

32. Kochem, K-H., W. Sohn, N. Hebel, K. Jung, and H. Ehrhardt, *J. Phys. B: At. Mol. Phys.*, 18, 4455, 1985. Integral cross sections are not provided. The differential cross sections are measured as a function of the angle, covering the range 15° to 105°. Integrating the differential cross sections is not attempted by the author, due to the inaccuracy that exists in extrapolation.

33. Buckman, S. J., M. T. Elford, and D. S. Newman, *J. Phys. B: At. Mol. Phys.*, 20, 5175, 1987.

34. Kanik, I., D. C. McCollum, and J. C. Nickel, *J. Phys. B: At. Mol. Opt. Phys.*, 22, 1225, 1989. Experimental differential cross sections are measured and integral elastic scattering cross sections are not provided.

35. Johnstone, W. M., N. J. Mason, and W. R. Newell, *J. Phys.B: At. Mol. Opt. Phys.*, 26, L147, 1993.

36. Gianturco, F. A. and T. Stoecklin, *Phys. Rev. A*, 29, 3933, 1996. Integral and differential cross sections have been theoretically calculated for the energy range 0 to 100 eV.

37. Tanaka, H., T. Ishikawa, T. Masai, T. Sagara, L. Boesten, M. Takekawa, Y. Itikawa, and M. Kimura, *Phys. Rev. A*, 57, 1798, 1998.

38. Gibson, J. C., M. A. Green, K. W. Trantham, S. J. Buckman, P. J. O. Teubner, and M. J. Brunger, *J. Phys. B: At. Mol. Opt. Phys.*, 32, 213, 1999.

39. Gianturco, F. A. and T. Stoecklin, *Phys. Rev. A*, 34, 1695, 2001. Differential cross sections have been theoretically calculated for the energy range 0 to 100 eV.

40. Lee, C., C. Winstead, and V. McKoy, *J. Chem. Phys.*, 111, 5056, 1999.

41. Pack, J. L., R. E. Voshall, and A. V. Phelps, *Phys. Rev.*, 127, 2084, 1962.

42. Hake, R. D., Jr., and A. V. Phelps, *Phys. Rev.*, 158, 70, 1967. This paper deals with the momentum transfer cross sections in O_2, CO, and CO_2. Vibrational cross sections have also been derived in the low energy range.

43. Statamovic, A. and G. J. Schulz, *Phys. Rev.*, 188, 213, 1969. Ratios of vibrational cross sections in the modes 010 and 100 are given in terms of the normal mode 001, but the magnitudes of the cross sections as a function of energy are not given.

44. Sanchez, L. and G. J. Schulz, *J. Chem. Phys.*, 58, 479, 1973.

45. Boness, M. J. W. and G. J. Schulz, *Phys. Rev. A*, 9, 1969, 1974.

46. Antoni, Th., K. Jung, H. Ehrhardt, and E. S. Chang, *J. Phys. B: At. Mol. Phys.*, 19, 1377, 1986.

47. Takekawa, M. and Y. Itikawa, *J. Phys. B: At. Mol. Opt. Phys.*, 29, 4227, 1996. Elastic scattering cross sections are calculated theoretically in the 3 to 60 eV range.
48. Kitajima, M., S. Watanabe, H. Tanaka, M. Takekawa, M. Kimura, and Y. Itikawa, *J. Phys. B: At. Mol. Opt. Phys.*, 34, 1929, 2001.
49. Cvejanović, S., J. Jureta, and D. Cvejanivić, *J. Phys. B: At. Mol. Phys.*, 18, 2541, 1985.
50. Green, M. A., P. J. O. Teubner, L. Campbell, M. J. Brunger, M. Hoshino, T. Ishikawa, M. Kitajima, H. Tanaka, Y. Itikawa, M. Kimura, and R. J. Buenker, *J. Phys. B: At. Mol. Opt. Phys.*, 35, 567, 2002.
51. Rapp, D. and P. Englander-Golden, *J. Chem. Phys.*, 43, 4081, 1965.
52. Adamczyk, B., A. J. H. Boerboom, and M. Lukasiewicz, *Int. J. Mass Spect.*, 9, 407, 1972.
53. Crowe, A. and J. W. McConkey, *J. Phys. B: At. Mol. Phys.*, 7, 349, 1974.
54. Straub, H. C., B. G. Lindsay, K. A. Smith, and R. F. Stebbings, *J. Chem Phys.*, 105, 4015, 1996; see also ref. 57.
55. Trajmar, S., D. F. Register, and A. Chutjian, *Phys. Rep.*, 97, 216, 1983.
56. Kimura, M., O. Sueoka, A. Hamada, M. Takekawa, Y. Itikawa, H. Tanaka, and L. Boesten, *J. Chem. Phys.*, 107, 6616, 1997.
57. Stebbings, R.F. and B. G. Lindsay, *J. Chem. Phys.*, 114, 4741, 2001.
58. Karwasz, G.P., R. S. Brusa, and A. Zecca, *Riv. Nuovo Cimento*, 24, 1, 2001.
59. Raju, G. G., *Dielectrics in Electric Fields*, Marcel Dekker, New York, 2003, p. 193.
60. Haddad, G.N. and M. T. Elford, *J. Phys. B: At. Mol. Phys.*, 12, L743, 1979.
61. Joshipura, K. N. and P. M. Patel, *Z. Phys. D*, 29, 269, 1994.
62. Read, F.H. and D. Andrick, *J. Phys. B: At. Mol. Phys.*, 4, 911, 1971.
63. Andrick, D. and F. H. Read, *J. Phys. B: At. Mol. Phys.*, 4, 389, 1971.
64. Pitchford, L., personal communication, 2003. See also www.siglo.com
65. Herzberg, G., *Molecular Spectra And Molecular Structure*, Van Nostrand, Princeton, NJ, 1950.
66. Herzenberg, A., *J. Phys. B: At. Mol. Phys.*, 1, 548, 1968.
67. Čadež, I., M. Tronc, and R. I. Hall, *J. Phys. B: At. Mol. Phys.*, 7, L132, 1974.
68. Crowe, A. and J. W. McConkey, *J. Phys. B: At. Mol. Phys.*, 7, 349, 1974.
69. Takekawa, M. and Y. Itikawa, *J. Phys. B: At. Mol. Opt. Phys.*, 31, 3245, 1998.
70. Takekawa, M. and Y. Itikawa, *J. Phys. B: At. Mol. Opt. Phys.*, 32, 4209, 1999.
71. Herzberg, G., *Molecular Spectra and Molecular Structure–III, Electronic Spectra and Electronic Structure of Polyatomic Molecules*, Van Nostrand, Princeton, NJ, 1965.
72. England, W.B., W. C. Ermler, and A. C. Wahl, *J. Chem. Phys.*, 66, 2336, 1977.
73. Hubin-Franskin, M-J., J. Delwiche, B. Leclerc, and D. Roy, *J. Phys. B: At. Mol. Phys.*, 21, 3211, 1988.
74. LeClair, L. R. and J. W. McConkey, *J. Phys. B: At. Mol. Opt. Phys.*, 27, 4039, 1994.
75. Strickland, D. J. and A. E. S. Green, *J. Geophys. Res.*, 74, 6415, 1969.
76. Ajello, J. M., *J. Chem. Phys.*, 55, 3169, 1971.
77. Orient, O.J. and S. K. Srivastava, *J. Phys. B: At. Mol. Phys.*, 20, 3923, 1987.
78. Freund, R. S., R. C. Wetzel, and R. J. Shul, *Phys. Rev. A*, 41, 5861, 1990. Gases studied are N_2, CO, CO_2.
79. Hwang, W., Y.-K. Kim, and M. E. Rudd, *J. Chem. Phys.*, 104, 2956, 1996.
80. Adamczyk, B., A. J. H. Boreboom, and M. Lukasiewicz, *Int. J. Mass Spectrom. Ion Phys.*, 9, 407, 1972.
81. Märk, T. D. and E. Hille, *J. Chem. Phys.*, 69, 2492, 1978.
82. Krishnakumar, E., *Int. J. Mass Spectrom. Ion Proc.*, 97, 283, 1990.

HYDROCARBON GASES

83. Zecca, A. G. Karwasz, R. S. Brusa, and C. Szmytkowski, *J. Phys. B: At. Mol. Opt. Phys.*, 24, 2747, 1991.
84. Karwasz, G. P., R. S. Brusa, and A. Zecca, *Riv. Nuovo Cimento*, 24, 1, 2001. Gases covered are acetylene (C_2H_2), ethene (C_2H_4), ethane (C_2H_6), and propane (C_3H_8).
85. Sohn, W., K.-H. Kochem, K.-M. Scheuerlein, K. Jung, and H. Ehrhardt, *J. Phys. B: At. Mol. Phys.*, 19, 3625, 1986.

86. Sakae, T., S. Sumiyoshi, E. Murakami, Y. Matsumoto, Y. Ishibashi, and A. Katase, *J. Phys. B: At. Mol. Opt. Phys.*, 22, 1385, 1989. Differential cross sections in CH_4, CF_4, and SF_6 are studied and integrated elastic scattering and momentum transfer cross sections are given.

87. Jones, R. K., *J. Chem. Phys.*, 82, 5424, 1985.

88. Ferch, J., B. Granitza and W. Raith, *J. Phys. B: At. Mol. Phys.*, 18, L445, 1985.

89. Floeder, K., D. Fromme, W. Raith, A. Schwab, and G. Sinapius, *J. Phys. B: At. Mol. Phys.*, 18, 3347, 1985.

90. Dababneh, M.S., Y. F. Hseih, W. E. Kauppila, Ch. K. Kwan, S. J. Smith, T. S. Stein, and M. N. Uddin, *Phys. Rev. A*, 38, 1207, 1988.

91. Lohmann, B. and S. J. Buckman, *J. Phys. B: At. Mol. Phys.*, 19, 2565, 1986.

92. Nishimura, H. and T. Sakae, *Jpn. J. Appl. Phys.*, 29, 1372, 1990.

93. Sueoka, O. and S. Mori, *J. Phys. B: At. Mol. Opt. Phys.*, 22, 963, 1989.

94. Cascella, M., R. Curik, and F. A. Gianturco, *J. Phys. B: At. Mol. Opt. Phys.*, 34, 705, 2001.

95. Alvarez-Pol, H., I. Duran, and R. Lorenzo, *J. Appl. Phys.*, 30, 2455, 1997.

96. Boesten, L. and H. Tanaka, *J. Phys. B: At. Mol. Opt. Phys.*, 24, 821, 1991.

97. Mapstone, B. and W. R. Newell, *J. Phys. B: At. Mol. Opt. Phys.*, 25, 491, 1992.

98. Bundschu, C. T., J. C. Gibson, R. J. Gulley, M. J. Brunger, S. J. Buckman, N. Sanna, and F. A. Gianturco, *J. Phys. B: At. Mol. Opt. Phys.*, 30, 2239, 1997.

99. Nakano, T., H. Toyoda, and H. Sugai, *Jpn J. Appl. Phys.*, 29, 2908, 1991.

100. Chatham, H., D. Hils, R. Robertson, and A. Gallagher, *J. Chem. Phys.*, 87, 1770, 1984.

101. Durić, N., I. Čadež, and M. Kurepa, *Int. J. Mass. Spectrom. Ion. Proc.*, 108, R1, 1991.

102. Tarnovsky, V., A. Levin, H. Deutsch, and K. Becker, *J. Phys. B: At. Mol. Opt. Phys.*, 29, 139, 1996.

103. Straub, H. C., D. Lin, B. G. Lindsay, K. A. Smith, and R. F. Stebbings, *J. Chem. Phys.*, 106 4430, 1997.

104. Tian, C. and C. Vidal, (a) *Chem. Phys.*, 222, 105, 1997; (b) *J. Phys. B: At. Mol. Opt. Phys.*, 31, 895, 1998.

105. Zheng S.-H. and S. K. Srivastava, *J. Phys. B: At. Mol. OPt. Phys.*, 29, 3235, 1996.

106. Nishimura, H. and H. Tawara, *J. Phys. B: At. Mol. Opt. Phys.*, 27, 2063, 1994.

107. (a) Kim, Y.-K., K. K. Inkuro, M. E. Rudd, D. S. Zucker, M. A. Zukker, J. S. Coursey, K. I. Olsen, and G. G. Wiersma 2001, Web site http://physics.nist.gov/PhysRefData/Contents.html; (b) Bart, M., P. W. Harland, J. E. Hudson, and C. Vallance, *Phys. Chem. Chem. Phys.*, 3, 800, 2001.

MERCURY VAPOR (Hg)

108. Dwarakanath, K. and Govinda Raju, *Ion Thrusters for Space Electric Propulsion*, Indian Space Research Report, 1978.

109. Jost, K. and B. Ohnemus, *Phys. Rev. A*, 19, 641, 1979.

110. Burrow, P. D., J. A. Michejda, and J. Comer, *J. Phys. B: At. Mol. Phys.*, 9, 3225, 1976.

111. Holtkamp, G., K. Jost, F. J. Peitzmann, and J. Kessler, *J. Phys. B: At. Mol. Phys.*, 20, 4543, 1987.

112. Peitzmann, F. J. and J. Kessler, *J. Phys. B: At. Mol. Opt. Phys.*, 23, 4005, 1990.

113. Rockwood, S. D., *Phys. Rev., A*, 8, 2348, 1973.

114. England, J. P. and M. T. Elford, *Aust. J. Phys.*, 44, 647, 1991.

115. McEachran, R. P. and M. T. Elford, *J. Phys. B: At. Mol. Opt. Phys.*, 36, 427, 2003.

116. Borst, W. L., *Phys. Rev.*, 181, 257, 1969.

117. Liu, J. and G. G. Raju, *J. Phys. D: Appl. Phys.*, 25, 167, 1992; 25, 465, 1992.

118. Srivastava, R., T. Zuo, R. P. McEachran, and A. D. Stauffer, *J. Phys. B: At. Mol. Opt. Phys.*, 26, 1025, 1993.

NITROUS OXIDE (N₂O)

119. Johnstone, W. M. and W. R. Newell, *J. Phys. B: At. Mol. Opt. Phys.*, 26, 129, 1993.

120. Schulz, G. J., *Phys. Rev.*, (a) 116, 1141, 1959; (b) 135, 988, 1964.

121. Brunger, M. J. and S. J. Buckman, *Phys. Rep.*, 357, 215, 2002.
122. Ch. K., Kwan, Y.-F. Hseih, W. E. Kauppilla, S. J. Smith, T. S. Stein, and M. N. Uddin, *Phys. Rev. Lett.*, 52, 1417, 1984.
123. (a) Szmytkowski, C., G. Karwasz, and Maciąg, *Chem. Phys. Lett.*, 107, 481, 1984; energy range 0.4 to 40 eV. (b) Szmytkowski, C., K. Maciąg, G. Karwasz, and D. Filipović, *J. Phys. B: At. Mol. Opt. Phys.*, 22, 525, 1989; the range is extended to 40 to 100 eV.
124. Shilin, X., Z. Fang, Y. Liqiang, Yu Changqing, and X. Kezun, *J. Phys. B: At. Mol. Opt. Phys.*, 30, 2867, 1997.
125. Ramsauer, C. and R. Kollath, *Ann. Phys.*, 7, 176, 1930.
126. Kitajima, M., Y. Sakamoto, S. Watanabe, T. Suzuki, T. Ishikawa, H. Tanaka, and M. Kimura, *Chem. Phys. Lett.*, 309, 415, 1999.
127. Sarpal, B. K., K. Pfingst, B. N. Nestmann, and S. D. Peyerimhoff, *J. Phys. B: At. Mol. Opt. Phys.*, 29, 857, 1996.
128. King, G. C., J. W. McConkey, F. W. Read, and B. J. Dobson, *J. Phys. B: At. Mol. Phys.*, 13, 4315, 1980.
129. Marinković, B., C. Szmytkowski, V. Pejčev, D. Filipović, and L. Vušković, *J. Phys. B: At. Mol. Phys.*, 19, 2365, 1986.
130. Michelin, S. E., T. Kroin, and M. T. Lee, *J. Phys. B: At. Mol. Opt. Phys.*, 29, 2115, 1996.
131. Pack, J. L., R. E. Voshall, and A. V. Phelps, *Phys. Rev.*, 127, 2084, 1962.
132. Singh, Y., *J. Phys. B: At. Mol. Phys.*, 3, 1222, 1970.
133. Azria, R., S. F. Wong, and G. J. Schulz, *Phys. Rev. A*, 11, 1309, 1975.
134. Andrick, L. and R. I. Hall, *J. Phys. B: At. Mol. Phys.*, 17, 2713, 1984.
135. Marinković, B., Cz. Szmytkowski, V. Pejčev, D. Filipović, and L. Vušković, *J. Phys. B: At. Mol. Opt. Phys.*, 19, 2365, 1986.
136. Hall, R. I., A. Chutjian, and S. Trajmar, *J. Phys. B: At. Mol. Phys.*, 6, L365, 1973.
137. Marinković, B., R. Panajotović, Z. D. Pešić, D. Filipović, Z. Felfi, and A. Z. Maezane, *J. Phys. B: At. Mol. Opt. Phys.*, 32, 1949, 1999.
138. Strakeljahn, G., J. Ferch, and W. Raith, *5th European Conference on Atomic and Molecular Physics*, Edinburgh, 1955, Abstracts p. 542. Citation and digitized values are taken from Sarpal et al.[127]
139. Iga, I., M. V. V. S. Rao, and S. K. Srivastava, *J. Geophys. Res.*, 101, 9261, 1996.
140. Kim, Y.-K., W. Hwang, N. M. Weinberger, M. A. Ali, and M. E. Rudd, *J. Chem. Phys.*, 106, 1026, 1997.
141. Allcock, G. and J. W. McConkey, *Chem. Phys.*, 34, 169, 1978.
142. Mason, N. J. and W. R. Newell, *J. Phys. B: At. Mol. Opt. Phys.*, 22, 2297, 1989.
143. LeClair, L. R., J. J. Corr, and J. W. McConkey, *J. Phys. B: At. Mol. Opt. Phys.*, 25, L647, 1992.
144. LeClair, L. R. and J. W. McConkey, *Chem. Phys.*, 99, 4566, 1993.

Ozone (O₃)

145. Eliasson, B. and U. Kogelschatz, *J. Phys. B: At. Mol. Phys.*, 19, 1241, 1986.
146. Gulley, R. J., T. A. Field, W. A. Steer, N. J. Mason, S. L. Lunt, J. P. Ziesel, and D. Field, *J. Phys. B: At. Mol. Opt. Phys.*, 31, 5197, 1998.
147. Joshipura, K. N., B. K. Antony, and M. Vinod Kumar, *J. Phys. B: At. Mol. Opy. Phys.*, 35, 4211, 2002.
148. Shyn, T. W. and C. J. Sweeney, *Phys. Rev. A*, 47, 2919, 1993.
149. Sarpal, B. K., B. M. Nestmann, and S. D. Peyerimhoff, *J. Phys. B: At. Mol. Opt. Phys.*, 31, 1333, 1998.
150. Bettega, M. H. F., M. T. do N. Varella, L. G. Ferreira and M. A. P. Lima, *J. Phys. B: At. Mol. Opt. Phys.*, 31, 4419, 1998.
151. Allan, M., K. R. Asmis, D. B. Popvic, M. Stepanovic, N. J. Mason, and J. A. Davies, (a) *J. Phys. B: At. Mol. Opt. Phys.*, 29, 3487, 1996; (b) *J. Phys. B: At. Mol. Opt. Phys.*, 29, 4727, 1996.

152. Walker, I. C., J. M. Gingell, N. J. Mason, and G. M. Marston, *J. Phys. B: At. Mol. Opt. Phys.*, 29, 4749, 1996. For an explanation of electron configuration in polyatomic molecules see Herzberg,[71] pp. 296–351.

153. Skalny, J. D., S. Matejcik, A. Kindler, A. Stamatovic, and T. D. Mark, *Chem. Phys. Lett.*, 255, 112, 1996.

154. Swanson, N. and R. J. Celotta, *Phys. Rev. Lett.*, 35, 783, 1975.

155. Johnstone, W. M., N. J. Mason, W. R. Newell, P. Biggs, G. Marston, and R. P. Wayne, *J. Phys. B: At. Mol. Opt. Phys.*, 25, 3873, 1992.

156. Allan, M., J. A. Davies, and N. J. Mason, *J. Chem. Phys.*, 105, 5665, 1996.

157. Sweeney, C. J. and T. W. Shyn, *Phys. Rev.*, 53, 1576, 1996.

158. Siegel, M. W., *Int. J. Mass Spectr.*, 44, 19, 1982.

159. Newson, K. A., S. Luc, S. D. Price, and N. J. Mason, *Int. J. Mass Spectr. Ion Proc.*, 148, 203, 1995.

Silane (SiH_4)

160. Wan, H.-X., J. H. Moore, and J. A. Tossell, *J. Chem. Phys.*, 91, 7340, 1989.

161. Zecca, A., G. P. Karwasz, and R. S. Brusa, *Phys. Rev.*, A, 45, 2777, 1992.

162. Sueoka, O., S. Mori, and A. Hamada, *J. Phys. B: At. Mol. Opt. Phys.*, 27, 1453, 1994.

163. Szmytkowski, C., P. Możejko, and G. Kasperski, *J. Phys. B: At. Mol. Opt. Phys.*, 30, 4363, 1997.

164. Jain, A. and D. J. Thompson, *J. Phys. B: At. Mol. Opt. Phys.*, 24, 1087, 1991.

165. Jiang, Y., J. Sun, and L. Wan, *Phys. Rev. A*, 52, 398, 1995.

166. Ohmori, Y., M. Shimozuma, and H. Tagashira, *J. Phys. D: Appl. Phys.*, 19, 1029, 1986.

167. Mathieson, K. J., P. G. Millican, I. C. Walker, and M. G. Curtis, *J. Chem. Soc., Faraday Trans. II*, 83, 1041, 1987.

168. Kurachi, M. and Y. Nakamura, *J. Phys. D: Appl. Phys.*, 22, 107, 1989.

169. Tanaka, H., L. Boesten, H. Sato, M. Kimura, M. A. Dillon, and D. Spence, *J. Phys. B: At. Mol. Opt. Phys.*, 23, 577, 1990.

170. Chatham, H., D. Hills, R. Robertson, and A. Gallagher, *J. Chem. Phys.*, 81, 1770, 1984.

171. Krishnakumar, E. and S. K. Srivastava, *Contrib. Plasma. Phys.*, 35, 395, 1995.

172. Basner, R., M. Schmidt, V. Tarnovsky, K. Becker, and H. Deutsch, *Int. J. Mass. Spectrom. Ion Proc.*, 171, 83, 1997.

173. Ali, M. A., Y.-K. Kim, W. Hwang, N. M. Weinberger, and M. E. Rudd, *J. Chem. Phys.*, 106, 9602, 1997.

Sulfur Hexafluoride (SF_6)

174. (a) Christophorou, L. G. and R. J. van Brunt, *IEEE Trans. Dielect. and Elec. Insul.*, 2, 952, 1995; (b) Christophorou, L. G. and J. K. Olthoff, *J. Phys. Chem. Ref. Data*, 29, 267, 2000; (c) Christophorou, L. G. and J. K. Olthoff, *Fundamental Electron Interactions with Plasma Processing Gases*, Kluwer Academic/Plenum Publishers, New York, 2004.

175. Karwasz, G. P., R. S. Brusa, and A. Zecca, *Riv. Nuovo Cimento*, 24, 93, 2001.

176. Holland, D. M. P., D. A. Shaw, A. Hopkirk, M. A. McDonald, P. Baltzer, L. Karlsson, M. Lundqvist, B. Wannberg, and W. W. von Niessen, *Chem. Phys.*, 192, 333, 1995.

177. Dehmer, J. L., J. Siegel, and D. Dill, *J. Chem. Phys.*, 69, 5205, 1978.

178. Dehmer, J. L., A. C. Parr, S. Wallace, and D. Dill, *Phys. Rev. A*, 26, 3283, 1982.

179. Trajmar, S. and A. Chutjian, *J. Phys. B: At. Mol. Phys.*, 10, 2943, 1977.

180. Hildenbrand, D. L., *J. Phys. Chem.*, 77, 897, 1973.

181. Ischenko, A. A., J. D. Ewbank, and L. Schäfer, *J. Phys. Chem.*, 98, 4287, 1994.

182. Chase, M. W., Jr., C. A. Davis, J. R. Downey, Jr., D. J. Frurip, R. A. McDonald, and A. N. Syverud, *J. Phys. Chem. Ref. Data*, 14, 1163, 1985.

183. Rohr, K., *J. Phys. B: At. Mol. Opt. Phys.*, 12, L185, 1979.

184. Kennerly, R. E., R. A. Bonham, and M. McMillan, *J. Chem. Phys.*, 70, 2039, 1979.

185. Ferch, J., W. Raith, and K. Schröder, *J. Phys. B: At. Mol. Phys.*, 15, L175, 1982.

186. Romanyuk, N. I., I. V. Chernyshova, and O. B. Shpenik, *Sov. Phys. Tech. Phys.*, 29, 1204, 1984.
187. Dababneh, M. S., Y.-F. Hseih, W. E. Kauppila, C. K. Kwan, S. J. Smith, T. S. Stein, and M. N. Uddin, *Phys. Rev. A*, 38, 1207, 1988.
188. Zecca, A., G. Karwasz, and R. S. Brusa, *Chem. Phys. Lett.*, 199, 423, 1992.
189. Wan, H.-X., J. H. Moore, J. K. Olthoff, and R. J. Brunt, *Plasma Chem. Plasma Process.*, 13, 1, 1993.
190. Kasperski, G., P. Możejko, and C. Szmytkowski, *Z. Phys. D*, 42, 187, 1997.
191. Hickam, W. M. and R. E. Fox, *J. Chem. Phys.*, 25, 642, 1956.
192. Chutjian, A., *Phys. Rev. Lett.*, 46, 1511, 1981.
193. Kellert, F. G., C. Higgs, K. A. Smith, G. F. Hildebrandt, F. B. Dunning, and R. F. Stebbings, *J. Chem. Phys.*, 72, 6312, 1980.
194. Kline, L. E., D. K. Davies, C. L. Chen, and P. J. Chantry, *J. Appl. Phys.*, 50, 6789, 1979.
195. Trajmar, S., D.F. Register, and A. Chutjian, *Phys. Rep.*, 97, 219, 1983.
196. Gianturco, F. A., R. R. Lucchese, and N. Sanna, *J. Chem. Phys.*, 102, 5743, 1995. Differential scattering cross sections in the range from 3 to 30 eV are given.
197. Srivastava, S. K., S. Trajmar, A. Chutjian, and W. Williams, *J. Chem. Phys.*, 64, 2767, 1976.
198. Johnstone, W. M. and W. R. Newell, *J. Phys. B: At. Mol. Opt. Phys.*, 24, 473, 1991.
199. Cho, H., R. J. Gulley, and S. J. Buckman, *J. Phys. B: At. Mol. Opt. Phys.*, 33, L309, 2000.
200. Phelps A. V., and R. J. van Brunt, *J. Appl. Phys.*, 64, 4269, 1988.
201. Liu, J. and Govinda Raju, *IEEE Trans. Plasma. Sci.*, 20, 515, 1992.
202. Randell, J., D. Field, S. L. Lunt, G. Mrotzek, and J. P. Ziesel, *J. Phys. B*, 25, 2899, 1992.
203. Benedict, M. G. and I. Gyemant, *Int. J. Quantum Chem.*, 13, 597, 1978.
204. Jiang, Y., J. Sun, and L. Wan, *Phys. Lett. A*, 231, 231, 1997.
205. Herzberg, G., *Molecular Spectra and Molecular Structure*, Van Nostrand, Princeton, NJ, 1967, p. 5.
206. Herzberg, G., *Molecular Spectra and Molecular Structure—III, Electronic Spectra and Electronic Structure of Polyatomic Molecules*, Van Nostrand, Princeton, NJ, 1965, p. 644.
207. Blanks, K. A., A. E. Tabor, and K. Becker, *J. Chem. Phys.*, 86, 4871, 1987.
208. Herzberg, G., *Molecular Spectra and Molecular Structure*, Van Nostrand, Princeton, NJ, 1967, p. 545.
209. Ito, M., M. Goto, H. Toyoda, and H. Sugai, *Contri. Plasma Phys.*, 35, 405, 1995.
210. Corr, J. J., M. A. Khakoo, and J. W. McConkey, *J. Phys. B: At. Mol. Phys.*, 20, 2597, 1987.
211. Asundi, R. K. and J. D. Craggs, *Proc. Phys. Soc.*, 83, 1964, 611.
212. Stanski, T. and B. Adamczyk, *Int. J. Mass Spectrom. Ion. Phys.*, 46, 31, 1983.
213. Margreiter, D., G. Walder, H. Deutsch, H. U. Poll, C. Winkler, K. Stephan, and T. D. Märk, *Int. J. Mass. Spectrom. Ion Processes*, 100, 143, 1990.
214. Margreiter, D., H. Deutsch, M. Schmidt, and T. D. Märk, *Int. J. Mass. Spectrom. Ion Processes*, 100, 157, 1990.
215. Rao, M. V. V. S. and S. K. Srivastava, *XX International Conference on the Physics of Electronic and Atomic Collisions, Scientific Program and Abstracts of Contributed Papers*, Vol. II, Vienna, Austria, 23–29 July 1997, Paper MO 151; cited by Christophorou and Olthoff.[174b]
216. Rejoub, R., D. R. Sieglaff, B. G. Lindsay and R. F. Stebbings, *J. Phys. B: At. Mol. Opt. Phys.*, 34, 1289, 2001.
217. Hwang, W., Y.-K. Kim, and M. E. Rudd, *J. Chem. Phys.*, 104, 2956, 1996. This paper gives important data such as molecular orbitals and binding energy for a number of molecules. Calculated ionization cross sections are given for 19 molecules.
218. Tarnovsky, V., H. Deutsch, S. Matt, T. D. Märk, R. Basner, M. Schimdt, and K. Becker, in *Gaseous Electronics VIII*, Ed. L. G. Christophorou and J. K. Olthoff, Plenum, New York, 1998, p. 3.
219. Kim, Y.-K. and M. E. Rudd, *Comments At. Mol. Phys.*, 34, 309, 1999.
220. Deutsch, H., K. Becker, S. Matt, and T. D. Mark, *Int. J. Mass. Spectr.*, 197, 37, 2000.

WATER VAPOR (H$_2$O)

221. Mansfield, P., *J. Phys. E: Sci. Instrum.*, 21, 18, 1988.

222. Rajapandian, S. and G. Raju, *J. Inst. Eng.* 53, EI 187, 1973.
223. Sueoka, O., S. Mori, and Y. Katayama, *J. Phys. B: At. Mol. Phys.*, 19, L373, 1986.
224. Zecca, A., G. Karwasz, S. Oss, R. Grisenti, and R. S. Brusa, *J. Phys. B: At. Mol. Phys.*, 20, L133, 1987.
225. Nishimura, H. and K. Yano, *J. Phys. Soc. Jpn*, 57, 1951, 1988.
226. Sağlam, Z. and N. Aktekin, *J. Phys. B: At. Mol. Opt. Phys.*, 23, 1529, 1990.
227. Sağlam, Z. and N. Aktekin, *J. Phys. B: At. Mol. Opt. Phys.*, 24, 3491.
228. Danjo, A. and H. Nishimura, *J. Phys. Soc. Jpn*, 54, 1224, 1985.
229. Katase, A., K. Ishibashi, Y. Matsumoto, T. Sakae, S. Maezono, E. Murakami, K. Watanabe, and H. Maki, *J. Phys. B: At. Mol. Phys.*, 19, 2715, 1986.
230. Shyn, T. W. and S. Y. Cho, *Phys. Rev. A*, 36, 5138, 1987.
231. Johnstone, W. M. and W. R. Newell, *J. Phys. B: At. Mol. Phys.*, 24, 3633, 1991.
232. Shimamura, I., *Sci. Papers Inst. Phys. Chem. Res (Jpn)*, 82, 1, 1989.
233. Gianturco, F. A. and D. G. Thompson, *J. Phys. B: At. Mol. Phys.*, 13, 613, 1980.
234. Jain, A. and D. G. Thompson, *J. Phys. B: At. Mol. Phys.*, 16, 2593, 1983.
235. Rescigno, T. N. and B. H. Lengsfield, *Z. Phys. D*, 24, 117, 1992.
236. Seng, G. and F. Linder, *J. Phys. B: At. Mol. Phys.*, 9, 2539, 1976.
237. Jung, K., T. H. Antoni, R. Müller, K. H. Kochem, and H. Ehrhardt, *J. Phys. B: At. Mol. Phys.*, 15, 3535, 1982.
238. Cvejanović, D., L. Andrić, and R. I. Hall, *J. Phys. B: At. Mol. Opt. Phys.*, 26, 2899, 1993.
239. El-Zein, A. A. A., M. J. Brunger, and W. R. Newell, *J. Phys. B: At. Mol. Opt. Phys.*, 33, 5033, 2000.
240. Wong, S. F. and G. J. Schulz, *Proc. 9th Int. Conf. on Physics of Electronic and Atomic Collisions*, Seattle, Abstracts University of Washington Press, Seattle, pp. 283–284.
241. Mark, T. D. and F. Egger, *Int. J. Mass. Spectr. Ion Phys.*, 20, 89, 1976.
242. Bolorizadeh, M. A. and M. E. Rudd, *Phys. Rev.*, 33, 882, 1986.
243. Orient, O. J. and S. K. Srivastava, *J. Phys. B: At. Mol. Phys.*, 20, 3923, 1987.
244. Djurić, N. L., I. M. Čadež, and M. V. Kurepa, *Int. J. Mass. Spectr. Ion Process.*, 83, R7, 1988.
245. Rao, M. V. V. S., I. Iga, and S. K. Srivastava, *J. Geophys. Res.*, 100, E6421, 1995.
246. Straub, H. C., B. G. Lindsay, K. A. Smith, and R. F. Stebbings, *J. Chem. Phys.*, 108, 109, 1998.

PLASMA PROCESSING GASES

247. Christophorou, L. G., J. K. Olthoff, and M. V. V. S. Rao, *J. Phys. Chem. Ref. Data*, 25, 1341, 1996. Gas studied is CF_4.
248. Christophorou, L. G., J. K. Olthoff, and M. V. V. S. Rao, *J. Phys. Chem. Ref. Data*, 28, 967, 1999. Update on gases, CF_4, C_2F_6, and C_3F_8.
249. Christophorou, L. G., J. K. Olthoff, and M. V. V. S. Rao, *J. Phys. Chem. Ref. Data*, 26, 1, 1997. Gas studied is CH_3F.
250. Christophorou, L. G., J. K. Olthoff, and Y. Wang, *J. Phys. Chem. Ref. Data*, 26, 1205, 1997. Gas studied is CCl_2F_2.
251. Christophorou, L. G. and J. K. Olthoff, *J. Phys. Chem. Ref. Data*, 27, 1, 1998. Gas studied is C_2F_6.
252. Christophorou, L. G., J. K. Olthoff, and M. V. V. S. Rao, *J. Phys. Chem. Ref. Data*, 27, 889, 1998. Gas studied is C_3F_8.
253. Sieglaff, D. R., R. Rejoub, B. G. Lindsay, and R. F. Stebbings, *J. Phys. B: At. Mol. Opt. Phys.*, 34, 799, 2001.
254. Torres, I., R. Martinez, and F. Castaño, *J. Phys. B: At. Mol. Opt. Phys.*, 35, 2423, 2002.
255. Torres, I., R. Martínez, M. N. Sánchez Rayo, J. A. Fernández and F. Castaño, *J. Chem. Phys.*, 115, 4041, 2001.
256. Nishimura, H., W. M. Huo, and M. A. Ali, *J. Chem. Phys.*, 110, 3811, 1999.
257. Sanabia, J. E., G. D. Cooper, J. A. Tossell, and J. H. Moore, *J. Chem. Phys.*, 108, 389, 1998.
258. Szmytkowski, Cz., P. Możeko, G. Kasperski, and E. Ptasińska-Denga, *J. Phys. B; At. Mol. Opt. Phys.*, 33, 15, 2000.

259. Takagi, T., L. Boesten, H. Tanaka, and M. A. Dillon, *J. Phys. B: At. Mol. Opt. Phys.*, 27, 5289, 1994.

260. Hayashi, M. and A. Niwa, in *Gaseous Dielectrics*, vol. V, ed. L. G. Christophorou and D. W. Boldin, Pergamon Press, New York, 1987, p. 27.

261. Pirgov, P. and B. Stefanov, *J. Appl. Phys. B: At. Mol. Opt. Phys.*, 23, 2879, 1990.

262. Winters, H. F. and M. Inokuti, *Phys. Rev. A*, 25, 1420, 1982.

263. Tanaka, H., Y. Tachibana, M. Kitajima, O. Sueoka, H. Takaki, A. Hamada, and M. Kimura, *Phys. Rev. A*, 59, 2006, 1999. The caption for Table II should be C_3F_8.

264. Herzberg, G., *Molecular Spectra and Molecular Structure—III, Electronic Spectra and Electronic Structure of Polyatomic Molecules*, Van Nostrand, Princeton, NJ, 1965, p. 4.

265. *CRC Handbook of Chemistry and Physics*, CRC, Boca Raton, FL.

266. Novak, J. P. and M. F. Fréchette, *J. Appl. Phys.*, 57, 4368, 1985.

267. Randell, J., J. P. Ziesel, S. L. Lunt, G. Mrotzek, and D. Field, *J. Phys. B: At. Mol. Opt. Phys.*, 26, 3423, 1993.

268. Jones, R. K., *J. Chem. Phys.*, 84, 813, 1986.

269. Underwood-Lemons, T., D. C. Winkler, J. A. Tossell, and J. H. Moore, *J. Chem. Phys.*, 100, 9117, 1994.

270. Zecca, A., G. P. Karwasz, and R. S. Brusa, *Phys. Rev. A*, 46, 3877, 1992.

271. Mann, A., and F. Linder, *J. Phys. B: At. Mol. Opt. Phys.*, 25, 1633, 1992.

OTHER GASES

272. Zecca, A., G. P. Karwasz, and R. S. Brusa, *Phys. Rev. A*, 45, 2777, 1992.

273. Alle, D. T., R. J. Gulley, S. J. Buckman, and M. J. Brunger, *J. Phys. B: At. Mol. Opt. Phys.*, 25, 1533, 1992.

274. Tarnovsky, V., H. Deutsch, and K. Becker, *Int. J. Mass. Spectr. Ion. Process*, 167/168, 69, 1997.

275. Szmytkowski, C., P. Możejko, and G. Kasperski, *J. Phys. B: At. Mol. Opt. Phys.*, 34, 605, 2001.

276. Dillon, M. A., L. Boesten, H. Tanaka, M. Kimura, and H. Sato, *J. Phys. B: At. Mol. Opt. Phys.*, 27, 1209, 1994.

277. Krishnakumar, E., S. K. Srivastava, and I. Iga, *Int. J. Mass. Spectrom. Ion Process.*, 103, 107, 1991.

278. Krishnakumar E., and S. K. Srivastava, *Contrib. Plasma Phys.*, 35, 395, 1995.

279. Możejko, P., G. Kasperski, and C. Szmytkowski, *J. Phys. B: At. Mol. Opt. Phys.*, 29, L571, 1996.

280. Dillon, M. A., L. Boesten, H. Tanaka, M. Kimura, and H. Sato, *J. Phys. B: At. Mol. Opt. Phys.*, 26, 3147, 1993.

281. Soejima, H. and Y. Nakamura, *J. Vac. Sci. Technol. A*, 11, 1161, 1993.

282. Ali, M. A., Y.-K. Kim, W. Hwang, N. M. Weinberger, and M. E. Rudd, *J. Chem. Phys.*, 106, 9602, 1997.

283. Szmytkowski, C., P. Możejko, and S. Kwitnewski, *J. Phys. B: At. Mol. Opt. Phys.*, 35, 1267, 2002.

284. Hart, M., P. W. Harland, J. E. Hudson, and C. Vallance, *Phys. Chem. Chem. Phys.*, 3, 800, 2001.

285. Hunter, S. R., L. G. Christophorou, D. L. McCorkle, I. Sauers, H. W. Ellis, and D. R. James, *J. Phys. B: At. Mol. Opt. Phys.*, 16, 573, 1983.

286. Szmytkowski, Cz. and K. Maciąg, *Chem. Phys. Lett.*, 129, 321, 1986.

287. Zecca, A., G. P. Karwasz, and R. S. Brusa, *Phys. Rev. A*, 45, 2777, 1992.

288. Jain, A. and K. L. Baluja, *Phys. Rev. A*, 45, 202, 1992.

289. Rao, M. V. V. S. and S. K. Srivastava, *J. Geophys. Res.*, 98, 13137, 1993.

290. Zecca, A., J. C. Nogueira, G. P. Karwasz, and R. S. Brusa, *J. Phys. B: At. Mol. Opt. Phys.*, 28, 477, 1995.

291. Curik, R., F. A. Gianturco, R. R. Lucchese, and N. Sanna, *J. Phys. B. At. Mol. Opt. Phys.*, 34, 59, 2001.

292. Lindsay, B. G., M. A. Mangan, H. C. Straub, and R. F. Stebbings, *J. Chem. Phys.*, 112, 9404, 2000.
293. Gulley, R. J. and S. J. Buckman, *J. Phys. B: At. Mol. Opt. Phys.*, 27, 1833, 1994.
294. Raj, D. and S. Tomar, *J. Phys. B: At. Mol. Opt. Phys.*, 30, 1989, 1997.
295. Basner, R., M. Schmidt, H. Deutsch, V. Tarnovsky, A. Levin, and K. Becker, *J. Chem. Phys.*, 103, 211, 1995.
296. Guillot, F., C. Dézarnaud-Dandine, M. Tronc, A. Modelli, A. Lisini, P. Declava, and G. Fronzoni, *Chem. Phys.*, 205, 359, 1996.
297. Cartwright, D. C., S. Trajmar, A. Chutjian, and S. Srivastava, *J. Chem. Phys.*, 79, 5483, 1983.
298. Compton, R. N., *J. Chem. Phys.*, 66, 4478, 1977.
299. Margreiter, D., H. Deutsch, M. Schmidt, and T. D. Märk, *Int. J. Mass Spectr. Ion Process.*, 100, 157, 1990.
300. Brusa, R. S., G. P. Karwasz, and A. Zecca, *Z. Phys. D*, 38, 279, 1996.
301. Phelps, A. V. and Z. Lj. Petrović, *Plasma Sources Sci. Tech.*, 8, R21, 1999.
302. Hudson, J. E., C. Vallance, M. Bart, and P. W. Harland, *J. Phys. B: At. Mol. Opt. Phys.*, 34, 3025, 2001.
303. (a) Raju, G. R. G. and R. Hackam, *J. Appl. Phys.* (USA), 52, 3912, 1981;
     (b) Raju, G. R. G. and R. Hackam, *J. Appl. Phys.* (USA), 53, 5557, 1982.
304. Rost, J. M. and T. Pattard, *Phys. Rev.*, 55, R5, 1997.
305. Szhuińska, M., P. V. Reeth, and G. Laricchia, *J. Phys. B: At. Mol. Opt. Phys.*, 35, 4059, 2002.
306. Rejoub, R., B. G. Linday and R. F. Stebbings, *Phys. Rev. A*, 65, 042713, 2002.
307. Dunning, F. B., *J. Phys. B: At. Mol. Opt. Phys.*, 28, 1645, 1995.
308. Huo, W. M. and Yong-Ki Kim, *IEEE Trans. Plasma. Sci.*, 27, 1225, 1999.
309. Itikawa, Y., *J. Phys. B: At. Mol. Opt. Phys.*, 37, R1, 2004.
310. Kurihara, M., Z. Lj Petrović, and T. Makabe, *J. Phys. D: Appl. Phys.*, 33, 2146, 2000.
311. De Pablos et al., *J. Phys. B*, 35, 865, 2002.
312. Davies , J. A., W. M. Johnstone, N. J. Mason, P. Biggs, and R. A. Wayne, *J. Phys. B: At. Mol. Opt. Phys.*, 26, L767, 1993.

6 Drift and Diffusion of Electrons—I

At the turn of the twentieth century the discovery of the electron and the advent of quantum mechanics as a new field of study gave birth to two schools for the study of electron–atom interactions. One school, led by Ramsauer and his colleagues, used beam techniques to measure scattering cross sections, which have been dealt with in earlier chapters. The other school, led by Townsend and his colleagues, adopted the swarm technique in which the electrons move in a gaseous medium under the influence of an applied electric field. The average properties of the electron swarm were studied and from these results the details of electron–atom collisions were theoretically deciphered. This chapter is devoted to the techniques adopted and the results obtained with regard to the diffusion and drift of electrons in gases. The emphasis will be on the data obtained, as they form the foundation on which the theory of electrical discharge, from corona to high density plasma, is developed. Theoretical treatment may be found in McDaniel,[1] Hasted,[2] and Huxley and Crompton.[3]

6.1 DEFINITIONS

As the electron undergoes collision, the velocity with which it travels between collisions is known as the random velocity and the velocity determined by the distance it drifts in the direction of the electric field is termed drift velocity. The random velocity is determined by its energy at the instant after collision and the random velocity vector adopts all directions. One therefore takes the root mean square of the magnitude and designates it as $<W>$. The drift velocity of electrons is denoted by W (m/s); the suffixes e and i will be added to denote electrons and ions only if necessary. The mobility μ is defined by the simple relationship

$$W = \mu E \tag{6.1}$$

where E is the electric field (V/m). The unit of μ is $V^{-1}m^2s^{-1}$. Reduced mobility (μ_0) refers to standard temperature (273 K) and pressure (101.3 kPa) according to

$$\mu_0 = \frac{N\mu}{2.69 \times 10^{25}} \tag{6.2}$$

where μ is the mobility measured at gas number density N.

The electron mean free path is λ (m). If it is assumed to be independent of energy it will have a constant value. The mean free path is an alternative way of quantifying the collision cross section (Q, m^2) according to

$$\lambda = \frac{1}{NQ_M} \tag{6.3}$$

where N is the number of gas neutrals (m^{-3}) at a given pressure p (Pa) and temperature T (K) and Q_M is the momentum transfer cross section (also called the diffusion cross section). For convenience, the equation for calculating N at a given p and T is repeated:

$$N = \frac{7.244 \times 10^{22}}{T(K)} \times p\,(Pa) \quad \text{per m}^3 \tag{6.4}$$

The measured quantities are E (V m^{-1}) and p (Pa). W is given as

$$W = \mu N \times \frac{E}{N} \quad \text{m s}^{-1} \tag{6.5}$$

In a drift distance d, the actual distance traversed is more, $d < W > /W$. The parameter E/N is used extensively in swarm studies and is expressed in units of Td (townsend), $1\,\text{Td} = 10^{-21}\,\text{V m}^2$, with N being determined from experimental gas pressure according to Equation 6.4.

The fraction of energy lost per collision (f) is determined by both $<W>$ and W. A simplified treatment[2] gives

$$f = 2.356 \frac{W^2}{<W>^2} \tag{6.6}$$

A constant mean free path implies that the momentum transfer cross section is independent of electron energy. This assumption is not realistic, and a rigorous treatment involves the calculation of energy distribution of electrons in the swarm. We shall consider energy distribution in a later section, after briefly referring to the method of measurement of drift velocity.

6.2 DRIFT AND DIFFUSION MEASUREMENT

In principle, the measurement of drift velocity and diffusion coefficient is simple. A pulse of electrons is released from the cathode and arrives at a grid to which a marker pulse is applied. The electrons that pass through the grid arrive at the anode and the time of arrival is measured. The method is simple in principle and the accuracy achieved is very good; however, to achieve an accuracy of 1 to 2% one should bear in mind the fact that the time of flight measured may not exactly coincide with the time interval between the center of mass of the electrons released at the first electrode and that of the electrons arriving at the second electrode. Both forward and backward diffusion at the electrodes distort the electron density profile. Further corrections to the time of transit are required due to longitudinal diffusion in the inter-electrode region.[4] Schlumbohm[5] has measured the current pulse due to an electron avalanche and the drift velocity can be derived by analyzing the pulse, up to $E/N = 1000$ Td. The width of the pulse is a measure of diffusion.

Historically, the first method developed for measuring the drift velocity used what is known as the Bradbury–Nielsen method.[6] Here the electrons are produced by an ultraviolet lamp or a suitable radioactive source. The grid consists of fine gold-plated wires to reduce contact potential. Alternate wires are connected to each other and to a radio-frequency voltage source.

A group of electrons passes through the grid during one half cycle and arrives at a second grid, constructed identically, to which the same radio-frequency voltage is applied. The two voltages are exactly in phase and, by varying the frequency, the current measured at the

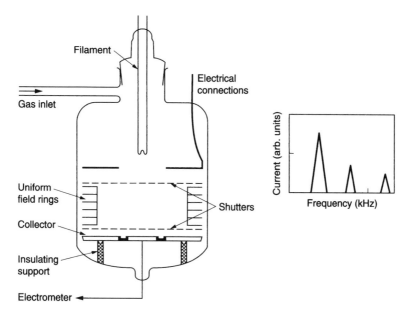

FIGURE 6.1 Schematic diagram of drift tube used for measurement of drift velocity in gases. The filament is the electron source and the drift distance is the spacing between the electrical shutters. The current, measured as a function of applied frequency, is shown on the right.

anode is obtained as pulses. The drift time is measured as the time interval between two successive peaks and the drift velocity is calculated from a knowledge of the drift distance. The usual precautions, such as ensuring the uniformity of the electric field and minimizing the contact potentials, are taken. The drift tube is constructed in such a way that it can be immersed in a constant temperature or cryogenic bath. Figure 6.1 shows a typical drift tube employed for the purpose.[7]

For the measurement of the lateral diffusion coefficient, the cathode is shaped in the form of annular rings separated from each other (Figure 6.2). Electrical contacts are made to the rings from outside the diffusion apparatus. Electrons released from a point source drift toward the anode and there is both radial and longitudinal diffusion of electrons. The ratio of current flowing to an annular ring to the total current involves all three swarm parameters (W, D_r, D_L) and given by[8] is

$$R = 1 - \left(\frac{h}{d'} - \frac{1}{\lambda_L h} + \frac{h}{\lambda_L d'\,2}\right)\left(\frac{h}{d'}\right)\exp[-\lambda_L(d' - h)] \tag{6.7}$$

where the following relationships hold (Figure 6.2):

$$\lambda_L = \frac{W}{2D_L}; \quad d' = b'^2 + h^2; \quad b' = \left(\frac{D_L}{D_r}\right)^{1/2} b \tag{6.8}$$

where D_L is the longitudinal diffusion coefficient.

Equation 6.7 has not been applied extensively, but the results of diffusion experiments have been analyzed using a simplified solution of the diffusion equation:

$$R = 1 - \frac{h}{d}\exp\left[-\frac{\lambda_T b^2}{d' + h}\right] \tag{6.9}$$

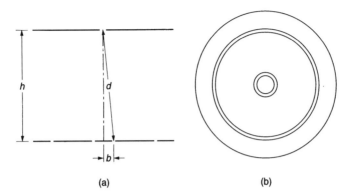

FIGURE 6.2 Schematic diagram of Townsend–Huxley apparatus for measurement of diffusion coefficient. Left: cross section; right: plan view of the anode, comprising three concentric rings. The annular electrodes are insulated from each other and electrical leads are made available for current measurement.

where $\lambda_T = W/2D_r$ and $d^2 = b^2 + h^2$. Equation 6.9 is known as the Huxley equation and assumes isotropic diffusion, which is rarely valid. However, the results obtained using this equation have been remarkably successful.[7]

The steady-state diffusion equation for electrons diffusing in a uniform electric field E (parallel to the z-axis) in a gas that does not form negative ions is given by

$$\nabla^2 n = \frac{\mu E}{D} \frac{\partial n}{\partial z} \tag{6.10}$$

Equation 6.9 is the original solution of Equation 6.10 as derived by Huxley and Bennett.[9] When this equation was applied to an electron-attaching gas the ratio (D/μ) obtained was very high when compared with those of Bailey;[10] Crompton et al.[11] recognized that the observed difference was due to the attachment process. The negative ions which were formed fell on the central electrode and hence the ratio of currents measured did not give the ratio of currents due to electron diffusion alone.

An additional factor that limits the accuracy of determining D/μ is the ionization that occurs when the reduced electric field E/N is sufficiently large. In the presence of ionization and attachment processes the diffusion equation becomes

$$\nabla^2 n = \frac{\mu E}{D} \left\{ \frac{\partial n}{\partial z} - (\alpha - \eta)n \right\} \tag{6.11}$$

where α and η are Townsend's first ionization and attachment coefficients respectively. Neglecting detachment, Huxley[12] and independently Lucas[13] found the solution of Equation 6.11. The solution is, however, too complicated to be of practical use, particularly since the attachment coefficient is not known to the desired accuracy.[14] A simpler solution is due to Francey,[15] who deduced the ratio of currents as

$$R = 1 - \left\{ 1 - \frac{Dmab^2}{2(\mu kTE - D)Bh^2} \right\} \exp\left(\frac{-2^{3/2}\mu E}{6B^{1/2}} \right) \frac{b^2}{h} \tag{6.12}$$

where m is the electron mass, $a = eE/m$, k is the Boltzmann constant, T the gas temperature, and

$$B = \frac{mM}{6} \left(\frac{a\lambda}{kT} \right)^2 \tag{6.13}$$

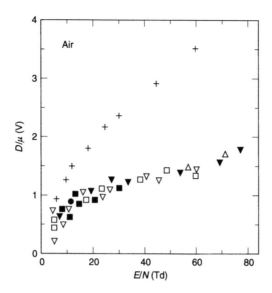

FIGURE 6.3 Values of D/μ in air as a function of E/N. ($\square$) Townsend and Tizard[17]; (∇) Crompton et al.[11]; ($\triangle$) Rees and Jory[18]; (+) Huxley and Crompton[3]; ($\blacktriangledown$), ($\blacksquare$), Rao and Raju[16] at (3.27, 6.53, 16.33, 22.86, 32.65) $\times 10^{22}\,\mathrm{m}^{-3}$ respectively.

M being the molecular mass and λ its mean free path. Typical values are: $m = 9.1091 \times 10^{-31}\,\mathrm{kg}$, $a = 1.7588 \times 10^{11} E\,\mathrm{m\,s}^{-2}$, $M = 46.5 \times 10^{-27}\,\mathrm{kg}$ (N_2), $\lambda = 4\sqrt{2} \times \lambda_{\mathrm{atom}} = 35.5 \times 10^{-8}\,\mathrm{m}$, and $kT = 3.98 \times 10^{-21} = 0.025\,\mathrm{eV}$, giving $B^{1/2} = 1.313 \times 10^{-3} E$ (E in $\mathrm{V\,m}^{-1}$). Equation 6.13 has been successfully applied to air, which is a moderately attaching gas, by Rao and Raju.[16] Figure 6.3 shows their results in comparison with previous results.[10,11,17,18]

The ratio of currents in a diffusion experiment varies, depending upon the processes that are operative in the swarm. These are:

1. Only drift and diffusion, with longitudinal diffusion neglected. This is the simplest of all situations and the current ratio is given by Equation 6.9.
2. Only drift and diffusion, but with longitudinal diffusion considered, Equation 6.7.
3. Drift, diffusion, ionization, and attachment. The solution of the diffusion equation, neglecting lateral diffusion coefficient for simplicity, is given by Naidu and Prasad.[19] The solution is generally complicated, inevitably requiring a computer for the analysis and requiring reliable data on ionization and attachment coefficients as a function of E/N, which limits the accuracy of derived ratios of D_r/μ. Naidu and Prasad[19] investigated several gases by adopting a cathode that was constructed out of a quartz disc that had thin gold plating on one side, the ultraviolet light was allowed to fall on the uncoated side. The current to the central disc of the collecting electrode was not considered in calculating D_r/μ, because the negative ions mostly fell on it in addition to the electrons.
4. The effect of secondary electrons was taken into account by Virr et al.[20] at low values of E/N and by Kontoleon et al.[21,22] and Lakshminarasimha and Lucas.[23]

It is not always possible to carry out (D/μ) measurements at exactly 293 K. The relationship between the ratios measured at temperature T (K) is given by[24]

$$\left(\frac{D}{\mu}\right)_{293} = \frac{293}{T}\left(\frac{D}{\mu}\right)_T \tag{6.14}$$

6.3   ELECTRON ENERGY DISTRIBUTION

The energy distribution of electrons in a swarm provides a complementary method of evaluating the swarm parameters by using the measured scattering cross section. A fuller discussion of the methods available is given elsewhere.[25] However, brief comments and definitions relating the energy distribution function to the drift velocity and diffusion coefficient are given here. The energy distribution is also used to derive the momentum transfer cross sections, particularly in the low energy region, by comparing the calculated swarm data with the measured data. The two aspects of the same issue will not be distinguished here and the description holds good for both applications.

The Boltzmann equation describes the effect of applied electric field and collisions on the distribution function $f(\mathbf{r}, \mathbf{v}, t)$. The function represents the number density of electrons, at time t, that lie in a differential volume centered at $\mathbf{r}$ in three-dimensional configuration space, and which have velocities centered at $\mathbf{v}$ in three-dimensional velocity space. The function is abbreviated as f, representing the number density of electrons in six-dimensional phase space. It is usually written in the form

$$\frac{\partial f}{dt} + \mathbf{v} \cdot \nabla_r f + \mathbf{a} \cdot \nabla_v f = \left(\frac{\partial f}{\partial t}\right)_{\text{coll}} \tag{6.15}$$

where ∇_r is the gradient operator in three-dimensional configuration space, ∇_v is the gradient operator in three-dimensional velocity space, $\mathbf{a}$ the acceleration produced by the electric field, and $\mathbf{v}$ the velocity. For mathematical expediency, the distribution function is expressed as

$$f(\mathbf{r}, \mathbf{v}, t) = f_0(\mathbf{r}, \mathbf{v}, t) + f_1(\mathbf{r}, \mathbf{v}, t) \tag{6.16}$$

The first term on the right side of Equation 6.16 is the steady-state distribution and the second term is a small perturbation imposed upon it.

Equation 6.15 is mathematically challenging and various approximations are resorted to in order to find the solution. One assumes that the function can be separated into spatially dependent density function $n(\mathbf{r}, t)$ and a spatially independent velocity distribution function $g(\mathbf{v}, t)$, that is

$$f(\mathbf{r}, \mathbf{v}, t) \equiv n(\mathbf{r}, t) g(\mathbf{v}, t) \tag{6.17}$$

In the time-independent simplification $g(\mathbf{v}, t) \equiv g(\mathbf{v})$. A further assumption is made that $g(\mathbf{r})$ may be expanded in spherical harmonics according to

$$g(v) = \sum_{j=0}^{n-1} g_j(v) P_j(\cos\theta) \tag{6.18}$$

where P_j is the jth order of the Legendre function. Considering only two terms in this expansion is referred to as the two-term approximation and is found to be satisfactory as long as the momentum transfer cross section does not vary rapidly with energy. Representing the steady-state energy distribution by f and changing the variable from $\mathbf{v}$ to ε, one rewrites the distribution function as

$$f(\varepsilon) = \sum_{j=0}^{n-1} f_j(\varepsilon) P_j(\cos\theta) \tag{6.19}$$

The time-independent Boltzmann equation, successfully employed by Phelps and his colleagues,[33] has the form

$$\frac{d}{d\varepsilon}\left(\frac{e^2E^2\varepsilon}{3NQ_M}\frac{df}{d\varepsilon}\right) + \frac{2m}{M}\frac{d}{d\varepsilon}\left(\varepsilon^2 NQ_M\left[f + kT\frac{df}{d\varepsilon}\right]\right)$$

$$+ \sum_j\left[(\varepsilon + \varepsilon_j)f(\varepsilon + \varepsilon_j)NQ_j(\varepsilon + \varepsilon_j) - \varepsilon f(\varepsilon)NQ_j(\varepsilon)\right]$$

$$+ \sum_j\left[(\varepsilon-\varepsilon_j)f(\varepsilon - \varepsilon_j)NQ_{-j}(\varepsilon-\varepsilon_j) - \varepsilon f(\varepsilon)NQ_{-j}(\varepsilon)\right] = 0 \qquad (6.20)$$

Once the solution has been found the following definitions apply to the drift velocity, radial diffusion coefficient, and rate coefficient for the jth inelastic process[26,27]:

$$W = -\frac{e}{3}\frac{E}{N}\left(\frac{2}{m}\right)^{1/2}\int_0^\infty \frac{\varepsilon}{Q_M(\varepsilon)}\frac{d}{d\varepsilon}f(\varepsilon)d\varepsilon \qquad (6.21)$$

$$D_T = \frac{1}{3N}\left(\frac{2}{m}\right)^{1/2}\int_0^\infty \frac{\varepsilon}{Q_M(\varepsilon)}f(\varepsilon)d\varepsilon \qquad (6.22)$$

$$k = \left(\frac{2}{m}\right)^{1/2}\int_0^\infty \varepsilon Q_j(\varepsilon)f(\varepsilon)d\varepsilon \qquad (6.23)$$

The distribution function is normalized to one (meaning the total number of electrons) so that the fraction of electrons having energy between ε and $\varepsilon + d\varepsilon$ is computed. Authors use one of two alternatives for normalization, according to

$$\int_0^\infty \varepsilon^{1/2}f(\varepsilon)d\varepsilon = 1 \quad \text{or} \quad \int_0^\infty F(\varepsilon)d\varepsilon = 1 \qquad (6.24)$$

Depending upon the normalization definition, Equations 6.21 to 6.23 take a somewhat modified expression. For example, compare Equation 1.121 with 6.23 for the same quantity, k. The numerical method of solution of Equation 6.20 and computation of swarm parameters shown in Equations 6.21 to 6.23 require extensive programming. A software is provided by Pitchford (www.siglo.com) for the two-term solution of the Boltzmann equation.

A second method of solving the Boltzmann equation is to use the Fourier expansion method, first used by Tagashira et al.[28] and applied to mercury vapor by Liu and Raju.[29] The distribution function is expressed as

$$\frac{\partial}{\partial t}\int_0^\varepsilon n(\varepsilon', z, t)\,d\varepsilon' = n_c + n_E + n_z \qquad (6.25)$$

where $n(\varepsilon', z, t)$ is the electron number density with ε, z, and t as the energy, space, and time variables respectively, n_c, n_E, and n_z are the change rate of electron number density due to collision, applied field, and gradient respectively. Equation 6.25 is a mathematical statement for the conservation of electron number density.

The solution of Equation 6.25 may be written in the form of a Fourier expansion[28]

$$n_s(\varepsilon, z, t) = e^{jsz}e^{-w(s)t}H_0(\varepsilon, s) \qquad (6.26)$$

where s is the parameter representing the Fourier component, and

$$w(s) = -w_0 + w_1(js) - w_2(js)^2 + w_3(js)^3 + \cdots \qquad (6.27)$$

$$H_0(\varepsilon, s) = f_0(\varepsilon) + f_1(\varepsilon)(js) + f_2(\varepsilon)(js)^2 + \cdots \qquad (6.28)$$

where w_n ($n = 0, 1, 2, 3, \ldots$) are constants. The details of the method of solution may be found in Liu and Raju.[29]

The Monte-Carlo simulation method also provides another means of determining the electron energy distribution. Different variations such as the mean free path, mean free time, and null collision methods have been described by Raju.[25] A comparison of the two methods, applied to the same gas, has been published by Liu and Raju.[30] Figure 6.4 shows the energy distribution as a function of E/N in molecular oxygen as calculated by the Boltzmann equation method.

The two-term solution of the Boltzmann equation is satisfactory for many applications in which one wishes to obtain the drift velocity and mean energy as a function of E/N. However, situations arise that require additional care in the computation process. The first one is when the cross section changes rapidly with increase in electron energy, as in the case of the right side of the Ramsauer–Townsend minimum. The second situation is when the momentum transfer cross section is small and the inelastic scattering cross section at low energy is significant. The latter situation has been investigated by Pitchford et al.[31] and Pitchford and Phelps,[32] who observed differences of up to ~8% in the transport coefficients at low E/N when a six-term expansion was used in place of the two terms. The effect of anisotropic scattering has been investigated by Phelps and Pitchford[33] in the higher range of E/N from 500 to 1500 Td, using the multiterm expansion.

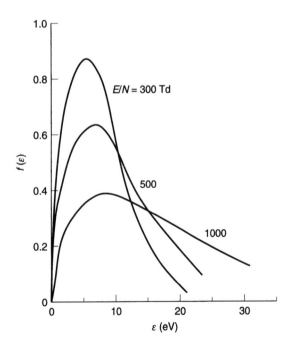

FIGURE 6.4 Electron energy distribution function in oxygen at several values of the parameter E/N. As E/N increases, the distribution spreads out with decreasing peak energy.[30]

6.4 APPROXIMATE METHODS

A number of approximate methods have been developed to calculate the drift velocity and the characteristic energy (D/μ) by using simplified energy distribution functions. The methods fall broadly into two categories: those that do not need the characteristic energy or mean energy and those that need this quantity as a starting point. One of the earliest distribution functions that belongs to the former kind is due to Druyvesteyn (reported by Huxley and Crompton[34]) whose derivation is

$$f(\varepsilon) = A\varepsilon^{1/2} \exp\left[-\frac{3m}{M}\left(\frac{2NQ_{M}\varepsilon}{E}\right)^{2} \right] \tag{6.29}$$

where E/N is the reduced electric field, m/M is the electron-to-atom (neutral) mass ratio, and A is the normalization constant. Note that the distribution falls off more rapidly than the Maxwellian distribution, due to the ε^2 term. Though the original derivation assumed that Q_M is independent of energy, one can find an approximate analytical expression to take into account the shape of the Q_M–ε curve. If we suppose that Q_M is proportional to a power of ε according to

$$Q_M = Q_0\varepsilon^n \tag{6.30}$$

one can calculate the swarm parameters using Equations 6.30, 6.21 to 6.23. In the case that the momentum transfer cross section varies as a function of ε and low energy inelastic losses (rotational and vibrational excitation) do not occur, as in rare gases, the energy distribution function is given by[4]

$$f(\varepsilon) = A\varepsilon^{1/2} \exp -\left\{ \int_{0}^{\varepsilon} \frac{d\varepsilon}{[(E/N)]^{2}e^{2}M[6m\varepsilon Q_{M}^{2}(\varepsilon)]^{-1}+kT} \right\} \tag{6.31}$$

where A is the normalizing factor according to

$$\int_{0}^{\infty} f(\varepsilon)d\varepsilon = 1 \tag{6.32}$$

Let us now consider the distribution function which can be calculated from a knowledge of the mean energy $\bar{\varepsilon}$. Morse et al.[35] have derived the function as

$$f(\varepsilon)d\varepsilon = A\frac{\varepsilon^{1/2}}{\bar{\varepsilon}^{3/2}}\exp\left[-B\left(\frac{\varepsilon}{\bar{\varepsilon}}\right)^{2(p+1)} \right] \tag{6.33}$$

in which $\bar{\varepsilon}$ is the mean energy and A and B are constants, defined by

$$A = 2(p+1)\left[\Gamma\frac{5}{4(p+1)} \right]^{3/2} \times \left[\Gamma\frac{3}{4(p+1)} \right]^{5/2} \tag{6.34}$$

$$B = \left[\Gamma\frac{5}{4(p+1)} \times \frac{1}{\Gamma\frac{3}{4(p+1)}} \right]^{2(p+1)} \tag{6.35}$$

In Equations 6.34 and 6.35 the symbol Γ stands for the gamma function, defined according to

$$\Gamma(n+1) = n! = 1 \cdot 2 \cdot 3 \cdots (n-1) \cdot n \tag{6.36}$$

Thus

$$\Gamma(n+1) = n\Gamma n \tag{6.37}$$

Some useful values of the Gamma function are:

$$\Gamma\tfrac{1}{4} = 3.6256; \quad \Gamma\tfrac{1}{3} = 2.6789; \quad \Gamma\tfrac{1}{2} = 1.7724; \quad \Gamma\tfrac{3}{4} = 1.2254$$

Values of the gamma function are tabulated in Abramowitz and Stegun.[36] The energy distribution function given by expression 6.33 is known as the *p*-set and is quite simple to use: $p = -1/2$ gives the Maxwellian distribution and $p = 0$ gives the Druyvesteyn function, the same as Equation 6.29 except that the latter expression is expressed in more fundamental quantities (m/M) of the atom. Brand and Kopainsky[37] adopted the Druyvestyn distribution function in calculating the sparking voltage of SF_6.

Equation 6.33 can be inserted into Equations 6.21 to 6.23 to obtain analytical expressions for the drift velocity and diffusion coefficient only if the quantity NQ_M can be expressed as an analytical function of ε. Further, not all possible values of p can be explored. Heylen[38,39] successfully derived an analytical equation for W and D assuming that $p = -1/2$, that is, the distribution is Maxwellian and the momentum transfer cross section in H_2, N_2, O_2, and air could be approximated according to

$$Q_M(\varepsilon) = Q_{M0}\varepsilon^n \exp(-s\varepsilon) \tag{6.38}$$

in which Q_{M0}, n, and s are adjustable constants. In the case of the same constants not satisfying Equation 6.38 over the entire range, as in N_2, O_2, and air, successive ranges were found and integration carried out. Since the resulting expression involves evaluating incomplete gamma functions the calculation becomes tedious but, with software now available for numerical integration, the use of Equation 6.33 with Equations 6.21 to 6.23 has become much simpler. Heylen's work[38] has shown that W and D_r/μ are relatively insensitive to small variations in the electron energy distributions when the cross section does not vary too rapidly with electron energy. Further, the relationship between (D_r/μ) and the mean energy $\bar{\varepsilon}$, given by

$$\bar{\varepsilon} = 1.5\frac{D}{\mu} \tag{6.39}$$

is applicable only for the Maxwellian distribution, and other distributions combined with variation of momentum transfer cross section with energy could give values in the range of 1.5 to 0.3 in Equation 6.39.

This method was further explored by Raju et al.[40–42] in several gases by approximating Q_M as

$$Q_M(\varepsilon) = Q_{M0}\varepsilon^n \exp(-g\varepsilon) \tag{6.40}$$

in which Q_{M0}, n and g are adjustable constants for the particular gas. Substituting Equation 6.40 in 6.21 and integrating, one obtains the expression for the drift velocity as

$$W = \frac{1.04E}{NQ_{M0}} \frac{\Gamma 2 - n}{\bar{\varepsilon}^{2.5}} \frac{1}{(1.5/\bar{\varepsilon} - g)^{2-t}} \tag{6.41}$$

Such expressions show the functional dependence of swarm parameters on E/N. In concluding this section we note that the diffusion coefficient D is not measurable as a single quantity and that measurements reported in the literature are of W and D_r/μ. Theoretical methods are required for calculating D_r and the longitudinal diffusion coefficient D_L.

6.5 DATA ON DRIFT AND DIFFUSION

The voluminous data on drift and diffusion of electrons in gases that are available in the literature are summarized in this chapter and Chapter 7. As already stated, there are three different methods available for accumulating this vast quantity of data, namely experiments, the Boltzmann distribution method, and the Monte Carlo method. We choose the data in the range of E/N of our interest from all three sources, assigning priority to experimental data. For each gas about ten citations are given and the recommended values are shown graphically or in tabular form. While we note that such comprehensive presentation is not available in any previous volume, as far as the author of the present volume is aware, the following sources have been used in addition to individual papers. Data up to 1974 are taken from Huxley and Crompton,[3] the bibliography of electron swarm data by Beaty et al.[43] and the report of Raju.[44]

6.5.1 AIR (DRY AND HUMID)

Drift velocity data in dry air are reported by: Hessanauer et al.[45] current pulse method; Rees,[46] data corrected for diffusion effects; Huxley and Crompton,[3] compilation; Roznerski and Leja,[47] experimental, using an electrical shutter method; Liu and Raju,[48] Monte Carlo analysis. Further, the author of the present volume has calculated the drift velocity and mean energy using BOLSIG software[49] with two-term expansion for synthetic air with a composed mixture of 78% N_2 + 21% O_2 + 1% CO_2. These data are shown in Figure 6.5. Table 6.1

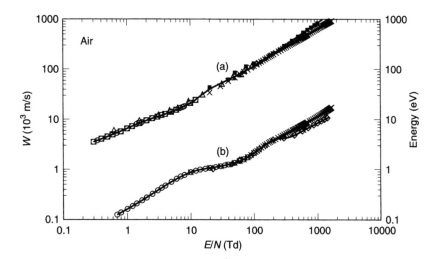

FIGURE 6.5 Drift velocity (a) and characteristic energy (b) of dry air. Drift velocity: ($\triangle$) Hessenauer et al.[45]; ($\square$) Rees[46]; ($\circ$) Huxley and Crompton[3]; ($\bullet$) Roznerski and Leja[47]; ($\blacksquare$) Liu and Raju[48]; ($\times$) calculated from BOLSIG[49]; (—) recommended. (D_T/μ): ($\circ$) Huxley and Crompton[3]; ($\diamond$) Lakshminarasimha and Lucas[51]; ($\times$) calculated from BOLSIG[49]; (—) recommended.

TABLE 6.1
Recommended Drift Velocity of Electrons in Dry Air

E/N	W	E/N	W	E/N	W	E/N	W
0.3	3.60	4	12.76	80	97.10	700	650.50
0.4	4.00	6	15.50	100	112.80	800	726.00
0.6	4.91	8	18.26	200	210.00	900	788.30
0.8	5.78	10	21.00	300	309.6	1000	859.90
1	6.58	20	40.00	400	383.6	1300	1103.40
2	9.52	40	62.50	500	496.2	2000	1566.5
3	11.36	60	77.60	600	556.70		

E/N in units of Td, W in units of 10^3 m/s.

shows recommended values. The drift velocity may be represented by the analytical equations[44]

$$W = -3.2\left(\frac{E}{N}\right)^4 + 92.6\left(\frac{E}{N}\right)^3 - 928.1\left(\frac{E}{N}\right)^2 + 5128.4\left(\frac{E}{N}\right) + 2184.3 \, (\text{m/s}) \quad (6.42)$$

for the range $0.3 \leq E/N \leq 12$, deviation limits: $+1.85\%$ at 1.2 Td, -2.64% at 0.35 Td; and

$$W = \left[-9 \times 10^{-5}\left(\frac{E}{N}\right)^2 + 0.94\left(\frac{E}{N}\right) + 24.61\right] \times 10^3 \, (\text{m/s}) \quad (6.43)$$

for the range $40 \leq E/N \leq 2000$, deviation limits: -4.3% at 100 Td, 4.9% at 500 Td, $R^2 = 0.9995$. An alternative representation for the mobility has been provided by Chen and Davidson[50]:

$$\mu_e N = 6.4421 \times 10^{19}\left(\frac{E}{N}\right)^2 - 3.0314 \times 10^{22}\left(\frac{E}{N}\right) + 5.1925 \times 10^{24} \quad (6.44)$$

with $R^2 = 0.9959$ for $39.9 \leq E/N \leq 206.1$ Td; and

$$\mu_e N = 9.4833 \times 10^{24}\left(\frac{E}{N}\right)^{-0.3321} \quad (6.45)$$

with $R^2 = 0.9971$ for $206.1 \leq E/N \leq 1000$ Td. It is noted here that, according to theory, the drift velocity increases linearly with E/N till the limit of energy equality with gas molecules is exceeded. The purpose of representing W by a polynomial is to provide data economically and the upper limit of E/N is not related to the above theory.

In addition to the references shown in Figure 6.5 the ratio D_r/μ has been measured by Lakshminarasimha and Lucas,[51] who developed techniques for analysis in the presence of secondary electrons. These data are shown in Figure 6.5 and Table 6.2.

The addition of small quantities of water has a complicated influence on the mobility of electrons. While at low values of E/N (~ 0.7 Td) the mobility decreases with increasing vapour content, above ~ 1 Td there is an increase of mobility. These effects are due to interaction of momentum transfer cross sections and their energy dependency. Table 6.3 shows the results of measurements of mobility with various concentrations of water.[52]

TABLE 6.2
Recommended D_r/μ in Dry Air

E/N	D_r/μ	E/N	D_r/μ	E/N	D_r/μ	E/N	D_r/μ
0.70	0.13	10.0	0.86	70.0	1.64	500	5.29
0.80	0.14	14.0	0.99	80.0	1.79	600	6.13
1.00	0.16	20.0	1.07	100	2.13	700	6.68
1.40	0.21	30.0	1.19	150	2.97	800	7.35
2.00	0.28	40.0	1.28	200	3.88	900	8.00
3.00	0.38	50.0	1.37	300	4.25	1000	8.59
5.00	0.57	60.0	1.50	400	4.50	1500	10.90

E/N in units of Td, D_r/μ in units of V.

TABLE 6.3
Mobility of Electrons in Humid Air

H_2O (%)	$N\mu = W/(E/N)$				
	$E/N = 0.7$	0.8	1.0	1.5	2.0
0.30	14.3	13.4	11.7	8.60	6.75
0.45	13.7	13.2	11.9	9.01	7.13
0.60	13.1	12.7	11.7	9.32	7.50
0.75	12.0	11.8	11.2	9.41	7.80
0.90	11.1	10.9	10.5	9.29	8.00
1.0	10.5	10.4	10.1	9.00	8.00
1.2	9.56	9.45	9.25	8.66	7.85
1.5	8.40	8.43	8.27	7.83	7.45

E/N in units of Td, $N\mu$ in units of $10^3 \, \text{m}\,\text{s}^{-1}\text{Td}^{-1}$
From Molloy, B. et al., *Aust. J. Phys.*, 28, 231, 1975.

Collision frequency is a parameter closely related to W. Figure 6.6 shows total electron–neutral collision frequencies v/N as a function of E/N, as calculated by the present author using the two-term BOLSIG software and the cross sections for synthetic air. The collision frequencies may be expressed by an empirical expression

$$\frac{v}{N} = \left[10^{-12} \left(\frac{E}{N}\right)^4 + 6 \times 10^{-9} \left(\frac{E}{N}\right)^3 - 8 \times 10^{-6} \left(\frac{E}{N}\right)^2 + 0.0044 \left(\frac{E}{N}\right) + 0.4433 \right] \times 10^{-13} \quad (6.46)$$

applicable in the range $10 \leq E/N \leq 800 \, \text{Td}$; the quality of fit between Equation 6.46 and the calculated values is also shown in Figure 6.6. The computed values are in very good agreement with the microwave data, $v/N = 1.6 \times 10^{-13} \, \text{m}^3 \, \text{s}^{-1}$.

6.5.2 ARGON

Drift velocity data in argon are provided by Bowe,[53] Wagner et al.,[54] Pack et al.,[55] Jager and Otto,[56] Brambring,[57] Huxley and Crompton,[3] Long (Jr) et al.,[58] Robertson,[59] Tagashira et al.,[60] Sakai et al.,[61] Christophorou et al.,[62] Kücükarpaci and Lucas,[63] Nakamura and Kurachi,[64] and Pack et al.[65] Combining the compilation of Huxley and Crompton[3] with the

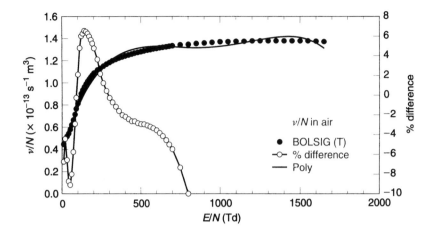

FIGURE 6.6 Calculated collision frequency in air as a function of E/N using BOLSIG[49] software for two-term solution of the Boltzmann equation. (●) ν/N (Raju, 2003, unpublished); (——) Equation 6.46 with $R^2 = 0.9832$; (○) percentage difference between theoretical results and values obtained from Equation 6.46. Positive values of the ordinate imply that theoretical values are higher.

experimental results of Nakamura and Kurachi[64] and those of Kücükarpaci and Lucas,[63] obtained both by experiment and by Monte Carlo simulation, one covers the range of $10^{-2} \leq E/N \leq 10^4$ Td, a range of six decades. Selected data, including the data obtained at 77 K, are shown in Figure 6.7 and Table 6.4.

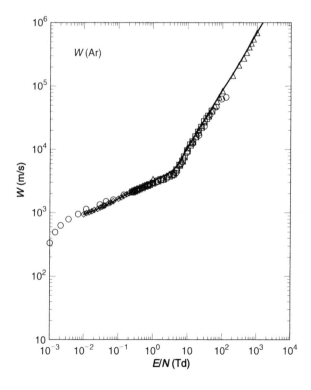

FIGURE 6.7 Experimental and theoretical values of W for electrons in argon. (□) Nakamura and Kurachi[64]; (△) Kücükarpaci and Lucas[63]; (◇) Huxley and Crompton[3]; (○) Pack et al.[65]; (——) Table 6.4.

TABLE 6.4
Drift Velocity in Argon at 293 K

E/N	W	D_r/μ	E/N	W	D_r/μ	E/N	W	D_r/μ
1×10^{-2}	0.935	0.45	0.4	2.39	—	40	34.4	7.46
1.2	0.972	—	0.5	2.52	—	50	41.2	6.91
1.4	1.005	—	0.60	2.63	—	60	49.1	6.44
1.7	1.047	—	0.70	2.73	—	70	58.4	6.19
2.0	1.084	0.8	0.80	2.81	—	80	68.0	6.08
2.5	1.144	—	1.0	2.95	4.78	100	85.5	6.05
3.0	1.205	—	2.0	3.44	6.35	200	149	5.93
3.5	1.252	—	3.0	3.81	7.36	300	217	5.47
4.0	1.294	1.05	4.0	4.12	8.04	400	282	5.40
5.0	1.368	—	5.0	4.75	8.24	500	351	5.83
6.0	1.437	—	6.0	5.64	8.29	600	424	6.07
7.0	1.500	1.45	7.0	6.54	8.29	700	497	6.26
8.0	1.556	—	8.0	7.61	8.18	800	568	6.61
0.1	1.654	1.85	10	9.56	7.70	1000	704	7.38
0.12	1.741	—	12	11.7	7.42	2000	1190	8.61
0.14	1.820	—	14	13.4	7.37	3000	1600	10.62
0.17	1.918	—	17	16.3	—	4000	1770	13.91
0.20	2.00	—	20	18.8	7.36	5000	1860	16.86
0.25	2.13	—	25	22.4	7.50	6000	2060	19.98
0.30	2.23	—	30	27.1	7.61			
0.35	2.31	2.8	35	31.1	7.60			

E/N in units of Td, W in $10^3 \, \mathrm{m \, s^{-1}}$, D_r/μ in volts. Values of D_r/μ for $E/N > 100$ Td are based on Lakshminarasimha and Lucas.[85]

The drift velocity in the range of $0.014 \leq E/N \leq 1$ may be represented by the analytical expressions[44]

$$W = 2943.9 \times \left(\frac{E}{N}\right)^{0.25} \text{m/s}; \quad R^2 = 0.999 \quad (0.01 \leq E/N \leq 4 \, \text{Td}) \tag{6.47}$$

$$W = 1381.6 \times \left(\frac{E}{N}\right)^{0.877} \text{m/s}; \quad R^2 = 0.9962 \quad (10 \leq E/N \leq 500 \, \text{Td}) \tag{6.48}$$

The quality of agreement between values calculated by these expressions and measured values is shown in Table 6.5. The lowest discrepancy between calculated values using expression 6.47 and those shown in Table 6.4 occurs at 0.01 Td (+0.09%) and the highest discrepancy at 0.025 Td (−2.33%).

The influence of the Ramsauer–Townsend minimum on the drift velocity has been theoretically investigated by Makabe and Mori.[66] A temperature dependence is observed at low values of E/N in the range $1 \times 10^{-4} \leq E/N \leq 1$ Td, with the drift velocity decreasing with decreasing temperature. Their contribution deserves more recognition, particularly since most theoretical studies prefer to choose higher values of E/N. Using a three-term solution of the Boltzmann equation, the authors examined the isotropic and directional (nonisotropic) contributions, justifying the conclusion that the two-term solution is applicable.

TABLE 6.5
Quality of Fit between Analytical Expressions and True Values

Equation	E/N range	Highest Discrepancy		Lowest Discrepancy	
		%	E/N	%	E/N
(6.47)	0.01–4.0	−2.33	0.025	−0.07	1.4
(6.48)	10–500	−8.82	10.0	−0.34	35

E/N in units of Td. Positive values mean that analytical values are lower.

Makabe and Mori[66] find a significant distortion of the distribution function due to the Ramsauer–Townsend minimum in the momentum transfer cross section. The number of electrons decreases with energy near the minimum, resulting in a double peak distribution in the range $25 \times 10^{-4} \leq E/N \leq 80 \times 10^{-4}$ Td. Figure 6.8 clearly demonstrates this effect. At low energies the distribution function becomes a function of the gas mean temperature. As a result both the drift velocity and the characteristic energy become sensitive to the temperature, in agreement with experimental observations. At low values the drift velocity depends on E/N according to a power law, before the slope decreases to a lower value at $\sim E/N = 40 \times 10^{-4}$ Td. For higher reduced electric fields (E/N) the temperature dependence of the drift velocity disappears (Figure 6.9).

The experimental results of Robertson[59] at 90 and 293 K provide experimental confirmation of this observation. Dezhen and Tengcal[67] have investigated theoretically the influence of the degree of ionization on the transport properties in a narrow range of $10 \leq E/N \leq 100$ Td and suggest that the electron energy distribution shifts towards the Maxwellian as the degree of ionization increases. This result needs to be confirmed since most of the previous studies have shown that the distribution in rare gases is far from Maxwellian.

The influence of the density on the drift velocity has been calculated by Atrazhev and Iakubov,[68] who considered the short and long range polarization fields; the drift velocity in argon increases with density at densities of the order of $N \simeq 10^{26}$ to 10^{27} m³, in conformity

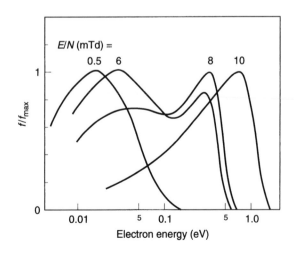

FIGURE 6.8 Influence of Ramsauer–Townsend minimum on the electron energy distribution in argon. Double peaks are observed in range $5 \times 10^{-4} \leq E/N \leq 8 \times 10^{-3}$ Td. Figure reproduced from Makabe, T. and T. Mori, *J. Phys. D*, 15, 1395, 1982. With permission of the Institute of Physics (IOP), U.K.

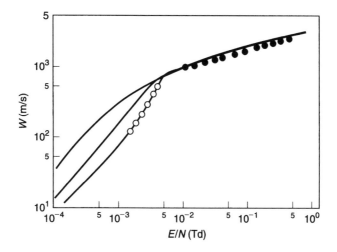

FIGURE 6.9 Influence of temperature on the drift vrlocity of electrons in argon. At low E/N the drift velocity increases with increasing temperature. Beyond a certain value of E/N the temperature does not influence the velocity. Theoretical: (----) 89.6 K; (—) 293 K; (—·—) 1000 K. Experimental: (○) Robertson,[59] 89.6 K; (●) Robertson.[59] Figure reproduced from Makabe, T. and T. Mori, *J. Phys. D*, 15, 1395, 1982. With permission of IOP, U.K.

with CH_4 but in contrast with He, N_2, and H_2 which show a decrease of drift velocity with increase of density. The results of Christophorou and McCorkle[69] provide experimental proof of this observation. On the other hand, Allen and Prew[70] find that higher densities in the range of 300 kPa (~3 atmospheres) to 10 MPa (~100 atmospheres) do not make an appreciable difference. Since the range of E/N considered by Atrazhev and Iakubov[68] is very low (~0.001 to 0.006 Td) compared with that of Allen and Prew[70] (0.5 to 6.0 Td), their different conclusions remain to be explained.

Argon is a favorite gas for studying, both theoretically and experimentally, the influence of mixtures on transport properties because of the deep Ramsauer–Townsend minimum in the momentum transfer cross section. The schematic variation of momentum transfer cross section of several gases as a function of electron energy is shown in Figure 6.10. Argon, silane

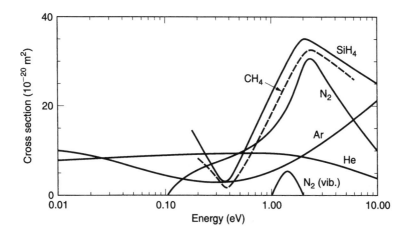

FIGURE 6.10 Schematic sketch of momentum transfer cross sections in several gases. Argon, silane and methane show the Ramsauer–Townsend effect, though argon has been shown on an expanded scale for clarity. Helium shows monotonic decline. N_2 has a significant vibrational excitation cross section.

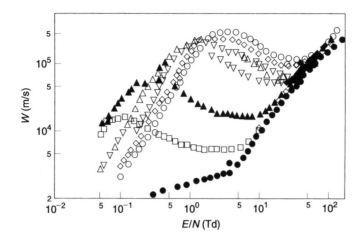

FIGURE 6.11 Measured drift velocity in mixtures of argon and methane. (•) 100% Ar; (○) 0.5%, (▲) 3%, (△) 25%, (▽) 50%, (◇) 75%, (○) 100% CH_4. Figure reproduced from de Urquijo, J. et al., *J. Phys. D*, 32, 1646, 1999. With permission of IOP, U.K.

(SiH_4), and methane (CH_4) show the Ramsauer–Townsend minimum, whereas nitrogen and helium do not. The rise of Q_M in argon, to the left of the minimum, is not as steep as in the other two gases. Nitrogen has a significant vibrational cross section in addition to a shape resonance.

Transport properties of mixtures of Ar + CO_2,[71] Ar + N_2,[72] Ar + CO,[58] Ar + SiH_4,[73] Ar + CH_4,[74,75] Ar + H_2,[76] Ar + SF_6,[77] Ar + Hg,[78] Ar + CF_4,[79] and Ar + c-C_4F_8[80] have been studied by several researchers. Addition of small amounts of molecular gas changes the drift velocity disproportionately. Further, de Urquijo et al.[74] experimentally observe (Figure 6.11) a decrease in the drift velocity in mixtures of Ar+CH_4 as E/N is increased, and the authors attribute this effect to the Ramsauer–Townsend minimum. Their results confirm the original findings of Foreman et al.[75] that the drift velocity in the low E/N region in mixtures of Ar + CH_4 is greater than in either pure gas; this effect is attributed to inelastic processes that occur in CH_4. A theoretical description of this phenomenon is given in Chapter 7.

D/μ values as a function of E/N may be used to derive momentum transfer cross sections at very low electron energies where experimental measurement of Q_M presents formidable difficulties. A detailed calculation of the energy distribution function is involved in such derivation, but the gross features of dependence of Q_M on the electron energy are revealed by the dependence of D/μ on E/N. This is attributable to the fact that D/μ is approximately equal to $2\bar{\varepsilon}/3$, where $\bar{\varepsilon}$ is the mean energy of a swarm which is, in turn, sensitive to variations of the different cross sections. At low enough values of E/N the electrons in the swarm are at thermal equilibrium with neutrals and one can put $D/\mu = kT/e$.

As E/N is increased, the average electron energy and D/μ rise well above this value. In simple cases where only a constant elastic scattering cross section is involved, D/μ increases linearly with E/N for D/μ values above kT/e. If, however, the elastic scattering cross section increases or decreases with electron energy, the D/μ values increase with E/N more rapidly than linearly. Similarly, the onset of an inelastic process is suggestive of a sudden decrease of the slope of the D/μ–E/N plot.[82]

Data on the characteristic energy defined by D_r/μ have been provided by Townsend and Bailey,[81] Warren and Parker,[82] Heylen and Lewis,[83] Engelhardt and Phelps,[84] Wagner et. al.,[54] Lakshminarasimha and Lucas,[85] Robertson,[86] Milloy et al.,[87] Kücükarpaci and Lucas,[88] and Al-Amin and Lucas.[89] Selected data are shown in Figure 6.12 and Table 6.4. The recommended values of D_r/μ for $E/N > 100\,Td$ are based on the experimental results of

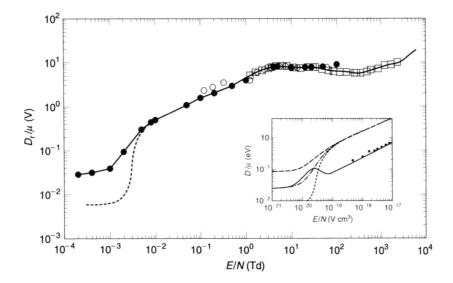

FIGURE 6.12 Characteristic energy for electrons in argon at 300 K. Experimental: (○) Townsend and Bailey[17b]; (- - -) Warren and Parker,[82] (77 K); (△) Lakshminarasimha and Lucas[85]; (□) Kücükarpaci and Lucas.[88] Theory: (—) Kücükarpaci and Lucas[88]; (●) Pack et al.[65] Inset shows the dependence of D_r/μ at low E/N on temperature, calculated by Makabe and Mori[66]; temperatures from top to bottom are 1000 K, 293 K, and 89.6 K(D_r/μ) and 293 K(D_L/μ). Experimental points in the inset: (◇) Milloy et al.,[87] at 294 K; (○) Wagner et al.[54] at 293 K.

Lakshminarasimha and Lucas,[85] who used a steady-state technique. The measurements were repeated by Al-Amin and Lucas,[89] using the time-of-flight technique; the latter method gives higher values.

The characteristic energy initially increases rapidly with E/N, reaching a plateau and increasing again slowly (Figure 6.12). The initial rapid rise and the subsequent flattening are due to the marked minimum in the momentum transfer cross section of argon as a function of energy.[82] An experimental detail to note here is the fact that measurements at very low E/N are accomplished by the use of much higher gas pressures, in the range of 100 to 300 kPa. The temperature dependence of D_r/μ has been studied theoretically by Makabe and Mori[66]; their results are shown as an inset in Figure 6.12. The ratio of longitudinal diffusion coefficient to mobility (D_L/μ) shown in Figure 6.7, will be discussed in the section on helium.

As stated earlier, D/μ values are approximately 66% of the mean energy, though the exact relationship is dependent on the energy distribution function. The theoretically calculated mean energy is given by Kücükarpaci and Lucas[63] at moderate and high values of E/N, following the earlier experimental investigations of Losee and Burch[90] and Makabe et al.[91] The experiment of Losee and Burch[90] has not received the attention it deserves, though it is one of the very few studies in which the electron energy distribution is measured. Figure 6.13 and Table 6.6 present the mean energy as a function of E/N. Losee and Burch[90] used two different distributions to calculate the mean energy and the values shown in Figure 6.12 are the average of the two energies at the same E/N.

Figure 6.13 also includes the mean energy of electrons in a mixture of Ar + H₂ as reported by Petrović et al.[76] Addition of hydrogen causes inelastic collision losses which reduce the mean energy dramatically and help to determine the momentum transfer cross section at low E/N more accurately by the swarm method. A more detailed discussion of these results is presented later.

Returning to Figure 6.7, there are two different diffusion coefficients, the radial diffusion coefficient (D_r) and the lateral diffusion coefficient (D_L), that are shown as a function of E/N.

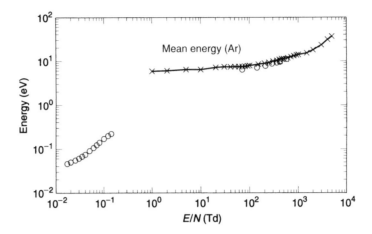

FIGURE 6.13 Mean energy of electron swarm in argon as a function of E/N. Experimental: (o) Losee and Burch[90]; (◇) Makabe et al.[91] Theory: (---) Kücükarpaci and Lucas,[88] overlapping the recommended values; (—×—) recommended. Lower curve: (o) Petrović et al.,[76] mean energy in Ar + 5.44% H_2, showing the dramatic reduction of $\bar{\varepsilon}$.

TABLE 6.6
Recommended Values of the Mean Energy of Electrons in Argon

E/N (Td)	$\bar{\varepsilon}$ (eV)	E/N (Td)	$\bar{\varepsilon}$ (eV)	E/N (Td)	$\bar{\varepsilon}$ (eV)
1	5.90	90	7.74	700	12.47
2	6.00	100	7.91	800	13.12
5	6.47	150	8.46	900	13.61
10	6.38	200	8.75	1000	13.94
20	7.14	250	9.38	1500	15.28
30	7.47	300	10.09	2000	18.09
40	7.44	350	10.59	3000	23.62
50	7.40	400	10.89	4000	31.42
60	7.40	450	11.09	5000	37.07
70	7.47	500	11.27		
80	7.58	600	11.80		

The ratio D_r/μ is measured by the Townsend method in which the electrons are emitted at a steady rate from a point source and drift under a steady electric field. They diffuse radially, spreading out, and the current collected by a cathode with annular rings yields the ratio D_r/μ. The mobility μ is measured by a time-of-flight method in which a pulse of electrons is formed and the speed of the center of gravity of the pulse gives the mobility.

Theoretical simulations of electron–atom collisions have provided details of energy losses in the collision processes. In rare gases the losses are due to elastic scattering, excitation, and ionization. The dominant process and the corresponding range of E/N are shown in Table 6.7. Data for other rare gases are also included in this table as reference to them will be made again. In argon, at low E/N, elastic scattering accounts for 85% of the energy loss. At intermediate E/N excitation processes dominate, the maximum of 95% occurring at ~ 40 Td. At high E/N over 95% of the energy is lost, due to ionization cross sections.[63]

The arriving pulse in the time-of-flight method was observed to have an increasing width, which was attributed to the diffusion of electrons.[92] The diffusion phenomenon was

TABLE 6.7
Dominant Energy Loss during Collision Process in Rare Gases

Gas	Elastic	Range of E/N (Td) Excitation	Ionization	Reference
Argon	< 8	$8 \leq E/N \leq 220$ (11.6 eV)	≥ 220	Kücükarpaci and Lucas[88]
Helium	< 15	$15 \leq E/N \leq 30$ (19.8 eV)	≥ 40	Kücükarpaci et al.[143]
		$30 \leq E/N \leq 40$ (21.5 eV)		
Krypton	< 7	$7 \leq E/N \leq 290$ (11.3 eV)	≥ 290	Kücükarpaci and Lucas[184]
Neon	< 3	$3 \leq E/N \leq 75$ (16.7 eV)	≥ 75	Kücükarpaci et al.[226]

Numbers in brackets are excitation thresholds.

recognized as longitudinal diffusion, in the direction of the electric field. The ratio of diffusion coefficient to mobility measured by the Townsend method was found to be different from those obtained by the time-of-flight method. Lowke and Parker[93,94] developed the theoretical method, based on the concept that the electron energy distribution is spatially dependent, for the calculation of the ratio D_L/μ.

The theory, applied to several gases, has shown that the ratio D_r/D_L is unity at low values of E/N where the electrons are in thermal equilibrium with gas neutrals. At higher E/N, this ratio is dependent on the collision cross section, being particularly sensitive to the gradient of the momentum transfer cross section as a function of energy. There is considerable difference between D_L/μ and D_r/μ, and the ratio D_r/D_L in argon is as large as seven, whereas in helium, hydrogen, and nitrogen it is about two. Pack et al.[65] suggest that the ratio D_L/μ reflects, more accurately than D_r/μ, the change of elastic scattering cross section with energy.

The electron–atom collision frequency of monoenergetic electrons is given by Equation 1.95 and the effective collision frequency in a swarm is given by Equation 1.98. A simple relationship between these two equations does not exist because of the complicated nature of the electron energy distribution. However, in a low temperature plasma one can make an approximation that the energy distribution is Maxwellian; the effective collision frequency (ν_{eff}) as a function of electron temperature (11,605 K = 1 eV) has been given by Baille et al.[95] for all the noble gases. The expressions for calculating ν_{eff} (in units of $\text{m}^3 \, \text{s}^{-1}$) as a function of the absolute electron temperature T_e are:

$$\text{Ar}: \quad \frac{\nu_{\text{eff}}}{N} = \begin{cases} (2.58 \times 10^{-12})T_e^{-0.96} + 2.25 \times 10^{-23}T_e^{2.29}; & (600 \leq T_e \leq 14,000 \, \text{K}) \\ 3.70 \times 10^{-14}T_e^{-0.315}; & (250 \leq T_e \leq 600 \, \text{K}) \end{cases} \quad (6.49)$$

$$\text{He}: \quad \frac{\nu_{\text{eff}}}{N} = \begin{cases} 3.53 \times 10^{-16}T_e^{0.562}; & (100 \leq T_e \leq 5000 \, \text{K}) \\ 1.08 \times 10^{-15}T_e^{0.425}; & (3000 \leq T_e \leq 30,000 \, \text{K}) \end{cases} \quad (6.50)$$

$$\text{Kr}: \quad \frac{\nu_{\text{eff}}}{N} = \begin{cases} (2.15 \times 10^{-9})T_e^{-1.67} + 1.41 \times 10^{-23}T_e^{2.4}; & (1300 \leq T_e \leq 25000 \, \text{K}) \\ 3.80 \times 10^{-13}T_e^{-0.45}; & (300 \leq T_e \leq 1000 \, \text{K}) \end{cases} \quad (6.51)$$

$$\text{Ne}: \quad \frac{\nu_{\text{eff}}}{N} = \begin{cases} (8.63 \times 10^{-18})T_e^{0.833}; & (100 \leq T_e \leq 5000 \, \text{K}) \\ 3.79 \times 10^{-17}T_e^{0.655}; & (3000 \leq T_e \leq 30,000 \, \text{K}) \end{cases} \quad (6.52)$$

$$\text{Xe}: \quad \frac{\nu_{\text{eff}}}{N} = \begin{cases} (2.13 \times 10^{-10})T_e^{-1.23} + 4.52 \times 10^{-26}T_e^{3.2}; & (1000 \leq T_e \leq 8000 \, \text{K}) \\ 1.34 \times 10^{-12}T_e^{-0.495}; & (250 \leq T_e \leq 1000 \, \text{K}) \end{cases} \quad (6.53)$$

As an example of application of the formulas 1.95, 1.98, and 6.49, let us substitute the following values: $\varepsilon = 1\,\mathrm{eV}$, $E/N = 1.8\,\mathrm{Td}$ (corresponding to $D/\mu = 0.67\,\mathrm{V}$), $W = 1.05 \times 10^3\,\mathrm{m/s}$, $T_e = 11{,}605\,\mathrm{K}$. The collision frequencies obtained are (in units of $\mathrm{m^3\,s^{-1}}$) 6.4×10^{-15}, 2.8×10^{-13}, and 4.6×10^{-14} respectively. The first value is for a monoenergetic beam, the second for swarm electrons, and the last for low temperature plasma with the assumption of Maxwellian distribution. A representative calculation using the Boltzmann equation gives $\nu_{\mathrm{eff}}/N \simeq 2 \times 10^{-13}\,\mathrm{m^3\,s^{-1}}$ in reasonable agreement with the second value above. Morgan et al.[96] have evaluated a collision frequency of $1.02 \times 10^{-13}\,\mathrm{m^3\,s^{-1}}$ in laser-induced breakdown, which corresponds to a mean energy of $\sim 4\,\mathrm{eV}$.

As a final remark for argon, Sakai et al.[97] have analyzed the electron avalanche growth by the Monte Carlo method and the Boltzmann equation method.[98] The latter study is notable for a new approach, the Fourier series method, for solving the Boltzmann equation.

6.5.3 Carbon Dioxide (CO$_2$)

Selected references for the electron drift velocity data in the range of $0.3 \le E/N \le 6200\,\mathrm{Td}$ are: Schlumbohm[99]; Elford[100]; Hake and Phelps[101]; Huxley and Crompton[3]; Saelee et al.[102]; Kücükarpaci and Lucas[103]; Haddad and Elford[104]; Elford and Haddad[105]; Rosnerski and Leja[47]; Hasegawa et al.[106]; and Hernández-Avila et al.[107] Hasegawa et al.[106] have the extended the earlier measurements, which were mostly up to 100 Td, to 1000 Td. There is a change of slope at $\sim 20\,\mathrm{Td}$, at which the mean energy is $0.7\,\mathrm{eV}$, and at this energy vibrational excitation is the dominant mechanism of energy loss (see Table 5.3).

The drift velocities measured by Hasegawa et al.[106] agree with those of Saelee et al.[102] and Elford.[100] The values of Roznerski and Leja[47] are about 25% lower than those of Hasegawa et al.[106] The theoretical results obtained by the Boltzmann equation method[101] are considerably higher, by as much as 70% for certain values of E/N. The Monte Carlo simulation method[103] gives good agreement with experimental data. Figure 6.14 and Table 6.8 present selected data with recommended drift velocity.

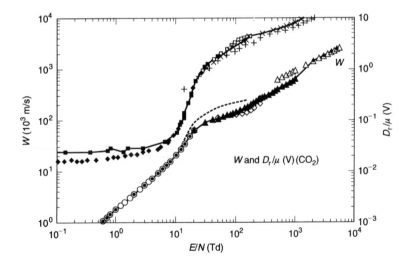

FIGURE 6.14 Drift velocity and D_r/μ of electrons in CO$_2$ as a function of E/N. Drift velocity W (bottom curve): experimental: ($\triangle$) Schlumbohm[99]; ($\times$) Elford[100]; ($\blacksquare$) Saelee et al.[102]; ($\diamond$) Roznerski and Leha[47]; ($\blacktriangle$) Hasegawa et al.[106] Theory: (- - -) Hake and Phelps[101]; ($+$) Kücükarpaci and Lucas.[103] Compilation: ($\circ$) Huxley and Crompton.[3] (—•—) Recommended. D_r/μ (top curve): ($\blacklozenge$) Warren and Parker,[82] 195 K; ($\blacksquare$) Warren and Parker,[82] 300 K; ($\square$) Rees[113]; ($\times$) Lakshminarasimha et al.[114]; ($+$) Kücükarpaci and Lucas[103]; (—) recommended.

TABLE 6.8
Recommended Drift Velocity (W) in CO_2

E/N	W	E/N	W	E/N	W	E/N	W
0.3	0.536	6.0	11.12	60	115.9	450	373.4
0.4	0.714	7.0	13.24	70	124.5	500	410.3
0.5	0.890	8.0	15.51	80	124.4	600	450.1
0.6	1.068	10.0	20.60	90	133.8	700	490.4
0.7	1.246	12.0	26.80	100	142.6	800	546.2
0.8	1.424	14.0	34.60	150	178.7	900	572.0
1.0	1.781	17.0	48.70	200	219.3	1000	663.0
2.0	3.56	20.0	63.20	250	266.0	2000	998
3.0	5.37	30	89.20	300	294.1	3000	1270
4.0	7.20	40	100.0	350	305.7	4000	1510
5.0	9.12	50	104.5	400	334.5	5000	1720

E/N in units of Td, W in 10^3 m/s.

The drift velocity may be expressed in analytical form by the following equations (E/N in units of Td):

$$W = 3.87 \times 10^{-3}\left(\frac{E}{N}\right)^4 + 3.41\left(\frac{E}{N}\right)^3 - 11.19\left(\frac{E}{N}\right)^2 + 1.79 \times 10^3\left(\frac{E}{N}\right) - 1.03 \text{ m/s} \tag{6.54}$$

$$W = -0.5\left(\frac{E}{N}\right)^4 + 21.2\left(\frac{E}{N}\right)^3 - 206.8\left(\frac{E}{N}\right)^2 + 2540.7\left(\frac{E}{N}\right) - 0.7552 \text{ m/s}; \quad R^2 = 0.9999 \tag{6.55}$$

$$W = 4 \times 10^{-4}\left(\frac{E}{N}\right)^3 - 0.8\left(\frac{E}{N}\right)^2 + 955.0\left(\frac{E}{N}\right) + 58.0 \times 10^3 \text{ m/s}; \quad R^2 = 0.9978 \tag{6.56}$$

$$W = 1.12 \times 10^4\left(\frac{E}{N}\right)^{0.591} \text{ m/s} \tag{6.57}$$

The range of applicability and the quality of fit of Equations 6.54 to 6.57 in representing the recommended drift velocity are shown in Table 6.9.

At low values of E/N (< 0.1 Td) the characteristic energy ($\varepsilon_k = eD/\mu$) is close to the thermal energy (kT) and the rotational excitations are relatively unimportant compared to the vibrational excitation. The threshold for the lowest vibrational state is 83 meV[101] and there is a relatively large cross section at threshold. The momentum transfer cross section varies as $\varepsilon^{-0.5}$ at relatively low electron energy. These features explain the experimentally observed fact that the drift velocity, at low values of E/N, does not depend on the temperature, unlike in argon.[55] Low energy electrons ($< kT$) elastically colliding with neutrals gain a small fraction of energy and the drift velocity therefore increases with increasing kT. If, however, energy is lost as a result of inelastic collisions the drift velocity tends to be independent of kT. Figure 6.15 shows the theoretically calculated energy loss processes as a function of E/N[103]; at low values of E/N vibrational excitation is the only loss mechanism.

TABLE 6.9

Applicability of Range and Maximum Difference of Analytical Representation of Drift Velocity

	Equation 6.54	Equation 6.55	Equation 6.56	Equation 6.57
Applicable E/N range	0.3–6.0	2.0–20.0	30–1000	450–6200
Max. % difference	−8.92	9.91	7.54	—
Corresponding E/N	4.0	20.0	1000	—
Min. % difference	0.03	0.05	−0.44	—
Corresponding E/N	5.0	5.0	200	—

A positive difference means that the calculated values are lower.

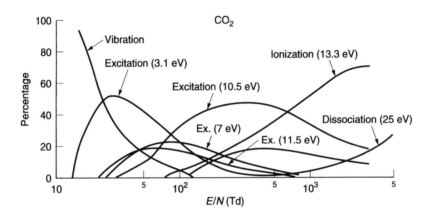

FIGURE 6.15 Energy loss processes in carbon dioxide as a function of E/N. Figure reproduced from Kücükarpaci, H. N. and J. Lucas, *J. Phys. D*, 12, 2123, 1979. With permission of IOP, U.K.

CO_2 has been suggested as a gas alternative to SF_6 and the high-pressure properties of the gas are of practical interest. Allen and Prew[108] have measured the drift velocity at gas pressures in the range of 0.24 to 1.7 MPa and find that the drift velocity decreases with increasing gas number density at constant E/N (Figure 6.16).

Theoretical studies of drift velocity of electrons in $CO_2 + Ar$ mixtures have been carried out by Uman and Warfield,[109] following the experimental measurements of English and Hanna[110] for low concentrations of CO_2. Lowke et al.[111] have calculated the electron transport coefficients in $CO_2 + N_2 + He$ mixtures. Figure 6.17 shows the variation of the total cross section as a function of energy to convey the complexity of the gas mixture.

The ratio D_r/μ has been studied by Warren and Parker,[112] Rees,[113] Lakshminarasimha et al.,[114] Roznerski and Mechlinska-Drewko,[115,116] and Kücükarpaci and Lucas,[103] covering the range $0.003 \leq E/N \leq 3000$ Td.

Figure 6.14 shows selected data for D_r/μ. The experimental data of Warren and Parker[112] up to 50 Td merge smoothly with those of Lakshminarasimha et al.[114] at $E/N = 50$ Td, and the data of the latter authors up to $E/N = 1500$ Td give the most accurate values. Figure 6.18 shows the ratio D_L/μ and the electron mean energy $\bar{\varepsilon}$ as determined by Hasegawa et al.[106] and Kücükarpaci and Lucas.[103] Both quantities increase smoothly with E/N.

The collision frequency for low energy thermal electrons is calculated using the method already described, Equations 1.95 and 1.98, as a function of electron energy and E/N

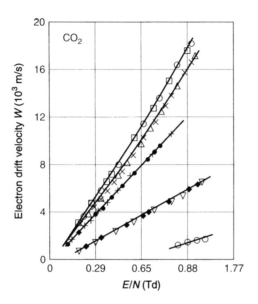

FIGURE 6.16 Drift velocity of electrons in CO_2 as a function of E/N at various gas number densities. Gas number densities and pressure are: (o, □) $N = 5.9 \times 10^{-25} \, m^{-3}$, $p = 237.9 \, kPa$; (×, △) $N = 19.5 \times 10^{25} \, m^{-3}$, $p = 757 \, kPa$; (●, +) $N = 26.4 \times 10^{25} \, m^{-3}$, $p = 1.01 \, MPa$; (■, ▽) $N = 35.7 \times 10^{25} \, m^{-3}$, $p = 1.33 \, MPa$, The drift velocity decreases with increasing gas pressure at constant E/N. Figure reproduced from Allen, N. L. and B. A. Prew, *J. Phys. B*, 3, 1113, 1970. With permission of IOP, U.K.

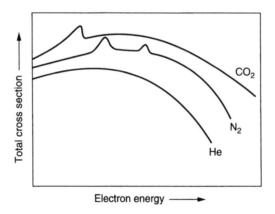

FIGURE 6.17 Schematic variation of total cross section as a function of energy of components of laser gas mixtures.

respectively. Argyropoulos and Casteel[117] give a variation of Equation 1.95 for collision frequency of thermal electrons as

$$\frac{\nu}{N} = \frac{4}{3} \left(\frac{8kT}{\pi m} \right)^{1/2} Q_M \tag{6.58}$$

The term within brackets is recognized as the mean thermal velocity (Equation 1.25) in terms of the absolute temperature T. For convenience, Equation 1.98 is repeated as

$$\frac{\nu}{N} = \frac{e}{m} \left[\frac{E}{N} \frac{1}{W} \right] \tag{6.59}$$

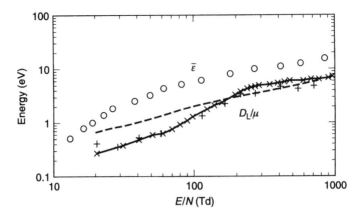

FIGURE 6.18 Mean energy $\bar{\varepsilon}$ and D_L/μ in CO_2. $\bar{\varepsilon}$: (o) Kücükarpaci and Lucas.[103] D_L/μ: (+) Saelee et al.[102]; (×) Hasegawa et al.,[106] which are also the recommended (—) values; (---) Kücükarpaci and Lucas.[103]

TABLE 6.10
Parameters for Calculation of Collision Frequency in CO_2

Temperature	Q_M	Temperature	Q_M	Temperature	Q_M
1000	30.4	5500	6.32	10,000	8.47
1500	20.0	6000	6.38	11,000	9.00
2000	14.4	6500	6.53	12,000	9.49
2500	11.1	7000	6.75	13,000	9.91
3000	9.10	7500	7.00	14,000	10.3
3500	7.85	8000	7.29	15,000	10.6
4000	7.07	8500	7.58	16,000	10.8
4500	6.61	9000	7.88	17,000	11.1
5000	6.38	9500	8.18	18,000	11.3

Temperature in units of K and Q_M in units of $10^{-20}\,m^2$. 11,605 K = 1 eV.
From Argyropoulos, G. S. and M. A. Casteel, *J. Appl. Phys.*, 41, 4162, 1990. With permission.

The link between Equations 6.58 and 6.59 is provided through the $\bar{\varepsilon}$–E/N plot shown in Figure 6.18. Table 6.10 shows the values of T and Q_M given by Argyropoulos and Casteel.[117] The method of calculation of collision frequency is demonstrated below, for clarity.

Let $T = 5000\,K$ ($= 0.43\,eV$), $Q_M = 6.38 \times 10^{-20}\,m^2$. The constants are $k = 1.381 \times 10^{-23}\,J/K$, $m = 9.109 \times 10^{-31}\,kg$. The calculated v/N, using Equation 6.58, is $3.74 \times 10^{-14}\,m^3/s$. E/N corresponding to 0.43 eV is 20 Td and $W = 6.5 \times 10^4\,m/s$. The collision frequency v/N, obtained from Equation 6.59, is $5.41 \times 10^{-14}\,m^3/s$. An updated cross section of $10.39 \times 10^{-20}\,m^2$ at 0.5 eV (Table 5.2) gives an improved agreement.

6.5.4 CARBON MONOXIDE (CO)

Selected references for data on drift velocity in CO are: Pack et al.[118]; Wagner et al.[119]; Huxley and Crompton[3]; Saelee and Lucas[120]; Roznerski and Leja[121]; and Nakamura.[122] The Monte Carlo simulation results of Saelee and Lucas[120] are considerably higher than their experimental ones and the results of Roznerski and Leja,[121] who provide just three data

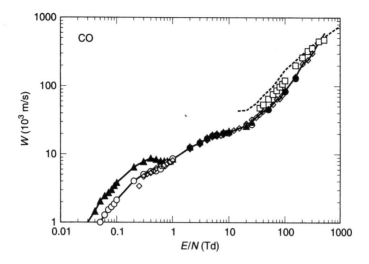

FIGURE 6.19 Drift velocity of electrons in CO as a function of E/N. Unless otherwise mentioned, the temperature is 293 K. (o) Huxley and Crompton[3]; (—▲—) Huxley and Crompton,[3] 77 K, same as Pack et al.[118]; (□) Saelee and Lucas[120]; (- - -) Saelee and Lucas[120]; (●) Roznerski and Leja[121]; (◇) Nakamura[122]; (—) recommended.

points, are also higher than the experimental results of Saelee and Lucas.[120] The experimental results of Nakamura[122] are in excellent agreement with those of Pack et al.[118] and join smoothly at 20 Td, extending up to 300 Td. More measurements are required at $E/N > 300$ Td. The measurements of Pack et al.[118] at 77 K are also shown in Figure 6.19. One interesting point to note is that the 77 K drift velocity lies above that at 300 K at low values of E/N, which is possibly due to the fact that rotational de-excitation also occurs at these low values of E/N.

Figure 6.19 and Table 6.11 show the recommended drift velocity, composed from data of Pack et al.,[118] for the range $0.04 \leq E/N \leq 20$ Td, and of Nakamura[122] for the range

TABLE 6.11
Recommended Drift Velocity as a Function of E/N in CO

E/N	W	E/N	W	E/N	W
0.04	0.744	0.8	7.522	40	42.61
0.05	1.003	1	8.620	50	46.68
0.06	1.280	2	12.301	60	54.56
0.07	1.520	3	14.778	70	64.81
0.08	1.693	4	16.778	80	65.74
0.09	1.873	5	18.329	90	74.73
0.1	2.125	6	19.265	100	87.22
0.2	4.026	8	19.977	150	139.4
0.3	5.039	10	20.164	200	211.9
0.4	5.572	20	26.134	250	240.6
0.5	5.693	25	31.97	300	303.3
0.6	6.035	30	34.59	400	436.6
0.7	6.725	35	38.77	500	583.6

Data at 400 and 500 Td are calculated from Equation 6.61. E/N in units of Td, W in 10^3 m/s.

TABLE 6.12
Quality of Fit between Recommended Values (Table 6.11) and Equations 6.60 and 6.61

Equation	E/N (Td) (Applicable Range)	Maximum Difference		Minimum Difference	
		%	E/N (Td)	%	E/N (Td)
6.60	0.6–10	8.75	0.8	−1.25	5
6.61	20–300	12.22	25	0.12	150

A positive difference means that the recommended value is higher.

$20 \leq E/N \leq 300$ Td. The drift velocity may be expressed by the analytical equations (in 10^3 m/s)

$$W = 0.049\left(\frac{E}{N}\right)^3 - 1.033\left(\frac{E}{N}\right)^2 + 7.3234\frac{E}{N} + 1.6415; \quad R^2 = 0.9883 \qquad (6.60)$$

$$W = 0.0008\left(\frac{E}{N}\right)^2 + 0.7495\left(\frac{E}{N}\right) + 8.8245; \quad R^2 = 0.9935 \qquad (6.61)$$

The range of applicability of Equations 6.60 and 6.61 and the quality of fit between the recommended and calculated drift velocity are shown in Table 6.12.

Selected references for diffusion studies (D_r/μ) are: Warren and Parker[112] at 77 K; Hake and Phelps[101]; Lowke and Parker[123]; Lakshminarasimha et al.[124]; Saelee and Lucas[125]; Roznerski and Mechlinska-Drewko[126]; and Al-Amin et al.[127] The data of Warren and Parker[112] show a sudden break at $D/\mu = 0.1$ V, attributed to vibrational excitation for which the threshold is 0.26 V.[128] The low energy computations of Hake and Phelps[101] at $\varepsilon_k < 1$ eV show that the dominant loss processes in CO may be categorized as follows: (a) rotational excitation, 0.00 to 0.06 eV; (b) direct vibrational excitation, 0.1 to 0.6 eV; (c) resonant vibrational excitation and electronic excitation, > 0.8 eV.

Selected references for (D_L/μ) are Wagner et al.,[129] Lowke and Parker,[123] and Saelee and Lucas.[120] Figure 6.20 shows both (D_r/μ) and (D_L/μ) as a function of E/N. Because values for 77 K and 293 K are included by several investigators the differences between radial and longitudinal diffusion coefficients are obscured. To alleviate this to some extent, the inset of Figure 6.20 shows the diffusion coefficients calculated by Lowke and Parker,[123] from which one can clearly see that (D_L/μ) is smaller than (D_r/μ).

Another point to note, which is common to all gases, is that the radial diffusion coefficient measured by the Townsend method is applicable at low values of E/N. As already stated, analysis of the radial distribution of electrons as a function of the gap separation gives the ratio D_r/μ by the use of the Townsend–Huxley equation[3]

$$\frac{D_r}{\mu} = \frac{E}{2u} \qquad (6.62)$$

where

$$u = -\ln\left[\frac{(1-F)R}{d}\right]\left[\frac{1}{(R-d)}\right] \qquad (6.63)$$

$$R = \left(d^2 + r^2\frac{D_L}{D_r}\right)^{1/2} \qquad (6.64)$$

F is the fractional current collected by the anode of radius r, and d is the gap separation.

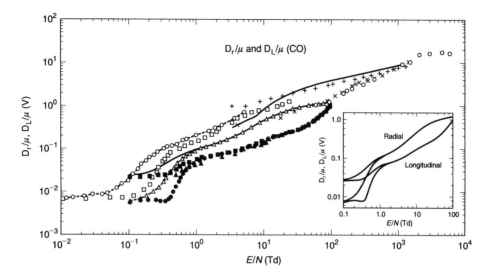

FIGURE 6.20 Radial and longitudinal diffusion coefficients of electrons in CO. Unless otherwise mentioned, the temperature is 293 K. Open symbols for radial and closed symbols for longitudinal diffusion. The letters E, T, C mean experimental, theoretical, and compiled respectively. Radial diffusion: (—○—) Warren and Parker,[112] 77 K, E; (□) Hake and Phelps,[101] T; (—△—) Lowke and Parker,[123] 77 K, T; (——) Lowke and Parker,[123] T; (×) Lakshminarasimha et al.,[124] E; (○) Al-Amin,[127] E. Longitudinal diffusion: (△) Wagner et al.,[129] E; (●) Lowke and Parker,[123] 77 K, T; (■) Lowke and Parker,[123] T; (—) Saelee and Lucas,[125] T. The inset shows details at low E/N.

For higher E/N the method has to be modified to overcome two sources of error. The first is the presence of secondary electrons caused by ion and photon bombardment of the cathode. These secondary electrons have a wider distribution across the anode than the primary electrons and hence give a false ratio of currents if not separated from the primary electrons. The second problem arises from the fact that the mean energy is high and the electrons are not in equilibrium till they fall through the potential that is equivalent to the mean energy expressed in volts. A minimum distance d_0 may be assigned for the electrons in the nonequilibrium region so that the effective gap length in the analysis is $d–d_0$.

Kontoleon and Lucas[130] developed a steady-state method that uses a method of analysis for separating the primary and secondary electrons. A second method, developed by Lucas and Kücükarpaci[131] uses the time-of-flight method and makes use of the principle that the times of arrival of secondary electrons due to photons and ions are different. The experimental results of Lakshminarasimha et al.,[124] shown in Figure 6.20, use the analysis developed by Kontoleon and Lucas.[130]

From microwave studies, Mentzoni[132] has deduced a constant electron–molecule collision frequency (v/N) of $3.1 \times 10^{-13} \, \mathrm{m^3 \, s^{-1}}$. The collision frequency, calculated from the formula of Argyropoulos and Casteel[117] as a function of electron energy, is shown in Table 6.13.

6.5.5 HELIUM (HE)

Selected references for data (both experimental and theoretical) on drift velocity over a wide range of E/N are: Bowe[133]; Phelps et al.[134]; Pack and Phelps[135]; Stern[136]; Anderson[137]; Lowke and Parker[138]; Hughes[139]; Blum et al.[140]; Huxley and Crompton[3]; Bartels[141]; Milloy and Crompton[142]; Kücükarpaci et al.[143]; Amies et al.[144]; and Pack et al.[145] Figure 6.21 shows the drift velocity; the figure also shows the ratios D_r/μ and D_L/μ, which will be commented

TABLE 6.13

Calculated Collision Frequency (v/N) for Electrons in CO Using Equation 6.58 of Argyropoulos and Casteel[117]

T	v/N	T	v/N	T	v/N
1000	2.78	5500	12.12	10,000	14.1
1500	3.98	6000	12.60	11,000	14.2
2000	5.19	6500	13.0	12,000	14.3
2500	6.42	7000	13.2	13,000	14.4
3000	7.67	7500	13.5	14,000	14.6
3500	8.82	8000	13.6	15,000	14.7
4000	9.90	8500	13.8	16,000	14.9
4500	10.8	9000	14.0	17,000	15.0
5000	11.5	9500	14.0	18,000	15.2

Temperature in units of K ($11,605 \, K = 1 \, eV$), v/N in units of $10^{-14} \, m^3 \, s^{-1}$.

upon below. Recommended drift velocity is shown in Table 6.14 and comprises data from Huxley and Crompton,[3] range 0.02 to 3.5 Td, average of the values of Kücükarpaci et al.[143] and Pack et al.,[145] range 4 to 200 Td and, for $E/N > 200$ Td, of Kücükarpaci et al.[143] The drift velocity at low values of E/N (< 6 mTd) is a linear function of E/N because the electrons are in thermal equilibrium with the gas atoms. As E/N increases, the electron temperature increases and the temperature dependence disappears.[135]

The drift velocity at 293 K may be expressed as an analytical function of E/N according to the following equations:

$$W = 4.87 \times 10^3 \left(\frac{E}{N}\right)^{0.559} \text{m/s}; \quad R^2 = 0.9960 \tag{6.65}$$

$$W = -5.79 \times 10^4 \left(\frac{E}{N}\right)^4 + 8.78 \times 10^4 \left(\frac{E}{N}\right)^3 - 5 \times 10^4 \left(\frac{E}{N}\right)^2 + 1.69 \times 10^4 \left(\frac{E}{N}\right) + 1.7 \times 10^2 \, \text{m/s} \tag{6.66}$$

$$W = 2268.0 \left(\frac{E}{N}\right) \text{m/s} \tag{6.67}$$

At 77 K:

$$W = \left[6.17 \times 10^5 \left(\frac{E}{N}\right)^3 - 1.93 \times 10^5 \left(\frac{E}{N}\right)^2 + 2.75 \times 10^4 \left(\frac{E}{N}\right) + 1.67 \times 10^4 \right] \times 10^{-2} \, \text{m/s} \tag{6.68}$$

The range of applicability and accuracy of Equations 6.65 to 6.68 in representing the drift velocity given in Table 6.14 are shown in Table 6.15. The drift velocity at 77 K is higher than that at 293 K for the same E/N till the two curves meet at 1.7 Td. This is because at low E/N the scattering cross section increases with increasing energy (Table 3.13), and therefore at the lower temperature the collision frequency is less, allowing the electron to build up energy and have higher drift velocity. Assuming that the momentum transfer cross section is independent of electron energy, a formula for the dependence of W on E/N has been derived from

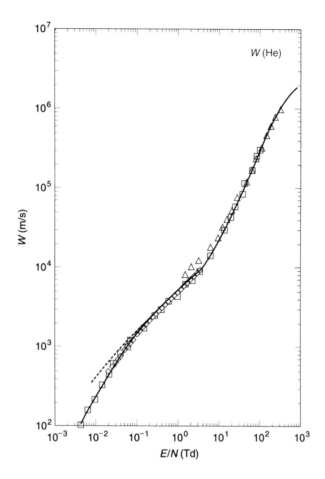

FIGURE 6.21 Drift velocity of electrons in helium as a function of E/N. Temperature is 293 K unless otherwise mentioned. ($\square$) Pack et al.[145]; ($\diamond$) Huxley and Crompton[3]; ($\triangle$) Kücükarpaci et al.[143]; (- - -) Huxley and Crompton[3]; 77 K; (—) Table 6.14.

theoretical considerations for the range of E/N from 10 to 70 Td,[134] but the equation is not easy to use from our point of view.

Drift velocity in mixtures of He + CH$_4$ has been measured by Foreman et al.[147] in the low E/N range of 0 to 5 Td. At certain concentrations of CH$_4$ inelastic scattering by methane causes a maximum in the drift velocity plotted as a function of E/N. As the methane concentration decreases, the drift velocity maximum decreases and occurs at lower E/N. Electron transport properties of He + SF$_6$ mixtures are measured by Xiao et al.[148] for limited mixture ratios.

We now turn our attention to the ratios D_L/μ and D_r/μ. As stated earlier, the former ratio is lower than the latter almost by a factor two up to 3 Td and, in the case of helium, both are increasing functions of E/N. Inelastic collisions do not occur at these values of E/N; Milloy[149] has calculated that inelastic collisions in helium do not set in below 8.5 Td. In the rising part of the curve the slope is approximately 45°. Warren and Parker[112] attribute this to a constant elastic scattering cross section as a function of electron energy.

The mean energy of electrons in the swarm is given by Kücükarpaci et al.[143] and Lymberopoulos and Schieber.[150] The latter authors use a variation of the Monte Carlo technique in which the starting point, unlike in the null collision technique,[151] is the Boltzmann equation. The cross sections employed for the study are shown in Table 6.16.

TABLE 6.14
Recommended Drift Velocities in Helium

E/N	W	E/N	W	E/N	W
0.02	0.453	1.0	4.85	40	99
0.03	0.631	2.0	6.86	50	134
0.04	0.786	3.0	8.51	60	159
0.05	0.922	4.0	10.07	70	176
0.06	1.044	5.0	11.52	80	203
0.07	1.156	6.0	13.13	100	285
0.08	1.259	7.0	15.03	200	534
0.1	1.446	8.0	18.77	300*	797
0.2	2.14	10.0	22.2	400*	965
0.3	2.65	15.0	32.6	500*	1147
0.4	3.08	20.0	49.2	600*	1380
0.5	3.44	25	68.1	700*	1642
0.6	3.77	30	77.8	800*	1729
0.8	4.35	35	87.5	1000*	1915

Untabulated values are obtained from digitizing the original publication and interpolating at specific E/N. E/N in units of Td, W in units of 10^3 m/s. * means unconfirmed by experiment.

TABLE 6.15
Range of Validity and Accuracy of Analytical Equations for Representing Drift Velocity in Helium

Equation	Range	Maximum Difference (%)		Minimum Difference (%)	
		Magnitude	E/N	Magnitude	E/N
6.65	0.03–7.0	−8.70	0.03	−0.41	1.0
6.66	0.025–0.6	−3.08	0.025	0.22	0.6
6.67	20–120	−7.49	40	0.80	120
6.68	0.014–1.7	−2.63	0.14	0.44	0.035

E/N in units of Td.[146] Positive differences mean recommended values are higher.

Figure 6.22 shows the electron energy distribution obtained by this method. At $E/N = 10$ Td the energy distribution falls rapidly due to the onset of inelastic collisions and, as the field increases, the tail of the distribution shifts to higher energies, becoming more like a Maxwellian distribution. The mean energy obtained by both studies is shown in Figure 6.23. At low E/N the values of Kücükarpaci et al.[143] appear to be too high when compared with those obtained by Lymberopoulos and Schieber.[150]

It is appropriate here to make some observations with regard to collision frequency in gases in general. A monoenergetic beam of electrons scattering in a medium of gas has a collision frequency that is dependent on the energy of the electrons. The relation of the collision frequency (ν/N) to electron energy is given by Equation 1.96 and we have already used this equation several times. In a swarm there is a distribution of electron energy and relating (ν/N) to ε is meaningless as far as the swarm properties are concerned. In this

TABLE 6.16
Simulation Parameters Adopted by Lymberopoulos and Schieber[150]

Collision Type	Cross Section (m²)	Energy Range (eV)	Energy Loss (eV)
Elastic	$Q_e = 8.5 \times 10^{-19}/(\varepsilon+10)^{1.1}$		0
Excitation	0	$0 \leq \varepsilon \leq 19.8$	0
$He + e \rightarrow He^* + e$	$Q_{ex} = 2.08 \times 10^{-22} (\varepsilon - 19.8)$	$19.8 \leq \varepsilon \leq 27.0$	19.8
Ionization	$Q_i = 1 \times 10^{-17}(\varepsilon - 24.5)* [(\varepsilon+50)(\varepsilon+300)^{1.2}]^{-1}$	$24.5 \leq \varepsilon$	24.5

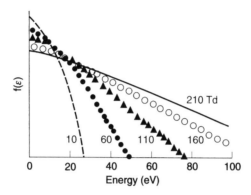

FIGURE 6.22 Electron energy distribution in helium as a function of E/N. At $E/N = 10$ Td the distribution falls sharply, due to the onset of inelastic collisions. As the electric field increases the tail of the distribution moves to higher energies, approaching a Maxwellian distribution.

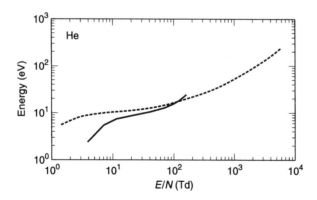

FIGURE 6.23 Mean energy as a function of E/N in helium. (- - -) Kücükarpaci et al.[143]; (—) Lymberopoulos and Schieber.[150]

situation one has to use Equation 1.98 to relate v/N to E/N. Though, at low energies, the energy of an electron in a beam may be approximated to the mean energy of a swarm there is no exact correspondence. One therefore uses the term "effective collision frequency" of a swarm and in helium this frequency, in the energy-independent region, is given as $v/N = 7.45 \times 10^{-14}\,m^3\,s^{-1}$,[152] which is in good agreement with the constant collision frequency (effective), $v/N = 7.20 \times 10^{-14}\,m^3\,s^{-1}$, given by Phelps et al.[134]

Frost and Phelps[183] have evaluated the momentum transfer cross section in He and, from their calculated drift velocity as a function of E/N, one calculates that $v/N = 7.90 \times 10^{-14}\,\mathrm{m^3\,s^{-1}}$ at $E/N = 10\,\mathrm{Td}$ and $W = 2.2 \times 10^4\,\mathrm{m/s}$. Capitelli et al.[153] have studied the change of collision frequency in RF bulk plasma in mixtures of He + CO. McColl et al.[154] give $v/N = 6.2 \times (10^{-14}$ to $10^{-16})\,\mathrm{m^3\,s^{-1}}$ from microwave-resonant-cavity discharges. In a recent study of terahertz time domain measurements of plasma characteristics, Jamison et al.[155] have measured the electron–neutral collision frequency as $v/N = 9.3 \times 10^{-12}\,\mathrm{m^3\,s^{-1}}$, but the authors consider the value as anomalous. From the shape of the total cross section as a function of energy the effective collision frequency in helium may be approximated to a constant value.

6.5.6 HYDROGEN (H_2) AND DEUTERIUM (D_2)

Selected references for data on drift velocity in hydrogen to cover a wide range of E/N are: Engelhardt and Phelps[156]; Schlumbohm[157]; Huxley and Crompton[3]; Bartels[158]; Snelson and Lucas[159]; Blevin et al.[160,161] Atrazhev and Yakubov[162]; Hunter[163]; Saelee and Lucas[164]; Hayashi[165]; and Roznerski and Leja.[166]

Deuterium, also called heavy hydrogen, is an isotope of hydrogen; it has an atomic number of one and an atomic weight of approximately two. Its nucleus contains one proton and one neutron. It is found in ordinary hydrogen at a concentration of 0.015% and is purified by distillation of hydrogen. It enters into all the chemical reactions that normal hydrogen enters into, D_2O (heavy water) being analogous to H_2O. Nuclear fusion of deuterium atoms releases enormous amounts of energy, and heavy water is used in nuclear reactors to control the reaction. Deuterium also finds applications as a tracer element in biochemical diagnosis.

The elastic scattering cross section of deuterium is identical to that of hydrogen. For example, Pack et al.[167] find that the momentum transfer cross section of D_2 in the energy range 0.003 to 0.05 eV is $Q_M^{-1} = (1.29 - 3.36\varepsilon) \times 10^{19}\,\mathrm{m^2}$, a difference of less than 1% from that of H_2. In the theoretical analysis of the two gases, it is customary to divide the energy range into three regions, in common with many molecular gases. The first region is covered by the electron having thermal energy to the onset of vibrational excitation (the onset of vibrational excitation is 0.516 and 0.360 eV in H_2 and D_2 respectively); in this region rotational excitation is the dominant loss mechanism. A detail that is often considered is the increase in rotational excitation cross section due to the polarization of the H_2 molecule. Inelastic collisions of the second kind are important in this range, and the importance is more significant in D_2 than in H_2.[156] The second region covers the energy range from the vibrational energy, giving significant importance to the dissociation (electronic excitation) energy.

In the third region, vibrational excitation, electronic excitation, and ionization occur. Figure 6.24 shows the schematic of the energy loss as a function of E/N. in hydrogen. Vibrational excitation is the dominant loss mechanism and this is true for D_2 also, though D_2 has lower loss than H_2 in this range. For higher values of E/N, electronic excitation is the dominant mechanism, followed by ionization collision and elastic collision losses. Over the entire range of E/N the vibrational excitation losses are higher for H_2 than for D_2. This observation is attributed partly to the fact that the quantum of energy for vibrational excitation is higher for H_2, and partly to the fact that the vibrational cross section in D_2 is lower. The loss due to elastic scattering in H_2 is higher than in D_2 and this is explained by the fact that the elastic collision loss is dependent on $2m/M$ and M is larger for the D_2 molecule. A significant conclusion from Figure 6.24 is that vibrational excitation is important in determining the degree of ionization.[156]

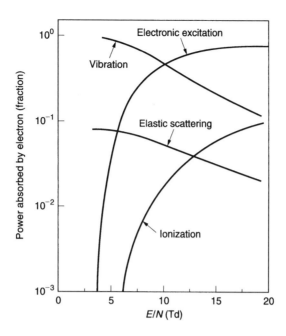

FIGURE 6.24 Schematic representation of power absorbed by the electron at low values of reduced electric field in molecular hydrogen. Note the dominant role of vibrational excitation.

With this background, we now consider the drift velocity in both gases. Data for drift velocity in deuterium are rather sparse and limited to $E/N \leq 125$ Td. Pack et al.[167] have measured the drift velocity by the pulsed drift tube method. The compilation of Huxley and Crompton[3] is based on a number of investigations of the Australian group. Roznerski et al.[168] have extended the measurements to 125 Td. Figure 6.25 shows the drift velocity in both gases at 77 K and 293 K. Tabulated values for 293 K are given in Table 6.17.

There is very good agreement between various results in the overlapping range of E/N in both gases. Further, the results merge smoothly when different ranges used by different investigators are plotted on the same graph. As an example, at $E/N = 250$ Td Schulumbohm[157] gives a value of 4.09×10^5 m/s against a value of 4.01×10^5 m/s reported by Roznerski and Leja.[166] Note that the ordinates of deuterium data in Figure 6.25 are multiplied by a factor of ten for clarity.

Analytical expressions for representation of drift velocity are as follows:

Hydrogen

$$W = -9.79 \times 10^5 \left(\frac{E}{N}\right)^4 + 5.90 \times 10^5 \left(\frac{E}{N}\right)^3 - 1.36 \times 10^5 \left(\frac{E}{N}\right)^2$$

$$+ 2.49 \times 10^4 \left(\frac{E}{N}\right) + 106.7 \, \text{m/s} \tag{6.69}$$

$$W = \left[-0.236 \left(\frac{E}{N}\right)^4 + 50.78 \left(\frac{E}{N}\right)^3 - 3.06 \times 10^3 \left(\frac{E}{N}\right)^2 + 1.58 \times 10^5 \left(\frac{E}{N}\right) \right.$$

$$\left. + 5.27 \times 10^5 \right] \times 10^{-2} \, \text{m/s} \tag{6.70}$$

$$W = -5 \times 10^{-8} \left(\frac{E}{N}\right)^4 = 5 \times 10^{-4} \left(\frac{E}{N}\right)^3 - 1.23 \left(\frac{E}{N}\right)^2 + 2052 \left(\frac{E}{N}\right) - 4.6 \times 10^4 \, \text{m/s} \tag{6.71}$$

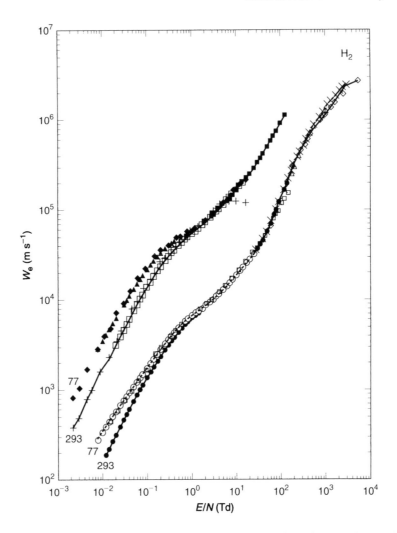

FIGURE 6.25 Drift velocity as a function of E/N in hydrogen and deuterium. Hydrogen (293 K): (◇) Schlumbohm[157]; (●) Huxley and Crompton[3]; (−✕−) Saelee and Lucas[164]; (△) Roznerski and Leja[166]; (—) recommended. Hydrogen (77 K): (○) Huxley and Crompton[3]; (−●−) Raju[44]; Deuterium (293 K): (+) Pack et al.[167]; (□) Huxley and Crompton[3]; (■) Roznerski and Leja[166]; (—) recommended. Deuterium (77 K): (◆) Pack et al.[167]; (▲) Huxley and Crompton.[3] Ordinates of deuterium points are multiplied by a factor of ten for the sake of clarity.

Deuterium

$$W = -1.13 \times 10^5 \left(\frac{E}{N}\right)^4 + 1.57 \times 10^5 \left(\frac{E}{N}\right)^3 - 8.7 \times 10^4 \left(\frac{E}{N}\right)^2$$
$$+ \, 2.78 \times 10^4 \left(\frac{E}{N}\right) + 88 \, \text{m/s} \tag{6.72}$$

$$W = -3.56 \times 10^3 \left(\frac{E}{N}\right)^4 + 1.35 \times 10^4 \left(\frac{E}{N}\right)^3 - 1.93 \times 10^4 \left(\frac{E}{N}\right)^2$$
$$+ \, 1.46 \times 10^4 \left(\frac{E}{N}\right) + 51.47 \, \text{m/s} \tag{6.73}$$

TABLE 6.17
Recommended Drift Velocity in Hydrogen and Deuterium

E/N	W H₂	W D₂	E/N	W H₂	W D₂	E/N	W H₂	W D₂
0.002		37.49	0.60	4820	4350	50	5.71(4)	4.96(5)
0.003		47.67	0.70	5240	4650	60	6.95(4)	5.85(5)
0.004		77.08	0.80	5610	4920	70	8.30(4)	6.54(5)
0.006		98.02	1.0	6.23(3)	5.38(3)	80	9.80(4)	7.47(5)
0.009		158.5	1.2	6.62(3)	5.79(3)	100	1.28(5)	9.20(5)
0.012	186.2	193.0	1.4	6.92(3)	6.19(3)	120	1.66(5)	1.05(6)
0.02	311.0	308.0	1.7	7.64(3)	6.74(3)	140	1.94(5)	
0.025	385.0	383.0	2.0	8.37(3)	7.27(3)	150	2.13(5)	
0.03	459.0	457.0	2.5	9.13(3)	8.10(3)	200	3.36(5)	
0.035	530.0	530	3.0	9.82(3)	8.88(3)	250	4.19(5)	
0.04	600.0	601.0	3.5	1.08(4)	9.61(3)	300	4.68(5)	
0.05	737.0	740.0	4.0	1.15(4)	1.02(4)	350	5.22(5)	
0.06	870.0	873.0	5.0	1.29(4)	1.12(4)	400	6.02(5)	
0.07	998.0	1002	6.0	1.42(4)	1.27(4)	500	7.25(5)	
0.08	1120	1130	7.0	1.54(4)	1.38(4)	600	8.30(5)	
0.10	1370	1335	8.0	1.65(4)	1.48(4)	700	9.23(5)	
0.12	1580	1553	10.0	1.87(4)	1.67(4)	800	9.94(5)	
0.14	1780	1759	12.0	2.07(4)	1.85(4)	900	1.04(6)	
0.17	2070	2046	14.0	2.27(4)	2.02(4)	1000	1.13(6)	
0.20	2350	2310	17.0	2.55(4)	2.28(4)	2000	1.80(6)	
0.25	2760	2700	20.0	2.81(4)	2.51(4)	2300	2.00(6)	
0.30	3130	3030	25.0	3.22(4)	2.90(4)	2650	2.35(6)	
0.35	3470	3340	30.0	3.66(4)	3.38(4)	5500	2.69(6)	
0.40	3790	3595	35.0	4.08(4)	3.72(4)			
0.50	4330	4017	40.0	4.54(4)	4.17(4)			

E/N in units of Td, W in m/s. Quantity $a(b)$ means $a \times 10^b$.

The range of applicability and the quality of fit with the recommended drift velocity are shown in Table 6.18. The quality of fit for H_2 at 293 K, particularly in the higher range 50 to 1000 Td, is less than satisfactory.

Drift velocity measurements in Ar–H_2 mixtures have been measured by Petrović et al.[169] with a view to determining the scattering length of electrons elastically scattered in argon (see Section 6.5.2). The scattering length A is defined according to $Q_T(0) = 4\pi A^2$, where $Q_T(0)$ is the total scattering cross section for electrons with zero energy. At zero energy the total cross section involves only the s-wave scattering and equals the momentum transfer cross section as determined by swarm experiments. The scattering length is a basic parameter in scattering theory that determines both the cross section and its first derivative.

To obtain the scattering length from the momentum transfer cross section one needs to make drift velocity measurements at low values of E/N. The minimum value of the electric field that can be applied is limited by the contact potentials, which may be of the order of 10 to 25 meV. To make measurements below 1 Td, therefore, one has to employ gas number densities that are larger than those at atmospheric pressure ($2.45 \times 10^{25}\,\mathrm{m}^{-3}$). The technique adopted by Petrović et al.[169] is to use a mixture of two gases; one of them, a small percentage of additive, serves the purpose of limiting the mean energy of electrons while the other gas, the majority constituent, determines the momentum transfer cross section.

TABLE 6.18
Range of Applicability and Quality of Fit with the Recommended Drift Velocity

Gas	H_2	H_2	H_2	D_2	D_2
Temperature (K)	77	293	293	77	293
Equation	6.69	6.70	6.71	6.72	6.73
Range (Td)	0.01–7	1.0–100	50–1000	0.01–0.5	1.4–14
Maximum deviation (%)	9.6	9.5	15.0	4.6	2.3
Corresponding E/N (Td)	0.25	1.0	100	0.01	5.0
Minimum deviation (%)	0.07	−0.08	0.82	0.09	−0.09
Corresponding E/N (Td)	0.1	70	300	0.08	10.0

Positive signs for differences mean that the recommended values are higher.

From these considerations, Petrović et al.[169] used 5.45% H_2 in argon; gas pressures were in the range of 33 to 77 kPa, E/N in the range of 17 to 140 mTd, and the mean energies derived were in the range 47 to 221 meV. Figure 6.26 shows the effect of adding a small quantity of H_2; the electron distribution is altered, drastically reducing the mean energy. Typically, the mean energy of argon at 0.02 eV is 1 eV (Figure 6.12) and the addition of a small quantity of hydrogen reduces it to 50 meV, a reduction of a factor of 20.

Since the MERT analysis for argon was not covered adequately in Chapter 3, additional details are provided here. The expression for momentum transfer is given by Equation 3.34, the four-term phase shift η_0 by Equation 3.52, η_1 and η_L by Equations 3.38 and 3.39 respectively. The result of the fit gives the following values for the constants: $D = 68\ 93\ a_0^3$, $F = -97\ a_0^4$, $A_1 = 8.69\ a_0^3$, $A = -1.459\ a_0$, where a_0 is the Bohr radius. From constant A the scattering length is obtained as 7.718×10^{-11} m as against the earlier value of 8.92×10^{-11} m determined by Pack et al.[170]

The dependence of drift velocity on density has been referred to earlier. The theory of electron motion in gases predicts that the drift velocity is dependent on E/N, on temperature, on the nature of the gas, but not on density. However, experimental observations of

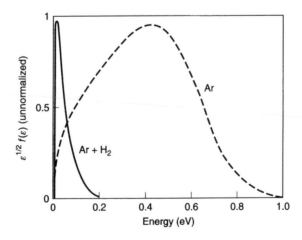

FIGURE 6.26 Electron energy distribution in argon (- - -) and argon + H_2 (—) at $E/N = 0.02$ Td. Addition of hydrogen alters the distribution drastically, with the mean energy decreasing by a factor of twenty as a result of low energy inelastic collisions. Figure reproduced from Petrović, Z. Lj. et al., *J. Phys. B*, 28, 3309, 1995. With permission of IOP, U.K.

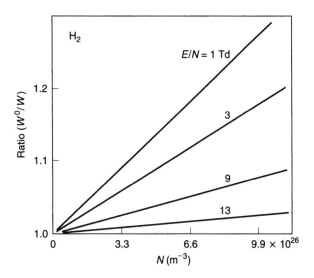

FIGURE 6.27 Drift velocity as a function of gas pressure at constant values of E/N. The reciprocal of drift velocity is a linear function of pressure. Note the order in which E/N changes.

Grünberg[171] have shown that the drift velocity decreases in both hydrogen and nitrogen. This author observed a variation of W of up to 30% in H_2 at neutral densities up to $10^{27} \, m^3$ (40 × atmospheric pressure). Crompton et al.[172] observed a density dependence of W in D_2 at 77 K.

Frommhold[173] has observed that the plot of W^{-1} is linear with N if E/N and T are kept constant, as shown in Figure 6.27. The straight lines at constant E/N satisfy the equation

$$W = W_0 \left(\frac{1}{1 + \lambda N} \right) \tag{6.74}$$

where W_0 is the zero density drift velocity (y-axis intercept) and λ is a constant having the dimension of m^3. It is a function of D/μ which is itself a function of E/N and T. λ decreases with increasing D/μ, necessitating that experiments should be carried out at densities of several atmospheric pressures to detect the effect.

Attachment of electrons to neutrals and subsequent detachment results in a lowering of the drift velocity. The lifetime of the negative ion, τ, depends on the detachment mechanism; if the detachment process is collisional in nature, τ will be longer. On the other hand, if the negative ion is a short-lived compound or resonant state, τ will be shorter. An electron that is attached and detached several times is delayed so that the average velocity, W, is lower and a decrease of W with pressure is observed. The zero-density drift velocity, W_0 in Equation 6.74, has the physical meaning that it is the velocity between two successive trapping collisions.

The drift velocity in terms of the lifetime of the ion is given by

$$W = W_0 \left(\frac{1}{1 + \nu_0 N \tau} \right) \tag{6.75}$$

in which ν_0 ($m^3 \, s^{-1}$) is the electron–neutral collision frequency. Comparing Equations 6.74 and 6.75, one derives

$$\lambda = \nu_0 \tau \tag{6.76}$$

From experimental data, Frommhold[173] has deduced that $\lambda/N = (0.2 \text{ to } 2.8) \times 10^{-28} \text{ m}^3$ for $E/N = 0.1$ to 12.0 Td. If one substitutes $N = 3.22 \times 10^{22} \text{ m}^3$, $\nu_0 = 10^{-13} \text{ s}^{-1} \text{ m}^3$, one gets $\tau = 10^{-15} \text{ s}$, which may be taken as evidence for a short-lived negative ion state.

The detachment process and the lifetime of the ion have the following relationship:

1. If the attachment and subsequent detachment of the electron is due to a two-body process, the ratio $W/W_0 < 1$ because $\nu_0 N \tau$ in Equation 6.75 is not equal to zero. There is no pressure dependence because the dependence of collision frequency (s^{-1}) and τ (s) cancel out. This case does not apply to hydrogen or nitrogen.
2. If the attachment is due to a three-body process and is followed by subsequent detachment, a pressure dependence of W will be observed. However, a three-body attachment process has never been observed in hydrogen or nitrogen.
3. A resonance phenomenon occurs, because the time required to impart energy to the neutral is of the same order of magnitude as the wavelength (divided by the orbital velocity) of the impacting electron. Resonance is independent of gas pressure except at very high pressures (> 100 atmospheres) at which the intermolecular distance is affected. The gas pressures shown in Figure 6.27 are not this high.
4. Frommhold[173] suggests, however, that resonance due to rotational states cannot be excluded. The energy for the Jth rotational level is given by

$$\varepsilon_J = J(J+1)B_0 \tag{6.77}$$

where B_0 is the rotational constant (7.56 meV for H_2, 3.77 meV for D_2[174] and $J = 0, 1, 2, \ldots$. Resonance occurs if the energy ε_J of an incoming electron lies between $(\varepsilon_{J+1} - \varepsilon_J)$ and $(\varepsilon_{J+2} - \varepsilon_J)$. The two levels are known as lower and upper levels respectively. In H_2, for $J = 0$, the lower resonance level is 15 meV and the upper level is 45 meV. Electrons in this energy range can cause rotational resonance. In D_2 the rotational levels of interest at 77 K are 7.5, 15, 23, 38, and 53 meV. Hence even thermal electrons (energy $= 6.6$ meV) can cause rotational resonance. As the energy increases, these resonance states become unimportant; this agrees with the experimental observation that, at high values of E/N, mean energy $(\bar{\varepsilon})$ is higher and density dependence of W is not observed.

It is appropriate to mention here that polarization of the neutral also affects the drift velocity because of interaction between an electron and the polarized neutral. This interaction ceases to be important beyond a certain radial distance between the interacting particles. This distance is a function of gas number density. However, the cross section for scattering decreases with increasing density, resulting in an increase of W. This situation applies to argon and methane.

The factors that should be considered when calculating energy gain or loss due to rotational excitation have been succinctly summarized by Engelhardt and Phelps.[156] Both H_2 and D_2 are nonpolar molecules and therefore the selection rule for rotational excitation is $\pm \Delta J = 2$. Let us suppose that rotational excitation occurs from the state $J = 0$ to $J = 2$. The energy levels for the states are given by Equation 6.77. The cross section $Q_{J \rightarrow J+2})$ for loss of energy sustained in rotationally exciting the molecule $J \rightarrow J+2$ (both H_2 and D_2) is given by

$$Q_{J \rightarrow J+2} = \left[\left(\frac{P_J}{P_r} \right) \exp\left(-\frac{\varepsilon_J}{kT} \right) \right] \sigma_{J \rightarrow J+2} \tag{6.78}$$

The expression in the square brackets is the fraction of the molecules in the Jth rotational level according to the Boltzmann equation. P_J is calculated from the formula

$$P_J = (2t+1)(t+a)(2J+1) \tag{6.79}$$

where t is the nuclear spin (1/2 for H_2, 1 for D_2) and

$$a = 0, \quad J \text{ even}$$
$$= 1, \quad J \text{ odd} \tag{6.80}$$

In Equation 6.78 the total population is

$$P_r = \sum_J p_J \exp\left(-\frac{\varepsilon_J}{kT}\right) \tag{6.81}$$

The cross section for an energy loss of

$$\varepsilon_J = (4J + 6)B_0 \tag{6.82}$$

is given by

$$\sigma_{J \to J+2}(\varepsilon) = \frac{(J+1)(J+2)}{(2J+3)(2J+1)} \left[1 - \frac{(4J+6)B_0}{\varepsilon} \right]^{1/2} \sigma_0 \tag{6.83}$$

where

$$\sigma_0 = \frac{8}{15} \pi \mu_q^2 a_0^2 \tag{6.84}$$

where μ_q is the quadrupole moment and a_0 the Bohr radius. The quadrupole moment for H_2 and D_2 is $0.62 \, ea_0^2 \, (= 2.78 \times 10^{-40} \, \text{C m}^2)$.

When transition occurs from $J \to J-2$ there is an energy gain of (using the notation of Engelhardt and Phelps[156])

$$\varepsilon_{-J} = (4J - 2)B_0 \tag{6.85}$$

and the cross section for the transition is given by

$$\sigma_{J \to J-2}(\varepsilon) = \frac{J(J-1)}{(2J-1)(2J+1)} \left[1 + \frac{(4J-2)B_0}{\varepsilon} \right]^{1/2} \sigma_0 \tag{6.86}$$

Applications of these equations to calculate the cross sections are demonstrated here by a numerical example. As already stated, the rotational constant for H_2 (B_0) is $7.56 \, \text{meV}$. The energy for excitation from $J=0 \to 2$ is, according to Equation 6.82, $45 \, \text{meV}$. This is the onset energy for inelastic collision in H_2 as experimentally observed by Randell.[175] The cross section with $J=0$ (σ_0) is obtained from Equation 6.84 as $\sigma_0 = 8\pi \times (0.529 \times 10^{-10})^2 = 2.69 \times 10^{-20} \, \text{m}^2$. At gas temperature of $300 \, \text{K}$ $(kT = 26 \, \text{meV})$ the ground state and the lower two rotational states are populated in the ratio $0.3:0.45:0.22$, according to Equation 6.78.[173] At $\varepsilon = 1 \, \text{eV}$ the cross section $\sigma_{0 \to 2}$ $(1 \, \text{eV})$ is given by Equation 6.83 as $1.8 \times 10^{-20} \, \text{m}^2$. The theoretical value is substantially higher than the value $(\sim 10^{-22} \, \text{m}^2)$ determined by Engelhardt and Phelps[156] at $77 \, \text{K}$.

Frommhold[173] has also considered the possibility of formation of dimers $(H_2)_2$ for an explanation of decrease of W with increasing number density. At higher pressures evidence exists for the presence of these dimers. Dimers are formed according to the reaction

$$H_2 + 2H_2 \to (H_2)_2 + H_2 \tag{6.87}$$

and the binding energy is quite small, $\sim 0.4\,\text{meV}$. A dimer hardly survives a collision if kT is much larger than the binding energy, and a reverse reaction will occur. The scattering cross section is now the sum of the partial cross sections of the monomer and the dimer and the latter is a function of the gas pressure. However, the theoretical concentration of $(H_2)_2$ dimer, $\sim 5.1 \times 10^{15}\,\text{m}^{-3}$, is considered to be at least an order of magnitude lower than that required to show an influence on the drift velocity. The experimentally determined concentration of the dimer is at least two orders of magnitude lower, adding further support to this conclusion.

We now consider the ratios D_r/μ and D_L/μ. Selected references are: Warren and Parker,[82] H_2 and D_2; Engelhardt and Phelps,[156] H_2 and D_2; Lowke and Parker[138]; Virr et al.[176]; Kontoleon et al.,[177] H_2; Saelee and Lucas,[164] H_2; Roznerski and Leja,[166] H_2; Al-Amin et al.[127]; Petrović and Crompton,[178] D_2; and Roznerski et al.,[168] D_2. These results are shown in Figures 6.28 and 6.29. Figure 6.28 shows that the ratio D_r/μ is dependent on the temperature at low values of E/N and rises rapidly from a near-constant value to a temperature-independent value, as theory shows. The values given by Huxley and Crompton[3] for E/N up to 212 agree very well with those of Kontoleon et al.[177] and merge smoothly at higher E/N. The ratios D_L/μ are lower than D_r/μ, as in other gases (note that the former has been multiplied by a factor of ten for improved presentation).

Data for deuterium are not as plentiful as for hydrogen, Figure 6.29. 77 K data reported by Huxley and Crompton[3] agree very well with those of Warren and Parker[82] and Engelhardt and Phelps.[156] The 293 K data of Huxley and Crompton[3] join smoothly with the more recent data of Roznerski et al.[168] up to 1000 Td. As far as the author knows, there is only a single theoretical study of D_L/μ, due to Lowke and Parker.[138] Note that the D_L/μ values in Figure 6.29 have been multiplied by a factor of ten for clarity of presentation.

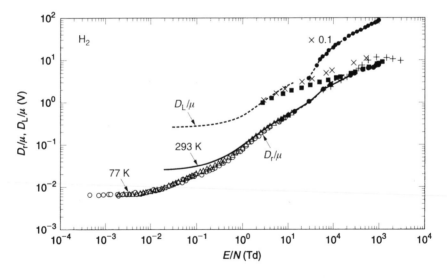

FIGURE 6.28 Ratios of radial diffusion coefficient to mobility (D_r/μ) and longitudinal diffusion coefficient to mobility (D_L/μ) in H_2 as a function of E/N. Unless otherwise specified, temperature is 293 K. D_r/μ: (○) Warren and Parker,[82] 77 K; (---) Engelhardt and Phelps,[156] 77 K; (×) Virr et al.[176]; (●) Kontoleon et al.[177]; (△) Huxley and Crompton,[3] 77 K; (—) Huxley and Crompton[3]; (■) Saelee and Lucas[164]; (+) Al-Amin et al.[127] D_L/μ: (---) Lowke and Parker[123]; (●) Saelee and Lucas.[164] The experimental results of Wagner et al.[119] (D_L/μ) have not been shown because they are in excellent agreement with those of Lowke and Parker.[138] Note that the ratio D_L/μ is multiplied by a factor of 10 for better presentation.

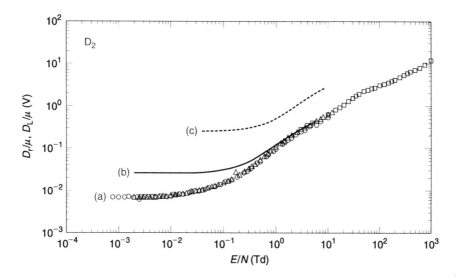

FIGURE 6.29 Ratios D_r/μ and D_L/μ as a functions of E/N in deuterium. (a): D_r/μ, 77 K: (o) Warren and Parker[82]; (- - -) Engelhardt and Phelps[156]; (Δ) Huxley and Crompton.[3] (b): D_r/μ, 293 K: (—) Huxley and Crompton[3]; ($\square$) Roznerski and Leja.[166] (c): D_L/μ, 77 K: (- - -) Lowke and Parker.[123] Ordinates of curve (c) have been multiplied by a factor of ten for better presentation.

The collision cross section in deuterium was not discussed in Chapter 4. The momentum transfer cross section is the same as in H_2, as demonstrated by Engelhardt et al.[156] However, the vibrational and rotational excitation thresholds and cross sections are different. The heavier isotopic molecule has lower vibrational frequency. In H_2 and D_2 they are 0.516 eV and 0.360 eV[156] respectively. The levels of the lighter isotope are always those of the heavier and the separation of energy for the same level between the isotopes increases with increasing level. The rotational constant for the isotopes is also different so that the energy levels are different as well. The separations between the same rotational state (J) for the isotopes are also different. When both vibration and rotation occur, the two isotope effects are added.[179]

The curve of rotational Q–ε for D_2 can lie above or below that of H_2, depending upon the rotational levels, whereas the cross section for vibrational excitation was found to be lower than that for H_2 over almost the entire energy range of 1 to 10 eV.[156] Buckman and Phelps[180] have proposed a set of cross sections for H_2 and D_2 that are stated to be consistent with transport equations. The collision frequency for momentum transfer and energy exchange is shown as a function of characteristic energy in Figure 1.14.

6.5.7 KRYPTON (KR)

Selected references for drift velocity data are : Bowe[181]; Pack et al.[182]; Frost and Phelps[183]; Huxley and Crompton[3]; Kücukarpaci and Lucas[184]; Al-Amin and Lucas[185]; Hunter et al.[186]; Suzuki et al.[187]; Pack et al.[188a]; and Nakamura and Naitoh.[188b] Figure 6.30 and Table 6.19 show selected data.

Since few publications have resulted since the paper of Pack et al.,[188a] we essentially reproduce their comments. The deviation of drift velocity data of Hunter et al.[186] from those of Pack et al.[182] in the range of $0.01 < E/N < 0.2$ is attributed to the trace impurity of 1 ppm in the latter's gas samples. Pack et al.[188a] suggest that the impurity is probably water vapour and/or hydrocarbons.

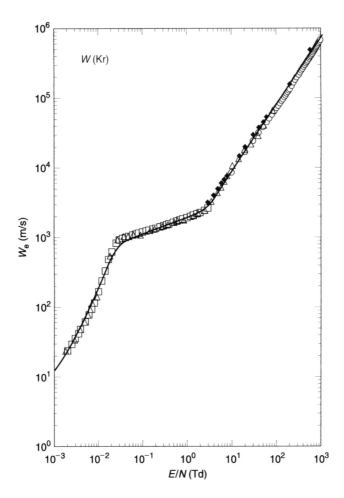

FIGURE 6.30 Experimental and theoretical transport parameters for electrons in krypton. (◇) Lucas et al.[184]; (□) Hunter et al.[186]; (○) computed by the author using BOLSIG software; (△) Pack et al.[188]; (—) Table 6.19.

TABLE 6.19
Experimental and Theoretical Drift Velocity in Krypton

E/N	W	E/N	W	E/N	W
0.001	12.00	0.4	1.57E+03	100	7.74E+04
0.002	23.70	0.7	1.78E+03	200	1.41E+05
0.004	48.30	1	1.94E+03	300	1.92E+05
0.007	95.00	2	2.30E+03	400	2.45E+05
0.01	166.8	4	3.38E+03	500	3.29E+05
0.02	625.00	7	5.75E+03	750	4.95E+05
0.04	992.00	10	8.75E+03	1000	7.34E+05
0.07	1.12E+03	20	1.69E+04	2000	1.05E+06
0.1	1.20E+03	40	3.16E+04	3000	1.35E+06
0.2	1.37E+03	70	4.85E+04	4000	1.52E+06

$0.002 \leq E/N \leq 2.0$ Td from Hunter et al.[186]; $2 < E/N \leq 100$ Td averaged from Figure 6.30; $E/N > 100$ Td digitized from Kücükarpaci and Lucas.[184] E/N in units of Td, and W in units of m s^{-1}.

The drift velocity may be represented by the analytical equation

$$W = 1087 \left(\frac{E}{N}\right)^{0.906} \text{m/s}; \quad R^2 = 0.9971; \quad 2 \leq E/N \leq 500 \,(\text{Td}) \qquad (6.88)$$

Pack et al.[182] have deduced the momentum transfer cross section (Q_M) in krypton from their drift velocity measurements at low energy ($\varepsilon \leq 0.09 \,\text{eV}$). The expressions are, depending upon the power of ε desired,

$$(Q_M)^{-1} = 3.68 \times 10^{18} + 4.05 \times 10^{20} \varepsilon^{3/2} \,\text{m}^{-2} \qquad (6.89)$$

$$(Q_M)^{-1} = 1.59 \times 10^{18} + 1.53 \times 10^{19} \varepsilon \,\text{m}^{-2} \qquad (6.90)$$

$$(Q_M)^{-1} = -5.22 \times 10^{18} + 6.92 \times 10^{19} \varepsilon^{1/2} \,\text{m}^{-2} \qquad (6.91)$$

The radial diffusion coefficient (D_r/μ) in krypton has been measured by Al-Amin and Lucas,[185] who covered very high values of E/N by using a drift distance of 20 cm. Such a large gap is found to be necessary to allow the electrons to reach equilibrium. The collecting electrode was 40 cm in diameter, huge by diffusion experiment standards. The experiment was carried out to obtain revised values for all the noble gases except xenon. Table 6.20 shows these data and one sees that the earlier results of Lakshminarasimha and Lucas[85] for helium and argon are higher by ~14 to 40%. Note that the values shown in this table appear to be different from the original ones because we have used interpolation to obtain D_r/μ at values of E/N that are multiples of ten, for reasons of economy of space.

The general behavior of the experimental and theoretical D_L/μ values is similar to that of argon. The time-of-flight method of determining the diffusion coefficients[185] has resulted in values that are higher by 14 to 23% than those of the steady-state method of Lakshminarasimha and Lucas[85]; compare Tables 6.20 and 6.4 for argon. According to the theory of Lowke and Parker,[138] the ratio D_L/D_r may have a value greater than unity if the momentum transfer frequency decreases with increasing electron energy. Such decrease occurs for both argon and krypton near the Ramsauer–Townsend minimum. The mean energy has been determined from Monte Carlo analysis.[184]

The effective collision frequency of electrons for momentum transfer is easily calculated with the help of Equation 1.98. Typical values are: $E/N = 0.01$ Td, $W = 143$ m/s, $\nu/N = 1.23 \times 10^{-14} \,\text{m}^3\,\text{s}^{-1}$. Dutt[189] has experimentally determined a value of $1.6 \times 10^{-14} \,\text{m}^3\,\text{s}^{-1}$ in high-frequency pulsed discharges.

6.5.8 METALLIC VAPORS

Metallic vapors such as cesium (Cs), mercury (Hg), sodium (Na), and thallium (Tl) have industrial applications and basic data on these vapors are of interest to industrial design engineers and to experimental and theoretical physicists. Cesium and sodium are used as seed materials in magnetohydrodynamics, an area of study in which high-temperature plasma is employed for generation of electrical power. Mercury vapor is extensively used in UV lamps and has potential in space thrusters.

6.5.8.1 Cesium (Cs)

Cesium (atomic number = 55, atomic weight = 139.9), discovered in 1860 by Bunsen and Kirchoff,[190] is one of the most reactive of the alkaline metals in the first column of the periodic table. Because of its heavy atom, the outer electron is far removed from the nucleus,

TABLE 6.20
Ratio D_r/μ by the Time-of-Flight Method[185] and Steady-State Method[85]

E/N	He 185	He 85	Ar 185	Ar 85	Ne 185	Kr 185
2	—	—	—	—	1.73	5.98
4	—	—	—	—	3.12	7.40
7	—	—	—	—	4.44	6.36
10	—	—	—	—	5.16	6.00
20	—	—	—	—	6.36	6.80
30	—	—	—	—	7.46	6.27
40	—	—	—	—	7.99	5.58
50	8.24	—	5.53	—	8.17	5.00
60	6.21	—	6.76	—	8.32	4.66
70	6.96	—	7.42	—	8.61	4.59
80	8.79	—	7.65	—	9.02	4.72
90	10.05	7.51	7.62	6.33	9.47	4.97
100	10.29	7.51	7.47	6.21	9.92	5.26
200	14.41	10.78	7.09	6.29	13.86	5.49
300	18.88	13.67	7.62	5.30	16.21	4.30
400	20.65	15.33	7.47	6.08	18.59	4.29
500	21.70	17.81	7.68	6.64	19.68	5.17
600	21.22	19.07	8.24	6.814	20.96	6.05
700	19.02	17.89	8.59	7.31	23.85	6.51
800	17.68	17.28	8.71	8.03	26.93	6.81
900	17.21	—	9.17	8.79	28.63	7.27
1000	16.97	—	10.07	9.43	28.58	7.95
1500	12.83	—	12.03	9.30*	24.27	11.00
2000	9.88	—	14.67	—	20.52	11.37
3000	8.06	—	16.26	—	17.50	13.65
4000	6.06	—	14.15	—	17.38	13.32
5000	6.92	—	13.71	—	—	—
5650	5.14	—	12.80	—	—	—

E/N in units of Td, D_r/μ in V. * means $E/N = 1250$ Td.

making it easier to detach for chemical reaction. Its melting point is 27°C, lower than that of all other alkalines; the only metal with a lower melting point is mercury, which is not alkaline. The dominant color emitted by the vapor is bright blue, which in fact was the beacon for its discovery. The frequency of the radiation emitted by cesium-133 is 9 GHz (precisely 9,192,631,770 s^{-1}); this, the only naturally occurring isotope of cesium, is used as a standard of time, defining one second.

Experimental measurements of transport parameters in metallic vapors provide a considerable challenge since higher temperatures need to be applied to achieve a satisfactory vapor pressure of the metal. At the high temperatures required to operate the drift tube (up to 1030°C) there will be increased leakage current due to condensation of the metal at lead-throughs, and considerable outgassing from the walls of the container. In order to overcome these problems Nakamura and Lucas[191a] have employed the heat tube technique that has been used by spectroscopists.[192]

Figure 6.31 shows the heat tube developed by Nakamura and Lucas.[191a] A stainless steel tube is constructed as a closed heat pipe with a capillary structure (a wick constructed from

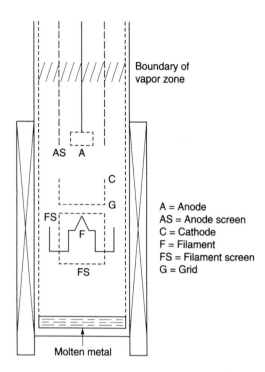

Boundary of vapor zone

AS   A

C

G

A = Anode
AS = Anode screen
C = Cathode
F = Filament
FS = Filament screen
G = Grid

FS

FS

Molten metal

FIGURE 6.31 Vertical heat tube developed by Nakamura and Lucas[191a] for the measurement of drift velocity of electrons in metallic vapors. Figure reproduced from Nakamura, T. and J. Lucas, *J. Phys. D*, 11, 325, 1978. With permission of IOP, U.K.

two layers of stainless steel mesh). The other end is water cooled and contains the electrical feed-throughs. The tube is heated at the closed end by a 1 kW heater; two radiation shields and a vacuum jacket are provided to prevent heat loss and corrosion of the outer walls by oxidation. The electrodes are made of nichrome wire so that they can be individually heated by dc to a slightly higher temperature than the vapor pressure temperature, thus preventing the condensation of vapor on the electrodes. The drift distance is 2 cm and over this distance the vapor pressure was found to be constant.

Selected references for drift velocity data of electrons in Cs are: Chanin and Steen[193]; Postma[194]; and Saelee and Lucas.[195] Figure 6.32 shows the transport parameters. The drift velocity is a power-law function of E/N. The earlier results of Chanin and Steen[193] are considerably higher than the subsequent data obtained by Saelee and Lucas.[195] The characteristic energy and the mean energy ($\bar{\varepsilon}$) are obtained by the Boltzmann distribution method. It is noted that D_r/μ is marginally higher than $\bar{\varepsilon}$ at the same E/N. If the electron energy distribution is Maxwellian, then $D_r/\mu = 0.67\,\bar{\varepsilon}$, Equation 6.39. Since this condition is not satisfied the electron energy distribution is non-Maxwellian, as is indeed found by the Boltzmann analysis. Saelee and Lucas[195] show that at the very low E/N investigated by them (1.41 Td) the electrons are in thermal equilibrium with gas molecules and that departure from a Maxwellian distribution occurs as the mean energy becomes larger.

The cross section data shown in Figure 6.33 are reproduced from Zecca et al.[196] The Cs atom shows the Ramsauer–Townsend minimum at 0.08 eV.[197] The agreement between the total collision cross section data of various workers is less than satisfactory. The references for cross sections shown in Figure 6.33 are as follows. Total collision cross section: Brode[198]; Visconti et al.[199]; Jaduszliwer and Chan.[200] Momentum transfer: Nighan and Postma[201]; Stefanov.[202] Elastic: Bartschat.[203] Excitation (2 p 6 ^{2}P): Chen and Gallagher.[204]; Ionization: Brink[205]; McFarland and Kinney[206]; Nygaard.[207]

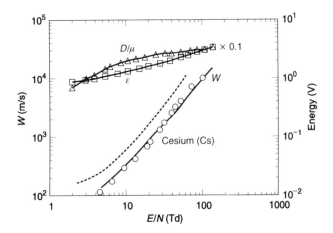

FIGURE 6.32 Transport parameters in cesium. Drift velocity W: (- - -) Chanin and Steen[193]; (○) Saelee and Lucas,[195] experimental; (—) Saelee and Lucas,[195] theoretical. D/μ (—△—) and $\bar{\varepsilon}$(—□—) are reported by Saelee and Lucas,[195] theoretical. Energy values are multiplied by a factor of ten for clarity.

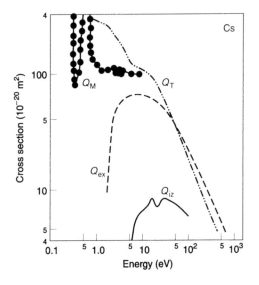

FIGURE 6.33 Scattering cross sections in cesium. The selected references are as follows. Total: Brode[198]; Visconti et al.[199]; Jaduszliwer and Chan.[200] Momentum transfer: Stefanov[202]; Nighan and Postma.[201] Elastic: Bartschat.[203] Excitation: Chen and Gallagher.[204] Ionization: Brink[205]; McFarland and Kinney[206]; Nygaard.[207] Compilation and discussion: Zecca et al.[196]

Using Equation 1.98 and the momentum transfer cross section, Nigham and Postma[201] have determined the effective collision frequency ν_m/N as $1 \times 10^{-12}\,\text{s}^{-1}\text{m}^3$ at 100 Td. The ratio ν_m/N remains essentially constant in the range of 40 to 400 Td. Constant ν_m/N suggests that W is a linear function of E/N (in terms of energy, Q_M varies as $\varepsilon^{-1/2}$), as is borne out by Figure 6.32.

6.5.8.2 Mercury (Hg)

Selected references for drift velocity data in mercury are: Killian[208]; Klarfeld[209]; McCutchen[210]; Kerzar and Weissglas[211]; Rockwood[212]; Judd[213]; Nakamura and Lucas[191a,b];

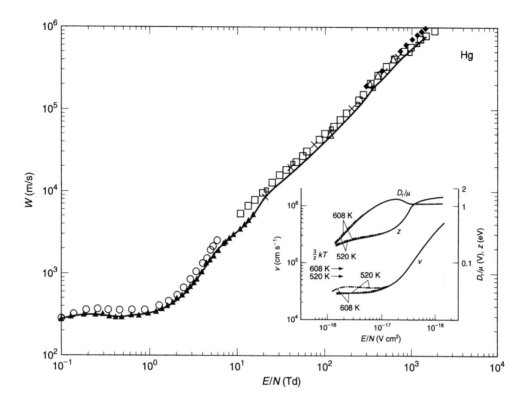

FIGURE 6.34 Drift velocity of electrons in mercury vapor. Letter E or T after citation denotes experimental or theoretical values. ($\triangle$) Klarfeld,[209] E; ($\circ$) McCutchen,[210] E; ($\blacktriangle$) Nakamura and Lucas,[191a,b] E, T; ($\blacklozenge$) Garamoon,[214] T; ($\times$) Rockwood,[212] T; ($\square$) Sakai et al.[215]; ($\square$) Liu and Raju,[216] T; (—) suggested. The inset shows both the measured and theoretical drift velocity, characteristic and mean energy at two temperatures. The dependence of W on temperature is attributed to the formation of mercury dimers (Hg_2) which possibly have higher collision cross section. Note that D_r/μ is higher than $\bar{\varepsilon}$, suggesting a non-Maxwellian energy distribution. Inset reproduced from Nakamura, T. and J. Lucas, *J. Phys. D*, 11, 337, 1978. With permission of IOP, U.K.

Garamoon and Abdelhaleem[214]; Sakai et al.[215]; and Liu and Raju.[216,217] Figure 6.34 shows these data. The experimental results (indicated by the letter E in the legend) do not extend beyond 20 Td, due to experimental difficulties. Higher E/N requires lower vapor pressure and therefore lower temperature, which is difficult to control accurately. The theoretical values, though they agree among themselves, are higher than the experimental values by 10 to 50% in the range 20 to 320 Td, with the theoretical values obtained by the Monte Carlo method or the Boltzmann method. The differences can be reconciled by resorting to the drift velocity as calculated by the Boltzmann equation method. Tagashira et al.[218] define four different drift velocities:

$$\left.\begin{array}{l} W_T = \text{time-of-flight (TOF) drift velocity} \\ W_P = \text{pulsed Townsend (PT) velocity} \\ W_S = \text{steady-state (SST) velocity} \\ W_d = \text{diffusion-modified velocity, Equation 6.102} \end{array}\right\} \qquad (6.92)$$

An analysis of the drift velocity using the Monte Carlo technique[216] and the Boltzmann equation[217] has been carried out, using the same collision cross section. The results of these

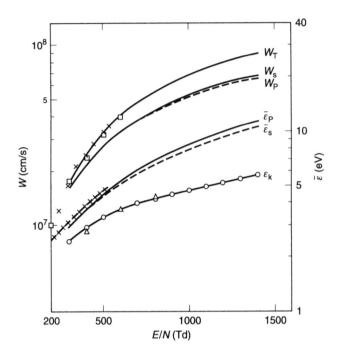

FIGURE 6.35 Transport parameters according to the definitions of Tagashira et al.[218] For the meaning of subscripts see Equations 6.92. ε_k is characteristic energy. Figure reproduced from Liu, J. and G. R. Govinda Raju, *Can J. Phys.*, 70, 216, 1992.

computations are shown in Figure 6.35. The Monte Carlo results give W_T and the steady-state velocity, which is identical to W_P, is lower. Further, the difference between W_T and W_S increases with increasing E/N. The suggested drift velocity is shown as a continuous line in Figure 6.34 and the Monte Carlo drift velocity is reduced here by 18% in the range of E/N from 20 to 320 Td.

The inset of Figure 6.34 shows the result of Nakamura and Lucas,[191b] who observed a temperature dependence of both W and D_r/μ. Since this dependence could not be attributed to the change in energy of the atom ($3kT/2 = 67.22$ and 78.59 meV at 520 K and 608 K respectively), which is small, the observed variation of W is attributed to the formation of Hg_2 dimer, which may have a much higher cross section than the monomer. The effects of temperature on W and D_r/μ are of the opposite kind. An increase of temperature lowers the W because of an increased concentration of the dimer, with its greater cross section. On the other hand, D_r/μ increases because the energy lost during collisions (proportional to m/M) is only 50%, due to the increase in M.

Figure 6.34 also shows the PT and SST mean energy and characteristic energy. There is no suitable definition of the mean energy according to TOF and it is difficult to compare this with the results of the Monte Carlo simulation as far as the mean energy is concerned. The characteristic energy D_r/μ agrees very well with that of Rockwood[212] for overlapping values of E/N. Sakai et al.[215] compare the SST drift velocity results with those of Rockwood[212] and report good agreement. It is noted that Sakai et al.[215] have included the effect of metastable atoms in the ionization process.[219]

The dominant states in the Hg discharge are the ground state (6 $^1S_0$) and the excited states, which are 6 $^3P_0$ (4.67 eV), 6 $^3P_1$ (4.89 eV), 6 $^3P_2$ (5.46 eV), and 6 $^1P_1$ (6.7 eV). Among these, 6 $^3P_0$ and 6 $^3P_2$ are metastable atoms with long life and play a significant role in the discharge process by generating secondary electrons.[220] The secondary electrons are produced by collision between excited states. Some of the possible processes are[220]

$$Hg(6\,^1S_0) + e \rightarrow Hg^+ + 2e \tag{6.93}$$

$$\left.\begin{array}{l} Hg(6\,^3P_0) + e \rightarrow Hg^+ + 2e \\ Hg(6\,^3P_1) + e \rightarrow Hg^+ + 2e \\ Hg(6\,^3P_2) + e \rightarrow Hg^+ + 2e \end{array}\right\} \tag{6.94}$$

$$\left.\begin{array}{l} Hg(6\,^3P_0) + Hg(6\,^3P_1) \rightarrow Hg_2^+ + e \\ Hg(6\,^3P_2) + Hg(6\,^3P_2) \rightarrow Hg^+ + Hg(6\,^1S_0) + e \end{array}\right\} \tag{6.95}$$

$$\left.\begin{array}{l} Hg(6\,^3P_1) + Hg(6\,^3P_2) \rightarrow Hg^* + Hg(6\,^1S_0) \\ Hg^* + e \rightarrow Hg^+ + 2e \end{array}\right\} \tag{6.96}$$

Reaction 6.93 is ionization of the ground state atom (ionization potential = 10.4 eV), 6.94 is ionization of excited states, 6.95 is ionization by collision between excited states, and 6.96 is ionization by collision between excited states, the asterisk showing excitation to a higher lying state and its subsequent ionization. In 6.95 the formation of the dimer ion (ionization potential = 9.46 eV) satisfies the energy conservation principle with heat of formation of 0.06 eV.[214]

Figure 6.36 shows the effective collision frequency obtained by Monte Carlo simulation. The momentum transfer collision frequency is nearly constant, having a value of

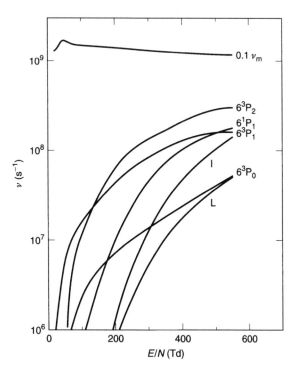

FIGURE 6.36 Collision frequency in mercury vapor as a function of E/N at $3.22 \times 10^{22}\,m^{-3}$ gas number density. L means lumped remaining excitation states, I means ionization collisions. Reproduced from Liu, J., Ph.D. Thesis, University of Windsor, 1993.

approximately $3.7 \times 10^{-13}\,\mathrm{m^3\,s^{-1}}$. The excitation collisions are predominantly due to the states shown. Nakamura and Lucas[191b] have shown that 90% of the energy lost in collisions is due to excitation collisions.

6.5.8.3   Sodium (Na) and Thallium (Tl)

Sodium has an atomic number of 11 and an atomic weight of 22.99. It reacts violently with water, sizzling on top of the water as it melts. It is far too reactive to be found in nature in its pure form. It was first isolated in 1807 by the British scientist Humphrey Davy. Sodium chloride is, of course, common salt, and sodium bicarbonate is found in every kitchen. Its characteristic spectroscopic color is yellow and it is used in highway lights because of its high luminous efficiency. Moreover, yellow light does not scatter as much as white light, making it suitable in foggy conditions.

Thallium (Tl) is the element that comes after Hg in the periodic table. It was discovered in 1861 by the British scientist William Crookes, who identified it by the brilliant green spectral line of its emitted light. It is obtained as a by-product of lead and zinc refining. It is highly toxic and possibly carcinogenic. Some thallium compounds change conductivity when exposed to infrared radiation, making them suitable for infrared detectors. One isotope of thallium is radioactive with a half life of 72 hours, making it suitable for medical diagnostic purposes. Thallium has an atomic number of 81 and an atomic weight of 204.38.

Figures 6.37 and 6.38 show the drift velocity of electrons in sodium and thallium respectively.[191a] In sodium W depends on the vapor pressure, with higher vapor pressure giving lower drift velocity at the same E/N. This behavior is attributed to the formation of dimers, the fraction of which increases with increasing temperature. The drift velocity is therefore determined at various temperatures at the same E/N and extrapolated to zero concentration of the dimer. These values are shown in Figure 6.37 by the broken line. The results shown in Figure 6.38 are the only measurements in thallium vapor.

Collision cross sections employed by Nakamura and Lucas[191b] are shown in Figure 6.39.

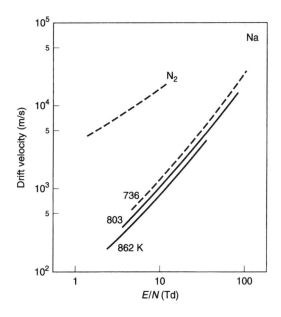

FIGURE 6.37 Drift velocity of electrons in sodium vapor. The pressure and the temperature are shown on the right. The curve of N_2 is included for comparison. Figure reproduced from Nakamura, T. and J. Lucas, *J. Phys. D*, 11, 325, 1978. With permission of IOP, U.K.

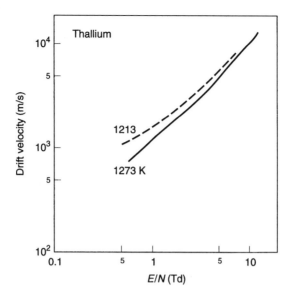

FIGURE 6.38 Drift velocity of electrons in thallium vapor. Figure reproduced from Nakamura, T. and J. Lucas, *J. Phys. D*, 11, 337, 1978. With permission of IOP, U.K.

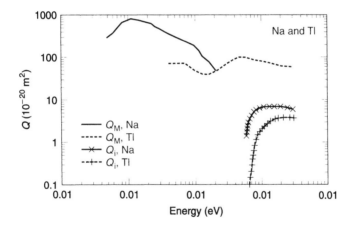

FIGURE 6.39 Scattering cross section in sodium and thallium employed by Nakamura and Lucas.[191b] Only momentum transfer and ionization cross sections are shown.

6.5.9 NEON (NE)

Selected references for drift velocity data in neon are: Bowe[221]; Pack and Phelps[222]; Anderson[223]; Sugawara and Chen[224]; Robertson[225]; Huxley and Crompton[3]; Kücükarpaci et al.[226]; Sakai et al.[227]; and Puech and Mizzi.[228] A large number of theoretical analyses prior to 1980 have been published which have not been cited in the present context.

Figure 6.40 and Table 6.22 show selected data of drift velocity at both 77 K and 300 K. The same figure contains data on diffusion coefficients and mean energy, which will be discussed below. Compared with helium and argon, the following differences are observed:

1. At 77 K the drift velocity in neon is marginally higher than that at 300 K. This is the converse of the temperature effect in argon.

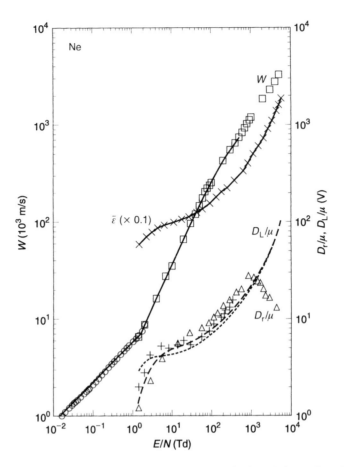

FIGURE 6.40 Transport parameters in neon as a function of E/N. Unless otherwise specified, the temperature of the gas is 300 K. W: (●) Huxley and Crompton,[3] 77 K; (○) Huxley and Crompton[3]; (□) Kücükarpaci et al.[226]; (—) recommended. D_r/μ: (- - -) Kücükarpaci et al.,[226] theory; (△) Al-Amin and Lucas.[185] D_L/μ: (+) Kücükarpaci et al.,[226] experimental; (- -) Kücükarpaci et al.,[226] theory. Mean energy: (-×-) Kücükarpaci et al.,[226] theory.

2. The difference at the two temperatures in neon is much smaller than that in argon.
3. At low values of E/N ($< \sim 10$ mTd) the log–log plot of W vs. E/N does not have a slope of 45°, in contrast to helium (Figure 6.21); a linear variation of W with E/N gives this slope. This is attributed to the fact that at low values of E/N the electron energy distribution in neon is determined more by the electric field than by the thermal motion of gas atoms.

In this context, Pack and Phelps[222] define an "energy relaxation time" given by

$$\tau = \frac{M}{2m\nu_{\text{eff}}} \tag{6.97}$$

where m and M are the masses of electron and atom respectively, and ν_{eff} is the effective electron–molecule collision frequency. Pack and Phelps[222] suggest an approximate value of

$$\tau = C\frac{3.22 \times 10^{16}}{N} \text{ s} \tag{6.98}$$

where N is the number of atoms at $300\,\text{K}$ (m^{-3}) and C has a value of 500, 100, and 5 for argon, neon, and helium respectively at low values of E/N. The significance of energy relaxation time is that the measurements should be carried out using experimental parameters that result in a drift time that is much greater than the relaxation time. Using equation 6.98, one gets a relaxation time of $\sim 1\,\mu\text{s}$ in neon, which is much lower than the drift time of $\sim 50\,\mu\text{s}$, as is desired. Approximate values of ν_{eff} according to Equation 6.97 are (3.7, 0.8, 0.004) $\times 10^{10}\,\text{s}^{-1}$ in argon, neon, and helium respectively. These values translate to (1.2, 0.25, 0.014) $\times 10^{-12}\,\text{m}^3\,\text{s}^{-1}$ respectively.

In the region of high E/N $(> 100\,\text{Td})$ experimental values of W have been provided by Kücükarpaci et al.[226] up to $\sim 400\,\text{Td}$. For higher values, the data shown are obtained from Monte Carlo simulation. The drift velocity (units of $10^3\,\text{m/s}$) may be expressed by the analytical function

$$W = 5.8321\left(\frac{E}{N}\right)^{0.433}; \quad R^2 = 0.9991 \tag{6.99}$$

$$W = 5.131\left(\frac{E}{N}\right)^{0.847} \tag{6.100}$$

$$W = 14.937\left(\frac{E}{N}\right)^{0.631}; \quad R^2 = 0.9992 \tag{6.101}$$

The range of applicability and the quality of agreement with the recommended values are shown in Table 6.21.

Sakai et al.[227] have theoretically evaluated the electron swarm properties in Penning mixtures of Ar + Ne, where the ratio of argon in the total concentration varied from 10^{-6} to 10^{-1}. The Penning effect will be considered in Chapter 8. The drift velocity calculated is corrected for diffusion according to

$$W_{\text{corr}} = W - \alpha D \tag{6.102}$$

where α is the ionization coefficient and D the diffusion coefficient. The mean energy is found to be considerably lower in mixtures with Penning additives than in pure neon. The diffusion-corrected drift velocity W_{corr}, however, changes only marginally for small percentages of argon added.

Figure 6.40 also shows the ratios (D_{r}/μ) and (D_{L}/μ) in neon. Of all the rare gases, measurements of these ratios in neon have been made only by Lucas and his group. There is only a single measurement of the former ratio by Al-Amin and Lucas,[185] and the decrease of D_{r}/μ at $E/N=1000\,\text{Td}$ is attributed to nonequilibrium effects. As in several of their

---

TABLE 6.21
Range of Applicability and Quality of Agreement of Analytical Expressions for W in Neon

Equation	Range	Maximum Difference (%)	E/N	Minimum Difference	E/N
6.99	0.017–1.7	+2.2	0.17	+0.05	0.7
6.100	4–100	+6.68	70	0.48	100
6.101	200–5000	+2.59	1000	0.58	800

E/N in units of Td. Positive differences mean that the recommended values are higher.

TABLE 6.22
Recommended Drift Velocity of Electrons in Neon

E/N	W	E/N	W	E/N	W
0.02	1.09	0.5	4.31	40	120
0.03	1.29	0.6	4.67	50	146
0.04	1.46	0.7	5.00	60	175
0.05	1.59	0.8	5.28	70	201
0.06	1.72	1	5.80	80	222
0.07	1.83	2	8.48	90	239
0.08	1.93	4	16.10	100	255
0.1	2.12	7	27.60	200	427
0.2	2.86	10	35.00	300	552
0.3	3.43	20	65.90	400	646
0.4	3.90	30	96.0	600	829

E/N in units of Td, W in 10^3 m/s.

theoretical works, Lucas and his group have derived the ratios for both elastic and isotropic scattering. The data taken in the present volume for other gases are those that come closer to experimental results. In neon the data are for elastic scattering and are identical to those for inelastic scattering up to $E/N \simeq 450$ Td. For higher E/N the isotropic scattering gives ratios that are a few percent higher. Again, there is only a single measurement of D_L/μ by Kücükarpaci et al.,[226] as shown.

The mean energy as a function of E/N is often a useful parameter for approximate calculations using the Maxwell or Druyvestyn electron energy distribution. Further, Kücükarpaci et al.[226] suggest that the electron swarm is in equilibrium with the electric field provided that the breakdown voltage is at least ten times as large as the mean energy expressed in volts (Figure 6.40).

6.5.10 NITROGEN (N₂)

As stated earlier, the nitrogen molecule, with 14 electrons, is iso-electronic with carbon monoxide. The two gases possess similarities in some respects and important dissimilarities. They have almost identical molecular mass, equilibrium nuclear distance, dissociation energy, and fundamental vibrational and rotational constants. There are similarities in the momentum transfer cross sections also, whereas the rotational and vibrational excitation cross sections are different in magnitude.

The nitrogen molecule has been one of the most widely investigated. Selected references for drift velocity data are: Bowe[229]; Pack and Phelps[230]; Heylen[231]; Engelhardt et al.[232]; Schlumbohm[233]; Huxley and Crompton[3]; Snelson and Lucas[234]; Raju and Rajapandian[235]; Raju and Gurumurthy[236]; Saelee et al.[238b]; Taniguchi et al.[237]; Kücükarpaci and Lucas[238a]; Tagashira et al.[239]; Raju and Hackam[240]; Novak and Frechette[241]; Roznerski and Leja[242]; Phelps and Pitchford[243]; Nakamura and Kurachi[244]; Nakamura[245]; Fletcher and Reid[246]; Liu and Raju[247,248]; Hasegawa et al.[249]; Roznerski[250]; and Tanaka.[251]

The drift tube employed in the studies of drift velocity uses grids or pulses of electrons for measurement of time. At high values of E/N ionization occurs and the additional electrons generated contribute to the current; this component must be eliminated by suitable correction to current ratios, as Lawson and Lucas[252] have demonstrated. An alternative strategy is to apply the Monte Carlo method in which the correction to the ionized component can be applied immediately after the simulated event.

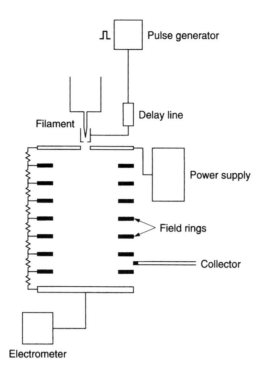

FIGURE 6.41 Experimental setup of Fletcher and Reid[246] for measuring transport parameters by the photon counting technique. The drift tube is provided with a movable electrode assembly to facilitate viewing by the photomultiplier, behind the collector, of different parts of the discharge region. The signal, after amplification, passes through a time-to-amplitude convertor and a multichannel analyzer before being recorded in the computer.

A novel technique has been employed in hydrogen and nitrogen for the measurement of transport properties, based on the principle that the photons generated in the collision region may be used as a probe. Blevin et al.[253] and Fletcher and Reid[254] have exploited this technique in hydrogen and nitrogen respectively. Their experimental arrangement is shown in Figure 6.41. The drift tube was provided with an indirectly heated oxide cathode as an electron source. The cathode was normally held at positive potential so that the electrons entered the drift region only when a negative potential, in the form of a pulse voltage (~30 ns), was applied to it. The cathode had a small aperture through which the electrons entered the drift region. This pulse was also used as the starting pulse for the time-to-amplitude converter.

In nitrogen, photons with a wavelength of 337 nm are generated by means of the C state $(^3\Pi_u)$,[246] the lifetime of the excited state being 35.6 ns. The photomultiplier used for counting was relatively insensitive to longer wavelengths whereas shorter wavelengths were filtered by the pyrex glass of which the tube was made. It is appropriate to mention here that photons are also generated by de-excitation of the E state into C state and subsequent decay of the C state to ground state. This process should be accounted for in the analysis of data to obtain the correct transport coefficients. The electrode assembly was mounted on a movable vacuum vernier so that any part of the drift tube could be viewed by the spatially fixed photomultiplier. The method was applied to nitrogen over the range of $50\,\text{Td} \leq E/N \leq 500\,\text{Td}$ in which ionization occurs.

Figure 6.42 shows selected data. The data of Pack and Phelps[230] are shown because they appertain to low-energy electrons for E/N values of 10^{-4} to 30 Td. The experimental data of Schlumbohm,[233] obtained by the current pulse method, is notable for the highest E/N at

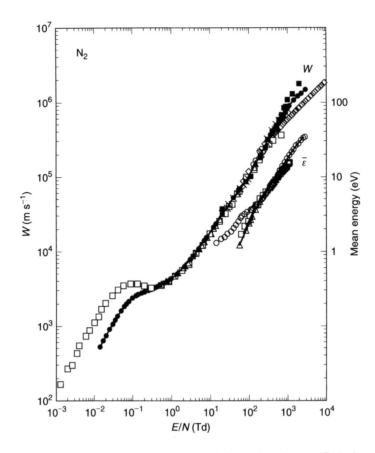

FIGURE 6.42 Drift velocity and mean energy in N_2. Unless otherwise specified, the temperature is 293 K. The letters E or T after citation mean experimental or theoretical respectively. W: ($\square$) Engelhardt et al.,[232] 77 K, T; ($\diamond$) Schlumbohm,[233] E; ($\bullet$) Huxley and Crompton,[3] E; ($\blacktriangle$) Snelson and Lucas,[234] E; ($\blacksquare$) Raju and Rajapandian,[235] E; ($+$) Saelee et al.,[238b] E; ($\circ$) Kücükarpaci and Lucas,[238a] T; Nakamura,[245] E; ($---$) Liu and Raju,[247] T; ($\blacksquare$) Liu and Raju,[248] T; ($\times$) Hasegawa et al.,[249] E; ($—$) recommended. Mean energy: ($\square$) Taniguchi et al.,[237] T; ($\circ$) Kücükarpaci and Lucas,[238a] T; ($\triangle$) Liu and Raju,[248] T; ($—$) recommended. The 77 K data of Huxley and Crompton[3] are obscured by points ($\square$) of Engelhardt et al.[232]

which measurements were made, 10,000 Td. The compilation of Huxley and Crompton[3] covers the range that is appropriate for many applications in the domain of plasma and gaseous electronics (0.01 to 240 Td). The groups of Lucas and Raju have independently carried out both theoretical and experimental investigations. Roznerski[242,250] has adopted the Bradbury–Nielsen double shutter method.

With regard to theoretical analysis, Engelhardt et al.[232] used a time- and space-independent Boltzmann equation (usually referred to as the Holstein form in the literature) to calculate the swarm parameters from derived self-consistent cross section data. They considered seven inelastic collisions in all, one of them being the ionization cross section and the remaining six considered as excitation cross sections (electronic and ro-vibrational). The range of E/N covered was also high, up to 1000 Td. Taniguchi et al.[237] studied the development of electron avalanche at E/N ranging from 56 to 1131 Td, taking into account the temporal ($\partial f/\partial t$) and spatial ($\partial f/\partial z$) variation of the distribution function. It is appropriate to mention here that detailed measurements of excitation cross sections have been published by the Jet Propulsion Laboratories group.[255–257] To include these data, Tagashira et al.[239]

presented a re-analysis which demonstrated that the excitation frequency for the respective electronic states lay within the experimental error of Cartwright.[256]

The Monte Carlo approach adopted by Kücükarpaci and Lucas[238] covers the range of E/N varying from 14 to 3000 Td. Eight vibrational excitations and four electronic excitation with threshold energies of 6.22, 7.39, 8.59, and 11.05 eV were used. The essential difference in cross section data required for the theory is somewhat different between the Boltzmann method and the Monte Carlo method. In the former method one uses the momentum transfer cross section whereas in the latter one uses the total cross section combined with the differential scattering cross section. These requirements usually result in a differing set of cross sections for theoretical purposes. The Boltzmann equation solution used by Novak and Frechette[241] was aimed at investigating the transport parameters in mixtures of $SF_6 + N_2$ and will be referred to again in the section on SF_6.

Liu and Raju[248] have investigated the dependence of the transport parameters on E/N in both methods by using two different sets of cross sections. In set 1 cross sections of Engelhardt et al.[232] are employed whereas, in set 2, the more accurate data that had become available were incorporated. For example, the results of Cartwright[256] were renormalized by Trajmar et al.[258] The review of Itikawa et al.[259] became available in 1986. The details of the two sets are shown in Table 6.23. The momentum transfer cross sections in set 1 are higher than those in set 2. The excitation cross section in set 1 is very large, $2.8 \times 10^{-20} m^2$, over a wide range of energy, $25.5 \leq \varepsilon \leq 150 \, eV$. The dissociation cross section of Spence and Schulz[260] is also included in set 2. A comparison of transport parameters obtained by both methods and with experimental data is given in Table 6.24.

With this background, the data available on drift velocity in N_2 are presented in Figure 6.42, which also includes mean energy data. The recommended values are shown in Table 6.25 and formulated by combining the results of Huxley and Crompton,[3] range 0.02 to 200 Td, Hasegawa et al.[249] 200 to 1000 Td, and Kücükarpaci and Lucas,[238a] range 1000 to 2500 Td. The drift velocity data may be expressed as analytical functions of E/N,[261] using the following expressions.

293 K

$$W = 2467.6 \left(\frac{E}{N} \right)^{0.836} \text{m/s}; \quad R^2 = 0.9975 \tag{6.103}$$

$$W = -3.84 \times 10^5 \left(\frac{E}{N} \right)^4 + 4.77 \times 10^5 \left(\frac{E}{N} \right)^3 - 2.08 \times 10^5 \left(\frac{E}{N} \right)^2$$
$$+ 3.99 \times 10^4 \frac{E}{N} + 34.36 \, \text{m/s} \tag{6.104}$$

$$W = -2.06 \times 10^{-4} \left(\frac{E}{N} \right)^4 + 9.11 \times 10^{-2} \left(\frac{E}{N} \right)^3 - 12.80 \left(\frac{E}{N} \right)^2$$
$$+ 1.60 \times 10^3 \left(\frac{E}{N} \right) + 3.02 \times 10^3 \, \text{m/s} \tag{6.105}$$

$$W = 5780 \left(\frac{E}{N} \right)^{0.708} \text{m/s}; \quad R^2 = 0.9867 \tag{6.106}$$

77 K

$$W = -30 \left(\frac{E}{N} \right)^3 + 305.9 \left(\frac{E}{N} \right)^2 + 784.2 \left(\frac{E}{N} \right) + 3.38 \times 10^3 \, \text{m/s} \tag{6.107}$$

TABLE 6.23
Collision Cross Section Data in N_2[247]

Type of Collision	Energy Range	Threshold Energy	Reference
Set 1			
Momentum transfer	$\varepsilon \leq 70$	0	Engelhardt et al.[232]
Set 2			
Momentum transfer	$\varepsilon \leq 1$		Engelhardt et al.[232]
	$1 \leq \varepsilon \leq 4$		Itikawa et al.[259]
	$\varepsilon > 4\,\mathrm{eV}$		Trajmar et al.[258]

Vibrational Excitation (Set 1)			Vibrational Excitation (Set 2)		
Energy	Energy loss	Peak Q_v	Energy	Onset	Peak Q_V
0.29	0.29	1.63	0.29	0.29	6.8
1.70	0.59	1.52	1.77	0.57	4.6
1.80	0.88	1.69	1.89	0.86	2.2
1.90	1.17	1.30	2.05	1.13	1.5
2.0	1.47	1.04	2.10	1.41	0.5
2.20	1.76	0.6	2.31	1.68	0.6
2.30	2.06	0.44	2.10	1.95	0.3
2.50	2.35	0.24	2.35	2.21	0.2
5.0	5.0	0.42	2.47	2.47	0.08
			2.73	2.73	0.04

Excitation Cross Section (Set 1)		Excitation Cross Section (Set 2)	
Energy Loss	Peak Q_{ex}	Energy Loss	Peak Q_{ex}
6.7	0.56	7.56	0.21
8.4	0.42	8.04	0.31
11.2	1.0	8.65	0.34
12.5	0.4	9.46	0.12
14.0	2.8	9.72	0.10
		9.26	0.34
		10.11	0.12
		11.10	0.38
		11.90	0.01
		12.30	0.06
		12.81	0.18
		13.11	0.08
		13.28	0.02
		13.4	0.2

Energy in units of eV, cross section in $10^{-20}\,\mathrm{m}^2$.

The range of applicability and quality of fit with the recommended values are shown in Table 6.26.

We shall now consider, briefly, the collision process. The low energy loss processes are those of rotational and vibrational excitation, with electronic excitation and ionization processes taking over as the energy of the electron increases. Figure 6.43 shows the processes that play a role as the mean energy increases. Figure 6.44 shows the energy losses as a function of E/N at higher values. In the range $E/N = 80$ to $600\,\mathrm{Td}$ excitation is the dominant loss mechanism and for $E/N > 600\,\mathrm{Td}$ the ionization process dominates.

TABLE 6.24
Comparison of Boltzmann Equation and Monte Carlo Methods in N_2 Using the Same Set of Scattering Cross Sections

E/N (Td)	Mean Energy $\bar{\varepsilon}$ (eV)		Drift Velocity W ($\times 10^5$/m/s)		
	Boltzmann	Monte Carlo	Boltzmann	Monte Carlo	Experimental
282	5.65	6.25	2.81	2.53	2.9
338	6.57	7.26	3.17	2.78	3.2
395	7.44	7.83	3.51	3.32	3.7
451	8.29	8.72	3.82	3.49	4.0
508	9.12	9.20	4.11	3.80	4.35

TABLE 6.25
Recommended Drift Velocity in N_2

E/N	W	E/N	W	E/N	W
0.02	7.33 (+02)	20.0	3.09 (+04)	300	3.21 (+05)
0.04	1.35 (+03)	30.0	4.17 (+04)	400	4.12 (+05)
0.07	1.99 (+03)	40.0	5.18 (+04)	500	5.00 (+05)
0.10	2.38 (+03)	50.0	6.09 (+04)	600	5.65 (+05)
0.20	2.87 (+03)	60.0	7.03 (+04)	700	6.21(+05)
0.40	3.30 (+03)	70.0	7.90 (+04)	800	6.59 (+05)
0.70	3.93 (+03)	80.0	8.74 (+04)	900	7.26 (+05)
1.00	4.43 (+03)	100	1.05 (+05)	1000	8.06 (+05)
2.0	5.98 (+03)	120	1.28 (+05)	1500	1.10 (+06)
4.0	9.33 (+03)	140	1.48 (+05)	2000	1.25 (+06)
7.0	1.41 (+04)	170	1.81 (+05)	2500	1.35 (+06)
10.0	1.84 (+04)	200	2.12 (+05)	3000	1.53 (+06)

E/N in units of Td, drift velocity in m/s. $a(b)$ means $a \times 10^b$.

TABLE 6.26
Range of Applicability and Quality of Fit of the Analytical Expressions for Drift Velocity in N_2

Equation	E/N Range	Maximum Difference (%)		Minimum Difference (%)	
		Magnitude	E/N	Magnitude	E/N
6.103	10–1500	10.91	500	−0.09	800
6.104	0.02–0.5	2.38	0.05	−0.02	0.1
6.105	0.6–200	−6.08	0.06	−0.06	25
6.106	210–3000	−9.41	3000	0.43	800
6.107	0.06–6.0	−7.81	0.35	−0.14	3.5

E/N in units of Td. A positive difference means that the recommended values are higher.

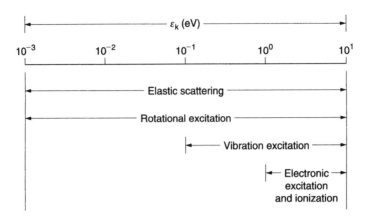

FIGURE 6.43 Collision processes in N_2 as the characteristic energy increases. At higher E/N ($\varepsilon_k \simeq 10\,eV$) electronic excitation and ionization dominate.

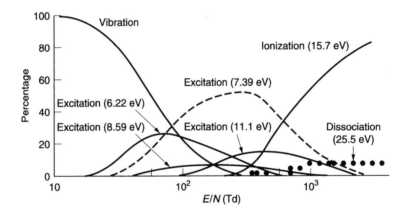

FIGURE 6.44 Percentage power lost by electrons in N_2 as a function of E/N. At low E/N ro-vibrational excitation dominates and at very high E/N ionization dominates. In the intermediate region excitation collisions followed by ionization are important. Figure reproduced from Kücükarpaci, H. N. and J. Lucas, *J. Phys. D*, 12, 2123, 1979. With permission of IOP, U.K.

There are many similarities between H_2 and N_2 in the transport of electrons; both are diatomic, nonpolar, and possess a quadrupole moment. We therefore refer to Equations 6.77 to 6.86 for discussion of rotational energy loss at low E/N (≤ 0.2 Td). The lowest rotational energy threshold in N_2 is 1.5 meV (rotational constant $B_0 = 0.25$ meV) and even at 77 K (where the thermal energy is 6.63 meV) there is a large number of rotationally excited molecules. In contrast with hydrogen, there are about 15 rotational states of significant population at 300 K. About 30 rotational resonances may be involved; the lifetimes of the resonances may be of the order of 10^{-14} s and their width about 30 meV. Another difference from hydrogen is the high density of these levels at thermal energies. To satisfy these requirements Engelhardt et al.[232] used a quadrupole moment of magnitude $4.66 \times 10^{-40}\,m^2$; see Equation 6.84 and the discussion following it.

Theoretical calculation of rotational excitation cross sections involves consideration of two effects. The first effect is that of the interaction of the incoming electron with the quadrupole molecule. This aspect has been discussed in connection with Equations 6.77 to 6.86. The second effect is the polarization of the molecule with which the electron interacts. Here we are concerned with the electronic polarizability and not the permanent dipole moment. A correction factor, greater than unity, has been suggested by Dalgarno and Moffett[262] that should be used to multiply the rotational excitation cross section. The factor is

$$f_R(\varepsilon) = 1 + \frac{P_\alpha(4\varepsilon - \varepsilon_J)}{\varepsilon^{1/2}} + \frac{9}{4}P_\alpha^2(2\varepsilon - \varepsilon_J) \qquad (6.108)$$

where

$$P_\alpha = \frac{4\pi^2 \varepsilon_0 \left(\alpha_\| - \alpha_\perp \right)}{24 \mu_q R^{1/2}} \qquad (6.109)$$

Here $\alpha_\|$ and $\alpha_\perp$ are the polarizabilities in the parallel and perpendicular directions, $\varepsilon_0 = 8.854 \times 10^{-12}$ F/m, μ_q = quadrupole moment, R = Rydberg constant (13.56 eV). To illustrate the use of Equations 6.108 and 6.109 we substitute the following values for N_2. Let $J = 4$ and $\varepsilon = 20$ meV. The energy for this rotational level, from Equation 6.77, is 5 meV ($B_0 = 0.25$ meV). Engelhardt et al.[232] have determined $\mu_q = 4.66 \times 10^{-40}$ C m², $\alpha_\| = 2.65 \times 10^{-40}$ F m² and $\alpha_\perp = 1.614 \times 10^{-40}$ F m². Substituting these values, we obtain from Equation 6.109, in a straightforward manner, $P_\alpha = 7.915 \times 10^{-3}$ and, from Equation 6.108, $f_R(\varepsilon) = 1.004$. This is a correction of 0.4%. The factor f_R increases with electron energy and at 2 eV it has a value of 1.04, only 4% more. The polarization effects are therefore negligible in N_2, which is the conclusion of Engelhardt et al.[232] However, Engelhardt and Phelps[156] calculate that the correction is more significant in H_2, between 10% and 30%.

Lowke et al.[263] have investigated the transport parameters in mixtures of CO_2, N_2, and He that are of interest in laser applications. Hasegawa et al.[264] have measured and theoretically calculated the drift velocity and (D_L/μ) in mixtures of $N_2 + CO_2$. Their results

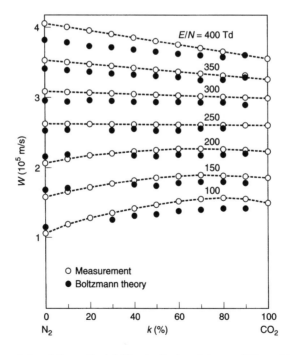

FIGURE 6.45 Experimental and theoretical drift velocity in mixtures of N_2 and CO_2. (o) measurement; (●) Boltzmann equation analysis. Note the nonlinear variation of W with increasing percentage of CO_2 at lower values of E/N. Figure reproduced from Hasegawa, H. et al., *J. Phys. D*, 29, 2664, 1996. With permission of IOP, U.K.

are shown in Figure 6.45. At E/N in the range of 100 to 200 Td, increasing the percentage content of CO_2 increases W and, for higher E/N, there is a decrease of W as the percentage content of CO_2 increases. W in pure N_2 is higher than in CO_2 in the range of 300 to 1000 Td. Grünberg[265a,b] measured the drift velocity at high gas pressures and found that there is a decrease in W in He, N_2, and H_2. Allen and Prew[266] have measured the influence of gas density on drift velocity in the range of 0.3 to 7 MPa ($N = 7.9 \times 10^{25}\,m^{-3}$ to $N = 1.8 \times 10^{27}\,m^{-3}$) and $E/N = 1$ to 5 Td. No significant change in W was observed (compare with Figure 6.16), in contradiction to the measurements of Grünberg[265a,b].

Selected data on the ratios D_r/μ and D_L/μ have been provided by the following, where the letter E or T denotes experiment or theory respectively. Warren and Parker,[267] E; Engelhardt et al.,[232] T; Wagner et al.,[54] E; Naidu and Prasad,[268] E; Lowke and Parker,[285] T; Huxley and Crompton,[3] E; Saelee et al.,[238b] E; Taniguchi et al.,[237] T; Fletcher and Reid[246]; Novak and Frechette[269]; Al-Amin et al.,[270] E; Wedding et al.[27] E; Nakamura,[245] E; Roznerski,[250] E; and Hasegawa et al.,[249] E. Figure 6.46 shows selected data. Mean energy has been evaluated by Taniguchi et al.,[237] T; Kücükarpaci and Lucas,[238a] T; and Liu and Raju,[247] T. These data are also shown in Figure 6.46.

The results of Roznerski[250] differ by about 10% from those of Naidu and Prasad[268] in the region of overlap. The theoretical results of Taniguchi et al.[237] and the experimental values of Al-Amin et al.[270] are higher by a few percent. With regard to the latter, at high values of E/N ionization occurs and it is not clear whether a correction has been applied. The recommended D_r/μ values, derived by combining the data of Huxley and Crompton[3], in the range $0.02 \leq E/N \leq 14$ Td, Roznerski[250] in the range $30 \leq E/N \leq 200$ Td, and the mean (visual) values of several data in the range $200 \leq E/N \leq 1100$ Td, are also shown. There are large differences, particularly at $E/N > 200$ Td, in the D_L/μ values between the three methods of

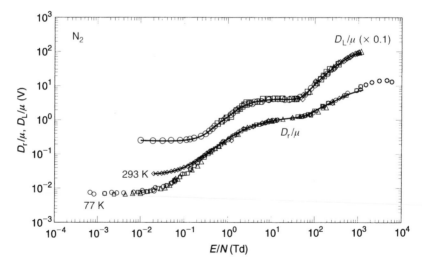

FIGURE 6.46 Ratios D_r/μ and D_L/μ in N_2 as a function of E/N. Unless otherwise mentioned, the temperature is 293 K. Letters E or T after the reference indicate experiment or theory respectively. D_r/μ: (○) Warren and Parker,[82] 77 K, E; (□) Huxley and Crompton,[3] 77 K, E; (Δ) Engelhardt et al.,[232] 77 K, T; (◇) Huxley and Crompton,[3] E; (×) Fletcher and Reid,[246] T; Al-Amin et al.,[270] E; (●) Roznerski,[250] E; (—) recommended. D_L/μ: (○) Lowke and Parker,[285] T; (□) Nakamura[245]; (▲) Roznerski,[250] E; (◆) Hasegawa et al.,[249] E. The experimental results of Wagner et al.[54] for D_L/μ are not shown as they agree very well with the theoretical results of Lowke and Parker.[285] Note that the values of D_L/μ are shown multiplied by the factor ten, for clarity.

evaluation: the experimental, Boltzmann equation, and Monte Carlo simulation. The recommended values of D_L/μ are obtained by combining the data of Lowke and Parker[285] in the range $0.02 \leq E/N \leq 100$ Td and of Hasegawa et al.[264] in the range $100 \leq E/N \leq 1000$ Td.

The mean energy is shown in Figure 6.42, with the references stated in the legend. The collision frequency is shown in Figures 1.14 to 1.16.

6.5.11 Oxygen (O_2)

We have sufficient theoretical knowledge to consider the transport properties in this gas straight away. Selected references for data are: Neilsen and Bradbury[272]; Schlumbohm[273]; Hake and Phelps[274]; Naidu and Prasad[275]; Huxley and Crompton[3]; Roznerski and Leja[276]; Al-Amin et al.[277]; Liu and Raju[278,279]; and Jeon and Nakamura.[280]

Figure 6.47 shows selected data with the recommended values, formulated by combining the values of Huxley and Crompton,[3] $0.6 \leq E/N \leq 20$ Td, with the experimental measurements of Al-Amin et al.[270] and Jeon and Nakamura,[280] $20 \leq E/N \leq 300$ Td. They do not differ by more than a few percent. We have therefore recommended the average values of Al-Amin et al.[270] and Jeon and Nakamura[280] in the overlapping ranges. Further, the results of Liu and Raju[278,279] are in better agreement with those of Al-Amin et al.[270] For higher values of $E/N > 800$ Td, the results of Liu and Raju[278] are recommended.

Recommended values of drift velocity are given in Table 6.27. The expressions of Schlumbohm[273] and Raju[261] are as follows.

Schlumbohm[273]

$$W = 2.13 \times 10^4 \sqrt{\frac{E}{N}} \, \text{m/s} \qquad (6.110)$$

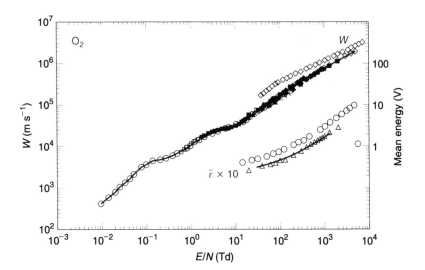

FIGURE 6.47 Drift velocity and mean energy in O_2 as a function of E/N. Letter E or T after citation means experimental or theoretical respectively. W: ($\diamond$) Schlumbohm,[273] E; ($\circ$) Hake and Phelps,[274] T; ($\square$) Huxley and Crompton,[3] E; ($\times$) Roznerski and Leja,[276] E; ($\blacklozenge$) Liu and Raju,[278] T; ($\blacksquare$) Liu and Raju,[279] T; ($\bullet$) Jeon and Nakamura,[280] E; (—) recommended. Mean energy: ($\circ$) Al-Amin et al.[277] E; (–) Liu and Raju,[278] T; ($\triangle$) Liu and Raju,[279] T.

TABLE 6.27
Recommended Drift Velocity in O_2

E/N	W	E/N	W	E/N	W	E/N	W
0.009	417.8	1	1.14 (04)	25	5.59 (04)	500	4.76 (05)
0.014	590.3	2	1.78 (04)	30	7.08 (04)	600	5.17 (05)
0.020	787.5	4	2.44 (04)	40	9.14 (04)	700	5.94 (05)
0.03	1.32 (03)	5	2.57 (04)	50	1.06 (05)	800	7.04 (05)
0.04	1.66 (03)	6	2.66 (04)	60	1.18 (05)	900	7.17 (05)
0.05	2.10 (03)	7	2.76 (04)	70	1.29 (05)	1000	7.71 (05)
0.08	3.13 (03)	8	2.87 (04)	80	1.44 (05)	1300	8.96 (05)
0.10	3.52 (03)	10	3.14 (04)	90	1.58 (05)	2000	1.19 (06)
0.20	4.71 (03)	12	3.46 (04)	100	1.72 (05)	3000	1.48 (06)
0.40	5.94 (03)	14	3.82 (04)	200	2.78 (05)	4000	1.70 (06)
0.6	7.20 (03)	17	4.36 (04)	300	3.55 (05)	5000	1.91 (06)
0.8	9.40 (03)	20	4.94 (04)	400	4.25 (05)		

E/N in units of Td, W in units of m/s. $a(b)$ means $a \times 10^b$.

Raju[261]

$$W = -1.53\left(\frac{E}{N}\right)^4 + 79.65\left(\frac{E}{N}\right)^3 - 1.33 \times 10^3 \left(\frac{E}{N}\right)^2 + 9.6 \times 10^3 \left(\frac{E}{N}\right) + 2.96 \times 10^3 \, \text{m/s}$$

(6.111)

$$W = 8.05 \times 10^3 \times \left(\frac{E}{N}\right)^{0.655} \, \text{m/s}; \quad R^2 = 0.9970$$

(6.112)

The range of applicability and quality of fit with the recommended values are shown in Table 6.28.

The low-energy inelastic processes operative in O_2 are rotational excitation (rotational constant $B_0 = 0.18 \, \text{meV}$), vibrational excitation (0.193 eV for $v = 0 \rightarrow v = 1$), electronic excitation of the a $^1\Delta_g$ and b $^1\Sigma_g^+$ with thresholds of 0.98 and 1.63 eV respectively, and the three-body attachment process.[281] Hake and Phelps[274] assumed a vibrational cross section as a series of spikes having cross sections of ~ 1 to $1.5 \times 10^{-22} \, \text{m}^2$ (see Figure 4.36). The vibrational excitation is possibly due to the short-lived negative ion with subsequent autodetachment. Very good agreement was obtained with the calculated and experimental drift velocity and characteristic energy.

TABLE 6.28
Range of Applicability and Quality of Fit of the Analytical Expressions

Equation	E/N Range	Maximum Difference		Minimum Difference	
		%	E/N	%	E/N
6.111	1.4–25.0	−9.78	20.0	−0.74	3.5
6.112	30–4000	8.70	800	0.24	60

E/N in units of Td. A positive difference means that the recommended values are higher.

A few comments with regard to the quadrupole moment of molecules are in order here. The diatomic molecule is visualized, for purposes of mathematical simplicity, as two points connected together by a weightless rigid rod. The atoms rotate about an axis that is perpendicular to the rod and the arrangement is known as a "rigid rotator." The possible energy states of such a rotator are quantized and only certain rotational frequencies are allowed. The rotational quantum number J gives approximately the angular momentum of the atom in units of $h/2\pi$. The rotational frequency increases approximately linearly with J.[282]

If the atoms that rotate are electrically charged, then the rotation of the atoms results, according to classical theory, in radiation of electromagnetic waves. This situation applies to electrical dipoles. The permanent dipole moment is due to the difference in electronegativity of the unlike atoms of the molecule and it lies in a direction of the perpendicular from the mass point to the axis of rotation. The emitted frequency due to rotation of the molecule is in the infrared region ($\sim 10^{12}$ to 10^{13} Hz). Absorption of the same frequency also occurs. In quantum mechanics, emission or absorption of radiation from the dipole occurs for a change in the rotational quantum number.

However, the elementary molecules H_2, N_2, O_2, F_2, Cl_2 do not show an infrared spectrum, though there is emission and absorption from these molecules at visible frequencies. This phenomenon is attributed to the fact that these molecules, though not possessing a dipole moment, possess a quadrupole moment which is smaller than the dipole moment by a factor of 10^{-8} to 10^{-10}, expressed in units of C m^2. The potential due to a quadrupole varies as r^{-3} whereas that due to a dipole varies as r^{-2}. Gerjuoy and Stein[156] derived the equation relating the rotational quantum number and the quadrupole moment to determine the rotational excitation cross section, Equation 6.86. An approximate value of the rotational excitation cross section may be obtained[274] with the help of Equation 6.86 by using an approximate quadrupole moment of 1.8 ea_0^2 ($\mu_q = 8.07 \times 10^{-40}$ C m^2).

The data on (D_r/μ) and (D_L/μ) are presented in Figure 6.48 and Table 6.29. Evaluation of D_r/μ in electron-attaching gases is complicated by the fact that at low values of E/N

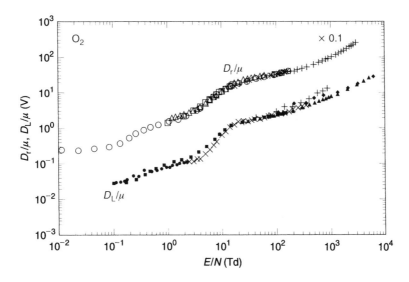

FIGURE 6.48 Ratios of radial and longitudinal diffusion coefficients to mobility as a function of E/N. Letters E or T after citation indicate experimental or theoretical respectively. D_r/μ: (○) Hake and Phelps,[274] T; (×) Naidu and Prasad,[275] E; (□) Fleming et al.,[284] E; (Δ) Huxley and Crompton,[3] E; (+) Al-Amin et al.,[277] T. D_r/μ values are multiplied by a factor of ten for clarity of presentation. D_L/μ: Schlumbohm,[273] E; (■) Lowke and Parker,[285] T; (●) Nelson and Davis,[286] E; (+) Al-Amin et al.[277] E; (×) Jeon and Nakamura,[280] E.

TABLE 6.29

Recommended Ratios of D_r/μ and D_L/μ in Oxygen

E/N (Td)	D_r/μ (V)	D_L/μ (V)	E/N (Td)	D_r/μ (V)	D_L/μ (V)
0.01	0.025	—	10	1.09	0.65
0.02	0.024	—	20	2.06	1.50
0.04	0.025	—	40	2.70	1.77
0.07	0.029	—	70	2.99	2.21
0.1	0.033	0.030	100	3.39	2.45
0.2	0.053	0.035	120	3.58	2.60
0.4	0.094	0.050	140	3.85	2.69
0.7	0.13	0.080	200	(4.05)	3.36
1	0.15	0.10	400	—	6.62
2	0.20	0.14	700	—	10.90
4	0.36	0.15	1000	—	—
7	0.73	0.33	2000	—	—

The number in brackets requires experimental confirmation.

attachment occurs and at high values of E/N both ionization and attachment occur. Naidu and Prasad[283] have used a solution for the ratio of current to the collecting electrode that requires the availability of both ionization and attachment coefficients as a function of E/N and an analysis using a computer. Secondary electrons also add to the complexity. For these reasons, the D_r/μ values show considerable deviation from each other.

Monte Carlo simulated results, shown in Figure 6.48,[277] are also subject to the method employed; whether the elastic and inelastic scattering cross sections are isotropic, or according to differential scattering. The results shown in the figure are for isotropic scattering of both elastic and inelastic processes because they are the most consistent with the measured ones. The measurements of Fleming et al.[284] are shown because they measured the ratio (D_r/μ) by removing the negative ions formed in O_2 by addition of small amounts of hydrogen, according to the process

$$O^- + H_2 \rightarrow H_2O + e \qquad (6.113)$$

Experimental data on (D_r/μ) for $E/N > 150\,\text{Td}$ are lacking.

Data on D_L/μ are reported by Lowke and Parker,[285] Nelson and Davis,[286] Price et al.,[287] Al-Amin et al.,[277] Liu and Raju,[278] and Jeon and Nakamura.[280] The results agree very well among themselves except at the lower range of $E/N < 15\,\text{Td}$. The accuracy of the time-of-flight measurements of D_L/μ decreases as the ratio E/N decreases (~ 18 to 30%). This is because the gas number densities are higher at low E/N and the half width of the emitted electron pulse from the cathode is comparable with the half pulse width produced by diffusion. Such conditions produce inaccurate D_L/μ values.[277]

This ratio rises rapidly at $\sim 5\,\text{Td}$, and the corresponding slow saturation of W at the same E/N indicates that the energy gain of electrons from the electric field starts to dominate over the energy loss due to vibrational excitations. Further, increase of W and the relatively slow increase of D_L/μ at $\sim 10\,\text{Td}$ are attributed to new inelastic energy losses, possibly due to excitation to the a $^1\Delta_g$ and b $^1\Sigma_g^+$ states and other higher electronic states. These losses are much more dominant over vibrational excitation losses[280] in the electron swarm. The inelastic collision frequency calculated by Hake and Phelps[274] also shows a flat region in the range of $E/N = 5$ to $10\,\text{Td}$ adding support to this observation.

6.5.12 XENON (XE)

Pack et al.[288] measured the drift velocity of slow electrons with the pulsed drift tube method. Huang and Freeman[289] have studied the influence of density, electric field, and temperature on the mobility. Other studies are due to Bowe,[290] Wagner et al.,[54] and Brooks et al.[291] Notable measurements are due to Hunter et al.,[292] who have improved the accuracy to ± 1.0 to 2.0%, though the range of E/N covered is narrow.

Pack et al.[293] have made a succinct analysis of the transport parameters of xenon in the range of $10^{-3} \leq E/N \leq 100$ Td. Their measurements of longitudinal diffusion coefficients from current waveforms and calculated values of W and D_r/μ are shown in Figure 6.49. To the knowledge of the author of the present volume there is only one set of tabulated values of W in this gas in the range 0.004 to 3.0 Td (Table 6.30). For the range $3 < E/N \leq 60$ the average of Brooks et al.[291] and Pack et al.[288] have been shown. For the higher range, $E/N > 100$ Td, the only data available are the experimental ones due to Makabe and Mori[294] and Xiao et al.[297]

In view of the limited number of publications since the paper of Pack et al.,[288] we essentially reproduce their comments. The drift velocity data of Brooks et al.[291] at $1 < E/N < 4$ Td are about 10% higher than than those of Pack et al.[288] Hunter et al.[292] suggest that the sample of Brooks et al.[291] was contaminated with halogen-containing compounds.

Puech and Mizzi[295] have calculated W and D_L/μ for the range 10^{-3} to 10^3 Td, Santos et al.[296] have employed the three-dimensional model Monte Carlo method for low E/N in the

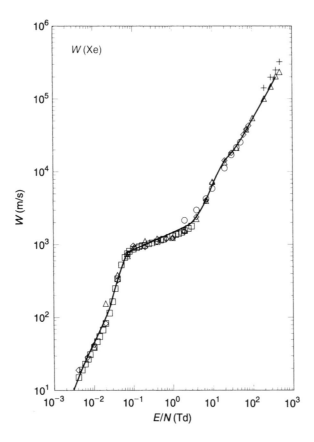

FIGURE 6.49 Drift velocity of electrons in Xe as a function of E/N. ($\square$) Hunter et al.[292]; ($\triangle$) Peuch and Mizzi[295]; (+) Xiao et al.[297]; ($\diamond$) Pack et al.[288]; (o) Brooks et al.[291]; (—) Table 6.30.

TABLE 6.30
Drift Velocity in Xenon

E/N (Td)	W (m/s)	E/N (Td)	W (m/s)	E/N (Td)	W (m/s)
0.002	6.00	0.70	1.20E+03	60	3.24E+04
0.004	15.20	1	1.31E+03	70	3.84E+04
0.007	26.80	2	1.55E+03	80	4.34E+04
0.010	38.70	4	2.38E+03	100	5.48E+04
0.020	82.00	7	3.99E+03	200	1.01E+05
0.040	336.00	10	6.93E+03	300	1.49E+05
0.070	743.00	20	1.42E+04	400	2.04E+05
0.100	858.00	30	1.81E+04	500	2.33E+05
0.20	981.00	40	2.15E+04		
0.40	1.11E+03	50	2.63E+04		

$0.002 \leq E/N \leq 80$ Td digitized from Pack et al.[288] $80 < E/N \leq 500$ Td digitized from Puech and Mizzi.[295]

3 to 16 Td range, and Xiao et al.[297] have conducted experimental measurements by the pulsed Townsend method in the range of E/N from 32.24 to 564.2 Td. They have also studied the transport parameters in mixtures of SF_6 and xenon; the drift velocity is the highest in pure xenon, lower in pure SF_6, and is intermediate in value for 50% Xe and 50% SF_6. Similar behavior is observed by Xiao et al.[298] in $SF_6 + He$ mixtures. This is possibly because an increasing percentage of xenon reduces the attachment coefficient and low-energy inelastic scattering losses that occur in molecular gases.

Xenon shows a pronounced Ramsauer–Townsend minimum (Figure 3.61). Makabe and Mori[294] have measured the drift velocity and the electron energy distribution function. The calculated energy distribution using the Boltzmann equation is compared with the measured distribution and reasonable agreement is obtained between the two.

The ratios (D_r/μ) and (D_L/μ) as a function of E/N are qualitatively similar to those in argon, which also shows a pronounced Ramsauer–Townsend minimum (Figure 6.50).

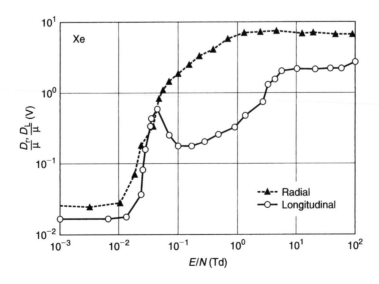

FIGURE 6.50 Ratio of radial and longitudinal diffusion coefficients to mobility in xenon.

Further, the ratio (D_L/μ) reflects more accurately the variation of the momentum transfer cross section as a function of energy.

Concluding remarks for this chapter will be presented at the end of Chapter 7.

REFERENCES

1. McDaniel, E. W., *Collision Phenomena in Ionized Gases*, John Wiley and Sons, New York, 1964.
2. Hasted, J. B., *Physics of Atomic Collisions*, American Elsevier, New York, 1972.
3. Huxley, L. G. H. and R. W. Crompton, *The Diffusion and Drift of Electrons in Gases*, John Wiley and Sons, New York, 1974.
4. Crompton, R. W., in *Advances in Atomic, Molecular, and Optical Physics*, ed. M. Inokuti, Academic Press, Boston, 1994, vol. 33, p. 119.
5. Schlumbohm, H., *Z. Phys.*, 182, 306, 1965.
6. Bradbury, N. E. and R. A. Nielsen, *Phys. Rev.*, 49, 388, 1936.
7. Crompton, R. W., in *Advances in Atomic, Molecular, and Optical Physics*, ed. M. Inokuti, Academic Press, Boston, 1994, vol. 33, p. 97.
8. Lowke, J. J., *Aust. J. Phys.*, 26, 469, 1973.
9. Huxley, L. G. H. and F. W. Bennett, *Phil. Mag.*, 30, 396, 1940.
10. Bailey, V. A., *Phil. Mag.*, 50, 825, 1925.
11. Crompton, R. W., L. G. H. Huxley, and D. J. Sutton, *Proc. Roy. Soc. A*, 215, 507, 1953.
12. Huxley, L. G. H., *Aust. J. Phys.*, 12, 171, 1959.
13. Lucas, J., *Int. J. Electron.*, 18, 419, 1965.
14. Raju, G. R. G., *Brit. J. Appl. Phys.*, 16, 279, 1965.
15. Francey, J. L. A., *J. Phys. B: At. Mol. Phys.*, 2, 680, 1969.
16. Rao, C. R. and G. R. Govinda Raju, *J. Phys. D: Appl. Phys.*, 4, 769, 1971.
17. (a) Townsend, J. S. and H. T. Tizard, *Proc. Roy. Soc. A*, 88, 336, 1913; (b) Townsend, J. S. and V. A. Bailey, *Phil. Mag.*, 43, 593, 1922.
18. Rees, J. A. and R. L. Jory, *Aust. J. Phys.*, 17, 307, 1964.
19. Naidu, M. S. and A. N. Prasad, *J. Phys. D: Appl. Phys.*, 2, 1431, 1969.
20. Virr, L. E., J. Lucas, and N. Kontoleon, *J. Phys. D: Appl. Phys.*, 5, 542, 1972.
21. Kontoleon, N., J. Lucas, and L. E. Virr, *J. Phys. D: Appl. Phys.*, 5, 956, 1972.
22. Kontoleon, N., J. Lucas, and L. E. Virr, *J. Phys. D: Appl. Phys.*, 6, 1237, 1973.
23. Lakshminarasimha, C. S. and J. Lucas, *J. Phys. D: Appl. Phys.*, 10, 313, 1977.
24. Fleming, I., D. R. Gray, and J. A. Rees, *J. Phys. D: Appl. Phys.*, 5, 291, 1972.
25. Raju, G. G., *Dielectrics in Electric Fields*, Marcel Dekker, New York, 2003.
26. Huxley, L. G. H. and R. W. Crompton, *The Diffusion and Drift of Electrons in Gases*, John Wiley and Sons, New York, 1974, p. 182.
27. Morrison, M. A., R. W. Crompton, B. C. Saha, and Z. Lj. Petrović, *Aust. J. Phys.*, 40, 239, 1987.
28. Tagashira, H., Y. Sakai, and S. Sakamoto, *J. Phys. D: Appl. Phys.*, 10, 1051, 1977.
29. Liu, J. and G. R. Govinda Raju, *Can. J. Phys.*, 70, 216, 1992. Also see J. Liu, Ph.D. thesis, University of Windsor, 1993.
30. Liu, J. and G. R. Govinda Raju, *Can. J. Phys.*, 71, 476, 1993.
31. Pitchford, L. C., S. V. O'Neil, and J. R. Rumble, Jr., *Phys. Rev. A.*, 23, 294, 1981.
32. Pitchford, L. C. and A. V. Phelps, *Phys. Rev. A*, 25, 540, 1982.
33. Phelps, A. V. and L. C. Pitchford, *Phys .Rev. A*, 31, 2932, 1985.
34. Huxley, L. G. H. and R. W. Crompton, *The Diffusion and Drift of Electrons in Gases*, John Wiley and Sons, New York, 1974, pp. 25 and 76.
35. Morse, B. M., W. P. Allis, and E. S. Lamar, *Phys. Rev.*, 48, 412, 1935.
36. Abramowitz, M. and I. A. Stegun, *Handbook of Mathematical Physics*, Dover, New York, 1970.
37. Brand, K. P. and J. Kopainsky, *Appl. Phys.*, 18, 321, 1979.
38. Heylen, A. E. D., *Proc. Phys. Soc.*, 79, 284, 1962.
39. Heylen, A. E. D., *Proc. Phys. Soc.*, 76, 779, 1960.
40. Raju, G. R. Govinda and R. Hackam, *J. Appl. Phys.*, 52, 3912, 1981.

41. Raju, G. R. Govinda and R. Hackam, *Proc. IEEE*, 69, 850, 1981.
42. Raju, G. R. Govinda and R. Hackam, *J. Appl. Phys.*, 53, 5547, 1982.
43. Beaty, E. C., J. Dutton, and L. C. Pitchford, *A Bibliography of Electron Swarm Data*, JILA Information Report No. 20, 1979.
44. Raju, A., *Basic Data on Gas Discharges*, internal report, University of Windsor, 1985.

AIR

45. Hessenauer, H., et al., *Z. Physik*, 204, 142, 1967.
46. Rees, J. A., *Aust. J. Phys.*, 26, 403, 1973.
47. Roznerski, W. and K. Leja, *J. Phys. D: Appl. Phys.*, 17, 279, 1984.
48. Liu, J. and Gorur G. Raju, *IEEE Trans. Elec. Insul.*, 28, 154, 1993.
49. Morgan, W. L., J. P. Boeuf and L. C. Pitchford, www. siglo-kinema.com, 1996.
50. Chen, J. and J. H. Davidson, *Plasma Chem. Plasma Process.* 23, 83, 2003.
51. Lakshminarasimha, C. S. and J. Lucas, *J. Phys. D: Appl. Phys.*, 10, 313, 1977.
52. Milloy, H. B., I. D. Reid, and R. W. Crompton, *Aust. J. Phys.*, 28, 231, 1975.

ARGON (AR)

53. Bowe, J. C., *Phys. Rev.*, 117, 1411, 1960.
54. Wagner, E. B., F. J. Davis, and G. S. Hurst, *J. Chem. Phys.*, 47, 3138, 1967.
55. Pack, J. L., R. E. Voshall, and A. V. Phelps, *Phys. Rev.*, 127, 2084, 1962.
56. Jager, G. and W. Otto, *Z. Phys.*, 169, 517, 1962.
57. Brambring, J., *Z. Phys.*, 179, 539, 1964.
58. Long, W. H., Jr., W. F. Bailey, and A. Garscadden, *Phys. Rev. A*, 13, 471, 1976.
59. Robertson, A. G., *Aust. J. Phys.*, 30, 39, 1977.
60. Tagashira, H., Y. Sakai, and S. Sakamoto, *J. Phys. D: Appl. Phys.*, 10, 1051, 1977.
61. Sakai, Y., Y. H. Tagashira, and S. Sakamoto, *J. Phys. D: Appl. Phys.*, 10, 1035, 1977.
62. Christophorou, L. G., D. L. McCorkle, D. W. Maxey, and J. G. Carter, *Nucl. Instr. Methods*, 163, 141, 1979.
63. Kücükarpaci, H. N. and J. Lucas, *J. Phys. D: Appl. Phys.*, 14, 2001, 1981.
64. Nakamura, Y. and M. Kurachi, *J. Phys. D: Appl. Phys.*, 21, 718, 1988.
65. Pack, J. L., R. E. Voshall, A. V. Phelps, and L. E. Kline, *J. Appl. Phys.*, 71, 5363, 1992.
66. Makabe, T. and T. Mori, *J. Phys. D: Appl. Phys.*, 15, 1395, 1982.
67. Dezhen, W. and Ma Tengcal, *J. Phys. D: Appl. Phys.*, 24, 1367, 1991.
68. Atrazhev, V. M. and I. T. Iakubov, *J. Phys. D: Appl. Phys.*, 10, 2155. 1977.
69. Christophorou, L.G. and D. L. McCorkle, *Chem. Phys. Lett.*, 42, 533, 1976.
70. Allen, N. L. and B. A. Prew, *J. Phys. B: At. Mol. Phys.*, 3, 1113, 1970.
71. English, W. H. and G. C. Hanna, *Can. J. Phys.*, 31, 768, 1935.
72. Long, W. H., Jr., W. F. Bailey, and A. Garscadden, *Phys. Rev.*, 13, 471, 1976.
73. Kurachi, M. and Y. Nakamura, *J. Phys. D: Appl. Phys.*, 21, 602, 1988.
74. de Urquijo, J., I. Alvarez, E. Basurto, and C. Cisnero, *J. Phys. D: Appl. Phys.*, 32, 1646, 1999.
75. Foreman, L., P. Kleban, L. D. Schmidt, and H. T. Davis, *Phys. Rev. A*, 23, 1553, 1981.
76. Petrović, Z. Lj., T. F. O'Malley, and R. W. Crompton, *J. Phys. B: At. Mol. Opt. Phys.*, 28, 3309, 1995.
77. Raju, Gorur G., in *Proceedings of the International Conference on Properties and Applications of Dielectric Materials* (ICPADM), Nagoya, Japan, 2003, p. 999.
78. Sawada, S., Y. Sakai, and H. Tagashira, *J. Phys. D: Appl. Phys.*, 22, 282, 1989.
79. Kurihara, M., Z. Lj. Petrović, and T. Makabe, *J. Phys. D: Appl. Phys.*, 33, 2146, 2000.
80. Yamaji, M. and Y. Nakamura, *J. Phys. D: Appl. Phys.*, 36, 640, 2003, with an erratum in *J. Phys. D: Appl. Phys.*, 37, 644, 2004.
81. Townsend, J. S. and V. A. Bailey, *Phil. Mag.*, 44, 1033, 1922.
82. Warren, R. W. and J. H. Parker, *Phys. Rev.*, 128, 2661, 1962.
83. Heylen, A. E. D. and T. J. Lewis, *Proc. Roy. Soc. A*, 271, 531, 1963.
84. Engelhardt, A. G. and A. V. Phelps, *Phys. Rev. A*, 133, 375, 1964.

85. Lakshminarasimha, C. S. and J. Lucas, *J. Phys. D: Appl. Phys.*, 10, 313, 1977.
86. Robertson, A. G., *Aust. J. Phys.*, 30, 39, 1977.
87. Milloy, H. B., R. W. Crompton, J. A. Rees, and A. G. Robertson, *Aust. J. Phys.*, 30, 61, 1977.
88. Kücükarpaci, H. N. and J. Lucas, *J. Phys. D: Appl. Phys.*, 14, 2001, 1981.
89. Al-Amin, S. A. J. and J. Lucas, *J. Phys. D: Appl. Phys.*, 20, 1590, 1987.
90. Losee, J. R. and D. S. Burch, *Phys. Rev. A*, 6, 1652, 1972.
91. Makabe, T., T. Goto, and T. Mori, *J. Phys. B: At. Mol. Phys.*, 10, 1781, 1977.
92. Wagner, E. B., F. J. Davis, and G. S. Hurst, *J. Chem. Phys.*, 47, 3138, 1967.
93. Parker, J. H. and J. L. Lowke, *Phys. Rev.*, 181, 290, 1969.
94. Lowke, J. L. and J. H. Parker, *Phys. Rev.*, 181, 302, 1969.
95. Baille, P., J. Chang, A. Claude, R. M. Hobson, G. L. Ogram, and A. W. Yau, *J. Phys. B: At. Mol. Phys.*, 14, 1485, 1981.
96. Morgan, F., L. R. Evans, and C. G. Morgan, *J. Phys. D: Appl. Phys.*, 4, 225, 1971.
97. Sakai, Y., H. Tagashira, and S. Sakamoto, *J. Phys. D: Appl. Phys.*, 10, 1035, 1977.
98. Tagashira, H., Y. Sakai, and S. Sakamoto, *J. Phys. D: Appl. Phys.*, 10, 1051, 1977.

Carbon Dioxide (CO_2)

99. Schlumbohm, H., *Z. Phys.*, 182, 317, 1965.
100. Elford, M. T., *Aust. J. Phys.*, 19, 629, 1966.
101. Hake, R. D. and A. V. Phelps, *Phys. Rev.*, 158, 70, 1967.
102. Saelee, H. T., J. Lucas, and J. W. Limbeek, *Solid State Electron Dev.*, 1, 111, 1977.
103. Kucukarpaci, H. N. and J. Lucas, *J. Phys. D: Appl. Phys.*, 12, 2123, 1979.
104. Haddad, G. N. and M. T. Elford, *J. Phys. B: At. Mol. Phys.*, 12, L743, 1979.
105. Elford, M. T. and G. N. Haddad, *Aust. J. Phys.*, 33, 517, 1980.
106. Hasegawa, H., H. Date, M. Shimozuma, K. Yoshida, and H. Tagashira, *J. Phys. D: Appl. Phys.*, 29, 2664, 1996.
107. Hernández-Ávila, J. L., E. Basurto, and J. de Urquijo, *J. Phys. D: Appl. Phys.*, 35, 2264, 2002. The authors have used the pulsed Townsend technique, which is also capable of yielding the ionization and attachment coefficients. The range of E/N covered is 2 to 400 Td and the gases investigated are SF_6, CO_2, and mixtures thereof.
108. Allen, N. L. and B. A. Prew, *J. Phys. B: At. Mol. Phys.*, 3, 1113, 1970.
109. Uman, M. A. and G. Warfield, *Phys. Rev.*, 120, 1542, 1960.
110. English, W. H. and G. C. Hanna, *Can. J. Phys.*, 31, 768, 1935.
111. Lowke, J. J., A. V. Phelps, and B. W. Irwin, *J. Appl. Phys.*, 44, 4664, 1973.
112. Warren, R. W. and J. H. Parker, *Phys. Rev.*, 128, 2661, 1962.
113. Rees, J. A., *Aust. J. Phys.*, 17, 462, 1964.
114. Lakshminarasimha, C. S., J. Lucas, and N. Kontoleon, *J. Appl. Phys. D: Appl. Phys.*, 7, 2545, 1974.
115. Roznerski, W. and J. Mechlinska-Drewko, *J. Phys. (Paris), Colloq. c7*, 40, 149, 1979.
116. Roznerski, W. and J. Mechlinska-Drewko, *Phys. Lett. A*, 70, 271, 1979.
117. Argyropoulos, G. S. and M. A. Casteel, *J. Appl. Phys.*, 41, 4162, 1970.

Carbon Monoxide (CO)

118. Pack, J. L., R. E. Voshall, and A. V. Phelps, *Phys. Rev.*, 127, 2084, 1962.
119. Wagner, E. B., F. J. Davis, and G. S. Hurst, *J. Chem. Phys.*, 47, 3138, 1967.
120. Saelee, H. T. and J. Lucas, *J. Phys. D: Appl. Phys.*, 10, 343, 1977.
121. Roznerski, W. and K. Leja, *J. Phys. D: Appl. Phys.*, 17, 279, 1984.
122. Nakamura, Y., *J. Phys. D: Appl. Phys.*, 20, 933, 1987.
123. Lowke, J. J. and J. H. Parker, Jr., *Phys. Rev.*, 181, 302, 1969. D_L/μ and D_r/μ for several gases are calculated and compared with measured values.
124. Lakshminarasimha, C. S., J. Lucas, and N. Kontoleon, *J. Appl. Phys. D: Appl. Phys.*, 7, 2545, 1974.
125. Saelee, H. T. and J. Lucas, *J. Phys. D: Appl. Phys.*, 10, 343, 1977.

126. Roznerski, W. and J. Mechlinska-Drewko, *Phys. Lett. A*, 70, 271, 1979. The lateral diffusion coefficient is measured.
127. Al-Amin, S. A. J., J. Lucas, and H. N. Kücükarpaci, *J. Phys. D: Appl. Phys.*, 18, 2007, 1985.
128. Herzberg, G., *Spectra of Diatomic Molecules*, Van Nostrand, Princeton. NJ, 1950, Appendix.
129. Wagner, E. B., F. J. Davis, and G. S. Hurst, *J. Chem. Phys.*, 47, 3138, 1967.
130. Kontoleon, W. and J. Lucas, *J. Phys. D: Appl. Phys.*, 5, 956, 1972.
131. Lucas, J. and H. N. Kücükarpaci, *J. Phys. D: Appl. Phys.*, 12, 703, 1979.
132. Mentzoni, M., *J. Phys. D: Appl. Phys.*, 6, 490, 1973.

HELIUM (HE)

133. Bowe, J. C., *Phys. Rev.*, 117, 1411, 1960.
134. Phelps, A. V., J. L. Pack, and L. S. Frost, *Phys. Rev.*, 117, 470, 1960.
135. Pack, J. L. and A. V. Phelps, *Phys. Rev.*, 121, 798, 1961.
136. Stern, R. A., *Proc. 6th International Conference on Phenomena in Ionized Gases*, Serma Paris, 1963, vol. 1, pp. 331–333.
137. Anderson, J. M., *Phys. Fluids*, 7, 1517, 1964.
138. Lowke, J. J. and J. H. Parker, Jr., *Phys. Rev.*, 181, 302, 1969.
139. Hughes, M. H., *J. Phys. B: At. Mol. Phys.*, 3, 1544, 1970.
140. Blum, W., K. Sochting, and U. Stierlin, *Phys. Rev. A*, 10, 491, 1974. The drift velocity of electrons generated by cosmic rays is measured using a spark chamber at low E/N, 0 to 2 Td.
141. Bartels, A., *Appl. Phys.*, 8, 59, 1975.
142. Milloy, H. B. and R. W. Crompton, *Phys. Rev. A.* 15, 1847, 1977.
143. Kücükarpaci, H. N., H. T. Saelee, and J. Lucas, *J. Phys. D: Appl. Phys.*, 14, 9, 1981.
144. Amies, B. W., J. Fletcher, and M. Sugawara, *J. Phys. D: Appl. Phys.*, 18, 2023, 1985.
145. Pack, J. L., R.E. Voshall, A. V. Phelps, and L. E. Kline, *J. Appl. Phys.*, 71, 5363, 1992.
146. Raju, A., internal report (unpublished), University of Windsor, 1985.
147. Foreman, L., P. Kleban, L. D. Schmidt, and H. T. Davis, *Phys. Rev. A*, 23, 1553, 1981.
148. Xiao, D. M., L. L. Zhu, and Y. Z. Chen, *J. Phys. D: Appl. Phys.*, 32, L18, 1999.
149. Milloy, H. B., *J. Phys. D: Appl. Phys.*, 8, L414, 1975.
150. Lymberopoulos, D. P. and J. D. Schieber, *Phys. Rev. E*, 50, 4911, 1994.
151. Liu, J. and Govinda Raju, *J. Franklin Inst.*, 329, 181, 1992.
152. Morgan, F., L. R. Evans, and C. Grey Morgan, *J. Phys. D: Appl. Phys.*, 4, 225, 1971.
153. Capitelli, M., R. Celiberto, C. Gorse, R. Winkler, and J. Wilhelm, *J. Appl. Phys.*, 62, 1987, 4398.
154. McColl, W., C. Brooks, and M. L. Brahe, *J. Appl. Phys.*, 74, 3724, 1993.
155. Jamison, S. P., Jingling Shen, D. R. Jones, R. C. Issac, B. Ersfeld, D. Clark, and D. A. Jaroszynski, *J. Appl. Phys.*, 93, 4334, 2003.

HYDROGEN AND DEUTERIUM

156. Engelhardt, A. G. and A. V. Phelps, *Phys. Rev.*, 131, 2115, 1963. Theoretical calculations of rotational excitation cross sections are given by Gerjuoy, E. and S. Stein, *Phys. Rev.*, 97, 1671, 1955; 98, 1848, 1955.
157. Schlumbohm, H., *Z. Physik*, 182, 317, 1965.
158. Bartels, A., *Appl. Phys.*, 8, 59, 1975.
159. Snelson, R. A. and J. Lucas, *Proc. IEE*, 122, 333, 1975.
160. Blevin, H. A., J. Fletcher, S. R. Hunter, and L. M. Marzec, *J. Phys. D: Appl. Phys.*, 9, 471, 1976.
161. Blevin, H. A., J. Fletcher, and S. R. Hunter, *J. Phys. D: Appl. Phys.*, 9, 1671, 1976.
162. Atrazhev, V. M. and I. T. Yakubov, *J. Phys. D: Appl. Phys.*, 10, 2155, 1977.
163. Hunter, S. R., *Aust. J. Phys.*, 30, 83, 1977.
164. Saelee, H. T. and J. Lucas, *J. Phys. D: Appl. Phys.*, 10, 343, 1977.
165. Hayashi, M., *J. Phys.* (Paris), *Colloq. c7*, 45, 1979.
166. Roznerski, W. and K. Leja, *J. Phys. D: Appl. Phys.*, 17, 279, 1984.
167. Pack, J. L., R. E. Voshall, and A. V. Phelps, *Phys. Rev.*, 127, 2084, 1962.

168. Roznerski, W., J. Mechlińska-Drewco, K. Leja,and Z. Lj. Petrović, *J. Phys. D: Appl. Phys.*, 27, 2060, 1994.
169. Petrović, Z. Lj., T. F. O'Malley, and R. W. Crompton, *J. Phys. B: At. Mol. Opt. Phys.*, 28, 3309, 1995.
170. Pack, J. L., R. E. Voshall, A. V. Phelps, and L. E. Kline, *J. Appl. Phys.*, 71, 5363, 1992.
171. Grünberg, R., *Z. Physik*, 204, 12, 1967.
172. Crompton, R. W., M. T. Elford, and A. I. McIntosh, *Aust. J. Phys.*, 21, 43, 1968.
173. Frommhold, L., *Phys. Rev.*, 172, 118, 1968.
174. Herzberg, G., *Molecular Spectra and Molecular Structure: Spectra of Diatomic Molecules*, Van Nostrand, Princeton, NJ, 1950.
175. Randell, J., S. L. Lunt, G. Mrotzek, J.-P. Ziesel, and D. Field, *J. Phys. D: At. Mol. Opt. Phys.*, 27, 2369, 1994.
176. Virr, L. E., J. Lucas, and N. Kontoleon, *J. Phys. D: Appl. Phys.*, 5, 542, 1972.
177. Kontoleon, N., J. Lucas, and V. E. Virr, *J. Phys. D: Appl. Phys.*, 6, 1237, 1973.
178. Petrović, Lj. and R. W. Crompton, *Aust. J. Phys.*, 42, 4999, 1989.
179. Herzberg, G., *Molecular Spectra and Molecular Structure: Spectra of Diatomic Molecules*, Van Nostrand, Princeton, NJ, 1950, pp. 142.
180. Buckman, S. J. and A. V. Phelps, *J. Chem. Phys.*, 82, 4999, 1985.

Krypton (Kr)

181. Bowe, J. C., *Phys. Rev.*, 117, 1411, 1960.
182. Pack, J. L., R. E. Voshall, and A. V. Phelps, *Phys. Rev.*, 127, 2084, 1962.
183. Frost, L. S. and A. V. Phelps, *Phys. Rev.*, 136, A1538, 1964.
184. Kücükarpaci, H. N. and J. Lucas, *J. Phys. D: Appl. Phys.*, 14, 2001, 1981.
185. Al-Amin, S. A. J. and J. Lucas, *J. Phys. D: Appl. Phys.*, 20, 1590, 1987.
186. Hunter, S. R., J. G. Carter, and L. G. Christophorou, *Phys. Rev. A*, 38, 5539, 1988.
187. Suzuki, M., T. Taniguchi, and H. Tagashira, *J. Phys. D: Appl. Phys.*, 22, 1848, 1989.
188. (a) Pack, J. L., R. E. Voshall, A. V. Phelps, and L. E. Kline, *J. Appl. Phys.*, 71, 5363, 1992; Nakamura, Y. and A. Naitoh (cited as ref. 42 of Pack et al.[188a])
189. Dutt, T. L., *J. Phys. B: At. Mol. Phys.*, 2, 234, 1969.

Metallic Vapors

Cesium (Cs)

190. Stwertka, A., *A Guide to the Revised Edition*, Oxford University Press, Oxford, 1998.
191. Nakamura, T. and J. Lucas, (a) *J. Phys D: Appl. Phys.*, 11, 325, 1978; (b) *J. Phys. D: Appl. Phys.*, 11, 337, 1978. Boltzmann analysis was used in the second paper and the momentum transfer cross section that was used to obtain agreement was higher by a factor of 2.8.
192. Vidal, C. R., *Physik uns. Zeit*, 3, 174, 1972.
193. Chanin, L. M. and R. D. Steen, *Phys. Rev.*, 136, A138, 1964.
194. Postma, A. J., *Physica* (Utrecht), 44, 38, 1969.
195. Saelee, H. T. and J. Lucas, *J. Phys. D: Appl. Phys.*, 12, 1275, 1979.
196. Zecca, A., G. P. Karwasz, and R. S. Brusa, *Riv. Nuovo Cimento*, 19, 1, 1996.
197. Crown, J. C. and A. Russek, *Phys. Rev. A*, 45, 197, 1992.
198. Brode, R. B., *Phys. Rev.*, 34, 673, 1929. Vapors studied are sodium, potassium, rubidium, and cesium. Brode's experimental setup has been shown in Figure 2.1 of this volume.
199. Visconti, P. J., J. A. Slevin, and K. Rubin, *Phys. Rev. A*, 3, 1310, 1971. Total cross sections are provided for cesium. rubidium, and potassium in the energy range 0 to 10 eV by the crossed beam atom recoil technique, which involves measuring the energy of scattered atoms.
200. Jaduszliwer, B. and Y. C. Chan, *Phys. Rev. A*, 45, 197, 1992. The atom recoil technique was adopted and the energy range covered was 2 to 18 eV.
201. Nighan, W. L. and A. J. Postma, *Phys. Rev. A*, 6, 2109, 1972. Electron drift velocity was analyzed to obtain the momentum transfer cross sections in the energy range 0 to 1 eV. The maximum cross section of $2500 \times 10^{-20}\,\text{m}^2$ is observed at 0.25 eV.

202. Stefanov, B., *Phys. Rev. A*, 22, 427, 1980. An interesting approach was adopted to derive the momentum transfer cross sections in the energy range 0.05 to 2 eV. Six quantities, all of which involve the momentum transfer cross sections, were considered. The six quantities are: (a) width of the electron cyclotron resonance; (b) attenuation of microwaves; (c) electrical conductivity in both equilibrium and nonequilibrium pure Cs or Ar–Cs plasmas; (d) electron thermal conductivity in a strong magnetic field; and (f) electron drift velocity. An algorithm was generated to obtain the best fit with the experimental data. The Ramsauer–Townsend minimum was observed at 0.278 eV and the minimum cross section was 72×10^{-20} m^2. Two maxima are observed at 0.15 eV and 0.45 eV. Tabulated results are given in terms of the velocity of the electron. The conversion factor is ε (eV) = $[W\,(10^7)$ cm s$^{-1}/5.93\}^2$.

203. Bartschat, K. J., *J. Phys. B: At. Mol. Opt. Phys.*, 26, 3595, 1993.

204. Chen, S. T. and A. C. Gallagher, *Phys. Rev. A*, 17, 551, 1978.

205. Brink, G. O., *Phys. Rev.*, 134, A345, 1964.

206. McFarland, R. H. and J. D. Kinney, *Phys. Rev.*, 137, A1058, 1965.

207. Nygaard, J. K., *J. Chem. Phys.*, 49, 1995, 1968.

Mercury (Hg)

208. Killian, T. J., *Phys. Rev.*, 35, 1238, 1930.

209. Klarfeld, B., *Tech. Phys.* (USSR), 5, 913, 1938.

210. McCutchen, C. W., *Phys. Rev.*, 112, 1848, 1958.

211. Kerzar, B. and P. Weissglas, *Electron. Lett.*, 1, 43, 1965.

212. Rockwood, S. D., *Phys. Rev. A*, 8, 2348, 1973.

213. Judd, O., *J. Appl. Phys.*, 47, 467, 1976.

214. Garamoon, A. A. and A. S. Abdelhaleem, *J. Phys. D: Appl. Phys.*, 12, 2181, 1979.

215. Sakai, Y., S. Sawada, and H. Tagashira, *J. Phys. D: Appl. Phys.*, 22, 276, 1989.

216. Liu, J. and G. R. Govinda Raju, *J. Phys. D: Appl. Phys.*, 25, 167, 1992.

217. Liu, J, and G. R. Govinda Raju, *Can. J. Phys.*, 71, 571, 1993.

218. Tagashira, H, Y. Sakai, and S. Sakamoto, *J. Phys. D: Appl. Phys.*, 10, 1051, 1977.

219. Liu, J., Ph.D. thesis, University of Windsor, 1993.

220. Vriens, L., R. A. J. Keijser, and F. A. S. Lighthart, *J. Appl. Phys.*, 49, 3807, 1978.

NEON (NE)

221. Bowe, J. C., *Phys. Rev.*, 117, 1411, 1960.

222. Pack, J. L. and A. V. Phelps, *Phys. Rev.* 121, 798, 1961.

223. Anderson, J. M., *Phys. Fluids*, 7, 1517, 1964.

224. Sugawara, M. and C. J. Chen, *J. Appl. Phys.*, 41, 3442, 1970.

225. Robertson, A. G., *J. Phys. B: At. Mol. Phys.*, 5, 648, 1972.

226. Kücükarpaci, H. N., H. T. Saelee, and J. Lucas, *J. Phys. D: Appl. Phys.*, 14, 9, 1981.

227. Sakai, Y., S. Sawada, and H. Tagashira, *J. Phys. D: Appl. Phys.*, 19, 1741, 1986.

228. Puech, V. and S. Mizzi, *J. Phys. D: Appl. Phys.*, 24, 1974, 1991.

NITROGEN

229. Bowe, J. C., *Phys. Rev.*, 117, 1411, 1960.

230. Pack, J. L. and A. V. Phelps, *Phys. Rev.*, 121, 798, 1961.

231. Heylen, A. E. D., *Proc. Phys. Soc.* (London), 79, 284, 1962.

232. Engelhardt, A. G., A. V. Phelps, and C. G. Risk, *Phys. Rev.*, 158, A1566, 1964.

233. Schlumbohm, H., *Z. Phys.*, 182, 317, 1965.

234. Snelson, R. A. and J. Lucas, *Proc. IEE*, 122, 107, 1975.

235. Raju, G. R. Govinda and S. Rajapandian, *Int. J. Electron.*, 40, 65, 1976.

236. Raju, G. R. Govinda and G. R. Gurumurthy, *IEEE Trans. Plasma Sci.*, 4, 241, 1976.

237. Taniguchi, T., H. Tagashira, and Y. Sakai, *J. Phys. D: Appl. Phys.*, 11, 1757, 1978.

238. (a) Kücükarpaci, H. N. and J. Lucas, *J. Phys. D: Appl. Phys.*, 12, 2123, 1979; (b) Saelee, H. T., J. Lucas, and J. W. Limbeck, *IEE Solid-state Electron Dev.*, 1, 111, 1977.
239. Tagashira, H., T. Taniguchi, and Y. Sakai, *J. Phys. D: Appl. Phys.*, 13, 235, 1980.
240. Raju, G. R. Govinda and R. Hackam, *J. Appl. Phys.*, 52, 3912, 1981.
241. Novak, J. P. and M. Frechette, *J. Appl. Phys.*, 55, 107, 1984.
242. Rosnerski, W. and K. Leja, *J. Phys. D; Appl. Phys.*, 17, 279, 1984.
243. Phelps, A. V. and L. C. Pitchford, *Phys. Rev.*, 31, 2932, 1985.
244. Nakamura, Y. and M. Kurachi, *J. Phys. D: Appl. Phys.*, 21, 718, 1988.
245. Nakamura, Y., *J. Phys. D: Appl. Phys.*, 20, 933, 1987.
246. Fletcher, J. and I. D. Reid, *J. Phys. D: Appl. Phys.*, 13, 2275, 1988.
247. Liu, J. and G. R. Govinda Raju, *J. Franklin Inst.*, 329, 181, 1992.
248. Liu, J. and G. R. Govinda Raju, *IEEE Trans. Elec. Insul.*, 28, 154, 1993.
249. Hasegawa, H., H. Date, M. Shimozuma, K. Yoshida, and H. Tagashira, *J. Phys. D: Appl. Phys.*, 29, 2664, 1996.
250. Roznerski, W., *J. Phys. D: Appl. Phys.*, 29, 614, 1996.
251. Tanaka, Y., *J. Phys. D: Appl. Phys.*, 37, 851, 2004. The Boltzmann energy distribution is calculated for N_2, O_2 and mixtures thereof in the temperature range of 300 to 3500 K.
252. Lawson, P. A. and J. Lucas, *Proc. Phys. Soc.*, 85, 177, 1965. Huxley and Crompton[3] (footnote, p. 381) comment that the procedure adopted by these authors contains a fallacy, though the final expression is correct.
253. Blevin, H. A., J. Fletcher, S. R. Hunter, and L. M. Marzec, *J. Phys. D: Appl. Phys*, 11, 2295, 1978. Earlier references for the photon technique can be found in this paper. The photons generated in a hydrogen avalanche are mostly in the ultraviolet region and to detect these photons by photomultiplier a layer of sodium salicylate having a conversion efficiency of 94% for radiation in the region 80 to 300 nm, to visible wavelength, was applied to the inside of the window.
254. Fletcher, J. and I. D. Reid, *J. Phys. D: Appl. Phys.*, 13, 2275, 1980.
255. Cartwright, D. C., S. Trajmar, A. Chutjian, and W. Williams, *Phys. Rev. A*, 16, 1041, 1977.
256. Cartwright, D. C., *J. Appl. Phys.*, 49, 3855, 1978.
257. Chutjian, A., D. C. Cartwright, and S. Trajmar, *Phys. Rev. A*, 16, 1052, 1977.
258. Trajmar, S., D. F. Register, and A. Chutjian, *Phys. Rep.*, 97, 219, 1983.
259. Itikawa, Y., M. Hayashi, A. Ichimura, K. Onda, K. Sakimoto, K. Takayanayi, M. Nakamura, H. Nishimura, and T. Takayanagi, *J. Phys. Chem. Ref. Data*, 15, 985, 1986.
260. Spence, D. and G. J. Schulz, *J Chem. Phys.*, 58, 1800, 1973.
261. Raju, A., *Basic Data on Gas Discharges*, Internal Report, University of Windsor, 1985. Analytical expressions for ion and electron drift velocities in N_2 are given by Morrow, R., and J. J. Lowke, *J. Phys. D: Appl. Phys.*, 14, 2027, 1981.
262. Dalgarno, A. and R. J. Moffett, *Indian Academy of Sciences Symposium on Collision Processes*, 1962. Unfortunately the original publication is not available and Equations 6.108 and 6.109 have been cited from Engelhardt and Phelps.[156]
263. Lowke, J. J., A. V. Phelps, and B. W. Irwin, *J. Appl. Phys.*, 44, 4664, 1973.
264. Hasegawa, H., H. Date, Y. Ohmori, P. L. G. Ventzek, M. Shimozuma, and H. Tagashira, *J. Phys. D: Appl. Phys.*, 31, 737, 1998.
265. Grünberg, R., (a) *Z. Naturforsch.* 23a, 1994, 1968; (b) *Z. Naturforsch.*, 24a, 1838, 1968.
266. Allen, N. L. and B. A. Prew, *J. Phys. B: At. Mol. Phys.*, 3, 1113, 1970.
267. Warren, R. W. and J. H. Parker, *Phys. Rev.*, 128, 2661, 1962.
268. Naidu, M. S. and A. N. Prasad, *Brit. J. Appl. Phys.*, 1, 763, 1968.
269. Novak, J. P. and M. Frechette, *J. Appl. Phys.*, 55, 107, 1984.
270. Al-Amin, S. A. J., J. Lucas, and H. N. Kücükarpaci, *J. Phys. D: Appl. Phys.*, 18, 2007, 1985.
271. Wedding, A. B., H. A. Blevin, and J. Fletcher, *J. Phys. D: Appl. Phys.*, 18, 2361, 1985.

OXYGEN (O_2)

272. Nielsen, R. A. and N. E. Bradbury, *Phys. Rev.*, 51, 69, 1937.
273. Schlumbohm, H., *Z. Phys.*, 182, 317, 1965.

274. Hake, R. D., Jr., and A. V. Phelps, *Phys. Rev.*, 158, 70, 1967.

275. Naidu, M. S. and A. N. Prasad, *J. Phys. D: Appl. Phys.*, 3, 957, 1970.

276. Roznerski, W. and K. Leja, *J. Phys. D: Appl. Phys.*, 17, 279, 1984.

277. Al-Amin, S. A. J., H. N. Kücükarpaci, and J. Lucas, *J. Phys. D: Appl. Phys.* 18, 1781, 1985.

278. Liu, J. and G. R. Govinda Raju, *Can. J. Phys.*, 70, 216, 1992.

279. Liu, J. and G. R. Govinda Raju, *IEEE Trans. Elec. Insul.*, 28, 154, 1993.

280. Jeon, B. and Y. Nakamura, *J. Phys. D: Appl. Phys.*, 31, 2145, 1998.

281. Chanin, L. M., A. V. Phelps, and M. A. Biondi, *Phys. Rev.*, 128, 1962, 219.

282. Herzberg, G., *Molecular Spectra and Molecular Structure*, Van Nostrand, New York, 1950.

283. Naidu, M. S. and A. N. Prasad, *J. Phys. D: Appl. Phys.*, 3, 957, 1970.

284. Fleming, I., A, D. R. Gray, and J. A. Rees, *J. Phys. D: Appl. Phys.* 5, 291, 1972.

285. Lowke, J. J. and J. H. Parker, *Phys. Rev.*, 181, 302, 1969.

286. Nelson, D. R. and F. J. Davis, *J. Chem. Phys.*, 57, 4079, 1972.

287. Price, D. A., J. Lucas, and J. L. Moruzzi, *J. Phys. D: Appl. Phys.*, 6, 1514, 1973.

XENON (XE)

288. Pack, J. L., R. E. Voshall, and A. V. Phelps, *Phys. Rev.*, 127, 2084, 1992.

289. Huang, S. S.-S. and G. R. Freeman, *J. Chem. Phys.*, 68, 1355, 1978.

290. Bowe, J. C., *Phys. Rev.*, 117, 1411, 1960.

291. Brooks, H. L., M. C. Cornell, J. Fletcher, L. M. Littlewood, and K. J. Nygaard, *J. Phys. D: Appl. Phys.*, 15, L51, 1982.

292. Hunter, S. R., J. G. Carter, and L. G. Christophorou, *Phys. Rev.*, 38, 5539, 1988.

293. Pack, J. L., R. E. Voshall, A. V. Phelps, and L. E. Kline, *J. Appl. Phys.*, 71, 5363, 1992.

294. Makabe, T. and T. Mori, *J. Phys. B: At. Mol. Phys.*, 11, 3785, 1978.

295. Puech, V. and S. Mizzi, *J. Phys. D: Appl. Phys.*, 24, 1974, 1991.

296. Santos, F. P., T. H. V. T. Dias, A. D. Stauffer, and C. A. N. Conde, *J. Phys. D: Appl. Phys.*, 27, 42, 1994.

297. Xiao, D. M., L. L. Zhu, and X. G. Li, *J. Phys. D: Appl. Phys.*, 33, L145, 2000.

298. Xiao, D. M., L. L. Zhu, and Y. Z. Chen, *J. Phys. D: Appl. Phys.*, 32, L18, 1999.

7 Drift and Diffusion of Electrons—II: Complex Molecules

We continue the discussion of drift and diffusion of electrons in gases that have complex molecules. A theoretical framework has been established so that we can present data in a compressed manner, wherever possible. The chapter begins with an introduction to the pulse method of measuring the drift velocity, as the technique still needs to be employed to study a large number of molecules for which data are incomplete or for which there is a need to update the older data with respect to accuracy and range of E/N to be covered. Drift velocity is measured by the use of the following techniques[1]:

1. Electrical shutter method. The method was introduced by Bradbury and Nielsen and was applied to a number of gases. A variation of the method is the double shutter method, in which two closely spaced shutters are placed near the cathode and anode. The distance between them is the drift distance.
2. Detection of the photon pulse that accompanies the electron pulse. The method has advantages at high values of E/N where secondary electron effects are a source of error.
3. Time-of-flight method and the pulsed Townsend technique. The technique is conceptually quite simple in that it follows from the definition of the drift velocity, that is the drift distance divided by the transit time. However, considerable efforts are required to ensure that back diffusion of electrons to the cathode at low voltages, and also diffusion, particularly at low gas pressures, are properly taken into account. The method has the advantage of measuring the ionization coefficient and lateral diffusion coefficient in addition to its applicability over a wide range of E/N. Recent users of this technique are Xiao et al.[2] and Hernández-Ávila and Urquijo,[3] who have made measurements up to 500 Td.
4. The method of arrival time distribution. In principle this method is a variation of the time-of-flight technique. Electrons are released from the cathode at known intervals and arrive at the anode as well separated groups. The time taken for the drift is a function of both W and D_L. This technique has been employed by Nakamura[4] and Hasegawa et al.[5]

7.1 CURRENT PULSE DUE TO AVALANCHE

Let us consider parallel plane electrodes that have been designed according to the well-known Rogowski profile to give uniform electric fields (Figure 7.1a). A pulse of light is allowed to fall on the cathode and the electrons released at the cathode drift toward the anode. As in the steady-state method, a resistor is placed in series with the collecting electrode and the voltage drop across the resistor due to the pulse is displayed on an oscilloscope, following suitable amplification. The equivalent circuit for this arrangement is shown in Figure 7.1a.

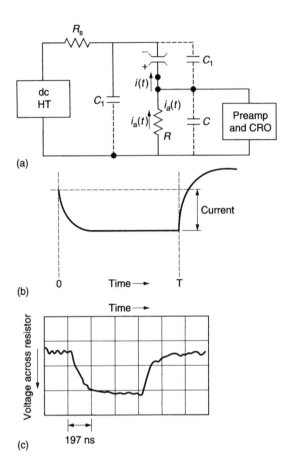

FIGURE 7.1 Schematic of the experimental setup for measurement of drift velocity by the pulsed Townsend method. (a) Circuit diagram. R is the measuring resistance and C is the input capacitance of the measuring system. The stray capacitances are lumped as C_1. (b) Theoretical current pulse. (c) Experimentally obtained pulse.[10]

In an ideal case, the voltage pulse that appears at the oscilloscope is a faithful reflection of the electron pulse that traverses the gap. However, stray capacitances are invariably present. These are[6]

- $C_s =$ stray capacitance of the cathode to ground;
- $C_g =$ capacitance of the gap;
- $C =$ stray capacitance of the input circuit and preamplifiers to ground.

The current $i_a(t)$ flowing through the input resistor is given by the equation

$$R(C + C_g)\frac{di_a(t)}{dt} + i_a(t) = i(t) \tag{7.1}$$

The solution of this equation gives $i_a(t)$ but one must first substitute an expression for $i(t)$, which is the current in the gap. This current is

$$i(t) = \frac{NeW}{d} \tag{7.2}$$

where N = number of electrons between the electrodes, e is the electronic charge, W the drift velocity, and d the gap distance. The solution of Equations 7.1 and 7.2 shows that the voltage across the measuring resistance initially rises according to

$$V_R = i_R R = R\frac{NeW}{d}\left[1 - \exp\left\{\frac{-t}{R(C+C_g)}\right\}\right] \tag{7.3}$$

and attains an approximately constant value of

$$V_R = \left|R\frac{NeW}{d}\right| \tag{7.4}$$

until $t = T$, where T is the electron transit time. For $t > T$ the voltage falls according to

$$V_R(t) = R\frac{NeW}{d}\exp\left\{\frac{-t}{R(C+C_g)}\right\} \tag{7.5}$$

The general shape of the electron pulse is shown in Figure 7.1b, together with an experimentally obtained oscillogram in nitrogen, Figure 7.1c. Note that the rise of current is toward the negative axis. It is assumed that the duration of the ultraviolet pulse is zero though in practice it is small enough to justify this assumption. The rise time 2.2 $R(C+C_g)$ of the input circuit of the preamplifier is much smaller than the electron transit time T.

The method is also applicable to measurement of the drift velocity of ions. The mobility of the electrons is much larger than that of ions, and after the electrons have reached the anode the positive and negative ions left behind arrive at the cathode and anode respectively. The method was first adopted by Hornbeck[8] in 1951. The detailed theory was given in 1964 by Raether,[9] who showed that in a real system the large initial electron peak could be eliminated by suitable design of the measurement circuit. The positive ion current will remain at reasonably constant amplitude if the measuring resistance and the capacitances associated with the input to the measurement circuit are appropriately chosen. The condition that is satisfied with such design is[10]

$$\alpha W + R(C+C_g) = 1 \tag{7.6}$$

provided also that $\exp(\alpha d) \gg 1$.

The rate of fall of the electron pulse (Figure 7.1b) is dependent on the electron diffusion and this must be considered before evaluating the drift velocity. If diffusion were insignificant, the electron transit time could be obtained by measuring the time from the point at which the voltage began to rise to the point at which it began to fall. This is the time of travel of electrons from their first release from the cathode to the arrival of the same electrons at the anode. If diffusion is appreciable, this method yields a drift time that is too small, since it would include, on the average, the effects of those electrons that happened to diffuse in the forward direction. As we have seen, the effect of diffusion is greatest at lower pressure and its effects should be considered.[11]

The theory outlined by Equation 7.5 is rudimentary, exposing the idea of the experimental method. The theory that includes the effects of longitudinal diffusion, ionization, and attachment has been considered by the pioneering group of Raether[9] and revived by Christophorou et al.[21] and more recently by de Urquijo and colleagues.[12] A summary of these developments is given below.

The voltage across the resistor R (Figure 7.1 (a)) is measured by two different methods. In the first, generally known as the "integrated method," the time constant of the preamplifier input is so chosen to satisfy the condition that $RC > T_e$ ($\sim 10^{-7}$ to 10^{-5} s) or T_+ ($\sim 10^{-4}$ to 10^{-2} s) or T_- ($\sim T_+$), where T_e, T_+, and T_- are the transit times of the electron, positive ion, and negative ion, respectively. If ionization occurs, then the condition for the integrated method is $RC > (\alpha W_e)^{-1}$, where α and W_e are the ionization coefficient (m^{-1}) and electron drift velocity respectively. The drift velocities of positive and negative ions are approximately the same. This is the method adopted by Hunter et al.,[11] who used $RC = 1$ s.

The second method, generally known as the "differentiated method," applies for the condition $RC < T_e$ if no ionization occurs or $RC < (\alpha W_e)^{-1}$ in the presence of ionization.[12] The expressions for the voltage across the measuring resistor and the observed wave forms are different for each case, as is shown in Figure 7.2.

7.1.1 ELECTRON CURRENT (INTEGRATING MODE)

Figure 7.2 shows (schematically) the current wave shape in a gas in which ionization and attachment occur. The voltage is measured in the integrating mode. The electron current begins to rise steeply and during a single transit time T_e the contribution of the positive and negative ions to the total current is negligibly small. A break will occur in the wave shape from which T_e can be determined. In practice a distinct break will become noticeable only at high N (low E/N, therefore T_e is long) and when α and η are small. T_+ or T_- is determined from the most significant rounding of the current waveform.

Hunter et al.[11] give the expression for the voltage across R_a as

$$V(T_e) = -\frac{eN_0}{C_0(\rho - \pi_L)d} \times \left[1 - \left(\frac{2\rho}{\rho + \lambda_L}\right) \exp[-(\rho - \lambda_L)d]\right] \quad (7.7)$$

where C_0 is the input capacitance, N_0 the initial number of electrons, d the gap length, and

$$\rho^2 = \lambda_L^2 + 2\lambda_L(\eta - \alpha) \quad (7.8)$$

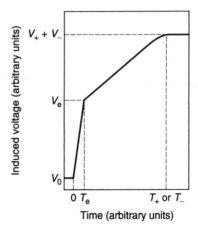

FIGURE 7.2 Voltage induced across the measuring resistor in the integrating mode ($RC \gg (\alpha W_+)^{-1}$. V_e is the voltage due to the electrons, V_+ and V_- are the voltages due to positive and negative ions respectively.

Here $\lambda_L = W/2D_L$ and D_L is the longitudinal diffusion coefficient. If diffusion is not significant, Equation 7.7 reduces to

$$V(T_e) = -\frac{eN_0}{C_0(\eta - \alpha)d}\{1 - \exp[(\alpha - \eta)d]\} \tag{7.9}$$

7.1.2 ELECTRON CURRENT (DIFFERENTIAL MODE)

Figure 7.3 shows the definitions of transit time and current due to each species of charge carrier. The current rises sharply and falls to zero at the electron transit time T_e. The positive ion current is relatively constant and falls to zero at T_+; the negative ion current decays with a time constant, essentially till $T_- \approx T_+$. The difference in time between T_+ and T_-, which is negligible in reality, is shown exaggerated for the sake of clarity.

Figure 7.4 shows the influence of the time constant of the measurement circuit on the voltage pulse. The voltage pulse is due to the total current, comprising the charge carrier component, $I_e + I_+ + I_-$, and the charging current $I_c = CdV/dt$. If $RC \ll (\alpha W_e)^{-1}$ then the voltage transient is a mirror of the electron current pulse as in Figure 7.1, with the positive ion current not being observable. The condition $RC \ll (\alpha W_e)^{-1}$ is realized by reducing R. The curve in Figure 7.4a applies to the case of $RC \ll (\alpha W_+)^{-1}$ and the voltage pulse is due to currents $I_+ + I_- + I_c$. The curve in Figure 7.4c applies to the case of $RC = (\alpha W_+)^{-1}$ and the curve in Figure 7.4b to the case of $RC \gg (\alpha W_+)^{-1}$. This is the integral mode adopted by Hunter et al.,[11] referred to previously.

Figure 7.5 shows the shape of the electron current in the differentiated pulse mode. The condition satisfied here is that

$$CR < (\alpha W_e)^{-1} \tag{7.10}$$

The expression for current has been derived by Brambring[13] and also used by Hernández-Ávila[14]

$$I_e(t) = P\{Q + R\} \tag{7.11}$$

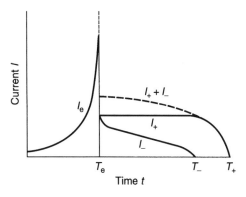

FIGURE 7.3 Current due to charge carriers in an avalanche. The electron transit time T_e is shown expanded for clarity. The negative ion transit time is also shown smaller for the same reason. The total ion current $(I_+ + I_-)$ is shown by the broken line. The initial spike due to electron current may be completely suppressed by a suitable choice of time constant for the measuring circuit.

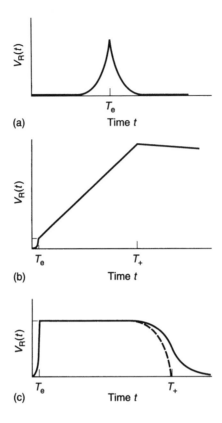

FIGURE 7.4 Theoretical voltage shapes across the measuring resistor for three different time constants. (a) Electron current. T_e is the electron transit time; $RC \ll (\alpha W_e)^{-1}$. The voltage pulse mirrors I_e because the capacitive current is negligible because of the small RC. Compare with Figure 7.5a. (b) Here $RC \gg (\alpha W_+)^{-1}$. The voltage pulse is due to electron avalanche. The electron component reaches its maximum at T_e. The ionic component rises linearly because the number of ions is constant. Compare with Figure 7.2. (c) Here, $RC = (\alpha W_+)^{-1}$. The electron current is integrated and the voltage reaches its maximum at T_e. Beyond T_e the current remains constant, falling to zero at $(T_e + T_+) \approx T_+$. Compare with Figure 7.5b.

where

$$P = \frac{N_0 e}{2T_e} \exp(\alpha W_e t) \tag{7.12}$$

$$Q = \left[1 - \phi\left(\frac{(W_e + \alpha D_L)t - d}{\sqrt{4 D_L t}}\right)\right] \tag{7.13}$$

$$R = \exp\left(\frac{W_e + \alpha D_L}{D_L} d\right)\left[\phi\left(\frac{(W_e + \alpha D_L)t + d}{\sqrt{4 D_L t}}\right) - 1\right] \tag{7.14}$$

In Equations 7.13 and 7.14 the function ϕ is the error function of argument x, defined by

$$\phi(x) = \frac{2}{\sqrt{\pi}} \int_0^x \exp(-u^2) du \tag{7.15}$$

All other symbols in Equations 7.12 to 7.14 have been defined.

Figure 7.5 is an actual electron pulse recorded by Hernández-Ávila et al.[14] in CO_2 along with computed current. Since the laser pulse has a finite width, instead of the theoretically assumed zero pulse width and other instrumental effects the rise time is not as sharp as the

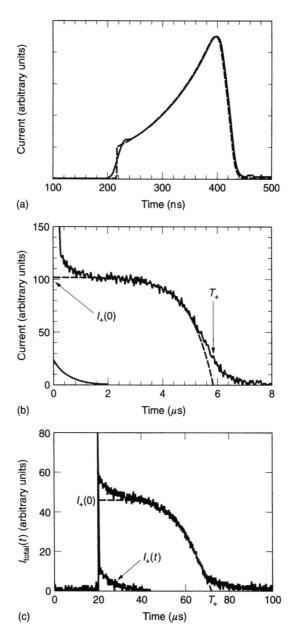

(a)

(b)

(c)

FIGURE 7.5 Experimental current wave shapes in gases. (a) Electron current pulse in CO_2, with the fitting curve according to Equation 7.11. The electron transit time is 197 ns. Conditions of measurement are: $E/N = 130$ Td, $N = 4.1 \times 10^{23}$ m^{-3}, $d = 3$ cm.[14] (b) Ion current pulse in methane for $E/N = 410$ Td, $N = 5.8 \times 10^{22}$ m^{-3}, $d = 2$ cm. Top curve is positive ion current and bottom curve is negative ion current.[12] (c) Ionic currents in 50:50 mixtures of $SF_6 + N_2$. Note the dominant negative ion current compared to the positive ion current. Negative ion transit time is shown as T_-.[15] With acknowledgment to the authors.

theoretical one. However, the pulse is remarkably free of noise, which increases the accuracy. The fall is affected by the longitudinal diffusion effects which, for constant E/N, become more apparent as the gas pressure is reduced. When diffusion is negligible or absent, Equation 7.11 reduces to

$$I_e(t) = \begin{cases} \dfrac{N_0 e}{T_e} \exp[(\alpha - \eta)W_e t] & 0 \leq t \leq T_e \\ \neq 0 & t > T_e \end{cases} \tag{7.16}$$

where T_e is the time at which the electron current goes to zero. Hernández-Ávila[14] obtained the initial values of α and W_e from Equation 7.16 and refined them by resorting to the complete solution given by Equation 7.11.

7.1.3 ION CURRENTS (DIFFERENTIAL MODE)

The positive ion current is[9]

$$I_+(t) = \frac{N_0 e}{T_+} \frac{\alpha}{(\alpha - \eta)} [\exp(\alpha - \eta)d - \exp(\alpha - \eta)W_+ t]; \quad T_e \leq t \leq T_+ + T_- \tag{7.17}$$

where $T_+ + T_- \approx 2T_+$ is the duration for which the ion current flows. The negative ion current is

$$I_-(t) = \frac{N_0 e}{T_-} \frac{\eta}{(\alpha - \eta)} \left\{ \exp[(\alpha - \eta)(d - W_- t)] - 1 \right\}; \quad T_e \leq t \leq T_- + T_e \tag{7.18}$$

The negative ion current decreases with the time constant

$$\tau_- = \frac{1}{(\alpha - \eta)W_-} \tag{7.19}$$

The negative ion current decreases slowly after the electrons have entered the anode. From Equations 7.16 to 7.18 the ratios of currents yield

$$\frac{\alpha}{\alpha - \eta} = \frac{I_+(T_e)T_+}{I_e(T_e)T_e} \tag{7.20}$$

$$\frac{\eta}{\alpha} = \frac{I_-(T_e)T_-}{I_+(T_e)T_+} \tag{7.21}$$

The transient voltage that appears across the measuring resistor is given by Raether[9] and is shown in Figure 7.4. The experimental wave shapes observed are shown in Figure 7.5. The exact expressions for voltage that apply to these conditions are given by Raether,[9] Hunter et al.,[11] and de Urquiho et al.[15]

7.2 ARRIVAL TIME SPECTRUM METHOD

This method, for the simultaneous measurement of drift velocity and longitudinal diffusion coefficient, operates on the principle that an electron leaving a cathode arrives at the anode after a certain time delay. If successive electrons are released at intervals longer than the drift time, observation of the arrival time of these electrons will yield an average drift velocity and the longitudinal diffusion coefficient. The modern version of this technique, as used by Yamaji and Nakamura,[16] is described with respect to Figure 7.6.

Initial electrons are generated by an ultraviolet light that falls on the back side of a metallized quartz disc. The electrons pass through a double shutter comprising two grids separated by a small distance (1 cm). In earlier measurements Nakamura[4] had used a hot filament to supply the electrons. The grids are meant to serve the purpose of relaxing the energy of the electrons so that they are in equilibrium with the electric field as they enter the drift region.

The two shutters are normally biased in such a way that the electrons are blocked there. This is accomplished by keeping the upper grid positive relative to the lower grid of shutter 1; the lower grid of shutter 2 is kept negative with respect to the upper grid. The voltages of the two pairs of grids are chosen such that no current flows to the collector while the other pair is unbiased. A typical bias pulse to the lower grid of S_2 cancels the blocking bias and electrons pass through shutters when the pulses are on. The pulse height to each shutter can be tuned to maximize the signal current to the collector.

The pulses to the two shutters have the same duration. The delay time of the second-shutter pulse relative to the first is slowly varied by a scan delay generator which feeds a ramp voltage, proportional to the delay time, to the horizontal input of a chart recorder. The collector current is proportional to the number of electrons arriving at the second shutter at a certain delay time after they started at the first shutter. The current is amplified

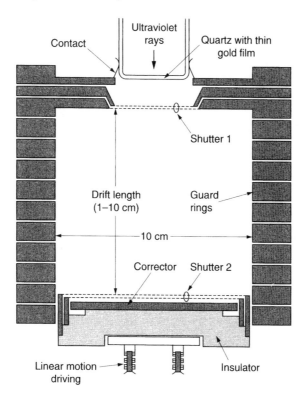

FIGURE 7.6 Setup for the measurement of transport parameters by the arrival time distribution method. A double grid separated by a small distance is employed at each end of the drift distance. Guard rings ensure a uniform electric field. Figure reproduced from Yamaji, M. and Y. Nakamura, *J. Phys. D*, 36, 640, 2003. With permission of IOP, U.K.

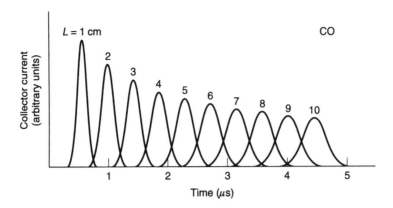

FIGURE 7.7 Arrival time spectrum of electron avalanche in CO for the measurement of W and ND_L. Reproduced from Nakamura, J., *J. Phys. D*, 20, 933, 1987. With permission of IOP, U.K.

and is recorded as a function of the delay time on the chart recorder. The arrival time spectrum obtained by Nakamura[4] in CO is shown in Figure 7.7. A plot of the peak time as a function of drift distance gives W and ND_L. Nakamura's[4] setup was designed such that the drift distance was variable up to 10 cm.

7.3 HYDROCARBON GASES

The hydrocarbons considered in this section have the general formula C_xH_y. Gases that belong to the category $y = 2x + 2$ are methane (CH_4), ethane (C_2H_6), propane (C_3H_8), n-butane (C_4H_{10}), and isobutane. Gases that belong to the category $y = 2x$ are ethene, also known as ethylene (C_2H_4), propene, also known as propylene (C_3H_6), and cyclopropane (C_3H_6). Acetylene (C_2H_2) belongs to the category $y = 2x - 2$ and has a triple bond. Table 7.1 shows the general classification of hydrocarbons and it is clear that an exhaustive treatment is not attempted. Selected physical properties of the gases are shown in Table 5.5.

A number of industrial applications of electrical discharges in hydrocarbons are being investigated. The following applications are being explored:

1. Hydrocarbon plasmas are employed for various applications such as hard coatings and polymer deposition. Nanosized particles are formed in the plasma; such plasmas are referred to as "dusty" plasmas in the literature.[17]
2. A low-temperature electrical discharge is an efficient means of decomposing hydrocarbons into harmless, or at least less toxic, compounds. For example, hydrocarbon contaminants are converted to CO_2, CO, and H_2O in electrical corona

TABLE 7.1
Four Types of Hydrocarbon

Type	Name	Formula	Saturation
Aliphatic	Alkane	C_nH_{2n+2}	Saturated
Open chain	Alkene	C_nH_{2n}	Unsaturated
	Alkyne	C_nH_{2n-2}	Unsaturated
Aromatic	—	—	Unsaturated

discharges. The electrons in such a discharge are accelerated to energies sufficient to create active radicals which decompose the gaseous contaminant.[18]

3. Methane is a major component of natural gas and is used as a fuel for industrial and residential heating. However, much more methane is produced or available than is used and it may be commercially advantageous to convert methane to higher hydrocarbons which can be readily transported in liquid form for possible use as a fuel or feedstock to produce industrial chemicals. A possible method for such conversion is to generate methyl ions and accelerate them to suitable energies would enable them to interact with neutral methane molecules on a surface.[19]

4. Laser writing on gold-containing hydrocarbon films for pattern generation or lithographic mask repair has been explored, using a discharge. The method, briefly, is to produce a radio-frequency capacitively coupled discharge using propane–argon gas in the discharge chamber. The powered electrode is a flat gold surface which constitutes the cathode, while a parallel grounded electrode (anode) bears a glass substrate. The gold atoms are sputtered off the cathode and on reaching the substrate become trapped in the polymer matrix produced simultaneously and uniformly on the substrate. The metal tracings are formed upon laser irradiation of the film at a particular wavelength. The laser generates local heating and decomposes the hydrocarbon matrix. The resistivity is decreased by about 10^{15}, reaching the conductivity levels of a metal.[20]

5. CH_4 has been suggested as a buffer gas in gas mixtures for use in externally sustained diffuse discharge opening switches. Opening times of the order of $\sim 20\,ns$ have been reported, with a possibility of further reduction by addition of a small amount of electronegative gas to sweep out the low energy electrons.[21,22]

Hydrocarbon gases exhibit a Ramsauer–Townsend minimum in the total cross section. Table 7.2 shows the energy and cross section at the minimum of selected hydrocarbons. The table also contains the value of the index n where, at thermal energies, the momentum transfer cross section is proportional to ε^{-n}

$$Q \propto \varepsilon^{-n} \tag{7.22}$$

and the mobility is constant. Both polarizability and dipole moment have a bearing on the shape of the ε–Q_M curve, and these two parameters are also shown. Increasing either the molecular polarizability α (compare propane and butane) or the permanent dipole moment μ (compare propane and propene) acts chiefly to shift the Q_M curve upwards. The magnitude of Q_M depends on the interaction potential between the electron and the molecule. This potential is of the basic form

$$V = -\left[\frac{\alpha e^2}{2r^4} + \frac{e\mu}{r^2}\cos\theta\right] \tag{7.23}$$

where r is the distance between the electron and the molecule, and θ the angle between the dipole axis and vector r. Each of the factors on the right side of Equation 7.23 may be corrected with extra factors, or shorter range terms may be added.[30]

The value of the index n in Equation 7.22 is close to unity for simple hydrocarbons (Table 7.2). Exceptions are cyclopropane ($n=0.2$) and ethene ($n=0.5$). The molecular structure and its relation to n are interesting. The spherical shape of the molecule and the absence of C–C bonds, as in methane, may have a significant role to play. From Table 7.2 one can discern that the Ramsauer–Townsend minimum occurs at the highest energies for

TABLE 7.2
Ramsauer–Townsend Minimum and Momentum Transfer Cross Sections in Selected Hydrocarbons

Gas	Formula	μ (D)	α	R--T Minimum (eV)	Minimum Q_T	n	Reference
(1)	(2)	(3)	(4)	(5)	(6)	(7)	(8)
Methane	CH_4	0	2.88	0.4	1.36	0.9	43
Ethane	C_2H_6	0	4.97	0.12	1.2	1.1	29
Propane	C_3H_8	0.084	6.9	0.14	2.8	1.2	30
Butane	C_4H_{10}	≤ 0.05	8.9	0.14	4.6	1.0	29
Acetylene	C_2H_2	0	3.71	0.2	1.9	—	44
Propene	C_3H_6	0.366	6.45	0.17	—	0.9	30
c-propane	C_3H_6	0	6.12	0.10	—	0.2	30

Column 3: μ is dipole moment in debyes; $1\,D = 3.3356 \times 10^{-30}\,C\,m$. Col. 4: $\alpha =$ polarizability in units of $10^{-40}\,F\,m^2$. Col. 6 in units of $10^{-20}\,m^2$, Col. 7: $Q_M \propto \varepsilon^{-n}$ ($0.03 \leq \varepsilon \leq 0.15\,eV$). The reference for columns 3, 4, and 7 is Gee and Freeman.[30]

the most sphere-like molecule (methane). and much lower (0.1 eV) for cyclopropane. These aspects have been further explored by Floriano et al.[23] in a number of low-density alkanes.

For a proper understanding of the transport properties the more important cross sections are the momentum transfer cross sections and the rotational and vibrational cross sections. These are presented in some detail. The cited references for selected hydrocarbons in the same study are: Bortner et al.[24]; Bowman and Gordon[25]; Cottrell et al.[26]; Huber[27]; Christophorou et al.[28]; McCorkle et al.[29]; and Gee and Freeman.[30]

7.3.1 METHANE (CH_4)

Methane is one of the more widely studied of all hydrocarbons as far the transport parameters are concerned. Selected references are: Schlumbohm,[31] current pulse method; Wagner et al.,[32] time-of-flight (TOF) method; Pollock[33]; Haddad,[34] TOF method; Al-Amin et al.,[35] TOF method; Hunter et al.,[36] pulsed Townsend method; Davies et al.[37]; Schmidt and Roncossek[38]; and Yoshida et al.,[39] drift tube method. These data are shown in Figure 7.8. Momentum transfer cross sections are shown as an inset to Figure 7.8 and listed in Table 7.3. Theoretical studies are due to Kline,[40] who used the Boltzmann two-term solution, Al-Amin et al.,[35] and, Ohmori et al.,[41] both using the Monte Carlo simulation method.

At low $E/N < 0.05$ Td the electrons are in thermal equilibrium with the gas molecules and the reduced mobility of the electrons remains approximately constant. This means that the mobility varies as $1/N$; the reduced mobility observed by Hunter et al.[36] is $(\mu N)_{th} = 3.12 \times 10^{25}\,V^{-1}\,m^{-1}\,s^{-1}$. This is in very good agreement with the thermal mobility of $(\mu N)_{th} = 3.3 \times 10^{25}\,V^{-1}\,m^{-1}\,s^{-1}$ measured by Cipollini et al.,[42] who made measurements at densities of the order of $10^{27}\,m^{-3}$ ($\sim$5 MPa) at 295 K. Additional references for the measurement of the Ramsauer minimum in methane are Ferch et al.[43] and Karwasz et al.[44]

In this region of low E/N the drift velocity is a linear function of the electric field, as is seen clearly from Figure 7.8. The thermal mobility is related to the temperature and the momentum transfer cross section according to[45]

$$\mu_{th} = \frac{2}{3}\left(\frac{2}{\pi m k T}\right)^{1/2} \frac{e}{N Q_M} \tag{7.24}$$

where e and m are the charge and mass of the electron respectively and Q_M is the momentum transfer cross section. The variation of μ_{th} with temperature and Q_M is decided by the energy dependence of Q_M. If T increases and Q_M decreases faster, then μ_{th} will increase with temperature. This is observed in argon, xenon, and methane, all of which have a pronounced Ramsauer–Townsend effect.

Methane exhibits the Ramsauer–Townsend effect at $\sim 0.36\,\text{eV}$,[46] followed by a broad maximum centered around $7\,\text{eV}$.[47] See the inset of Figure 7.8 and Table 7.2 for details. The Ramsauer–Townsend minimum in the total cross section and the large vibrational excitation cross sections at energies below $1\,\text{eV}$[48] cause a pronounced negative slope in the

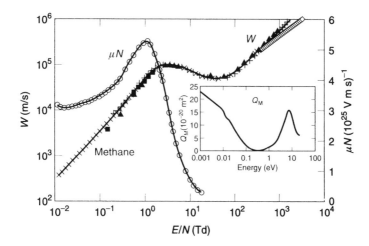

FIGURE 7.8 Drift velocity W and density-normalized mobility μN of electrons in methane. The pronounced negative slope in W is due to the Ramsauer–Townsend minimum. The peak occurs at 3 to 3.5 Td. ($\circ$) Schlumbohm[31]; ($\blacksquare$) Wagner et al.[32]; ($\blacktriangle$) Al-Amin et al.[35]; ($\times$) Hunter et al.[36]; ($+$) Yoshida et al.[38]; (—) recommended W, same as Davies et al.[37]; μN curve is due to Hunter et al.[36] The inset shows the momentum transfer cross sections given in Table 7.3.

TABLE 7.3
Momentum Transfer Cross Sections in CH_4

Energy	Q_M	Energy	Q_M	Energy	Q_M
0.000	50.0	0.600	0.953	40.0	3.26
0.010	40.0	0.800	1.40	60.0	2.13
0.015	24.4	1.00	1.90	80.0	1.67
0.020	17.0	1.50	3.10	100.0	1.43
0.030	10.2	2.00	4.40	150	1.07
0.040	7.30	3.00	7.20	200	0.900
0.060	4.60	4.00	10.5	300	0.714
0.080	2.90	6.00	14.7	400	0.612
0.100	2.00	8.00	16.1	600	0.496
0.150	0.990	10.00	15.5	800	0.430
0.200	0.700	15.0	12.6	1000	0.384
0.300	0.560	20.0	8.80	3000	0.120
0.400	0.650	30.0	4.76	10000	0.0360

Energy in units of eV, and cross section in units of $10^{-20}\,\text{m}^2$.

W–E/N plots, a theoretical discussion of which is provided in the next section. Hunter et al.[36] observed that the drift velocity was dependent on pressure, both at low E/N (<10 Td) and high E/N (>100 Td) values. This effect is attributed to the longitudinal diffusion coefficient (ND_L) and corrections were applied.

The longitudinal diffusion coefficient has a maximum at 2 Td (Figure 7.9). At higher E/N, >100 Td, gas ionization becomes prominent and the measured values were corrected by Hunter et al.[36] These authors state an uncertainty of $\pm2\%$ in the drift velocity when no ionization or attachment occurs, and $\pm5\%$ when these processes are present. Attachment in CH_4 has been shown to be a true effect by using the Boltzmann equation method. The drift velocity shown in Table 7.4 is formulated by the results of Hunter et al.[36] up to 300 Td and by Yoshida et al.[39] in the range $300 \leq E/N \leq 2000$ Td.

The drift velocity in selected hydrocarbons measured by Gee and Freeman[30] is shown in Figure 7.9. The linear relationship at low values of E/N where the electrons are in thermal equilibrium has been commented upon already.

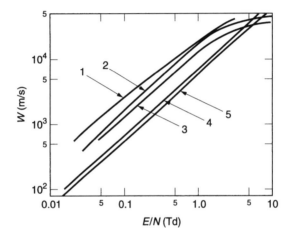

FIGURE 7.9 Drift velocity of electrons in selected hydrocarbon gases as a function of E/N near 300 K. Curve 1, cycloproane; 2, propane, 3, n-butane; 4, i-butane; 5, propene.

TABLE 7.4
Drift Velocity of Electrons in CH_4

E/N	W	E/N	W	E/N	W
0.012	380.0	2.0	9.10 (4)	300	2.18 (5)
0.017	530.0	4.0	10.1 (4)	350	2.16 (5)
0.02	630.0	6.0	9.22 (4)	400	2.47 (5)
0.03	940.0	8.0	8.25 (4)	450	2.66 (5)
0.04	1.30 (3)	10.0	7.40 (4)	500	2.90 (5)
0.05	1.60 (3)	20.0	5.79 (4)	600	3.85 (5)
0.06	1.99 (3)	40.0	5.35 (4)	700	4.25 (5)
0.1	3.50 (3)	60.0	5.58 (4)	800	4.68 (5)
0.2	7.62 (3)	80.0	6.23 (4)	900	5.25 (5)
0.4	1.75 (4)	100	7.21 (4)	1000	5.74 (5)
0.6	2.93 (4)	160	1.10 (5)	1500	8.96 (5)
1.0	5.26 (4)	200	1.36 (5)	2000	1.14 (6)

E/N in units of Td, W in units of m/s. $a(b)$ means $a \times 10^b$.

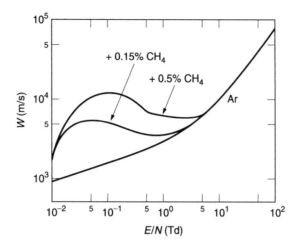

FIGURE 7.10 Schematic variation of drift velocity in argon and mixtures of Ar + CH4 to illustrate the change in magnitude and shape of the curve with the addition of small percentages of CH4.

CH_4 has often been used as a test gas in the derivation of scattering cross sections from measured transport parameters. To increase the accuracy and test uniqueness of the derived cross sections an additional set of parameters is required. In such cases swarm parameters are measured by adding small amounts of CH_4 to a noble gas, usually argon. The small percentage does not change the scattering cross sections appreciably, yet the drift velocity changes drastically due to the inelastic vibrational excitations that occur. The change therefore occurs at low E/N where the mean energy should be of the same order of magnitude as the energy at which these inelastic collisions occur. Figure 7.10 shows such a situation in mixtures of argon and 0.1% and 0.5% CH_4. Pure argon has a lower drift velocity, which increases more or less monotonically. The mixtures show much larger drift velocity and a negative slope at low values of E/N. A theoretical discussion of this aspect is given in the section that follows.

Data on ratios D_r/μ and D_L/μ in CH_4 are provided by: Schlumbohm[49]; Wagner et al.[32]; Lakshminarasimha and Lucas[50]; Al-Amin et al.[35]; Haddad[51]; Millican and Walker[52]; Schmidt and Roncossek[38]; and Yoshida et al.[39] These data are shown in Figure 7.11. The D_r/μ results of Haddad[51] and of Schmidt and Roncossek[38] agree very well over the overlapping ranges of E/N. For higher values of E/N ($>12\,\mathrm{Td}$) only the data of Lakshminarasimha and Lucas[50] are available.

The ratio D_L/μ shows greater disparity at low values of E/N, particularly among the earlier results. It is noted that the time-of-flight technique is not particularly suited for diffusion measurements at low values of E/N. This is because at low E/N one has to use higher gas pressures and the half-width of the electron pulse from the cathode becomes comparable to the increase of half-width of the electron pulse due to diffusion.[35] Again, the best results are those of Schmidt and Roncossek[38] for low values of E/N ($<12\,\mathrm{Td}$) and Yoshida et al.[39] The latter authors measured the ratio by the arrival spectra method over a wide range of E/N, 0.2 to 2000 Td, and their data form an excellent set of values. For lower E/N the results of Schmidt and Roncossek[38] are equally accurate.

7.3.2 ETHANE (C_2H_6)

Transport parameter data are given by: Heylen[53]; Cottrell et al.[26]; McCorkle et al.[29]; Schmidt and Roncossek[38]; and Shishikura et al.[54] These data are compiled in Figure 7.12. The inset

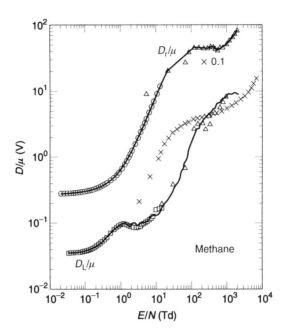

FIGURE 7.11 Ratios D_r/μ and D_L/μ in methane. D_r/μ: ($\triangle$) Lakshminarasimha and Lucas[50]; ($\square$) Haddad[51]; ($\circ$) Schmidt and Roncossek[38]; (—) visual best fitting. The values have been shown with a multiplying factor of ten for clarity. D_L/μ: ($\times$) Schlumbohm[31]; ($\triangle$) Al-Amin et al.[35]; ($\square$) Schmidt and Roncossek[38]; (—) Yoshida et al.[39]

of the figure also shows the momentum transfer cross sections derived by Shishikura et al.[54] from the drift velocity studies. Only one measurement is available[54] for E/N greater than 20 Td. The drift velocity data agree very well among themselves in the overlapping range of E/N. The negative slope in the W curve appears to be less pronounced, but this is due to the logarithmic scale.

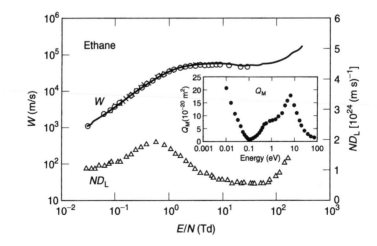

FIGURE 7.12 Transport parameters in ethane. W: ($\times$) Cottrell et al.[26]; ($\circ$) McCorkle et al.[29]; ($\square$) Schmidt and Roncossek[38]; (—) Shishikura et al.[54] ND_L: ($\blacktriangle$) Shishikura et al.[54] Inset shows the momentum transfer cross section.

7.3.3 Selected Hydrocarbons

Figures 7.13 and 7.14 show the drift velocity and diffusion coefficients in selected hydrocarbons measured by McCorkle et al.[29] and Schmidt and Roncossek.[38] For propane, the drift velocity measurements of the two groups agree very well. The mean energy $\bar{\varepsilon}$ given by McCorkle et al.[29] has been converted into the ratio D_r/μ, assuming a Maxwellian distribution. Again, good agreement is seen between the two references cited for ethane and propane.

7.4 NITROGEN COMPOUNDS

We consider ammonia (NH_3), nitric oxide (NO), nitrous oxide (N_2O), and nitrogen dioxide (NO_2) in this section. Ammonia molecules are present in the interstellar medium and the gas is preferentially used as a source of nitrogen in plasma deposition techniques.[55] It has a

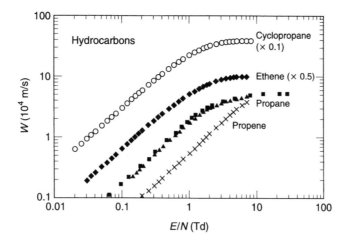

FIGURE 7.13 Drift velocity of electrons in selected hydrocarbon gases. Propane: (■) McCorkle et al.[29]; (▲) Schmidt and Roncossek.[38] All other gases: Schmidt and Roncossek.[38]

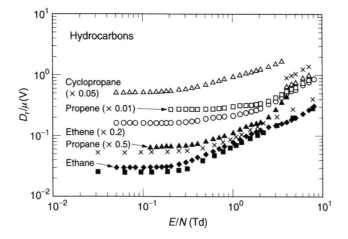

FIGURE 7.14 Ratio D_r/μ in selected hydrocarbon gases. Ethane: (■) McCorkle et al.[29]; (♦) Smith and Roncossek.[38] Propane: (×) McCorkle et al.[29]; (♦) Smith and Roncossek.[38] (○), (□), and (△) are ethene, propene, and cyclopropane respectively, Schmidt and Roncossek.[38]

TABLE 7.5
Selected Properties of Nitrogen Compounds

Name	Formula	At. Wt.	Electrons	α	μ (D)
Ammonia	NH_3	17	10	2.51	1.47
Nitric oxide	NO	30	15	1.89	0.148
Nitrous oxide	N_2O	44	22	3.37	0.18
Nitrogen dioxide	NO_2	46	23	3.37	0.32

α in units of $10^{-40}\,F\,m^2$. 1 debye $= 3.3356 \times 10^{-30}\,C\,m$.

dissociation potential of 4.48 eV. Nitric oxide is a minor but significant gas in the atmosphere. It is classified as an open shell molecule. Nitrous oxide has lasing properties due to vibrational and rotational transitions that are excited by an electrical discharge. Simultaneously it also undergoes dissociation, requiring rapid replacement by flowing gas. Table 7.5 gives selected properties.

7.4.1 AMMONIA

Brief comments are provided with regard to cross sections in the gas. The low energy momentum transfer cross section has been derived by Pack et al.[56] Q_M is represented as a function of energy by three alternative equations in the energy range 0.005 to 0.08 eV. Their analytical function is

$$Q_M = \left(4.93 \times 10^{17} \varepsilon^{1/2} + 1.51 \times 10^{19} \varepsilon^{3/2}\right)^{-1} m^2 \qquad (7.25)$$

The cross section increases steeply toward zero energy, reaching values of the order of $10^{-16}\,m^2$,[57] mainly as a result of inelastic rotational excitation, and reaching $1/\varepsilon$ as zero energy is approached. Electron-attaching gases, both nonpolar and polar (examples are SF_6 and NH_3 respectively), show an increasing cross section as zero energy is approached, but the mechanisms are different. Electron–dipole scattering is due to long-range forces whereas scattering with attaching target is due to relatively strong but short-range forces. For higher energies, in the range 0.1 to 30 eV, an averaged value of Q_M has been taken from the analysis of Alle et al.[58] There is paucity of data on Q_M at higher electron energies.

The total cross section is made up by combining the measurements of Sueoka et al.,[59] Szmytkowski et al.,[60] and Zecca et al.[61] to cover the range 1.5 to 1000 eV. The most noticeable feature is the very broad hump centered around 10 eV, which is attributed to shape resonance (short-lived negative ion formation). NH_3 is isoelectronic with H_2O and CH_4, and these molecules also show a broad maximum at $\sim$10 eV. Figure 7.15 shows the cross sections in NH_3 and NO_2; the latter is made up by combining the results of Szmytkowski et al.,[62] at 0.6 to 220 eV and Zecca et. al.,[63] at 100 to 1000 eV.

Drift velocity data in these gases are shown in Figure 7.16. To the author's knowledge there has been no measurement of W in ammonia since those of Pack et al.,[56] and these are limited to low values of E/N at which the drift velocity is a linear function of E/N. Earlier measurements are due to Nielsen and Bradbury.[65a] Yousfi and Benabdessadok[65] have used the hydrodynamic version of the Boltzmann equation to calculate the swarm parameters and their paper may be used to supplement the cross section data shown in Figure 7.15.

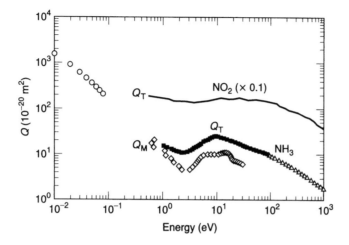

FIGURE 7.15 Scattering cross sections in ammonia and nitrogen dioxide. Ammonia: Q_M: (o) Pack et al.[53]; (◇) Alle et al.[58] Q_T: (■) Szmytkowski et al.[60]; (▲) Zecca et al.[61] NO₂: Q_T: (—) Szmytkowski et al.[62] and Zecca et al.[63] The cross sections of NO2 have been multiplied by ten for clarity of presentation.

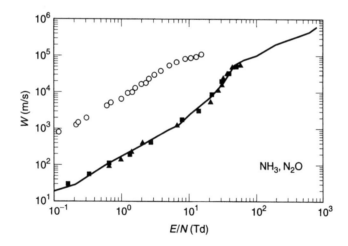

FIGURE 7.16 Drift velocity in ammonia and nitrous oxide. NH₃: (▲) Nielsen and Bradbury[64a]; (■) Pack et al.[56]; (—) Yousfi and Benabdessadok.[65] N₂O: (o) Pack et al.[56] More recent measurements by Yoshida et al.[67b] are given in Section 9.3.5.

They employed four rotational excitation, four vibrational cross sections with thresholds 0.12, 0.201, 0.414, and 0.427 eV. The dissociative excitation cross sections employed were specific to the reactions

$$e + NH_3 \rightarrow NH + H_2 \quad (3.9\,eV) \tag{7.26}$$

$$e + NH_3 \rightarrow NH_2 + H \quad (5.6\,eV) \tag{7.27}$$

$$e + NH_3 \rightarrow NH + 2H \quad (8.6\,eV) \tag{7.28}$$

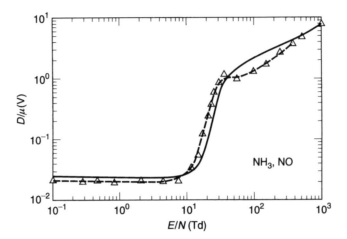

FIGURE 7.17 Radial and lateral diffusion coefficients in ammonia and nitric oxide. NH$_3$: (—) D_r/μ, Yousfi and Benabdessadok[65]; (–△–) D_L/μ, Yousfi and Benabdessadok.[65]; NO: (♦) Lakshminarasimha and Lucas.[67a]

7.4.2 NO, N$_2$O, AND NO$_2$

There has been a single drift velocity measurement in NO by Parkes and Sugden[66] since the measurement of Bailey and Somerville.[64b] Meager data exist for other gases, as shown in Figures 7.16 and 7.17. A point to note is that the radial diffusion to mobility ratios reported by Lakshminarasimha and Lucas[67a] are experimental, based on use of the steady-state method. A comprehensive set of data has been provided by Yoshida et al.[67b] for N$_2$O, which is presented in Section 9.3.5.

7.5 PLASMA INDUSTRIAL GASES

In this section we consider gases that have one or more halogen atoms, in order of increasing molecular weight. The compilations of Christophorou and colleagues are invaluable in this respect and this section is based mainly on their analyses. Table 7.6 shows selected references. As an aid to understanding the drift and diffusion properties, the momentum transfer cross sections are compiled in Figure 7.18. Figure 7.19 shows the drift velocity in a number of these gases.

7.5.1 TRIFLUOROMETHANE (CHF$_3$)

Trifluoromethane is used in plasma processing and, since it does not possess a chlorine atom, it is not destructive of ozone. The molecule is polar with a permanent dipole moment of 1.65 D (5.49 × 10^{-30} C m). As a consequence of the dipole moment it has a strong rotational spectrum. The electronic polarizability is variously reported to be between 3.0 and 4.0 × 10^{-40} F m^2. The energies of the six fundamental vibrational modes, ν_1, ν_2, ν_3, ν_4, ν_5, and ν_6, are 0.3763, 0.1415, 0.0864, 0.1708, 0.1435, and 0.063 eV respectively. The lowest excitation potential is 10.92 eV. As in CF$_4$, the molecules of CHF$_3$ predominantly dissociate or predissociate. Its ionization potential is 14.8 eV.

Christophorou et al.,[68] derived a reduced drift mobility of electrons in the gas, $\mu N = 5.71 \times 10^{25}$ (V m s)$^{-1}$ at low values of E/N (below 3.69 Td). At this value the electrons are in thermal equilibrium with gas molecules and W varies linearly with E/N. More recent measurements of Wang et al.[70] in the range $0.40 \leq E/N \leq 80$ Td have confirmed this reduced

TABLE 7.6
Selected References for Transport Parameters in Halocarbons

		Gas			
Formula	Name	W	D_r/μ	D_L/μ	Reference
CHF_3	Trifluoromethane				69
$M = 70$	(Fluoroform)	•	•		70
$Z = 34$			•		71
					72
CF_4	Tetrafluoromethane	•	•		73
$M = 88$					74
$Z = 42$		•			75
					76
					77
		•	•	•	78
		•			79*
CCl_2F_2	Dichlorodifluoromethane				80, 81
$M = 120.9$					
$Z = 58$					
C_2F_6	Perfluoroethane	•	•		73
$M = 138$					76
$Z = 66$			•	•	82
C_3F_8	Perfluoropropane	•	•		80
$M = 188$					
$Z = 90$					
$c\text{-}C_4F_8$	Octafluorocyclobutane	•	•		83
$M = 200$	(Perfluorocyclobutane)	•			84
$Z = 96$					
		•			85
		•			86
		•		•	87*

See references at end of chapter for mixture composition and brief comments. An asterisk (∗) denotes a gas mixture. M = molecular weight. Z = number of electrons.

mobility value. Momentum transfer cross sections are not available and therefore the total scattering cross section has been shown in Figure 7.8.

The dipole moment of a molecule is related to the momentum transfer cross section, according to the approximate theory of Altschuler,[88] by

$$Q_M = 1.72 \times 10^{-8} \frac{\mu_d^2}{W^2} \, m^2 \tag{7.29}$$

where μ_d is the dipole moment in units of Debye (1 Debye = 3.3356×10^{-30} C m) and W is the drift velocity in m/s. The equation has been shown to be accurate to $\pm 50\%$ in several gases by Christophorou and Christodoulides[80] in the range of 0.6 to 4.1 D.

The total scattering cross section for thermal and near-thermal energy scattering[89] is approximately equal to

$$Q_T = \frac{A_1}{W^2} \tag{7.30}$$

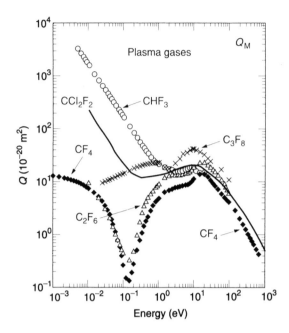

FIGURE 7.18 Momentum transfer cross sections of plasma industrial gases. (○) CHF$_3$;. total scattering cross section has been shown due to paucity of data on Q_M; (◆) CF$_4$; (—) CCl$_2$F$_2$; (△) C$_2$F$_6$; (×) C$_3$F$_8$, where the cross section in the energy range 0.025 to 1.0 eV is for total scattering.

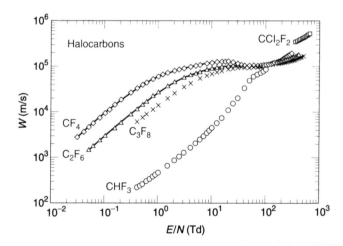

FIGURE 7.19 Drift velocity in halocarbons as a function of E/N. (●) CHF$_3$; (—●—) CF$_4$; (—□—) CCl$_2$F$_2$, where the values have been multiplied by a factor of two for clarity's sake; (—△—) C$_2$F$_6$; (—×—) C$_3$F$_8$. At low electron energy CHF$_3$ has the highest cross section and the lowest drift velocity of these gases. The same observation holds good in reverse: CF$_4$ has the lowest cross section and the highest drift velocity. (The author thanks de Urquiho for providing tabulated data on CHF$_3$.)

where A_1 is a constant having a value of $7.7574 \times 10^5 \, S^{-1}$. S is the slope of the line of W vs. E/N,

$$S = W\frac{N}{E} = \mu N \tag{7.31}$$

for $E/N < 3.9$ Td. As an example of application of this method to CHF_3 we substitute the following values: $E/N = 3.11$ Td, $W = 1.77 \times 10^3$ m/s, $\mu_d = 1.65$ D, obtaining 1.48×10^{-18} m^2 at 0.09 eV, which is in very good agreement with Figure 7.18. Due to dipole interaction, the momentum transfer cross section increases with decreasing W (energy), as Figure 7.18 shows. The large Q_M at low electron energy increases the number of collisions and the drift velocity is much lower than in other gases, as is shown in Figure 7.19.

Figure 7.20 shows the influence of nonpolar argon in modifying the drift velocity of the strongly polar gas, as observed by Wang et al.[70] The curve of pure argon has been added by the author of the present volume for discussion purposes. In pure argon W is a linear function of E/N in the range $0.01 \le E/N \le 3.5$ Td, and the slope begins to increase at higher values. In pure CHF_3, also, W increases with E/N linearly at low values, as already mentioned, but more rapidly at $E/N > 10$ Td. In the low E/N region argon has higher W, and in the high E/N region CHF_3 has higher W, the crossover occurring at ~45 Td. Neither argon (Figure 6.7) nor CHF_3 (Figure 7.19) exhibit a negative slope in the W–E/N plots. However, a small percentage of CHF_3 in argon results in a qualitative change of the W–E/N plots in the following way:

1. At low E/N (< 10 Td) even a small volume fraction of CHF_3 (0.1%) in argon increases W by a factor of ten or more. As an example, at $E/N = 5$ Td, W in pure argon is 4.75×10^3 m/s, in pure CHF_3 2.13×10^3 m/s, in 95% Ar + 5% CHF_3 it is 34.6×10^3 m/s.
2. The negative slope that appears in mixtures is seen at all mixture ratios investigated (0.1 to 100% CHF_3)
3. As the CHF_3 fraction increases, the peak of the W–E/N plot shifts towards higher E/N and the peak magnitude becomes larger.

Wang et al.[70] provide the following explanation for the observed phenomena. If one assumes that the local maxima of W are the result of electrons being scattered by CHF_3 (principally through inelastic vibrational excitation of the CHF_3 molecules) into the energy region where the electron scattering cross section in argon has a minimum (~0.23 eV, Table 3.4), the value of $(E/N)_{max}$ is the value of E/N at which the mean energy is 0.23 eV. Calculations of Clark et al.[90] confirm this argument.

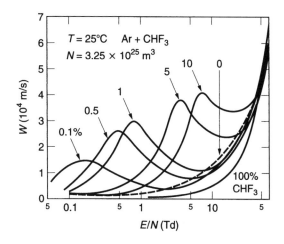

FIGURE 7.20 Drift velocity in mixtures of Ar and CHF_3[70]. The volumetric percentages of CHF_3 are shown. The dashed line is 100% argon.

7.5.2 TETRAFLUOROMETHANE (CF₄)

Compounds having the general formula C_nF_{2n+2} ($n = 1$ to 4) belong to the class known as perfluoroalkanes. Tetrafluoromethane (CF_4) is highly attaching and can be mixed with SF_6 at temperatures below $-50°$ C for high power electrical equipment such as circuit breakers.[91] The essential features of the scattering cross sections are as follows: The Ramsauer–Townsend minimum occurs at 0.15 eV with a cross section of 0.13×10^{-20} m²; corresponding values of the Ramsauer–Townsend minimum in argon are 0.25 eV and 0.09×10^{-20} m² (Table 3.4). There is a broad maximum between 10 and 20 eV. Vibrational excitation shows two maxima, near threshold and at 8 eV. The attachment cross section also has a maximum near 8 eV electron energy.

The drift velocity in the gas is independent of pressure, unlike D_r/μ.[73] The measurements of Hunter et al.[76] using the pulsed Townsend method cover the range of 0.03 to 300 Td. Note the negative slope of the W–E/N plot (Figure 7.19) in the 30 to 140 Td range. CF_4 exhibits a pronounced Ramsauer–Townsend effect (Figure 5.30), and the negative slope is probably due to this since the D_r/μ values in the 30 to 140 Td range are between of 0.5 and 5.0 eV; in this energy range the momentum transfer cross section and the vibrational cross sections vary rapidly with electron energy, though in opposite directions. Argon too shows a relatively flat region in which W increases slowly with E/N (Figure 6.7).

Another distinct feature of the W–E/N plots in mixtures of $CF_4 + Ar$ are the pronounced peaks that occur with a small percentage of CF_4 in the region of 2.0 to 5.0 Td (Figure 7.21) in a manner very similar to $Ar + CHF_3$. Since the ratio D_r/μ is ∼0.03 to 0.04 eV in CF_4 at these values of E/N, the observed effect is possibly due to the Ramsauer–Townsend minimum in argon. The inelastic collisions that occur to the right side of the Ramsauer–Townsend minimum of CF_4 push the electrons to the region where the minimum occurs

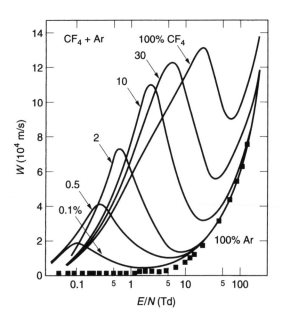

FIGURE 7.21 Electron drift velocity in mixtures of argon and CF_4. Both gases show a distinct Ramsauer–Townsend effect. The drift velocity is much lower in argon than in CF_4 at the same E/N. However, CF_4 has vibrational excitation and the change in the shape of the curves with increasing percentage of CF_4 in the mixture is attributed to this effect. Figure reproduced from Hunter, S. R. et al., *J. Appl. Phys.*, 58, 3001, 1985. With permission of AIP, USA.

in argon.[92] The peaks become broader as the percentage of CF_4 in the mixture increases, and this is attributed to the rapid increase of vibrational excitation losses.

Transport parameters in mixtures of gases are capable of yielding rich information dividends because one can choose different gases for examination. Rare gases such as helium and neon do not have either a Ramsauer–Townsend minimum or low energy inelastic losses due to rotational and vibrational excitation. Argon, krypton, and xenon show the R–T minimum but not low energy inelastic scattering cross sections. Diatomic molecules such as H_2, N_2, and O_2 have rich low energy inelastic scattering processes, but they are nonpolar. They possess a quadrupole moment, whereas SF_6 and CF_4 have neither dipole nor quadrupole moment. CO has all these attributes, namely the dipole moment, quadrupole moment, low energy inelastic scattering losses, and electron attachment.

By selectively choosing the mixture composition for both experimental and theoretical investigations one can hope to decipher the role of each of these processes in modifying the transport parameters. We have already referred to the contributions of Hunter et al.[92] who demonstrated that the drift velocity could be increased by pushing the electrons, through collisions, into the region of the R–T minimum. We therefore investigate the behavior of the mixtures in some detail.

The drift velocity may be written approximately as[93,94]

$$W = \left(\frac{2e}{m} \frac{E}{N} \frac{1}{v_T(\bar{\varepsilon})} \Lambda^{1/2} \right)^{1/2} \tag{7.32}$$

where

$$\Lambda = \frac{1}{v_T(\bar{\varepsilon})} \left(\frac{2m}{M} \bar{\varepsilon} v_{el}(\bar{\varepsilon}) + \sum_k \varepsilon_k v_k(\bar{\varepsilon}) \right) \tag{7.33}$$

and e and m are the electron charge and mass respectively. Λ is roughly proportional to the energy exchange during an electron–molecule collision. v_T, v_{el}, and v_k are the effective total, elastic, and kth inelastic scattering frequency with threshold energy ε_k, respectively. M is the rest mass of the neutral molecule, and $\bar{\varepsilon}$ is the mean energy of the electron.

The second term in Equation 7.33 applies to inelastic collisions; CF_4 does not have dipole moment or quadrupole moment so we need to consider only the vibrational and electronic excitations. Note that, if one neglects the inelastic collisions, the collision frequency for elastic scattering is approximately equal to the total collision frequency and the energy exchange equation simplifies to

$$\Lambda = \frac{2m}{M} \bar{\varepsilon} \tag{7.34}$$

as expected according to Equation 1.65.

Equation 7.32 shows that the drift velocity increases if the total energy losses increase, that is, if Λ increases. W also increases if v_T decreases. Obviously, introducing the rotational and vibrational excitations increases Λ; that is, the second term in Equation 7.33 will change from zero to a positive quantity.

Again, from theoretical considerations,[92] it has been demonstrated that the conductivity will have a negative differential coefficient (NDC) if two conditions are satisfied:

1. A molecular gas, or a mixture of gases with at least one component being molecular, will exhibit the NDC over a range of E/N values when the gas possesses a resonant-type inelastic loss process which peaks at energies just above the threshold and then either remains constant or decreases with increasing electron energy.

2. The total momentum transfer cross section rapidly increases with electron energy as on the right side of the R–T minimum. This results in an increase of effective momentum transfer frequency and a decrease of inelastic collision frequency. Data shown in Figure 7.11 is explained on the basis of these theoretical considerations.

7.5.3 DICHLORODIFLUOROMETHANE (CCl_2F_2)

Dichlorofluoromethane was used as a refrigerant till it was banned due to its adverse impact on the ozone layer. It is used as a plasma processing gas. In the atmosphere it releases chlorine atoms due to solar photolysis, and chlorine is an ozone-depleting molecule. The collision cross section has been presented in Chapter 5. It is a molecule with 58 electrons and is moderately polar (0.55 D) and moderately attaching, with an electron affinity of 0.4 ± 0.3 eV. The molecule has nine nondegenerate fundamental vibration modes, ν_1 to ν_9. The ionization potential is 12.3 eV.

The momentum transfer cross section shown in Figure 7.18 has a broad maximum in the vicinity of 10 eV due to negative ion resonance. It does not show any structure, in contrast with the total scattering cross section (Figure 5.33), and the observed structures in the latter are due to negative ion formation at 0.6 and 3.9 eV. The minimum observed at 0.04 to 0.06 eV in Q_T is interpreted as a Ramsauer–Townsend minimum, though an energy value of 0.5 ev has also been quoted for the minimum. Again, this is not observed in Q_M. The molecules's dipole moment results in there being many rotational levels.

There has been only a single measurement of drift velocity at higher values of E/N,[80] $350 \le E/N \le 640$ Td. The drift velocity shown in Figure 7.19 has been multiplied by a factor of two for the sake of clarity.

7.5.4 HEXAFLUOROETHANE (C_2F_6)

This is a factory-made gas that has wide applications in the aluminum industry, semiconductor industry, plasma chemistry, pulsed power switching, and etching. It is a greenhouse gas with a very long retention time in the atmosphere, reportedly 100 centuries! It is, however, destroyed by collision with O^+. It has no permanent electric dipole moment and has a theoretical electronic polarizability ranging from 5.1 to 7.2×10^{-40} F m^2, depending on the method of calculation.

The momentum transfer cross sections shown in Figure 7.18 show a Ramsauer–Townsend minimum at 0.15 eV with a cross section of 0.32×10^{-20} m^2. The drift velocity shown in Figure 7.19 also shows a negative differential conductivity (decrease in W with increasing E/N), as in CF_4.

7.5.5 PERFLUOROPROPANE (C_3F_8)

The perfluoropropane molecule is nonpolar and electron attaching. The total scattering cross section shows a Ramsauer–Townsend minimum but the momentum transfer cross section below 1.5 eV is not available. The drift velocity data of Hunter et al.[92] and Naidu et al.[83] differ considerably; the latter investigators did not find any dependence of W on E/N, in contrast with the former researchers. The observed dependence of W is attributed to the pressure-dependent attachment coefficients, the attachment process slows down the electrons considerably, to different extents.

We now turn our attention to the ratios D_r/μ and D_L/μ in these plasma processing gases. The data are rather sparse, partly because the steady-state method is not satisfactory, due to excessive attachment at lower values of E/N, and the negative ions fall on the central electrode. Over a considerable range of E/N, the available data exist only for CF_4 and C_2F_6. These data are shown in Figure 7.22.

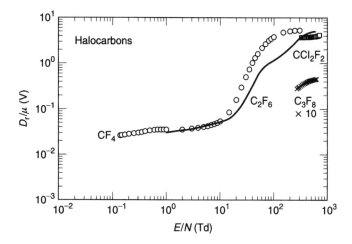

FIGURE 7.22 Radial diffusion coefficients in halocarbons. The values of C_3F_8 are multiplied by a factor of 0.1 for clarity of presentation.

7.5.6 SILANE (SIH$_4$)

The importance of silane in the semiconductor industry has already been referred to in Chapter 5. Cottrell and Walker[95] measured the drift velocity, Pollock[96] measured the drift velocity and characteristic energy, a set of collision cross sections was derived by Ohmori et al.,[97] and Millican and Walker[98] measured the characteristic energy. Kurachi and Nakamura[99] measured the drift velocity in mixtures of argon and silane, observing negative differential conductivity. The peak of W shifts towards higher E/N as the percentage of silane in the mixture is increased. The data for 100% argon was provided by the same authors in a separate publication.[100] [Nakamura and Kurachi, 1988]. More recently, Shimada et al.[101] have published a Boltzmann equation analysis of electron swarms. The transport parameters calculated by them are shown in Figure 7.23. Silane exhibits negative differential conductivity

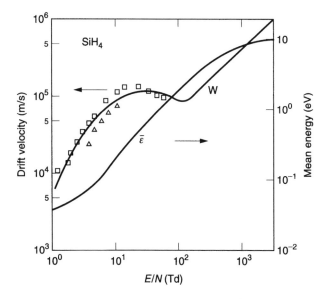

FIGURE 7.23 Drift velocity (W) and mean energy ($\bar{\varepsilon}$) in silane. Drift velocity: ($\triangle$) Cottrell and Walker[95]; ($\square$) Pollock[96]; (—) Boltzmann equation analysis.

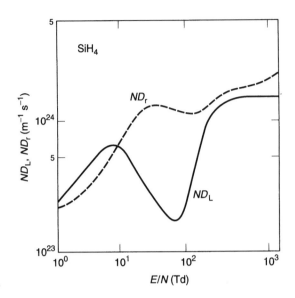

FIGURE 7.24 Longitudinal and radial normalized diffusion coefficients in silane.

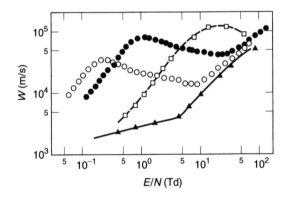

FIGURE 7.25 Drift velocity of electrons in mixtures of $Ar + SiH_4$ as a function of E/N. (▲) 100% Ar, Nakamura and Kurachi[100]; (□) 100% SiH_4, Pollock[96]; (●) 5.04% SiH_4; (○) 0.501% SiH4.

(Figure 7.24), and even small percentages of the gas in argon as buffer demonstrate the same effect (Figure 7.25). Table 7.7 shows the characteristic energy for the range of E/N in which experimental data are available.

7.6 SULFUR HEXAFLUORIDE (SF_6)

The transport parameters in this gas have been thoroughly analyzed by Christophorou and Olthoff[102] and only a few publications have appeared since their review. [103] We therefore substantially reproduce their analysis.

Drift velocity data for SF_6 have been provided by the following selected authors: Harris and Jones[104]; Sangi [105]; Naidu and Prasad [106]; Branston [107]; de Urquijo-Carmona[108]; Aschwanden[109]; Lisovskiy and Yegorenkov[110]; Nakamura[111]; and Xiao et al.[112, 113] Theoretical studies are due to Dincer and Raju[114b]; Novak and Fréchette[115]; Phelps and Van Brunt[116]; Itoh et al.[117]; and Pinheiro and Loureiro.[118]

TABLE 7.7
Characteristic Energy in Silane

E/N (Td)	D_r/μ (V)	E/N (Td)	D_r/μ (V)	E/N (Td)	D_r/μ (V)
1.0	0.0298	7	0.056	40	0.700
1.2	0.031	8	0.066	50	0.853
1.4	0.031	10	0.089	60	1.01
1.7	0.031	12	0.109	70	1.16
2.0	0.031	14	0.145	80	1.25
2.5	0.032	17	0.201	100	1.49
3.0	0.034	20	0.266	120	1.53
4.0	0.037	25	0.363	140	1.62
5.0	0.043	30	0.482	170	1.71
6.0	0.048	35	0.587		

Reproduced from Millican, P. G. and I. C. Walker, *J. Phys. D*, 20, 193, 1987. With permission of the Institute of Physics (IOP), U.K.

Figures 7.26, 7.27, and Table 7.8 show the drift velocity data and Figure 7.28 shows the energy loss in various types of collisions in the gas. The reported uncertainties range from <1% to 15%, as evidenced by the uncertainty of ±15% in the measurements of Harris and Jones,[104] ±5% claimed by Naidu and Prasad,[106] and <0.5% uncertainty claimed by Aschwanden.[109] In the E/N range 250 to 600 Td the data are more consistent than at lower E/N. A possible explanation is that at lower E/N the attachment process needs to be accounted for and the coefficient for this process cannot be less than a few percent.

The drift velocity may be analytically expressed as

$$W = 4349.1\left(\frac{E}{N}\right)^{0.666} \text{ m/s}; \quad R^2 = 0.9947 \tag{7.35}$$

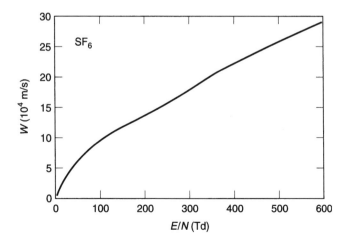

FIGURE 7.26 Drift velocity of electrons in SF_6 as a function of E/N. Measurements are those of Harris and Jones[104]; Sangi[105]; de Urquiho[108]; Aschwanden[109]; Nakamura[111]; Xiao et al.[112] Individual points are not shown, for clarity's sake.

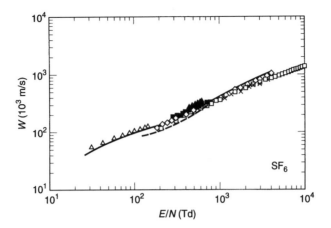

FIGURE 7.27 Drift velocity in SF6 for the complete range. ($\triangle$) Harris and Jones[104]; ($\blacklozenge$) Sangi[105]; ($\blacktriangle$) Naidu and Prasad[106]; ($+$) de Urquiho[108]; ($\diamond$) Aschwanden[109]; ($\blacksquare$) Nakamura[111]; ($- - -$) Xiao et al.[112]; ($\times$) Lisovskiy and Yegorenkov[110]; ($—$) recommended by Christophorou and Olthoff[102]; ($\square$) Raju (2001, unpublished).

TABLE 7.8
Drift Velocity of Electrons in SF$_6$

E/N (Td)	W (10^4 m/s)	E/N (Td)	W (10^4 m/s)	E/N (Td)	W (10^4 m/s)
25	4.1	450	24.6	950	41.8
50	6.8	500	26.4	1000	43.4
100	10.2	550	28.2	1500	58.3
150	12.1	600	30.0	2000	71.7
200	13.5	650	31.7	2500	83.9
250	15.6	700	33.4	3000	95.4
275	17.0	750	35.1	3500	106.3
300	18.3	800	36.8	4000	116.8
350	20.5	850	38.5		
400	22.6	900	40.1		

Values recommended and suggested by Christophorou, L. G. and J. K. Olthoff, *J. Phys. Chem. Ref. Data.*, 29, 267, 2000. With permission of the authors.

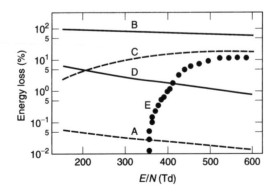

FIGURE 7.28 Energy loss in collisions in SF$_6$ as a function of E/N. (A) elastic collisions; (B) vibrational plus electronic excitation; (C) ionization; (D) attachment; (E) drift current. Note the equal energy loss due to ionization and attachment at $\sim$200 Td.[118]

The quality of fit with the recommended drift velocity is shown in Table 7.9.

The ratio D_r/μ has been measured by Naidu and Prasad[106] and Maller and Naidu.[119] Values suggested by Christophorou and Olthoff[102] are given in Table 7.10.

Considerable experimental and theoretical work has been carried out in mixtures containing SF_6. Table 7.11 shows selected references.

The effective collision frequency (ν_{eff}) as a function of E/N in SF_6, and in mixtures of SF_6 and argon, has been calculated by Raju,[125] using the two-term Boltzmann distribution code. Figure 7.29 shows the results. The calculations do not extend to the lower values of E/N,

---

TABLE 7.9
Quality of Fit of Equation 7.35 with the Recommended Values

E/N range	Maximum Difference	E/N	Minimum Difference	E/N
25–4000	−9.79%	200	−0.09%	950

A negative difference means that recommended values are lower. E/N in units of Td.

TABLE 7.10
D_r/μ Ratio as a Function of E/N

E/N (Td)	D_r/μ (V)	E/N (Td)	D_r/μ (V)	E/N (Td)	D_r/μ (V)
365	4.86	500	5.21	650	5.82
400	4.99	550	5.32	700	6.22
450	5.13	600	5.52	725	6.44

Values suggested by Christophorou, L. G. and J. K. Olthoff, *J. Phys. Chem. Ref. Data,* 29, 267, 2000. With permission of AIP, USA.

TABLE 7.11
Selected References for Transport Properties in Mixtures of Gases with SF_6

Additive	T or E	Reference
N_2	T	Raju and Hackam[114a]
N_2, O_2, Ne	T	Phelps and Van Brunt[116]
He	E	Xiao et al.[112]
Ne	E	Xiao et al.[120]
CO_2	E	de Urquiho et al.[121]
Xe	E	Xiao et al.[122]
He	E	de Urquijo et al.[15]
CO_2	E	Xiao et al.[123]
CO_2	E	Hernández-Ávila et al.[124]
N_2	E	Hernández-Ávila and de Urquijo[103]
Ar	T	Raju[125]

T or E means theory or experimental.

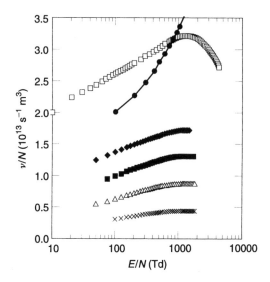

FIGURE 7.29 Effective collision frequency in mixtures of SF_6 and argon.[125] Argon percentage by volume: (□) 100; (◆) 80; (■) 60; (△) 40; (×) 20; (—●—) 0%.

so only the rising part of the ν_{eff}–E/N curve is seen. With increasing total cross section with energy the effective collision frequency increases up to ~1000 Td, at which the mean energy is in the vicinity of ~13 eV (Figure 6.13), and Q_T reaches a maximum value at this energy (Figure 3.1). For higher electron energies Q_T decreases and this is reflected by a decrease of ν_{eff}. With addition of a small amount of SF_6 the electrons lose energy by vibrational excitation and the effective collision frequency decreases. With further increase of the SF_6 additive, ν_{eff} continues to decrease.

However, surprisingly, pure SF_6 exhibits an effective collision frequency that is higher than those of the mixtures but lower than that of pure argon. Only at high vaues of E/N (~1000 Td) do the curves of pure argon and SF_6 cross over and the latter has a higher ν_{eff}, consistent with the broad concept that larger molecules have larger Q_T. A possible explanation for the fact that the ν_{eff} in 100% SF_6 is higher than that in mixtures of SF_6 and Ar is that even a small argon content lowers the cross section of SF_6 disproportionately.

7.7 WATER VAPOR (H_2O AND D_2O)

Drift velocity measurements have been carried out by Pack et al.[126]; Lowke and Rees[127]; Wilson et al.[128]; Ryzko[129, 130]; Crompton et al.[131]; and Elford.[132] Theoretical analyses have been carried out by Ness and Robson[133] and by Yousfi and Benabdessadok.[134] Figure 7.30 shows these data; a comparison with heavy water (D_2O) is included, to the extent of available information.

The variation of drift velocity with E/N is linear below $E/N = 15$ Td for H_2O. The same comment applies to D_2O, although the values are slightly higher than those for H_2O. A possible explanation for the higher W has been provided by Wilson et al.[128] A high drift velocity is favored for a small collision cross section (Equation 1.98) and a large fractional energy loss due to inelastic collisions (Equation 1.99). Electrons in many deuterated gases have a lower W than in hydrated gases (for example, in CD_4 and CH_4[135]) and the explanation was that the vibrational excitation losses in deuterated gases were lower than those in hydrated gases. However, the mean energy of electrons in D_2O is higher than in H_2O in the range of E/N from 20 to 50 Td. The momentum transfer cross section is inversely

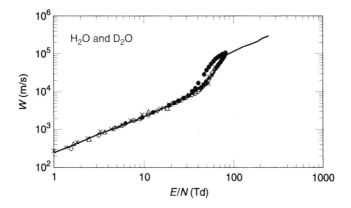

FIGURE 7.30 Drift velocity in H_2O and D_2O as a function of E/N. H_2O: ($\triangle$) Pack et al.[126]; ($\times$) Lowke and Rees[127]; Wilson et al.[128]; Yousfi and Benabdessadok.[134] D_2O: ($\bullet$) Wilson et al.[128] The drift velocity varies linearly below $E/N = 15$ Td and the electrons are in thermal equilibrium with molecules, having a Maxwellian distribution. For higher E/N the distribution becomes non-Maxwellian.

proportional to the energy for polar gases and therefore the momentum transfer cross section in the latter vapor will be higher than in D_2O (approximately 4% more at low electron energy). This results in a lower drift velocity in H_2O.

Below $E/N = 15$ Td the electrons are in thermal equilibrium, having a Maxwellian distribution. For higher E/N, the electron energy distribution departs from that of the Maxwellian, as is also confirmed by the measured D_r/μ ratio.

Figure 7.31 displays the ratios D_r/μ and D_L/μ in H_2O. Experimental values are those of Crompton et al.[136] and Wilson et al.[128] Theoretical values in H_2O vapor are due to Lowke and Parker,[137] and Yousfi and Benabdessadok.[134] There is good agreement between theory and experiment. The experimental values of the two ratios appear to be converging to the same value at approximately 90 Td. For E/N greater than 15 Td the ratio D_L/μ is greater than the ratio D_r/μ. The measured value of D_L/μ extrapolates to the thermal energy of 0.026 eV in accordance with the Einstein relation that predicts kT/e, where T is 300 K. The characteristic energy remains thermal for both H_2O and D_2O up to ~15 Td. A rapid increase in D_L/μ is observed above 15 Td for D_2O and above 30 Td for H_2O.

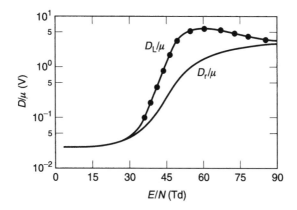

FIGURE 7.31 Ratios D_r/μ and D_L/μ in H_2O as a function of E/N. The D_L/μ ratio extrapolates to 0.026 eV (thermal energy). The diffusion coefficients converge at ~90 Td.

A large difference is observed in D_L/μ for H_2O and D_2O in the range of E/N from 15 to 60 Td, the values for D_2O being as much as 10 times larger than those for H_2O. For E/N greater than 60 Td the D_L/μ values for D_2O fall significantly below those for H_2O.

The parameters W, D_r/μ, and D_L/μ relate to electron motion in a gas as a function of E/N. Klots and Nelson[138] have derived a simple relationship between these parameters as

$$\frac{D_L}{D_r} = \frac{d(\ln W)}{d[\ln(E/N)]} \tag{7.36}$$

Using the measured W and D_L, one can calculate the ratio on the left-hand side of Equation 7.36 and hence determine ND_r and D_r/μ.

7.8 MISCELLANEOUS GASES

We shall consider a few selected gases in which data are scanty or have not been obtained since 1960.

7.8.1 HALOGENS

Chlorine, bromine, fluorine, and iodine belong to this group and occupy the last but one column in the periodic table. The significance of this placement is that, with one electron more, the elements acquire the same stable configuration as the rare gas atoms. The elements are highly electronegative and compounds with one or more halogen atoms, such as SF_6, CF_4, CCl_2F_2, become highly electron attaching. Due to the presence of the halogen atom in the molecule the bond becomes polar, but the molecule itself may or may not possess a dipole moment. Thus hydrogen chloride (HCl) and CCl_2F_2 are polar, but SF_6 and CF_4 are nonpolar. The interest in these gases is due to the scope that exists in increasing the dielectric strength of a gas by forming compounds. Hydrogen fluoride and rare gas fluoride lasers have added impetus for the fundamental studies in halogens.[139] A summary is provided below in the order of increasing number of electrons in the molecule.

7.8.1.1 Fluorine (F_2)

The fluorine atom has nine electrons with the configuration $1s^2 2s^2 2p^5$. The dissociation energy is $<2.75\,eV$ and the vibrational excitation threshold is $0.11\,eV$.[140] The cross section at low electron energy is mainly due to the dissociative electron attachment process, as determined by Chantry[141] from beam studies and by McCorkle et al.[139] from swarm unfolding. The dissociative attachment is

$$e + F_2 \rightarrow F_2^{-*} \rightarrow F^-(1S_0) + F(^2P) \tag{7.37}$$

The second step in the above reaction may also result in vibrational excitation according to

$$F_2^{-*} \rightarrow F_2^*(v = 1,2) + e \tag{7.38}$$

The cross section at 6 meV has been determined as $2 \times 10^{-18}\,m^2$ by Chutjian and Alajajian[142]; it decreases rapidly to $1 \times 10^{-20}\,m^2$ at 0.5 eV.

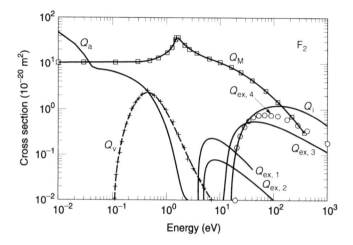

FIGURE 7.32 Scattering cross sections in fluorine as a function of electron energy. Note the very large attachment cross section (Q_a), which is even larger than the momentum transfer cross section (Q_M) at low energies.

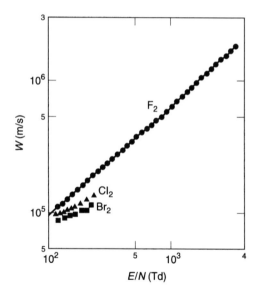

FIGURE 7.33 Drift velocity in fluorine: (——) Boltzmann analysis; ($\bullet$) Monte Carlo analysis. Data for Cl_2 and Br_2 are due to Bailey and Healy.[146]

There are no experimental data available for W in fluorine. Nygaard et al.[143] have measured the drift velocity in mixtures of helium and fluorine. McCorkle et al.[139] have measured W with fluorine in the buffer gases N_2 and Ar. Hayashi and Nimura[144] have calculated the swarm parameters by both the Monte Carlo method and Boltzmann analysis. The cross section data of F_2 are shown in Figure 7.32. Excitation cross sections, vibrational and attachment cross sections have been included. Figures 7.33 and 7.34 show the swarm parameters. Limited data available for chlorine and bromine are also included for the purpose of comparison.

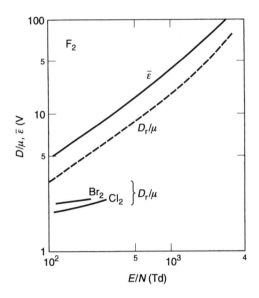

FIGURE 7.34 Characteristic energy (D_r/μ) and mean energy $(\bar{\varepsilon})$ in fluorine. (D_r/μ) data for Cl_2 and Br_2 are due to Bailey and Healy,[146] and Bailey et al.,[147] respectively.

7.8.1.2 Chlorine (Cl_2)

Chlorine is a plasma processing gas because the discharging generates chlorine atoms which etch the silicon surface.[145] Chlorine has 17 electrons with the configuration $1s^2 2s^2 2p^6 3s^2 3p^5$. The drift velocity was measured by Bailey and Healy[146] in 1935, a time when the method was figured out by the use of grids. The dissociation energy is 2.479 eV and the vibrational energy is 0.69 meV. Chlorine has an electron affinity of 2.45 eV and an ionization energy threshold of 11.50 eV. Christophorou and Olthoff[145] have reviewed the literature on the Cl_2 molecule, and the cross section set provided by them is shown in Figure 7.35.

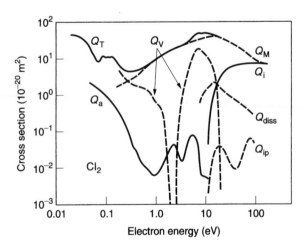

FIGURE 7.35 Scattering cross sections in Cl_2 as a function of electron energy. Q_{ip} = cross section for ion pair formation. Figure reproduced from Christophorou, J. G. and J. K. Olthoff, *J. Phys. Chem. Ref. Data*, 28, 132, 1999. With permission of the authors.

7.8.1.3 Bromine (Br₂)

Bromine has 35 electrons with the configuration $1s^2 2s^2 2p^6 3s^2 3p^6 3d^{10} 4s^2 4p^5$. The drift velocity was measured by Bailey et al.[147]

7.8.1.4 Iodine (I₂)

Iodine has 53 electrons with the configuration $1s^2 2s^2 2p^6 3s^2 3p^6 3d^{10} 4s^2 4p^6 4d^{10} 5s^2 5p^5$. Drift velocity measurements are due to Healy.[148]

7.9 CONCLUDING REMARKS

A comprehensive analysis of swarm parameter data in the more frequently encountered gases has been provided with sufficient, but not overwhelming, theoretical explanations and interpretations. The theoretical aspects of these parameters lie in the solution of Boltzmann equations and determination of the electron energy distribution functions. The number of collision cross sections and their accuracy, to be considered for the solution, depends on the degree of sophistication desired.

Having determined the electron energy distribution, one proceeds through the definitions of Equations 6.21 to 6.23 The swarm parameters are then converted into the desired format in a relatively simple matter. In this respect Equation 7.36 has not been used or tested extensively and its variations (for example, for different distributions of the s-set) may be extremely useful in complex gases in which not all the cross sections are yet readily available. Monte Carlo simulation also provides a desirable choice, particularly at high values of E/N at which the ionization process sets in. In the presence of attachment the Monte Carlo method is less satisfactory, and it is noted here that the first application of this method to an attaching gas was made by Raju and Dincer.[149] A discussion on the relative advantages of Boltzmann analysis and Monte Carlo method may be found in Raju.[150]

The steady-state method of measuring the drift velocity and D_r/μ has the advantage that the initial electron current can be measured without ambiguity (within certain constraints) but has the disadvantage that ionization and attachment processes are difficult to account for. In the time-of-flight method a narrow pulse of electrons is injected into a uniform electric field region containing the neutral gas at a certain number density. After injection, the pulse will drift under the action of the electric field, kept uniform by a set of grids maintained at appropriate potentials. As the electrons drift towards the anode, they spread by diffusion.

The initial electron energy distribution does not, in general, match the steady-state distribution. The distribution will change and this is reflected by a change in the rate of drift and diffusion as the pulse moves away from the source. Most of the change occurs during the relaxation distance, given by Parker and Lowke[151] as

$$d_\varepsilon \simeq \frac{MeE}{2m^2 v^2} \tag{7.39}$$

where m and M are the masses of electron and neutral respectively, e the electronic charge, and v the electron–neutral collision frequency. This equation assumes that only elastic scattering occurs. By substituting the values for H_2, $M/m = 3672.4$, $E = 500\,V/m$, and $v = 1 \times 10^{10}\,s^{-1}$, one gets the relaxation distance as 0.16 cm and the drift distance d may be kept sufficiently large to satisfy the condition, $d_\varepsilon \ll d$. However, the relaxation increases with the square of the collision frequency and it is easy to have this condition violated. The time-of-flight solution given by Parker and Lowke[151] removes this difficulty.

The improvement introduced by Nakamura[4] in having a large variation in drift distance ($\sim$10 cm) is stated to be very effective in the removal of end effects such as the energy relaxation of electrons, the irregularity of the electric field close to the shutters, the diffusion effects, and the width of the shutter pulse. Nakamura has used the same solution for anode pulse current as that of Parker and Lowke.[151]

New data on transport parameters to be acquired has been indicated in the two chapters. The group led by de Urquiho is making valuable contributions to the experimental measurements. Data on nitrogen compounds (particularly NO), ozone, hydrogen sulphide, and halogens is still very meager or nonexistent. All these gases have implications from environmental and safety considerations and the need for experimental data is likely to be felt more in the near future.

Data on collision frequency are also meager, except in common gases. In using the collision frequency to understand the various phenomena in gaseous electronics one should exercise extreme caution since this quantity is often expressed or evaluated for the particular set of conditions and type of discharge. It is expressed as a function of electron energy and therefore it has a range of values in a swarm, even though the external parameters such as the electric field and gas number density are held constant. In such cases one adopts the effective collision frequency, which is applicable at a given value of E/N for that particular type of discharge. At low electron energy the effective collision frequency is often calculated for a Maxwellian energy distribution, but other distributions result in different values.

The calculation of collision frequency from Equations 1.98 and 1.99 has the distinct advantage of separating the effects of elastic and inelastic collisions, as has been demonstrated by Phelps and his group of researchers.[152] These equations also show that a high drift velocity is favored for a small collision cross section (Equation 1.98) and a large fractional energy loss due to inelastic collisions (Equation 1.99).

The anisotropy of the diffusion phenomenon manifests itself in the fact that the lateral and radial diffusion coefficients are different. The theory of Lowke and Parker[151] and Skullerud[153] has established that this is because the electron energy distribution is spatially dependent. There does not seem to be a simple relationship between the effective collision frequency and the ratio D_L/D_r. Braglia[154] has noted that the effective collision frequency, if expressed as $\propto \varepsilon^n$, leads to this ratio decreasing with increasing value of the index n. For example, the theory of Parker and Lowke[151] yields a D_L/D_r ratio that varies from 4.00 to 0.29 as n is increased from $-3/8$ to 1.[154] Further, different authors obtain widely ranging D_L/D_r ratios for the same value of n. For example, for $n = -0.25$, the D_L/D_r ratio calculated by Parker and Lowke,[151] Skullerud,[155] and Lucas and Saelee[156] is 1.95, 2.02, and 1.49 respectively. For $n = 0$, that is a constant collision frequency as in helium, these authors obtain a ratio of one.

The diffusion coefficients, or rather the ratios D_r/μ and D_L/μ, usually span three different ranges of E/N. In the first range, below a certain E/N, the electrons are in thermal equilibrium with the molecules and Einstein's equation $D_r/\mu = kT$ holds good. The ratio D_r/D_L is one in this range. The drift velocity increases linearly with E/N and the momentum transfer cross section is inversely proportional to the electron energy. The second range begins with approximately the E/N at which vibrational excitations first set in and extends up to the electronic excitation thresholds. In several molecular gases the latter process occurs simultaneously with dissociation. Rotational excitations are usually strong with polar molecules, resulting from long-range electron–dipole interactions. In the third region all inelastic processes, including ionization, occur and appropriate solutions have been devised for each range so that experimental data can be properly analyzed. These developments have resulted in an added advantage that the transport parameters measured are a combination of W, ND_L, D_r/μ, D_L/μ, and α (first ionization coefficient).

Sulfur hexafluoride occupies an important place in view of its technological importance. In view of continued research on plasma and arc processes in this gas, relevant electron–molecule properties are summarized in Table 7.12.

Sulfur dioxide has an important role in determining the climatic conditions of the planet; it has a cooling effect which is of the same order of magnitude as the warming effect of CO_2. SO_2 and O_3 have structural similarities.[157] Removal of sulfur dioxide from flue gases of power plants is an important environmental problem[158] and several techniques are being explored. Yet very little information is available on the drift and diffusion of electrons in the gas.

A comparison of some physical properties of selected triatomic molecules is given in Table 7.13.

TABLE 7.12

Selected Electron–Molecule Interaction Properties of SF_6

Attachment

$e + SF_6 \rightarrow SF_6^-$	Threshold energy $(\varepsilon_{th}) = 0$
$e + SF_6 \rightarrow SF_5^- + F$	$\varepsilon_{th} = 0.1\,eV$
$e + SF_4 \rightarrow SF_3^- + F$	$\varepsilon_{th} = 0.7\,eV$
$e + SF_4 \rightarrow SF_3 + F-$	$\varepsilon_{th} = 0.2\,eV$
$e + F_2 \rightarrow F + F-$	

Dissociation

$e + SF_6 \rightarrow SF_5 + F + e$	$\varepsilon_{th} \approx 9.8\,eV$
$e + SF_5 \rightarrow SF_4 + F + e$	
$e + SF_4 \rightarrow SF_3 + F + e$	

Detachment

$e + F^- \rightarrow F + 2e$
$SF_6^{-*} \rightarrow SF_6 + e$

Ionization

$e + SF_6 \rightarrow SF_5^+ + F + 2e$	$\varepsilon_{th} = 15.9\,eV$
$e + SF_6 \rightarrow SF_4^- + F_2 + 2e$	$\varepsilon_{th} = 19.6\,eV$
$e + SF_6 \rightarrow SF_3^- + F_3 + F + 2e$	$\varepsilon_{th} = 19.8\,eV$
$e + SF_4 \rightarrow SF_3^- + F + 2e$	$\varepsilon_{th} = 12.7\,eV$
$e + F_2 \rightarrow F_2^- + 2e$	$\varepsilon_{th} = 15.8\,eV$

TABLE 7.13

Properties of Selected Triatomic Molecules

Gas	No. of Electrons	α_e ($10^{-40}\,F\,m^2$)	Bond Length (nm)	Bond Angle (°)	Dipole Moment (D)
H_2O	10	1.61	0.096	104.5	1.87
CO_2	22	3.24	0.116	180	0
N_2O	22	3.37	0.12	—	0.167
NO_2	23	3.37	0.12	134	0.316
O_3	24	3.57	0.13	117	0.53
SO_2	32	4.76	0.143	119	1.63

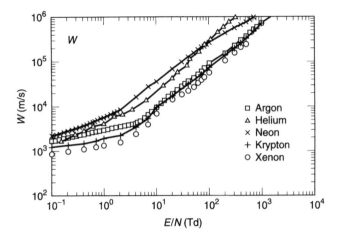

FIGURE 7.36 Comparison of drift velocities in rare gases. With the exception of helium, the drift velocity decreases with larger atoms because of the increased number of collisions.

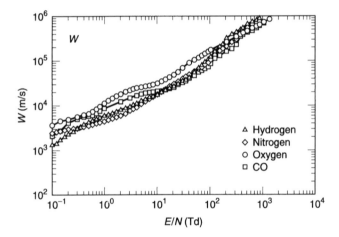

FIGURE 7.37 Comparison of drift velocities in diatomic gases.

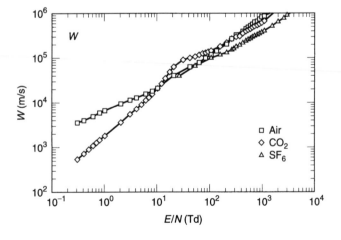

FIGURE 7.38 Comparison of drift velocities in polyatomic gases.

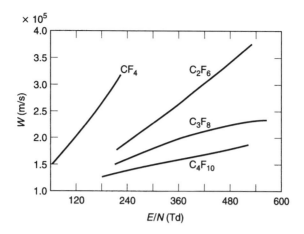

FIGURE 7.39 Experimental drift velocity in fluorocarbon gases. Figure reproduced from Naidu and Prasad.[73] With permission of IOP, U.K.

One often desires to refer to relative values of drift velocity as a function of E/N in various gases for the purpose of choosing candidate gases for a given purpose, be it theoretical or experimental. Since the paradigm adopted in the present volume so far does not facilitate such easy comparison, Figures 7.36 to 7.39 have been provided, using the data already presented.

REFERENCES

1. Huxley, L. G. H. and R. W. Crompton, *Drift and Diffusion of Electrons in Gases*, John Wiley and Sons, New York, 1974.
2. Xiao, D. M., L. L. Zhu, and Y. Z. Chen, *J. Phys. D: Appl. Phys.*, 32, L-18, 1999.
3. Hernández-Ávila, J. L. and J. de Urquijo, *J. Phys. D: Appl. Phys.*, 36, L51, 2003.
4. Nakamura, Y., *J. Phys. D: Appl. Phys.*, 20, 933, 1987.
5. Hasegawa, H., H. Date, M. Shimizuma, K. Yoshida, and H. Tagashira, *J. Phys. D: Appl. Phys.*, 29, 2664, 1996.
6. Allen, N. L. and B. A. Prew, *J. Phys. B: Atom. Mol. Phys.*, 3, 1113, 1970.
7. Bowe, J. C., *Phys. Rev.*, 117, 1411, 1960.
8. Hornbeck, J. A., *Phys. Rev.*, 83, 374, 1951.
9. Raether, H., *Electron Avalanches and Breakdown in Gases*, Butterworths, London, 1964.
10. Allen, N. L. and B. A. Prew, *J. Phys. B: Atom. Mol. Phys.*, 3, 1127, 1970.
11. Hunter, S. R., J. G. Carter, and L. G. Christophorou, *J. Appl. Phys.*, 60, 24, 1986.
12. de Urquijo, J., C. A. Arriga, C. Cisneros, and I. Alvarez, *J. Phys. D: Appl. Phys.*, 32, 41, 1999. A misprint in the equation for negative ion current has been corrected in a subsequent publication.[15]
13. Brambring, J., *Z. Phys.*, 179, 532, 1964.
14. Hernández-Ávila, J. L., *J. Phys. D: Appl. Phys.*, 35, 2264, 2002.
15. de Urquijo, J., E. Basurto, and J. L. Hernández-Ávila, *J. Phys. D: Appl. Phys.*, 34, 2151, 2001.
16. Yamaji M., and Y. Nakamura, *J. Phys. D: Appl. Phys.*, 36, 640, 2003.

Hydrocarbon Gases

17. Deschenaux Ch., A. Affolter, D. Magni, Ch. Hollenstein, and P. Foyet, *J. Phys. D: Appl. Phys.*, 32, 1876, 1999.
18. Jaworek, A., A. Krupa, and T. Czech, *J. Phys. D: Appl. Phys.*, 29, 2439, 1996.
19. Morgan, B. L., R. W. Airey, W. M. Sackinger, and V. A. Kamath, *J. Phys. D: Appl. Phys.*, 29, 1587, 1996.

20. Cambril, E., J. Akinnifesi, J. M. Enjalbert, and B. Despax, *J. Phys. D: Appl. Phys.*, 26, 149, 1993.
21. Christophorou, L. G., S. R. Hunter, J. G. Carter, S. M. Spyrou, and V. K. Lakdawala, *Proceedings of the fourth IEEE Pulsed Power Conference*, ed. T. H. Martin and M. F. Rose, Albuquerque, N M, June 6–8, 1983.
22. Hunter, S. R., J. G. Carter, L. G. Christophorou, and V. K. Lakdawala, in *Gaseous Dielectrics IV*, ed. L. G. Christophorou and M. O. Pace, Pergamon Press, New York, 1984.
23. Floriano, M. A., N. Gee, and G. R. Freeman, *J. Chem. Phys.*, 84, 6799, 1986.
24. Bortner, T. E., G. S. Hurst, and W. G. Stone, *Rev. Sci. Instr.*, 28, 103, 1957.
25. Bowman, C. R. and D. E. Gordon, *J. Chem. Phys.*, 46, 1878, 1967.
26. Cottrell, T. L., W. J. Pollock, and I. C. Walker, *Trans. Faraday Soc.*, 64, 2260, 1968.
27. Huber, B., *Z. Naturforsch.*, A24, 578, 1969.
28. Christophorou, L. G., R. P. Blaustein, and D. Pittmann, *Chem. Phys. Lett.*, 18, 509, 1973.
29. McCorkle, D. L., L. G. Christophorou, D. V. Maxey, and J. G. Carter, *J. Phys. B: At. Mol. Phys.*, 11, 3067, 1978.
30. Gee, N. and G. R. Freeman, *J. Chem. Phys.*, 78, 1951, 1983.
31. Schlumbohm, H., *Z. Phys.*, 182, 317, 1965.

Valuable data for drift velocity of electrons at high values of E/N are given according to the expression

$$W = C(E/N)^p \, \mathrm{m/s}$$

where C and p are constants. The measured values for several molecules are shown below:

Gas	Formula	C $(\mathrm{ms^{-1}\,Td^{-p}})$	p	E/N range (Td)
Oxygen	O_2	2.13×10^4	0.5	300–24,000
Nitrogen	N_2	1.87×10^4	0.5	360–9000
Acetone	C_2H_6CO	9.24×10^3	0.5	600–12,000
Benzene	C_6H_6	9.58×10^3	0.5	600–9000
Carbon dioxide	CO_2	8.08×10^3	0.591	450–6000
Methane	CH_4	2.45×10^3	0.758	360–3000
Diethyl ether	$(C_2H_5)_2O$	5.04×10^3	0.555	600–15,000
Methylal	$CH_2(OCH_3)_2$	3.07×10^3	0.635	300–9000

32. Wagner, E. B., F. J. Davis, and G. S. Hurst, *J. Chem. Phys.*, 47, 3138, 1967.
33. Pollock, W. J., *Trans. Faraday Soc.*, 64, 2919, 1968.
34. Haddad, G. N., *Aust. J. Phys.*, 38, 677, 1985.
35. Al-Amin, S. A. J., H. N. Kücükarpaci, and J. Lucas, *J. Phys. D: Appl. Phys.*, 18, 1781, 1985.
36. Hunter, S. R., J. G. Carter, and L. G. Christophorou, *J. Appl. Phys.*, 60, 24, 1986.
37. Davies, D. K., L. E. Kline, and W. E. Bies, *J. Appl. Phys.*, 65, 3311, 1989.
38. Schmidt B. and M. Roncossek, Aust. J. Phys., 45, 351, 1992.
39. Yoshida, K., T. Ohshima, Y. Ohmori, H. Ohuchi, and H. Tagashira, *J. Phys. D.:* 29, 1209, 1996.
40. Kline, L. E., *IEEE Trans. Plasma. Sci.*, PS-10, 224, 1982.
41. Ohmori, Y., K. Kitamori, M. Shimozuma, and H. Tagashira, *J. Phys. D: Appl. Phys.*, 19, 437, 1986.
42. Cipollini, N. E., R. A. Holroyd, and M. Nishikawa, *J. Chem. Phys.*, 67, 4636, 1977.
43. Ferch, F., B. Granitza and W. Raith, *J. Phys. B: At. Mol. Phys.*, 18, L445, 1985.
44. Karwasz, G. P., R. S. Brusa, and A. Zecca, *Riv. Del Nuovo Cimento*, 24, 1, 2001.
45. Cohen, M. H. and J. Lekner, *Phys. Rev.*, 158, 305, 1967.
46. Schmidt, B., *J. Phys. B: At. Mol. Opt. Phys.*, 24, 4809, 1991.
47. Tanaka, H., T. Okada, L. Boesten, T. Suzuki, T. Yamamoto, and M. Kubo, *J. Phys. B: At. Mol. Phys.*, 15, 3305, 1982.

48. Bordage, M. C., P. Ségur, and A. Chouki, *J. Appl. Phys.*, 80, 1325, 1996. A theoretical evaluation of cross sections of CF_4 by the swarm unfolding technique is presented.

49. Schlumbohm, H., *Z. Phys.*, 182, 306, 1965. Schlumbohm did not realize that he was measuring longitudinal diffusion, which was discovered in 1967, and reported his measurements as D_r/μ.

50. Lakshminarasimha, C. S. and J. Lucas, *J. Phys. D: Appl. Phys.*, 10, 313, 1977.

51. Haddad, G. N., *Aust. J. Phys.*, 38, 677, 1985.

52. Millican, P. G. and I. C. Walker, *J. Phys. D: Appl. Phys.*, 20, 193, 1987.

53. Heylen, A. E. D., *Proc. Phys. Soc.*, 80, 1109, 1962.

54. Shishikura, Y., K. Asano, and Y. Nakamura, *J. Phys. D: Appl. Phys.*, 30, 1610, 1997.

NITROGEN COMPOUNDS

55. Karwasz, G. P., R. S. Brusa, and A. Zecca, *Riv. Nuovo Cimento*, 24, 1, 2001.

56. Pack, J. L., R. E. Voshall, and A. V. Phelps, *Phys. Rev.*, 127, 2084, 1962.

57. Ling, X., M. T. Frey, K. A. Smith, and F. B. Dunning, *Phys. Rev. A.*, 48, 1252, 1993.

58. Alle, D. T., R. J. Gulley, S. J. Buckman, and M. J. Brunger, *J. Phys. B: At. Mol. Opt. Phys.*, 25, 1533, 1992.

59. Sueoka, O., S. Mori, and Y. Katayama, *J. Phys. B: At. Mol. Phys.*, 20, 3237, 1987.

60. Szmytkowski, Cz., K. Maciąg, G. Karwasz, and D. Filipović, *J. Phys. B: At. Mol. Opt. Phys.*, 22, 525, 1989.

61. Zecca, A., G. P. Karwasz, and R. S. Brusa, *Phys. Rev. A*, 45, 2777, 1992.

62. Szmytkowski, Cz., K. Maciąg, and A. M. Krzysztofowicz, *Chem. Phys. Lett.*, 190, 141, 1992.

63. Zecca, A., J. C. Nogueira, G. P. Karwasz, and R. S. Brusa, *J. Phys. B: At. Mol. Opt. Phys.*, 28, 477, 1995.

64. Nielsen R. A., and N. E. Bradbury, *Phys. Rev.*, 51, 69, 1937; (b) Bailey, V. A. and J. M. Somerville, *Phil. Mag.*, 17, 1169, 1934.

65. Yousfi, M. and M. D. Benabdessadok, *J. Appl. Phys.*, 80, 6619, 1996.

66. Parkes D. A., and T. M. Sugden, *J. Chem. Soc. Faraday. Trans.*, 68, 600, 1972.

67. (a) Lakshminarasimha C. S., and J. Lucas, *J. Phys. D: Appl. Phys.*, 10, 313, 1977; (b) Yoshida, K. et al., *J. Phys. D: Appl. Phys.*, 32, 862, 1999.

PLASMA PROCESSING GASES

68. Christophorou, L. G., D. R. James, and R. A. Mathis, *J. Phys. D: Appl. Phys.*, 14, 675, 1981. Dielectric gas mixtures with one component being electron attaching and the other component dipolar are investigated from the point of view of increasing the sparking voltage. The electron–dipole interaction reduces the energy of the electron considerably, thus increasing the sparking voltage. Attaching gases considered are c-C_4F_8 and SF_6. The polar gas considered is CHF_3. Since the present volume was completed a book has been published: Christophorou, L. G., and J. K. Olthoff, *Fundamental Electron Interactions with Plasma Processing Gases*. Kluwer Academic/Plenum Publishers, NY, 2004.

69. Christophorou, L. G., J. K. Olthoff, and M. V. V. S. Rao, *J. Phys. Chem. Ref. Data*, 26, 1, 1997. Data on the gas CHF_3 are analyzed.

70. Wang, Y., L. G. Christophorou, J. K. Olthoff, and J. K. Verbrugge, in *Gaseous Dielectrics VIII*, ed. L. G. Christophorou and J. K. Olthoff, Plenum, New York, 1998, p. 39. See also Wang, Y., L. G. Christophorou, and J. K. Verbrugge, *Chem. Phys. Lett.*, 304, 303, 1999.

71. Christophorou L. G. and J. K. Olthoff, *J. Phys. Chem. Ref. Data*, 28, 967, 1999. An update of complete data on CF_4, CHF_3, C_2F_6, and C_3F_8 is published. Drift velocity data on CHF_3 are taken from this publication.

72. de Urquiho, J., E. Basurto, C. Cisnero, and I. Alvarez, *Proc. 1999 Int. Conf. Physics of Ionized Gases*, Warsaw, Poland, 1999.

73. Naidu, M. S. and A. N. Prasad, *J. Phys. D: Appl. Phys.*, 5, 983, 1972. Gases studied are CF_4, C_2F_6, C_3F_8, and C_4F_{10}, belonging to the class known as perfluoroalkenes.

74. Lakshminarasimha, C. S., J. Lucas, and D. A. Price, *Proc. Inst. Elec. Eng.* 120, 1044, 1973.

75. Christophorou, L. G., D. L. McCorkle, D. W. Maxey, and J. G. Carter, *Nucl. Instrum. Meth.*, 163, 141, 1979. Mixtures studied were $CF_4 + Ar$, $CF_4 + Ne$, $CF_4 + Kr$, $CF_4 + He$, for particle detectors.

76. Hunter, S. R., J. G. Carter, and L. G. Christophorou, *Phys. Rev., A*, 38, 58, 1988. Drift velocity of C_3F_8 is studied.

77. Schmidt, B. and M. Polenz, *Nucl. Instrum. Meth., Phys. Res. A*, 273, 488, 1988.

78. Christophorou, L. G., J. K. Olthoff, and M. V. V. S. Rao, *J. Phys. Chem. Ref. Data*, 25, 1341, 1996. A complete compilation of scattering cross sections and transport parameters of CF_4 is presented.

79. de Urquijo, J., E. Basurto, and J. L. Hernández-Ávila, *J. Phys. D: Appl. Phys.*, 36, 3132, 2003. Mixtures of SF_6 and CF_4 have been studied in the range of 60 to 520 Td. Addition of SF_6 to CF_4 is observed to marginally change W and ND_L but the change in ionization and attachment coefficient is more substantial.

80. Naidu, M. S. and A. N. Prasad, *J. Phys. D: Appl. Phys.*, 2, 1431, 1969.

81. Christophorou, L. G., J. K. Olthoff, and Y. Wang, *J. Phys. Chem. Ref. Data*, 26, 1205, 1997. A complete compilation of scattering cross sections and transport parameters of CCl_2F_2 is presented.

82. Christophorou, L. G. and J. K. Olthoff, J. Phys. Chem. Ref. Data, 27, 1, 1998. A complete compilation of scattering cross sections and transport parameters of C_2F_6 is presented.

83. Naidu, M. S., A. N. Prasad, and J. D. Craggs, *J. Phys. D: Appl. Phys.*, 5, 741, 1972.

84. Novak, J. P., and M. F. Fréchette, *Gaseous Dielectrics–Vol. IV*, ed. L. G. Christophorou and D. W. Bouldin, Pergamon Press, Oxford, 1984, p. 34.

85. Wen, C. and J. M. Wetzer, *Proc. 9th Int. Conf. on Gas Discharges and their Applications*, GD 88, Venice, 1988, p. 367.

86. de Urquijo, J. and E. Basurto, *J. Phys. D: Appl. Phys.*, 34, 1352, 2001.

87. Yamaji M., and Y. Nakamura, *J. Phys. D: Appl. Phys.*, 36, 640, 2003. Mixtures of $c-C_4F_8$ + Ar have also been studied in the range of E/N 1 to 100 Td with 0.468% and 4.91% of $c-C_4F_8$.

88. Altschuler, S., *Phys. Rev.*, 107, 114, 1957.

89. Christophorou, L. G. and A. A. Christodoulides, *J. Phys. B: At. Mol. Phys.*, 2, 71, 1969.

90. Clark, J. D., B. W. Wright, J. D. Wrbanek, and A. Garscadden, in *Gaseous Dielectrics VIII*, ed. L. G. Christophorou and J. K. Olthoff, Kluwer Academic/Plenum Press, New York, 1998, p. 23.

91. Pradayrol, C., A. M. Casanovas, A. Hermoune, and J. Casanovas, *J. Phys. D: Appl. Phys.*, 29, 1941, 1996.

92. Hunter, S. R., J. G. Carter, and L. G. Christophorou, *J. Appl. Phys.*, 58, 3001, 1985. A large number of hydrocarbon and fluorocarbon gases are discussed in this paper, with many data.

93. Robson, R. E., *Aust. J. Phys.*, 37, 35, 1984.

94. Bordage, M. C., P. Ségur, and A. Chouki, *J. Appl. Phys.*, 80, 1325, 1996. A theoretical evaluation of cross sections of CF_4 by the swarm unfolding technique is presented.

SILANE (SiH$_4$)

95. Cottrell, T. L. and I. C. Walker, *Trans. Faraday. Soc.*, 61, 1583, 1965.

96. Pollock, W. J., *Trans. Faraday. Soc.*, 64, 2919, 1968.

97. Ohmori, Y., M. Shimozuma, and H. Tagashira, *J. Phys. D: Appl. Phys.*, 19, 1029, 1986.

98. Millican, P. G. and I. C. Walker, *J. Phys. D: Appl. Phys.*, 20, 193, 1987.

99. Kurachi, M. and Y. Nakamura, *J. Phys. D: Appl. Phys.*, 21, 602, 1988.

100. Nakamura, Y. and M. Kurachi, *J. Phys. D: Appl. Phys.*, 21, 718, 1988.

101. Shimada, T., Y. Nakamura. Z. Lj. Petrović, and T. Makabe, *J. Phys. D: Appl. Phys.*, 36, 1936, 2003.

SULFUR HEXAFLUORIDE (SF$_6$)

102. Christophorou, L. G. and J. K. Olthoff, *J. Phys. Chem. Ref. Data*, 29, 267, 2000.

103. Hernández-Ávila, J. L. and J. de Urquijo, *J. Phys. D: Appl. Phys.*, 36, L51, 2003.

104. Harris, F. M. and G. J. Jones, *J. Phys. D: Appl. Phys.*, 4, 1536, 1971.

105. Sangi, B., Ph.D. thesis, University of Manchester, 1971, as quoted in Christophoroi and Olthoff[102] as their reference 299.

106. Naidu, M. S. and A. N. Prasad, *J. Phys. D: Appl. Phys.*, 5, 1090, 1972.
107. Branston, D. W., Ph.D. thesis, University of Manchester, 1973, as quoted in Christophoroi and Olthoff[102] as their reference 301.
108. de Urquiho-Carmona, J., Ph.D. thesis, University of Manchester, 1980, as quoted in Christophoroi and Olthoff[102] as their reference 302.
109. Aschwanden, Th., in *Gaseous Dieiectrics IV*, ed. L. G. Christophorou and M. O. Pace, Pergamon, New York, 1984, p. 24.
110. Lisovskiy, V. A. and V. D. Yegorenkov, *J. Phys. D: Appl. Phys.*, 32, 2645, 1999.
111. Nakamura, Y., *J. Phys. D: Appl. Phys.*, 21, 67, 1988.
112. Xiao, D. M., H. L. Liu, and Y. Z. Chen, *J. Appl. Phys.*, 86, 6611, 1999.
113. Xiao, D. M., H. L. Liu, and Y. Z. Chen, *J. Phys. D: Appl. Phys.*, 32, L18, 1999. Experimental swarm properties of SF_6 and SF_6 + He are provided in this paper.
114. (a) Raju, G. R. Govinda and R. Hackam, *J. Appl. Phys.*, 52, 3912, 198;. (b) Dincer, M. S. and G. R. Govinda Raju, *J. Appl. Phys.*, 54, 6311, 1983.
115. Novak, J. P. and M. F. Fréchette, *J. Appl. Phys.*, 55, 107, 1984.
116. Phelps, A. V. and R. J. Van Brunt, *J. Appl. Phys.*, 64, 4269, 1988.
117. Itoh, H., Y. Ohmori, M. Kawaguchi, Y. Miura, Y. Nakao, and H. Tagashira, *J. Phys. D: Appl. Phys.*, 23, 415, 1990.
118. Pinheiro, M. J. and J. Loureiro, *J. Phys. D: Appl. Phys.*, 35, 3077, 2002. Boltzmann equation analysis has been applied to mixtures of SF_6 with helium, xenon, CO_2, and N_2.
119. Maller, V. N. and M. S. Naidu, *IEEE Trans. Plasma Sci.*, 3, 205, 1975.
120. Xiao, D. M., H. L. Liu, and Y. Z. Chen, *J. Appl. Phys.*, 86, 6611, 1999.
121. de Urquiho, J., J. Cisneros, C. Alvarez, and E. Basurto, *Proc. 24th Int. Conf. on Phenomena in Ionized Gases*, Warsaw, Poland, vol. 1. ed. P. Pisarczyk, T. Pisarczyk, and J. Wolowski, Warsaw Institute of Plasma Physics and Laser Microfusion, 1999, p. 67.
122. Xiao, D. M., L. L. Liu, and X. G. Li, *J. Phys. D: Appl. Phys.*, 33, L145, 2000.
123. Xiao, D. M., X. G. Liu, and X. Xu, *J. Phys. D: Appl. Phys.*, 34, L133, 2001.
124. Hernández-Ávila, J. L., E. Basurto, and de Urquijo, *J. Phys. D: Appl. Phys.*, 35, 2264, 2002.
125. Raju, G., *Proc. 7th Int. Conf. Properties and Applications of Dielectric Materials*, Nagoya, June 1–5, 2003, pp. 999–1002.

WATER VAPOR

126. Pack, J. L., R. E. Voshall, and A. V. Phelps, *Phys. Rev.*, 127, 2084, 1962.
127. Lowke, J. J. and J. A. Rees, *Aust. J. Phys.*, 16, 447, 1963.
128. Wilson, J. F., F. J. Davis, D. R. Nelson, R. N. Compton, and O. H. Crawford, *J. Chem. Phys.*, 62, 4204, 1975. This paper provides data on drift velocity in water vapor and heavy water.
129. Ryzko, H., *Proc. Phys. Soc. Lond.*, 85, 1283, 1965.
130. Ryzko, H., *Ark. Fys.*, 32, 1, 1966.
131. Crompton, R. W., M. T. Elford, and R. L. Jory, *Aust. J. Phys.*, 20, 369, 1967.
132. Elford, M. T., *Int. Conf. on Phenomena in Ionized Gases*, ed. W. T. Williams, University College of Swansea Press, Swansea, 1987.
133. Ness K. F., and R. E. Robson, *Phys. Rev. A*, 38, 1446, 1988.
134. Yousfi M., and M. D. Benabdessadok, *J. Appl. Phys.*, 80, 6619, 1996.
135. Cottrell, T. L. and I. C. Walker, *Trans. Farad. Soc.*, 61, 1585, 1965.
136. Crompton, R. W., M. T. Elford, and R. L. Jory, *Aust. J. Phys.*, 20, 1967, 1965.
137. Lowke, J. J. and J. H. Parker Jr., *Phys. Rev.*, 181, 302, 1969.
138. Klots, C. E. and D. R. Nelson, *Bull. Am. Phys. Soc.*, 15, 424, 1970, cited by Wilson et al.[128] as their reference 28.

HALOGENS

139. McCorkle, D. L., L. G. Christophorou, A. A. Christodoulides, and L. Pichiarella, *J. Chem. Phys.*, 85, 1966, 1986.
140. Herzberg, G., *Molecular Spectra and Molecular Structure*, Van Nostrand, Princeton, NJ, 1950.

141. Chantry, P. J., *Applied Atomic Collision Physics, vol. 3: Gas lasers*, ed. E. W. McDaniel and W. L. Nighan, Academic Press, New York, 1982, p. 15, cited by Hayashi and Nimura[144] as their reference 1.

142. Chutjian, A. and S. H. Alajajian, *Phys. Rev. A*, 35, 4512, 1987.

143. Nygaard, K. J., J. Fletcher, S. R. Hunter, and S. R. Foltyn, *Appl. Phys. Lett.*, 32, 612, 1978.

144. Hayashi, M. and T. Nimura, *J. Appl. Phys.*, 54, 4879, 1983.

145. Christophorou, J. G. and J. K. Olthoff, *J. Phys. Chem. Ref. Data*, 28, 132, 1999.

146. Bailey, V. A. and R. H. Healy, *Phil. Mag.*, 19, 725, 1935.

147. Bailey, J. E., R. E. B. Makinson, and J. M. Somerville, *Phil. Mag.*, 24, 177, 1937.

148. Healy, R. H., *Phil. Mag.*, 26, 940, 1938.

Concluding Remarks

149. Raju, Gorur G., *Dielectrics in Electric Fields*, Marcel Dekker, New York, 2003.

150. Raju, G. R. Govinda and M. S. Dincer, *J. Appl. Phys.*, 53, 8562, 1982.

151. Parker, J. H. and J. J. Lowke, *Phys. Rev.*, 181, 290, 1969.

152. Pack, J. L., R. E. Voshall, and A. V. Phelps, *Phys. Rev.*, 127, 2084, 1962.

153. Skullerud, H. R., *J. Phys. B: At. Mol. Phys.*, 2, 696, 1969.

154. Braglia, G. L., *Physica*, 92C, 91, 1977.

155. Skullerud, H. R., *Aust. J. Phys.*, 27, 195, 1974.

156. Lucas, J. and H. T. Saelee, *J. Phys. D: Appl. Phys.*, 8, 640, 1975.

157. Zecca, A., J. C. Nogueira, G. P. Karwasz, and R. S. Brusa, *J. Phys. B: At. Mol. Opt. Phys.*, 28, 477, 1995.

158. Novoselov, Y. N., G. A. Mesyats, and D. L. Kuznetsov, *J. Phys. D: Appl. Phys.*, 34, 1248, 2001.

8 Ionization Coefficients—I. Nonelectron-Attaching Gases

Ionization of gas neutrals by electrons is fundamental to the understanding of any type of discharge because this process is the precursor of subsequent developments of current buildup in the discharge. Irrespective of the discharge parameters, such as the applied electric field, gap length, spatial and temporal variation of the electric field, gas number density, etc., one has to consider the ionization process as the first step in the discharge buildup process.

The ionization processes may be broadly divided into primary and secondary processes, the latter being further subdivided into processes that occur within the gaseous medium and those that occur at the electrode surface and walls of the container. Secondary processes that occur within the gas are predominantly photoionization and Penning ionization in a limited number of situations. Secondary processes that occur at the cathode are due to photoelectric action, positive ion and metastable bombardment, and so on. The primary ionization process and the rate or coefficient that quantifies the ionization are our main concerns, though secondary processes will not be completely excluded. We aim to briefly discuss the ionization process and provide data for ionization coefficients as a function of the parameter E/N.

8.1 DISCHARGE DEVELOPMENT

We begin in the traditional way by introducing the development of discharge. Consider a pair of parallel electrodes situated in a container of gas with number density $N\,m^{-3}$ and a continuously variable, uniform electric field E applied. The cathode supplies the initial number of electrons from a source such as a UV lamp. The current–voltage characteristic has the distinct features shown in Figure 8.1.

The current–voltage relationship does not follow Ohm's law. The region below the first current maximum (A–B) is known as the dark discharge and in this region a source of electrons is required to generate the current. It is also called the Townsend discharge region. In the region (B–C) of negative resistance a self-sustaining current is generated and an increase of current is associated with a drop in voltage. At a current of $\sim 10\,\mu A$ the discharge turns into a glow and is maintained by a constant voltage. When the current in the normal glow discharge reaches $\sim 10\,mA$ the voltage across the discharge gradually increases, reaching a limiting value. The region of increasing voltage is called the abnormal glow. A further increase in current ($> 1\,A$) will turn the discharge into an arc. Primary and secondary ionization are the most dominant features in every region of the discharge.

8.2 CURRENT GROWTH IN UNIFORM FIELDS

We develop an equation for the growth of current in an electric field E due to electrons released from the cathode. The energy gained during a free path is $eE\lambda$; if this energy exceeds

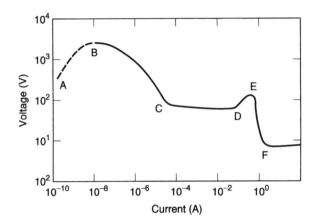

FIGURE 8.1 Current–voltage relationship of a low pressure discharge. A–B, Townsend dark discharge; B–C, self-sustained discharge; C–D, normal glow; D–E, abnormal glow; E–F, arc.

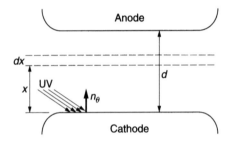

FIGURE 8.2 Current growth in an electric field. n_0 is the number of initial electrons released from the cathode due to the external agency (UV).

the ionization potential then ionization occurs, resulting in two electrons and a positive ion. Each electron in turn will then drift and the process of gaining energy is repeated till the next ionizing collision occurs. The current buildup due to the ionization process was first derived by Townsend as

$$i = i_0 \exp \alpha d \tag{8.1}$$

in which i_0 is the initial photoelectric current released from the cathode and α is called Townsend's first ionization coefficient. The spatial growth of current is even more rapid than that given by Equation 8.1, due to secondary processes, and we shall derive the equation for the current growth. It is assumed that the secondary processes are predominantly due to the action of positive ions and photons, generated in the primary avalanche, on the cathode. The secondary ionization coefficient, defined as the number of secondary electrons produced by a primary electron, is denoted by γ.

Let n_0 be the number of electrons released from the cathode due to an external source. Let n_0' be the number of electrons due to all sources and n the number of electrons arriving at the anode (Figure 8.2). The number of electrons released from the cathode due to the secondary effects is $(n_0' - n_0)$. The number of electrons generated within the gap due to ionization by collision is $(n - n_0')$. Therefore

$$(n_0' - n_0) = (n - n_0')\gamma \tag{8.2}$$

or

$$n_0' = \frac{n_0 + \gamma n}{1 + \gamma} \qquad (8.3)$$

Since

$$n = n_0' e^{\alpha d} \qquad (8.4)$$

substitution of Equation 8.3 into 8.4 yields

$$n = \frac{n_0 + \gamma n}{1 + \gamma} \exp \alpha d \qquad (8.5)$$

Rearranging terms, one gets

$$n(1 + \gamma - \gamma e^{\alpha d}) = n_0 e^{\alpha d} \qquad (8.6)$$

which may be simplified, after converting the number of electrons to current, as

$$i = \frac{i_0 e^{\alpha d}}{1 - \gamma(e^{\alpha d} - 1)} \qquad (8.7)$$

Equation 8.7 has been extensively used in the description of low pressure discharges in both uniform and nonuniform electric fields. Its applicability is the basis on which the ionization coefficients α and γ are determined from steady-state current growth experiments. We briefly refer to two other fundamental processes that are relevant to current growth expressions.

1. *Electron attachment coefficient (η)*. The number of electrons that get attached per meter length of drift is usually denoted by the symbol η (m^{-1}). The current growth expression is given as[1]

$$i = i_0 \frac{\alpha \exp[(\alpha - \eta)d - 1] - \eta}{(\alpha - \eta)} \qquad (8.8)$$

The current growth shown in Equation 8.8 assumes that the secondary ionization coefficient is zero. The contribution of the secondary electrons to the current may be taken into account by using the current growth equation

$$i = i_0 \frac{\left\{ \dfrac{\alpha}{\alpha - \eta} \exp(\alpha - \eta)d - \dfrac{\eta}{\alpha - \eta} \right\}}{1 - \left\{ \dfrac{\gamma \alpha}{(\alpha - \eta)} [\exp(\alpha d) - 1] \right\}} \qquad (8.9)$$

2. *Electron detachment coefficient (δ)*. Electron detachment is an additional process that occurs in electron attaching gases and the coefficient δ is defined analogously to α and η. The current growth equation in the presence of this process has been given by Dutton[2]; see Equations 9.16 to 9.20.

Attachment and detachment coefficients will be dealt with in Chapter 9. For the present we consider only nonattaching gases.

8.2.1 MEASUREMENT OF α AND γ

Basically, there are three techniques for the measurement of the coefficients α and γ: (1) current pulse method; (2) avalanche statistics method; and (3) steady-state method.

1. The first method has already been explained in Section 7.1 and Equations 7.11 to 7.14. The method has the advantage of providing data on W and α simultaneously, though determining these values from the current shape is not straightforward. The steady-state method has the advantage of separating the initial photoelectric current and therefore the amplification that occurs due to both primary and secondary processes may be obtained unambiguously. However, the disadvantage is that the parameters for experiment should be chosen judiciously to enable the separation of the two coefficients.
2. The avalanche statistics method relies on the fact that each electron liberated at the cathode multiplies into an avalanche in which the multiplication process is a statistical quantity. The probability $q(n, d)\delta n$ that an electron starting from one electron at the cathode reaches the number between n and $n + \delta n$ is given by[3]

$$q(n, d) = \frac{1}{\bar{n}}\exp\left(-\frac{n}{\bar{n}}\right) \tag{8.10}$$

where $\bar{n}$, the mean avalanche size, is given by

$$\bar{n} = \exp(\alpha d) \tag{8.11}$$

The distribution function $q(n,d)$ may be obtained experimentally by recording the magnitudes of a sufficiently large number of avalanches, using a pulse height analyzer.

3. Basically, current growth experiments using the steady-state method adopt one of two variations: the distance variation technique and the N variation technique. In the distance variation technique N is kept constant; the gap length between the cathode and the anode is made variable and the current is measured at each gap setting, keeping E constant. In the pressure variation technique the gap distance is kept constant and the gas pressure is varied; the current is measured at each pressure setting, keeping E/N constant. The distance variation technique has been used more extensively than the pressure variation technique. For the pressure variation technique Equation 8.7 is rewritten as

$$i = i_0 \frac{\exp\left(\frac{\alpha}{N} \times Nd\right)}{1 - \gamma\left[\exp\left(\frac{\alpha}{N} \times Nd\right) - 1\right]} \tag{8.12}$$

A schematic block diagram of the experimental apparatus is shown in Figure 8.3.

8.2.1.1 Ionization Chamber

The ionization chamber is usually constructed of glass or stainless steel. Glass chambers tend to be fragile, and assembly and dismantling of the chamber for the purpose of cleaning or changing electrodes becomes cumbersome. For a large electrode system glass chambers are not at all suitable. Stainless steel chambers are more rugged and the availability of

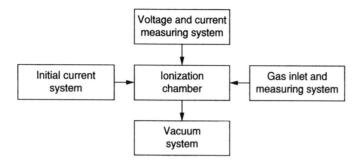

FIGURE 8.3 Schematic diagram of the experimental setup for the measurement of ionization coefficients.

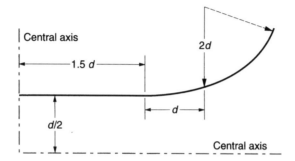

FIGURE 8.4 Practical uniform electric field electrode. d is the maximum gap length.

demountable and bakable components makes them the preferred choice. The chamber should be designed for high vacuum operation and provided with at least two viewing windows. Bakable systems use either copper or gold "O" rings; the latter is obviously more expensive and for "one-time use" only. Copper O rings have the advantage that they can be "refinished," permitting two or three times repeated usage.

The electrodes are designed to have a uniform electric field. This profile is referred to as the Rogowski or Harrison[4] profile (Figure 8.4). Holes are drilled in the anode to facilitate irradiation by ultraviolet light for the supply of initial electrons; the effect of the hole size on the electric field has been determined by Pearson and Harrison.[5] In a typical setup Raju and Dincer[6] have used 0.5 mm diameter holes, with at least twice that distance between adjacent holes in the anode. A rule-of-thumb for the profile is that the flat region of the electrode should be at least twice the maximum distance to be used and the radius of curvature should be the same as the flat region. The distance to the chamber walls should be at least equal to the diameter of the electrode.

One of the electrodes is usually made movable in the distance variation technique and this is accomplished by a vacuum-movable micrometer assembly. In the pressure variation technique, unless specially desired, the gap length remains fixed. This technique assumes that the ionization coefficients are independent of N and depend only on the parameter E/N. However, it is desirable to have experimental verification of this assumption in every gas studied and one should use the pressure variation technique at two or three gap distances.

8.2.1.2 Vacuum System

The ionization chamber is evacuated using high vacuum techniques.[7] These comprise of a backing pump giving ~0.01 torr (1 Pa), a sorption pump giving ~10^{-4} torr (10 mPa), and an ionization pump. The system is bakable up to at least 250°C, giving an ultimate vacuum of ~1 μPa. A sorption pump is preferred for the second stage in place of an oil diffusion pump with or without a baffle because of the uncertainty in eliminating back streaming of oil vapor. The ultimate vacuum attained before filling the gas is less important than the leak rate when the ionization chamber is cut off from the pumps because the rise of the impurity level adds to any impurity in the supply gas. A typical leak rate to aim for is $0.001 \, \text{Pa/m}^3 \, \text{s}$.

8.2.1.3 Gas Handling System

The gas is admitted into the chamber by a separately pumped vacuum line. The gas pressure is measured by a capacitive manometer which is also bakable. Usually about 60 minutes are required for the gas to come into equilibrium with the chamber temperature and pressure readings should be taken just before the start of the current measurements. In the N variation technique it may be advantageous to have a larger chamber attached to the experimental chamber to act as a reservoir that has already attained the chamber temperature.

8.2.1.4 Irradiation System

The initial photoelectric current is obtained by a mercury vapor lamp or, less frequently, an α-particle source. The ultraviolet light of ~254.3 nm wavelength corresponds to 4.86 eV and the work function of an ultraclean gold surface is 4.85 eV. Two major problems with photoelectron emission are the oxidation of the surface and photoelectric emission fatigue, both of which reduce the initial number of electrons. The former is reduced by choosing a surface that is less liable to oxidize within 3 to 4 hours, which is the duration of a typical experiment; gold and platinum coatings are the best, copper and aluminum are the least desirable. Further, running a glow discharge in the same gas at ~1 torr (100 Pa) at a current of ~100 μA before current measurements begin helps to obtain a stable emission surface. Photoelectric fatigue is monitored by repeatedly measuring i_0 at intervals of about 60 minutes.

A second method, which uses the same principle of photoelectric emission, is to coat a quartz window with a thin layer (~10 nm) of gold by vapor deposition.[8] The method has the advantage of reducing field distortion due to holes in the anode, but it suffers from the disadvantage of using a fragile component.

High energy particles have also been used to generate initial electrons, but the method has the disadvantage that ionization in the gas may also contribute to the total current. In addition, it is preferable to ensure that photoelectrons released from the cathode have as low an energy as possible, which is not satisfied if high energy particles are used.

8.2.1.5 Voltage and Current Measurement

Steady-state currents greater than 10^{-12} A may be measured to an accuracy of ±1% using an electrometer or channel multiplier. Highly stabilized direct voltages may be measured to a much better accuracy using high stability resistor dividers and a balancing bridge at the low-voltage arm end. The technique of accurately measuring the voltage is the same from low voltages up to 100 kV. A relatively recent reference to the steady-state measurement of the ionization coefficient is made by Yoshida et al.,[9] whose experimental setup is shown in Figure 8.5.

8.3 FUNCTIONAL DEPENDENCE OF α/N ON E/N

The quantity α/N is usually referred to as the reduced ionization coefficient and is dependent on the parameter E/N only, in the absence of attachment and multiple ionization,

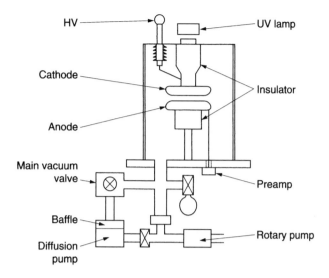

FIGURE 8.5 Experimental setup for the measurement of Townsend's ionization coefficients. Figure reproduced from Yoshida, K. et al., *J. Phys. D*, 28, 2478, 1995. With permission of the Institute of Physics (IOP), U.K.

according to the semiempirical equation

$$\frac{\alpha}{N} = F \exp - \left(G \frac{N}{E} \right) \tag{8.13}$$

in which F (m^2) and G (V m^2) are constants specific to the gas.[10] A modified form of Equation 8.13 is given by

$$\frac{\alpha}{N} = F \exp - \left(K \varepsilon_i \frac{N}{E} \right) \tag{8.14}$$

in which

$$G = K \varepsilon_i^{\text{eff}} \tag{8.15}$$

where $\varepsilon_i^{\text{eff}}$ is the effective ionization potential. The effective ionization potential calculated this way is always higher than the ionization potential of the atom or molecule because the latter is measured by using monoenergy beam or photoionization techniques.

A compilation of values of the constants in Equation 8.13 is given by Raju.[1] There have been other similar expressions with different constants in the literature; for example, the reduced ionization coefficient for atomic gases is also represented by an equation of the types

$$\frac{\alpha}{N} = C \exp \left(-\frac{D}{(E/N)^{1/2}} \right) \tag{8.16}$$

Figure 8.6 shows a plot of α/N as a function of E/N, drawn using Equation 8.13 for a hypothetical gas, with values $F = 4.22 \times 10^{-20}$ m^2 and $G = 7.30 \times 10^{-19}$ V m^2.

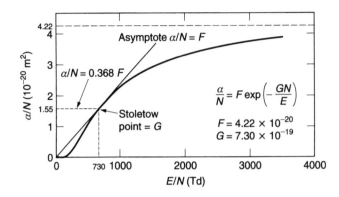

FIGURE 8.6 Calculated α/N values of argon according to Equation 8.13 with $F=4.22 \times 10^{-20} \, \text{m}^2$ and $G=7.30 \times 10^{-19} \, \text{V m}^2$. The tangent to the curve from the origin intersects at the Stoletow point, which has the co-ordinates $(E/N=G; \, \alpha/N=0.368 \, F)$. The asymptotic value of α/N is F.

An inflexion point is seen in the curve and a tangent to the curve from the origin intersects the curve at a value of $E/N=G$ called the Stoletow point. At this value of E/N the reduced ionization coefficient is $\alpha/N=0.368 \, F$. These results can be proved by differentiating Equation 8.13 with respect to N and equating the differential to zero to obtain the maximum α.[1]

The constants F and G have the units of cross section and E/N respectively. Intuitively one expects these constants to be related to the scattering cross section and ionization potential of the gas, but theoretical efforts to find such a correlation have been less than satisfactory. The coefficient G has been related to the ionization potential ε_i according to

$$G = F\varepsilon_i \tag{8.17}$$

Equation 8.17 has not been found to hold for many gases and the ratio G/F has been found to have different values for different gases. Lewis[11] has suggested that G is given by $\sqrt{3}$ times the area beneath the square root of the product of the elastic and inelastic electron–molecule collision cross sections between the onset of excitation and ionization. Mathematically, this is expressed as

$$G = \sqrt{3} \int_{\varepsilon_{ex}}^{\varepsilon_i} \sqrt{Q_{el} Q_{inel}} \, d\varepsilon \tag{8.18}$$

where Q_{el} and Q_{inel} are the elastic and inelastic scattering cross sections respectively. The lower and upper limits of the integration are the excitation and ionization potentials of the gas respectively. At the time the suggestion was made by Lewis[11] available inelastic cross section data were rather scanty and Equation 8.18 has still not been tested for many gases. Verification of this equation with available literature would be a fruitful area of investigation.

Often, the parameter $\eta=\alpha/E$ is used to denote the ionization efficiency, defined as the number of ionizations per volt. Equation 8.13 then modifies to

$$\eta = \frac{\alpha}{E} = \frac{F}{E/N} \exp\left(-\frac{GN}{E}\right) \tag{8.19}$$

Differentiating Equation 8.19 with respect to E/N, one gets

$$\frac{d\eta}{d(E/N)} = \frac{FG}{(E/N)^3}\exp\left(-\frac{GN}{E}\right) - \frac{F}{(E/N)^2}\exp\left(-\frac{GN}{E}\right) = 0 \qquad (8.20)$$

Simplification yields

$$G = E/N \qquad (8.21)$$

as shown in Figure 8.6.

A qualitative explanation of Figure 8.6 may be provided by considering the ionization cross sections as a function of electron energy. The cross section Q_i does not increase monotonically with electron energy, but reaches a maximum and then begins to decline for a further increase in energy. Electrons of very high energy do not lose all of their energy in an ionizing collision and the increase of α/N begins to fall back even though E/N has increased. The left-over excess energy of the electron adds to the energy gain during the next free path, or is delivered to the anode.

The effects of multiple ionization and attachment on the shape of the α/N–E/N curve has been analyzed by Swamy and Harrison.[12] In order to understand these effects, the true ionization coefficient is defined by α_T and the apparent ionization coefficient is then $\alpha = (\alpha_T - \eta)$; the growth equation (8.8) then becomes

$$i = i_0 \frac{\alpha_T \exp[(\alpha_T - \eta)d - 1]}{(\alpha_T - \eta)} - \frac{\eta}{(\alpha_T - \eta)} \qquad (8.22)$$

If the attachment coefficient is negligible compared with the ionization coefficient, that is $\eta \ll \alpha_T$, Equation 8.22 simplifies to

$$i = i_0 \left[\frac{\alpha_T \exp\{(\alpha_T - \eta)d\}}{\alpha_T - \eta}\right] \qquad (8.23)$$

The current growth curves give the apparent ionization coefficients.

The curve of $\ln (\alpha_T/N) - N/E$ is a straight line (Equation 8.13) when attachment and multiple ionizations are absent. Electron attachment is stronger at low values of E/N and multiple ionization becomes more evident at higher values of E/N where the mean energy approaches the second ionization potential. The curve of $\ln(\alpha/N)$ vs. N/E therefore tends to curve upwards at low values of N/E due to multiple ionization, whereas it tends to curve downwards at high values of N/E. The effect of attachment in several gases has been calculated by Swamy and Harrison[12] as shown in Figures 8.7 and 8.8. Table 8.1 shows the relevant data. While the curves of $\ln(\alpha_T/N)$ vs. N/E are straight lines, the curves of $\ln(\alpha/N)$ vs. N/E, where $\alpha = (\alpha_T - \eta)$, droop downward at higher values of N/E.

8.4 SPACE CHARGE EFFECTS

In the steady-state technique buildup of space charge modifies the electric field and the growth of avalanche may be different from that in a uniform electric field. Further, the buildup of space charge plays a central role in the transition of an electron avalanche into a streamer. It is important to realize that the electrons, due to their higher drift velocity, get absorbed into the anode and the positive ions, still remaining in the gap, are the space

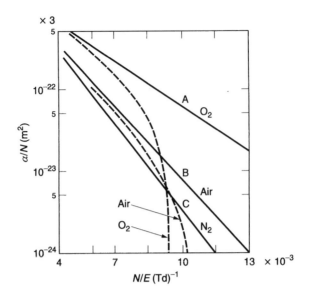

FIGURE 8.7 Semilogarithmic plot of α/N (broken curves) and α_T/N (full lines) as functions of N/E for (A) O_2; (B) air; (C) N_2. Figure and table reproduced from Swamy, M. N. and J. A. Harrison, *Brit. J. Appl. Phys.*, 2, 1437, 1969. With permission of IOP, U.K.

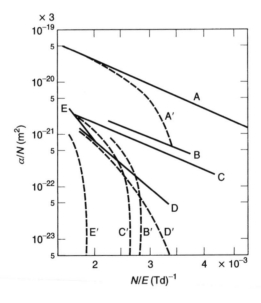

FIGURE 8.8 Semilogarithmic plot of α/N (curved broken line) and α_T/N (full line) as functions of N/E in several attaching gases. For designations see Table 8.1. Figure and table reproduced from Swamy, M. N. and J. A. Harrison, 2, 1437, 1969. With permission of IOP, U.K.

charge creating species. In electron attaching gases the negative ions also contribute to the creation of the space charge field.

An electron moving in an electric field produces $\exp(\alpha d)$ electrons at a distance d from the starting point. Let the charge of these electrons be denoted by Q,

$$Q = e \times \exp(\alpha d) \tag{8.24}$$

TABLE 8.1
Constants F and G in Electron-Attaching Gases: Chlorine (Cl_2); Sulfur Hexafluoride (SF_6); Freon-12 (CCl_2F_2); Silicon Tetrachloride ($SiCl_4$); CF_3SF_5

Gas	Range of E/N	F	G	G/F	Reference	Designation
Cl_2	217–465	5.43(−20)	1.00(−18)	18.59	Bozin and Goodyear[13]	A, A'
SF_6	279–465	4.22(−20)	9.25(−19)	21.93	Bhalla and Craggs[14]	B, B'
CCl_2F_2	68–248	3.26(−20)	9.81(−19)	30.10	Harrison and Geballe[15]	C, C'
$SiCl_4$	310–620	1.28(−19)	2.08(−18)	16.28	Swamy and Harrison[12]	D, D'
CF_3SF_5	542–698	5.50(−19)	2.84(−18)	5.16	Harrison and Geballe[15]	E, E'
O_2	93–155	6.3(−21)	3.65(−19)	57.67	Prasad and Craggs[16]	
CO_2	81–155	8.23(−21)	4.78(−19)	57.48	Bhalla and Craggs[17]	
CO	112–174	1.91(−20)	8.45(−19)	44.22	Bhalla and Craggs[18]	
Air	310–2480	4.66(−20)	1.13(−18)	24.25	Rao and Raju[20]	

E/N in units of Td, F in units of m^2, G in units of $(V\ m^2)$ $a(b)$ means $a \times 10^b$. Designation refers to Figure 8.8. Table reproduced from Swamy, M. N. and J. A. Harrison, *Brit. J. Appl. Phys.*, 2, 1437, 1969. With permission of the Institute of Physics (IOP), U.K.

The electric field due to these electrons as a congregate at a distance $\bar{r}$, assuming a spherical distribution, is given by

$$E_r = \frac{Q}{4\pi\varepsilon_0 \bar{r}^2} \tag{8.25}$$

where ε_0 is the permittivity of free space ($= 8.854 \times 10^{-12}\,\text{F/m}$) and $\bar{r}$ is the mean radial length, given by Equation 1.100 as

$$\bar{r} = \sqrt{3Dt} \tag{8.26}$$

The time t may be expressed as

$$t = \frac{d}{W} = \frac{d}{\mu E} \tag{8.27}$$

Substituting Equation 8.27 into 8.26, one gets

$$\bar{r} = \left(3D\frac{d}{\mu E}\right)^{1/2} \tag{8.28}$$

Substituting Equation 8.28 into 8.25, one gets, for the radial electric field due to the avalanche, the expression

$$E_r = \frac{Q}{4\pi\varepsilon_0}\frac{\mu E}{3dD} \tag{8.29}$$

The ratio D/μ is the characteristic energy, equal to

$$\frac{D}{\mu} = \frac{kT}{e} \tag{8.30}$$

The mean energy is given by (assuming a Maxwellian distribution of electron energy)

$$\bar{\varepsilon} = \frac{3}{2}\frac{D}{\mu} \tag{8.31}$$

Substituting Equation 8.31 into 8.29, one gets

$$E_r = \frac{Q}{4\pi\varepsilon_0}\frac{1}{2\bar{\varepsilon}d}E \tag{8.32}$$

The ratio E_r/E signifies the field distortion. Q may be calculated for a given distortion as a percentage of the applied field.

To show the method of applying Equation 8.32, we substitute the following values for N_2: let $d = 0.02$ m, $E = 3$ MV/m at a gas pressure of 101 kPa. The ratio $E/N = 122$ Td and the corresponding mean energy is 3.75 eV (Figure 6.46). Substituting these values in Equation 8.32, one gets $Q = 1.7 \times 10^{-11}$ C, which is equivalent to $\sim 10^8$ electrons for 10% electric field distortion. It should be remembered that the number of electrons multiplies exponentially and the field distortion increases rapidly for marginally higher E/N under the same conditions. Further, if one assumes that the ionic space charge is spherical in shape, the electric field will be reduced at the anode and enhanced at the cathode, departing from the uniformity of the field. Such an analysis has been provided by Raju[1] for the electric field at both the anode and the cathode. A current density of 0.01 to 0.1 A m^{-2} is the threshold for the onset of space charge effects.

We can approach the calculation of the space charge from a different angle. Let a number N_0 of initial electrons be released from the cathode and let N' be the ion density at the head of the electron avalanche. Assuming that the charge is uniformly spread in a spherical shape of radius r, the radial electric field may be calculated by applying the relationship

$$\hat{D} = \varepsilon_0 E \tag{8.33}$$

where $\hat{D}$ is the electric flux density (C/m^2). The surface charge density is given by

$$D = \frac{eN'}{4\pi r^2}\left(\frac{4\pi r^3}{3}\right) = \frac{N're}{3} \tag{8.34}$$

Since N' is the ion density, it is approximately equal to

$$N' = \frac{N_0\alpha\,e^{\alpha d}}{\pi r^2} \tag{8.35}$$

Substituting Equations 8.34 and 8.35 into 8.33, one gets the radial field as

$$E = eN_0\frac{\alpha e^{\alpha d}}{3\pi\varepsilon_0 r} \tag{8.36}$$

The space-charge field is created by a larger number of initial electrons (N_0) or larger multiplication ($e^{\alpha d}$). The former is important at low gas pressure discharges and the latter at high pressures (>0.2 MPa) where the influence of single avalanche dominates. If the electron emitting area is 0.01 m^2, then the initial current due to N_0 electrons is about 10^{-8} A in low pressure discharges. In high pressure discharges the number of electrons at the avalanche head is $\sim 10^8$, which causes appreciable space charge distortion.

8.5 BREAKDOWN IN UNIFORM FIELDS

Reverting to Equation 8.7 for current growth, one finds that the current tends to infinity if the denominator becomes zero and the current in the gap is then limited by the external resistance in the circuit only. The point of operation, B (Figure 8.1), has been reached and the current becomes self-sustaining due to the regenerative mechanisms. The condition for breakdown is expressed as

$$1 - \gamma\left[e^{\alpha d} - 1\right] = 0 \quad \text{or, approximately,} \quad \gamma e^{\alpha d} = 1 \tag{8.37}$$

The physical interpretation is that each primary electron released from the cathode generates a secondary electron at the cathode. Using the first part of Equation 8.37, the sparking distance d_s is expressed as

$$d_s = \frac{1}{\alpha} \ln\left(1 + \frac{1}{\gamma}\right) \tag{8.38}$$

We have already discussed the functional dependence of α/N on E/N according to Equation 8.13. For convenience, one can express

$$\frac{\alpha}{N} = f\left(\frac{E}{N}\right) \tag{8.39}$$

If we assume, for now, that γ is also a function of E/N, say $\phi(E/N)$, then Equation 8.38 may be expressed as

$$d_s = \frac{1}{Nf(E/N)} \ln\left[1 + \frac{1}{\phi(E/N)}\right] \tag{8.40}$$

Since we are considering a uniform electric field without space charges, we can substitute $V_s = E_s/d$ and Equation 8.40 becomes, dropping the subscript of d_s,

$$Nd = \frac{1}{f(V_s/Nd)} \ln\left[1 + \frac{1}{\phi(V_s/Nd)}\right] \tag{8.41}$$

This equation means that the sparking voltage is a function of E/N only and does not depend on the individual values of N and d. This law is known as the Paschen law and it has been verified in a number of gases.[19]

By substituting Equation 8.13 in 8.38 and rearranging terms, the sparking voltage as a function of Nd is obtained as

$$V_s = \frac{GNd}{\ln\left(\dfrac{FNd}{\ln(1 + 1/\gamma)}\right)} \tag{8.42}$$

To obtain the minimum sparking voltage, Equation 8.42 is differentiated and equated to zero. Accordingly, one gets $Nd_{\min}$ as

$$Nd_{\min} = \frac{e}{F} \ln\left(1 + \frac{1}{\gamma}\right) = \frac{2.718}{F} \ln\left(1 + \frac{1}{\gamma}\right) \tag{8.43}$$

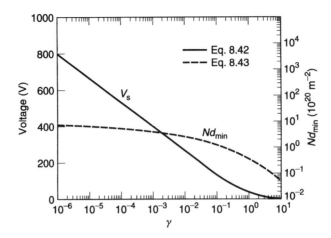

FIGURE 8.9 Plot of Equations 8.43 and 8.44 as a function of γ in dry air. Both equations are satisfied at the same value of γ.

Substitution of Equation 8.43 into 8.42 simplifies the minimum sparking voltage to

$$V_{s,min} = 2.718 \frac{G}{F} \ln\left(1 + \frac{1}{\gamma}\right) \tag{8.44}$$

As an example of application of Equations 8.42 and 8.43, we substitute the values of air, $F = 5.16 \times 10^{-20}\,m^2$, $G = 1.10 \times 10^{-18}\,V\,m^2$.[20] A value of $\gamma = 1.5 \times 10^{-3}$ gives $Nd_{min} = 3.43 \times 10^{20}\,m^{-2}$ and $V_{s,min} = 377\,V$, which compares well the experimental value of $380\,V$.[21] Figure 8.9 shows the curves as a function of γ; the same value of γ should satisfy both equations.

It should be remembered that the secondary ionization coefficient is a notoriously non-reproducible quantity since it depends upon the surface conditions of the cathode. The factors that change γ are the metal of which the cathode is made, the degree of smoothness, the oxidation layer, whether the cathode has been exposed to a spark discharge, and so on. Even under ultrapure and high vacuum conditions a single spark discharge changes the surface considerably, changing the coefficient γ by an order of magnitude. Since the Paschen minimum voltage depends on γ, $V_{s,min}$ also changes, though not to the same extent. It can be reproduced within a few volts of uncertainty. Figure 8.10 shows measured γ in dry air for several experimental conditions.

Two additional factors come into effect in the determination of secondary electron yields and are not always taken into account. The first factor is that the electron has to travel a certain minimum distance, d_0, in the field direction before attaining equilibrium with the electric field. The second factor is the "electron escape fraction" (f_{es}) that depends upon the local field.[22] The current growth equation that includes these factors is given by

$$i = \frac{i_0 f_{es} \exp[\alpha(d - d_0)]}{1 - \gamma\{\exp[\alpha_e(d - d_0)] - 1\}} \tag{8.45}$$

The electron yield per Ar^+ ion is discussed by Phelps and Petrović.[22] The data are collected from Schade,[23] Kruithof,[24] Kachikas and Fisher,[25] Menes,[26] Golden and Fisher,[27] Guseva,[28] Klyarfel'd et al.,[29] Heylen,[30] Pace and Parker,[31] Bhasavanich and Parker,[32] Auday et al.,[33] and Petrović and Phelps.[34] The experimental conditions included in the citation of these

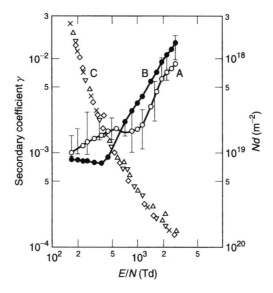

FIGURE 8.10 Secondary ionization coefficient in dry air for a gold-plated cathode. Curve A, before running a glow discharge; curve B, after running a glow discharge. The plot of Nd vs. E/N (curve C) is shown to display the range of parameters used. Symbols indicate the different gap lengths as delineated in the original publication.[20]

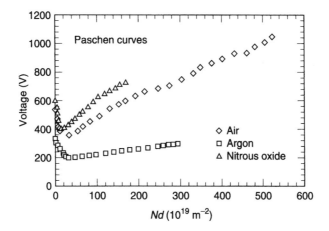

FIGURE 8.11 Paschen minima in several gases, measured by Raju and Saeed (2003, unpublished).

references at the end of the chapter are reproduced from Phelps and Petrović.[34] The secondary electron yield initially decreases up to $\sim$200 Td and, after reaching a minimum, increases again. The high values on the left of the minimum are attributed to the large VUV emission during collision of electrons with argon atoms. The large values of γ_{eff} at $E/N \geq 300$ Td are attributed to the combination of increasing electron yield per ion and the ionization produced by Ar^+ ions and fast atom collisions with Ar. The relative contributions of various secondary processes to the secondary emission are shown in Figure 8.11.

Typical values of the Paschen minimum voltage and corresponding $(Nd)_{min}$ are shown in Table 8.2. Figure 8.11 shows the result of recent measurements in several gases, using high vacuum techniques.

TABLE 8.2
Paschen Minimum Voltage in Gases

Gas	V_{smin} (V)	Nd_{min} ($10^{20}\,m^{-2}$)	Gas	V_{smin} (V)	Nd_{min} ($10^{20}\,m^{-2}$)
Nitrogen	270	2.2	He	161	2.10
Air	330	1.83	CO_2	418	1.64
O_2	435	2.25	Ar	245	1.90
H_2	275	3.7	N_2O	420	1.61

It is appropriate at this juncture to refer to the idea of relating the dielectric strength of a gas to its molecular properties. The matter was explored with reasonable success in hydrocarbons, which are nonattaching, by Devins and Sharbaugh[35] and explored further by Vijh.[36] A large number of gases show a rough empirical correlation between dielectric strength and boiling point. Qualitatively, it was argued that a high boiling point for a gas was characteristic of a large molecule and therefore a large collision cross section. The attachment cross section is, of course, a molecular property, and larger attachment cross sections result in higher dielectric strength. However, in the absence of these data for many molecules with several halogen atoms, a simpler approach would be advantageous.

On the basis of these reasonings, Vijh[36–38] has approached the problem on several fronts: the heat of atomization per mole,[37] boiling point,[38] and molecular weight.[36] From the Paschen law,

$$V_s = f(Nd) \tag{8.42}$$

for a given gap length d, one can write

$$V \propto N \tag{8.46}$$

According to the Avagadro hypothesis, equal numbers of molecules are contained in equal volumes of all gases under the same conditions. Therefore, higher molecular weight gases have higher density and one can write

$$V \propto \text{molecular weight} \tag{8.47}$$

Table 8.3 and Figure 8.12 shows this kind of correlation for 40 gases, though the data fall into two groups. Class A gases have higher dielectric strength than class B gases. Class A gases have a –CN group or an –SO fragment and do not contain a hydrogen atom. These observations possibly provide a clue to realization of higher dielectric strength.

Returning to the Paschen law, the ratio E/N at the Paschen minimum is high, between 700 and 1000 Td, and therefore the ionization coefficient we are dealing with is the true ionization coefficient. However, as the number density increases the sparking voltage increases and the ratio E/N becomes smaller. At these number densities, $\sim 10^{25}\,m^{-3}$, the breakdown is not so much controlled by the cathode as by the photoionization process in the gas. The phenomenon becomes more controlled by the space charge development; the criterion for breakdown has been derived as[2]

$$x_c = \frac{1}{\alpha}\left[\ln\frac{8\pi\varepsilon_0\bar{\varepsilon}}{e} + \ln x_c + \ln\frac{E_r}{E}\right] \tag{8.48}$$

where $\bar{\varepsilon}$ is the mean energy, which is a function of E/N, and x_c is known as the critical avalanche length, defined as the length of the avalanche at the head of which the radial

TABLE 8.3
Relative Electric Strengths (RES) at 0.1 MPa and Molecular Weights of Dielectric Gases

Number	Gas	RES	Mol. Wt.	B.P. (°C)	Class
1	He	0.15	4	−269	B
2	Ne	0.25	20	−246	B
3	H_2	0.50	2	−253	B
4	CH_3NH_2	0.80	31	6.8	B
5	CO_2	0.88	44	−28.5	B
6	N_2	1.0	28	−196	B
7	C_2H_5Cl	1.00	64	12.8	B
8	CH_4	1.00	16	−161.4	B
9	CH_3CHO	1.01	44	19.8	B
10	CF_4	1.01	88	−182	B
11	CO	1.02	28	−191.6	B
12	S_2Cl_2	1.02	135	137.1	B
13	CH_2ClF	1.03	68	—	B
14	$(CH_3)_2NH$	1.04	45	7.4	B
15	$C_2H_5NH_2$	1.06	45	16.51	B
16	C_2H_2	1.10	26	−83.5	B
17	SO_2	1.3	64	−10	B
18	$CHCl_2F$	1.33	103	8.9	B
19	CF_3Br	1.35	149	—	B
20	$CHClF_2$	1.40	86	−40.8	B
21	$CClF_3$	1.43	104	−81	B
22	Cl_2	1.55	71	−34.6	B
23	C_2F_6	1.82	138	−78	B
24	C_3F_8	2.19	188	−37	B
25	CH_3I	2.20	142	42.4	B
26	C_2ClF_5	2.3	154	−38.7	B
27	CCl_2F_2	2.42	121	−30	B
28	SF_6	2.5	146	−64	B
29	SOF_2	2.50	86	—	A
30	$C_2Cl_2F_4$	2.52	171	3.8	B
31	ClO_3F	2.73	102	—	A
32	C_4F_8	2.8	200	−6	B
33	C_4F_{10}	3.08	238	−2	B
34	CF_3CN	3.5	95	−63	A
35	CCl_3F	3.50	137	24.1	A
36	$CHCl_3F$	4.2	139	61.3	A
37	C_2F_5CN	4.5	145	−35	A
38	C_3F_7CN	5.5	195	1	A
39	C_5F_8	5.5	212	—	A
40	CCl_4	6.33	154	76.7	A

Numbers in Figure 8.12 correspond to the first column. Rearranged by the author in order of increasing RES: RES $1 = 30.54\,kV/cm$.
Table reproduced from Vijh, A.K., *IEEE Trans. Elec. Insul.*, EI-12, 313, 1977. With permission © 1977, IEEE.

field approaches having the same magnitude as the applied field. The equation is known as Meek's criterion for space charge dominated breakdown.

The critical avalanche length can be smaller or larger than the gap length. If smaller, a streamer type of breakdown occurs; if not, the breakdown will be controlled the secondary processes. One of the primary requirements for the development of a space charge field is that ionization should occur ahead of the electron avalanche by photoionization.

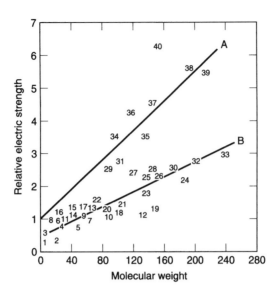

FIGURE 8.12 Relative electric strength as a function of molecular weight in gases. The numbers on the plot correspond to those in Table 8.3.

In gases in which photoionization occurs by production of energetic photons, or mixtures of gases in which the excitation potential of one component is higher than the ionization potential of the second, breakdown occurs by the streamer mechanism if the gas pressure and gap length are such that Equation 8.48 is satisfied. Details of streamer theory and its application to several gases may be found in Meek and Craggs.[2]

8.6 MULTIPLICATION IN NONUNIFORM FIELDS

We shall now devote our attention to multiplication in nonuniform electric fields that exist between dissimilar electrodes. A concentric wire–cylinder, concentric spheres, sphere–plane electrodes are examples of dissimilar electrodes.

The current growth in a nonuniform electric field in the absence of secondary effects is given by

$$i = i_0 \exp \int_0^d \alpha dx \tag{8.49}$$

If the secondary effects are included, the current growth equation in a nonuniform electric field, analogous to Equation 8.7, is

$$i = i_0 \frac{\exp \int_0^d \alpha dx}{1 - \gamma \left[\exp \int_0^d \alpha dx - 1 \right]} \tag{8.50}$$

In Equations 8.49 and 8.50 the first ionization coefficient, α, is obtained from uniform electric field measurements as discussed in Section 8.2.1. The question is whether the uniform-field values of α give the true multiplication in nonuniform electric fields. This question is answered, in principle, by measuring experimentally the current multiplication in the nonuniform electric field geometry and calculating the same by using

Equation 8.49. However, there are several practical limitations to obtaining an unambiguous answer.

The factors that should be taken into account in calculation are the spatial variation of the electric field and the true value of α that one should use in the calculation. The electric field variation is calculated from electrostatics in well-defined geometries such as coaxial cylinders or concentric spherical electrodes. Of course, this method assumes that there are no edge effects, an assumption that is almost invariably inapplicable unless elaborate numerical computations are resorted to. With regard to the true value of α the differences between the results of various workers lead to further uncertainty. To alleviate this difficulty, Shallal and Harrison[39] have measured the ionization coefficient in hydrogen in both uniform and nonuniform fields, assuming that the same gas purity and the technique cancel out the uncertainty.

The experimental difficulties in measuring the ionization currents in nonuniform fields also include the edge effects of the electrodes. In concentric cylinder geometry the cylinders should end bell shaped.[39] In sphere–plane geometry the distance to the walls should be large enough to avoid distortion of the electric field due to nearby earthed objects. To determine the current multiplication in nonuniform fields accurately, several data points are required at different Nd but at the same E/N; this is achieved only by the N variation method because a change of electrode spacings will mean a new set of electrodes with a different distribution of the electric field in the inter-electrode geometry. The pressure variation method can be adopted only in the case of gases in which the ionization and attachment coefficients are not pressure dependent.

Measurement of the initial photoelectric current from the cathode also presents several problems when compared with uniform field studies. In the pressure variation technique the initial photoelectric current decreases more rapidly with increasing pressure, due to higher absroption of the ultraviolet light by the gas, than with the distance variation technique. Gosseries[40] and Raju[41] have described methods of analysis that do not require the knowledge of i_0. Shallal and Harrison[39] have shown a method of taking back-diffusion into account in the analysis.

There have been very few studies of current growth experiments in gases, for these reasons. Fisher and Weissler[42] used confocal paraboloids in hydrogen and concluded that the use of α from uniform field studies was invalid in nonuniform fields. Morton[43] used concentric cylinder geometry in hydrogen; Johnson[44] used the same geometry in air and hydrogen. All of these investigators concluded that the observed ionization was greater or smaller than that given by uniform electric field integration, provided that the variation of electric field intensity was more than 2% over an electron mean free path.

Golden and Fisher[45] made a reanalysis of these measurements and Shallal and Harrison[39] measured the ionization currents in both uniform and nonuniform electric fields using the same apparatus (except for the electrode geometry) and same gas sample. They concluded that there is agreement between measured and calculated values of ionization currents in nonuniform electric fields even when the electric field changes markedly over an electron mean free path. Liu and Raju[46,47] have made theoretical calculations using Monte Carlo simulation techniques, and concluded that the ionization coefficients are higher than uniform field values where the electric field from cathode to anode is decreasing (the smaller electrode is the cathode). On the contrary, the ionization coefficients in nonuniform fields are lower than the uniform field values where the electric field from the cathode to anode is increasing.

There has been no measurement of ionization currents in nonuniform fields by the use of the current pulse technique. There is a critical need for experimental data in electron attaching gases.

The breakdown criterion, or corona inception voltage if the field is sufficiently divergent, is given by equating the denominator of Equation 8.50 to zero, i.e.

$$\gamma\left[\exp\int_{d_0}^{d}\alpha dx - 1\right] = 1 \tag{8.51}$$

Application of Equations 8.51 and 8.13 to a given geometry gives the sparking voltage as a function of Nd. Some useful formulas are collected below.

8.6.1 ELECTRIC FIELD ALONG THE AXIS OF CONFOCAL PARABOLOIDS

The smaller electrode is a point paraboloid and the larger electrode is a hollow paraboloid. The electric field along the axis of a pair of paraboloids is given by Fisher and Weissler[48] as

$$E = \frac{V}{\ln(f/F)}\frac{1}{x+f} \tag{8.52}$$

where f and F are the focal lengths of the point and plate paraboloids, and x is the axial distance from the point.

8.6.2 RADIAL ELECTRIC FIELD IN A COAXIAL CYLINDRICAL GEOMETRY

The radial electric field in a coaxial cylindrical geometry is given by Raju and Gurumurthy[49] as

$$E_r = \frac{V}{R\ln(R_2/R_1)} = \frac{V}{R\ln(n)} \tag{8.53}$$

where R_1 and R_2 are the radii of the inner and outer cylinders respectively, $n = R_2/R_1$.
The criterion for self-sustaining discharge is

$$\int_{R_1}^{R_2}\left(\frac{\alpha}{N}\right)Ndr = \ln\left(1+\frac{1}{\gamma}\right) \tag{8.54}$$

Combining Equations 8.53, 8.13, and 8.54, one obtains the sparking voltage as

$$\frac{F}{G}\frac{V_s}{\ln(n)}\left[\exp\left(-\frac{GNR_1\ln(n)}{V_s}\right) - \exp\left(-\frac{GNnR_1\ln(n)}{V_s}\right)\right] = \ln\left(1+\frac{1}{\gamma}\right) \tag{8.55}$$

Substituting

$$n = \frac{R_2}{R_1} \quad \text{and therefore } R_1 = \frac{(R_2 - R_1)}{(n-1)} \tag{8.56}$$

Equation 8.55 may be written as

$$\frac{FV_s}{G\ln(n)}\left[\exp\left(-\frac{GN(R_2-R_1)\ln(n)}{V_s(n-1)}\right) - \exp\left(-\frac{GN(R_2-R_1)n\ln(n)}{(n-1)V_s}\right)\right]$$
$$= \ln(1+1/\gamma) \tag{8.57}$$

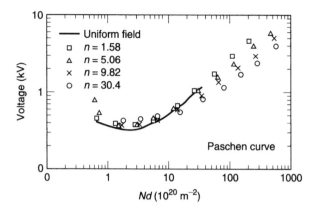

FIGURE 8.13 Sparking voltages in nitrogen as a function of Nd. Here $d = (R_2 - R_1)$ and $n = R_2/R_1$. On the left of the Paschen minimum, increasing n at the same d increases the sparking voltage. On the right of the minimum the converse is true.[21]

The reason for arranging Equation 8.57 in this format is to provide a methodology for comparing the sparking voltage as a function of $(R_2 - R_1)N$ which is analogous to the quantity Nd in uniform electric fields. Figure 8.13 shows the Paschen curves for several ratios of n and uniform field in nitrogen for positive polarity of the smaller electrode. Similar data for the negative polarity of the central electrode in dry air are given by Raju and Gurumurthy.[49]

We note the following effect with regard to the Paschen minimum in nonuniform electric fields. On the right side of the minimum the uniform field has the highest sparking voltage whereas increasing nonuniformity (increasing n) decreases the voltage. The reason is that increasing n increases the electric field, and hence E/N, at the central electrode much more than at the outer electrode. At the left of the Paschen minimum, however, increasing the ratio n increases the sparking voltage at the same E/N. This observation is attributed to the shape of the α/N–E/N curve, which has a maximum for each gas. On the right of the Paschen minimum the uniform electric field has the highest dielectric strength of any geometry for the same Nd. On the left, on the other hand, the uniform field has the lowest dielectric strength.

To compare the Paschen minimum one differentiates Equation 8.57 with respect to Nd and equates it to zero,

$$\frac{\partial V_s}{\partial (Nd)} = 0 \tag{8.58}$$

and one substitutes the value of $(Nd)_{min}$ to get the sparking voltage as

$$V_{s,min} = \frac{G}{F} \ln\left(1 + \frac{1}{\gamma}\right) \frac{\ln(n)}{n - 1} n^{n/(n-1)} \tag{8.59}$$

The ratio of the minimum voltage in a nonuniform field to that in a uniform field is obtained by dividing Equation 8.59 by Equation 8.44,

$$\frac{V_{s,min}(\text{nonuniform})}{V_{s,min}(\text{uniform})} = \frac{\ln(n)}{2.718(n - 1)} (n)^{n(n-1)^{-1}} \tag{8.60}$$

The minimum sparking voltage in a nonuniform field is higher than that in a uniform field. For $n = 1.58$ the ratio is 1.009 whereas for $n = 30$ it is 1.456.

8.6.3 CONCENTRIC SPHERE–HEMISPHERE

The electric field between concentric sphere and hemisphere of radius R_1 for the inner sphere and R_2 for the outer sphere is given by Raju and Hackam[50] as

$$E = \frac{R_1 R_2}{r^2 (R_2 - R_1)} V \tag{8.61}$$

Substitution of Equations 8.61 and 8.13 in Equation 8.54 yields the condition for self-sustaining discharge as

$$\ln\left(1 + \frac{1}{\gamma}\right) = \pi^{1/2} FN \left\{ \frac{\operatorname{erf}(K^{1/2} R_2) - \operatorname{erf}(K^{1/2} R_1)}{2K^{1/2}} \right\} \tag{8.62}$$

where

$$K = GN \frac{(R_2 - R_1)}{V_s R_1 R_2} \tag{8.63}$$

It is noted that, for a value of R_2 greater than a critical value, the outer sphere has negligible influence on V_s because intense ionization takes place near the inner spherical electrode.

8.6.4 SPHERE–PLANE

The electric field along the axis of a sphere–plane gap is given by Peek[51] as

$$E_x = V \frac{2\partial [\partial^2 (f+1) + x^2 (f-1)]}{[\partial^2 (f+1) - x^2 (f-1)]^2} \tag{8.64}$$

where E_x = electric field strength at a distance x from the plane along the perpendicular from the center of the high voltage sphere to the plane, V = voltage applied between the electrodes, and f = a constant, defined by

$$f = \frac{1}{4}\left[(2p-1) + \sqrt{(2p-1)^2 + 8} \right] \tag{8.65}$$

where $p = 1 + (\partial/R)$; ∂ = gap length, R = radius of the sphere. By inserting $x = \partial$ one gets the field strength on the surface of the sphere.

8.6.5 SPHERE–SPHERE

Equation 8.64 may easily be modified for a sphere–sphere gap with equal radii by assuming that the mid plane is at zero potential.

8.7 RECOMBINATION

Recombination of charge carriers is an electron loss mechanism and recombination may take place in the volume or at a surface. We restrict ourselves to the former case. Photo-ionization occurs when photons having sufficient energy strike an atom and two charge carriers, an electron and a positive ion, result. Ionization by an electron results in three charge carriers, two electrons and a positive ion. Photo-ionization involves negligible momentum transfer and reversing it requires particles that have virtually zero velocity. Reversing electron-impact ionization takes three particles to come together to conserve energy and momentum. This mechanism (often referred to as collisional radiative recombination) is represented by

$$e + e + X^+ \rightarrow X + e \text{ (higher energy)}$$

The volume recombination is facilitated by electron attachment[52]

$$e + XY \rightarrow XY^-$$
$$XY^- + A^+ \rightarrow XY + A$$

Radiative recombination (reversal of photo-ionization) occurs according to

$$e + X^+ \rightarrow X + h\nu$$

The recombination rate coefficient is expressed in units of $m^3 s^{-1}$. The three-body collisional recombination rate coefficient is dependent on the electron number density and the energy of the electrons. It increases with the number density and decreases with the energy. A representative rate coefficient in helium is $10^{-15} m^3 s^{-1}$ at an electron number density of $10^{16} m^{-3}$ and at a temperature of 300 K.[53] In view of the larger electron densities and low electron energies prevailing in plasmas, recombination is a considerably more important process in plasmas than in self-maintained discharges. In weakly ionized gases and Earth's atmosphere, a process known as dissociative recombination is likely to be more significant.[54] The dissociative recombination cross section in gases is of the order $\sim 10^{-20} m^2$, although it could be as large as $\sim 10^{-17} m^2$.[55]

8.8 DATA ON IONIZATION COEFFICIENTS

8.8.1 RARE GASES

We now present data on measured Townsend's first coefficient in gases in the range of E/N covering from onset to 1000 Td, supplementing the data from theoretical studies where experimental data are not available. Experimental data are so copious that one has to restrict the number of authors cited. Tabulated values are given in Appendix 4. Note that, for economy of space, data are sometimes interpolated at desired values of E/N. Limited data on recombination are also presented as the process represents an electron loss mechanism. The compilation of references up to the year 1979 has been made available by Beaty et al.[56]

8.8.1.1 Argon (Ar)

Experimental measurements of α/N in argon have been presented by: Kruithof[57]; Davies and Milne[58]; Heylen[59]; Lakshminarasimha and Lucas[76b] Abdulla et al.[60]; Kruithof and Penning,[61] mixtures of Ar + Ne; Heylen,[62] mixtures of argon + methane (CH_4) and argon + propane (C_3H_8); and Yamane,[63] mixtures of several hydrocarbons and rare gases. Analytical expressions have been given by Ward,[64] and Phelps and Petrović.[22]

The measurement of α/N in rare gases is notoriously difficult due to the disproportionately large influence of impurities, including the Penning effect (see below). Kruithof and

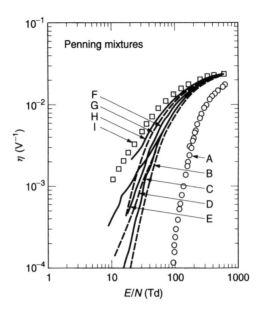

FIGURE 8.14 Primary ionization coefficient as a function of reduced field in argon, ethane, and mixtures thereof. (A) C_2H_6; (B) 99.995% Ar; mixtures with Ar (B); (C) 0.000 03% C_2H_6; (D) 0.0003% C_2H_6; (E) 99.5% Ar; mixture with Ar (E); (F) 3% C_2H_6; (G) 0.3% C_2H_6; (H) 0.03% C_2H_6; (I) 0.003% C_2H_6. Note the order of increase of η with increasing percentage of C_2H_6.[59]

Penning,[61] and Heylen[59] have devoted particular attention to this aspect, employing a bakable system, high purity gases, and chemically stable electrodes such as gold. Figure 8.15 shows the values given by Phelps and Petrović[22] according to their equation

$$\frac{\alpha}{N} = 1.1 \times 10^{-22} \exp\left[\frac{-72N}{E}\right] + 5.5 \times 10^{-21} \exp\left[\frac{-187N}{E}\right]$$

$$+ 3.2 \times 10^{-20} \exp\left[\frac{-700N}{E}\right] - 1.5 \times 10^{-20} \exp\left[\frac{-10^4 N}{E}\right] \tag{8.66}$$

where E/N is expressed in units of Td. Tabulated values are given in Appendix 4. A further analysis is given in Chapter 11 (Figure 11.10).

When one gas is mixed in increasing proportion with another gas, it is usually observed that certain swarm properties change from the first to the second in a fairly monotonic way. However, this behavior is not conformed to when the metastable excitation level of one gas (ε_{me}) exceeds the ionization potential (ε_i) of the second gas. Two conditions should be satisfied for Penning ionization to occur. The first one is that the energy for excitation of one species should be higher than the ionization potential of the other species. The second condition is that the excited level should survive long enough in the gap for a collision with a neutral atom or molecule to occur. The average time to diffuse to the cathode from a distance x is given by x^2/D, where D is the diffusion coefficient in m²/s, and a collision should occur within this time.

Penning and Addinck[65] observed that the sparking potential of neon decreased with minute traces of argon. Kruithof and Penning[61] observed that the first ionization coefficient increased out of proportion for small quantities of argon additive. This behavior was first observed by Penning and his colleagues and the effect is usually referred to as the Penning

TABLE 8.4
Penning Effect Cross Section in Selected Gases, in Units of $10^{-20}\,m^2$

A*		Energy (A*) (eV)	B					
			Ar (15.759)	Kr (13.999)	Xe (12.130)	Ne (21.564)	N_2	O_2
He	$2\,^3S$	19.82	7	8.8	12		5.3	14
	$2\,^1S$	20.62	25	34	40		11	25
	$2\,^1P$	21.22	86	89	84		69	69
	$3\,^1P$	23.09	56	50	73			
Ne	$^3P_2$	16.619	3.1	1	12		6	1.7
	$^1P_1$	16.848						4
	$^3P_0$	16.716						2.8
	$^3P_1$	16.671						2.8
Ar	$^3P_2$	11.55		1				1.2
	$^1P_1$	11.83						1.8
	$^3P_0$	11.72						1.8
	$^3P_1$	11.623						1.6
Kr	$^3P_2$	9.915						1.9
	$^1P_1$	10.644						1.6
	$^3P_0$	10.563						1.8
	$^3P_1$	10.033						2.2

A* and B refer to Equation 8.67. Data taken from Grigoriev and Meilikhov.[70] Numbers in brackets are energy (eV). Row: ionization potential; column: excitation potential. With permission of CRC Press.

effect in the literature. Lagushenko and Maya[66] and Sakai et al.[67] have theoretically calculated the Penning effect in small quantities of argon in neon. The Penning effect is described by the reaction

$$A^* + B \rightarrow A + B^+ + e \qquad (8.67)$$

The Penning effect is not restricted to rare gas mixtures; it has been observed in helium + hydrogen and in neon + hydrogen mixtures with considerable intensity.[68] Argon–ethane mixtures have also exhibited the Penning effect[62]; an optimum concentration of 0.66% of ethane increased the ionization coefficient by a factor of 40 from that in pure argon. Figure 8.14 shows the results for selected percentages of ethane in the mixture. The measurements of LeBlanc and Devins[69] in ethane at higher values of E/N are also included for comparison. Table 8.4 shows the cross section for Penning ionization with selected gas mixtures.[70]

8.8.1.2 Helium (He)

Selected references for the reduced first ionization coefficients in helium are: Davies et al.[71]; Chanin and Rork,[72] 9 to 900 Td; Dutton and Rees[73] at low values of E/N (9 to 15 Td); Cavelleri[74]; Lakshminarasimha et al.[75]; and Kücükarpaci et al.,[76] simulation. Chanin and Rork[72] used a bakable system with cataphoretically purified helium and considered the nonequilibrium distance in the analysis.

At higher E/N ($> 200\,Td$) the results of Lakshminarasimha et al.[75] are lower by a few percent, but this is offset by the fact that Chanin and Rork[72] have given tabulated values.

TABLE 8.5
Quality of Fit of Equation 8.68

Range of applicable E/N	60–900 Td
Maximum deviation	−8.39%
Corresponding E/N	60 Td
Minimum deviation	−0.03%
Corresponding E/N	150 Td
R^2	0.9959

A positive difference means that the tabulated values are higher.

The latter values are shown in Figure 8.14 and Appendix 4. An analytical expression to represent reduced ionization coefficients as a function of E/N is

$$\frac{\alpha}{N} = \left[-9 \times 10^{-7} \left(\frac{E}{N} \right)^2 + 0.0016 \left(\frac{E}{N} \right) - 0.0287 \right] \times 10^{-20}\, \mathrm{m}^2 \tag{8.68}$$

The quality of fit with the tabulated values is shown in Table 8.5.

8.8.1.3 Krypton (Kr)

Selected references for the reduced first ionization coefficient in krypton are: Kruithof[77]; and Heylen[78]; and Specht et al.[79] The gas samples used by Kruithof[77] had a small amount (<0.05%) impurity of xenon, and it was not expected that the influence of the contaminant on α/N was important because the ionization potential of neon is higher than the excitation potential of xenon (no Penning effect). The theoretically simulated results of Kücükarpaci and Lucas[80] agree well with those of Kruithof[77] in the range of E/N shown.

8.8.1.4 Neon (Ne)

Selected references for the reduced first ionization coefficient in neon are: Kruithof,[81] 6 to 1100 Td; Chanin and Rork[82]; De Hoog and Kasdorp[83], Willis and Morgan,[84] 25 to 250 Td; Dutton et al.[85]; Buursen et al.[86]; Bhattacharya,[87] 30 to 600 Td. Simulation results are given by Kücükarpaci et al.[88]

The analytical expressions for α/N as a function of E/N are:

$$\frac{\alpha}{N} = \left[-8 \times 10^{-8} \left(\frac{E}{N} \right)^3 + 2 \times 10^{-5} \left(\frac{E}{N} \right)^2 + 0.0002 \left(\frac{E}{N} \right) - 0.0054 \right] \times 10^{-20}\, \mathrm{m}^2 \tag{8.69}$$

$$\frac{\alpha}{N} = \left[-9 \times 10^{-11} \left(\frac{E}{N} \right)^3 - 7 \times 10^{-7} \left(\frac{E}{N} \right)^2 + 0.0018 \left(\frac{E}{N} \right) - 0.029 \right] \times 10^{-20}\, \mathrm{m}^2 \tag{8.70}$$

The range of applicability and quality of fit are shown in Table 8.6.

8.8.1.5 Xenon (Xe)

Selected references for data on reduced ionization coefficients are: Kruithof[89] 6 to 1100 Td; Burkley and Sexton[90]; Bhattacharya,[91] 30 to 600 Td, Makabe and Mori[92]; and Specht et al.[93]

TABLE 8.6

Range of Applicability and Quality of Fit of Analytical Expressions for α/N of Krypton

Equation	8.69	8.70
Range of E/N (Td)	20–150	150–1000
Maximum deviation	5.19%	−4.21%
Corresponding E/N (Td)	150	150
Minimum deviation	−0.31%	−0.53%
Corresponding E/N (Td)	35	250
R^2	0.9999	0.9992

Positive values indicate that values shown in Table 1 of Appendix 4 are higher.

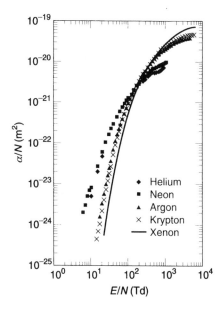

FIGURE 8.15 α/N as a function of E/N in rare gases. Tabulated values are shown in Appendix 4.

Figure 8.15 shows the data obtained by Kruithof[89] and Appendix 4 gives the tabulated values. The comments for krypton with regard to Kruithof's data apply equally to xenon. Analytical expressions for representing α/N as a function of E/N are

$$\frac{\alpha}{N} = \left[-10^{-8}\left(\frac{E}{N}\right)^3 + 10^{-5}\left(\frac{E}{N}\right)^2 - 0.0004\left(\frac{E}{N}\right) - 0.0022 \right] \times 10^{-20}\,\text{m}^2 \tag{8.71}$$

$$\frac{\alpha}{N} = \left[3 \times 10^{-11}\left(\frac{E}{N}\right)^3 - 5 \times 10^{-7}\left(\frac{E}{N}\right)^2 + 0.0033\left(\frac{E}{N}\right) - 0.235 \right] \times 10^{-20}\,\text{m}^2 \tag{8.72}$$

The range of applicability and quality of fit with the tabulated data in Appendix 4 are shown in Table 8.7.

TABLE 8.7
Range of Applicability and Quality of Fit of Analytical Expressions for α/N as a Function of E/N in Xenon with Tabulated Values of Appendix 4

Equation	8.71	8.72
Range of E/N (Td)	60–450	450–4000
Maximum difference	10.63%	−12.29%
Corresponding E/N (Td)	200	450
Minimum difference	0.24%	−0.51%
Corresponding E/N (Td)	60	2000
R^2	0.9996	0.9987

A positive difference means that the tabulated value is higher.

8.9 MOLECULAR GASES (NONATTACHING)

We shall now consider the data of reduced ionization coefficients as a function of E/N in selected, nonattaching, molecular gases.

8.9.1 HYDROCARBON GASES

8.9.1.1 Methane (CH₄)

Methane is one of the gases in which several measurements have been carried out over a wide range of E/N. Selected references are: Heylen[94]; Davies and Jones[95]; Schlumbohm[96]; Cookson and Ward[97]; Cookson et al.[98]; Vidal et al.[99]; Heylen[100a]; Hunter et al.[101]; and Davies et al.[102] Analyses of ionization currents prior to 1986 were carried out on the basis that CH_4 was purely a nonattaching gas. However, Hunter et al.[101] reported, using the pulse technique, that it is a weakly attaching gas. This result was subsequently questioned by Davies et al.,[102] who found no evidence for attachment in the drift tube measurements. The reduced ionization coefficients measured by Davies et al.[102] are in good agreement with the results of Heylen,[94] the latter being marginally higher. If the attachment process observed by Hunter et al.[101] is true, the true ionization coefficients measured before 1986 must be identified with the effective ionization coefficients. Figure 8.16 and Table 2 of Appendix 4 present these data.

The reduced ionization coefficient in methane as a function of E/N satisfies the equation[102]

$$\frac{\alpha}{N} = 2.51 \times 10^{-20} \exp\left[-\frac{624N}{E}\right] \text{m}^2 \qquad (8.73)$$

where E/N is expressed in units of Td. The reduced attachment coefficients are shown in Table 8.8, with possible collisional detachment occurring under swarm conditions.[102]

8.9.1.2 Other Hydrocarbons

Heylen[94, 100a,b] has carried out a series of careful measurements in several hydrocarbons to determine the applicability of Equation 8.13 to hydrocarbons and the relation of the molecular parameters F and G to the molecular weight. Townsend's α measurements have been reported by Le Blanc and Devins[103] for the series methane to n-hexane. Table 8.9

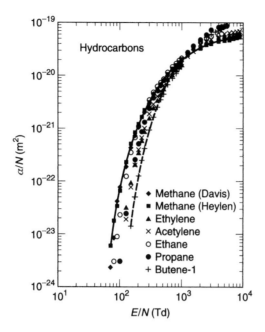

FIGURE 8.16 Reduced ionization coefficients in hydrocarbons as a function of E/N. Except for methane ($\blacklozenge$) all other measurements are due to Heylen: methane, ethylene, and acetylene[94]; ethane, propane, and butane[100a]; ethylene, propylene, and butene-1.[100b] Curves for heavy and complex hydrocarbons intersect those for lighter gases; the former have lower coefficients below the crossover E/N and higher coefficients above the crossover E/N. Tabulated values are given in Table 2 of Appendix 4.

TABLE 8.8
Theoretical Reduced Attachment Coefficient in Methane[102]

E/N (Td)	η/N (10^{-22} m^2)	E/N (Td)	η/N (10^{-22} m^2)
50	0.018	300	0.149
100	0.148	350	0.140
150	0.191	400	0.131
200	0.180	450	0.119
250	0.162	500	0.110

---

shows the constants determined by them, using the ionization current method. The following conclusions are drawn from these investigations.

1. The gas constants derived from ionization current measurements agree with those from sparking potential measurements, provided both cover the same range of E/N.
2. The molecular constants F and G increase with molecular weight and degree of unsaturation. Gases with triple bond have higher values than those with double bond (compare the values of acetylene and ethylene).
3. The secondary ionization coefficient, γ, is small ($< 10^{-4}$) in hydrocarbons and this fact can be made use of to suppress this coefficient in other gases, enabling current

TABLE 8.9
Gas Constants in Hydrocarbon Gases

Gas	$F (\times 10^{-20} \, m^2)$	G (Td)	E/N range (Td)	Reference
Methane	2.28	599	82–405	Cookson et al.[98]
Ethane	3.5	677	141–532	Heylen[94]
Ethylene	2.7	840	121–434	Heylen[94]
Acetylene	4.2	939	132–804	Heylen[94]
Propane	6.5	973	100–1889	Heylen[100a]
Butane	7.6	1136	166–1889	Heylen[100a]
Propylene	3.7	981	127–818	Heylen[100b]
Butene-1	5.4	1198	149–1120	Heylen[100b]

measurements to be made for larger multiplication ($e^{\alpha d} > 10^6$) to investigate the effects of space charge.

4. In general, heavy and complex hydrocarbons are better conductors (larger α/N) at higher E/N and better insulators at lower E/N.

Figure 8.16 and Table 2 of Appendix 4 show measured α/N as a function of E/N.

8.9.2 HYDROGEN AND NITROGEN

Hydrogen and nitrogen are nonpolar diatomic gases that have been studied quite extensively. We also include the data for deuterium in this section.

8.9.2.1 Hydrogen (H₂)

Selected references for measurements of α/N in H_2 are: Rose[104]; Folkard and Haydon[105]; Haydon and Robertson[106]; Chanin and Rork[107]; Barna et al.[108]; Golden et al.[109]; Haydon and Stock[110]; Tagashira and Lucas,[111] who derived a complicated equation to account for the loss of photons from the gap; Shimozuma et al.[112]; Blevin et al.,[113] photon flux method; Shallal and Harrison[114]; and Lakshminarasimha et al.[115]. Theoretical calculations are due to: Engelhardt and Phelps[116]; and Saelee and Lucas.[117]

Rose[104] determined the values of the gas constants in Equation 8.13 as

$$F = 1.44 \times 10^{-20} \, m^2; \quad G = 391.4 \, Td$$

Rose's[104] results were questioned for apparent neglect of secondary ionization and uncertainty of equilibrium of electron energy with applied voltage. However, Chanin and Rork[107] and Golden et al.,[109] who repeated the measurements, obtained values in agreement with those of Rose[104] up to 1500 Td. The measured values of Shallal and Harrison[114] above 600 Td fall between those of Rose[104] and Haydon and Stock.[110]

The experimental setup of Rose[104] was more sophisticated than that of many investigations that followed, such as purity of gas, high vacuum system, etc. Thus the values shown in Table 3 of Appendix 4 include those of Rose[104] as well as those given by Shallal and Harrison.[114] Figure 8.17 shows the reduced ionization coefficients as a function of E/N in all three gases: hydrogen, deuterium, and nitrogen.

With regard to γ, Fletcher and Blevin[118] have determined from measurement of photon flux in a Townsend discharge that significant secondary electrons are released by H^+ ions

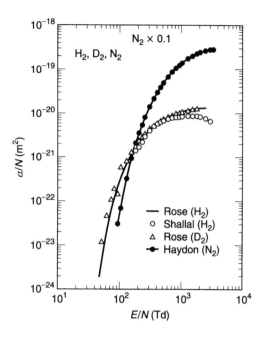

FIGURE 8.17 Townsend's first ionization coefficients in H_2, D_2 and N_2. Ordinates of N_2 have been multiplied by a factor of 10 for better presentation. Tabulated values are given in Table 4 of Appendix 4.

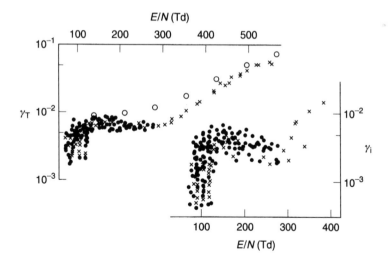

FIGURE 8.18 Secondary ionization coefficient in hydrogen with gold-plated electrodes. (o) Folkard and Haydon.[105] Figure reproduced from Shimozuma, M. et al., *J. Phys. D*, 10, 1671, 1977. With permission of IOP, U.K.

only above $E/N > 220\,\text{Td}$ in hydrogen and $E/N > 480\,\text{Td}$ in nitrogen. A rapid rise in the secondary ionization coefficient in these gases is observed above the stated E/N.

Figure 8.18 shows the total ionization coefficient (γ_T) and the contribution due to positive ions (γ_i) as a function of E/N, determined from pre-breakdown current measurements.[112] For $E/N < 110\,\text{Td}$ γ_i falls rapidly with decreasing E/N, suggesting that at low E/N the secondary action is mainly due to the photons. γ_T also decreases but is found

TABLE 8.10

Ionization Cross Sections Q_i in D_2 as a Function of Electron Impact Energy

Energy	Q_i Cowling and Fletcher[122a]	Rapp[122b]	Energy	Q_i Cowling and Fletcher[122a]	Rapp[122b]
16	0.034	0.034	60	0.977	0.977
17	0.097	0.104	70	0.986	0.981
18	0.159	0.173	80	0.981	0.974
19	0.218	0.239	90	0.964	0.958
20	0.273	0.300	100	0.946	0.939
21	0.325	0.355	125	0.890	0.877
22	0.378	0.404	150	0.833	0.813
23	0.423	0.454	175	0.781	—
24	0.470	0.498	200	0.735	0.716
25	0.515	0.537	250	0.661	0.638
30	0.700	0.699	300	0.604	0.576
35	0.809	—	350	0.553	0.523
40	0.876	0.876	400	0.510	0.482
45	0.917	—	450	0.469	0.446
50	0.951	0.950	500	0.431	0.414

Energy in units of eV, cross sections in units of $10^{-20}\,m^2$.

to be pressure dependent. For $110 \leq E/N \leq 310$ Td γ_i and γ_T do not vary much and, of course, $\gamma_i < \gamma_T$ because of partial contributions from photons (γ_p). For $E/N > 300$ Td both γ_T and γ_i increase again, due to an increased positive ion contribution.

8.9.2.2 Deuterium (D_2)

Selected references are: Rose[119]; Barna et al.[120]; Rork and Chanin[121]; and Cowling and Fletcher.[122] The latter investigations are notable because they measured both ionization cross sections and reduced ionization coefficients, removing the uncertainty of gas samples and measurement of parameters such as voltage, current, pressure and so on. The results of Cowling and Fletcher[122] agree with those of Rose[119] in the range of E/N from 100 to 400 Td. However, Rose's[119] results are increasingly higher as E/N increases above 400 Td. Since the ionization cross section of deuterium was not given in Chapter 4, it is presented in Table 8.10.

8.9.2.3 Nitrogen (N_2)

Selected references are: Masch[123]; De Bitteto and Fischer[124]; Cookson et al.[125]; Daniel et al.[126]; Kontoleon et al.[127]; Maller and Naidu[128]; Raju and Rajapandiyan[129]; Haydon and Williams[130]; and Wedding et al.[131] Table 3 of Appendix 4 gives the values of α/N due to Haydon and Williams.[130] High pressure breakdown occurs at low values of E/N and one often seeks the ionization coefficients at these low values. Figure 8.19 and Table 8.11 give α/N at these high gas number densities. Daniel et al.[126] and, Farish and Tedford[132] have concluded that the Paschen law is applicable to nitrogen at high values of the product of gas pressure and gap distance ($0.5 \times 10^{24} \leq Nd \leq 6.0 \times 10^{24}\,m^{-2}$).

Theoretical investigations are due to: Engelhardt et al.[133]; Taniguchi et al.[134]; Raju and Gurumurthy[135]; Kücükarpaci and Lucas[136]; Phelps and Pitchford[137]; and Liu and Raju.[138]

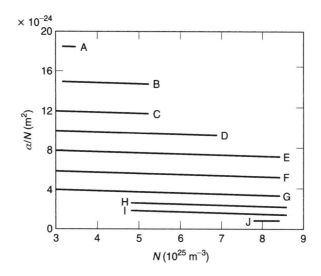

FIGURE 8.19 Ionization coefficients in nitrogen as a function of gas number density at low values of E/N. Letters A–J refer to column 1 of Table 8.11. Figure reproduced from Daniel, T.N. et al., *J. Phys. D*, 2, 1559, 1969. With permission of IOP, U.K.

TABLE 8.11
Values of α/N in Nitrogen at High Gas Number Density and Low E/N[126]

Curve	Pressure (MPa)	E/N (Td)	α/N (10^{-24} m^2)
A	0.13	112	18.62
B	0.13–0.20	109	15.13
C	0.13–0.20	106	12.18
D	0.13–0.26	103	9.83
E	0.13–0.33	100	7.76
F	0.13–0.33	97.0	5.83
G	0.13–0.33	93.9	4.39
H	0.20–0.33	90.9	3.21
I	0.20–0.13	87.9	2.31
J	0.33	84.8	1.54

Curve letters refer to Figure 8.19.

8.10 OTHER GASES (NONATTACHING)

8.10.1 MERCURY VAPOR

Selected references for measurement of α/N in mercury vapor are: Badareu and Bratescu[139]; Davies and Smith[140]; Overton and Davies.[141] Theoretical calculations are published by Rockwood[142] and Liu and Raju.[143] Figure 8.20 shows the data of Davies and Smith[140] and Overton and Davies.[141] The α/N values of the two measurements agree for $E/N \leq$ 2700 Td. For higher values of E/N the results of Overton and Davies[141] are progressively larger than those of Davies and Smith.[140] The differences at such high values of E/N are likely to be due to nonequilibrium effects. Figure 8.20 and Table 3 of Appendix 4 show the α/N values.

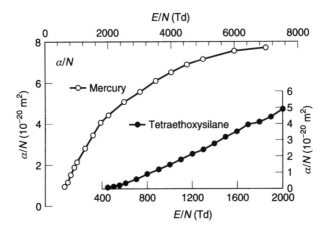

FIGURE 8.20 α/N as a function of E/N in mercury vapor[141] and tetraethoxysilane.[145]

TABLE 8.12
Values of α/N as a Function of E/N in Tetraethoxysilane.[145] Digitized and Interpolated from Figure 8.20

E/N	α/N	E/N	α/N	E/N	α/N
450	0.0847	900	1.17	1500	3.16
500	0.142	1000	1.47	1600	3.50
550	0.211	1100	1.78	1700	3.94
600	0.335	1200	2.13	1800	4.10
700	0.578	1300	2.39	1900	4.41
800	0.897	1400	2.76	2000	4.90

E/N in units of Td, α/N in units of $10^{-20}\,m^2$.

The gas constants for mercury vapor are given by Davies and Smith[140] as $F = 8.47 \times 10^{-20}\,m^2$ and $G = 1414\,Td$ (E/N in Equation 8.13 is presently in units of Td) in the range of E/N from 707 to 5090 Td. The ionization coefficients in the vapor are much higher than those of other gases since the ionization potential of the vapor is considerably lower. As an example, the maximum α/N in the vapor is $6.78 \times 10^{-20}\,m^2$, compared with the maximum in H_2, $8.72 \times 10^{-21}\,m^2$, an order of magnitude less.

The secondary ionization coefficient in the vapor shows a peak at $\sim560\,Td$. Overton and Davies[141] attribute this effect to the following possible secondary mechanisms, in addition to the impact of positive ions on the cathode:

1. The effect of delayed resonance radiation;
2. Collision-induced radiation from metastables;
3. The destruction of metastables at the cathode.

The Paschen minimum occurs in the vapor in the range from 287.0 to 296.0 volts, with larger gap lengths showing higher minimum voltage. This effect is attributed to the loss of active particles (photons and metastables) from the electrode space.[144]

8.10.2 Tetraethoxysilane

Tetraethoxysilane is a new plasma-processing gas that has the potential of being used for deposition of silicon dioxide thin films which serve as insulators in semiconductor chips.[145] The structure of the molecule is $Si(-O-CH_2-CH_3)_4$. It is liquid at room temperature, with vapor pressure of 0.47 torr. Yoshida et al.[145] have made the only reported measurements of ionization coefficients in the gas. An experimental fact to note is that, like measurements with vapors, condensation at the chamber walls should be prevented by heating the walls to a temperature close to that of equilibrium with vapor pressure. The current growth curves did not exhibit any evidence for electron attachment. Figure 8.20 and Table 8.12 show the ionization coefficients as a function of E/N.

REFERENCES

1. Raju, Gorur G., *Dielectrics in Electric Fields*, Marcel Dekker, New York, 2003.
2. Dutton, J., in *Electrical Breakdown in Gases*, ed. J. M. Meek and J. D. Craggs, John Wiley & Sons, New York, 1978, ch. 3.
3. Cookson, A. H., B. W. Ward, and T. J. Lewis, *Brit. J. Appl. Phys.*, 17, 891, 1966.
4. Pearson, J. S. and J. A. Harrison, *J. Phys. D: Appl. Phys.*, 2, 77, 1969.
5. Pearson, J. S. and J. A. Harrison, *J. Phys. D: Appl. Phys.*, 2, 1583, 1969.
6. Raju, G. R. Govinda, and M. S. Dincer, *J. Appl. Phys.*, 53, 8562, 1982.
7. Dushman, S., *Scientific Foundations of Vacuum Techniques*, 2nd edition, John Wiley & Sons, New York, 1962.
8. Hunter, S. R., J. G. Carter, and L. G. Christophorou, *J. Appl. Phys.*, 60, 24, 1986.
9. Yoshida, K., T. Ohshima, H. Ohuchi, and H. Tagashira, *J. Phys. D: Appl. Phys.*, 28, 2478, 1995.
10. In the literature on ionization coefficients published before 1975 the units frequently used were: E/p (V cm^{-1} torr^{-1}) and α/p (cm^{-1} torr^{-1}). The reduced ionization coefficient was expressed as

$$\frac{\alpha}{p} = A \exp - \left(\frac{Bp}{E}\right)$$

The conversion factors for F and G are:

$$F = \frac{A}{N_0} \times 100 \, (\text{m}^2); \quad G = \frac{B}{N_0} \times 100 \, (\text{V m}^2)$$

where $N_0 = 3.22 \times 10^{22}$ m^{-3} torr^{-1} at 293 K and 3.54×10^{22} m^{-3} torr^{-1} at 273 K. If E/N is expressed in units of Td in Equation 8.13, then $G = 3.1B$. If E/N is expressed in units of Td according to Equation 8.16, then $C = F$ and $D = B\sqrt{3.1}$ or $B\sqrt{2.82}$, according to whether $T = 293$ K or 273 K respectively.
11. Lewis, T. J., *Proc. Roy. Soc. A*, 244, 166, 1958.
12. Swamy, M. Narayana, and J. A. Harrison, *Brit. J. Appl. Phys.*, 2, 1437, 1969.
13. Božin, S. E. and C. C. Goodyear, *Brit. J. Appl. Phys.*, 18, 49, 1967.
14. Bhalla, M. S. and J. D. Craggs, *Proc. Phys. Soc.*, 80, 151, 1962.
15. Harrison, M. A. and R. Geballe, *Phys. Rev.*, 91, 1, 1953.
16. Prasad, A. N. and J. D. Craggs, *Proc. Phys. Soc.*, 77, 385, 1961.
17. Bhalla, M. S. and J. D. Craggs, *Proc. Phys. Soc.*, 76, 369, 1960.
18. Bhalla, M. S. and J. D. Craggs, *Proc. Phys. Soc.*, 78, 438, 1961.
19. Meek, J. M. and J. D. Craggs, *Electrical Breakdown in Gases*, Clarendon Press, Oxford, 1953.
20. Rao, C. Raja, and G. R. Govinda Raju, *J. Phys. D: Appl. Phys.*, 4, 494, 1971.
21. Gurumurthy, G. R. and G. R. Govinda Raju, *IEEE Trans. Electr. Insul.* EI-12, 3325, 1977.

22. Phelps, A. V. and Z. Lj. Petrović, *Plasma Sources Sci. Technol.*, 8, R-21, 1999.

23. Schade, R., *Z. Phys.*, 108, 353, 1938. "Tubes baked at 770 K and nickel electrodes heated to bright red. Electrode spacing from 0.3 to 5 cm."

24. Kruithoff, A. A., *Physica*, 7, 519, 1940. "Tubes baked with hydrogen at 670 K and then under vacuum at 740 K. These thermally grounded large area Cu cathodes were heavily sputtered ($\sim 10^{19}$ ions per cycle) with a Ne discharge at $2\,\mathrm{mA\,cm^{-2}}$, but at unspecified pressure and voltage. Earlier results with argon in which copper cathode of the same experimental tube was cleaned by sputtering and was not heated to higher temperatures also showed low γ_i values (0.009 to 0.04). The breakdown data were obtained with $d = 1$ cm. See Kruithoff, A. A. and F. M. Penning, *Physica*, 3, 515, 1936."

25. Kachikas, G. A. and L. H. Fisher, *Phys. Rev.*, 91, 775, 1953. "Electrode (brass) spacings were 0.3 to 3 cm."

26. Menes, M., *Phys. Rev.*, 116, 481, 1959. "System baked at 620 K and residual pressure of 3×10^{-9} torr. Probable purity of gas supply is 0.9999. Electrode spacing 0.3 cm to 1 cm."

27. Golden, D. E. and L. H. Fisher, *Phys. Rev.*, 123, 1079, 1961. "Electrodes (Ni) exposed to (a) a 10^{-6} A discharge for several days and (b) allowing an intermittent spark discharge to pass between the electrodes for about 50 hours."

28. Guseva, L. G., in *Investigations into Electrical Discharges in Gases*, ed. B. N. Klyarfel'd, Pergamon, New York, 1964, p. 1.

29. Klyarfel'd, B. N., L. G. Guseva, and A. S. Pokrovskaya-Soboleva, *Zh. Tekh. Fiz.*, 36, 704, 1966; English translation: *Sov. Phys.—Tech. Phys.*, 11, 520, 1966.

30. Heylen, A. E. D., *J. Phys. D: Appl. Phys.*, 1, 179, 1968. "Baking system at 470 K produced pressure of 10^{-7} torr ($\sim 10^{-5}$ Pa). Typical electrode spacings at breakdown were 0.5 cm. Electrodes exposed to ethane from mixture experiments."

31. Pace, J. D. and A. B. Parker, *J. Phys. D: Appl. Phys.*, 6, 1525, 1973. "Electropolished and chemically cleaned stainless steel electrodes separated by 0.3 to 2.0 cm. System baked to yield 10^{-8} torr ($\sim 10^{-6}$ Pa)."

32. Bhasavanich, D. and A. B. Parker, *Proc. Roy. Soc. A*, 358, 385, 1977.

33. Auday, G., Ph. Guillot, J. Galy, and H. Brunet, *J. Appl. Phys.*, 83, 5917, 1998. "Tube heated to 240° C for 24 hours. Electrode diameter 5 cm, $d = 0.2$ to 1 cm."

34. Petrović, Z. Lj. and A. V. Phelps, *Phys. Rev. E.*, 56, 5920, 1997. "Cathode chemically cleaned, electropolished, and gold plated. Mild bakeout of system gave rate of rise of better than $10^{-4}\,\mathrm{torr\,h^{-1}}$ ($\sim 10^{-2}\,\mathrm{Pa\,h^{-1}}$)."

35. Devins, J. C. and A. H. Sharbaugh, *Electrotechnology*, p. 104, Feb. 1961.

36. Vijh, A. K., *IEEE Trans. Elec. Insul.*, EI-12, 313, 1977.

37. Vijh, A. K., *J. Mater Sci.*, 11, 784, 1976.

38. Vijh, A. K., *J. Mater Sci.*, 11, 1374, 1976.

39. Shallal, M. A. and J. A. Harrison, *J. Phys. D: Appl. Phys.*, 4, 1550, 1971.

40. Gosseries, A., *Physica*, 6, 458, 1939.

41. Raju, G. R. Govinda, *Brit. J. Appl. Phys.*, 16, 279, 1965.

42. Fisher, L. H. and G. L. Weissler, *Phys. Rev.*, 66, 95, 1944.

43. Morton, P. L., *Phys. Rev.*, 70, 358, 1946.

44. Johnson, G. W., *Phys. Rev.*, 73, 284, 1948.

45. Golden, D. E. and L. H. Fisher, *Phys. Rev.*, 139, A613, 1965.

46. Liu, J. and G. R. Govinda Raju, *IEEE Trans. Plasma Sci.*, 20, 515, 1992.

47. Liu, J. and G. R. Govinda Raju, *IEEE Trans. Elec. Insul.*, 28, 154, 1993.

48. Fisher, L. H. and G. L. Weissler, *Phys. Rev.*, 66, 95, 1944.

49. Raju, G. R. Govinda and G. R. Gurumurthy, *IEEE Trans. Electr. Insul.*, EI-12, 325, 1977.

50. Raju, G.R. Govinda and R. Hackam, *Proc. IEE*, 120, 927, 1973.

51. Peek, F. W., *Dielectric Phenomena in High Voltage Engineering*, McGraw Hill, New York, 1929.

RECOMBINATION

52. Braithwaite, N. St J., *Plasma Sources Sci. Technol.*, 9, 517, 2000.

53. Boulmer, J., F. Devos, J. Stevefelt, and J.-F. Delpech, *Phys. Rev. A*, 15, 1502, 1977.

54. Hasted, J. B., *Physics of Atomic Collisions*, American Elsevier, New York, 1972, p. 427.
55. Hasted, J. B., *Physics of Atomic Collisions*, American Elsevier, New York, 1972, pp. 15–16.
56. Beaty, E. C., J. Dutton, and L. C. Pitchford, Joint Institute for Laboratory Astrophysics (JILA) Information Center Report No. 20, 1979.

Argon

57. Kruithof, A. A., *Physica*, 7, 519, 1940. Measurements have been reported for argon, neon, krypton, and helium. These measurements have been considered to be the benchmark for rare gases.
58. Davies, D. E. and J. G. C. Milne, *Brit. J. Appl. Phys.*, 10, 301, 1959.
59. Heylen, A. E. D., *Brit. J. Appl. Phys.*, 1, 179, 1968.
60. Abdulla, R. R., J. Dutton, and A. W. Williams, *J. Phys.* (Paris) *Colloq.*, C7, 40, 73, 1979.
61. Kruithof, A. A. and F. M. Penning, *Physica*, 4, 430, 1937.
62. Heylen, A. E. D., *Int. J. Electron.*, 24, 165, 1968.
63. Yamane, M., *J. Phys. Soc. Jpn*, 15, 1076, 1960. Mixtures studied were of argon with C_2H_4, C_3H_6, C_4H_8, CH_4, C_2H_6, C_3H_8, C_4H_{10}, C_5H_2, CH_3OH, C_2H_5OH, n-C_3H_7OH, Kr, Xe.
64. Ward, A. L., *J. Appl. Phys.*, 33, 2789, 1962.
65. Penning, F. M. and C. C. J. Addinck, *Physica* , 1, 1007, 1934.
66. Lagushenko, R. and J. Maya, (a) *J. Appl. Phys.*, 54, 2255, 1983; (b) *J. Appl. Phys.*, 55, 3293, 1984. In the first paper normalized current density (j/p^2) in Penning mixtures has been measured. In the second paper the ionization coefficient has been calculated for Penning mixtures. This paper is interesting because of the analytical expressions for various quantities given. These are:

Elastic scattering cross section, Q_{el} (He):

$$6.16 \times 10^{-20} \frac{1}{\left[1 + \left(\dfrac{\varepsilon}{15}\right)^{1.5}\right]} \text{ m}^2$$

Ionization cross section, Q_i (He):

$$Q_i = \frac{19.8 \times 10^{-20}}{\varepsilon} \ln\left[1 + \frac{0.045(\varepsilon - 24.6)}{70}\right] \text{ m}^2$$

Elastic scattering cross section, Q_{el} (Ne):

$$Q_{el} = 1.51 \times 10^{-20} \frac{\varepsilon^{1/3}}{[1 + \varepsilon/40]^{1.4}} \text{ m}^2$$

Ionization cross section, Q_i (Ne):

$$Q_i = \frac{70 \times 10^{-20}}{\varepsilon} \ln\left[1 + \frac{0.047(\varepsilon - 21.6)\varepsilon}{250}\right] \text{ m}^2$$

Reduced first ionization coefficient, α/N (Ne):
where

$$\frac{\alpha}{N} = C \exp\left[-D\left(\frac{N}{E}\right)^{1/2}\right] \text{ m}^2$$

$C = 2.55 \times 10^{-20}\,\text{m}^2$ and $D = 9.47 \times 10^{-10}\,(\text{V}\,\text{m}^2)^{1/2}$.

Drift velocity of electrons, W (Ne):

$$W = \frac{E}{N} \frac{P}{\left[1 + Q\sqrt{E/N}\right]}\ \text{ms}^{-1}$$

where $P = 3.87 \times 10^3\,\text{m s}^{-1}\,\text{Td}^{-1}$, $Q = 0.085\,(\text{Td})^{-1/2}$, and E/N is in units of Td.

Elastic scattering cross section, Q_{el} (Ar):

$$Q_{el} = \begin{cases} 1.41 \times 10^{-20}\varepsilon\,(\varepsilon \leq 11.5) \\ 284 \times 10^{-20} \times \dfrac{1}{(\varepsilon + 6)}\,(\varepsilon > 11.5) \end{cases}\ \text{m}^2$$

Ionization cross section, Q_i (Ar):

$$Q_i = \frac{141 \times 10^{-20}}{\varepsilon}\ln\left[1 + \frac{0.50 - (\varepsilon - 15.9)}{500}\varepsilon\right]\text{m}^2$$

Drift velocity, W (Ar):

$$W = \frac{E}{N} \frac{P}{\left[1 + Q\sqrt{E/N}\right]}\ \text{m s}^{-1}$$

where $P = 1.26 \times 10^3\,\text{m s}^{-1}\,\text{Td}^{-1}$, $Q = 0.085\,(\text{Td})^{-1/2}$, and E/N is in units of Td. At $E/N = 300$ Td the calculated drift velocity is $1.53 \times 10^5\,\text{m s}^{-1}$ and the value shown in Table 6.4 is $2.17 \times 10^5\,\text{m s}^{-1}$, a difference of 30%.

Excitation cross section for the rare gases:

$$Q_{ex} = \frac{2.82 \times 10^{-21}}{\varepsilon}A_0 \times \ln\left[1 + \frac{[B_0 + B_1(\varepsilon - \varepsilon_1) + B_2(\varepsilon - \varepsilon_1)^2]\varepsilon}{A_0}\right]\text{m}^2$$

where the parameters have the following values:

Gas	A_0 (eV)	B_0 (eV)	B_1 (eV)	B_2 (eV)	ε
He	10	0.13	0.08	0.0	20
Ne	12	0.10	0.0	0.0090	16.5
Ar	38	0.0	0.13	0.073	11.5

67. Sakai, Y., S. Sawada, and H. Tagashira, *J. Phys. D: Appl. Phys.*, (a) 19, 1741, 1986; (b) 19, 2393, 1986.
68. Chanin, L. M. and G. D. Rork, *Phys. Rev.*, 135, A71, 1964.
69. LeBlanc, O. H. and J. C. Devins, *Nature* (London), 188, 219, 1960.
70. Eletskii, A. V. in *Handbook of Physical Quantities*, ed. I. S. Grigoriev and Z. Meilikhov, CRC Press, Boca Raton, 1999, ch. 20.

HELIUM

71. Davies, D. K., F. Llewellyn-Jones, and C. G. Morgan, *Proc. Phys. Soc. Lond.*, 80, 898, 1962.
72. Chanin, L. M. and G. D. Rork, *Phys. Rev.*, 133, A1005, 1964.
73. Dutton, J. and D. B. Rees, *Brit. J. Appl. Phys.*, 18, 309, 1965.
74. Cavelleri, G., *Phys. Rev.*, 179, 186, 1969.
75. Lakshminarasimha, C. S., J. Lucas, and R. A. Nelson, *Proc. Inst. Electr. Eng.*, 122, 1162, 1975.
76a. Kücükarpaci, H. N., H. T. Saelee, and J. Lucas, *J. Phys. D: Appl. Phys.*, 14, 9, 1981. *W* and D/μ are measured in He and Ne.
76b. Lakshminarasimha, C. S., and J. Lucas, *J. Phys. D: Appl. Phys.*, 10, 313, 1977. D_r/μ and α/N are measured in He, Ar, air, CH_4 and NO.

KRYPTON

77. Kruithof, A. A., *Physica*, 7, 519, 1940.
78. Heylen, A. E. D., *Int. J. Electron.* 31, 19, 1971.
79. Specht, L. T., S. A. Lawton, and T. A. Temple, *J. Appl. Phys.*, 51, 166, 1980.
80. Kücükarpaci, H. N. and J. Lucas, *J. Phys. D: Appl. Phys.*, 14, 2001, 1981.

NEON

81. Kruithof, A. A., *Physica*, 7, 519, 1940.
82. Chanin, L. M. and G. D. Rork, *Phys. Rev.*, 132, 2547, 1963.
83. De Hoog, F. J. and J. Kasdorp, *Physica*, 34, 63, 1967.
84. Willis, B. A. and C. G. Morgan, *J. Phys. D: Appl. Phys.*, 1, 1219, 1968.
85. Dutton, J., M. H. Hughes, and B. C. Tan, *J. Phys. B: At. Mol. Phys.*, 2, 890, 1969.
86. Buursen, C. G. J., F. J. De Hoog, and H. Van Montfort, *Physica*, 60, 244, 1972.
87. Bhattacharya, A. K., *Phys. Rev. A*, 13, 1219, 1976.
88. Kücükarpaci, H. N., H. T. Saelee, and J. Lucas, *J. Phys. D: Appl. Phys.*, 14, 2001, 1981.

XENON

89. Kruithof, A. A., *Physica*, 7, 519, 1940.
90. Burkley, C. J. and M. C. Sexton, *Brit. J. Appl. Phys.* 18, 443, 1967.
91. Bhattacharya, A. K., *Phys. Rev. A*, 13, 1219, 1976.
92. Makabe, T. and T. Mori, *J. Phys. B: At. Mol. Phys.*, 11, 3785, 1978.
93. Specht, L. T., S. A. Lawton, and T. A. Temple, *J. Appl. Phys.*, 51, 166, 1980.

HYDROCARBON GASES

94. Heylen, A. E. D., *J. Chem. Phys.*, 38, 765, 1963.
95. Davies, D. K. and E. Jones, *Proc. Phys. Soc. Lond.*, 82, 537, 1963.
96. Schlumbohm, H., *Z. Phys.*, 184, 492, 1965.

Measurements of α/N by the current pulse method are fitted to Equation (8.13) and the constants are shown below.

Gas	Formula	E/N range (Td)	F (10^{-20} m^2)	G (Td)
Oxygen	O_2	210–900	2.39	633.0
Carbon dioxide	CO^2	225–450	1.48	567.6
Methane	CH_4	270–375	2.17	597.4
Acetone	$(CH_3)_2CO$	300–1200	4.44	1122
Benzene	C_6H_6	480–1950	8.08	1870
Diethyl ether	$(C_2H_5)_2O$	270–2100	7.61	1209
Methylal	$CH_2(OCH_3)_2$	270–3000	9.22	1122

97. Cookson, A. H. and B. L. Ward, *Electron. Lett.*, 1, 83, 1965.

98. Cookson, A. H., B. W. Ward, and T. J. Lewis, *Brit. J. Appl. Phys.*, 17, 891, 1966.

99. Vidal, G., J. Lacaze, and J. Maurel, *J. Phys. D: Appl. Phys.*, 7, 1684, 1974.

100. Heylen A. E. D., *Int. J. Electron.*, (a) 39, 653, 1975; (b) 11, 367, 1978.

101. Hunter, S. R., J. G. Carter, and L. G. Christophorou, *J. Appl. Phys.*, 60, 24, 1986.

102. Davies, D. K., L. E. Kline, and W. E. Bies, *J. Appl. Phys.*, 65, 3311, 1989.

103. Leblanc, O. H., Jr., and J. C. Devins, *Nature*, 188, 288, 1960.

HYDROGEN AND NITROGEN

104. Rose, D. J., *Phys. Rev.*, 104, 273, 1956.

105. Folkard M. A. and S. C. Haydon, *Aust. J. Phys.*, 24, 519, 1971.

106. Haydon, S. C. and A. G. Robertson, *Proc. Phys. Soc.* (London), 78, 92, 1961.

107. Chanin L. M., and G. D. Rork, *Phys. Rev.*, 132, 2547, 1963.

108. Barna, S. F., D. Edelson, and K. B. McAffee, *J. Appl. Phys.*, 35, 2781, 1964.

109. Golden, D. E., H. Nakano, and L. H. Fisher, *Phys. Rev.*, 138, 1613, 1965.

110. Haydon, S. C. and H. M. P. Stock, *Aust. J. Phys.*, 19, 795, 1966.

111. Tagashira, H. and J. Lucas, *J. Phys. D: Appl. Phys.*, 2, 867, 1969.

112. Shimozuma, M., Y. Sakai, H. Tagashira, and S. Sakamoto, *J. Phys. D: Appl. Phys.*, 10, 1671, 1977.

113. Blevin, H. A., J. Fletcher, and S. R. Hunter, *J. Phys. D: Appl. Phys.*, 11, 2295, 1978.

114. Shallal, M. A. and J. A. Harrison, *J. Phys. D: Appl. Phys.*, 4, 1550, 1971. Ionization currents were measured in both uniform and nonuniform electric fields between 300 and 3300 Td.

115. Lakshminarasimha, C. S., J. Lucas, and R. A. Snelson, *Proc. Inst. Electr. Eng.*, 122, 1162, 975.

116. Engelhardt, A. G. and A. V. Phelps, *Phys. Rev.*, 131, 2115, 1964. Analysis of H_2 and D_2 was carried out in the range of E/N from 0.001 to 100 Td.

117. Saelee, H. T. and J. Lucas, *J. Phys. D: Appl. Phys.*, 10, 343, 1977.

118. Fletcher J., and H. A. Blevin, *J. Phys. D: Appl. Phys.*, 14, 27, 1981.

119. Rose, D. J., *Phys. Rev.*, 104, 273, 1956.

120. Barna, S. F., D. Edelson, and K. B. McAffee, *J. Appl. Phys.*, 35, 2781, 1964.

121. Rork, G. D. and L. M. Chanin, *J. Appl. Phys.*, 35, 2801, 1964.

122a. Cowling, I. R. and J. Fletcher, *J. Phys. B: At. Mol. Phys.*, 6, 665, 1973. Both ionization cross sections and reduced ionization coefficients in H_2 and D_2 are measured.

122b. Rapp, D., and P. Englander-Golden, *J. Chem. Phys.*, 43, 1464, 1965.

123. Masch, K., *Arch. Electrotech.*, 26, 587, 1932.

124. De Bitetto, D. J. and L. H. Fisher, *Phys. Rev.*, 104, 1213, 1956. High pd (gas pressure × gap length) values in the range of 133 to 110 Pa m were employed to measure α/N in both hydrogen and nitrogen in the range of 42 to 66 Td and 90 to 135 Td respectively.

125. Cookson, A. H., B. W. Ward, and T. J. Lewis, *J. Phys. D: Appl. Phys.*, 17, 891, 1966. The ionization coefficients in methane and hydrogen were measured by the avalanche statistics method in the E/N range from 70 to 600 Td.

126. Daniel, T. N., J. Dutton, and F. M. Harris, *J. Phys. D: Appl. Phys.*, 2, 1559, 1969. Accurate sparking potentials in nitrogen at high values of Nd (1 to 6×10^{24} m^{-2}) are reported in this paper.

127. Kontoleon, N., J. Lucas, and L. E. Virr, *J. Phys. D: Appl. Phys.*, 6, 1237, 1973.

128. Maller, V. N. and M. S. Naidu, *J. Phys. D: Appl. Phys.*, 7, 1406, 1974.

129. Raju, G. R. Govinda and S. Rajapandiyan, *Int. J. Electron.*, 40, 65, 1976.

130. Haydon, S. C. and O. M. Williams, *J. Phys. D: Appl. Phys.*, 9, 523, 1976. Both spatial and temporal current measurements are reported in nitrogen with tabulated values of α/N (cm^2) in the range of E/N from 85 to 3400 Td.

131. Wedding, A. B., H. A. Blevin, and J. Fletcher, *J. Phys. D: Appl. Phys.*, 18, 2361, 1985.

132. Farish, O. and D. J. Tedford, *J. Phys. D: Appl. Phys.*, 2, 1555, 1969.

133. Engelhardt, A. G., A. V. Phelps, and C. G. Risk, *Phys. Rev.*, 135, 1566, 1964.

134. Taniguchi, T., H. Tagashira, and Y. Sakai, *J. Phys. D: Appl. Phys.*, 11, 1757, 1978.

135. Raju, G. R. Govinda and G. R. Gurumurthy, *Int. J. Electron.*, 44, 714, 1978.

136. Kücükarpaci, H. N. and J. Lucas, *J. Phys. D: Appl. Phys.*, 12, 2123, 1979.
137. Phelps, A. V. and L. C. Pitchford, *Phys. Rev.*, 31, 2932, 1985.
138. Liu, J. and G. R. Govinda Raju, *J. Franklin Inst.*, 329, 181, 1992.

MERCURY VAPOR

139. Badareu, E. and G. G. Bratescu, *Bull. Soc. Roum. Fiz.*, 42, 82, 1942.
140. Davies D. E., and D. Smith, *Brit. J. Appl. Phys.*, 16, 697, 1965.
141. Overton, G. D. N. and D. E. Davies, *Brit. J. Appl. Phys.*, 1, 881, 1968.
142. Rockwood, S. D., *Phys. Rev. A*, 8, 2348, 1973.
143. Liu, J. and G. R. Govinda Raju, *J. Phys. D: Appl. Phys.*, 25, 167, 1992.
144. Overton, G. D. N., D. Smith, and D. E. Davies, *Brit. J. Appl. Phys.*, 16, 731, 1965.
145. Yoshida, K., T. Ohshima, H. Ohuchi, and H. Tagashira, *J. Phys. D: Appl. Phys.*, 28, 2478, 1995.

9 Ionization and Attachment Coefficients—II. Electron-Attaching Gases

Electron attachment is a loss mechanism that depletes electrons from the ionization region. We shall consider the attachment coefficients in addition to the ionization coefficients; the processes by which attachment occurs are not treated exhaustively, in view of the volume of literature available, though enough description will be included for the reader to understand the role of electron attachment in the ionization growth of an avalanche.

9.1 ATTACHMENT PROCESSES

Electron attachment may be a two-body or three-body process. In the two-body process, the formation of a negative ion by collision of an electron with a diatomic molecule may be explained by considering the formation of a molecule. Consider two isolated atoms that are situated so far apart that they do not influence one another. As they come closer together the outer electrons repel each other, but the electrons of one atom are attracted to the nucleus of the other. If they are too close, of course, their nuclei repel each other. At a certain critical distance between their nuclei the energy is a minimum and the molecule will be at ground state with minimum energy at the particular bond length, say X_2. For shorter or longer bond lengths the energy of the molecule is not minimum and the molecule will not have a stable configuration.

When an electron with a finite energy impacts the molecule and a negative ion is formed (X_2^-), the energy of the electron must be accounted for. The binding energy of the extra electron to the molecule is known as the electron affinity. If the negative ion is stable then the electron affinity is defined as the energy difference between the ground state of the neutral and the lowest state of the corresponding negative ion. The higher the electron affinity of the atom, the greater is the attachment. For example, halogen atoms have electron affinities in the range 3.1 to 3.5 eV, compared to 0.08 eV for helium which is not attaching.[1] The oxygen atom has an electron affinity of 1.46 eV and is moderately attaching. Atoms and molecules have different electron affinities; for example, O_2 and O_3 have electron affinities of 0.44 and 2.10 eV respectively.

The minimum energy of the negative ion is mostly higher than the energy of the molecule and the bond length has a higher value. If we assume that the lifetime of the negative ion is short compared to the time of the nuclear vibration time the X_2^- ion dissociates into X and X^- in time $\sim 10^{-13}$ s. The incoming electron must have precisely the exact energy for the single-bond distance of the molecule. This requirement translates into a peak in the cross section–energy $(Q_a - \varepsilon)$ curve and a resonance is said to occur. The resonance energy for a large number of molecules is in the 1–4 eV range.

Direct attachment is represented by

$$AB + e \rightarrow AB^-$$ (9.1)

TABLE 9.1
Electron Affinity (ε_A), Attachment Peak Energy (ε_p), and Total Cross Section at Peak for Negative Ion Formation in Molecular Gases[2]

Gas	ε_A	ε_p (eV)	Q_a (10^{-20} m^2)
O_2	0.44	6.5	1.407×10^{-2}
CO		9.9	2.023×10^{-3}
NO	0.024	8.15	1.117×10^{-2}
CO_2		8.1	4.283×10^{-3}
N_2O	~1.465	2.2	8.602×10^{-2}
SF_6	1.06^{163}	~0.1	2.146

The dissociative attachment is usually represented by

$$e + AB \rightarrow (AB^-) \rightarrow A + B^- \tag{9.2}$$

Ion pair production is another type of two-body process that results in electron attachment. The reaction is represented by

$$e + AB \rightarrow (AB^*) + e \rightarrow A^+ + B^- + e \tag{9.3}$$

Three-body processes of electron attachment require the presence of a third particle that may or may not be the same as the parent atoms. These reactions are:

$$e + AB \rightarrow (AB^*)^-; (AB^*)^- + M \rightarrow AB^- + M \tag{9.4}$$

$$e + AB + M \rightarrow AB^- + M \tag{9.5}$$

Electron affinity and total attachment cross section peaks of selected molecules are shown in Table 9.1. The cross section shown is for combined dissociative attachment and ion pair production (reactions 9.2 and 9.3). The cross section for SF_6 is at least two orders of magnitude higher than most of the others and the electron affinity is low; possibly the resonance peak is very narrow and the width of the electron beam partially contributes to the observed energy.[2]

The experimental cross sections of Rapp and Briglia[2] shown in Table 9.1 were obtained using a monoenergetic beam of electrons. Under swarm conditions the independent variable is the parameter E/N, related to the mean energy, and the rate constant for dissociative attachment is given in Table 9.2.[3] Again, note the large difference in the rate constant between molecules with halogen atoms and those without. The rate constant of dissociative attachment in gases as a function of mean energy is given in Appendix 5.

The usefulness of Table 9.2 is demonstrated by substituting typical values for SF_6 (say): $E/N = 100$ Td, $K_a = 7 \times 10^{-16}$ m^{3}s^{-1}, W_e (from Table 7.8) $= 10^5$ m s^{-1}, giving a reduced attachment coefficient (η/N) of 70×10^{-22} m^2, which agrees with Figure 9.26.

9.2 CURRENT GROWTH IN ATTACHING GASES

A large amount of data on attachment coefficients has been acquired in electron-attaching gases, using the current growth method. Let the attachment coefficient per meter length of drift be denoted by η and the reduced attachment coefficient by η/N (m^2). The true ionization

TABLE 9.2
Rate Constant ($10^{-16}\,\mathrm{m^3 s^{-1}}$) of Dissociative Attachment to the Molecule in an Electric Field

E/N (Td)	SF$_6$	CF$_4$	C$_2$F$_6$	C$_3$F$_8$	C$_4$F$_{10}$	CCl$_2$F$_2$	H$_2$O	N$_2$O	HCl	HBr	HI
2								6.5(−6)	1.6	34	53
5								1.2(−4)	2.6		26
10								0.001			
50	6.3	0.39					0.017				
100	7	0.68					0.8				
150	7.8	0.68					1.0				
200	7.3	0.64					0.78				
250	6.8	0.5	2.7	3.0	3.3						
300	6.8	0.46	2.6	2.4	2.3	4.6					
400	6.7		1.5	1.8	1.0	4.6					
500	6.5		0.65	1.0	0.6	3.9					
600	5.8					2.5					

$a(b)$ means $a \times 10^b$.
Reproduced from Grigoriev, I. S. and Z. Meilikhov, *Handbook of Physical Quantities*, CRC Press, Boca Raton, 1999. With permission.

coefficient is represented by α_T (m^{-1}) and the apparent ionization coefficient is then $(\alpha_T - \eta)$. For convenience, we drop the subscript in the latter symbol and show the apparent ionization coefficient simply as $(\alpha - \eta)$.

The current growth equation in an electron-attaching gas in the presence of secondary ionization at constant E/N is given by

$$I = I_0 \frac{\left(\dfrac{\alpha}{\alpha - \eta} \exp(\alpha - \eta)d - \dfrac{\eta}{\alpha - \eta} \right)}{1 - \left\{ \dfrac{\gamma \alpha}{\alpha - \eta} \right\} \{ \exp(\alpha - \eta)d - 1 \}} \tag{9.6}$$

where I_0 is the initial photoelectric current due to the external source and d is the gap distance. In the absence of the secondary ionization coefficient, Equation 9.6 simplifies to

$$I = I_0 \left(\frac{\alpha}{\alpha - \eta} \exp(\alpha - \eta)d - \frac{\eta}{\alpha - \eta} \right) \tag{9.7}$$

The unknown status of I_0 limits the accuracy with which the three coefficients α, η, γ can be evaluated from current growth experiments. Raju[4] has suggested a method that does not require I_0 in the analysis and thus improves accuracy. The method is an improvement over the Gosseries method[5] that, also, does not require the knowledge of I_0 but is limited to nonattaching gases. The method is outlined as follows. Let

$$A = I_0 \frac{\alpha}{\alpha - \eta}; \qquad B = I_0 \frac{\eta}{\alpha - \eta}$$

so that

$$A - B = I_0; \qquad \frac{A}{B} = \frac{\alpha}{\eta}$$

where A and B are constants for a given E/N as are also the parameters α, η, and I_0. Equation 9.7 may now be written as

$$I_d = A \exp(\alpha - \eta)d - B \tag{9.8}$$

where I_d denotes the ionization current at gap length d. Hence

$$A = \frac{I_d + B}{\exp(\alpha - \eta)d} \tag{9.9}$$

On the basis of Equation 9.8 one can write the current at gap $(d + \Delta d)$ as

$$I_{d+\Delta d} = A \exp(\alpha - \eta)(d + \Delta d) - B \tag{9.10}$$

Substituting for A from Equation 9.9, one gets

$$I_{d+\Delta d} = \{\exp(\alpha - \eta)\Delta d\}I_d + B\{\exp(\alpha - \eta)\Delta d - 1\} \tag{9.11}$$

If the intervals Δd, at which current measurements are made, are kept constant, a plot of $I_{d+\Delta d}$ against I_d will be a straight line and one can apply the least-squares method to find the best slope and intercept of such a plot. The slope and the intercept give the quantities

$$\text{slope} = \exp(\alpha - \eta)\Delta d; \quad \text{intercept} = B\{\exp(\alpha - \eta)\Delta d - 1\}$$

from which α and η are evaluated. Note that I_0 does not need to be evaluated. This is a distinct advantage because I_0 is a notoriously difficult quantity to measure and to maintain at a constant value, particularly in electron-attaching gases.

The breakdown criterion in electron-attaching gases is obtained by equating the denominator of Equation 9.6 to zero, that is,

$$1 - \left\{\frac{\gamma\alpha}{\alpha - \eta}\right\}\{\exp(\alpha - \eta)d - 1\} = 0 \tag{9.12}$$

At high gas number densities the breakdown criterion in electron-attaching gases is

$$\alpha - \eta - \alpha\gamma\{\exp(\alpha - \eta)d - 1\} = 0 \tag{9.13}$$

If γ is very small and the last term is negligible in comparison with the apparent ionization coefficient, the breakdown criterion will simply be

$$\frac{\alpha}{N} = \frac{\eta}{N} \tag{9.14}$$

Equation 9.14 is often misinterpreted as the criterion for breakdown. An accurate meaning is that in the limit of the left side becoming equal to the right side current growth will not occur. However, when the limit is exceeded in the positive direction, that is, when

$$\frac{\alpha + \Delta\alpha}{N} > \frac{\eta}{N} \tag{9.15}$$

the current multiplies extremely rapidly due to the exponential dependence of α/N on E/N, Equation 8.11. Equation 9.15 is also the basis for the often stated criterion that a limiting E/N exists below which breakdown does not occur in an electron-attaching gas.

For the sake of completeness, we provide the complete current growth equation in the presence of all processes—primary ionization (coefficient α), secondary ionization (γ), attachment (η), and detachment (δ)—as[6]

$$\frac{I}{I_0} = C\left(Ae^{\lambda_1 d} - Be^{\lambda_2 d}\right)\left[1 - \gamma\left\{\frac{\lambda_1 + \Delta}{\lambda_2}\left(e^{\lambda_1 d} - 1\right) - \frac{\lambda_2 + \Delta}{\lambda_2}\left(e^{\lambda_2 d} - 1\right)\right\}\right]^{-1} \tag{9.16}$$

where

$$\lambda_1 = \frac{1}{2}\left[\alpha - a - \Delta + \left\{(\alpha - a - \Delta)^2 + 4\alpha\Delta\right\}^{1/2}\right] \tag{9.17}$$

$$\lambda_2 = \frac{1}{2}\left[\alpha - a - \Delta - \left\{(\alpha - a - \Delta)^2 + 4\alpha\Delta\right\}^{1/2}\right] \tag{9.18}$$

$$A = \frac{(a + \Delta + \lambda_1)}{(\lambda_1 - \lambda_2)} \tag{9.19}$$

and

$$B = \frac{(a + \Delta + \lambda_2)}{(\lambda_1 - \lambda_2)} \tag{9.20}$$

where a and Δ are parameters related to constants A and B, as shown.

The breakdown criterion is then

$$1 - \gamma\left\{\frac{\lambda_1 + \Delta}{\lambda_1}\left(e^{\lambda_1 d_s} - 1\right) - \frac{\lambda_2 + \Delta}{\lambda_2}\left(e^{\lambda_2 d_s} - 1\right)\right\} = 0 \tag{9.21}$$

From Equations 9.17 and 9.18 it may be seen that λ_1 and λ_2 are positive and negative quantities respectively. Also, since $\lambda_1 d_s$ and $\lambda_2 d_s$ are large (~ 10), it follows that

$$e^{\lambda_1 d_s} \gg e^{\lambda_2 d_s} \tag{9.22}$$

and thus Equation 9.21 can be simplified to

$$1 - \frac{\gamma}{\lambda}\left(e^{\lambda_1 d_s}\right) = 0 \tag{9.23}$$

where

$$\lambda = \frac{\lambda_1}{\lambda_1 + \Delta} \tag{9.24}$$

Thus, if λ_1 and γ are known, by substituting them into Equation 9.23 the sparking distance d_s can be calculated, and hence V_s.

9.3 IONIZATION AND ATTACHMENT COEFFICIENTS

The methods of measuring attachment coefficients are:

1. Ionization current method.
2. Bradbury-type filter method, in which the electrons are allowed to drift through a known distance and the number lost through attachment is measured.[7]
3. Current pulse method.

All three methods have been described previously. Data on effective ionization coefficient $(\alpha - \eta)$ and attachment coefficient will be presented in the following sections.

9.3.1 DRY AND HUMID AIR

Formation of negative ions in dry air is entirely due to its oxygen component if one ignores trace impurities such as CO and CO_2. Since attachment in oxygen will be dealt with in a separate section, we include only brief comments here.

The attachment processes in dry air may be attributed to[8] (1) dissociative attachment, (2) ion pair formation, (3) three-body attachment, and (4) radiative attachment. Dissociative and radiative attachments are dependent on N in the same way as the ionization process. The three body process depends upon the square of the number density, N^2. In a mixture of gases in which one component is attaching and the other nonattaching, such as air, the three-body process may occur in two different ways, depending upon whether the third body is attaching or nonattaching. In the case of air the reactions are[11]

$$\left. \begin{array}{l} e + O_2 \rightarrow \left(O_2^-\right)^* \\ \left(O_2^-\right)^* + O_2 \rightarrow O_2^- + O_2^* \end{array} \right\} \qquad (9.25)$$

$$\left. \begin{array}{l} e + O_2 \rightarrow \left(O_2^-\right)^* \\ \left(O_2^-\right)^* + N_2 \rightarrow O_2^- + N_2^* \end{array} \right\} \qquad (9.26)$$

The third body in reaction 9.25 is an oxygen molecule and in 9.26 it is a nitrogen molecule. Dutton et al.[11] suggest a three-body attachment coefficient of $6 \times 10^{-24}\,m^2$ at $E/N = 99$ Td, which corresponds to a breakdown of 1 cm in air at atmospheric pressure. The effective ionization coefficient at the same conditions, for comparison, is $7 \times 10^{-23}\,m^2$.

The process of ion pair formation (reaction 9.3) does not remove the electron from the swarm and therefore it does not affect the primary ionization coefficient.

The attachment coefficient observed in air is lower than that expected on the basis of partial content of oxygen. Moruzzi and Price[14] suggest that this is because there is a fast detachment process, that is, $O^- + N_2^* \rightarrow$ products $+ e$. In humid air H_2O molecules also attach electrons and the attachment coefficient is then dependent on the partial vapor pressure of water. A sole investigation to cover the range of E/N from threshold to 1000 Td is not available and one has to combine the results of several investigators. The selected data for this purpose are given below.

Ionization coefficients and reduced attachment coefficients (η/N) have been measured by Prasad,[9] Dutton et al.,[10,11] Frommhold,[12] Ryzko,[13] and Moruzzi and Price.[14] The effective ionization coefficient in mercury-free dry air has been measured by Bhiday et al.[15] and Rao and Raju[17a] in the range of $E/N = 140$ to 2850 Td. In this range attachment coefficients are in-significant compared with α_T. Theoretical analysis is due to Liu and Raju.[16]

TABLE 9.3
Ionization and Attachment Coefficients in Dry Air at Low Values of E/N

E/N	α/N	η/N	E/N	α/N	η/N
80		1.45(−3)			
95	0.0001	1.6(−3)	130	0.0086	
100	0.0002		135	0.0105	
105	0.0009		140	0.0126	2.1(−3)
110	0.0020	1.80(−3)	150	0.0174	
120	0.0047		160	0.0224	
125	0.0066	1.98(−3)	170	0.0278	

E/N in units of Td, coefficients in units of $10^{-20}\,\mathrm{m}^2$. $a(b)$ means $a \times 10^b$.

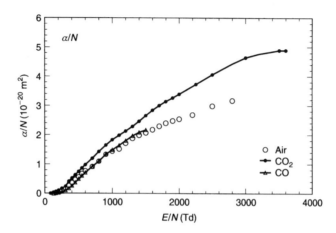

FIGURE 9.1 Reduced ionization coefficients as a function of E/N: ($\circ$) air, Moruzzi and Price,[14] Rao and Raju[17a]; ($-\bullet-$) CO_2, Bhalla and Craggs[38]; ($-\triangle-$) CO, Bhalla and Craggs,[31] and Lakshminarasimha et al.[33] Tabulated values are shown in Appendix 6, Table 1.

The best value of sparking potential for dry air at atmospheric pressure (101.3 kPa) and 1 cm gap length at 293 K is 31.55 kV. This corresponds to $E/N = 95$ Td. One often desires to find the attachment and effective ionization coefficient in the vicinity of these conditions; Table 9.3 gives these values at close intervals.[9,11] The attachment cross section derived by Prasad[9] is $3 \times 10^{-24}\,\mathrm{m}^2$ at an electron energy of 3 eV, relatively weakly attaching.

The recommended values of effective ionization coefficients are shown in Figure 9.1 and Table 1 of Appendix 6. They are derived by combining the measured values of Rao and Raju[17a] and Moruzzi and Price.[14] Table 9.4 gives accurate values of gas constants in Townsend's equation (8.11), obtained by least-squares curve fitting.[17]

With regard to attachment, we shall consider both humid air and water vapor simultaneously. The operative collision processes in H_2O are[18]

$$e + H_2O \rightarrow OH + H^-; \quad \text{appearence potential of 5.7 eV} \tag{9.27}$$

$$e + H_2O \rightarrow 2H + O^-; \quad \text{appearance potential of 7.3 eV} \tag{9.28}$$

$$e + H_2O \rightarrow H_2O^{-*} \rightarrow 2H + O^-; \quad \text{dissociation potential of 5.16 eV} \tag{9.29}$$

$$e + H_2O \rightarrow H + H^+ + O^- + e; \quad \text{appearance potential of 20.8 eV} \tag{9.30}$$

$$e + H_2O \rightarrow H^- + O^- + 2e; \quad \text{appearance potential of 34.3 eV} \tag{9.31}$$

TABLE 9.4
Gas Constants in Dry Air According to Equation 8.11

E/N range (Td)	167–372	372–608	608–1686	1686–3035
F ($\times 10^{-20}\,\text{m}^2$)	2.59	3.30	5.11	5.53
G (Td)	848	956	1228	1288
% max. deviation	±2.1%	±2.0%	±2.25%	±0.7%

Percentage maximum deviation is with respect to measured values.

Energy considerations rule out reactions 9.30 and 9.31 because the swarm does not have electrons that possess energy substantially extending to this high value. The remaining reactions fall into the category of dissociative attachment, $e + AB \rightarrow A + B^-$. The attachment cross section for H_2O is in the range of 4 to $13 \times 10^{-23}\,\text{m}^2$ for mean energies of the swarm in the range 1 to 3 eV.

Attachment in humid air and water vapor has been measured by: Kuffel[19]; Prasad and Craggs[18]; Crompton et al.[20]; Ryzko[13]; Parr and Moruzzi[21]; and Risbud and Naidu.[22] The attachment coefficients are shown as a function of E/N in Figure 9.2 and, for comparison, the attachment coefficients in dry and humid air are also included. The attachment coefficient in water vapor initially decreases as the value of E/N rises from near zero; this effect is attributed to attachment to clusters of water molecules that occurs. The formation of clusters decreases with increasing E/N, resulting in decreasing η/p. At $E/N \sim 40$ Td there is a sudden increase in the coefficient, rising up to $E/N \sim 150$ Td. This effect is attributed to a combination of dissociative attachment (reaction 9.27) and ion–molecule reactions of the form[19]

$$\left.\begin{array}{c} H^- + H_2O \rightarrow OH^- + H_2 \\ \\ \text{or} \\ \\ O^- + H_2O \rightarrow OH^- + OH \end{array}\right\} \tag{9.32}$$

where the ions O^- and H^- are formed by single collision of electrons with the H_2O molecule.

At this juncture, a few comments are in order about the ionization frequency (ν_i) in air. The ionization frequency is the number of ionizing collisions per second between electrons and gas molecules and is related to the ionization coefficient according to

$$\frac{\alpha}{N} = \frac{\nu_i}{WN} \tag{9.33}$$

where W is the drift velocity of electrons. The basic methods of measuring ν_i are:

1. Measurement of α/N and W for substitution in Equation 9.33.
2. The diffusion controlled method. Here a CW rf field is applied across a diffusion container, and breakdown is characterized by

$$\nu_i = \frac{D}{\Lambda^2} \tag{9.34}$$

where D is the diffusion coefficient and Λ is the characteristic diffusion length.[23]
3. The formative time lag method. In this method the formative time lag is measured and, by knowing certain parameters *a priori*, it can be related to ν_i.

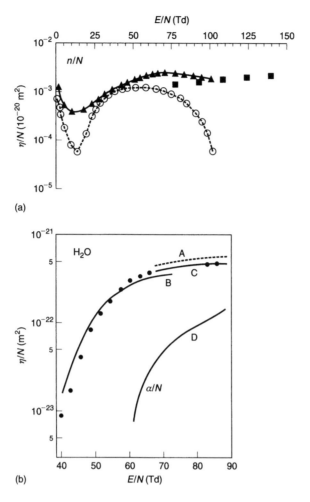

FIGURE 9.2 Attachment coefficients in (a) dry and humid air: ($-\circ-$) dry air, Kuffel[19]; ($\blacksquare$) dry air, Prasad[9]; ($-\bullet-$) humid air, Kuffel.[19] (b) H_2O vapor: (A) Crompton et al.[20]; (B) Kuffel[19]; (C and D) Ryzko[13]; ($\bullet$) Parr and Moruzzi.[21] Note that the attachment coefficients are larger than the ionization coefficients by a factor of 5 to 10 and that no crossover occurs. Figure reproduced from the Institute of Physics (IOP), U.K. With permission.

4. Measurement of the intensity of the photoemission that is related to the electron density.[24] Mentzoni[24] has measured $v_i/N = 6.2 \times 10^{-16} \, m^3 s^{-1}$ using the luminosity method at an equivalent dc $E/N = 320$ Td. Substituting for W from Table 6.1, one gets a value of α/N that is within 10% of the value shown in Figure 9.1.

9.3.2 CARBON DIOXIDE AND CARBON MONOXIDE

CO_2 and CO are both moderately electron-attaching gases. We shall consider each gas separately.

9.3.2.1 Carbon Dioxide (CO_2)

The dissociative attachment process is represented by

$$e + CO_2 \rightarrow CO + O^- \tag{9.35}$$

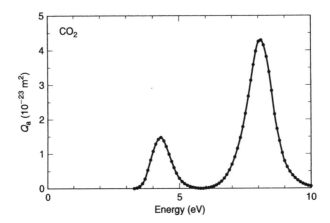

FIGURE 9.3 Total cross section for negative ion formation in carbon dioxide.[28] Both peaks are due to dissociative attachment but with different kinetic energy.[28]

This reaction is rapidly followed by a three-body process according to the reaction[25]

$$O^- + CO_2 + CO_2 \rightarrow CO_3^- + CO_2 \qquad (9.36)$$

and increases with N^2, as expected for a three-body process. The reaction rate decreases from 1×10^{-39} to 1×10^{-41} as E/N is increased from 10 to 160 Td; the coefficients for the process vary from $\mu/N^2 = 210 \times 10^{-44}$ at 61 Td to $\mu/N^2 = 7.9 \times 10^{-44}$ at 152 Td.[26] Recall that the coefficient and rate are related through the drift velocity of the ion. In contrast to oxygen, CO_2 does not show a significant detachment process.[27]

The attachment cross section derived from swarm studies has a peak of 5 to 7×10^{-24} m² in the energy range of 3 to 4 eV, depending upon the electron energy distribution used for the analysis. The total cross section for attachment has been measured by Rapp and Briglia,[28] whose measurements show two peaks, the first at 4.3 eV with a magnitude of 1.48×10^{-23} m² and the second at 8.1 eV with a magnitude of 4.3×10^{-23} m². The observed onset energy for attachment was 3.3 eV. Both the peaks are due to dissociative attachment but differ in kinetic energies involved,[29] the smaller peak having a magnitude of ~35% of the larger one Figure 9.3.

Ionization and attachment coefficients have been measured by: Schlumbohm[30,32]; Bhalla and Craggs[31]; Lakshminarasimha et al.[33]; Conti and Williams[34]; Alger and Rees[26], and Davies.[35] Theoretical studies are due to Hake and Phelps[36] and to Sakai et al.[37] The criterion $\alpha = \eta$ is satisfied at $E/N = 46.6$ Td.[35]

Figure 9.3 shows the total negative ion cross section and Figure 9.4 shows the attachment coefficients in CO_2. There is good agreement between the results of several investigators, partly due to the fact that they used the same current growth method. The effective ionization coefficients are shown in Figure 9.1 and the gas constants are shown in Table 8.1.

9.3.2.2 Carbon Monoxide (CO)

Attachment processes in CO are[38]

$$CO + e \rightarrow C + O^- \qquad (9.37)$$

$$CO + e \rightarrow C^+ + O^- + e \qquad (9.38)$$

$$CO + e \rightarrow C^- + O^+ + e \qquad (9.39)$$

$$CO + e \rightarrow (CO^-)^* \qquad (9.40)$$

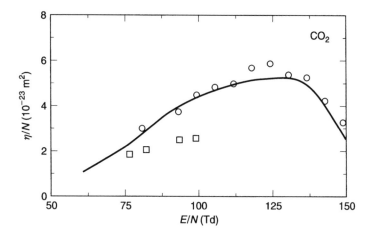

FIGURE 9.4 Reduced attachment coefficients in CO_2 as a function of E/N: (○) Bhalla and Craggs[31]; (—) Alger and Rees[26]; (□) Davies.[35]

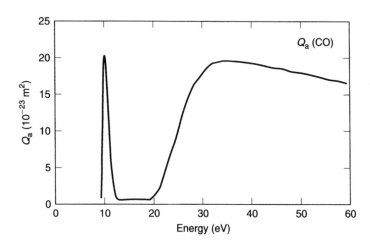

FIGURE 9.5 Total cross section for formation of negative ions in CO.[41] The lower energy peak is due to dissociative attachment and the higher energy peak is due to formation of an ion pair.

Reaction 9.37 is dissociative attachment with an appearance potential of 9.65 eV,[39] reactions 9.38 and 9.39 are ion pair formation at 21 and 23 eV respectively, and reaction 9.40 is the excited state of the negative ion. The electron affinity of O^- in CO is 1.6 eV.[40]

The total cross section for negative ion production has been measured by Rapp and Briglia[41] and is shown in Figure 9.5. The first peak corresponds to dissociative attachment and the second peak, at higher energy, corresponds to ion pair formation. The peak cross section for dissociative attachment is about 50% of that in CO_2. Destruction of negative ions occurs through the autodetachment reaction

$$CO + O^- \rightarrow CO_2 + e \tag{9.41}$$

The attachment coefficient as measured by the ionization current method has been controversial, possibly because of electron detachment. Chatterton and Craggs,[42] using the Bradbury filter method, did not observe any attachment in CO. Moruzzi and Phelps[43] employed a drift tube coupled to a mass spectrometer and failed to detect any negative ions. Further, they detected, during the study of O^- formation in O_2, that the signal strength decreased with the addition of as little as 0.4% CO. They explained their results on the basis that the lifetime of the O^- ions produced by dissociative attachment in CO was very short.

Parr and Moruzzi,[44a] in an attempt to resolve the question, adopted three methods: (a) The Townsend current method, (b) a pulsed uniform electric field drift tube with a charge-integrating system, and (c) a steady uniform-field drift tube with mass spectrometer sampling. Methods (a) and (b) did not reveal any attachment. However, CO stored in a gas cylinder for a year revealed attachment by method (b); this observation was attributed to carbonyl compounds of iron, as determined by method (c). Davies and Williams[45] did not observe any attachment in the ionization current method. The available evidence points to the conclusion that attachment occurs in CO but, for reasons to be clarified, does not influence the current growth. So we present the reduced ionization coefficients in Table 1 of Appendix 6.

Reduced ionization has been measured by Bhalla and Craggs,[38] Price and Moruzzi,[44b] and Lakshminarasimha et al.[33] Figure 9.5 shows the attachment coefficients and Figure 9.1 includes the ionization coefficients for this gas. Theoretical results are derived by Saelee and Lucas[46] and by Land.[47]

9.3.3 FREON-12 (CCl_2F_2)

A review of electron-molecule interaction in Freon-12 has been published by Christophorou et al.[48a] We summarize their review, as very few publications have appeared since. Scattering cross section data have been provided in Chapter 5 and transport parameters in Chapter 7.

CCl_2F_2 is a member of a class of compounds called chlorofluorocarbons (CFCs) and, as in all other compounds of this group, impact of low energy electrons yields an abundance of fragment Cl^- ions. A competition exists for the formation of fragment Cl^- or F^- ions and the favored reaction is

$$CCl_2F_2 + e \rightarrow (CCl_2F_2^-)^* \rightarrow CClF_2 + Cl^- + 0.4\,eV \tag{9.42}$$

according to Kiendler et al.[49] The reaction is exothermic (release of energy) whereas the channel for formation of F^- is endothermic (absorption of energy) by about 1.3 eV, due to the strong CF bond. F^- can be produced and observed only at higher energies.

McCorkle et al.[50] and Christophorou et al.[51] used the swarm unfolding technique to determine the total attachment cross sections, using N_2 as the buffer gas. Electron beam measurement of the total attachment cross section was carried out by Pejčev et al.,[52] who found that the negative ion states were at 0.0, 0.6, and 3.5 eV. Hayashi[53] provided data from swarm studies and Chutjian and Alajajian[54] determined the zero energy attachment cross section as $1.0 \times 10^{-19}\,m^2$. A mass spectrometer study of dissociative attachment was carried out by Underwood-Lemons et al.[55] Kiendler et al.[49] measured the dissociative attachment cross section, using the high-resolution crossed beam technique. These cross sections, along with those recommended by Christophorou et al.,[48a] are shown in Figure 9.6.

Density-reduced attachment coefficients have been measured by Harrison and Geballe[56]; Moruzzi[57]; Boyd et al.[58]; Maller and Naidu[60,61]; Siddagangappa et al.[62], Rao and Raju[63]; and Frechette[64]. Figure 9.7 shows α/N and η/N values recommended by Christophorou et al.[48a] Tabulated values of α/N and η/N are shown in Table 9.5. The attachment rate constant shows a strong dependence on the temperature ($\sim T_e^{-1/2}$) with an average value of $15.5 \times 10^{-16}\,m^3s^{-1}$ at 298 K.[48a] The limiting value of E/N is 355 Td, below which breakdown does not occur. Figure 9.8 shows additional data in selected fluorocarbons, for comparison.

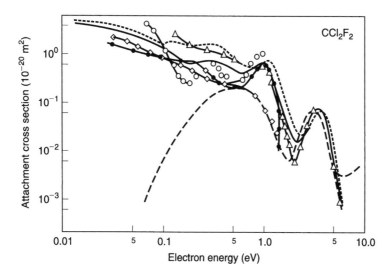

FIGURE 9.6 Electron attachment cross sections in CCl_2F_2. (—♦—) Christophorou et al.[51]; (—△—) Pejčev et al.[52]; (—○—) McCorkle et al.[50]; (– – –) Hayashi[53]; (——) Underwood-Lemons et al.[55]; (··◇··); Kiendler et al.[49] The cross sections recommended by Christophorou et al.[48a] are shown by the heavy line (——). Figure reproduced from Christophorou, L. G. et al.[48] *J. Phys. Chem. Ref. Data*, 26, 1205, 1997. With permission of the authors.

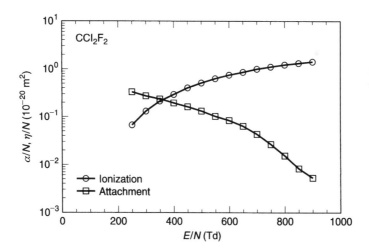

FIGURE 9.7 Density-reduced ionization and attachment coefficients in CCl_2F_2 as a function of E/N. Figure reproduced from IOP, U.K. With permission.

9.3.4 HALOGENS

We have already stated that the halogens have a high electron affinity, which is reflected by the fact that gases that have one or more atoms belonging to this group have a higher dielectric strength. The molecules are highly reactive and, to measure the attachment properties, one needs to use a buffer gas. These gases are discussed in order of increasing atomic weight.

TABLE 9.5
Coefficients α/N and η/N in CCl_2F_2 as a Function of E/N

E/N	α/N	η/N	E/N	α/N	η/N
250	0.066	0.33	800	1.18	0.015
300	0.13	0.27	850	1.27	0.008
350	0.21	0.23	900	1.36	0.005
400	0.29	0.19	950	1.46	
450	0.40	0.16	1000	1.56	
500	0.50	0.13	1250	2.08	
550	0.61	0.10	1500	2.65	
600	0.73	0.082	2000	3.85	
650	0.84	0.062	2500	5.15	
700	0.96	0.042	3000	6.51	
750	1.07	0.026			

Coefficients in units of $10^{-20}\,m^2$, E/N in units of Td.
Reproduced from Christophorou, L. G. et al. *J. Phys. Chem. Ref. Data*, 26, 1205, 1997.
With permission.

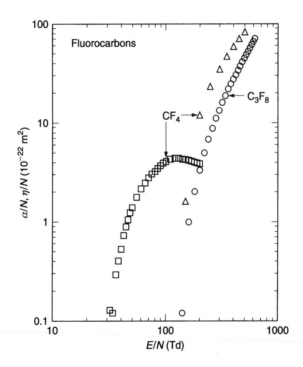

FIGURE 9.8 Density-reduced ionization and attachment coefficients in selected fluorocarbons: ($\circ$) C_3F_8 (α_T)[48b]; ($\square$) CF_4 (η/N)[48c]; ($\triangle$) CF_4 (α_{eff}/N).[48c] Reproduced from IOP, U.K. With permission.

9.3.4.1 Fluorine (F_2)

Dissociative attachment proceeds according to the formation of a short-lived negative ion (shape resonance), the reaction being

$$F_2 + e \rightarrow (F_2^-)^* \rightarrow F^- + F \tag{9.43}$$

TABLE 9.6
Attachment Rate Constant of Thermal Electrons in Molecular Halogens

Molecule	Electron Affinity (atom) (eV)	Electron Affinity (molecule) (eV)	Electron Energy (eV)	Rate Constant[74a] k (m^3 s^{-1})
F$_2$	3.40	3.08	0.043	7.5×10^{-15}
Cl$_2$	3.62	2.38	0.025	3.1×10^{-16}
Br$_2$	3.36	2.55	0.025	0.82×10^{-18}
I$_2$	3.06	2.55		

Reactions similar to Equation 9.43 also occur in other molecular halogens[65]:

$$\left.\begin{array}{l} Cl_2 + e \rightarrow Cl^- + Cl \\ Br_2 + e \rightarrow Br^- + Br \end{array}\right\} \qquad (9.44)$$

The mechanism of attachment is due to thermal electrons, the cross section increasing with decreasing energy. The thermal attachment rate constants for the three halogens[74a] are given in Table 9.6. The potential energy diagrams of the ground states of F$_2$ and F$_2^-$ cross very close to the equilibrium distance of the neutral molecule.[66] This behavior suggests that a peak of the attachment cross section exists near zero energy. The dissociative attachment rate of F$_2$ measured by Schneider and Brau,[74b] 7.0×10^{-15} m^3 s^{-1} at 1 eV mean energy, agrees very well with the value shown in the table.

Ion pair formation also occurs with an onset potential of 15.8 eV according to the reaction[67]

$$F_2 + e \rightarrow F^- + F^+ + e \qquad (9.45)$$

Figure 9.9 shows the ionization efficiency curve measured with a mass spectrometer[67]; the distinct peak seen at ~1 eV is typical of shape resonance. However, the steeply rising cross section at low energy is not found by Chantry[68] or McCorkle et al.[70a] There is reason to believe that the DeCorpo et al.[67] cross section is an experimental artefact. F$_2$ is a highly reactive gas and the gas pressure fluctuates considerably during the course of an experiment. McCorkle et al.[70a] derived the low-energy dissociative attachment cross section by an unfolding procedure and obtained a cross section of ~1.5×10^{-19} m^2 near thermal energy. Though this cross section is lower than that determined by Chantry,[68] 4×10^{-19} m^2, there is tolerable agreement between the two investigations.

The ionization potential of the gas is 17.442 eV. The only theoretically calculated values for the effective ionization and attachment coefficients are those reported by Hayashi and Nimura.[69] These are shown in Figure 9.10. The theoretically predicted $(E/N)_{lim}$ is 213 Td.

9.3.4.2 Chlorine (Cl$_2$)

Electron interaction with chlorine molecules has been reviewed by Christophorou and Olthoff[70b] and their recommended or suggested coefficients are summarized.

There are four negative ion states with energies of 0.0, 2.5, 5.5, and 6 eV. At low energies (~2 eV) dissociative attachment occurs with zero threshold potential. The electron affinity of the chlorine atom (3.69 eV) is higher than the dissociation energy of the molecule (2.479 eV). Ionization occurs at 11.5 eV and ion-pair formation occurs at 11.9 eV. Attachment cross sections are shown in Figure 7.37. Below 10 eV dissociative attachment is the only

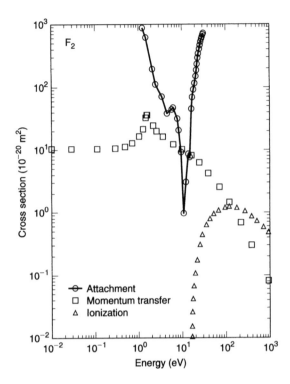

FIGURE 9.9 Cross sections in F_2. The apparently unusual shape of the attachment cross section is due to the fact that there is very high thermal attachment and dissociative attachment up to 16 eV. The increase of attachment cross section above 16 eV is due to ion pair formation.

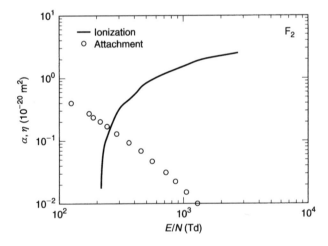

FIGURE 9.10 Theoretical effective ionization and attachment coefficients in fluorine.[69]

mechanism, whereas above this energy ion-pair formation has the dominant cross section. Tabulated values of $(\alpha - \eta)/N$ and η/N are shown in Table 9.7.

9.3.4.3 Bromine (Br_2)

The dissociation energy of Br_2 is 1.9 eV and the electron affinity of the atom is 3.5 eV. The electron affinity of Br_2 is much lower, probably ~1.5 eV.[71] Bromine has two isotopes and,

TABLE 9.7
Attachment and Effective Coefficients in Cl_2

E/N (Td)	$(\alpha - \eta)/N$ $(10^{-20}\,m^2)$	η/N $(10^{-20}\,m^2)$
215	−0.185	0.253
225		0.244
250	−0.107	0.223
275		0.200
300	0.0021	0.176
325		0.156
350	0.18	0.137
375		0.119
400	0.337	0.100
425		0.0814
450	0.509	0.0626
500	0.691	
550	0.868	
600	1.037	
650	1.203	
700	1.369	
750	1.534	

Reproduced from Christophorou, L. G. and J. K. Olthoff, *J. Phys. Chem. Ref. Data*, 28, 131, 1999. With permission.

in an attempt to detect a third one, Blewett[72] studied the formation of positive and negative ions by slow electrons, using a mass spectrometer. Both Br^- and Br_2^- ions were observed in relatively large quantities. The abundance of Br^- had a peak at about 2.8 eV, and the ion possessed considerable kinetic energy. This fact is usually observed by plotting the shift of the peak with electron energy. The attachment and ionization processes in the gas are as follows.

1. *Dissociative attachment:*

$$Br_2 + e \rightarrow Br + Br^- \tag{9.46}$$

The dissociation energy is 1. 97 eV. At electron energies below 10 eV this is the only process that contributes to negative ion formation.

2. *Ion pair formation:*

$$e + Br_2 \rightarrow Br^+ + Br^- + e \tag{9.47}$$

The threshold energy for this process is 10.41 eV.

3. *Three-body collision:*

$$e + Br_2 + M \rightarrow Br_2^- + M \tag{9.48}$$

Direct attachment of an electron to the bromine molecule during a binary collision is unlikely to occur.

4. *Molecular ionization:*

$$Br_2 + e \rightarrow Br_2^+ + e + e \tag{9.49}$$

The ionization potential is 11.80 eV.

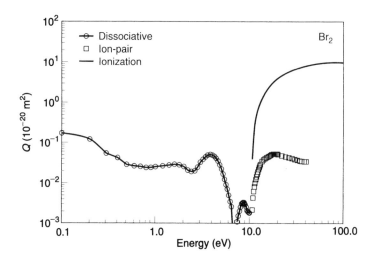

FIGURE 9.11 Attachment and ionization cross sections in molecular bromine: (—○—) dissociative attachment; (□) ion-pair production; (—) ionization.[73] The zero-energy dissociative attachment cross section is approximately the same as that at 0.1 eV, 17.74×10^{-20} m^2.

5. *Dissociative ionization:*

$$e + Br_2 \rightarrow Br^+ + Br + e + e \tag{9.50}$$

A comprehensive investigation of ionization and attachment cross sections has been carried out by Kurepa et al.[73] for electron impact energies between 0 and 100 eV. Since the beam technique was employed, the cross section measured was for total ionization or for total negative ion formation. However, some discrimination of the individual contributions is feasible because of the differences in the threshold energy for each process. The ion-pair formation process occurs at electron energies higher than the lowest positive ion threshold energy. This implies that the negative ions collected for energies lower than that required for ion-pair formation are due to the dissociative attachment process only.

Figure 9.11 shows the various cross sections as a function of electron impact energy. Though Kurepa et al.[73] split up the total cross section into its partial components we present the total cross section only. The attachment cross section curve showed four distinct maxima, though a detailed convolution process revealed two additional attachment processes. The first maximum, at zero energy, is attributed to the direct capture process forming Br$^-$ molecular ions. The cross section at zero energy is more than an order of magnitude lower than that in fluorine.

The rate constant for dissociative attachment has been measured by Sides et al.,[74] and is two orders of magnitude lower than that in F$_2$ (Table 9.6). Ionization and attachment coefficients have been measured by Bailey et al.[75] in their diffusion apparatus, Razzak and Goodyear[71] by the current growth method, and Frank et al.[76] The effective ionization and attachment coefficients are shown in Table 9.8. The attachment coefficient is observed to be dependent on the gas number density; this effect is attributed to the fact that excited negative ions are produced according to the reaction

$$Br_2 + e \rightarrow Br_2^{-*} \tag{9.51}$$

followed by collisional stabilization, which is a three-body process. $(E/N)_{lim}$ occurs at 285 Td.

TABLE 9.8
Ionization and Attachment Coefficients in Bromine[71]

E/N (Td)	α/N	η/N	$(\alpha - \eta)/N$
240	0.0557	0.158	−0.102
250	0.0715	0.151	−0.0791
260	0.0874	0.142	−0.0551
270	0.104	0.135	−0.0309
280	0.123	0.130	−0.0069
290	0.143	0.127	0.0169
300	0.162	0.120	0.0420
310	0.180	0.112	0.0683
320	0.200	0.105	0.0945
330	0.220	0.100	0.120
340	0.240	0.0939	0.146
350	0.261	0.0862	0.174
360	0.278	0.0778	0.202
370	0.300	0.0698	0.228
380	0.317	0.0633	0.254
390	0.337	0.0576	0.280
400	0.357	0.0515	0.310
410	0.378	0.0455	
420	0.401	0.0395	0.361
430	0.425	0.0335	
440			0.431
460			0.544
480			0.640
500			0.705
550			0.847
600			1.024
650			1.187

Coefficients in units of $10^{-20} \, m^2$. $N = 6.45 \times 10^{22} \, m^{-3}$.

9.3.4.4 Iodine (I_2)

Only fragmentary data are available in iodine. Hanstrop and Gustaffson[77] determined the electron affinity as 3.059 eV by the photodetachment method. The method depends on radiating the negative ion with a fixed-frequency laser beam that has energy greater than the electron affinity. The kinetic energy of the released electron is measured, employing an energy analyzer. Buchdahl[78] measured the attachment cross section, using the total ionization tube method, and Healy[79] derived the cross section from the swarm method. The dissociative attachment cross section, derived by Biondi and Fox,[80] shows a maximum at zero electron energy and falls to half value at ~0.01 eV. Figure 9.12 shows the attachment cross sections up to 10 eV electron impact energy. The dissociative attachment, as in O_2 and N_2O,[81] depends on the temperature.[82] Further references for attachment cross sections are Truby[83a,b] and Shipsey.[84] There are no publications of ionization coefficients in I_2. Development of a spark discharge in iodine vapor has been studied by Lewis and Woolsey.[85]

9.3.5 Nitrogen Compounds

The gases considered in this section are ammonia (NH_3), nitric oxide (NO), nitrous oxide (N_2O), and nitrogen dioxide (NO_2). The ionization cross sections, attachment cross sections, and some pertinent data are given in Table 9.9. Figure 9.13 shows the ionization cross sections for these four gases.

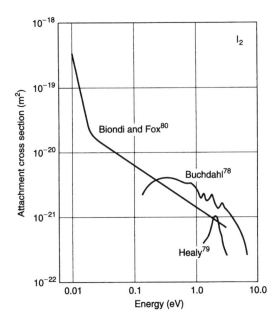

FIGURE 9.12 Attachment cross sections in iodine.

TABLE 9.9
Selected Data for Nitrogen Compounds

Parameter	Gas			
	NH_3	NO	N_2O	NO_2
Dipole moment (D)	1.30	0.16	0.167	0.29
Electron affinity (eV)	1.21	0.02	1.46	2.27
Dissociation energy (eV)	4.77	6.39	1.78	3.09
Ionization potential (eV)	10.18	9.26	12.89	9.59
Reference for Q_a	86	101a	101b	129
Reference for Q_i	87	88	89	90

The ionization cross section for NO has already been given in Table 4.28, and that for N_2O in Figure 5.17. Note that the more recent cross sections of Lindsay et al.[88] for NO, which agree excellently with those of Rapp and Englander-Golden,[89] have been plotted in Figure 9.13.

9.3.5.1 Ammonia (NH_3)

The dissociative attachment products and the minimum energy required for the formation of negative ions are[86]

$$e + NH_3 \rightarrow \left.\begin{array}{ll} H^- + NH_2 & (3.76\,eV) \\ H^- + NH_2{}^* & (5.03\,eV) \\ H + NH_2^- & (3.3\,eV) \\ H + NH_2^-{}^* & (5.78\,eV) \\ H^* + NH_2^- & (13.50\,eV) \end{array}\right\} \qquad (9.52)$$

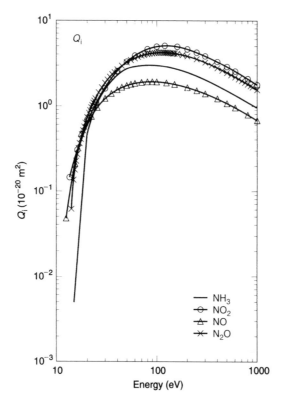

FIGURE 9.13 Ionization cross sections in nitrogen compounds: (—) NH_3, Rao and Srivastava[87]; (—○—) NO_2, Lindsay et al.[88]; (—△—) NO, Lindsay et al.[88]; (—×—) N_2O, Rapp and Englander-Golden[89]; for tabulated values see Table 4.28.

Sharp and Dowell[86] measured the total cross sections for negative ion formation; two peaks were observed, one at 5.65 eV and the second at 10.5 eV. Both peaks contained the ions H^- and NH_2^-. The cross sections for the formation of H^- and NH_2^- are $1.4 \times 10^{-22}\,m^2$ and $1.4 \times 10^{-22}\,m^2$ respectively. Stricklett and Burrow[91] found that the dissociative attachment cross section was a series of peaks in the 5.0 to 6.0 eV range, attributed to vibrational excitation quanta; the cross section in this energy range resembles that of NO (Figure 4.44). Tronc et al.[92] have also confirmed the existence of this series of peaks.

Ionization and attachment coefficients in NH_3 have been measured by Bailey and Duncanson,[93] Bradbury,[94] Parr and Moruzzi,[95] and Risbud and Naidu.[96] Attachment does not occur at low values of E/N (< 25 Td), and in this respect it resembles the behavior of water vapor. Attachment sets in at $E/N > 27$ Td and the process suggested is the two-body attachment (dissociative) that increases rapidly with increasing E/N. The measured attachment coefficients are shown in Figure 9.14. Calculated dissociative attachment rate coefficients using the drift velocity data of Pack et al.[97] are in the range of 10^{-18} to $10^{-16}\,m^3 s^{-1}$.

9.3.5.2 Nitric Oxide (NO)

Rapp and Briglia[98] have measured the total cross sections for negative ion formation in the energy range from 6.7 to 13.0 eV, observing a broad peak for O^- formation. The reaction is simple, viz.

$$e + NO \rightarrow N + O^- \tag{9.53}$$

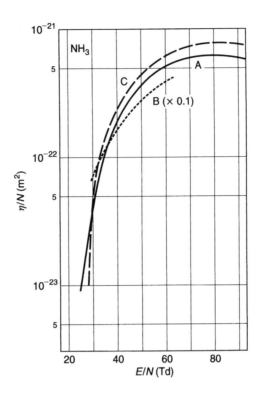

FIGURE 9.14 Density-reduced attachment coefficients in NH_3 as a function of E/N: (A) Bailey and Duncanson[93]; (B) Bradbury[94] note the multiplyimg factor; (C) Parr and Moruzzi.[95] Reproduced from IOP, U.K. With permission.

The threshold for negative ion formation was observed[98] as 6.6 eV whereas Chantry[99] observed a threshold of 7.42 eV. Though the peak in the cross section observed by Chantry[99] is higher than that reported by Rapp and Briglia,[98] the shape of the cross section curve is similar. A point to note is that the nitrogen atom is not in the ground state; it has the ^{2}D excited state.[100] Krishnakumar and Srivastava[101a] observed a peak cross section which is about 50% higher than that seen by Rapp and Briglia[98] (Figure 9.15), in addition to an additional channel for ion pair formation,

$$e + NO \rightarrow N^+ + O^- + e \qquad (9.54)$$

the threshold energy being 19.4 eV. Three-body attachment occurs according to

$$e + NO + M \rightarrow NO^- + M \qquad (9.55)$$

Data on attachment coefficients in NO are not available except for those of Bradbury,[102] who obtained attachment probabilities in the range from 0.1 to 1.5 cm^{-1}. The attachment coefficient decreased with increasing E/N from 0 to 15 Td and depended upon the pressure at the same E/N; higher pressures yielded higher coefficients with a linear dependence. A three-body collision process was postulated. Nitric oxide forms dimers easily with an energy of attraction between two molecules being 0.05 eV, facilitating the formation of NO^- ions.

Bradbury[103] suggested that direct attachment of type AB^- is favored in the case of molecules that do not have a resultant orbital angular momentum or spin equal to zero in the

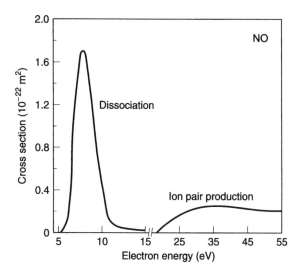

FIGURE 9.15 Cross section for formation of negative ions in NO by dissociation ($<15\,\mathrm{eV}$) and ion pair oroduction ($>19.4\,\mathrm{eV}$). Reproduced from IOP, U.K. With permission.

ground state. The extra electron should have a peg to hang its coat, so to speak; a path through which the orbital angular momentum and spin of the atom to which the electron attaches could be changed. Molecules that have the $^1\Sigma$ configuration ($^1S_0$ configuration of the atom) do not form stable negative ions (examples are H_2, N_2, CO, Cl_2, Br_2, I_2, HCl). O_2 and NO, which have $^3\Sigma$ and $^2\Pi$ configuration respectively, have negative ions of the parent molecule.

Although the formation of NO^- has been established, its presence is not readily detected because of collisional detachment in the process

$$NO^- + M \rightarrow NO + e + M \qquad (9.56)$$

where M is a neutral species such as He, Ne, H_2, CO, NO, CO_2, N_2O, and NH_3. To observe NO^- the ion should be collisionally stabilized, probably with trace amounts of impurity. Parkes and Sugden[104] have measured the drift velocity and attachment coefficient, using a pulsed drift tube, and their results are shown in Figure 9.16.

It is instructive to compare the results of the two investigations nearly forty years apart. At $E/N = 0.8\,\mathrm{Td}$ Parkes and Sugden[104] obtained $\eta/N^2 = 10^{-36}\,\mathrm{cm}^5$ at $N = 5.25 \times 10^{18}\,\mathrm{cm}^{-3}$, which gives $\eta = 2500\,\mathrm{m}^{-1}$. Bradbury[102] measured the same coefficient at a lower pressure and observed that it was proportional to N^2. Extrapolating to the number density used by Parkes and Sugden,[104] one obtains an attachment coefficient of $5000\,\mathrm{m}^{-1}$ which is about 100% larger. However, at $E/N = 10\,\mathrm{Td}$ this difference reduces to $\sim 20\%$, partly because of neglect of detachment by Parkes and Sugden.[104]

McFarland et al.[105] found that the detachment rate constant could be fitted to the Arrhenius equation

$$K_d = A \exp\left(-\frac{\varepsilon_d}{kT}\right) \qquad (9.57)$$

where ε_d is the activation energy in eV, T the absolute temperature, and the preexponential factor A is in units of $\mathrm{m}^3\mathrm{s}^{-1}$. The same type of equation may also be used for attachment rate, except that A will have a different value and ε_d is equal to the electron affinity, 0.02 eV for NO. Depending upon the detacher, A will have different values, as shown in Table 9.10.

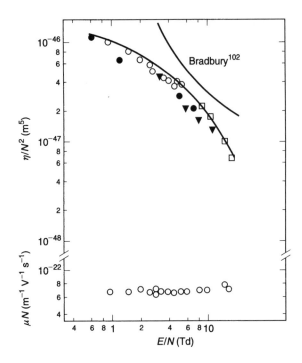

FIGURE 9.16 η/N^2 (top) and μN (bottom) as a function of E/N in NO.[104] The symbols are experimental points obtained at various gas number densities in the range of 2.2 to $5.25 \times 10^{24}\,m^3$ at 293 and 361 K. They are not delineated, for the sake of clarity. Reduced mobility is shown in the lower part of the figure.

TABLE 9.10
Three-Body Attachment (k_a) and Detachment Rates (k_d) in NO in Units of $m^3 s^{-1}$ [105]

M	ε_d (meV)	A (m³/s)	k_d (285 K) (m³/s)	k_a (300 K) m⁶/s
Ne	98.0	1.47(−18)	2.9(−20)	3.42(−45)
He	74.8	5.23(−17)	2.4(−19)	2.97(−43)
H_2	82.8	6.29(−18)	2.3(−19)	2.62(−44)
CO	45.5	3.46(−18)	5.0(−19)	6.06(−44)
NO	105.1	3.28(−16)	5.0(−18)	5.80(−43)
CO_2	81.2	2.2(−16)	8.3(−18)	9.76(−43)
N_2O	107.2	4.26(−16)	5.1(−18)	6.95(−43)
NH_3	59.0	2.32(−16)	2.0(−17)	2.42(−42)

The attachment process is shown in Equation 9.55 and detachment in 9.56. $a(b)$ means $a \times 10^b$. Note the unit of k_a, due to $(1/N^2)$ dependence.

9.3.5.3 Nitrous Oxide (N_2O)

Dissociative attachment occurs according to the reaction

$$N_2O + e \rightarrow N_2 + O^- \tag{9.58}$$

The reaction produces one nitrogen molecule per electron but experimentally the nitrogen yield has been observed to be larger. Several reactions involving O^- ions and NO_2 molecules have been suggested,[106] and it is beyond the scope of our treatment to consider these ion–molecule reactions and the rate constants. We therefore restrict ourselves to the total attachment and ionization coefficients.

Electron attachment to N_2O was studied by Schulz,[107] Curran and Fox,[108] and Bardsley.[109] The total cross section for formation of negative ions was measured by Rapp and Briglia[110] in the electron energy range from 0 to 55 eV. Thermal energy attachment does not occur (only a second-order effect has been observed)[111] due to the $^1\Sigma$ configuration of the ground state.[112] Mass spectrometer studies, however, showed an O^- signal at zero electron energy; Chantry[113] resolved the discrepancy by measuring the O^- signal at various temperatures in the range from 160 to 1040 K for electrons of energy less than 4 eV. At room temperature he observed a single peak with a maximum at 2.3 eV and a small shoulder at around 1 eV. At higher temperatures the low-energy signal increased dramatically, finally resulting in a pronounced peak near 0 eV. The temperature dependence was attributed to two different states of N_2O^-: a $^2\Sigma$ state at a vertical energy of 2.3 eV, and a $^2\Pi$ state with a vertical energy of 1.0 eV.[114]

The essential features of attachment cross sections are that a prominent peak of 8.60×10^{-22} m^2 occurs at 2.2 eV due to dissociative attachment, with a continuum beyond 15 eV due to ion-pair formation (Figure 9.17). Additional peaks have been reported: Curran and Fox,[108] and Schulz[107] observed a small peak at 0.7 eV; Paulson[115] observed smaller peaks at 8 and 13 eV.

Rapp and Briglia[110] obtained two additional peaks in the 5 to 15 eV range. The differences between various results in the low-energy region are attributed to the differences in source temperature, as expounded by Chantry.[113] Krishnakumar and Srivastava[101b] observed additional peaks at 5.40, 8.10, and 13.20 eV before the onset of the continuum. Brüning et al.[114] obtained a peak at 0.55 eV which is also smaller than the 2.3 eV resonance peak.

Returning to swarm measurements, attachment coefficients have been measured by Bailey and Rudd[116]; Bradbury and Tatel[117]; Phelps and Voshall[118]; Parkes[119]; Warman et al.[120]; Dutton et al.[121]; Hayashi et al.[122]; and Yoshida et al.[123] The latter authors have employed the double shutter method with arrival time spectra analysis, providing

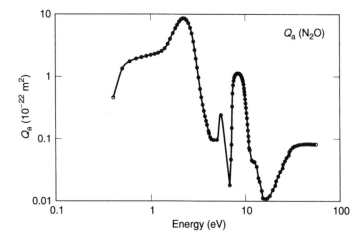

FIGURE 9.17 Total cross section for negative ion formation in N_2O. The peak at 2.3 eV is due to dissociative attachment and the continuum above 15 eV is due to ion-pair formation. The intermediate peaks at 5.5 and 9 eV are probably due to impurity O^-/O_2 and O^-/CO_2 ions, according to Krishnakumar and Srivastava.[101b]

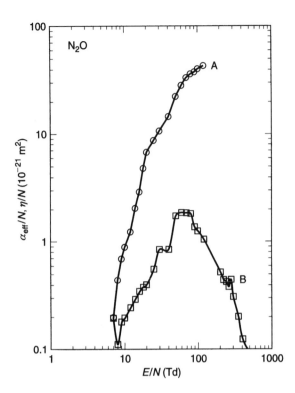

FIGURE 9.18 Density-reduced attachment and ionization coefficients in N_2O: (A) α_{eff}/N; (B) η/N.[123] Reproduced from IOP, U.K. With permission.

one of the most comprehensive sets of parameters for this gas. Figure 9.18 shows the α_{eff}/N and η/N measured by these authors. Table 9.11 shows the swarm parameters, including the drift velocity (W), density-normalized diffusion coefficient (ND_1), and the characteristic energy.

9.3.5.4   Nitrogen Dioxide and Sulfur Dioxide (NO_2 and SO_2)

NO_2 and SO_2 are both atmospheric pollutants. They are agents of acid rain, SO_2 more than NO_2. SO_2 has been reported to have a cooling effect, opposite to the warming effect of CO_2. SO_2 has been observed to be the main constituent of the atmosphere of Jupiter's satellite Io; it is also abundant in the atmosphere of Venus.[124]

From the chemical point of view, nitrogen dioxide and sulfur dioxide present some similarities with carbon dioxide since they are oxides in which the carbon atom (at. wt. = 12) has been replaced by a heavier atom of nitrogen (at. wt. = 14) or sulfur (at. wt. = 32). On the other hand, in contrast with CO_2, NO_2 is a bent molecule with a dipole moment of 0.316 D (bond angle 134°). Nitrogen and sulfur dioxide are also similar to O_3 in which an oxygen atom has been replaced with a nitrogen or sulfur atom, either of which is electronegative. In the case of NO_2 the replacement atom is lighter than O, whereas in the case of SO_2 the replacement is heavier. These considerations provide interesting aspects for study of these molecules.[125] A comparison of selected properties is given in Table 9.12.

We shall first provide a summary of scattering cross sections, which are required for theoretical studies as well as for understanding the collision physics, as they were not included in the earlier treatment.

TABLE 9.11
Electron Transport Coefficients as a Function of E/N in N_2O

E/N (Td)	α_{eff}/N (10^{-21} m^2)	η/N (10^{-21} m^2)	W (10^4 m/s)	ND_L (10^{24} m^{-1}s^{-1})	D_L/μ (eV)
7		0.193	5.85	0.746	0.0893
10		0.193	7.39	0.791	0.107
12		0.241	7.98	0.774	0.116
14		0.287	8.34	0.723	0.121
16		0.342	8.47	0.684	0.129
18		0.374	8.64	0.676	0.141
20		0.394	8.60	0.733	0.170
25		0.550	8.74	0.618	0.177
30		0.833	8.78	0.588	0.201
40		0.837	9.02	0.536	0.238
50		1.71	9.46	0.583	3.07
60		1.84	9.77	0.482	2.96
70		1.82	10.5	0.795	5.28
80		1.79	11.1	0.831	5.98
90		1.35	11.9	1.17	8.79
100		1.23	12.6	1.27	1.00
120		1.03	14.3	1.67	1.40
150			17.0	2.13	1.88
200	−0.0546	0.514	20.1	2.97	2.96
220	0.196	0.436	21.7	2.94	2.99
240	0.437	0.417	23.2	2.67	2.76
260	0.687	0.372	24.8	2.57	2.70
280	0.882	0.438	25.8	2.94	3.19
300	1.22	0.302	27.4	2.65	2.91
350	2.02	0.196	30.3	2.95	3.41
400	2.88	0.121	33.5	2.85	3.41
500	4.77	0.0877	39.4	2.93	3.72
600	6.68		44.7	3.10	4.16
700	8.66		49.9	3.50	4.90
800	10.6		55.2	4.03	5.85
1000	14.4		64.7	4.87	7.53
1500	22.1		85.9	6.84	11.9
2000	28.0		104	7.08	13.6
2500	33.1		121	7.75	16.0
3000	35.9		135	10.7	23.9
3500	37.3		153	—	—
4000	39.9		165	—	—
5000	42.7		193	8.55	22.1

Table reproduced from Yoshida, K. et al., *J. Phys. D*, 32, 862, 1999. With permission of IOP, U.K.

TABLE 9.12
Selected Properties of CO_2, NO_2, and SO_2 Molecules[124]

Molecule	Number of Electrons	Valence Electrons	Bond Length (nm)	Bond Angle (°)	Dipole Moment (D)
CO_2	22	14	0.1162	180	0
NO_2	23	15	0.1195	134	0.316
SO_2	32	18	0.1432	119	1.63

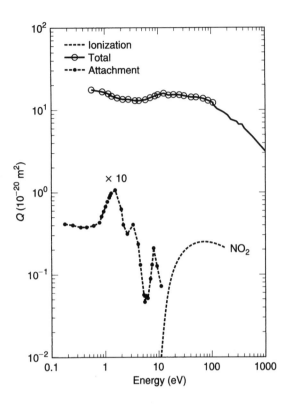

FIGURE 9.19 Scattering cross sections in NO_2: ($—○—$) total, Szmytkowski et al.[126]; ($—$) total, Zecca et al.[124]; ($—•—$) attachment, Rangwala et al.[129]; ($— —$) Ionization, Stephan et al.[128] Note the multiplying factor for the attachment cross section.

9.3.5.4.1　Nitrogen Dioxide (NO_2)

The total cross section has been measured by Szmytkowski et al.[126] in the electron impact energy range from 0.6 to 220 eV and by Zecca et al.[124] in the range range from 100 to 4000 eV. In the overlapping region there is good agreement between the two investigations. The rise in the low-energy Q_T near 0.6 eV has been attributed to a direct scattering process. A local maximum of $15 \times 10^{-20}\,m^2$ at 11.0 eV has been observed.[127] Ionization cross sections have been measured by Stephan et al.[128] These cross sections are shown in Figure 9.19. The high-energy cross sections above 300 eV have been interpreted by Zecca et al.,[124] using a "double Yukawa potential," as

$$V(r) = \frac{V_1}{r}\exp\left(-\frac{r}{a_1}\right) + \frac{V_2}{r}\exp\left(-\frac{r}{a_2}\right) \tag{9.59}$$

where V_i and a_i are the depth and radial extension of the so-called Yukawa well.

　　Disssociative attachment cross sections have been measured by Abouaf et al.[129a] and Rangwala et al.[129b] The dissociative attachment processes are[129a]

$$e + NO_2 \rightarrow O^- + NO \quad (1.61\,eV) \tag{9.60}$$

$$e + NO_2 \rightarrow NO^- + O \quad (3.11\,eV) \tag{9.61}$$

$$e + NO_2 \rightarrow O_2^- + N \quad (4.03\,eV) \tag{9.62}$$

The yields for the ions reach peaks at 1.5 eV, 3.2 eV, and 8.4 eV respectively (Figure 9.19) in the ratio of 1000:3:10. The attachment cross section at zero energy occurs by collisional stabilization through the three-body process[130]

$$
\left.
\begin{aligned}
e + NO_2 &\rightarrow \left(NO_2^-\right)^* \\
\left(NO_2^-\right)^* + M &\rightarrow NO_2^- + M
\end{aligned}
\right\}
\tag{9.63}
$$

The reaction rate for (9.63) depends upon the species M. Shimamori and Hotta[130] obtained rates ranging from 0.5 to $1.3 \times 10^{-14}\, m^3/s$ for several gases (He, Ar, Xe, N_2, CO_2, and n-C_4H_{10}).

Meager data exist on ionization and attachment coefficients, experimental or theoretical. Bradbury and Tatel[131] and Puckett et al.[132] measured attachment in flowing afterglows. The rate coefficient for NO_2^- was determined as $1.4 \times 10^{-17}\, m^3/s$. Data on ionization and attachment coefficients in the gas are needed.

9.3.5.4.2 Sulfur Dioxide (SO_2)

Measurements of total cross section in the low-energy range up to 10 eV show considerable discrepancies between several results as reported.[127] Figure 9.20 shows the measurements of Zecca et al.[124] The cross sections are considerably larger than those of CO_2 and NO_2, due to the large molecule. The main features of Q_T are a peak at ∼1 eV and a broad maximum at ∼10 eV, characteristic of shape resonance.

The energy for ionization of the molecule is 12.5 eV,[133] and the total ionization cross sections (fragments with appearance potentials are: SO^+, 16.5 eV; S^+, 16.5 eV; and O^+, 23.5 eV) measured by Orient and Srivastava[133] up to 200 eV energy are also shown in Figure 9.20. The range from 200 to 250 eV is covered by Čadež et al.[134] Again, one sees that the ionization cross sections are much larger than those of CO_2 and NO_2.

Dissociative attachment cross sections have been measured by Krishnakumar et al.[135] whose results are not in agreement with the results of Čadež et al.[134] A large number of

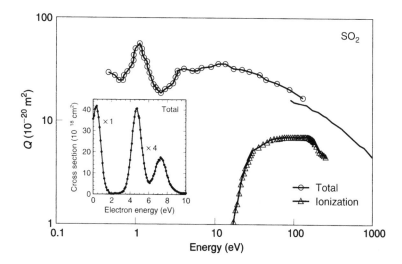

FIGURE 9.20 Scattering cross sections in SO_2: (— ○ —) total, Zecca et al.,[124] Gdansk laboratory; (—) total, Zecca et al.,[124] Trent laboratory; (— △ —) ionization cross section, Orient and Srivastava.[133] The inset shows the total attachment cross section.

negative ions of species O^-, S^-, and SO^- are formed by different routes. The total attachment cross section, shown in the inset of Figure 9.20, has three prominent peaks, at 1.0 eV, 4.7 eV, and 7.5 eV. The latter two peaks are lower in the ratio of 100:25:8. Tabulated values of attachment cross sections as a function of electron energy are given in Table 1, Appendix 6.

Ionization and attachment coefficients have been measured by Bradbury and Tatel[131]; Schlumbohm[136]; Bouby et al.[137]; Rademacher et al.[138]; and Moruzzi and Lakdawala.[139] The last mentioned authors have measured the attachment coefficients for $3 < E/N < 240$ Td, using the pulsed drift tube technique. For $E/N < 10$ Td the attachment process is of the three-body type whereas for $E/N > 10$ Td dissociative attachment occurs. The dissociative process is

$$SO_2 + e \rightarrow O^- + SO$$

or

$$SO_2 + e \rightarrow SO^- + O$$

Schlumbohm[136] suggested that the attachment coefficient is of the type

$$\left(\frac{\eta}{N}\right)_{total} = \left[\left(\frac{\eta}{N}\right) + \left(\frac{\bar{\eta}}{N^2}\right)N\right] \tag{9.64}$$

where $\eta/N = 1.24 \times 10^{-22}$ m^2 and $\bar{\eta}/N^2 = 2.89 \times 10^{-47}$ m^5. Tabulated values of density-reduced attachment coefficients as a function of E/N are given in Table 2 of Appendix 6. Three types of particle were detected, O^- ion, SO^- ion, and $(SO_2)_2$ cluster, with reduced mobilities of 0.69, 0.62, and 0.55 cm^{2}V^{-1}s^{-1} respectively.

9.3.6 OXYGEN (O_2)

Oxygen is one of the most extensively studied gases as far as the attachment cross sections are concerned because of its technological importance. Dissociative and three-body attachment processes occur according to

$$e + O_2 \rightarrow O^- + O \tag{9.65}$$

$$e + O_2 \rightarrow (O_2^-)^*; \quad (O_2^-)^* + M \rightarrow O_2^- + M \tag{9.66}$$

Ion-pair production occurs according to

$$e + O_2 \rightarrow O^+ + O^- + e \tag{9.67}$$

A very small cross section ($\sim 1 \times 10^{-28}$ m^2) has been estimated[140] from photodetachment of O_2^-;[140] in view of the small cross section we shall not discuss it further.

There is broad agreement on the dissociative attachment cross section as shown in Figure 9.21.[141a, b] The cross section has a relatively broad peak of 1.41×10^{-22} m^2 with its maximum at 6.7 eV; this peak is interpreted as a resonance peak. Rapp and Briglia[142] measured the cross section for total negative ion production in the range of energy from 4 to 55 eV. The increase of attachment cross section above 17 eV is attributed to ion-pair production, reaction 9.67. Attachment of low-energy electrons to the molecule is by three-body collision, reaction 9.66 and the attachment coefficients are pressure dependent. The earlier results of Chanin et al.[143a,b] are shown in Figure 1.17; Figure 9.22 shows the results

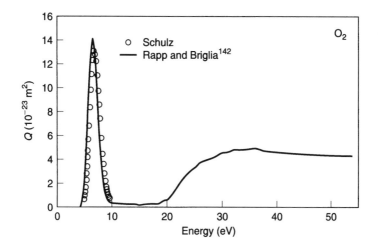

FIGURE 9.21 Cross section for electron attachment in O_2. The 6.7 eV peak is due to dissociative attachment and the high energy continuum is due to ion-pair formation.

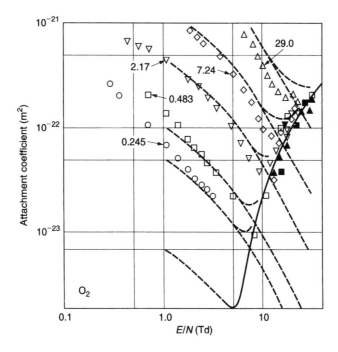

FIGURE 9.22 Electron attachment coefficient (η/N) as a function of E/N at low values of E/N. The left sides of the curves show a three-body attachment process below $E/N = 10$ Td; the density-reduced attachment coefficient is dependent on N. N is as shown ($\times 10^{24}\,\text{m}^{-3}$); note the increase of η/N with increasing N. Left-branch open symbols[143a, b]; (–□–□–both branches) Hunter et al.[144]; (▲) Chatterton and Craggs [143c]; (▼) Rees,[143d] agreeing with Herreng[143e]; (▽, right branch); (■) Kuffel [143f] (– – –) Taniguchi et al.[439] computed.

obtained by Hunter et al.,[144] who compare their measurements with those of Chanin et al.[143] and Grünberg.[145]

Density-reduced ionization and attachment coefficients have been measured or calculated by a number of authors. Selected references for the density-reduced coefficient are: Masch[146]; Bradbury[147]; Geballe and Harrison[148]; Chanin et al.[143a,b]; Schlumbohm[149]; Prasad and

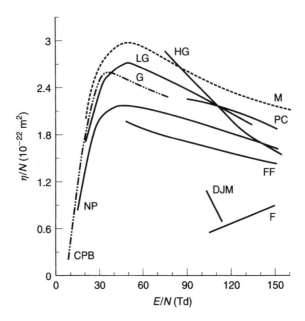

FIGURE 9.23 Density-reduced attachment coefficient as a function of E/N for O_2. The letters refer to: (HG) Geballe and Harrison[148]; (PC) Prasad and Craggs[150]; (CPB) Chanin et al.(143a); (DJM) Dutton et al.[151]; (F) Frommhold[153]; (FF) Freely and Fischer[152]; (G) Grünberg[145]; (NP) Naidu and Prasad[154]; (M) Masek et al.[157]; (LG) Liu and Raju.[160] Figure reproduced from Liu, J., Ph.D. Thesis, University of Windsor, Ontario, 1993.

Craggs[150]; Dutton et al.[151]; Freely and Fischer[152]; Frommhold[153]; Grünberg[145]; Naidu and Prasad[154]; Price et al.[155]; Gurumurthy and Raju[156]; and Masek et al.[157] Theoretical analyses are due to Hake and Phelps[158]; Lucas et al.[159]; Liu and Raju[160,161]; and Liu.[162] Figures 9.23 and 9.24 and Table 9.13 present these data. A large number of ion–molecule reactions occur, but these are beyond the scope of the present volume.

The values of the gas constants F and G have been evaluated by Gurumurthy and Raju[156] from their measurements, as shown in Table 9.14.

9.3.7 SULFUR HEXAFLUORIDE (SF$_6$)

As stated previously, the cross sections and the measurements of density-reduced coefficients have been thoroughly reviewed by Christophorou and Olthoff.[163] We essentially reproduce their recommended attachment cross sections and transport coefficients as a limited number of publications have appeared since their review.

The attachment processes in SF$_6$ are shown in Table 9.15. Process 9.68 is thermal attachment to the entire SF$_6$ molecule,[164] where the impact initially forms a metastable, (SF$_6^-$)*, which is subsequently stabilized by collision or radiation. The zero-energy cross section for direct attachment, SF$_6^-$, is very large, $\sim 8 \times 10^{-16} \, \text{m}^2$, and decreases according to ε^{-n} in the energy range 0.1 meV to 0.1 eV. For electron energies greater than 0.1 eV the attachment cross section decreases even faster, reaching a value of $\sim 3 \times 10^{-22} \, \text{m}^2$, a six order of magnitude decrease.

Dissociative attachment for the formation of SF$_5^-$ occurs at a low threshold energy of 0.1 eV with a small peak at 0.38 eV. Fragment SF$_3^-$ requires a slightly higher threshold energy of 0.7 eV compared with the threshold energy of 0.2 eV for the formation of

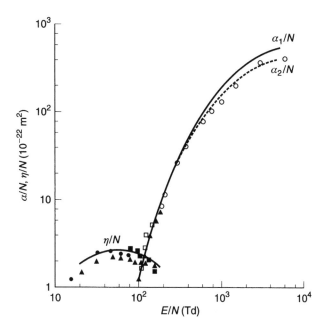

FIGURE 9.24 Reduced ionization and attachment coefficients as a function of E/N for O_2. α/N: (○) Schlumbohm[149]; (■) Naidu and Prasad[154]; (▲) Price et al.[155] η/N: (△) Geballe and Harrison[148]; (•) Grünberg[145]; [■] Naidu and Prasad[154]; (——) Liu and Raju,[160] corrected for diffusion; (– –) Liu and Raju,[160] uncorrected for diffusion. Figure reproduced from Liu, J., Ph.D. Thesis, University of Windsor, 1993.

TABLE 9.13
Density-Reduced Ionization Coefficients α/N, and Attachment Coefficients η/N as a Function of E/N[161]

E/N	α/N	η/N	E/N	α/N
20		1.8	400	48.9
30		2.4	500	71.5
40		2.6	600	93.2
50		2.7	700	114.2
60		2.7	800	133.1
70	0.10	2.5	900	151.1
80	0.30	2.5	1000	167.7
90	0.50	2.5	1300	210.3
100	0.70	2.3	2000	287.4
130	2.2	2.1	3000	358.2
200	9.2	2.1	4000	392.2
300	27.0		5000	396.0

α/N and η/N in units of $10^{-22}\,m^2$, E/N in units of Td.

---

neutral SF_3. Table 9.16 gives the cross sections recommended or suggested by Christophorou and Olthoff,[163] and Figure 9.25 shows the cross sections.

Selected references for the reduced attachment coefficients are: Geballe and Harrison[165]; Bhalla and Craggs[166]; McAfee and Edelson[167]; Boyd and Crichton[168]; Teich and Sang[169]; Maller and Naidu[170]; Kline et al.[171]; Raju and Dincer[172]; Shimozuma et al.[173];

TABLE 9.14
Gas Constants in O_2[156]

E/N Range (Td)	F (m^2)	G (Td)	Deviation
257–642	1.97×10^{-20}	646	±2.5%
642–1026	3.87×10^{-20}	1066	±2.5%

Deviation is calculated from measured values.

TABLE 9.15
Threshold Energy and Attachment Processes in SF_6

Threshold energy (eV)	Process	
0	$e + SF_6 \rightarrow SF_6^-$	(9.68)
0.1	$e + SF_6 \rightarrow SF_5^- + F$	(9.69)
2.8	$e + SF_6 \rightarrow SF_4^- + F_2$	(9.70)
0.7	$e + SF_4 \rightarrow SF_3^- + F$	(9.71)
0.2	$e + SF_4 \rightarrow SF_3 + F^-$	(9.72)
1.64	$e + F_2 \rightarrow F + F^-$	(9.73)

Neutral dissociation energies are given in Table 5.16.

TABLE 9.16
Attachment Cross Sections for SF_6 as a Function of Electron Energy and Cross Section

Energy (eV)	Attachment cross section (10^{-20} m^2)						
	Total	SF_6^-	SF_5^-	SF_4^-	SF_3^-	F_2^-	F^-
0.0001	7617	7617					
0.0002	5283	5283					
0.0003	4284	4284					
0.0004	3692	3692					
0.0005	3280	3280					
0.0006	2968	2968					
0.0007	2724	2724					
0.0008	2529	2529					
0.0009	2369	2369					
0.001	2237	2237					
0.002	1511	1511					
0.003	1202	1202					
0.004	993	993					
0.005	859	859					
0.006	760	760					
0.007	683	683					
0.008	621	621					
0.009	569	569					
0.010	526	526					

(Continued)

TABLE 9.16
Continued

Energy (eV)	Attachment cross section (10^{-20} m^2)						
	Total	SF_6^-	SF_5^-	SF_4^-	SF_3^-	F_2^-	F^-
0.015	383	383					
0.020	304	304					
0.025	257	257					
0.030	221	221					
0.035	190	190					
0.040	171	171					
0.045	149	149					
0.050	132	132					
0.060	109	109					
0.070	92.7	92.7					
0.080	82.9	82.9					
0.090	74.3	74.3					
0.10	51.4	49.5	1.85				
0.12	32.9	30.8	2.09				
0.14	20.2	17.8	2.36				
0.15	16.7	14.2	2.48				
0.16	13.1	10.5	2.61				
0.18	8.72	5.85	2.87				
0.20	6.01	2.86	3.15				
0.22	4.69	1.24	3.45				
0.25	4.38	0.52	3.86				
0.28	4.40	0.25	4.15				
0.30	4.40	0.16	4.24				
0.35	4.12	0.05	4.07				
0.40	3.46	0.01	3.45				
0.45	2.75		2.75				
0.50	2.15		2.15				
0.60	1.25		1.25				
0.70	0.722		0.72				
0.80	0.416		0.42				
0.90	0.245		0.25				
1.00	0.147		0.15				
1.20	0.060		0.060				
1.50	0.020		0.020			30(−6)	
2.00	0.0043					80(−6)	220(−6)
2.25	0.0020					75(−6)	750(−6)
2.50	0.0019					60(−6)	1490(−6)
3.0	0.0010					26(−6)	980(−6)
3.5	0.0018			84(−6)		26(−6)	1710(−6)
4.0	0.0092			350(−6)		265(−6)	8560(−6)
4.5	0.0290			1440(−6)		707(−6)	26900(−6)
5.0	0.0514			4570(−6)		489(−6)	46300(−6)
5.5	0.0493			5280(−6)		155(−6)	43900(−6)
6.0	0.0317			3940(−6)		38(−6)	27800(−6)
6.5	0.0162			2510(−6)		19(−6)	13700(−6)
7.0	0.0088			1300(−6)		15(−6)	7490(−6)
7.5	0.0066			460(−6)		15(−6)	6150(−6)
8.0	0.0099			84(−6)	14(−6)	18(−6)	9770(−6)
8.5	0.0143			32(−6)	46(−6)	31(−6)	1420(−5)
9.0	0.0159				150(−6)	65(−6)	1570(−5)
9.5	0.0144				330(−6)	104(−6)	1390(−5)

(Continued)

TABLE 9.16
Continued

Energy (eV)	Attachment cross section (10^{-20} m^2)						
	Total	SF$_6^-$	SF$_5^-$	SF$_4^-$	SF$_3^-$	F$_2^-$	F$^-$
10.0	0.0120			SF$_2^-$	510(−6)	200(−6)	1130(−5)
10.5	0.0142			3.7(−6)	640(−6)	503(−6)	1310(−5)
11.0	0.0227			14.0(−6)	750(−6)	910(−6)	2100(−5)
11.5	0.0252			42.0(−6)	710(−6)	954(−6)	2350(−5)
12.0	0.0206			76.0(−6)	490(−6)	615(−6)	1950(−5)
12.5	0.0128			106.0(−6)	260(−6)	274(−6)	1220(−5)
13.0	0.0066			87.0(−6)	110(−6)	111(−6)	6290(−6)
13.5	0.0041			44.0(−6)	31(−6)	50(−6)	4000(−6)
14.0	0.0035			17.0(−6)	8(−6)	32(−6)	3400(−6)
15.0	0.0030			2.8(−6)		12(−6)	3000(−6)

Reproduced from Christophorou, L. G. and J. K. Olthoff, *J. Phys. Chem. Ref. Data*, 29, 267, 2000. With permission of the authors.
$a(b)$ means $a \times 10^b$. Note the placement of SF$_2^-$ cross sections.

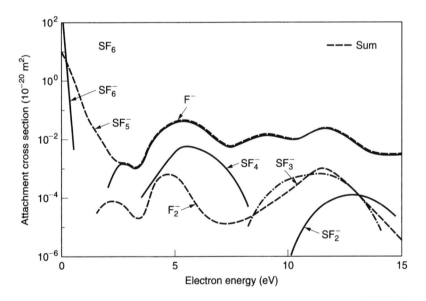

FIGURE 9.25 Direct and dissociative attachment cross sections of SF$_6$ as a function of energy. Figure reproduced from Christophorou, L. G. and J. K. Olthoff, *J. Phys. Chem. Ref. Data*, 29, 267, 2000. With permission of the authors.

Siddagangappa et al.[174,176]; Aschwanden[175]; Siddagangappa et al.[176]; Fréchette[177]; Urquijo-Carmona et al.[178]; Hayashi and Wang[179]; Hasegawa[180]; and Qiu and Xiao.[181] These results are shown in Figure 9.26 and Table 9.17. Selected references in addition to those already cited with reference to η/N are: Itoh et al.[182]; and Xiao et al.[183] These data are shown in Figure 9.27 and Table 9.18.

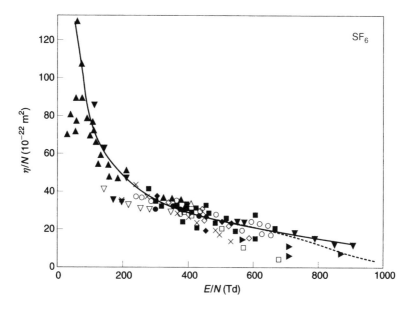

FIGURE 9.26 Density-reduced attachment coefficients as a function of E/N for SF$_6$: (○) Geballe and Harrison[165]; (•) Bhalla and Craggs[166]; (■) Teich and Sangi[169]; (□) Maller and Naidu[170]; (◆) Kline et al.[171]; (◇) Raju and Dincer[172]; (*) Shimozuma et al.[173]; (+) Siddagangappa et al.[174]; (▼) Aschwanden[175]; (▽) Siddagangappa et al.[176]; (×) Frechette[177]; (○) de Urquiho-Carmona[178]; (△) Hayashi[179]; (□) Hasegawa et al.[180]; (□) Qiu and Xiao[181]; (– – –) Christophorou and Olthoff[163] $\alpha/N - (\alpha-\eta)/N$; (—) suggested η/N.[163] Figure reproduced from Christophorou, L. G. and J. K. Olthoff, *J. Phys. Chem. Ref. Data*, 29, 267, 2000. With permission of the authors.

TABLE 9.17
Recommended Values of η/N as a Function of E/N[163]

E/N (Td)	η/N (10^{-22} m^2)	E/N (Td)	η/N (10^{-22} m^2)	E/N (Td)	η/N (10^{-22} m^2)
75	126	250	40.1	600	20.7
100	87	300	34.8	650	18.6
125	68	350	31.6	700	16.4
150	58.0	400	28.3	750	14.3
175	51.7	450	26.3	800	12.2
200	46.8	500	24.5	850	10.2
225	43.3	550	22.6	900	8.06

Analytical expressions for the ionization and attachment coefficients have been obtained by Raju and Hackam[184] by assuming a Maxwellian electron energy distribution on the reasoning that the low energy inelastic scattering cross sections push the distribution to the Maxwellian type. The ionization and attachment coefficient are defined by

$$\frac{\alpha}{N} = \frac{1}{W} \int_{\varepsilon_i}^{\infty} \varepsilon^{1/2} Q_i(\varepsilon) F(\varepsilon) d\varepsilon \tag{9.74}$$

$$\frac{\eta}{N} = \frac{1}{W} \int_{0}^{\infty} \varepsilon^{1/2} Q_a(\varepsilon) F(\varepsilon) d\varepsilon \tag{9.75}$$

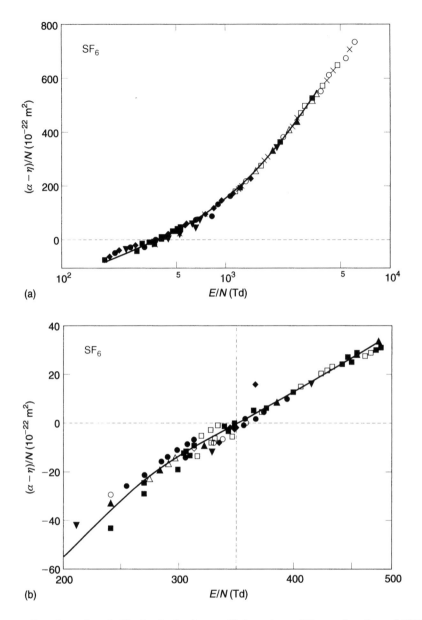

FIGURE 9.27 Density-reduced effective ionization coefficients $(\alpha - \eta/N)$ as a function of E/N for SF_6: (●) Geballe and Harrison[165]; (▲) Bhalla and Craggs[166]; (■) Boyd and Crichton[168]; (♦) Teich and Sangi[169]; (▼) Itoh et al.[182]; (○) Kline et al.[171]; (△) Raju and Dincer[172]; (□) Shimozuma et al.[173]; (◇) Aschwanden[175]; (▽) Fréchette[177]; (+) Hasegawa et al.[180]; (×) Qiu and Xiao[181]; (*) Xiao et al.[183]; (—) recommended by Christophorou and Olthoff.[163] The point of intersection of the curve with the dotted line gives $(E/N)_{lim}$: (b) shows a close-up of this point. Figure reproduced from Christophorou, L. G. and J. K. Olthoff, *J. Phys. Chem. Ref. Data*, 29, 267, 2000. With permission of the authors.

in which Q_i and Q_a are the ionization and attachment cross sections respectively. Note that the lower limit of integration in Equation 9.74 is ε_i, the ionization potential. By expressing Q_a and Q_i as analytical expressions of ε and substituting the Maxwellian equation for $F(\varepsilon)$, the derived coefficients have been shown to give very good agreement with the measured ones. Raju and Hackam[185] have extended these calculations to other industrially important gases (CCl_2F_2, c-C_4F_8, $CBrClF_2$, C_4F_6).

TABLE 9.18

Density-Reduced Effective Ionization Coefficients for SF$_6$ Recommended by Christophorou and Olthoff[163]

E/N (Td)	$(\alpha-\eta)/N$ (10^{-22} m^2)	E/N (Td)	$(\alpha-\eta)/N$ (10^{-22} m^2)	E/N (Td)	$(\alpha-\eta)/N$ (10^{-22} m^2)
200	−55.3	600	63.8	1000	154
250	−32.8	650	75.2	1250	204
300	−16.1	700	87.0	1500	250
350	−2.43	750	98.8	2000	338
400	10.9	800	110	2500	413
450	25.8	850	122	3000	478
500	39.3	900	132	3500	531
550	51.9	950	143	4000	578

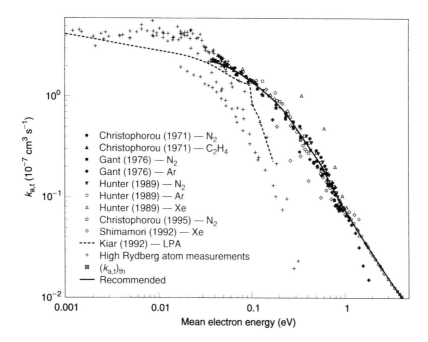

FIGURE 9.28 Total attachment rate constants as a function of mean energy for SF$_6$. Buffer gases for swarm measurements are as shown. LPA means laser photoelectron attachment. (●, ▲) Christophorou et al.[186]; (■, ◆) Gant[187]; (▼, ○, △) Hunter et al.[189]; (◇) Shimamori et al.[191]; (□) Christophorou and Datskos[190]; (− −) Klar et al.[192]; (+) high Rydberg atom measurements; (⊗) $k_{(a,t)th}$; (—) recommended by Christophorou and Olthoff.[163] Figure reproduced from Christophorou, L. G. and J. K. Olthoff, *J. Phys. Chem. Ref. Data*, 29, 267, 2000. With permission of the authors.

The attachment rate constants K_a at room temperature and low electron mean energies are measured by using a buffer gas, usually N$_2$,[186–190] Ar,[187,189] Xe,[189,191] or C$_2$H$_4$.[186] While many of the measurements are carried out using the swarm technique of measuring the attachment coefficient, other techniques employed are laser photoelectron attachment (LPA)[192] and high-Rydberg-atom measurements.[193] The attachment rate constants shown in Figure 9.28 are plotted as a function of mean energy and the differences in K_a in various buffer gases are due to the electron energy distribution, which affects the mean energy.

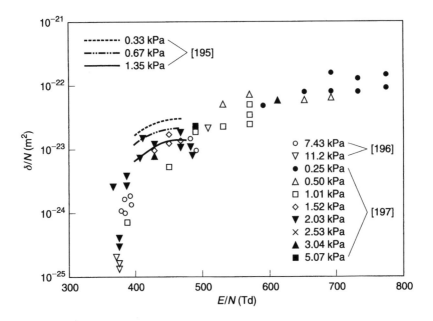

FIGURE 9.29 Detachment coefficients for SF_6 as a function of E/N at the gas pressures shown in the legend. Reference numbers are given in square brackets. Figure reproduced from Christophorou, L. G. and J. K. Olthoff, *J. Phys. Chem. Data*, 29, 267, 2000. With permission of the authors.

Detachment from ions occurs according to[194]

$$SF_6^- + SF_6 \rightarrow SF_6 + SF_6 + e \tag{9.76}$$

$$e + F^- \rightarrow F + 2e \tag{9.77}$$

$$SF_6^{-*} \rightarrow SF_6 + e \tag{9.78}$$

Reactions 9.76 and 9.77 are collisional whereas the last reaction is known as autodetachment. The cross sections for these processes are quite high.[163] The detachment coefficients have been measured by O'Neill and Craggs,[195] Hansen et al.,[196] and Hilmert and Schmidt.[197] The detachment threshold is also quite high, $\sim$360 Td, and the detachment coefficient δ/N is a function of gas number density, indicating three-body processes. Figure 9.29 shows measured detachment coefficients.

9.3.8 SELECTED INDUSTRIAL GASES

A large number of industrial gases belonging to the category of perfluorocarbon compounds (PFCs) have been investigated by Christodoulides et al.[198] and Pai et al.[199] using the swarm technique. The perfluorinated carbons that belong to the saturated group are CF_4, C_2F_6, C_3F_8, and n-C_4F_{10}. Attachment to these molecules has been studied by Davis et al.[200] and Fessenden and Bansal.[201] The attachment rate coefficients (ηW) are measured as a function of mean energy by the technique described in Section 2.5. The comprehensive studies have addressed the following issues: the magnitude of thermal attachment rate, the magnitude and energy dependence of the electron attachment rate and cross section, the position of the negative ion resonance states below $\sim$1.4 eV, the possible reaction mechanisms for formation and dissociation of the parent negative ions, and the effect of the molecular structure on

TABLE 9.19

Formulas and Physical Constants of Compounds Studied by Christodoulides et al.[198]

Perfluorocarbon	Molecular Formula	N_{CF} in N_2 (m^{-3})	N_{CF} in Ar (m^{-3})	B.P. (°C)	F.P. (°C)
Perfluorocyclobutene	c-C$_4$F$_6$	$(1.4–28) \times 10^{18}$	$(1.4–4.7) \times 10^{19}$	0–3	−60
Perfluoro-2-butyne	2-C$_4$F$_6$	$(2–4) \times 10^{18}$	$(2.5–3) \times 10^{18}$	(−24)–(−25)	−117.4
Perfluoro-1,3-butadiene	1,3-C$_4$F$_6$	$(2–3) \times 10^{18}$	$(3–4.5) \times 10^{18}$	6.0	−132
Perfluorocyclobutane	c-C$_4$F$_8$	$(2.5–5.0) \times 10^{18}$	$(2.5–4.0) \times 10^{18}$	(−4)–(6)	−40
Perfluoro-2-butene	2-C$_4$F$_8$	$(2.5–4.0) \times 10^{18}$	$(2.5–5.0) \times 10^{18}$	0.4-3.0	−129

The buffer gases used were nitrogen and argon with number densities in the range from 1.5 to 6×10^{25} m^{-3}. N_{CF} is the partial number density of the compound. B.P. = boiling point; F.P. = freezing point.

TABLE 9.20

Formulas and Physical Constants of Compounds Studied by Pai et al.[199]

Perfluorocarbon	Molecular Formula	N_{CF} in N_2 (m^{-3})	N_{CF} in Ar (m^{-3})	B.P. (°C)	F.P. (°C)
Perfluorocyclopentene	c-C$_5$F$_8$	$(1.5–15.0) \times 10^{18}$	$(2.8–5.5) \times 10^{18}$	25–35	
Perfluorocyclohexene	c-C$_6$F$_{10}$	$(3–24) \times 10^{17}$	30×10^{18}	9.68	
Perfluoro-1,2-dimethylcyclobutane	c-C$_6$F$_{12}$	$(1.3–4.0) \times 10^{18}$	$(1.8–3.2) \times 10^{18}$	54	
Perfluorotoluene	C$_7$F$_8$	$(4.5–15.0) \times 10^{18}$	15×10^{18}	80.5–103	−98
Perfluoro-1,3-dimethylcyclohexane	C$_8$F$_{16}$	$(1.8–3.2) \times 10^{18}$	$(2.4–5.5) \times 10^{18}$	100.2	−55

The buffer gases used were nitrogen and argon with number densities in the range from 1.5 to 6×10^{25} m^{-3}. N_{CF} is the partial number density of the compound. B.P. = boiling point; F.P. = freezing point.

the electron capture processes at low energies. Limitations of space preclude presentation of results and conclusions on all of these essential issues.

Relevant physical constants of compounds studied by these authors are reproduced in Tables 9.19 and 9.20. The tables show the buffer gases used and their pressure, which is relatively high, in the range of 75 to 150 kPa, to keep the mean energy low. The partial pressure of the gas under investigation was kept very low so that the electron energy distribution remained characteristic of the carrier gas. The significant points to note are:

1. The attachment cross section increases with decreasing electron energy for all gases shown in Table 9.19 except for c-C$_4$F$_8$, which reaches a maximum at zero electron energy.
2. Some gases exhibit a well defined second peak at ~0.02 to 0.03 eV. A third peak at ~0.8 eV is also discernible for some gases.
3. The attachment cross sections at electron energies greater than 0.4 eV are much larger than that for SF$_6$.

The attachment rates at thermal energy are several orders of magnitudes smaller for saturated open-chain PFC molecules than for cyclic PFC molecules.

Turning our attention to ionization and attachment coefficients, measurements and analyses have been carried out by the following: Tetrafluoromethane (CF$_4$): Bozin and Goodyear,[202] Lakshminarasimha et al.,[203] Dutton et al.,[204] and Christophorou and Olthoff;[48d] Perfluoroalkene (C$_4$F$_{10}$): Naidu and Prasad.[205]

9.4 CONCLUDING REMARKS

The review presented shows that the task of measuring the ionization and attachment coefficients has been accomplished in common diatomic gases, at least in a broad sense. In view of the technological importance of SF_6, voluminous literature has been published both on fundamental parameters and on discharge phenomena.[48d, 206–209] Considerable efforts are required to achieve the same degree of completeness in triatomic and polyatomic gases, such as SO_2. NO_2, N_2O, as in diatomic molecules. NO and O_3 also stand out for lack of reliable coefficients. Future measurements of ionization coefficients in vapors of carbon compounds will help us to understand better the relationship between the molecular structure and the coefficients.

REFERENCES

1. Hasted, J. B., *Physics of Atomic Collisions*, American Elsevier, New York, 1972, p. 451.
2. Rapp, D. and D. D. Briglia, *J. Chem. Phys.*, 43, 1480, 1965.
3. Grigoriev, I. S. and Z. Meilikhov, *Handbook of Physical Quantities*, CRC Press, Boca Raton, 1999, p. 506.
4. Raju, G. R. Govinda, *Brit. J. Appl. Phys.*, 16, 279, 1965.
5. Gosseries, A., *Physica*, 6, 458, 1939.
6. Daniel, T. N., J. Dutton, and F. M. Harris, *J. Phys. D: Appl. Phys.*, 2, 1559, 1969.
7. Kuffel, E., *Proc. Phys. Soc.* (London), 74, 297, 1959.

AIR

8. Dutton, J., F. M. Harris, and F. Llewellyn-Jones, *Proc. Phys. Soc.* (London), 82, 581, 1963.
9. Prasad, A. N., *Proc. Phys. Soc.* (London), 74, 33, 1959.
10. Dutton, J., F. M. Harris, and F. Llewellyn-Jones, *Proc. Phys. Soc.* (London), 78, 569, 1961.
11. Dutton, J., F. M. Harris, and F. Llewellyn-Jones, *Proc. Phys. Soc.* (London), 82, 581, 1963.
12. Frommhold, L., *Fortschr. Phys.*, 12, 597, 1964.
13. Ryzko, H., *Arkiv Fysik.*, 32, 1, 1966.
14. Moruzzi, J. L. and D. A. Price, *J. Phys. D: Appl. Phys.*, 7, 1434, 1974.
15. Bhiday, M. R., A. S. Paithankar, and B. L. Sharda, *J. Phys. D: Appl. Phys.*, 3, 943, 1970.
16. Liu, J. and G. R. Govinda Raju, *IEEE Trans. Elec. Insul.*, 28, 154, 1993.
17. (a) Rao, C. Raja, and G. R. Govinda Raju, *J. Phys. D: Appl. Phys.*, 4, 494, 1971; (b) Rajapandyan, S. and G. R. Govinda Raju, *J. Phys. D: Appl. Phys.*, 5, 16, 1972.
18. Prasad, A. N. and J. D. Craggs, *Proc. Phys. Soc.* (London), 76, 223, 1960.
19. Kuffel, E., *Proc. Phys. Soc.* (London), 74, 297, 1959.
20. Crompton, R. W., J. A. Rees, and R. L. Jory, *Aust. J. Phys.*, 18, 541, 1965.
21. Parr, J. E. and J. L. Moruzzi, *J. Phys. D: Appl. Phys.*, 5, 514, 1972.
22. Risbud, A. V. and M. S. Naidu, *J. Phys.* (Paris) *Colloq.*, c7, 40, 77, 1979.
23. Brown, S. C., *Basic Data of Plasma Physics*, M.I.T. Press, Cambridge, MA., 1966, p. 251.
24. Mentzoni, M. H., *J. Appl. Phys.*, 41, 1960, 1970.

CO₂ AND CO

25. Moruzzi, J. L. and A. V. Phelps, *J. Chem. Phys.*, 45, 4617, 1966.
26. Alger, S. R. and J. A. Rees, *J. Phys. D: Appl. Phys.*, 9, 2359, 1976.
27. Frommhold, L., *Fortschr. Phys.*, 12, 597, 1964.
28. Rapp, D. and D. D. Briglia, *J. Chem. Phys.*, 43, 1480, 1965.
29. Asundi, R. K., J. D. Craggs, and M. V. Kurepa, *Proc. Phys. Soc.* (London), 82, 967, 1963.
30. Schlumbohm, H., *Z. Angew. Phys.*, 11, 156, 1959.
31. Bhalla, M. S. and J. D. Craggs, *Proc. Phys. Soc.* (London), 76, 369, 1960.

32. Schlumbohm, H., *Z. Phys.*, 184, 492, 1965.
33. Lakshminarasimha, C. S., J. Lucas, and N. Kontoleon, *J. Phys. D: Appl. Phys.*, 7, 2545, 1974.
34. Conti, V. J. and A. W. Williams, *J. Phys. D: Appl. Phys.*, 8, 2198, 1975.
35. Davies, D. K., *J. Appl. Phys.*, 49, 127, 1978. Ionization and attachment coefficients in pure CO_2 and CO_2:N_2:He (1:2:3) mixtures were studied.
36. Hake, R. D. and A. V. Phelps, *Phys. Rev.*, 158, 70, 1967.
37. Sakai, Y., S. Kaneko, H. Tagashira, and S. Sakamoto, *J. Phys. D: Appl. Phys.*, 12, 23, 1979.
38. Bhalla, M. S. and J. D. Craggs, *Proc. Phys. Soc.* (London), 78, 438, 1961.
39. Chantry, P. J., *Phys. Rev.*, 172, 125, 1968.
40. Schulz, G. J., *Phys. Rev.*, 128, 178, 1962.
41. Rapp, D. and D. D. Briglia, *J. Chem. Phys.*, 43, 1480, 1965.
42. Chatterton, P. A. and J. D. Craggs, *Proc. Phys. Soc.*, 85, 355, 1965.
43. Moruzzi, J. L. and A. V. Phelps, *J. Chem. Phys.*, 45, 4617, 1966.
44. (a) Parr, J. E. and J. L. Moruzzi, *Proc. 10th Int. Conf. Phenomena in Ionized Gases*, Oxford, 1971, Donald Parsons, Oxford, p. 8, cited by Davies and Williams[45]; (b) Price, D. A. and J. L. Moruzzi, *J. Phys. D: Appl. Phys.*, 6, L17, 1973.
45. Davies, G. H. L. and A. W. Williams, *J. Phys. D: Appl. Phys.*, 8, 2198, 1975.
46. Saelee, H. T. and J. Lucas, *J. Phys. D: Appl. Phys.*, 10, 343, 1977.
47. Land, J. E., *J. Appl. Phys.*, 49, 5716, 1978.

FREON-12

48. (a) Christophorou, L. G., J. K. Olthoff, and Y. Wang, *J. Phys. Chem. Ref. Data*, 26, 1205, 1997. This reference deals with CCl_2F_2. (b) Christophorou, L. G. and J. K. Olthoff, *J. Phys. Chem. Ref. Data*, 27, 889, 1998. This study refers to C_3F_8. (c) Christophorou, L. G., J. K. Olthoff, and M. V. V. S. Rao, *J. Phys. Chem. Ref. Data*, 25, 1341, 1996. This study refers to CF_4. (d) Christophorou, L. G. and J. K. Olthoff, *Fundamental Electron Interactions with Plasma Processing Gases*, Kluwer Academic/Plenum Publishers, New York, 2004.
49. Kiendler, A., S. Matejcik, J. D. Skalny, A. Stamatovic, and T. D. Märk, *J. Phys. B: At. Mol. Phys.*, B29, 6217, 1996.
50. McCorkle, D. L., A. A. Christodoulides, L. G. Christophorou, and I. Szamrej, *J. Chem. Phys.*, 72, 4049, 1980.
51. Christophorou, L. G., D. L. McCorkle, and D. Pittman, *J. Chem. Phys.*, 60, 1183, 1974.
52. Pejčev, V. M., M. V. Kurepa, and I. M. Čadež, *Chem. Phys. Lett.*, 63, 301, 1979.
53. Hayashi, M., in *Swarm Studies and Inelastic Electron–Molecule Collisions*, ed. L. C. Pitchford, B. V. McKoy, A. Chutjian, and S. Trajmar, Springer, New York, 1987, p. 167.
54. Chutjian, A. and S. H. Alajajian, *J. Phys. B: At. Mol. Phys.*, 20, 839, 1987.
55. Underwood-Lemons, T., T. J. Gergel, and J. H. Moore, *J. Chem. Phys.*, 102, 119, 1995.
56. Harrison, M. A. and R. Geballe, *Phys. Rev.*, 91, 1, 1953.
57. Moruzzi, J. L., *Brit. J. Appl. Phys.*, 14, 938, 1963.
58. Boyd, H. A., G. C. Crichton, and T. Munk Nielsen, *IEEE Conf. Publ.* 70, 426, 1970.
59. Rao, C. Raja and G. R. Govinda Raju, *Int. J. Electron.*, 35, 49, 1973.
60. Maller, V. N. and M. S. Naidu, *Third IEEE Conf. Gas Discharges*, London, 1974, IEEE, Torbridge, 1974, p. 409.
61. Muller, V. N. and M. S. Naidu, *IEEE Trans. Plasma Sci.*, PS-3, 49, 1975.
62. Siddagangappa, M. C., C. S. Lakshminarasimha, and M. S. Naidu, *J. Phys. D: Appl. Phys.*, 16, 763, 1983.
63. Rao, C. Raja and G. R. Govinda Raju, *Int. J. Electron.*, 35, 49, 1973.
64. Frechette, M. F., *J. Appl. Phys.*, 61, 5254, 1987.
65. Sides, G. D., T. E. Tiernan, and R. J. Hanrahan, *J. Chem. Phys.*, 65, 1966, 1976.

HALOGENS

66. Rescigno, T. N. and C. F. Bender, *J. Phys. B: At. Mol. Phys.*, 9, L329, 1976.
67. DeCorpo, J. J., R. P. Steiger, J. J. Franklin, and J. L. Margrave, *J. Chem. Phys.*, 53, 936, 1970.

68. Chantry, P. J., in *Applied Atomic Collision Physics*, ed. H. S. W. Massey, E. W. McDaniel, and B. Bederson, Academic Press, New York, 1982, vol. 3, p. 35.

69. Hayashi, M. and T. Nimura, *J. Appl. Phys.*, 54, 4879, 1983.

70. (a) McCorkle, D. L., L. G. Christophorou, A. A. Christodoulides, and L. Pichiarella, *J. Chem. Phys.*, 85, 1968, 1986. This paper presents electron attachment cross sections in F_2 measured by the swarm unfolding technique. (b) Christophorou, L. G. and J. K. Olthoff, *J. Phys. Chem. Ref. Data*, 28, 131, 1999. This a review of data on Cl_2.

71. Razzak, S. A. A. and C. C. Goodyear, *J. Phys. D: Appl. Phys.* 2, 1577, 1969.

72. Blewett, J. P., *Phys. Rev.*, 49, 900, 1936.

73. Kurepa, M. V., D. S. Babić, and D. S. Belić, *J. Phys. B: At. Mol. Phys.*, 14, 375, 1981.

74. (a) Sides, G. D., T. E. Tiernan, and R. J. Hanrahan, *J. Chem. Phys.*, 65, 1966, 1976. (b) Schneider, B. I. and C. A. Brau, *Appl. Phys. Lett.*, 33, 569, 1978.

75. Bailey, J. E., R. E. B. Makinson, and J. M. Somerville, *Phil. Mag.*, 24, 177, 1937.

76. Frank, H., M. Neiger, and H.-P. Popp, *Z. Naturforschung*, A25, 1553, 1970.

77. Hanstrop, D. and M. Gustaffson, *J. Phys. B: At. Mol. Opt. Phys.*, 25, 1773, 1992.

78. Buchdahl, R., *J. Chem. Phys.*, 9, 146, 1941.

79. Healy, R. H., *Phil. Mag.*, 26, 940, 1938.

80. Biondi, M. A. and R. E. Fox, *Phys. Rev.*, 109, 2012, 1958.

81. Chantry, P. J., *J. Chem. Phys.*, 51, 3369, 1969.

82. Birtwistle, D. T. and A. Modinos, *J. Phys. B: At. Mol. Phys.*, 11, 2949, 1978.

83. Truby, F. K., (a) *Phys. Rev.*, 172, 24, 1968; (b) *Phys. Rev.*, 188, 508, 1969.

84. Shipsey, E. J., *J. Chem. Phys.*, 52, 2274, 1970.

85. Lewis, D. B. and G. A. Woolsey, *J. Phys. D: Appl. Phys.*, 14, 1445, 1981.

NITROGEN COMPOUNDS

Ammonia (NH$_3$)

86. Sharp, T. E. and J. T. Dowell, *J. Chem. Phys.*, 50, 3024, 1969.

87. Rao, M. V. V. S. and S. K. Srivastava, *J. Phys. B: At. Mol. Opt. Phys.*, 25, 2175, 1992.

88. Lindsay, B. G., M. A. Mangan, H. C. Straub, and R. F. Stebbings, *J. Chem. Phys.*, 112, 9404, 2000.

89. Rapp, D. and P. Englander-Golden, *J. Chem. Phys.*, 43, 1464, 1965.

90. Lindsay, B. G., M. A. Mangan, H. C. Straub, and R. F. Stebbings, *J. Chem. Phys.*, 112, 303, 2000.

91. Stricklett, K. L. and P. D. Burrow, *J. Phys. B: At. Mol. Phys.*, 19, 4241, 1986.

92. Tronc, M., R. Azria, and M. B. Afra, *J. Phys. B: At. Mol. Phys.*, 21, 2497, 1988.

93. Bailey, V. A. and W. E. Duncanson, *Phil. Mag.*, 10, 145, 1930.

94. Bradbury, N. E., *J. Chem. Phys.*, 2, 827, 1934.

95. Parr, J. E. and J. L. Moruzzi, *J. Phys. D: Appl. Phys.*, 5, 514, 1972.

96. Risbud, A. V. and M. S. Naidu, *J. Phys.*, (Paris) *Colloq.*, c7, 40, 77, 1979.

97. Pack, J. L., R. E. Voshall, and A. V. Phelps, *Phys. Rev.*, 127, 2084, 1962.

Nitric Oxide (NO)

98. Rapp, D. and D. D. Briglia, *J. Chem. Phys.*, 43, 1480, 1965.

99. Chantry, P. J., *Phys. Rev.*, 172, 125, 1968.

100. Van Brunt, R. J. and L. J. Kieffer, *Phys. Rev. A*, 10, 1633, 1974.

101. Krishnakumar, E. and S. K. Srivastava, (a) *J. Phys. B: At. Mol. Opt. Phys.*, 21, L607, 1988. This paper gives the dissociative attachment cross sections in NO. (b) *Phys. Rev. A*, 41, 2445, 1990. This paper gives the dissociative attachment in N_2O.

102. Bradbury, N. E., *J. Chem. Phys.*, 2, 827, 1934.

103. Bradbury, N. E., *J. Chem. Phys.*, 2, 840, 1934.

104. Parkes, D. A. and T. M. Sugden, *J. Chem. Soc., Faraday Trans. II*, 68, 600, 1972.

105. McFarland, M., D. B. Dunkin, F. C. Fehsenfeld, A. L. Schmeltekopf, and F. E. Ferguson, *J. Chem. Phys.*, 56, 2358, 1972.

Nitrous Oxide (N₂O)

106. Parkes, D. A., *J. Chem. Soc., Faraday Trans. I*, 68, 2103, 1972.
107. Schulz, G. J., *J. Chem. Phys.*, 34, 1590, 1961.
108. Curran, R. K., and R. E. Fox, *J. Chem. Phys.*, 34, 1590, 1961.
109. Bardsley, J. N., *J. Chem. Phys.*, 51, 3384, 1969.
110. Rapp, D. and D. D. Briglia, *J. Chem. Phys.*, 43, 1480, 1965.
111. Moruzzi, J. L. and J. T. Dakin, *J. Chem. Phys.*, 49, 5000, 1968.
112. Herzberg, G., *Molecular Spectra and Molecular Structure, III. Electronic Spectra and Electronic Structure of Polyatomic Molecules*, Van Nostrand, Princeton, NJ, 1967, p. 596.
113. Chantry, P. J., *J. Chem. Phys.*, 43, 3369, 1969.
114. Brüning, F., S. Matejcik, E. Illenberger, Y. Chu, G. Senn, D. Muigg, G. Denifl, and T. D. Märk, *Chem. Phys. Lett.*, 292, 177, 1998.
115. Paulson, J. F., *Adv. Chem. Ser.*, 58, 28, 1966.
116. Bailey, V. A. and J. B. Rudd, *Phil. Mag.*, 14, 1033, 1932.
117. Bradbury, N. E. and H. E. Tatel, *J. Chem. Phys.*, 2, 835, 1934.
118. Phelps, A. V. and R. E. Voshall, *J. Chem. Phys.*, 49, 3246, 1968.
119. Parkes, D. A., *J. Chem. Soc., Faraday Trans. I*, 68, 2103, 1972.
120. Warman, J. M., R. W. Fessenden, and G. Bakale, *J. Chem. Phys.*, 57, 2702, 1972.
121. Dutton, J., F. M. Harris, and D. B. Hughes, *J. Phys. B: At. Mol. Phys.*, 8, 313, 1975.
122. Hayashi, M., M. Ohoka, and A. Niwa, *18th Int. Conf. Phenomena in Ionized Gases*, Swansea, ed. W. T. Williams, pp. 14–15, 1987.
123. Yoshida, K., N. Sasaki, H. Ohuchi, H. Hasegawa, M. Shimozuma, and H. Tagashira, *J. Phys. D: Appl. Phys.*, 32, 862, 1999.

Nitrogen Dioxide (NO₂) and Sulfur Dioxide (SO₂)

124. Zecca, A., J. C. Nogueira, G. P. Karwasz, and R. S. Brusa, *J. Phys. B: At. Mol. Opt. Phys.*, 28, 477, 1995.
125. Curik, R., F. A. Gianturco, R. R. Lucchese, and N. Sanna, *J. Phys. B: At. Mol. Opt. Phys.* 34, 59, 2000.
126. Szmytkowski, Cz., K. Maciąg, and A. M. Krzysztofowicz, *Chem. Phys. Lett.*, 190, 141, 1992.
127. Karwasz, G. P., R. S. Brusa, and A. Zecca, *Riv. Nuovo Cimento*, 24, 1, 2001.
128. Stephan, K., H. Helm, Y. B. Kim, G. Seykora, J. Ramier, M. Grössl, E. Märk, and T. D. Mark, *J. Chem. Phys.*, 73, 303, 1980.
129. (a) Abouaf, R., R. Paineau and F. Fiquet-Fayard, *J. Phys. B: At. Mol. Phys.*, 9, 303, 1976; (b) Rangwala, S. A., E. Krishnakumar, and S. V. K. Kumar, *XXI Int. Conf. Physics of Electronic and Atomic Collisions*, Sendai, 1999, ed. Y. Itikawa et al., Abstract p. 342.
130. Shimamori, H. and H. Hotta, *J. Chem. Phys.*, 85, 887, 1986.
131. Bradbury, N. E. and H. E. Tatel, *J. Chem. Phys.*, 2, 835, 1934.
132. Puckett, L. J., M. D. Kregel, and M. W. Teague, *Phys. Rev. A*, 4, 1659, 1971.
133. Orient, O. J. and S. K. Srivastava, *J. Chem. Phys.*, 80, 140, 1984.
134. Čadež, I. M., V. M. Pejčev, and M. V. Kurepa, *J. Phys. D: Appl. Phys.*, 16, 305, 1983.
135. Krishnakumar, E., S. V. K. Kumar, S. A. Rangwala, and S. K. Mitra, *Phys. Rev. A*, 56, 1945, 1997.
136. Schlumbohm, H., *Z. Phys.* 166, 192, 1962.
137. Bouby, L., F. Fiquet-Fayard, and C. Bodero, *Int. J. Mass Spectr. Ion Phys.*, 7, 415, 1971.
138. Rademacher, J., L. G. Christophorou, and R. P. Blaunstein, *J. Chem. Soc., Faraday Trans. II*, 71, 1212, 1975.
139. Moruzzi, J. L. and V. K. Lakdawala, *J. Phys. (Paris) Colloq.*, c7, 40, 11, 1979.

OXYGEN

140. Branscomb, L. M., in *Atomic and Molecular Processes*, ed. D. R. Bates, Academic Press, New York, 1962, p. 100.

141. (a) Itikawa, Y., A. Ichimura, K. Onda, K. Sakimoto, K. Takayanagi, Y. Hatano, M. Hayashi, H. Nishimura and S. Tsurubuchi, *J. Phys. Chem. Ref. Data*, 18, 23, 1989. (b) Schulz, G. J. *Phys. Rev.*, 128, 178, 1962.
142. Rapp, D. and D. D. Briglia, *J. Chem. Phys.*, 43, 1480, 1965.
143. (a) Chanin, L. M., A. V. Phelps, and M. A. Biondi, *Phys. Rev.*, 128, 219, 1962; (b) Chanin, L.M., A.V. Phelps, and M. A. Biondi, *Phys. Rev. Lett.*, 2, 344, 1959; (c) Chatterton, P. A. and J. D. Craggs, *J. Electron. Control.*, 11, 425, 1971; (d) Rees, J. A., *Aust. J. Phys.*, 18, 41, 1965; (e) Herreng, P., *Cahiers Phys.*, 38, 1, 1952; (f) Kuffel, E., *Proc. Phys. Soc.*, 74, 297, 1959; (g) Taniguchi, T., H. Tagashira, I. Okada, and Y. Sakai, *J. Phys. D: Appl. Phys.*, 11, 2281, 1978.
144. Hunter, S. R., J. G. Carter, and L. G. Christophorou, *J. Appl. Phys.*, 60, 24, 1986.
145. Grünberg, R., *Z. Naturforsch.* 24a, 1039, 1969.
146. Masch, K., *Arch. Electrotech.*, 26, 587, 1932.
147. Bradbury, N. E., *Phys. Rev.*, 44, 883, 1933.
148. Geballe, R. and M. A. Harrison, *Phys. Rev.*, 85, 372, 1952.
149. Schlumbohm, H., *Z. Angew. Phys.*, 11, 156, 1959.
150. Prasad, A. N. and J. D. Craggs, *Proc. Phys. Soc.* (London), 77, 385, 1961.
151. Dutton, J., F. Llewellyn-Jones, and G. B. Morgan, *Nature*, 198, 680, 1963.
152. Freely, J. B. and L. H. Fischer, *Phys. Rev.*, 133, A304, 1964.
153. Frommhold, I., *Fortschr. Physik*, 12, 597, 1964.
154. Naidu, M. S. and A. N. Prasad, *J. Phys. D: Appl. Phys.*, 3, 957, 1970.
155. Price, D. A., J. Lucas, and J. L. Moruzzi, *J. Phys. D: Appl. Phys.*, 6, 1514, 1973.
156. Gurumurthy, G. R. and G. R. Govinda Raju, *IEEE Trans. Plasma Sci.*, 3, 131, 1975.
157. Masek, K., L. Laska, and T. Ruzicka, *J. Phys. D: Appl. Phys.*, 10, L125, 1977.
158. Hake, R. D. and A. V. Phelps, *Phys. Rev.*, 158, 70, 1967.
159. Lucas, J., D. A. Price, and J. L. Moruzzi, *J. Phys. D: Appl. Phys.*, 6, 1503, 1973.
160. Liu, J. and G. R. Govinda Raju, *Can. J. Phys.*, 70, 216, 1992.
161. Liu, J. and G. R. Govinda Raju, *IEEE Trans. Elec. Insul.*, 28, 154, 1993.
162. Liu, J., Ph.D. Thesis, University of Windsor, 1993.

Sulfur Hexafluoride

163. Christophorou, L. G. and J. K. Olthoff, *J. Phys. Chem. Ref. Data*, 29, 267, 2000.
164. Harland, P. W. and J. C. Thynne, *J. Phys. Chem.*, 75, 3517, 1971.
165. Geballe, R. and M. A. Harrison, as reported in *Basic Processes of Gaseous Electronics*, Loeb, L. B., University of California Press, Berkeley, CA, 1965, chapter 5, p. 415. Also see *Phys. Rev.* 91, 1, 1953.
166. Bhalla, M. S. and J. D. Craggs, *Proc. Phys. Soc.* (London), 80, 151, 1962.
167. McAfee, K. B., Jr., and D. Edelson, *Proc. Phys. Soc.* (London), 81, 382, 1963.
168. Boyd, H. A. and G. C. Crichton, *Proc. IEE*, 118, 1872, 1971.
169. Teich, T. H. and R. Sangi, *Proc. 1st Int. Symp. on High Voltage Engineering*, Munich, 1972, ed. F. Heidbromer, vol. 1. p. 391.
170. Maller, V. N. and M. S. Naidu, *Proc. IEE*, 123, 107, 1976.
171. Kline, L. E., D. K. Davies, C. L. Chen, and P. J. Chantry, *J. Appl. Phys.*, 50, 6789, 1979.
172. Raju, G. R. Govinda and M. S. Dincer, *J. Appl. Phys.*, 53, 8562, 1982.
173. Shimozuma, M., H. Itoh, and H. Tagashira, *J. Phys. D: Appl. Phys.*, 15, 2443, 1982.
174. Siddagangappa, M. C., C. S. Lakshminarasimha, and M. S. Naidu, *J. Phys. D: Appl. Phys.*, 15, L83, 1982.
175. Aschwanden, Th., in *Gaseous Dielectrics IV*, ed. L. G. Christophorou and M. O. Pace, Pergamon Press, New York, 1984, p. 24.
176. Siddagangappa, M. C., C. S. Lakshminarasimha, and M. S. Naidu, in *Gaseous Dielectrics IV*, ed. L. G. Christophorou and M. O. Pace, Pergamon Press, New York, 1984, p. 49.
177. Fréchette, M. F., *J. Appl. Phys.*, 59, 3684, 1986.
178. de Urquijo-Carmona, J., I. Alvarez, and C. Cisneros, *J. Phys. D: Appl. Phys.*, 19, L207, 1986.
179. Hayashi, M. and G. Wang, cited by Christophorou and Olthoff[193] as their reference number 182. (Unpublished and private communication, 1999.)

180. Hasegawa, H., A. Taneda, K. Murai, M. Shimozuma, and H. Tagashira, *J. Phys. D: Appl. Phys.*, 21, 1745, 1988.
181. Qiu, Y. and D. M. Xiao, *J. Phys. D: Appl. Phys.*, 27, 2663, 1994.
182. Itoh, H., M. Shimozuma, H. Tagashira, and S. Sakamoto, *J. Phys. D: Appl. Phys.*, 12, 1979, 2167.
183. Xiao, D. M., H. L. Liu, and Y. Z. Chen, *J. Appl. Phys.*, 86, 6611, 1999.
184. Raju, G. R. Govinda and R. Hackam, *J. Appl. Phys.*, 52, 3912, 1981.
185. Raju, G. R. Govinda and R. Hackam, *J. Appl. Phys.*, 53, 5557, 1982.
186. Christophorou, L. G., D. L. McCorkle, and J. G. Carter, *J. Chem. Phys.*, 54, 253, 1971; 57, 2228, 1972.
187. Gant, K. S., Ph.D. Dissertation, University of Tennessee, 1976.
188. Lakdawala, V. K. and J. L. Moruzzi, *J. Phys. D: Appl. Phys.*, 13, 1439, 1980.
189. Hunter, S. R., J. G. Carter, and L. G. Christophorou, *J. Chem. Phys.*, 90, 4879, 1989.
190. Christophorou, L. G. and P. G. Datskos, *Int. J. Mass. Spectr., Ion Process.*, 149/150, 59, 1995.
191. Shimamori, H., Y. Tatsumi, Y. Ogawa, and T. Sunagawa, *J. Chem. Phys.*, 97, 6335, 1992.
192. Klar, D., M.-W. Ruf, and H. Hotop, *Aust. J. Phys.*, 45, 263, 1992.
193. Christophorou, L. G. and J. K. Olthoff, *J. Phys. Chem. Ref. Data*, 29, 267, 2000 (see page 312 for several references to this technique).
194. Picard, A., G. Turban and B. Grolleau, *J. Phys. D: Appl. Phys.*, 19, 991, 1986.
195. O'Neill, B. C. and J. D. Craggs, *J. Phys. B: At. Mol. Phys.*, 6, 2634, 1973.
196. Hansen, D., H. Jungblut, and W. F. Schmidt, *J. Phys. D: Appl. Phys.*, 16, 1623, 1983.
197. Hilmert, H. and W. F. Schmidt, *J. Phys. D: Appl. Phys.*, 24, 915, 1991.

INDUSTRIAL GASES

198. Christodoulides, A. A., L. G. Christophorou, R. Y. Pai, and C. M. Tung, *J. Chem. Phys.*, 70, 1156, 1979.
199. Pai, R. Y., L. G. Christophorou, and A. A. Christodoulides, *J. Chem. Phys.*, 70, 1169, 1979.
200. Davis, F. J., R. N. Crompton, and D. R. Nelson, *J. Chem. Phys.*, 59, 2324, 1973.
201. Fessenden, R. W. and K. W. Bansal, *J. Chem. Phys.*, 53, 3468, 1970.
202. Bozin, S. E. and C. C. Goodyear, Brit. *J. Appl. Phys.*, 1, 327, 1968.
203. Lakshminarasimha, C. S., J. Lucas, and R. A. Snelson, *Proc. Inst. Electr. Engrs.*, 122, 1162, 1975.
204. Dutton, J., A. Goodings, A. K. Lucas, and A. W. Williams, *J. Phys. D: Appl. Phys.*, 20, 1322, 1987.
205. Naidu, M. S. and A. N. Prasad, *J. Phys. D: Appl. Phys.*, 5, 983, 1972.
206. Raju, G. R. Govinda and J. Liu, *IEEE Trans. Dielectr. and Electr. Insul.*, 2, 1004, 1995.
207. Raju, G. R. Govinda and J. Liu, *IEEE Trans. Dielectr. and Electr. Insul.*, 2, 1015, 1995.
208. Christophorou, L. G. and R. J. van Brunt, *IEEE Trans. Dielectr. and Electr. Insul.*, 2, 952, 1995.
209. M. F. Fréchette, D. Roberge, and R. Y. Larocque, *IEEE Trans. Dielectr. and Electr. Insul.*, 2, 925, 1995.

10 High Voltage Phenomena

We change our focus to the discharge phenomena that occur under high voltages in various gaseous media. For the purpose of this chapter the discussion is limited to sparking phenomena. It is well known that, given a uniform electric field, the breakdown voltage of a gap between smooth, clean electrodes spaced one cm apart in nitrogen at atmospheric pressure (101.3 kPa) is approximately 30.4 kV and that the breakdown voltage of a gap in the range from 0.1 to 5 cm shows a linear relationship with the gap length. Any deviation from linearity may be attributed to nonuniformity of the electrodes or to electrodes that are not tolerably clean and lack mechanical polish. Humidity also contributes to an increase in observed breakdown voltage, as discussed in connection with attachment processes in water vapor.

For our present purpose we define high voltage breakdown as that occurring above 100 kV, with no restriction as to the electrode geometry, gas pressure, or gap length. The definition is purely arbitrary, to serve the purpose of limiting the scope of discussion and providing pointers for further study. At atmospheric pressure such high voltage breakdown is of extreme importance in high voltage transmission studies; phase conductor-to-conductor spacings, conductor-to-tower and ground wire clearances are some of the examples. Clearances to be provided for arcing horns protecting insulator strings used on transmission towers at 400 kV and higher and other types of protective gaps are additional examples. In the case of high pressure breakdown we restrict ourselves to sulfur hexafluoride (SF_6), quoting other gases, mainly nitrogen and air, for comparison purposes.

The chapter begins with a brief description of breakdown in alternating voltages of power frequency (50 or 60 Hz), followed by the considerations that apply when dealing with lightning impulse and switching impulse voltages. A brief description of how these voltages are generated in the testing laboratories of industries and universities is provided before we present information on breakdown. One should be cautious not to apply the available experimental data indiscriminately as the sparking voltage is a function of a number of parameters such as the shape of the electrodes, polarity of the applied voltage in the case of direct and impulse voltages, formation of corona, presence of nearby earthed objects, etc.

10.1 TYPES OF VOLTAGE

As stated above, one encounters the following types of voltage in industrial applications:

1. Direct voltages of either polarity.
2. Alternating voltages (50 or 60 Hz).
3. Lightning impulse voltages (standard wave shape: 1.2/50 µs; see Section 10.4) of either polarity. These voltages arise in a power system following due to a direct or indirect lightning stroke. Very fast (~0.1 µs) rising impulse voltages are a special category.

This chapter is dedicated to Mr. Fritz Roy Perry (1902–1976) — always a perfect gentleman in every way.

4. Switching impulses of either polarity (standard wave shape: 200/2000 µs). These voltages arise from switching operations in a power system, such as opening or closing a switch. When a switch opens in a system it is possible for the voltage to attain a value that is theoretically twice the peak value of the system voltage. A discharged capacitor can also attain a similar voltage if energized at the peak of the system voltage cycle. There are other circumstances in which voltages far in excess of these can arise.[1]

5. Slowly varying (~0.1 s) high voltages.

6. Voltages with oscillatory components.

7. Impulse voltages superimposed on the ac voltage. This type of voltage arises due to a transient on an energized phase conductor. The transient voltage wave shape, its polarity, and the angle of superposition on the ac voltage lead to a bewildering array of sparking phenomena which cannot be classified systematically in a brief discussion of the present type.

10.2 HIGH DIRECT VOLTAGE GENERATION

High direct voltages are generated by simple rectification of alternating voltage (ac) to direct voltage (dc). The principle of the single-phase half-wave rectifier is illustrated in Figure 10.1.[2] The high voltage transformer transforms the input voltage to the desired value and the voltage across the reservoir capacitor C builds up to the maximum voltage $+V_{max}$ of the alternating high voltage $V_{alt}(t)$ while the rectifier R is conducting. The leakage reactance of the transformer and the internal impedance of the rectifier are neglected, which will be the case as long as the voltage across the load V is less than $V_{alt}(t)$ for the polarity of R assumed. The voltage at terminal b oscillates between $\pm V_{max}$ and the voltage across the load R_L remains constant at $+V_{max}$ during no-load conditions ($R_L = \infty$) but falls to V_{min} during the

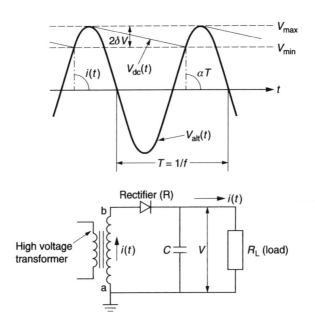

FIGURE 10.1 Single-phase half-wave rectifier with reservoir capacitance C. Top: voltage and current. T is the period and αT is the period of conduction. Bottom: the circuit: ab is the high voltage winding of the step-up transformer, R the rectifier, R_L the load resistance, and V the output direct voltage.

cycle if R_L starts conducting. The rectifier should withstand a peak inverse voltage of $2V_{max}$. If one defines the ripple voltage, in the customary way, as

$$\delta V = 0.5(V_{max} - V_{min}) \tag{10.1}$$

where V_{min} is as shown in Figure 10.1, the value of the direct voltage is the arithmetic mean of V_{max} and V_{min},

$$\bar{V} = \frac{V_{max} + V_{min}}{2} \tag{10.2}$$

and the ripple factor is defined as the ratio $\delta V / \bar{V}$.

Half-wave rectifiers have been built up to the megavolt range[3] by extending the high voltage transformer and using a selenium-type solid-state rectifier. To ensure voltage distribution across the rectifier, parallel capacitors with capacitance lower than C are used in a sectionalized mode. Since the peak inverse voltage is twice the output voltage, the size of the rectifier becomes enormous ($>10\,\text{m}$). The problem of saturation of the high voltage transformer, particularly if the amplitude of the direct current is comparable to the nominal current of the transformer, is a serious disadvantage.

Cascade circuits have the advantage that the transformers, rectifiers, and capacitor units are subject to only a fraction of the total output voltage. Of several standard circuits available,[4] the Cockcroft–Walton[5] circuit has prevailed for its simplicity and ease of operation. The principle of operation of the circuit is presented in Figure 10.2.

The circuit has a total of n stages with two capacitors and a rectifier array for each stage, resulting in a total voltage of $2nV_{max}$ volts. Focusing attention to a segment of the circuit, $0–1'–0'–0$, one recognizes that it is a half-wave rectifier as described previously. Capacitor C_1' charges to $+V_{max}$ if the transformer voltage has reached $-V_{max}$. If C_1 has not been charged the potential difference between $1'$ and 1 will turn the rectifier array R_1 to the conducting

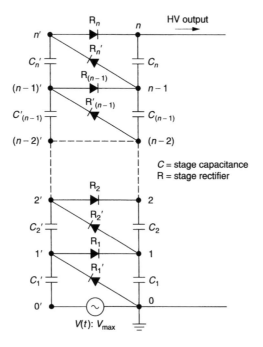

FIGURE 10.2 Cockcroft and Walton circuit for generation of high direct voltages.[2]

mode; the potential of node 1 will reach $+2V_{max}$ if the voltage $V(t)$ attains the maximum value, $+V_{max}$. The segment $1'-1-0$ is therefore a half-wave rectifier in which the voltage across the rectifier array R_1' can be assumed to be the alternating voltage source. The current through R_1 that charges C_1 is provided by $V(t)$ and the capacitor C_1' and not by R'_1.

The potential of node $1'$ oscillates between zero and $+2V_{max}$ and therefore the potential of capacitor C_2' will also rise to $+2V_{max}$. The next oscillation of voltage from $-V_{max}$ to $+V_{max}$ will force the rectifier R_2 to conduct and charge the capacitor C_2 to $+2V_{max}$. Kuffel and Zaengl[2] provide a succinct summary of the operation of all stages of the circuit:

1. The potentials at nodes $1', 2', 3' \ldots n'$ oscillate due to oscillation of $V(t)$.
2. The potentials at nodes $1, 2, 3 \ldots$ remain steady with respect to ground.
3. The voltages across all capacitors are of dc type, the magnitude of which is $2V_{max}$ across each stage except for capacitor C_1' which is stressed with V_{max} only.
4. Each rectifier array $R_1, R_1', R_2, R_2', \ldots R_n, R_n'$ is stressed to a voltage $2V_{max}$ or twice the peak voltage.
5. The output voltage is $2nV_{max}$.

High voltage dc generators operate silently even when charged to the fullest voltage, demanding extremely tight security arrangements for the protection of operating personnel.

A second type of high voltage generator for the generation of high dc voltages, known as the Van de Graaff generator, operates on the principle of generation and deposition of charge carriers. Figure 10.3 demonstrates the principle.

Electrical charges are produced by a set of corona points on to a moving belt and the charges are collected by a collector and deposited on an insulated high voltage terminal. The entire equipment may be placed in an enclosure filled with an insulating gas such as sulfur hexafluoride (SF_6) or Freon-12 (CCl_2F_2). This type of generator suffers from the disadvantage that the current that may be drawn is low, $\sim$500 μA, and is hence mostly employed for particle acceleration rather than sparkover studies. Voltages up to 25 MV have been generated by tandem connection. A variation of the principle is to dispense with

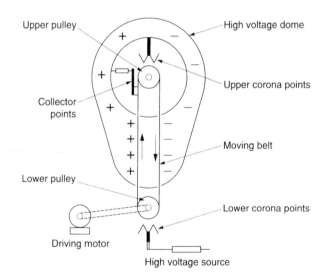

FIGURE 10.3 Principle of the operation of the Van de Graaff generator. Electrical charges are produced by a set of corona points onto a moving belt and the charges are collected by a collector and deposited on an insulated high voltage terminal.

the belt drive, which tends to vibrate at high speeds and is subject to frequent mechanical failures. Featuring an insulated stator separated from an insulated rotor by a small air gap, this arrangement is known as the Felici generator and eliminates the problem of vibration.

10.3 HIGH ALTERNATING VOLTAGE GENERATION

The demand for high alternating voltages comes mainly from the electrical power industry for testing equipment such as transformers, lightning arresters, insulator strings, etc. The single-phase step-up transformer, with the primary energized from an alternating current generator in the range of 2.2 to 11 kV, is the most common type of generator used for this purpose. The primary voltage is fed from an auxiliary motor–generator set designed to remove harmonics and the primary of the transformer is designed for twice the intended voltage to suppress electrical noise due to corona and incipient discharges. One of the problems frequently encountered with cascaded units is the reduction of the output voltage when the current drawn is significant, $\sim$1A in the output circuit. Often, the primary is composed of two identical windings with a facility for connection in either a series or a parallel configuration to improve the regulation.

Several such transformers are connected in cascade to increase the range of output voltage above $\sim$250 kV. Figure 10.4 shows the diagram of connection; the primary of the first transformer is energized by the alternator, as stated above. The primary of the second transformer is energized by an appropriate voltage tap from the secondary winding of the first transformer, or a tertiary winding. The output voltage with respect to ground will now be twice that of a single-phase transformer. It is worthwhile to point out that the voltage transformation of each transformer in the cascaded unit is exactly the same, though the voltage with respect to ground increases with the number of transformers in the cascade.

Since the tank (reference potential) of the second transformer is at the same potential as the output voltage of the first transformer, the former should be insulated from the ground with an insulating support that is mechanically strong enough to withstand the entire weight of the second unit in the cascade. The third unit in the cascade, similarly, should be insulated to a voltage of $2V_{max}$ with respect to the ground, where V_{max} is the secondary voltage of each transformer. One of the requirements of the design is that the secondary voltages are in phase so that they add up linearly to give an output voltage of nV_{max}, where n is the number of cascaded units.

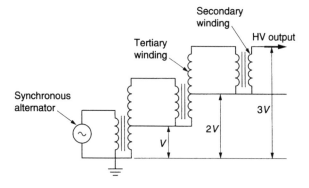

FIGURE 10.4 Schematic diagram of connection of cascaded transformers for generation of high alternating voltages. A tertiary winding on the secondary side of the first transformer provides the primary voltage of the second transformer. For a three-transformer cascaded unit the output voltage is $3V$.

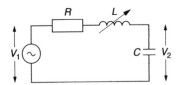

FIGURE 10.5 Equivalent circuit of a series resonant circuit. R and L are the resistance and inductance of the transformer, C the capacitance of the load.

Series resonance circuits may also be employed to generate high alternating voltages, though historically the instantaneous voltage reached high values (20 to 50 times V_1) due to accidental resonance occurring.[2] The beast has been tamed and series resonance circuits have been successfully employed to deliver high current (6 A) at 800 kV.

An equivalent circuit for series resonance is shown in Figure 10.5, where the resistance R and inductance L account for the series resistance and inductance of the primary and secondary windings respectively, and C is the capacitance of the test object. The shunt impedance of the transformer is very high so that it can be neglected for practical considerations. The resonance condition is given by[12a]

$$V_2 = V_1 \frac{1}{1 - \omega^2 LC} \tag{10.3}$$

As the resonance condition is approached ($\omega^2 LC \rightarrow 1$), $V_2 \gg V_1$ and the transformer ratio does not bear any relation to the number of turns on the primary and secondary in the traditional sense.

10.4 HIGH IMPULSE VOLTAGE GENERATION

Impulse voltages are required to test power equipment for its ability to withstand lightning or switching impulses. Although the wave shape of the natural lightning voltage is a statistically varying parameter, international standards specify a standard wave shape of 1.2/50 μs as representative of lightning impulse. The wave shapes standardized by the International Electrotechnical Commission (IEC)[6] are:

1. *Lightning impulse*: 1.2/50 μs, that is the wave has a rise time, to its peak value, of 1.2 μs and decays to 50% of the peak in 50 μs. The tolerances are ±30% for the rise time and ±20% for the tail time. Exact definitions of the front and tail times are somewhat more rigorous.[2]
2. *Switching impulse*: 250/2500 μs, that is the wave has a rise time of 250 μs and decays to 50% of the peak in 2500 μs. The tolerances are ±20% for the rise time and ±60% for the tail time.

Figure 10.6 shows the two wave shapes.

It is appropriate to consider the analytical representation of an impulse wave before describing the actual circuit employed to generate such impulses. One of the most frequently used analytical expressions is known as the double exponential,

$$e = E(\exp[at] - \exp[-bt]) \tag{10.4}$$

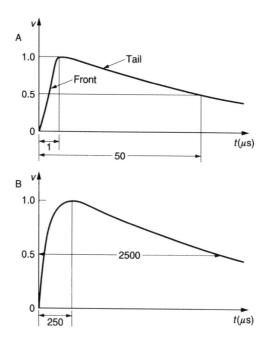

FIGURE 10.6 (A) $1.2/50\,\mu s$ lightning impulse; (B) $250/2500\,\mu s$ switching impulse.

in which E, a, and b are arbitrary constants and e is the impulse voltage or current. The values of the constants for selected wave shapes are given by Bewley[7] and are shown in Figure 10.7.

The generation of high impulse voltages is accomplished by employing the Marx circuit, which operates on the principle of charging a number of capacitors in parallel simultaneously and discharging them, at a given instant, in series. Figure 10.8 shows the essential components of a high voltage impulse generator. These are:

1. *Stage capacitors.* A number of capacitors are connected in parallel, each having the same voltage rating and capacitance. The rated voltage of each capacitor is in the range of 100 to 200 kV and the capacitance is of the order of 0.5 μF. They are stacked in one or more columns to provide mechanical stability and insulation with respect to ground, as the voltage builds up progressively from the first to the last capacitor.

2. *The charging unit.* Charging of the capacitors is accomplished by a transformer–rectifier combination with a high-resistance potential divider connected in parallel for the purpose of measuring and monitoring the charging voltage, V, which is continually variable. The dc output of the transformer–rectifier combination is grounded by a remotely operated switch for safety purposes. This switch is opened to charge the capacitors just before the test begins, after the test area has been cleared of all persons and extraneous objects. The charging of the first stage and subsequent stages is through high resistors (R_1 in Figure 10.8) so as to provide a relatively long charging time, approximately 30 s to 1 minute.

3. *Switching the capacitors in series.* This is accomplished by spark gaps, which are connected between stages. The voltage that appears on the first spark gap is the charging voltage and the voltage on the subsequent gaps is higher, erroneously thought to be $2V$, $3V$, etc. for other gaps. The setting of the second and subsequent gaps need only be slightly larger than that of the first gap for consistent operation.

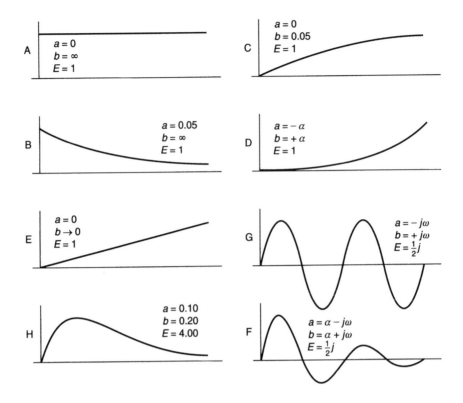

FIGURE 10.7 Wave shapes represented by Equation 10.4. The wave shapes are: (A) infinite rectangular wave; (B) simple exponential wave; (C) convex rising front; (D) concave rising front; (E) linear front wave; (F) damped sinusoidal wave; (G) sinusoidal wave; (H) impulse wave. From Bewley, L. V., *Travelling Waves on Transmission Systems*, John Wiley & Sons, New York, 1951.

Often one inserts a small resistance ($\sim$1 kΩ) R_2 in series with each gap to limit the current and also to serve as a distributed wave-shaping resistance. A distributed resistance improves the efficiency of the impulse generator.

The sphere gaps are illuminated from a source of ultraviolet light so that the statistical time lag is negligible at the instant of firing the generator by closing the series switch. The gaps are mounted on a column that can be rotated by remote control to increase the gap setting as the charging voltage is increased to obtain higher impulse voltages. Alternatively, enclosing the sphere gaps in compressed nitrogen or sulfur hexafluoride (SF_6) and varying the pressure to obtain different sparking voltages have also been tried.

4. *Triggering the generator.* The capacitors are switched into a series combination by triggering the first gap, which in fact is a triggered gap with three electrodes. The trigger electrode is concentric but insulated from the grounded sphere; sparkover occurs between the trigger electrode and the grounded sphere when a trigger voltage of the order of 2 to 10 kV is applied. This spark will initiate a spark across the main sphere gap. which in turn will cause the remaining gaps to sparkover.

5. *Wave-shaping elements.* The front and tail times of the impulse are controlled by the series resistance R_s, C_p, and R_p. When the last gap fires the charge in the capacitors C flows into C_p through the series resistace R_s and the time constant of this circuit determines the wave front duration. Upon reaching the peak voltage, both C_p and the n capacitors in series discharge into the ground through the

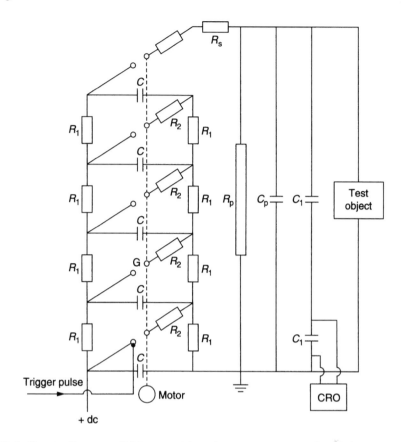

FIGURE 10.8 Circuit diagram of Marx-type impulse generator. Details of the trigger pulse and charging voltage source are not shown. C is the stage capacitance, R_1 the charging resistance, R_2 the gap series resistance, R_s the wave front control resistance, R_p the wave tail control resistance, C_p the load capacitance, C_1 and C_2 the potential divider capacitances, G the sphere gap, CRO the cathode ray oscilloscope. Details of the measuring circuit are not shown.

wave tail control resistance R_p and the time constant of this circuit determines the duration of the wave tail. For long wave fronts R_s will have a relatively larger value and it is often found that by inserting a resistance R_2 in series with each spark gap the wave shape becomes smoother and as already stated, the efficiency of the generator will be higher.

The capacitance of the test object adds to C_p and does not require further discussion except that too large a capacitance of the test object requires that the capacitance on the generator side should be larger for improved efficiency. Capacitances of typical loads are shown in Table 10.1.

6. *Potential divider.* The output high voltage is measured by a potential divider which may be of resistive, capacitive, or voltage grading type. The last-mentioned device consists of a number of capacitors in parallel with a resistor chain which are connected in series to form the high-voltage arm of the divider. The high-voltage arm of a capacitive divider will have low capacitance ($\sim$10 to 100 pF) to provide low loading of the measuring circuit, which is designed to withstand the voltage. The low-voltage arm of the divider is a low resistance (for the resistive divider) or a high capacitance (for the capacitive divider), as the case may be.

TABLE 10.1
Typical Capacitance Values

Test Object	Typical Capacitance
Standard sphere gap	10–20 pF
Post or suspension insulators	~10 pF
Gas-insulated HV power cable	~50 pF/m
Oil-paper-insulated HV power cable	200–250 pF/m
Bushings	100–1000 pF
Potential transformers	200–500 pF
Power transformers (<1 MVA)	~1000 pF
Power transformers (>1 MVA)	~1000–10,000 pF

10.5 IONIZATION IN ALTERNATING FIELDS

It is appropriate to consider at this juncture the ionization phenomena that lead to electrical breakdown of a gaseous medium in power-frequency electric fields in the light of discussion of phenomena in direct fields. Since a separate chapter has been devoted to high-frequency fields (≥ 1 kHz), we consider only the power-frequency fields, 50 or 60 Hz, occasionally 400 Hz. To illustrate the nature of the problem we restrict ourselves to uniform electric fields.

The motion of electrons in the gaseous medium between the electrodes is unaffected by the alternating field because the time required for the electrons to drift from cathode to anode is much smaller than that required for the electric field to change sign. At 60 Hz the latter time is 8.33 ms and an electron having a drift velocity of 10^4 m/s is able to travel 83 m; gap lengths of this magnitude are not encountered in practice except in the case of lightning discharge. However, the ion drift velocity is much smaller and the effect of alternating voltage cannot be ignored if the ion transit time and the half period of the voltage cycle are of the same order of magnitude. To illustrate this aspect, the motion of ions in alternating fields is explained below[8] and illustrated in Figure 10.9.

(a) The voltage is at the maximum at $t = 0$ and the electron avalanche has been built up, depending upon the magnitude of the applied field. As stated above, the time of formation of the avalanche and the electron drift time to the anode are negligibly small and we need to consider the ion drift only. Because of the exponential nature of the electron multiplication we concentrate on the bulk of the ions situated at the head of the avalanche.

(b) Though the applied field is decreasing, the ions move towards the cathode and continue to do so till the field becomes zero at $\omega t = \pi/2$.

(c) The ions have reached their closest position to the cathode. If the cathode is situated at the extreme position the ions are absorbed and the gap is cleared of all charge carriers. The next avalanche starts at the cathode as the field increases in the opposite direction and the effect of the alternating field is not felt.

(d) If the cathode is situated further away than the extreme position of the ionic slab, as shown here, reversal of the electric field reverses the direction of travel of the ions.

(e) The new direction is maintained till $\omega t = 3\pi/2$, at which instant the field is again zero.

(f) The second cycle now begins, repeating the situation at (a).

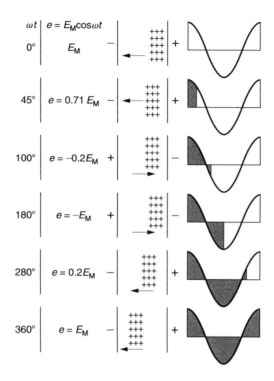

FIGURE 10.9 Ion motion in a gap in alternating electric fields. The instaneous polarity of the electrodes and the lapsed time on the alternating cycle of the electric field are shown. Arrows indicate the direction of motion of the bulk of the ions in the avalanche head. The hatched region shows the time lapsed.

It is clear that the influence of the alternating field is related to the mobility of the ions and the gap length in addition to the magnitude of the electric field applied. A simple derivation establishes the relationship between these parameters.[8] Let the alternating electric field (V/m) be represented by

$$e = E_{max} \cos(\omega t) \tag{10.5}$$

where $\omega = 2\pi f$ is the angular frequency corresponding to the frequency f (Hz). The positive ions have a mobility of μ ($V^{-1} m^2 s^{-1}$) and move a distance of

$$x = W^+ t = \int_0^{T/4} \mu e dt \tag{10.6}$$

where T is the period of one cycle ($1/f$) of the alternating electric field. Substituting Equation 10.5 into 10.6 and integrating, one gets

$$x = \frac{\mu E_{max}}{\omega} = \frac{\mu E_{max}}{2\pi f} \tag{10.7}$$

To appreciate the relationship between the frequency and the gap length we substitute $\mu = 1.83 \times 10^{-4} V^{-1} m^2 s^{-1}$ (N_2^+ ions in N_2) and $E = 3 \times 10^6$ V/m for nitrogen at one atmospheric pressure, to get $x = 1.45$ m. This means that for gap lengths less than ~ 1.5 m the influence of alternating field may be neglected. Similarly, if we substitute $x = 0.1$ m to find the frequency at which such an effect is discernible, this frequency is obtained as ~ 1 kHz or higher. A minor point that should not be overlooked is that the critical frequency is

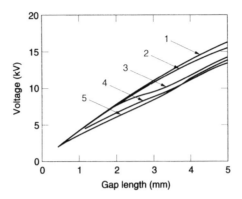

FIGURE 10.10 Breakdown voltage of a sphere gap in atmospheric air as a function of gap length for direct and alternating voltages.[8]

Curve	1	2	3	4	5
Frequency (kHz)	0–10	140	310	880	1450

twice the value given by Equation 10.7, because the time to clear the charge is related to half the cycle. Denoting the critical frequency by f_c, one obtains

$$f_c = \frac{\mu E_{max}}{\pi x} \tag{10.8}$$

The breakdown mechanism will be influenced by the ionic charge if $f > f_c$ and the sparkover voltage will be different from the value with steady-state voltage. The breakdown mechanism is usually referred to as mobility controlled because the secondary effects at the cathode are nonexistent. Another type of breakdown, usually referred to as diffusion controlled, arises when the frequency is so high that the electron is not able to reach the anode prior to voltage reversal. The frequency range for the process is the microwave region of the electromagnetic spectrum. A further increase in frequency leads to optical wavelengths; breakdown by laser irradiation occurs by a different mechanism.

Figure 10.10 shows the influence of frequency on the breakdown voltage of a sphere gap in atmospheric air at low gap lengths. As the gap length decreases, the frequency at which the ac sparking voltage becomes lower than the dc sparking voltage increases, in accordance with Equation 10.8.

The discussion presented assumes that the electric field is uniform or nearly so, and that only positive ions are formed in the inter-electrode region. However, in a nonuniform electric field and in electron-attaching gases, with the formation of negative ions, the breakdown mechanism is complicated. The breakdown voltage in the positive half-cycle and the negative half-cycle can be substantially different, even at lower values of the gap lengths than those given by Equation 10.8. Figure 10.11[9] clearly shows the polarity dependence of the breakdown voltage with alternating voltage. A special point to note is that the breakdown voltage during the positive half cycle is higher than that during the negative half cycle, which is contrary to observations of direct voltage.

10.6 SPARKING VOLTAGES

Several factors influence the sparking voltage of a gap. Primary factors are: (1) type of gas—the major distinction is between electron attaching and nonattaching; (2) pressure of gas; (3) shape of electrodes; (4) gap length; (5) type of voltage; as listed in Section 10.4 and

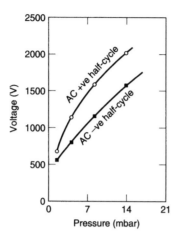

FIGURE 10.11 Breakdown voltage of a wire–cylinder gap as a function of gas pressure in Freon-12 (CCl_2F_2). Electrode radii are 0.063 and 1.25 cm.[9]

(6) polarity of the high voltage electrode. In addition there are secondary factors such as the purity of the gas, smoothness of the electrodes, in high-voltage experiments the clearance to nearby grounded objects, occurrence of corona, etc. A wealth of data exists in the literature and we restrict ourselves to selected data in dry air and sulfur hexafluoride. Exhaustive presentation and detailed discussion of mechanisms will not be attempted.

10.6.1 ATMOSPHERIC AIR

10.6.1.1 Direct Voltages

The sparking voltage of uniform-field gaps in atmospheric air as a function of the product Nd is shown in Figure 10.12. The results are compiled from: Gurumurthy and Raju,[10] Meek and Craggs,[9] and Kuffel and Zaengl.[2] The Paschen law is found to hold true on both sides of the Paschen minimum up to $\sim 3 \times$ atmospheric pressure ($Nd = 10^{26}\,m^{-2}$). The sparking voltage may be analytically represented by the equation.[11]

$$V_s = 1.36 \times 10^{-8}\left[\sqrt{Nd}\right] + 9.96 \times 10^{-20} Nd \tag{10.9}$$

As an example of application of Equation 10.9, the number of molecules at atmospheric pressure ($p = 760$ torr $= 1.01$ bar $= 101.3$ kPa) at 293 K is $N = 2.45 \times 10^{25}\,m^{-3}$, $d = 10^{-2}\,m$, and the sparking voltage is obtained as 31.13 kV. In high voltage applications measurements carried out in the laboratory (pressure $= p$, temperature $= T$) are corrected to standard pressure (1 bar) and temperature (293 K) by applying a correction factor (δ) calculated according to

$$\delta = \frac{p}{760}\frac{293}{273 + T}; \quad p \text{ in torr} \tag{10.10}$$

$$\delta = \frac{p}{101.3 \times 10^3}\frac{293}{273 + T}; \quad p \text{ in Pa} \tag{10.11}$$

$$\delta = \frac{p}{1.01}\frac{293}{273 + T}; \quad p \text{ in bar} \tag{10.12}$$

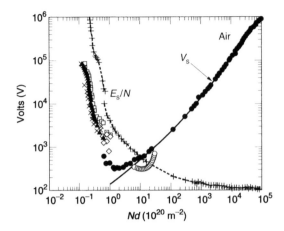

FIGURE 10.12 Sparking voltage in air as a function of Nd in uniform electric fields. (o) Gurumurthy and Raju[10]; ($\triangle$) Meek and Craggs,[9] 2 mm gap length; ($\square$) *Ibid.*, 2.9 mm gap length; ($\diamond$) *Ibid.*, 5 mm gap length; ($\blacktriangle$) *Ibid.*, 8 mm gap length; ($\times$) *Ibid.*, 9.8 mm; ($\bullet$) Kuffel and Zaengl,[2] (———) Equation 10.9. Paschen's law holds good on either side of the Paschen minimum. The arrow indicates the sparking voltage of a 1 cm gap at atmospheric pressure, 31.14 kV. The curve labeled E_s/N shows the reduced electric field in units of Td as a function of Nd.

Lower pressures and higher temperatures than the standard conditions defined above yield lower breakdown voltages. The effect of humidity is to increase the breakdown voltage, particularly at voltages higher than 100 kV. The saturated vapor pressure of water at 293 K is 20 torr and the absolute humidity (h, g/m³) is related to the partial vapor pressure (p_v, torr) according to[2]

$$h = p_v \frac{288.8}{T(\text{K})} \qquad (10.13)$$

The increase of breakdown voltage is between 0.2% and 0.35% per g/m³ of humidity.

For gas pressures greater than ~3 atmospheres the breakdown field becomes successively lower due to field emission and thermionic emission from micro-tips on the surface of the cathode. Field emission is given by the Fowler–Nordheim equation[12]

$$J = \frac{e^3 E^2}{8\pi h \phi} \exp\left\{ -\frac{4}{3} \left(\frac{2m}{\hbar^2} \right)^{\frac{1}{2}} \frac{(\phi - E_F)^{\frac{3}{2}}}{eE} \right\} \qquad (10.14)$$

where J is the current density in A/m², e the electronic charge in coulomb, E the electric field in V/m, h Planck's constant in eV, ϕ the work function in eV, and m the electron rest mass.

The thermionic emission equation, also known as the Richardson–Dushman equation, is given as

$$J = (1 - R)B_0 T^2 \exp\left(-\frac{\phi}{kT} \right); \qquad B_0 = \frac{4\pi emk^2}{h^2} \qquad (10.15)$$

where T is the temperature in K, and R is the reflection coefficient of the electron at the surface. The value of R depends on the surface conditions.

The field-assisted thermionic emission equation, also known as the Schottky equation, is given as

$$J = B_0 T^2 \exp\left[-\frac{(\phi - \beta_s E^{1/2})}{kT}\right]; \quad \beta_s = \left[\frac{e^3}{4\pi\varepsilon_0}\right]^{1/2} \qquad (10.16)$$

where β_s, called the Schottky coefficient, is related to the permittivity of free space, ε_0 (8.854×10^{-12} F/m).

10.6.1.2 Alternating Voltages

Alternating breakdown voltages of large air gaps in nonuniform electric fields are shown in Figures 10.13 and 10.14. Both figures show that there is a critical gap length, for each geometry of electrodes, below which corona cannot be separated from the breakdown voltage. This critical gap length increases with increasing diameter of the sphere, due to the fact that a certain nonuniformity must prevail before corona inception occurs. For the 100 cm diameter sphere (Figure 10.14) corona inception does not occur even at the highest gap length of 3 m. Feser[13] gives the following empirical equations for calculating the 50 Hz breakdown voltage:

$$
\left.
\begin{aligned}
V_s &= 25 + 4.55d &&\text{for rod–plane gaps} \\
V_s &= 10 + 5.25d &&\text{for rod–rod gaps}
\end{aligned}
\right\} \qquad (10.17)
$$

where V_s is in kV and the gap length d in cm. Such geometry is used to produce chopped waves that decay to zero almost instantaneously and are used in impulse testing of

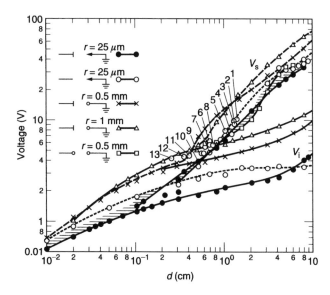

FIGURE 10.13 Corona inception and breakdown voltage (RMS values) at 50 Hz frequency as a function of gap length in a variety of nonuniform field gaps in atmospheric air. Points 1 to 13 refer to gaps that occur in printed circuits. A critical distance is required before the corona inception and breakdown voltages can be separated. From Meek, J. M. and J. D. Craggs, *Electrical Breakdown in Gases*, John Wiley & Sons, New York, 1978.

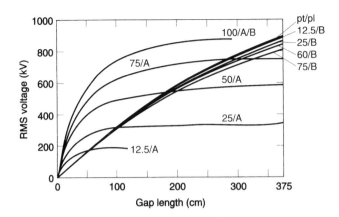

FIGURE 10.14 Corona inception and breakdown voltage (RMS values), at 50 Hz frequency, of sphere–plane gaps in air at atmospheric pressure up to one million volts. Sphere diameter is varied from 12.5 cm to 100 cm. A and B refer to corona inception and breakdown voltages respectively. A critical gap length is required below which the corona inception voltage and the breakdown voltage cannot be separated. The 100 cm diameter sphere–plane gap does not show corona before breakdown even at the highest voltage. The curve marked "pt/pl" refers to a point–plane gap, plotted for comparison. Note that the breakdown voltage of a sphere–plane gap approaches that of the point–plane gap asymptotically.[10]

power transformers. Meek and Craggs[14] give the following equations for much larger gap lengths, $1 \leq d \leq 9$ m:

$$\left. \begin{array}{ll} V_s = 0.0798 + 0.4779d - 0.0334d^2 + 0.0007d^3; & \text{rod–plane gaps} \\ V_s = -0.990 + 0.6794d - 0.405d^2; & \text{rod–rod gaps} \end{array} \right\} \qquad (10.18)$$

where V_s is in MV. Other formulas are available in Meek and Craggs.[9]

10.6.1.3 Lightning Impulse Voltages

Impulse breakdown voltages are measured as a probability. Ten voltage applications are made at the same peak voltage, and 50% breakdown voltage is defined as the voltage that causes breakdown during five out of ten applications. International specifications[15] recommends applying ten impulses at successively higher voltage, each differing from the previous level by 2 to 3%. The breakdown probability is determined at each voltage level and the 50% probability is determined from the curve of voltage peak vs. probability assuming a Gaussian cumulative distribution. Application of an air density factor is recommended for impulse voltages. A test setup for automated measurements of impulse breakdown voltage has been described by Venkatesh and Naidu.[16] The influence of humidity on the breakdown voltage of medium-length rod–plane gaps stressed by positive impulse voltages has been measured by Mikropoulos and Stassinopoulos.[17]

Figure 10.15 shows the positive and negative impulse voltage characteristics of rod–plane gaps as a function of gap distance at various rise times in the range from 2 to 12 μs. For comparison, 60 Hz peak voltage has also been included. The following points are noted with regard to Figure 10.15:

1. With increasing gap length the impulse breakdown voltages increase monotonically for both polarities.

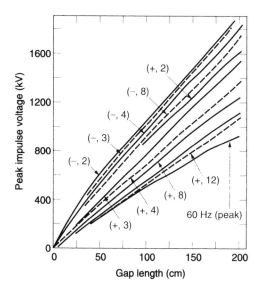

FIGURE 10.15 Positive and negative impulse voltage characteristics of rod–plane gaps. Brackets show polarity and rise time in microseconds. With increasing wave front the 50% breakdown voltage decreases for the same gap length. Further, for the same gap length and same rise time the negative impulse breakdown voltages are higher than the positive impulse voltages. Both the 60 Hz crest voltage for the rod–plane gap and the 50% standard lightning impulse (1.2/50 μs) voltage for 2 m diameter spheres are shown, for comparison.[9]

2. For the same polarity and same gap length the 50% breakdown voltage increases with decreasing wave front of the impulse.
3. For the same gap length and same rise time negative impulse voltages are higher than the positive impulse voltages.
4. The 60 Hz crest voltage for the rod–plane gaps is the lowest and the positive (or negative) impulse breakdown voltage for near-uniform field geometry is the highest in the envelope.

Figure 10.16 shows the influence of gap length, polarity, and electrode geometry on the impulse breakdown voltage of large air gaps in air at atmospheric pressure and at 60% relative humidity (11 g/m³).

The monotonic increase of breakdown voltage with standard impulse is characteristic of gaps in which formation of corona precedes breakdown. However, formation of corona is predicated on the nonuniformity of the electric field, and one can increase or decrease the electric field at the same gap length by using spheres of various diameters. Increasing the sphere diameter reduces or eliminates the nonuniformity whereas decreasing it has the opposite effect. Such measurements have been carried out by the author for sphere–plane gaps using spheres of various diameters in the range from 6.25 to 50 cm, with impulse voltages of both polarities, up to 2.5 million volts (Figure 10.17). The main features of Figure 10.17 are:

1. The sparking voltage for the point–plane gap increases monotonically with gap length for both polarity impulses. The negative polarity impulse breakdown voltage is higher than the positive impulse voltage.
2. For sphere–plane gaps the breakdown voltage characteristic as a function of gap length depends upon the sphere diameter, provided that all other parameters such as the gap length, polarity, etc. are kept the same.

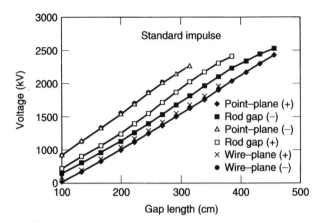

FIGURE 10.16 Standard impulse breakdown voltage of nonuniform field geometry as a function of gap length for both positive and negative polarity. Humidity correction has been applied (relative humidity 60%, absolute humidity 11 g/m^3). Data taken from Meek and Craggs.[9]

3. Corona inception voltage is dependent upon the radius of the high-voltage sphere and the gap length. The ratio of gap length to sphere radius should be sufficiently large ($\sim \geq 4$) for corona inception to occur prior to complete breakdown of the gap. Negative impulse corona inception voltages are lower than positive inception voltages for the same gap length and sphere radius.
4. Negative impulse breakdown voltages are higher than those due to positive impulse provided the gap length and sphere radius are kept the same.
5. For each sphere radius two different critical gap lengths exist. The first critical length is the gap length at which corona inception occurs, as already stated. The second critical gap length is that at which the negative impulse breakdown voltage increases in a step-like manner, increasing linearly with gap length beyond the critical gap length. This increase is seen in Figure 10.17 at 1.25 m gap length for the 50 cm diameter sphere (curve marked 8).

Many of these results could be explained, at least qualitatively, by the nonuniformity of the electric field, the polarity of the high-voltage electrode, and the formation of positive and negative ion space charges and their relative position within the gap and relative distance to the cathode.

10.6.1.4 Switching Impulse Voltages

Switching impulses are generated in the power system by a switching operation such as connecting or disconnecting a transmission line. In the laboratory, they are generated mostly using a Marx-type impulse generator. The specified wave shape for the study of switching impulses is 200/2000 μs. Data on breakdown voltages of sphere–rod and sphere–sphere geometry for small gap lengths (2.5 to 15 cm) for both lightning and switching impulses are given by Gourgoulis et al.[18]

Extensive investigations were carried out by various researchers between 1965 and 1985 to measure and explain the influence of wave front duration on the sparking voltage of large air gaps, particularly with reference to transmission of electrical power at high voltages.[19,20] The switching impulse strength of an air gap for a sparkover distance depends strongly on the geometrical characteristics of the electrodes. Efforts to determine the effects of gap (electric field) characteristics on the switching impulse strength in a general way have

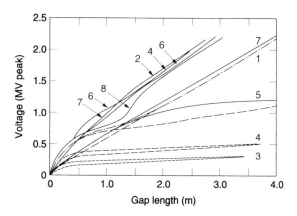

FIGURE 10.17 50% corona inception and breakdown voltage of sphere–plane gaps with standard impulse as a function of gap length for various sphere diameters. The electrode geometry is: 1 and 2, point–plane gap, positive and negative polarity respectively, shown for the purpose of comparison; 3 and 4, 6.25 cm and 12.5 diameter sphere–plane gap; 5, 50 cm diameter sphere–plane gap. Corona inception voltage increases in order of sphere diameter and negative inception voltage is lower than positive inception voltage for the same gap length and sphere diameter. 6, 7, 8, negative impulse breakdown voltage for 6.25, 12.5, and 50 cm diameter sphere–plane gap.

been only partially successful. Paris[21] reported extensive studies on the influence of the switching impulse on the breakdown voltage of a number of electrode geometries, air gap clearances, and insulator strings. The geometries were combinations of electrodes with the following shapes:

1. *High-voltage electrode*: Square rod, 1 cm^2 in cross section, for conductor bundle arranged horizontally.
2. *Earth electrode*: Plane, rod, tower structure, window, conductor.

In addition, characteristics of insulator strings through air gaps either in vertical or V-string formation were studied. Gap lengths in the range from 2 to 6 m, voltages having peak value up to 2400 kV of either polarity, with wave shape 120/4000 μs, were employed under both dry and wet (simulated rain) conditions. Figure 10.18 shows selected data from studies of Paris.[21]

The switching impulse breakdown voltages in Figure 10.18 show that the negative impulse breakdown voltage is greater than the positive impulse breakdown voltage. The height of the rod electrode above the ground plane is found to change the characteristics; in the latter case one observes that the negative polarity breakdown voltage is lower than the positive breakdown voltage, an observation that is contrary to the case in many nonuniform field geometries. Paris[21] observed that the 50% breakdown voltage for the rod–plane gap with positive polarity switching impulse may be expressed by a simple relationship

$$V_{50\%} = 500 \, d^{0.6} \tag{10.19}$$

where V is expressed in kilovolts and d in meters. The 50% breakdown voltage of other geometries may be expressed by using a factor k, the value of which is shown in Table 10.2 Figure 10.19 is provided to show the relevant electrode structures.

To demonstrate the application of Table 10.2, we obtain the 50% dry flashover of a conductor–window arrangement for a 3.5 m gap length as

$$V_{50\%} = 1.20 \times 500 \, (3.5)^{0.6} = 1060 \, \text{kV}$$

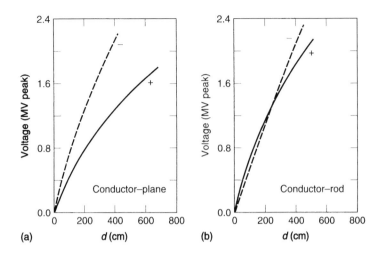

FIGURE 10.18 (a) Switching impulse breakdown voltage of four conductor bundle–plane gaps under dry conditions.[21] (b) Switching impulse breakdown voltage of four conductor bundle–rod gaps under dry conditions.[21] The rod has a vertical height of 3 m.

TABLE 10.2
50% Flashover Strength of Air Gaps According to $V_{50\%} = k \times 500\, d^{0.6}$

Electrodes		Factor k	
Impulsed	**Grounded**	**Without String**	**With String**
Rod	Plane	1.00	1.00
Rod	Structure (under)	1.05	
Conductor	Plane	1.15	
Conductor	Window	1.20	1.15
Conductor	Structure (under)	1.30	
Rod	Rod ($h = 3$ m)	1.30	
Conductor	Structure (over and laterally)	1.35	1.30
Rod	Rod ($h = 6$ m)	1.40	1.30
Conductor	Rope	1.40	
Conductor	Rod ($h = 3$ m, under)	1.65	
Conductor	Crossarm end		1.50
Conductor	Rod ($h = 6$ m, under)	1.90	
Conductor	Rod (over)	1.90	1.75

$h =$ distance from ground plane, $d =$ gap length.

Table 10.3 shows the tabulated values of 50% flashover voltages of rod–plane and rod–rod gaps for switching impulses of both polarities.[20]

Figure 10.20 shows the influence of the wave front on the critical flashover voltage of rod–rod gaps for impulses of both polarities. The influence of polarity is clearly seen. For positive impulse (Figure 10.20a) the sparkover voltage decreases with increasing wave front duration up to ~750 µs. For further increase in the wave front the sparkover voltage reaches a minimum and then begins to rise for wave fronts longer than 1000 µs (not shown). At 60 Hz (17 ms), of course, the flashover voltage (peak value) is higher than that of the standard lightning impulse.

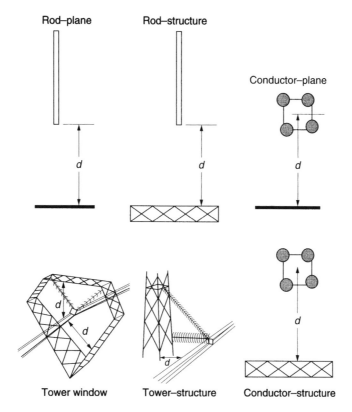

FIGURE 10.19 Gap configurations for Table 10.2.[21]

For the negative polarity impulses (Figure 10.20b) a minimum is not seen for the rod–rod gaps; the flashover voltage increases with increasing wave front duration. It is appropriate to remark here that the influence of longer wave front duration is very marginal, if at all, in the case of uniform field gaps. Sphere gap calibration tables[6] therefore give the same critical flashover voltage for positive lightning and switching impulses up to ~2.5 MV. A second point to note is that, for design purposes, the lowest flashover strength is the deciding value and at very high voltages one measures only the positive polarity flashover voltages.

10.6.2 Sulfur Hexafluoride (SF$_6$)

10.6.2.1 Breakdown Voltage

Voluminous data have been published on the breakdown voltages of this gas, which is extensively used now in enclosed gas-insulated equipment. We restrict ourselves to compressed SF$_6$ and particle-initiated breakdown as they are of practical interest in high voltage equipments. Extensive references have been compiled by Cookson[22] in a review article. In a series of articles Wiegart et al.[23–25] have considered in homogeneous field breakdown in gas-insulated substations. The mechanism of breakdown is pictorially represented as shown in Figure 10.21.

Dakin et al.[26] provided data for the ac uniform electric field breakdown voltage up to 20 MPa mm for the product of gas pressure and gap length (pd). Figure 10.22 shows the applicability of the Paschen law in uniform fields with alternating voltages. Reasonable agreement exists in the literature that this law holds true in the gas at very high pressures provided extreme precautions are taken in respect of electrode preparation, cleanliness, etc.[27]

TABLE 10.3
50% Switching Impulse (T_f/T_t) Flashover of Rod–Plane and Rod–Rod Gaps in Air[20]

<table>
<tr><td colspan="6" align="center">Rod–Plane Gaps</td></tr>
<tr><td colspan="3" align="center">Positive Polarity</td><td colspan="3" align="center">Negative Polarity</td></tr>
<tr><td>d (m)</td><td>$V_{50\%}$ (kV)</td><td>T_f (μs)</td><td>d (m)</td><td>$V_{50\%}$ (kV)</td><td>T_f (μs)</td></tr>
<tr><td>5.0</td><td>1340</td><td>180</td><td>1.5</td><td>1500</td><td>180</td></tr>
<tr><td>7.0</td><td>1650</td><td>180</td><td>2.0</td><td>1800</td><td>180</td></tr>
<tr><td>8.0</td><td>1790</td><td>180</td><td>2.5</td><td>2080</td><td>180</td></tr>
<tr><td>9.0</td><td>1910</td><td>180</td><td>3.0</td><td>2320</td><td>180</td></tr>
<tr><td>10.0</td><td>2050</td><td>180</td><td>3.5</td><td>2480</td><td>180</td></tr>
<tr><td>11.0</td><td>2100</td><td>180</td><td>4.0</td><td>2620</td><td>180</td></tr>
<tr><td>12.0</td><td>2180</td><td>180</td><td></td><td></td><td></td></tr>
<tr><td>13.0</td><td>2300</td><td>180</td><td></td><td></td><td></td></tr>
<tr><td>9.0</td><td>2200</td><td>80</td><td>3.0</td><td>2040</td><td>80</td></tr>
<tr><td>9.0</td><td>2070</td><td>110</td><td>3.0</td><td>2060</td><td>110</td></tr>
<tr><td>9.0</td><td>1950</td><td>180</td><td>3.0</td><td>2300</td><td>180</td></tr>
<tr><td>9.0</td><td>1930</td><td>250</td><td>3.0</td><td>2340</td><td>250</td></tr>
<tr><td>9.0</td><td>1900</td><td>550</td><td>3.0</td><td>2340</td><td>550</td></tr>
<tr><td colspan="6" align="center">Rod–Rod Gaps</td></tr>
<tr><td>4.0</td><td>1690</td><td>180</td><td>1.5</td><td>1110</td><td>180</td></tr>
<tr><td>5.0</td><td>1940</td><td>180</td><td>2.0</td><td>1450</td><td>180</td></tr>
<tr><td>6.0</td><td>2180</td><td>180</td><td>2.5</td><td>1790</td><td>180</td></tr>
<tr><td>7.0</td><td>2490</td><td>180</td><td>3.0</td><td>2080</td><td>180</td></tr>
<tr><td>7.5</td><td>2490</td><td>180</td><td>3.5</td><td>2330</td><td>180</td></tr>
<tr><td></td><td></td><td></td><td>4.0</td><td>2620</td><td>180</td></tr>
<tr><td></td><td></td><td></td><td>4.5</td><td>2780</td><td>180</td></tr>
<tr><td></td><td></td><td></td><td>5.0</td><td>2970</td><td>180</td></tr>
<tr><td>5.5</td><td>2150</td><td>80</td><td>3.0</td><td>1940</td><td>80</td></tr>
<tr><td>5.5</td><td>2100</td><td>110</td><td>3.0</td><td>1960</td><td>110</td></tr>
<tr><td>5.5</td><td>2080</td><td>180</td><td>3.0</td><td>2020</td><td>180</td></tr>
<tr><td>5.5</td><td>2050</td><td>250</td><td>3.0</td><td>2030</td><td>250</td></tr>
<tr><td>5.5</td><td>2050</td><td>550</td><td>3.0</td><td>2040</td><td>550</td></tr>
</table>

From Watanabe, Y., *IEEE Trans. Power Power Apparat. Syst.*, PAS-86, 948, 1967. With permission © 1967, IEEE.

At high gas pressures and high voltages the electric field intensification at the tips of microprojections assumes an increasingly important role in initiating breakdown; in this respect there is some similarity with breakdown in high vacuum.[28,29] Applying a dielectric coating to one of the electrodes, particularly the high-voltage electrode in a nonuniform field, increases the breakdown voltage above the Paschen law value.[30,31]

The corona inception and breakdown voltages in the gas depend, of course, on a number of factors like the electrode geometry, voltage polarity, and the ratio of gap length to the dimension of the high-voltage electrode. Figure 10.23 shows the representative results of corona inception and breakdown voltage for a rod–plane gap for both polarities of the direct voltage. Negative corona inception voltage is lower than that for positive polarity; this behavior is common to all electron-attaching gases. A peak in the positive voltage breakdown occurs at approximately 1.5 bar gas pressure due to a phenomenon usually referred to as "corona stabilization" in the literature.

Figure 10.24 shows the effects of corona stabilization in rod–plane geometry with alternating voltages.[32] The impulse breakdown voltage of the 1 mm diameter rod–plane gap

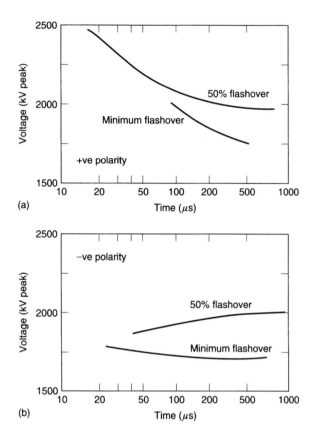

FIGURE 10.20 Minimum and 50% flashover strength of rod–rod gap as a function of time to crest or flashover: (a) positive polarity; (b) negative polarity.[20]

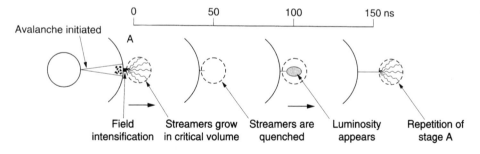

FIGURE 10.21 Pictorial development of discharge in nonuniform fields with the time scale shown, according to Wiegart et al.[23] With permission © 1988, IEEE.

(curve 5) has been plotted for comparison. A point to note is that the breakdown voltage with ac (curve 1) is higher than that with impulse voltage over the entire pressure range.

During the formation of corona prior to breakdown, electrons and positive ions are generated in the high field region close to the rod electrode. The electrons are quickly (~0.1 μs) absorbed by the positive electrode whereas the positive ions have to drift through a larger distance to the cathode. Since the mobility of positive ions is 100 to 1000 times lower than that of the electrons, the ions are still present in the gap, effectively reducing the electric field at the anode. Subsequent avalanches have to grow in a reduced electric field, requiring a higher voltage to cause breakdown. It was thought that corona

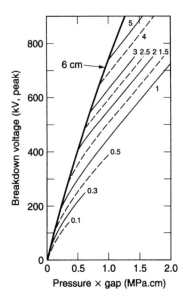

FIGURE 10.22 Ac breakdown voltage of SF_6 as a function of pd.[26]

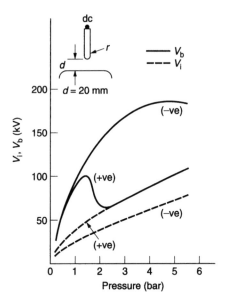

FIGURE 10.23 Corona inception and breakdown voltage in SF_6 as a function of gas pressure. The radius of the rod is 1 mm and gap length 2 cm. V_b = breakdown voltage, V_i = corona inception voltage. Note the pronounced corona stabilization effect.[2]

stabilization occurred only in SF_6. Due to corona stabilization, the corona disappears, leading to breakdown at higher pressures. Lakshminarasimha and Raju[33] have, however, shown that the phenomenon occurs in compressed air as well.

10.6.2.2 Particle-Initiated Breakdown

Particle-initiated breakdown in enclosures that use SF_6 for insulation is a concern because the full insulating capability of the gas is not realized.[34,35] A loose particle

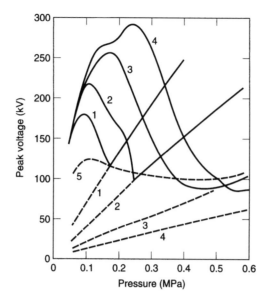

FIGURE 10.24 Corona stabilization in rod–plane gap with alternating voltages.[32] Curves 1 to 4, varying rod diameter from 1 to 10 mm, gap length 60 mm. (– – –) Ac corona onset; (———) ac breakdown. Curve 5, 1 mm diameter rod, 60 mm gap length impulse breakdown (1.2/2200 µs).

acquires electrical charge and, in the high electric field that exists within the enclosure, the electrostatic force is large enough to counteract the gravitational force. The bouncing particle crosses the inter-electrode gap over a period of several cycles.[36–38] As the particle approaches the other electrode it discharges just before impact, initiating the breakdown process. The mechanism of breakdown is different for direct voltage as there is no bouncing and initiation of discharge probably occurs by ionization at the particle tip.

Theoretical aspects of the motion of a particle in an electric field and its dynamic behavior under the action of various forces aiding and inhibiting the motion of the particle have been considered by Felici,[39] Asano et al.,[40] Indira and Ramu,[41] and, more recently, again by Indira and Ramu.[42] The surface charge acquired by a prolate ellipsoidal particle in a nearly homogeneous 3D field with possible axial symmetry has been calculated, making certain assumptions. The potential at any point between the two electrodes is a function of coordinate axis and the dimension of the ellipsoid. The induced charge on the surface of the conducting particle can be obtained as

$$Q = -8\pi\varepsilon_0 A \tag{10.20}$$

where ε_0 = permittivity of free space (8.854×10^{-12} F/m) and A is the elliptic integral in three dimensions, taken over zero and infinity on the variable θ,

$$A = -\varphi_0 \int_0^\infty \frac{d\theta}{\sqrt{(a^2 + \theta) + (b^2 + \theta) + (c^2 + \theta)}} \tag{10.21}$$

where a, b, c are the semi-axes of the ellipsoidal particle. Assuming that the potential of the particle is the same as that of the point in its absence, the surface charge density (ρ) is given by

$$\rho = \frac{Q}{4\pi abc} \frac{1}{\sqrt{\dfrac{X^2}{a^4} + \dfrac{Y^2}{b^4} + \dfrac{Z^2}{c^4}}} \tag{10.22}$$

The induced surface charge on the semi-ellipsoidal particle lying in quadrature to the electric field direction is given by

$$Q = 2\pi\varepsilon_0 rlE \tag{10.23}$$

where l is the length of the particle, r the radius of the particle, and E the electric field seen by the particle.

The motion of the particle under an alternating electric field is, of course, more complex than under a constant field, since the particle interacts with a changing field. In a coaxial cylindrical geometry the particle executes a series of bounces, as stated previously. The maximum charge acquired by the particle is determined by the applied voltage and the loss of charge due to corona or pulseless discharge.

The lift-off is the electric field required to set the particle in motion towards the opposite electrode,

$$E_L = \sqrt{\frac{\rho g r}{1.43\varepsilon_0}} \tag{10.24}$$

Equation 10.24, due to Indira and Ramu,[42] should be compared with the earlier derivation of Asano et al.,[40]

$$E_L = 808\sqrt{r\rho} \quad \text{for half cylinder} \tag{10.25}$$

$$E_L = 880\sqrt{r\rho} \quad \text{for full cylinder} \tag{10.26}$$

The force on the particle is given by

$$F = EQ \tag{10.27}$$

The equation of motion for a particle of mass m is written as

$$m\frac{d^2x}{dt^2} = F - mg \tag{10.28}$$

where g is the acceleration due to gravity. Taking into account the viscous drag (proportional to the square of the velocity) in the course of motion toward the region of maximum field, Equation 10.28 is modified to

$$m\frac{d^2x}{dt^2} = F - mg - C\left(\frac{dx}{dt}\right)^2 \tag{10.29}$$

Substituting Equation 10.27 into 10.29, Indira and Ramu[41] derive

$$m\frac{d^2x}{dt^2} = 2\pi rl\varepsilon_0 E^2 - mg - k\left(\frac{dx}{dt}\right)^2 \tag{10.30}$$

Figure 10.25 shows the calculated lift-off field for two different materials of the particle, with comparisons with the earlier derivations.

The deleterious effects of free particles may be mitigated to certain extent by adopting dielectric coating of the enclosure. Felici[30] has derived the expression for the

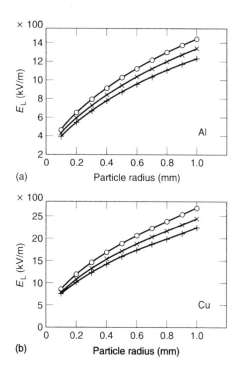

FIGURE 10.25 Calculated lift-off field as a function of particle size: (a) aluminum; (b) copper. (o) Felici's model[39]; (+) Indira and Ramu[42]; (×) Asano et al.[40]; full cylinder model.

lifting field for a wire particle as

$$E_{\mathrm{L}} = \left[\ln\frac{2l}{r_{\mathrm{w}}} - 1\right]\left[\frac{r_{\mathrm{w}}^2 \rho_{\mathrm{w}} g}{\varepsilon_0 l\left(\ln\left[\frac{l}{r_{\mathrm{w}}}\right] - 0.5\right)}\right]^{1/2} \tag{10.31}$$

where l is the length of the wire, r_{w} the radius, and ρ_{w} the particle material density. This equation has been put to the test theoretically by Prakash et al.[43] with a dielectric-coated enclosure; the reduction in the lifting field realized is shown in Figure 10.26, where the maximum height reached by the particle is reduced by a factor of five.

10.6.3 VOLT–TIME CHARACTERISTICS

The time to breakdown (t_{b}) of a gas with applied impulse voltage consists of three components: t_0, the time to reach the peak value or the critical sparkover voltage, as the case may be; t_s, the statistical time lag, defined as the time required for an initiating electron to appear at a suitable location in the gap; and t_{f}, the formative time, defined as the time required to complete the breakdown process.

$$t_{\mathrm{b}} = t_0 + t_s + t_{\mathrm{f}} \tag{10.32}$$

The component t_0 is measured by firing the first stage of the impulse generator by a trigatron or three-electrode gap built into the impulse generator. High pressure gas and

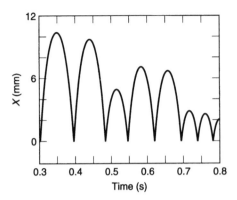

FIGURE 10.26 Free particle motion in Gas Insulated Substation (GIS). Anodized enclosure: movement of an aluminum wire particle (0.45 mm dia/6.4 mm long) with time for an applied voltage of 0.9 per unit (pu) in a 70/190 mm GIS/Gas Insulated Transmission Line (GITL) system.[43]

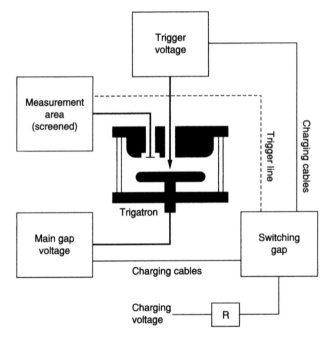

FIGURE 10.27 Trigatron gap and associated circuit for time lag measurements.[48] With permission © 1997, IEEE.

vacuum trigatrons[44–47] may also be employed for the triggering purpose, as shown in Figure 10.27.[48] The sweep of the oscilloscope is initiated by the trigger pulse and the wave shape is recorded by employing potential dividers specially designed for high frequency response. Fast-rise transients are generated by special techniques.[49–51] A simple method of obtaining fast rising transients with the Marx type of impulse generator is to use a series gap at the output end of the impulse generator, as demonstrated by Qiu et al.[52] The initial electrons may be generated from the processes of: (1) field emission from protrusion tips on the surface of the cathode; (2) electron detachment from the negative ions in the gas; or (3) emission from the cathode surface due to ultraviolet light or other forms of radiation.

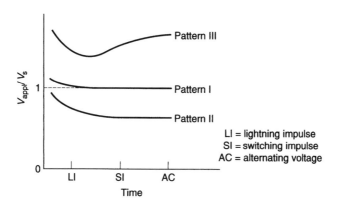

FIGURE 10.28 Classification of V–t characteristics in SF_6 gas. V_{appl} = applied voltage, V_s = threshold breakdown voltage. From Izeki, N., *Electra*, July 1985, p. 41.

The statistical time lag and the formative time lag are separated from the measured time lag by the application of the Laue equation[12b]

$$n = n_0 \exp\left(-\frac{t}{t_s}\right) \tag{10.33}$$

where t_s is the mean statistical time lag, n_0 the number of breakdowns out of n that have a time lag longer than t. The time corresponding to n/n_0 gives the formative time lag t_f. A plot of voltage as a function of time lag is known as the volt–time characteristic and is used by high-voltage engineers to design power systems. Equipment using self-restoring insulation (gas) should spark over at an earlier time than that employing nonself-restoring insulation. The experimental technique involves measuring the time lag for a large number of impulse applications (50 to 100) at the same voltage to determine t_f according to Equation 10.33.

The volt–time characteristics may generally be divided into three patterns,[53] as shown in Figure 10.28. The breakdown voltage is given by the streamer theory as

$$V_s = 89 \, pdu\left(1 + \frac{0.175}{\sqrt{pr}}\right) \text{ kV} \tag{10.34}$$

where p is the gas pressure (bar), d the gap length (cm), u the field utilization factor, defined as the ratio of the average to the maximum field, and r the radius of curvature of the electrode tip (cm). The ordinate in the figure is the ratio of the applied voltage to the threshold voltage V_s. Table 10.4 summarizes the patterns observed, with relevant details such as the electrode radius, gas pressure, and so on.

Theoretical formulation of volt–time characteristics in uniform and slightly nonuniform electric fields is based upon an idea generally known as the equal area criterion. Darveniza and Vlastos[54] define a disruptive effect (DE) of a discharge as

$$DE = \int_{t_0}^{t} [V(t) - V_0]^k dt \tag{10.35}$$

where $V(t)$ is the applied voltage, V_0 the onset voltage, t_0 the time on the wave front when $V(t)$ exceeds V_0, and k an empirical voltage-dependent constant. If $k = 1$, then

TABLE 10.4
Selected Classification of V–t Characteristics[53]

Configuration	Electrode Material	Irradiation	Radius r (mm)	Gap Length d (mm)	Factor u	Surface Area s (cm²)	Positive Pressure				Negative (bar)			
							1	2	4	6	1	2	4	6
Sphere–sphere	Stainless steel	No	75	10	0.956	1.5	I	II	II	II	I	III	II	II
			75	20	0.915		I	II	II	II	I	III	II	II
			75	40	0.820	23	I				I			
			75	80	0.621		I				II			
Hemisphere–hemisphere	Stainless steel	No	15	10	0.787	0.2	I	I	II	II	I	III	II	II
			15	20	0.606		I	I	II		I	I	II	
			15	40			I						III	
			15	160			I						III	
	Stainless steel	No	5	10	0.481	0.2	I	I	II		I	I	II	II
			5	40	0.481		III	I	I	I	III	III	III	II
			5	160	0.045		III	I	I	I	III	III	III	III

Configuration	Electrode Material	Irradiation	Radius r (mm)	Gap Length d (mm)	Factor u	Surface Area s (cm²)	Pressure (bar)			Pressure (bar)		
							1	3.5	11	1	3.5	11
Coaxial cylinders	Aluminum	No	16	32	0.549	322	II	II	II	II	II	II
			25	23	0.709	785	II	II	II	II	II	II

Configuration	Electrode Material	Irradiation	Radius r (mm)	Gap Length d (mm)	Factor u	Surface Area s (cm²)	Pressure (bar)			Pressure (bar)		
							1	3	5	1	3	5
Sphere–plane	Stainless steel	Yes	75	10	0.917	5.7	I	I	I	I	I	I
			75	30	0.780	18.8	I	I		I	I	II
			75	50	0.675	34.2	I			I		
	Stainless steel	Yes	10	10	0.569	1.1	I	I	I	I	I	I
			10	30	0.293	3.4	I	I	I	I	I	II
			10	50	0.200	4.2	I	I	I	I	I	II
	Stainless steel	No	25	25	0.667		I	II		I	II	
			25	50	0.455		I	II		I	II	

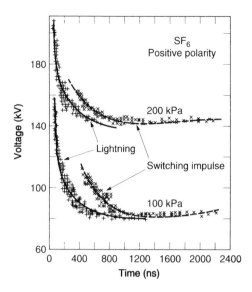

FIGURE 10.29 Voltage–time characteristics of coaxial cylinder geometry with positive polarity impulse. Steep-fronted impulse, 50 ns.[53]

Equation 10.35 is known as the equal area criterion. Data provided by Pfeiffer[55] and Ishikawa et al.[56] appear to conform to this criterion. However, there is some doubt as to precise measurement of time lags with steep-fronted impulses,[57] particularly since the curve rises steeply as the time becomes shorter than 200 ns (see Figure 10.29 and Qiu et. al.[52]). Thalji and Nelson[58] observe a transition in the volt–time characteristics of divergent field gaps with a 0.0075/80 µs pulse, and suggest that the transition may be due to corona stabilization.

Volt–time characteristics show the following properties:

1. With increasing pressure there is a tendency to change from pattern I to II, as Table 10.4 shows.
2. With increasing surface area of the cathode the critical sparkover voltage decreases since a larger number of electrons is released into the gap and the $V–t$ curve tends to shift from pattern I to II.
3. The formative time lag becomes shorter for larger over-voltage, smaller gap length, and positive polarity.
4. The time to breakdown varies from 10^{-4} to 10^4 hours, depending upon the electric field strength and gas pressure.[59]

REFERENCES

1. Greenwood, A., *Electrical Transients in Power Systems*, John Wiley and Sons, New York, 1991.
2. Kuffel, E. and W. S. Zaengl, *High Voltage Engineering*, 2nd ed., Pergamon Press, Oxford, 1984.
3. Prinz, H., Feuer, Blitz and Funke, F. Bruckman-Verlag, Munich, 1965.
4. Craggs, J. D. and J. M. Meek, *High Voltage Laboratory Techniques*, Butterworth, London, 1954.
5. Cockcroft, J. D. and E. T. S. Walton, *Proc. Roy. Soc. Lond. A*, 136, 619, 1932.
6. International Electrotechnical Commission (IEC), *High Voltage Test Techniques, Part 2: Test Procedures*, Publication 60–2, 1973.
7. Bewley, L. V., *Traveling Waves on Transmission Systems*, John Wiley and Sons, New York, 1951.

8. Nasser, E., *Fundamentals of Gaseous Ionization and Plasma Electronics*, Wiley-Interscience, New York, 1971. Also see (b) Cobine, J. D., *Gaseous Conductors*, Dover Publications, New York, 1958.

9. Meek, J. M. and J. D. Craggs, *Electrical Breakdown in Gases*, Clarendon Press, Oxford, 1953. For Figure 10.11 see page 104. For Figure 10.15 see page 324.

10. Gurumurthy, G. R. and G. R. Govinda Raju, *IEEE Trans. Electr. Insul.*, EI-12, 3325, 1977.

11. Dakin, T. W., *Electra*, 32, 61, 1974.

12. Raju, G., (a) *Fundamentals of Electric Circuits*, IBH Prakashana, Bangalore, India, 1980; (b) *Dielectrics in Electric Fields*, Marcel Dekker, New York, 2003.

13. Feser, K., Energ. Technik, 22, 319, 1970.

14. Meek, J. M. and J. D. Craggs, *Electrical Breakdown in Gases*, John Wiley and Sons, New York, 1978.

15. International Electrotechnical Commission (IEC), *High Voltage Test Techniques. Part I: General Definitions and Requirements*, Publication 60–1, 1989.

16. Venkatesh, S. K. and M. S. Naidu, *Eleventh International Symposium on High Voltage Engineering*, Conf. Pub. No. 467, Vol. 1, 1999, p. 23.

17. Mikropoulos, P. N. and C. A. Stassinopoulos, *IEE Proc. Sci. Meas. Technol.*, 141, 407, 1994.

18. Gourgoulis, D. E., P. N. Mikropoulos, and C. A. Stassinopoulos, *IEE Proc. Sci. Meas. Technol.*, 144, 11, 1997.

19. Phillips, T. A., L. M. Robertson, A. F. Rohlfs, and R. L. Thompson, *IEEE Trans. Power Apparat. Syst.*, PAS-86 933, 1967.

20. Watanabe, Y., *IEEE Trans. Power Apparat. Syst.*, PAS-86, 948, 1967.

21. Paris, L., *IEEE Trans. Power Apparat. Syst.*, PAS-86, 936, 1967.

22. Cookson, A., *Proc. 3rd Int. Conf. Properties and Applications of Dielectric Materials*, Tokyo, July 8–12, 1991, p. 369.

23. Wiegart, N., L. Niemeyer, F. Pinnekamp, J. Kindersberger, R. Morrow, W. Zaengel, M. Zwicky, I. Gallimberti, and S. A. Boggs, *IEEE Trans. Power Deliv.*, PWRD-3, 923, 1988.

24. Wiegart, N., L. Niemeyer, F. Pinnekamp, W. Boeck, J. Kindersberger, R. Morrow, W. Zaengl, M. Zwicky, I. Gallimberti, and S. A. Boggs, *IEEE Trans. Power Deliv.*, PWRD-3, 931, 1988.

25. Wiegart, N., L. Niemeyer, F. Pinnekamp, W. Boeck, J. Kindersberger, R. Morrow, W. Zaengl, M. Zwicky, I. Gallimberti, and S. A. Boggs, *IEEE Trans. Power Deliv.*, PWRD-3, 939, 1988.

26. Dakin, T. W. et al., *Electra*, 32, 64, 1974.

27. Baumgartner, E., *ETZ-A*, 98, 369, 1977.

28. Pedersen, A., *IEEE Trans.*, PAS-94, 721, 1989.

29. McAllister, I. W. and G. C. Crichton, *IEEE Trans. Electr. Insul.*, EI-24, 325, 1989.

30. Hing, D. Chee and K. D. Srivastava, *IEEE Trans. Electr. Insul.*, EI-10, 119, 1975.

31. Morgan, J. D. and M. Abdellah, *IEEE Trans. Electr. Insul.*, EI-23, 467, 1988.

32. Zwicky, M., *IEEE Trans. Electr. Insul.*, EI-22, 317, 1987.

33. LakshmiNarasimha, C. S. and G. R. Govinda Raju, *Elect. Lett.*, 3, 460, 1967.

34. Cookson, A. H., *Proc. IEE*, 128, 303, 1981.

35. Takuma, T., *IEEE Trans. Electr. Insul.*, EI-21, 855, 1986.

36. Anis, H. and K. D. Srivastava, *IEEE Trans. Electr. Insul.*, EI-16, 327, 1981.

37. Morcos, M. M., H. Anis, and K. D. Srivastava, *IEEE Trans. Electr. Insul.*, EI-24, 825, 1979.

38. Rizk, F. A. M., C. Masetti, and R. P. Comsa, *IEEE Trans.*, PAS-98, 825, 1979.

39. Felici, N. J., *Rev. Cen. Electr.*, 75, 1145, 1966.

40. Asano, K., K. Anno, and Y. Higashiyang, *IEEE Trans. Ind. and Appl.*, 33, May–June, 1997.

41. Indira, M. S. and T. S. Ramu, *Conference Record of the 1998 IEEE International Symposium on Electrical Insulation*, Arlington, VA, June 7–10, 1998.

42. Indira, M. S. and T. S. Ramu, *IEEE Trans. on Dielectr. Electr. Insul.*, 7, 247, 2000.

43. Prakash, K. S., K. D. Srivastava, and M. M. Morcos, *IEEE Trans. Dielectr. Electr. Insul.*, 4, 344, 1997.

44. Raju, G. R., Govinda, R. Hackam, and F. A. Benson, *J. Appl. Phys.* (USA), 47, 1310, 1976.

45. Raju, G. R., Govinda, R. Hackam, and F. A. Benson, *J. Appl. Phys.* (USA), 48, 1101, 1977.

46. Raju, G. R., Govinda, R. Hackam, and F. A. Benson, *Int. J. Electron.*, 42, 185, 1977.

47. Raju, G. R., Govinda, R. Hackam, and F.A. Benson, *Proc. IEE*, 124, 828, 1977.

48. MacGregor, S. J., F. A. Tuema, S. M. Turnbull, and O. Farish, *IEEE Trans. Plasma Sci.*, 25, 118, 1997.
49. Mankowski, J., J. Dickens, and M. Kristiansen, *IEEE Trans. Plasma Sci.*, 26, 874, 1998.
50. Yashima, M., H. Goshima, H. Fujinami, E. Oshita, T. Kuwahara, K. Tanaka, Y. Kawakita, Y. Miyai, and M. Hakoda, *High Voltage Engineering Symposium*, 22–27 Aug. 1999, Conf. Publ. No. 467, p. 394.
51. Baum, C. E. and J. M. Lehr, *IEEE Trans. Plasma Sci.*, 30, 1712, 2002.
52. Qiu, Y., C. Y. Lu, and M. Zhang, *Conference Record of the IEEE International Symposium on Electrical Insulation*, Baltimore, MD, USA, June 7–10, 1992, p. 310.
53. Izeki, N., *Electra*, July, 41, 1985.
54. Darveniza, M. and A. E. Vlastos, *IEEE Trans. Electr. Insul.*, 23, 373, 1988.
55. Pfeiffer, W., in *Gaseous Dielectrics IV*, ed. L. G. Christophorou and M. O. Pace, Pergamon Press, New York, 1984, pp. 323–334.
56. Ishikawa, M., T. Hattori, M. Honda, H. Aoyagi, and K. Terasaka, in *Gaseous Dielectrics III*, ed. L. G. Christophorou, Pergamon Press, New York, 1982, pp. 285–392.
57. Tekietsadik, K. and J. C. Campbell, IEE Proc. Sci. Meas. Technol., 143, 270, 1996.
58. Thalji, J. Y. and J. K. Nelson, *Conference Record of the IEEE International Symposium on Electrical Insulation*, Baltimore, MD, USA, June 7–10, 1992, p. 314.
59. Nitta, T. and H. Kuwahara, *IEEE PES 1980 Summer Meeting*, Paper no. 80SM703-9, 1980.

11 Ionization in $E \times B$ Fields

Ionization and breakdown in gases in crossed electric and magnetic fields is still only a moderately explored area of research, relative to the volume of literature on other areas of gaseous electronics. The potential industrial uses of an electrical discharge where the magnetic field is perpendicular to the electric field are at least as great as those of other manifestations of discharge phenomena. The author is aware of just a single volume published on this topic, as far back as 1949.[1] This has been followed by an excellent review, again the only one known to the author, by Heylen,[2] covering the work done till 1980. Inevitably, we refer very frequently to this review in presenting this chapter.

The chapter begins with the motion of an electron in a crossed magnetic field (often abbreviated as XF) and then considers the collision processes with neutrals in the presence of crossed fields. Ionization coefficients, breakdown, energy distribution, and time lags are dealt with. Brief comments on corona inception and sparking in nonuniform electric fields with crossed magnetic fields are included, and the chapter concludes with comments on research still to be accomplished.

11.1 LIST OF SYMBOLS

Table 11.1 lists the symbols used in this chapter.

TABLE 11.1
List of Chapter Symbols

A	Regeneration factor	
A_T	Gas constant (p in torr)	$\text{cm}^{-1}\,\text{torr}^{-1}$
B	Magnetic flux density	tesla
B_T	Gas constant (p in torr)	$\text{V}\,\text{cm}^{-1}\,\text{torr}^{-1}$
B/N	Reduced magnetic field	$\text{T}\,\text{m}^3$
C_T	Gas constant	m^2
D	Diffusion coefficient	$\text{m}^2\,\text{s}^{-1}$
D_T	Gas constant	$(\text{Td})^{1/2}$
d	Gap length	m
d_1	Diameter of inner electrode	m
d_2	Diameter of outer electrode	m
E	Electric field	$\text{V}\,\text{m}^{-1}$
E_{eff}	Effective field intensity	$\text{V}\,\text{m}^{-1}$
$E_{\text{av,eff}}$	Averaged effective field intensity in $E \times B$ fields	$\text{V}\,\text{m}^{-1}$
EREF	Equivalent reduced electric field	$\text{V}\,\text{m}^2$, Td
EIDC	Equivalent increased density concept	m^{-3}
E/N	Reduced electric field	$\text{V}\,\text{m}^2$, Td
e	Electrical charge of electron	C

(Continued)

TABLE 11.1
Continued

F	Force	N
F_T	Gas constant (N in m^{-3})	m^2
$F(\varepsilon)$	Energy distribution function	
G_T	Gas constant	Td
H	Magnetic field strength	$A\,m^{-1}$
I	Current	A
I_B, I_{XF}	Current in B and $E \times B$ fields	A
I_0	Initial photoelectric current	A
L	Mean free path at 1 torr	m
l, l'	Electron free path	m
M_F	Multiplication factor	
m	Mass of electron	kg
N	Number of neutrals per unit volume	m^{-3}
N_e	Equivalent number density	m^{-3}
P_c	Probability of collision	m^{-1}
p	Gas pressure	torr, Pa
p_e	Equivalent gas pressure	torr, Pa
$Q_i(\varepsilon)$	Ionization cross section	m^2
$Q_M(\varepsilon)$	Momentum transfer cross section	m^2
$Q_T(\varepsilon)$	Total scattering cross section	m^2
q	Charge	C
R_1, r_1, R_2, r_2	Radii of electrodes	m
t_f	Formative time lag	s
u	Mean velocity of agitation of electrons	$m\,s^{-1}$
v	Velocity	$m\,s^{-1}$
V	Voltage	V
V_c	Corona inception voltage	V
$V_{s,\,min}$	Minimum sparking voltage	V
$V_{c,\,XF}$	Corona inception in crossed fields	V
V_s	Sparking voltage	V
$V_{s\,min,\,XF}$	Minimum sparking voltage in crossed fields	V
V_{XF}	Sparking voltage in crossed fields	V
W	Drift velocity	$m\,s^{-1}$
W_e	Drift velocity in equivalent electric field	$m\,s^{-1}$
W_p	Perpendicular drift velocity	$m\,s^{-1}$
W_T	Transverse (along E) drift velocity	$m\,s^{-1}$
W_{XF}	Net drift velocity in crossed fields	$m\,s^{-1}$
α	Townsend's first ionization coefficient	m^{-1}
α_{XF}	First ionization coefficient in $E \times B$ crossed fields	m^{-1}
γ	Townsend's second ionization coefficient	
γ_{XF}	γ in crossed fields	
ε	Electron energy	eV
ε_i	Ionization potential	eV
ε_e	Threshold for electronic excitation	eV
ε_0	Permittivity of free space	$F\,m^{-1}$
η	Ionization efficiency $= \alpha/E$	V^{-1}
η_{XF}	Ionization efficiency in $E \times B$ crossed fields	V^{-1}
μ	Mobility	$V^{-1}\,m^2\,s^{-1}$
$\nu = 1/\tau$	Collision frequency	s^{-1}
$\nu_0 = \nu/N$	Collision frequency at unit density	$m^3\,s^{-1}$
τ	Collision time	s
ω	Cyclotron frequency ($= eB/m$)	$rad\,s^{-1}$

11.2 BRIEF HISTORICAL NOTE

In the presence of crossed fields, electrons describe cycloidal paths and experience a larger number of collisions than they do in electric field only. Historically, Townsend[3a] noted the deflection of an electron avalanche by a magnetic field and obtained the transverse drift velocity. Wehrli[3b] observed the influence of a magnetic field on Townsend's ionization coefficient α. On the assumption of inelastic collisions between electrons and gas molecules, and with constant free path for electrons of all energies, Wehrli[3b] derived an expression for the distance traveled in crossed fields by the electron in the electric field direction as

$$l' = l\left(1 - \frac{ceH^2l}{8 \times 10^8 Em}\right) \tag{11.1}$$

where c is a constant for converting the older units to SI units ($\sim 10^3$). When $H = 0$, of course, there is no change in the free path and $l = l'$. The component of the free path in the direction of E only causes a change in the energy gained. Wehrli[3b] stated that the effect of a crossed magnetic field is equivalent to an increase of gas pressure from p to p_e according to

$$p_e = \frac{p}{\left(1 - \dfrac{ceH^2l}{8 \times 10^8 Em}\right)} \tag{11.2}$$

Valle[4] introduced Equation 11.2 into the expression for α/N,

$$\frac{\alpha}{N} = F_T \exp\left[\frac{-G_T N}{E}\right] \tag{11.3}$$

and obtained a relation between α_{XF} and H. Both Wehrli[3b] and Valle[4] ignored the influence of the magnetic field on γ. The agreement between the measured sparking voltage in crossed fields and that calculated using this theory was poor. Somerville,[5] followed independently by Haefer,[6] reasoned that the neglect of the distribution of free paths about the mean was the cause of the observed discrepancy. Somerville[5] derived an expression for α_{XF} assuming that the ionizing probability was unity for all electrons having energy greater than the ionization potential,

$$\frac{\alpha}{N} = \frac{F_T \sinh(a_1/2l)}{\phi(1/a_1)\sinh(a_1/2l)}\left[1 - \frac{4G_T l}{Ea_1}\right] \tag{11.4}$$

where a_1 is the complete cycloidal path described by an electron starting from rest,

$$a_1 = \frac{8 \times 10^8 Em}{eH^2} \tag{11.5}$$

and

$$\phi(x) = \coth\left(\frac{1}{2x}\right) - 2x \tag{11.6}$$

Somerville realized that, under certain conditions, the maximum energy gained by an electron over a cycloidal path was less than that required for ionization. He further suggested that in this situation the theory should take into account the possibility of an electron gaining sufficient energy over several free paths to cause ionization.

The various theories summarized above approached the problem by considering the effect of the magnetic field on the motion of individual electrons. Blevin and Haydon[7] considered the effect of a magnetic field on the bulk properties of an electron avalanche such as the energy distribution, mean energy, and drift velocity. The equivalent pressure concept, derived by them, is given by

$$p_e = p\left[1 + c\left(\frac{B}{p}\right)^2\right]^{1/2} \tag{11.7}$$

where

$$c = \left(\frac{e}{m}\frac{L}{u}\right)^2 \tag{11.8}$$

Since the gas number density has replaced pressure in gas discharge literature we rename, without claiming originality, the equivalent pressure concept the equivalent increased density concept (EIDC). According to EIDC, then,

$$N_e = N\left[1 + c\left(\frac{B}{N}\right)^2\right]^{1/2} \tag{11.9}$$

An alternative expression for N_e is

$$N_e = N\left[1 + \frac{(w/N)^2}{(v/N)^2}\right]^{1/2} \tag{11.10}$$

The assumptions made by Blevin and Haydon[7] in their derivation are:

1. The energy distribution of electrons is Maxwellian.
2. The electron molecule collision frequency is constant and independent of E/N and B/N.
3. The magnetic field does not alter the form of energy distribution.

According to EIDC, the magnetic field effectively increases the gas number density from N to N_e; the reduced ionization coefficient is expressed as

$$\left(\frac{\alpha}{N}\right)_{XF} = \frac{N_e}{N}\left(\frac{\alpha}{N}\right)_{0, E/N_e} \tag{11.11}$$

As is shown later, Heylen[2] has laid a solid foundation by developing the effective reduced electric field concept (EREF). Theoretical aspects of energy distribution in magnetic fields are discussed by Allis[8] in a classical paper.

11.3 ELECTRON MOTION IN VACUUM IN $E \times B$ FIELDS

The motion of a charge carrier situated in crossed fields[9] is obtained through the Lorentz equation for force,

$$\mathbf{F} = e[\mathbf{E} + (\mathbf{v} \times \mathbf{B})] \tag{11.12}$$

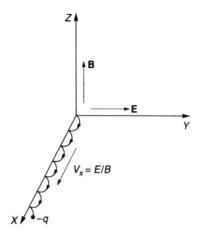

FIGURE 11.1 Motion of a charge in $E \times B$ fields. The crossed field drift velocity is E/B.

Choosing the coordinate axes as shown in Figure 11.1, the components of Equation 11.12 may be written as

$$m\frac{dv_x}{dt} = q(v \times B)_x = qv_y B \tag{11.13}$$

$$m\frac{dv_y}{dt} = q\left[E + (v \times B)_y\right] = qE - qv_x B \tag{11.14}$$

$$m\frac{dv_z}{dt} = 0 \tag{11.15}$$

According to Equation 11.15 the particle does not experience any force along the Z-axis and the motion is confined to the XY-plane. The angular velocity or the gyrofrequency of the circular motion is given by

$$\omega = \frac{qB}{m} \tag{11.16}$$

Substituting Equation 11.16 into 11.13, one gets

$$\frac{dv_x}{dt} = \omega v_y = \omega\frac{dy}{dt} \tag{11.17}$$

Integrating Equation 11.17, one obtains

$$v_x = \omega y + C_1 \tag{11.18}$$

where C_1 is the constant of integration. If one assumes a positive charge initially at rest at the origin of the coordinate axes, then $C_1 = 0$ and Equation 11.18 becomes

$$v_x = \omega y \tag{11.19}$$

Substituting this value of the velocity component in the X-direction into Equation 11.14, one gets, noting that $v_y = dy/dt$,

$$\frac{d^2y}{dt^2} + \omega^2 y = \frac{Eq}{m} \tag{11.20}$$

This is the equation for a simple harmonic oscillator subject to a constant force. The solution is

$$y = \frac{Eq}{m\omega^2} + C_2 \cos \omega t + C_3 \sin \omega t \tag{11.21}$$

where C_2 and C_3 are arbitrary constants of integration.

The Y-component of the velocity is given by

$$v_y = \frac{dy}{dt} = -\omega C_2 \sin \omega t + \omega C_3 \cos \omega t \tag{11.22}$$

The constants of integration are determined by the boundary conditions. As already stated, for a positive charge initially at rest at the origin we have $t = 0$, $v_x = 0$, and $v_y = 0$, resulting in $C_3 = 0$. Equation 11.19 then yields

$$C_2 = -\frac{Eq}{m\omega^2} \tag{11.23}$$

Equation 11.21 now simplifies to

$$y = \frac{Eq}{m\omega^2}(1 - \cos \omega t) \tag{11.24}$$

Inserting this into Equation 11.19 and integrating, the displacement in the X-direction is given by

$$x = \frac{Eq}{m\omega^2}(\omega t - \sin \omega t) \tag{11.25}$$

Equations 11.25 and 11.24 are the parametric equations of a cycloid [$x = (at - a \sin t)$, $y = a - a \cos t$], as shown in Figure 11.1. By differentiating these equations one obtains the velocities as

$$v_x = \frac{E}{B}(1 - \cos \omega t) \tag{11.26}$$

$$v_y = \frac{E}{B} \sin \omega t \tag{11.27}$$

The average velocity in the Y-direction is zero, as shown in Figure 11.1. The average velocity in the X-direction is $W_{XF} = E/B \, (m/s)$. Figure 11.2 shows the motion of a positive and a negative charge in $E \times B$ fields.

11.4 EFFECTIVE REDUCED ELECTRIC FIELD (EREF)

The equivalent pressure concept, because of its limitations, has been replaced by the effective reduced electric field (EREF) as first proved by Heylen[10] [1965] and independently by Gurumurthy and Raju.[11] The EREF is defined as that electric field in a zero magnetic field that will keep the mean electron energy the same as that in the $E \times B$ fields under

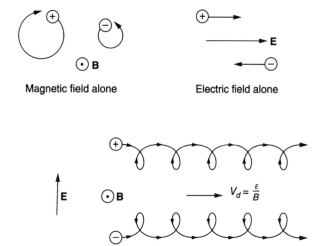

Magnetic field alone   Electric field alone

Crossed electric and magnetic fields

FIGURE 11.2 Motion of an electrical charge in $E \times B$ fields. The direction of the magnetic field is shown by the circle and dot within, out of the plane of the page. Figure reproduced from Roth, J. R., *Industrial Plasma Engineering*, Institute of Physics (IOP), 1995. With permission of IOP, U.K.

consideration. The influence of a magnetic field perpendicular to the electric field may be visualized as a reduction in the electric field according to

$$\left(\frac{E}{N}\right)_e = \frac{E}{N}\cos\theta \tag{11.28}$$

where the quantity θ is related to the magnetic flux density according to

$$\theta = \cos^{-1}\left[\left\{1 + \left(\frac{e}{m}\frac{1}{v_0}\frac{B}{N}\right)^2\right\}^{1/2}\right] \tag{11.29}$$

In terms of the electron neutral collision frequency, Equation 11.28 is expressed as[2]

$$\left(\frac{E}{N}\right)_e = \frac{E}{N}\frac{(v/N)}{\left[(v/N)^2 + (\omega/N)^2\right]^{1/2}} \tag{11.30}$$

Comparing Equation 11.30 with 11.10, one observes the similarity, but the EREF concept is based on a more solid theoretical basis[8] and there is no need to make too restrictive assumptions, except that the effective electron–molecule collision frequency should remain constant. In fact there is evidence that in SF_6 even this assumption is unnecessary because the EREF concept holds true even though the collision frequency is a function of $(E/N)_e$. An engineering approach to the influence of the crossed magnetic field is to visualize that the avalanche, due to incessant collisions of electrons with neutrals, is tilted at an angle θ to the direction of the electric field (Figure 11.3). The angle of the tilt is given by

$$\tan\theta = \frac{W_p}{W_T} \tag{11.31}$$

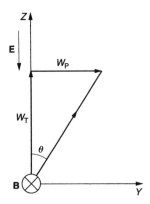

FIGURE 11.3 The electron avalanche is tilted with respect to the direction of the electric field. W_T and W_P are transverse and perpendicular drift velocities, respectively.

In terms of the magnetic field strength the angle of deflection is alternatively expressed as[2]

$$\tan \theta = \frac{e}{m} \frac{B}{N} \left(\frac{N}{v}\right) \tag{11.32}$$

The net drift velocity[52] is given by

$$W_{\mathrm{XF}} = \sqrt{W_T^2 + W_p^2} \tag{11.33}$$

To show the magnitude of the angle of tilt we substitute typical values in Equation 11.32: $B = 0.04$ T, $N = 3.22 \times 10^{23}\,\mathrm{m}^{-3}$, $v = 10^{10}\,\mathrm{s}^{-1}$, to get $\theta = 35°$. Substantial deflection occurs even at moderate magnetic field strengths as long as the number density is not very high (less than $\sim 10^{25}\,\mathrm{m}^{-3}$).

Assuming a Maxwellian distribution and that the collision frequency is independent of electron energy, Allis[8] and Heylen[12] have derived that

$$W_T = \frac{e}{m} \frac{v}{v^2 + \omega^2} E \tag{11.34}$$

and

$$W_p = \frac{e}{m} \frac{\omega}{v^2 + \omega^2} \tag{11.35}$$

In dealing with discharges in $E \times B$ fields one encounters two different frequencies, the cyclotron frequency ω and the elastic scattering frequency v. The ratio of these two frequencies forms an important parameter. It is customary to define the magnetic field strength according to

$$\left.\begin{aligned}
\frac{\omega}{v} &= \frac{e}{m} \frac{B}{v} \ll 1 \quad \text{weak magnetic field} \\[2ex]
\frac{\omega}{v} &= \frac{e}{m} \frac{B}{v} \approx 1 \quad \text{moderate magnetic field} \\[2ex]
\frac{\omega}{v} &= \frac{e}{m} \frac{B}{v} \gg 1 \quad \text{strong magnetic field}
\end{aligned}\right\} \tag{11.36}$$

To appreciate the magnitude of the quantities involved in this Equation one uses Equation 1.100 and momentum transfer data in H_2, given in Table 4.8, and calculates the

TABLE 11.2
Moderate Magnetic Flux Density (Condition 2 of Equation 11.36) at $N = 3.22 \times 10^{22}$ m^{-3} for Various Electron Energies in H$_2$

Energy (eV)	Q_M ($\times 10^{-20}$ m^2)	v/N ($\times 10^{-13}$ m^{3}s^{-1})	B/N ($\times 10^{-23}$ T m^3)	B (T)
0.1	10.50	0.197	0.011	0.0036
1.0	17.40	1.03	0.059	0.0190
10.0	6.80	1.28	0.073	0.0234

Momentum transfer cross sections (Q_M) are taken from Table 4.8.

magnetic flux density to satisfy the condition $\omega/v = 1$. Table 11.2 gives values of moderate flux density for various electron energies.

11.5   EXPERIMENTAL SETUP

One of the major concerns in accurate determination of sparking voltages and ionization coefficients in crossed fields is the large diffusion that occurs in the $E \times B$ fields, particularly since one uses only low gas number densities, as referred to above. Two alternative solutions have been sought to reduce or eliminate errors due to loss of electrons from the discharge region. The first one is to adopt ski-shaped electrodes so that the longer dimension is aligned in the $E \times B$ direction,[13] as shown in Figure 11.4. In this instance the magnetic field was formed by an electromagnet capable of producing a uniform flux density of 0.75 tesla in a region of approximately 16 cm^3. Gap lengths up to 1.2 cm were employed. The advantage of the system was that several gap lengths could be investigated without dismantling the system, though complex maneuvering was involved in moving the electromagnet into position.

As an alternative strategy, Gurumurthy and Raju[11] adopted a coaxial geometry to overcome errors due to diffusion (Figure 11.5). The advantage was that the electromagnet was an integral part of the equipment without requiring manipulation, but the disadvantage was that the gap length could not be varied. This demanded a pressure variation technique for the measurement of ionization coefficients.

11.6   IONIZATION COEFFICIENTS

The crossed magnetic field influences the coefficients α and γ because of changes that occur in fundamental properties such as the energy distribution, drift velocity, mean energy, etc. The current growth in crossed fields is given by[11]

$$I_{XF} = I_0 \frac{\exp\left[\left(\frac{\alpha}{N}\right)_{XF} \times Nd\right]}{1 - \gamma_{XF}\left\{\exp\left[\left(\frac{\alpha}{N}\right)_{XF} \times Nd\right] - 1\right\}} \tag{11.37}$$

Recalling that the ionization current in an electric field only is given by

$$I = I_0 \frac{\exp\left(\frac{\alpha}{N} \times Nd\right)}{1 - \gamma\left\{\exp\left(\frac{\alpha}{N} \times Nd\right) - 1\right\}} \tag{11.38}$$

the ratio of I_{XF}/I can be measured with and without the magnetic field.

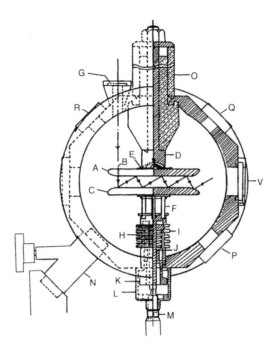

FIGURE 11.4 Experimental set up for $E \times B$ field studies with ski-shaped electrodes.[13] A: anode; B: slit; C: cathode; D: anode holder; E: anode lining-up platform; F: quartz rods; G: quartz window; H: stainless steel windows; I: spring; J: cathode guiding spindle; K: knurled nut; L: cathode holder; M: micrometer; N: butterfly valve; P: current-measuring flange; Q: vacuum-measuring flange; R: pressure-measuring flange; V: viewing window.

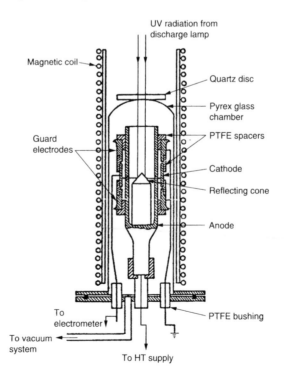

FIGURE 11.5 Schematic of the experimental setup for $E \times B$ field studies with coaxial electrodes.[11]

Assuming that the secondary ionization coefficient remains the same in crossed fields ($\gamma = \gamma_{XF}$), the ionization coefficient is obtained as

$$\left(\frac{\alpha}{N}\right)_{XF} = \frac{1}{Nd} \ln \frac{1}{\dfrac{\gamma}{(1+\gamma)} + \dfrac{I}{I_{XF}(1+\gamma)}[P]} \tag{11.39}$$

where

$$P = \frac{1 - \gamma\left\{\left[\exp\left(\dfrac{\alpha}{N} \times Nd\right)\right] - 1\right\}}{\exp\left(\dfrac{\alpha}{N} \times Nd\right)} \tag{11.40}$$

In the case of negligible γ, Equations 11.39 and 11.40 simplify to

$$\left(\frac{\alpha}{N}\right)_{XF} = \frac{1}{Nd} \times \ln\left[\frac{I_{XF}}{I} \exp\left(\frac{\alpha}{N} \times Nd\right)\right] \tag{11.41}$$

The first ionization coefficient (α_{XF}) is expressed, in terms of the EIDC (Equation 11.10), as

$$\left(\frac{\alpha}{N}\right)_{XF} = \frac{N_e}{N}\left(\frac{\alpha}{N}\right)_{0, E/N_e} \tag{11.42}$$

where the ratio N_e/N is given by Equation 11.10. The familiar Townsend's equation (8.13) is modified to

$$\left(\frac{\alpha}{N}\right)_{XF} = A_T y \exp\left(-\frac{B_T N y}{E}\right) \tag{11.43}$$

where

$$y = \frac{N_e}{N} \tag{11.44}$$

In a given experiment the quantities measured are N, E, B, and α_{XF}. The equivalent density is calculated with the help of Equation 11.10 from a knowledge of v/N, and Equation 11.42 is verified.

Heylen and Dargan[14] have shown that the expression corresponding to Equation 11.43, in terms of EREF, is given by

$$\left(\frac{\alpha}{N}\right)_{XF} = F_T \sec\theta\left\{\exp -\frac{G_T}{(E/N)\sec\theta}\right\} \tag{11.45}$$

Often it is useful to evaluate the ionization efficiency ($\eta = \alpha/E$). The relationship, in terms of ionization efficiency in crossed fields, is given by

$$\eta_{XF} = \eta_e \tag{11.46}$$

That is, the ionization efficiency in crossed fields is equivalent to that in an effective reduced electric field without the magnetic field.

Since all required quantities have been defined, we can now proceed to present experimental data on ionization coefficients in crossed fields in several gases.

11.7 EXPERIMENTAL DATA

Measured data are presented in this section.

11.7.1 AIR

There have been just two measurements of $(\alpha/N)_{XF}$ in air, those of Bhiday et al.[15] and Gurumurthy and Raju.[11] These data are shown in Tables 11.3 and 11.4, respectively. It is useful to recall that in electric fields only, the α/N values reported by Bhiday et al.[15] are in very good agreement with the results of Rao and Raju;[16] see Appendix 6 for tabulated values. Figures 11.6 and 11.7 show alternative ways of presenting the measured $(\alpha/N)_{XF}$.

From the data provided in Tables 11.3 and 11.4 one sees that $(\alpha/N)_{XF}$ decreases with increasing B/N at the same value of E/N. Further, at the same B/N, $(\alpha/N)_{XF}$ increases with

TABLE 11.3
First Ionization Coefficients in Air in Crossed Fields[15]

E/N (Td)	B/N = 0 α/N (10^{-20} m²)	B/N = 0.3 × 10^{-23} T m³ $(\alpha/N)_{XF}$ (10^{-20} m²)
600	0.725	—
900	1.33	0.786
1200	1.73	1.50
1500	2.08	2.04
1800	2.33	2.75

TABLE 11.4
$(\alpha/N)_{XF}$ in Crossed Fields in Air. Averaged Values at B/N=0 Are Taken from Appendix 6

B/N E/N	0	0.029	0.039	0.048	0.063	0.075	0.096	0.114	0.150	0.18
400	0.368	0.322	0.313	0.271	0.198					
500	0.549	0.529	0.504	0.497	0.470	0.428	0.286			
600	0.760			0.699	0.675	0.639	0.556			
700	0.908				0.872	0.874	0.792	0.754	0.572	0.526
800	1.091					1.11	1.04	1.01	0.851	0.680
900	1.33						1.28	1.23	1.11	0.955
1000	1.43						1.54	1.46	1.35	1.25
1100	1.51							1.68	1.59	1.50
1200	1.71							1.85	1.80	1.72
1300	1.87							2.06	2.00	1.98
1400	1.98							2.30	2.22	2.18
1500								2.52	2.44	2.39
1600								2.66	2.69	

Table values are $(\alpha/N)_{XF}$ (10^{-20} m²).

E/N in units of Td and B/N in units of 10^{-23} T m³. Interpolated by the author from the data of Gurumurthy and Raju.[11] © IEEE

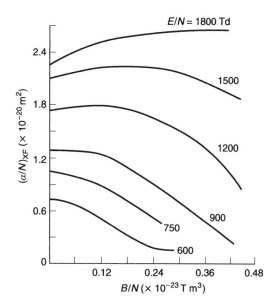

FIGURE 11.6 $(\alpha/N)_{XF}$ in dry air, measured by Bhiday et al.[15]

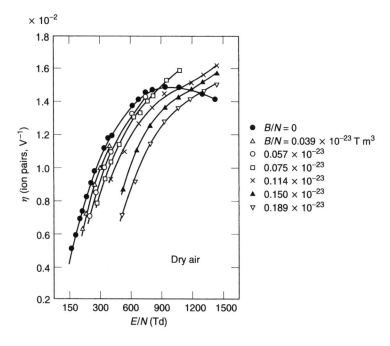

FIGURE 11.7 Ionization efficiency in crossed fields in dry air. Note the decrease in η with increasing B/N at the same value of E/N.[11] Reproduced with permission of IOP, U.K.

increasing E/N. Both observations are consistent with the theory given. From the measured ionization coefficients one can evaluate the collision frequency at unit density, ν_0. Since a separate section has been devoted to this topic we just remark on the representative value. Observed values of ν_0 lie between 10^{-13} and $10^{-12}\,\mathrm{s^{-1}\,m^3}$, increasing with magnetic field at the same value of E/N. Bhiday et al.[15] obtain values of 2 to $5 \times 10^{-13}\,\mathrm{s^{-1}\,m^3}$.

11.7.2 ARGON

The only experimental study of $(\alpha/N)_{XF}$ in argon is due to Heylen,[17] whose values cover a limited range, $120 \leq E/N \leq 500$ Td (Figure 11.8). Tabulated data digitized from the figure are shown in Table 11.5. The recapture of electrons at the cathode has been observed to be greater in the gas, particularly at higher B/N, as is evident from the fact that the ionization

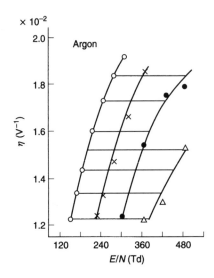

FIGURE 11.8 Ionization efficiency in argon in crossed fields as a function of reduced electric field. The intersection of horizontal lines with the curves give E/N to maintain the same $(E/N)_e$.[17] B/N: ($\circ$) 0; ($\times$) 0.15; ($\bullet$) 0.3; ($\triangle$) 0.45; in units of 10^{-23} Tm^3.

TABLE 11.5
Ionization Efficiency in Argon in Crossed Fields[17]

E/N	B/N = 0	$(\eta)_{XF} \times 10^{-2}$ 0.15 × 10^{-23}	0.30 × 10^{-23}	0.45 × 10^{-23}
150	1.24			
170	1.35			
190	1.45			
210	1.57			
230	1.66	1.25		
250	1.73	1.36		
270	1.81	1.46		
290	1.87	1.56		
310	1.92	1.66	1.29	
330		1.74	1.39	
350		1.80	1.48	
370		1.86	1.55	1.23
400			1.65	1.30
425			1.72	1.36
450			1.77	1.43
475			1.82	1.50
500				1.58

B/N in units of T m^3, E/N in units of Td, and $(\eta)_{XF}$ in V^{-1}.

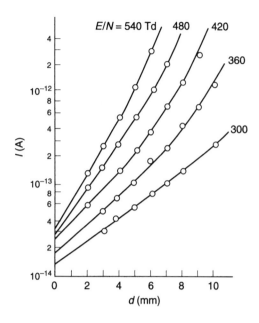

FIGURE 11.9 Ionization current growth in argon in crossed fields at various reduced electric fields. The currents at zero gap length (I_0) do not converge to the same point on the ordinate, because of back diffusion. Back diffusion increases with decreasing E/N, with greater reduction in I_0.[17]

currents intercept the ordinate at different points; the lower the E/N, the lower is I_0 because recapture is higher (Figure 11.9). A similar effect was observed by Bernstein,[18] who used coaxial electrodes. According to Equation 11.46, if one draws a horizontal line at a given value of η in electric fields only (e.g. Figure 11.7), the horizontal line will intersect the ionization efficiency curves at E/N in the presence of a magnetic field. The corresponding collision frequency may then be evaluated.

We have already shown that Townsend's semiempirical equation (8.13) for α/N fits the molecular gases much better than the rare gases. Ward[19] has given an alternative expression that fits rare gases much better,

$$\frac{\alpha}{N} = C_T \exp\left(-\frac{D_T}{(E/N)^{1/2}}\right) \tag{11.47}$$

The α/N values shown in Appendix 4 are plotted in Figure 11.10 and the constants C_T and D_T evaluated by the least-squares method are shown in Table 11.6. In crossed fields, Heylen[20] has verified that Equation 11.47 modified according to EREF, holds true as

$$\left(\frac{\alpha}{N}\right)_{XF} = C_T \sec\theta \exp\left[-\frac{D_T N \sec\theta}{E}\right]^{1/2} \tag{11.48}$$

11.7.3 HYDROGEN

Hydrogen has been relatively more investigated, mostly by the Australian group, as far as $(\alpha/N)_{XF}$ is concerned. Bernstein[18] made measurements extending to magnetic field strengths up to 0.8 T. The work done up to 1980 has been summarized by Heylen;[2] selected references to the early work are: Haydon and Robertson;[21] Fletcher and Haydon.[22] Tabulated values of

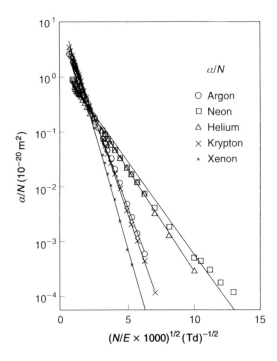

FIGURE 11.10 Reduced ionization coefficients in rare gases. See Appendix 4 for tabulated values.

TABLE 11.6
Constants in Ward's Analytical Expression (Equation 11.47) for (α/N) as a Function of E/N

Gas	C_T ($10^{-20}\,m^2$)	D_T $(Td)^{1/2}$	E/N range (Td)	R^2
Argon	7.885	44.7	30–2000	0.9991
Helium	1.70	75.9	10–900	0.9993
Krypton	10.59	48.0	25–2000	0.9996
Neon	1.69	23.7	20–1100	0.9917
Xenon	18.12	60.33	25–2000	0.9996

$(\alpha/N)_{XF}$ in hydrogen are given in Tables 11.7 and 11.8, the latter extending up to 1.2×10^{-23} T m^3. Representative curves are shown in Figure 11.11. Agreement with Berntein's[18] measurements is adequate.

11.7.4 NITROGEN

Bagnall and Haydon,[23] Fletcher and Haydon,[22] Raju and Rajapandian,[24] and Heylen and Dargan[14] have measured $(\alpha/N)_{XF}$ in N$_2$. Table 11.9 and Figure 11.12 show these data. Heylen and Dargan[14] report gas constants $F_T = 3.73 \times 10^{-20}$ m^2 and $G_T = 1038$ Td in the range $300 \leq E/N \leq 1800$ Td in electric fields only.

With magnetic field added, Equation 11.48 is found to hold good, with the slope of the $(\alpha/N)_{XF}$–N/E curve increasing with increasing B/N as predicted by theory. A point to note is that the application of the magnetic field reduces $(\alpha/N)_{XF}$ if the reduced electric field is

TABLE 11.7
Reduced Ionization Coefficients in H_2 in $E \times B$ Fields[22]

E/N (Td)	B/N (10^{-23} T m^3)	$(\alpha/N)_{XF}$ 10^{-20} m^2	E/N (Td)	B/N (10^{-23} T m^3)	$(\alpha/N)_{XF}$ (10^{-20} m^2)
150	0	0.10	525	0.03	0.733
	0.03	0.073		0.06	0.689
	0.06	0.039	600	0	0.733
225	0	0.245		0.03	0.795
300	0	0.382		0.06	0.783
	0.03	0.37		0.15	0.491
	0.06	0.25		0.30	0.205
	0.15	0.087	750	0	0.770
360	0	0.494		0.03	0.876
	0.03	0.450		0.06	0.932
	0.06	0.363		0.15	0.721
420	0	0.587		0.30	0.394
	0.03	0.590	900	0	0.783
	0.06	0.503		0.03	0.922
450	0	0.621		0.06	1.15
	0.03	0.640		0.30	0.634
	0.06	0.559			
	0.15	0.289			
	0.30	0.065			

TABLE 11.8
Reduced Ionization Coefficients $(\alpha/N)_{XF}$ in H_2 in $E \times B$ Fields[18]

E/N(Td)	B/N ($\times 10^{-23}$ T m^2)				
	0.075	0.15	0.3	0.6	1.2
150	0.034				
187.5	0.065	0.012			
225	0.118	0.038			
262.5	0.180	0.053			
300	0.255	0.009	0.012 (60%)		
375	0.404	0.186	0.037		
450		0.298	0.065		
525		0.450	0.124		
600		0.609	0.199		
675		0.745			
750			0.419	0.062 (60%)	
900			0.699	0.118	
1050				0.248	
1200				0.388	
1350				0.575	
1500				0.808	0.165
1800				1.12	0.028

Units are 10^{-20} m^2. Stated accuracy is between 5 and 25% for all measurements except for two values, as shown.

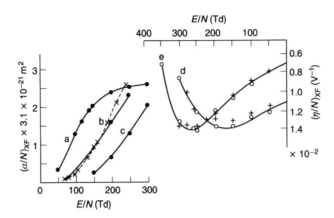

FIGURE 11.11 Reduced ionization coefficient and ionization efficiency in hydrogen as a function of E/N. B/N values, in units of 10^{-23} T m^3, are: (a) 0; (b) 0.15, dashed portion;[18] (c) 0.3; (d) 0; (e) 0.06.[22]

TABLE 11.9
Reduced Ionization Coefficients in N_2 in $E \times B$ Fields[22]

E/N (Td)	B/N $(10^{-23}$ T m$^3)$	$(\alpha/N)_{XF}$ 10^{-20} m^2	E/N (Td)	B/N $(10^{-23}$ T m$^3)$	$(\alpha/N)_{XF}$ $(10^{-20}$ m$^2)$
225	0	0.046	975	0	1.35
300	0	0.13	1050	0	1.44
360	0	0.24	1200	0	1.65
420	0	0.36		0.225	1.40
450	0	0.38		0.3	1.17
	0.15	0.31		0.375	0.89
600	0	0.65		0.45	0.75
	0.075	0.59		0.525	0.59
	0.15	0.41	1500	0	2.04
	0.225	0.26		0.6	0.34
675	0	0.79	1800	0	2.34
750	0	0.96		0.6	1.31
825	0	1.10		0.75	0.89
900	0	1.23			

kept constant, as shown by curves a–c of Figure 11.11. If, however, $(E/N)_e$ is held constant in the presence of the magnetic field, the ionization efficiency is lower, as shown by Figure 11.12, as long as the peak of the η–$(E/N)_e$ curve has not been reached. Since $\eta = \alpha/E$, an effective reduction of E should increase η. An explanation has been provided by Heylen and Dargan,[14] who theoretically calculated the ionization efficiency using the familiar integrals

$$\eta_{XF} = \left(\frac{2e}{m}\right)^{1/2} \frac{1}{W_T(E/N)} \int_{\varepsilon_i}^{\infty} N_0 Q_i \varepsilon^{1/2} F(\varepsilon) d\varepsilon \qquad (11.49)$$

$$\eta_e = \left(\frac{2e}{m}\right)^{1/2} \frac{1}{W(E/N)_e} \int_{\varepsilon_i}^{\infty} N_0 Q_i \varepsilon^{1/2} F(\varepsilon) d\varepsilon \qquad (11.50)$$

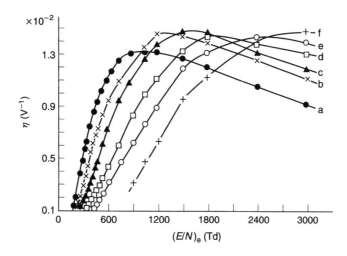

FIGURE 11.12 Ionization efficiency in nitrogen in $E \times B$ fields as a function of reduced electric field. B/N, in units of 10^{-23} T m^3: (a) 0; (b) 0.09; (c) 0.15; (d) 0.21; (e) 0.30; (f) 0.45.[14]

TABLE 11.10
Gas Constants in O$_2$[11]

E/N range (Td)	F_T ($\times 10^{-20}$ m^2)	G_T (Td)	Max. Deviation
240–620	1.94	625	±2.5%
620–990	3.9	1031	±2.5%

Using somewhat involved reasoning, the variation of η_{XF} with $(E/N)_e$ is explained[14] from a consideration of the energy distribution.

11.7.5 OXYGEN

There is just one publication of measurement of $(\alpha/N)_{XF}$ in this gas, by Gurumurthy and Raju.[11] The electrode geometry employed was coaxial cylinders with the density variation technique, in the range $0.039 \times 10^{-23} \le B/N \le 0.172 \times 10^{-23}$ T m^3, $309 \le E/N \le 993$ Td. The density-reduced ionization coefficients in zero magnetic field were measured to determine the accuracy of measurement and to verify that there was no density dependence. At low values of E/N three-body attachment is a possible process, introducing errors. A lower limit of $E/N = 240$ Td was determined, to be free of this error. Good agreement was obtained with the measurements of Schlumbohm[25] and of Thomas Betts and Davies.[26] The gas constants determined from (α/N) are shown in Table 11.10.

Table 11.11 shows the measured $(\alpha/N)_{XF}$. Figure 11.13 shows the ionization efficiency as a function of E/N; we have already commented upon the increase of η_{XF} with increasing E/N and its decrease with increasing B/N. An interesting feature observed in O$_2$ with regard to the gas constants, Equation 11.48, is that they do not change by the same amount, sec θ (Figure 11.14). The reasons for this peculiarity are not clear. Originally it was proposed, on the basis of the fact that the exponential term is related to the cross section,[27] that the ionization cross section might be influenced by the presence of a magnetic field. However, this aspect remains to be clarified.

TABLE 11.11
$(\alpha/N)_{XF}$ in O_2 in Units of $10^{-20}\,\mathrm{m}^2$

E/N	B/N = 0	0.039	0.057	0.078	0.112	0.153	0.192
250	0.159						
300	0.232	0.202					
350	0.323	0.277					
400	0.401	0.384					
450	0.476	0.471	0.415	0.334			
500	0.557		0.508	0.427			
550	0.635		0.579	0.514			
600	0.700		0.651	0.601			
650	0.781		0.731	0.693	0.486		
700	0.859		0.820	0.795	0.618	0.282	0.272
750	0.956		0.919	0.909	0.743	0.434	0.369
800	1.10			1.02	0.844	0.570	0.414
850	1.15			1.13	0.938	0.636	0.450
900	1.21			1.22	1.05	0.738	0.523
950	1.29			1.28	1.17	0.878	0.683

E/N in units of Td and B/N in units of 10^{-23} T m^3. Data are taken from Gurumurthy and Raju,[11] interpolated by the author.

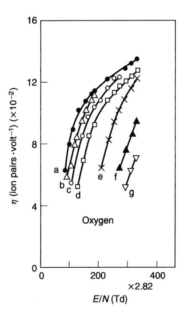

FIGURE 11.13 η_{XF} as a function of E/N in oxygen. Reduced magnetic field, B/N in units of 10^{-23} T m^3 are: (a) 0; (b) 0.039; (c) 0.057; (d) 0.078; (e) 0.114; (f) 0.153; (g) 0.192.[11]

11.8 SECONDARY IONIZATION COEFFICIENT

So far we have constrained our treatment to the assumption, mainly for the sake of simplicity, that the secondary ionization coefficient γ remains constant, with or without a magnetic field. This assumption, of course, is only approximate because even an elementary consideration shows that factors such as back diffusion at the cathode may have appreciable

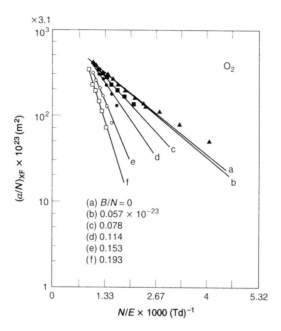

FIGURE 11.14 Reduced ionization coefficients as a function of N/E.[11]

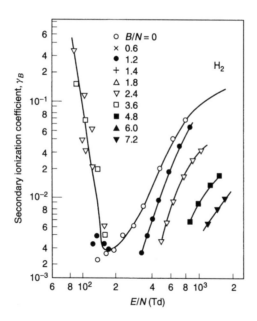

FIGURE 11.15 Secondary ionization coefficient in H_2 as a function of E/N at various magnetic field strengths. The decrease in γ_B with increasing B/N at the same E/N (>200 Td) is noticeable. E/N values to the left of $E/N = 180$ Td are of the equivalent reduced electric field $(E/N)_e$.[2]

influence. However, positive ions, because of their heavier mass, are not significantly affected by the magnetic field. Figure 11.15 is presented to show that caution is required in accepting this assumption.

Sen and Ghosh[28] provide a formula for γ as a function of E/N in crossed fields for an electrodeless discharge. A further reference to this aspect of discharge is included in Section 11.10.

11.9 SPARKING POTENTIALS

We shall consider uniform and nonuniform electric fields separately.

1.9.1 UNIFORM ELECTRIC FIELDS

The sparking criterion is given by the denominator of Equation 11.37 equated to zero. Expressing $(\alpha/N)_{XF}$ as a function of E/N according to Equation 11.45, one obtains in a straightforward way the equation for sparking voltage as

$$V_{s,XF} = \frac{GNd\sec\theta}{\ln(Nd\sec\theta) + \ln\left\{\dfrac{F}{\ln(1+1/\gamma)}\right\}} \tag{11.51}$$

where θ, the angle of deflection of the electron avalanche, is given by Equation 11.32. Equation 11.51 shows that the sparking potential in crossed fields is a function of Nd and B/N only, assuming that γ remains constant with and without the magnetic field, viz.,

$$V_{s,XF} = f(Nd, B/N) \tag{11.52}$$

Applying the trigonometric relationship

$$\sec\theta = \left(1 + \tan^2\theta\right)$$

to Equation 11.32, it is easy to show that

$$\sec\theta = \left\{1 + \left(\frac{e}{m}\frac{1}{v_0}\frac{B}{N}\right)^2\right\}^{1/2} \tag{11.53}$$

To examine the Paschen law, one plots the sparking voltage as a function of Nd; if the law holds good, then all points at various combinations of N and d fall on the same curve. In crossed fields, one plots $V_{s,XF}$ as a function of Nd_e, where

$$Nd_e = Nd\sec\theta \tag{11.54}$$

The most useful relationship between the sparking potentials with and without a magnetic field is given by Raju and Mokashi[29] as

$$\frac{v}{N} = \frac{e}{m}\frac{B}{N}\left[\frac{1}{\sqrt{[V_s/(V_s)_{XF}]^2 - 1}}\right] \tag{11.55}$$

from which v/N may be calculated.

The measured sparking potentials at various magnetic field strengths are shown in Figure 11.16a.[2] From these data the sparking potentials as a function of Nd_e are derived as shown in Figure 11.16b; the Paschen law is verified for crossed fields also. The equivalent increased density concept postulates the same result as well.

It is instructive to familiarize ourselves with the units of presentation of data (Figure 11.16) in terms of measured quantities. The point shown in Figure 11.16a

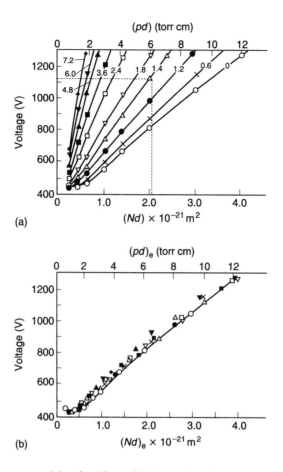

FIGURE 11.16 Sparking potentials of uniform field gaps in nitrogen with crossed magnetic field. (a) Measured data. Numbers attached to curves are reduced magnetic field (B/N) in units of 10^{-24} T m^{-3}. Dotted lines show the increase in Nd with magnetic field at the same sparking potential. (b) Replotted as a function of $(Nd)_e$ according to Equation 11.54. Paschen's law is observed to hold well. Note the increase in sparking voltage with the application of the magnetic field, shown by the values of Nd lying to the right of the Paschen minimum.[2]

corresponds to $Nd = 2.05 \times 10^{21}$ m^2 having a sparking potential of 1120 volts. If we assume a gap length of 0.008 m (8 mm), the number density is 2.56×10^{23} m^3. The reduced magnetic field is $B/N = 1.4 \times 10^{-24}$ T m^{-3} and therefore the applied magnetic field strength is 0.358 tesla (3580 gauss).

It is noted that the sparking potential for all the curves shown in Figure 11.16 is higher with a magnetic field than without. The effect of the crossed magnetic field on the sparking potential is, of course, dependent upon the value of Nd. If Nd is to the right of the Paschen minimum the crossed magnetic field increases the sparking voltage; on the other hand, if Nd is to the left of the Paschen minimum the crossed magnetic field decreases the sparking potential within certain limits. The proof of this statement is given below.

The criterion for a self-sustained discharge derived from Townsend's current growth equation is given by (Chapter 8)

$$\gamma \left[\exp\left(\frac{\alpha}{N} Nd\right) - 1 \right] = 1 \qquad (11.56)$$

which may be rewritten as

$$\frac{\alpha}{N} = \frac{1}{Nd}\ln\left(\frac{1+\gamma}{\gamma}\right)$$

(11.57)

In terms of the electric field, E_s at sparking (Equation 11.57) may be expressed as

$$\frac{\alpha}{E_s} = \frac{1}{E_s d}\ln\left(\frac{1+\gamma}{\gamma}\right)$$

(11.58)

In the presence of a magnetic field Equation 11.58 becomes

$$\left(\frac{\alpha}{E_s}\right)_{XF} = \frac{1}{(E_s)_{XF}d}\ln\left(\frac{1}{1+\gamma}\right)$$

(11.59)

Combining Equations 11.58 and 11.59, the sparking voltage in crossed fields becomes, in terms of that without the magnetic field,[11]

$$(V_s)_{XF} = \left(\frac{\alpha}{E}\right)\left(\frac{\alpha}{E}\right)_{XF}^{-1} V_s$$

(11.60)

This relationship shows that the sparking voltage in crossed fields will be higher or lower than that without the magnetic field. Now, turning one's attention to the type of data given in Figure 11.12, one sees a turning point at which the ratio of the reduced ionization coefficients is smaller or greater than unity. In other words, the sparking voltage in crossed fields is higher than that without the magnetic field, to the right side of the Paschen minimum, but $(V_s)_{XF}$ is lower than V_s to the left side of the minimum. Figure 11.17 shows the situation schematically . Examples may be found in Raju and Gurumurthy.[30] The fact that a crossed magnetic field lowers the sparking potentials below Paschen minimum conditions has important implications for spacecraft in the ionosphere.[31,70]

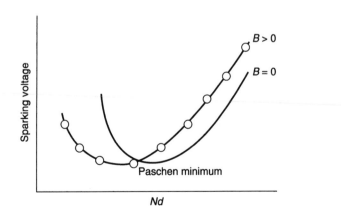

FIGURE 11.17 Schematic diagram showing the influence of a crossed magnetic field on the sparking voltage. For *Nd* less than the Paschen minimum, the crossed magnetic field decreases the sparking voltage.

11.9.2 NONUNIFORM ELECTPIC FIELDS

Paschen's law is the result of application of the similarity principle to uniform electric fields. In nonuniform electric fields the electric field within the inter-electrode region does not remain the same for all combinations of N and d, even though the product Nd is kept constant. In coaxial cylinder geometry, for example, the gap length is $(R_2 - R_1)$ and the breakdown condition is expressed by Equation 8.57 as a function of $(R_2 - R_1)$ and the ratio R_2/R_1.[32] Following a similar procedure in crossed fields, Raju and Gurumurthy[30] derived the expression

$$\frac{F_T}{G_T} \frac{(V_s)_{XF}}{\ln(n)} \left\{ \exp\left[-\frac{G_T y (R_2 - R_1) \ln(n)}{(n-1)(V_s)_{XF}} \right] - \exp\left[-\frac{G_T n y (R_2 - R_1) \ln(n)}{(n-1)(V_s)_{XF}} \right] \right\} = \ln\left(1 + \frac{1}{\gamma_B} \right)$$

(11.61)

If γ_B does not vary with gas pressure in the region of minimum breakdown voltage, the value of the sparking potential will be more sensitive to variation of α_{XF} and the right side of Equation 11.61 may be treated as a constant. Differentiating this expression with respect to B and setting

$$\frac{\partial (V_s)_{XF}}{\partial B} = 0$$

(11.62)

the condition to be satisfied for minimum $(V_s)_{XF}$ is obtained as

$$N(R_2 - R_1) = \frac{1}{F_T y} \frac{\ln(n) \ln(1 + 1/\gamma_B)}{\left\{ \exp[-\ln(n)/(n-1)] - \exp[-n\ln(n)/(n-1)] \right\}}$$

(11.63)

$$(V_s)_{XF\,min} = \frac{F_T}{G_T} \frac{\ln(n) \ln(1 + 1/\gamma_B)}{\left\{ \exp[-\ln(n)/(n-1)] - \exp[-n\ln(n)/(n-1)] \right\}}$$

(11.64)

Equation 11.64 shows that the minimum sparking voltage in crossed fields is independent of B and depends on constants F_T, G_T, and n only. This is contrary to what is observed in uniform electric fields[13] and needs further investigation. Letting $n \to 1$ in Equations 11.63 and 11.64 allows conditions approximating to uniform fields to be obtained. Experimental results for n varying from 1.58 to 30.4 have been obtained by Raju and Gurumurthy[30] to verify the theory just outlined in dry air and nitrogen for low and moderate values of B/N.

Rasmussen et al.[33] provided a succinct summary of the EDIC and EREF up to 1969, and showed that the two concepts are not exactly identical in the case of electric fields only. In the latter case one considers the ionization per unit length whereas in the equivalent $E \times B$ situation one considers ionization per unit of time (v_i). The two are related by $\alpha W = v_i$ and one can calculate the ionization rate. A comparison with experimental current growth rate in H_2 by Barbian and Rasmussen[34] showed qualitative agreement. It is useful to note here that the temporal growth of current, related to currents in the range of 0.1 to 100 A and space charge conditions which we have not considered, prevailed.

Raju and Rajapandiyan[35,36] have further examined nonuniform field ionization in crossed fields. The corona in a wire cylinder geometry in crossed fields has been studied by Gurumurthy and Raju.[37] Dincer and Gokmen[38] report a strong dependence of the breakdown voltage on the magnetic field strengths near the Paschen minimum in SF_6 (Figure 11.18). The very strong magnetic fields used by them (up to 1.2 T), resulting in $B/N = (0.35 \text{ to } 4.2) \times 10^{-23}$ T m^3, may possibly explain the discrepancy.

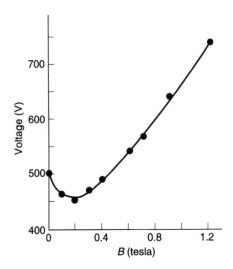

FIGURE 11.18 Variation of sparking voltage as a function of magnetic field strength in SF_6 in a coaxial cylindrical geometry.[38] Figure reproduced with permission of IOP, U.K.

11.10   TIME LAGS IN $E \times B$ CROSSED FIELDS

It is well known that the total time lag for breakdown consists of statistical and formative time components. The method of separating them has been described by Raju.[39] In this section we consider the influence of a crossed magnetic field on the formative lag. Deutsch[40] measured the time lags in air and hydrogen in crossed fields at number densities lower than those corresponding to the Paschen minimum. The formative time lags were observed to decrease with transverse magnetic field, then to reach a minimum, and finally to increase to infinity. These results were explained on the basis of EIDC as proposed by Blevin and Haydon.[7]

When the gyrofrequency for electrons is much greater than the electron molecule collision frequency, the theory proposed for formative time lags is analogous to the streamer mechanism theory of mid-gap breakdown. Kunkel and Sherwood[41] and Sherwood and Kunkel[42] suggest that, if the estimated avalanche crossing time T_c is greater than the breakdown time T_B, secondary processes do not play a significant role in the breakdown processes. For this condition of $T_c > T_B$ a nonlinear dependence of the breakdown time upon $1/N$ is observed. Formative time lags in hydrogen in a coaxial system of electrodes with impulses of either polarity have been reported by Rasmussen et al.[33]

While describing the operation of "GAMITRON," an experimental device for interrupting direct current in a power system, Lutz and Hoffman[43] have derived an expression for the formative time. The theory proposed refers to low gas number density and magnetic field strengths of the order of 10^{-2} T. The formative time is calculated by suggesting a mechanism analogous to the streamer theory of breakdown. Stock et al.[44] have also measured time lags in molecular hydrogen at gas number densities of 10^{16} to 10^{19} m^{-3}, where the electron mean free path is much longer than the gap length.

Raju and Mokashi[29] have measured the formative time lags in a number of gases in crossed fields. Figures 11.19 and 11.20 show the results in nitrogen and air at several gas number densities and magnetic field strengths. The hatched region in Figure 11.19 shows the envelope corresponding to upper and lower limits of the magnetic flux densities employed. In the absence of a magnetic field the formative time lag (t_f) decreases with increasing applied voltage, as expected. The influence of the magnetic field on the time lag depends upon the

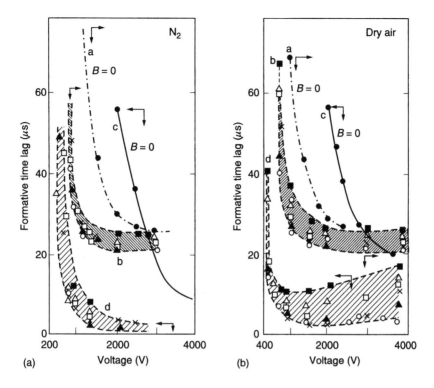

FIGURE 11.19 Formative time lags in nitrogen in $E \times B$ fields below the Paschen minimum. (A): N_2; (a) $N = 3.22 \times 10^{22}$ m^{-3}, $B = 0$; (b) same number density but $B > 0$; (c) $N = 1.6 \times 10^{22}$ m^{-3}, $B = 0$; (d) same number density as (c) but $B > 0$. Symbols for magnetic flux density in units of $\times 10^{-4}$ T: (●) $B = 0$; (■) $B = 85$; (△) $B = 93$; (□) $B = 100$; (×) $B = 108$; (▲) $B = 116$; (○) $B = 125$. (B): dry air; symbols as for N_2.[29]

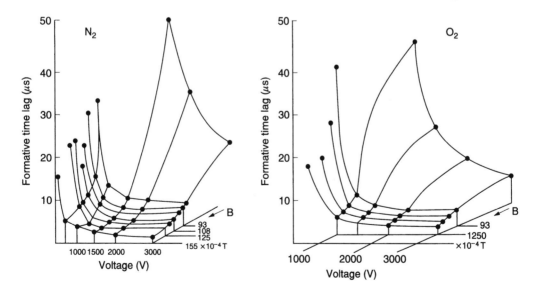

FIGURE 11.20 Composite diagram of formative time lags as a function of voltage and magnetic flux.[29]

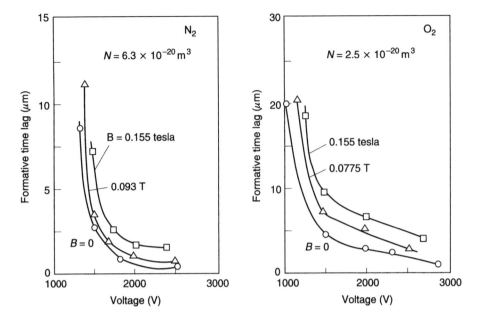

FIGURE 11.21 Formative time lags in N_2 and O_2 as a function of voltage at various magnetic flux densities. Note the increase of time lag with applied magnetic field.[29]

Nd value. If it is below the Paschen minimum, the magnetic field decreases the formative time, provided other parameters, N, d, and V, are held constant. If the Nd value is greater than that corresponding to the Paschen minimum, as shown in Figure 11.21, the magnetic field increases the time lag. This result is entirely consistent with the change of sparking voltages with magnetic field strength.

Raju and Mokashi[29] demonstrate that the streamer theory adopted by Lutz and Hoffman,[43] for low pressures does not agree with the observed time lags. The average ionizing collision time is expressed as

$$\tau = \frac{B}{Q_i NE} \tag{11.65}$$

Since Q_i is a function of electron energy, the average collision time may be calculated by substituting an average ionizng collision cross section, which may be found by approximating Q_i as

$$Q_i = Q_0(\varepsilon - \varepsilon_i) \tag{11.66}$$

in which Q_0 is a constant. Since electrons suffer collisions, their energy is limited to approximately $3\varepsilon_i$ and the average collision cross section is expressed as

$$Q_{i\,av} = \int_{\varepsilon_i}^{3\varepsilon_i} Q_0(\varepsilon - \varepsilon_i)d\varepsilon \approx Q_0\varepsilon_i \tag{11.67}$$

The ionization cross sections measured by Rapp and Englander-Golden in N_2 may be fitted to Equation 11.66 to yield $Q_0 = 6.9 \times 10^{-22}\,m^2\,eV^{-1}$ and hence an average collision cross section of $1.1 \times 10^{-20} \, m^2$. The calculated average collision time according to Equation 11.65 at an applied voltage of 810 V and a magnetic flux density of 0.0085 T is 9×10^{-8} s. Lutz and Hoffman[43] assume that a streamer type of breakdown occurs when the number of electrons reaches 10^{12} and the formative time lag is $t_f = 27 \, \tau$. Accordingly, the formative time for the example quoted above is obtained as 2.4 μs whereas the measured time lag is 20 μs.

Recognizing the need for a unified theory for formative time, on either side of the Paschen minimum and with or without a magnetic field, Raju has developed a unified theory using Heylen's[45] approach in electric fields only as a starting point. The relative contributions of photoelectric action (γ_{ph}) and positive ions (γ_i) at the cathode to the total secondary ionization coefficient (γ_T) are considered The regenerative mechanisms have cycle times (τ_{ph}, τ_i) and, at the low gas number densities we deal with, both actions are present. An effective cycle time may be defined as

$$\tau_{eff} = \frac{\gamma_i}{\gamma_T} \tau_i + \frac{\gamma_{ph}}{\gamma_T} \tau_{ph} \tag{11.68}$$

The formative time is obtained as

$$t_f = \frac{A}{A-1} \tau_{eff} \left[\ln \left\{ \frac{(A-1)M}{\exp(\alpha d)} + 1 \right\} \right] \tag{11.69}$$

where

$$A = \gamma_T \left(e^{\alpha d} - 1 \right) \tag{11.70}$$

and M is the current multiplication factor required for the spark to occur.

The general validity of Equation 11.69 has been established,[46,47] so that one can extend the theory to applied magnetic flux densities. The crossed field time lag is expressed as

$$t_{XF} = \frac{A_{XF}}{(A_{XF}-1)} \tau_{eff, XF} \left[\ln \left\{ \frac{A_{XF}-1}{\exp(\alpha_{XF}d)} + 1 \right\} \right] \tag{11.71}$$

in which the quantities with suffix XF are functions of E/N and B/N.

Applying this theory, Raju and Mokashi[29] have calculated time lags which show good agreement with the experimental results. In subsequent refinements to the theory, Raju[48] has included the effect of recapture of secondary electrons released at the cathode on the spatial and temporal growth of current. The fraction recaptured increases with increasing magnetic field. Secondary electrons liberated at the cathode by positive ions have thermal energy whereas those liberated by photoelectric action have considerably larger energy. Therefore the recapture applies to positive ion components only. The fraction recaptured is[5]

$$f = \exp - \left[\frac{m}{e} \frac{E}{N} \left(\frac{N}{B} \right)^2 P_c \right] \tag{11.72}$$

The calculated temporal growth of current is shown in Figure 11.22 and the influence of magnetic field is clearly seen. With increasing magnetic field, current growth increases up to a certain magnetic field strength. For higher field strengths the current multiplication decreases (compare curves 5 and 6) and, for very high field strengths (curve 7), breakdown is not possible at all. This condition is shown as "cutoff."

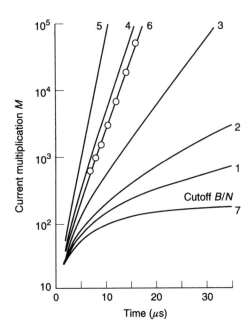

FIGURE 11.22 Temporal growth of current in crossed fields at a constant voltage of 950 V and constant $N = 3.32 \times 10^{21}$ m^{-3}. Flux densities in units of 10^{-4} are: curve (1) $B = 0$; (2) 5; (3) 10; (4) 20; (5) 100; (6) 1000; (7) 1200.[49]

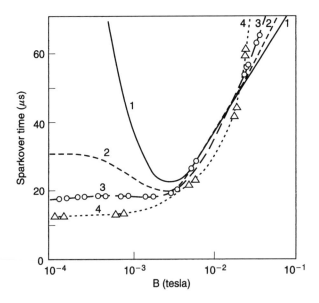

FIGURE 11.23 Sparkover time in O_2 as a function of B at various constant voltages at $N = 3.32 \times 10^{21}$ m^3. Curve (1) 950 V; (2) 1100 V; (3) 1300 V; (4) 2000 V.[48]

The time lags shown in Figure 11.23[49] are consistent with current growth curves: the formative time decreases with increasing magnetic flux density, reaches a minimum, and a further increase in B increases the time lag. The qualitative similarity with Figure 11.18 is good, notwithstanding that gases and other conditions are different.

11.11 COMPUTATIONAL METHODS

As stated earlier, the theoretical foundation for the drift of electrons in crossed fields was laid by Allis.[8] Bernstein[50] measured the drift velocity and diffusion in H_2 and D_2 in strong magnetic fields (Equation 11.36) and adopted a simplified Boltzmann equation to explain the results. Pearson and Kunkel[51] and Engelhardt and Phelps[52] used the Boltzmann equation to calculate the mobility and diffusion coefficients in strong magnetic fields.

Heylen and Bunting,[53] followed by Bunting and Heylen,[54] theoretically calculated the transverse and perpendicular drift velocities by assuming Maxwellian electron energy distribution. As expected from the theory of EREF, the ratio of W_e to $(E/N)_e$ remains constant for a given mean energy. Further, the increase in E/N required to keep the mean energy constant with application of a magnetic field was derived.

Raju and Gurumurthy[55] adopted the method of Engelhardt and Phelps[52] to calculate the transverse and perpendicular drift velocities in nitrogen at low values of E/N (3 to 30 Td) where ionization does not occur. As stated earlier, the effect of the magnetic field is to shift the high-energy electrons to lower energies and thereby lower the mean energy. In the EREF concept the mean energy is a function of $(E/N)_e$ irrespective of the combination of (E/N) and (B/N) to give that $(E/N)_e$. The energy distribution is far from Maxwellian and the energy distribution in crossed fields lowers the transverse drift velocity, the reduction being greater the larger the value of B/N.

This research was extended to higher values of E/N (60 to 600 Td),[56] at which ionization occurs, but where losses due to rotational excitation and collisions of the second kind are negligible.[57] The transverse drift velocity decreases with B/N at a given value of E/N, but the perpendicular drift velocity increases with B/N, reaching a peak value, and a further increase in B/N results in lower W_p. Calculated and measured ionization coefficients in crossed fields show good agreement in N_2.

Monte Carlo simulation methods provide a powerful technique for mapping the development of a discharge through the tracing of electron motion in crossed fields. The method, and its relative merits with regard to the Boltzmann solution method, have been described by Raju.[39] Dincer and Raju[58] extended their simulation in electric fields only to magnetic fields;[59] this was the first application of the simulation to crossed fields. $(\alpha/N)_{XF}$ and $(\eta/N)_{XF}$ are evaluated for $540 \leq E/N \leq 900$ Td and for $B/N \leq 0.18 \times 10^{-23}$ T m^3.

The mean energy is reduced in the presence of the magnetic field and therefore the density-reduced attachment coefficient increases, with a lowering of $(\alpha/N)_{XF}$. The $(E/N)_{lim}$ which determines the breakdown criterion is therefore related to the magnetic field, which may be visualized as having a critical value. The critical magnetic field and the corresponding values of the coefficients are shown in Table 11.12. The Monte Carlo method has also been applied to N_2[60] and compared with the Boltzmann equation method of Raju and Gurumurthy.[56]

TABLE 11.12
Critical Reduced Magnetic Flux Density for Breakdown in SF$_6$[59]

E/N (Td)	B/N (10^{-23} T m^3)	$(\alpha/N)_{XF} = (\eta/N)_{XF}$ (10^{-20} m^2)
540	0.07	0.56
630	0.09	0.62
720	0.12	0.81
810	0.13	0.87
900	0.15	0.97

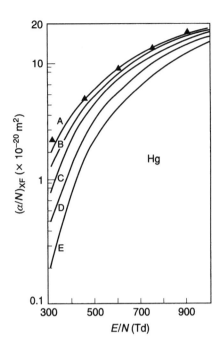

FIGURE 11.24 Calculated ionization coefficients in mercury vapor in crossed fields as a function of E/N at various magnetic flux densities. Curve (A) $B/N = 0$; (B) $B/N = 0.05 \times 10^{-23}$ T m³; (C) 0.1; (D) 0.15; (E) 0.2. (▲) Overton and Davies[63] in electric fields only. Figure reproduced from Liu, J. and G. R. Govinda Raju, *J. Phys. D*, 25, 485, 1992. With permission of IOP, U.K.

A further investigation of electron transport coefficients in mercury vapor without[61] and with crossed fields has been carried out by Liu and Raju.[62] The calculated ionization coefficients in crossed fields are shown in Figure 11.24 and compared with previous data[63] in electric fields only. Figure 11.25 shows these data replotted as a function of $(E/N)_e$; one observes the excellent quality of agreement with the EREF concept.

Gas breakdown in crossed fields has been investigated by Cho,[64] with emphasis on conditions prevailing in the ionosphere, using the Monte Carlo technique. The Townsend ionization coefficients of argon are calculated on the basis of the EREF concept, and breakdown voltages in the range of 200 to 10,000 volts are obtained at gas number densities less than 10^{19} m^{-3}.

In a recent contribution, Raju[65] has investigated the swarm properties of argon + SF$_6$ mixtures using the Boltzmann code developed by Siglo Kinema.[66] The derived ionization and attachment coefficients in this mixture, with increasing SF$_6$ content, are shown in Figure 11.26. With greater concentration of SF$_6$ in the mixture α/N decreases and η/N increases. The crossover point gives $(E/N)_{\text{lim.}}$. The calculations are extended to the $E \times B$ situation and the derived effective collision frequencies are shown in Figure 11.27.[67]

The effective collision frequencies in argon are relatively independent of E/N and may be represented by an equation of the type

$$\frac{v}{N} = \left[1.1 - 6 \times 10^{-5} \times \frac{E}{N} \right] \times 10^{-13} \tag{11.73}$$

where E/N is expressed in Td and v/N in s^{-1} m^3. As an example, at $E/N = 100 \times 10^{-21}$ V m^2 (100 Td) and $N = 3.22 \times 10^{22}$ m^{-3} (133.3 Pa), the collision frequency is obtained as 3.5×10^9 s^{-1}, which agrees very well with the literature value.[13]

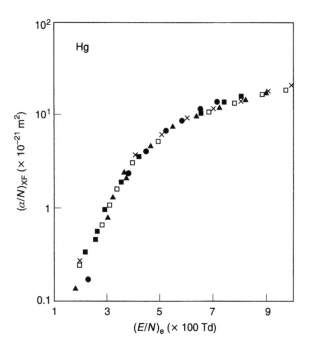

FIGURE 11.25 $(\alpha/N)_{XF}$ replotted as a function of $(E/N)_e$. Symbols are the same as in Figure 11.24. (▲) Overton and Davies[63] in electric fields only. Figure reproduced from Liu, J. and G. R. Govinda Raju, *J. Phys. D*, 25, 485, 1992. With permission of IOP, U.K.

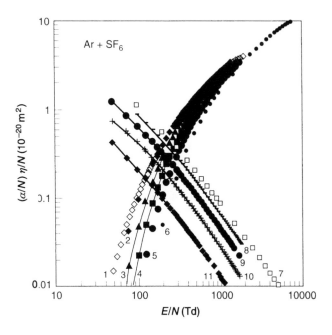

FIGURE 11.26 Ionization and attachment coefficients in mixtures of Ar + SF_6 as a function of E/N. α/N: (1) 100% Ar; (2) 80%; (3) 60%; (4) 40%; (5) 20%; (6) 0%. η/N: (7) 100% SF_6; (8) 80%; (9) 60% (10) 40%; (11) 20%.[65]

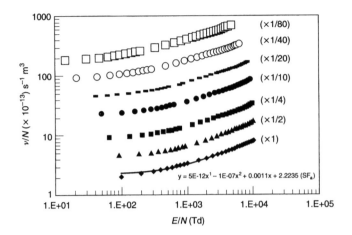

FIGURE 11.27 Effective collision frequency in SF_6 in $E \times B$ fields. Each curve is shifted as shown, for clarity of presentation. (◆) $B/N = 0$; (▲) 0.06×10^{-21} T m^{-3}; (■) 0.15; (●) 0.23; (---) 0.30; (○) 0.61; (□) 0.91; (———) Equation 11.74.[67]

The effective collision frequency in SF_6 is given by

$$\frac{\nu}{N} = \left[5 \times 10^{-12} \left(\frac{E}{N} \right)^3 - 1 \times 10^{-7} \left(\frac{E}{N} \right)^2 + 1.1 \times 10^{-13} \left(\frac{E}{N} \right) + 2.2 \right] \times 10^{-13} \ (s^{-1}m^3)(E/N \geq 300\,Td)$$

(11.74)

The mean energy obtained by the simulation at $E/N = 300$ Td is 7.52 eV. The momentum transfer cross section corresponding to this energy is 14.85×10^{-20} m^2 (Table 5.13) and, substituting these values in Equation 1.100, one obtains $\nu/N = 2.5 \times 10^{-13}$ s^{-1} m^3. Substituting $E/N = 300$ Td in Equation 11.74, one obtains 2.4×10^{-13} s^{-1} m^3, with excellent agreement. The influence of a crossed magnetic field on (ν/N), as seen from Figure 11.27, is to reduce the electric field and an interesting aspect in this gas (SF_6) is that the effective collision frequency is independent of $(E/N)_e$ from 100 to 10^4 Td. The constant effective frequency is 2.5×10^{-13} s^{-1} m^3.

11.12 EFFECTIVE COLLISION FREQUENCY

Reverting to Equation 11.32, the influence of the crossed magnetic field is expressed in terms of the fundamental electron–neutral collision frequency at unit density (ν/N). This quantity is a function of the electron energy and may be calculated from the collision cross sections. However, in a discharge there are electrons of all energies (within an upper limit) and, even if one knows the electron energy distribution, calculation of ν/N becomes a hopeless academic endeavor. Further, in crossed magnetic fields the electron energy distribution is assumed to remain constant, casting uncertainty on the calculated values. To address these difficulties an effective collision frequency at unit density has been assumed, remaining constant and dependent only upon $(E/N)_e$. Experimentally deduced values in several gases have largely supported this approach.

The effective collision frequency is determined, as the preceding discussion has revealed, by using one or more of the following approaches.

A. To measure the ionization currents with and without magnetic fields and then apply the EREF concept. The method is attractive as it quantifies the collision process at its most basic level; however, it is time and labor consuming.

B. Make breakdown measurements with and without magnetic fields and relate them through the collision frequency, making the necessary assumptions such as that the secondary ionization remains the same. A variation of the method is to decrease the gap length in a magnetic field so as to give the same sparking voltage as in a zero magnetic field, or, alternatively, to decrease the density to the same effect.[68]

C. An improvement to this method is that adopted by Dargan and Heylen.[13] The reduction in gap length that is required to keep the sparking voltage in crossed fields the same as that without is experimentally determined. The method has yielded accurate collision frequencies over a range of E/N and B/N.

D. Energy distribution method.

E. Simulation.[69]

Table 11.13 shows the selected values of ν/N in several gases. The agreement obtained by various methods is the best in N_2 and the worst in H_2. Whether this disagreement is another facet of disagreement between swarm evaluated and theoretically evaluated (also from beam experiments) vibrational excitation cross sections, as discussed in Chapter 4 (Figure 4.20),

TABLE 11.13

Selected Effective Collision Frequency in Gases in $E \times B$ Fields with Appropriate Parameters. Only Crossed Field Studies Are Considered

Gas	Method	$(E/N)_e$ (Td)	E/N (Td)	B/N $(10^{-23}\,\mathrm{T\,m^3})$	ν/N $(\times 10^{-13}\,\mathrm{s^{-1}\,m^3})$	Reference
Air	A		300–1800		2.0–5.2	15
	A	312–1040	366–1460	0.039–0.19	**1.1–3.5**	11
	B	1330–2310	1602–2718	0.078–0.31	2.4–5.9	11
Ar	B		30–540	0–0.84	**1.0**	13
	E	10–10⁴			4.88–13.4ᵃ	64
C_2H_6	B		240–750	0–2.1	**4.7**	13
H_2	A	4–200			0.96–1.9ᵇ	50
	D	80–170			**1.68**	52
	A		300–750		0.78	21
	A	81.5–252	150–900	0–0.3	1.2–0.78	22
	B		90–330	0–0.9	0.93	13
Hg	E	10–1000	10–1000	0–0.2	4.8–3.7	62
N_2	A	310–750	750–1200	0.15–0.42	1.96–2.64	23
	A	294–648	450–1200	0.15–0.75	2.3–2.6ᶜ	22
	B		450–600	0–0.9	2.5	13
	D		0.06–3000	0.0003–0.3	2.5	54
	A	180–660	240–1500	0.09–4.5	1.9–2.8ᵈ	14
	A		450–1080	0.025–1.94	**2.4**	24
	E		240–600	0–0.15	**2.1**	60
O_2	D		0.06–3000	0.0003–0.3	1.37–2.1ᵉ	54
	A	252–855	309–993	0.039–0.192	0.92–2.4	11
	B	1320–2340	1910–2810	0.06–0.3	1.13–6.6	11
SF_6	D		100–10000	0–0.91	**2.5**	65

Bold type indicates recommended ν/N.

ᵃThe author suggests an approximate constant value of $5 \times 10^{-13}\ \mathrm{m^3\,s^{-1}}$.

ᵇBernstein suggested that this effective collision frequency was lower than that obtained by other methods.

ᶜBoth parallel plane and coaxial geometry were investigated.

ᵈThe effective collision frequency increases with increasing $(E/N)_e$.

ᵉMaxwellian energy distribution was assumed.

remains to be resolved. The disagreement in air is also less than satisfactory and needs clarification.

11.13 CONCLUDING REMARKS

Compared with many other areas of gaseous electronics, the development of discharge in crossed magnetic and electric fields is less studied and provides scope for further research. Experimental data in a large number of gases, particularly in rare gases, are required to further verify the EREF concept. Drift velocity and diffusion measurements have been carried out only in H_2 and D_2, in strong magnetic fields, and to collect these data in other gases is a fruitful area of research. Further theoretical analyses for the remaining gases should be carried out to complete the task, at least in broad scope.

REFERENCES

1. Guthrie, A. and R. K. Wakerling, *The Characteristics of Electrical Discharges in Magnetic Fields*, McGraw-Hill, New York, 1949.
2. Heylen, A. E. D., *IEE Proc.*, 127, Pt. A, 221, 1980.
3. (a) Townsend, J. S., *Proc. Roy. Soc.*, A86, 571, 1912; (b) Wehrli, M. *Ann. Phys.* (Germany), 69, 285, 1922.
4. Valle, G., *Nuovo Cimento*, 7, 174, 1950.
5. Somerville, J. M., *Proc. Phys. Soc.*, B65, 620, 1952.
6. Haefer, R., *Acta Phys. Aust.*, 7, 52, 1953.
7. Blevin, H. A. and S. C. Haydon, *Aust. J. Phys.*, 11, 18, 1958.
8. Allis, W. P., *Handbuch der Physik* (Berlin), 21, 383, 1956.
9. Roth, J. R., *Industrial Plasma Engineering*, Institute of Physics, Bristol, 1995.
10. Heylen, A. E. D., *Brit. J. Appl. Phys.*, 16, 1151, 1965.
11. Gurumurthy, G. R. and G. R. Govinda Raju, *IEEE Trans. Plasma Sci.*, PS-3, 131, 1975.
12. Heylen, A. E. D., *Brit. J. Appl. Phys.*, 16, 1151, 1965.
13. Dargan, C. L. and A. E. D. Heylen, *Proc. IEE*, 115, 1034, 1968.
14. Heylen, A. E. D. and C. L. Dargan, *Int. J. Electron.*, 35, 433, 1973.
15. Bhiday, M. R., A. S. Paithankar, and B. L. Sharda, *J. Phys. D: Appl. Phys.*, 3, 943, 1970.
16. Rao, C. Raja and G. R. Govinda Raju, *J. Phys. D: Appl. Phys.*, 4, 494, 1971.
17. Heylen, A. E. D., *Int. J. Electron.*, 48, 129, 1980.
18. Bernstein, M. J., *Phys. Rev.*, 127, 342, 1962.
19. Ward, A. L., *Phys. Rev.*, 112, 1852, 1958.
20. Heylen, A. E. D., *Proc. IEE*, 126, 215, 1979.
21. Haydon, S. C. and A. G. Robertson, *Proc. Phys. Soc.*, 82, 79, 1963.
22. Fletcher, J. and S. C. Haydon, *Aust. J. Phys.*, 19, 615, 1966.
23. Bagnall, F. T. and S. C. Haydon, *Aust. J. Phys.*, 18, 227, 1965.
24. Raju, G. R. Govinda and S. Rajapandian, *Int. J. Electron*, 40, 65, 1976.
25. Schlumbohm, H., *Z. Phys.*, 184, 492, 1965.
26. Thomas Betts, A. and D. E. Davies, *J. Phys. D: Appl. Phys.*, 2, 213, 1969.
27. Lewis, T. J., *Proc. Roy. Soc.*, A244, 166, 1958.
28. Sen, S. N. and A. K. Ghosh, *Proc. Phys. Soc.*, 79, 293, 1962.
29. Raju, G. R. Govinda and A. D. Mokashi, *IEEE Trans. Electr. Insul.* EI-18, 436, 1983.
30. Raju, G. R. Govinda and G. R. Gurumurthy, *IEEE Trans. Electr. Insul.*, EI-12, 325, 1977.
31. Cho, M., *J. Phys. D: Appl. Phys.*, 26, 1398, 1993.
32. Heissen, A. *IEEE Trans. Plasma Sci.*, PS-4, 129, 1976.
33. Rasmussen, C. E., E. P. Barbian, and J. Kistemaker, *Plasma Phys.*, 11, 183, 1969.
34. Barbian, E. P. and C. E. Rasmussen, *Plasma Phys.*, 11, 197, 1969.
35. Raju, G. R. Govinda and S. Rajapandian, *IEEE Trans. Electr. Insul.*, EI-11, 1, 1976.
36. Raju, G. R. Govinda and S. Rajapandiyan, *Int. J. Electron.*, 40, 393, 1976.

37. Gurumurthy, G. R. and G. R. Govinda Raju, *Int. J. Electron.*, 46, 497, 1979.
38. Dincer, M. S. and A. Gokmen, *J. Phys. D: Appl. Phys.*, 25, 942, 1992.
39. Raju, Gorur G., *Dielectrics in Electric Fields*, Marcel Dekker, 2003.
40. Deutsch, F., *Proc. 7th International Conference on Phenomena in Ionized Gases*, 1965, p. 313.
41. Kunkel, W. B. and A. R. Sherwood, *Proc. 6th International Conference on Phenomena in Ionized Gases*, 1967, p. 177.
42. Sherwood, A. R. and W. B. Kunkel, *J. Appl. Phys.*, 39, 2343, 1968.
43. Lutz, M. A. and G. A. Hoffman, *IEEE Trans. Plasma Sci.*, PS-2, 11, 1974.
44. Stock, H. M. P., H. A. Blevin, and J. Fletcher, *J. Phys. D: Appl. Phys.*, 7, 635, 1974.
45. Heylen, A. E. D., *Proc. IEE*, 120, 1565, 1973.
46. Raju, G. R. Govinda, *J. Appl. Phys.*, 54, 6754 1983.
47. Raju, G. R. Govinda, *IEEE Trans.*, EI-19, 141, 1984.
48. Raju, G. R. Govinda, *IEEE Trans.*, EI-22, 287, 1987.
49. Raju, G. R. Govinda, *IEEE Trans.*, EI-22, 297, 1987.
50. Bernstein, M. J., *Phys. Rev.*, 127, 335, 1962.
51. Pearson, G. A. and W. B. Kunkel, *Phys. Rev.*, 130, 864, 1963.
52. Engelhardt, A. G. and A. V. Phelps, *Phys. Rev.*, 131, 2115, 1963.
53. Heylen, A. E. D. and K. A. Bunting, *Int. J. Electron.*, 27, 1, 1969.
54. Bunting, K. A. and A. E. D. Heylen, *Int. J. Electron.*, 31, 9, 1971.
55. Raju, G. R. Govinda and G. R. Gurumurthy, *IEEE Trans. Plasma Sci.*, PS-4, 241, 1976.
56. Raju, G. R. Govinda and G. R. Gurumurthy, *Int. J. Electron.*, 44, 355, 1978.
57. Lucas, J., *Int. J. Electron.*, 27, 201, 1969.
58. Dincer, M. S. and G. R. Govinda Raju, *J. Appl. Phys.*, 54, 6311, 1983.
59. Raju, G. R. Govinda and M. S. Dincer, *Proc. IEEE*, 73, 939, 1985.
60. Raju, G. R. Govinda and M. S. Dincer, *IEEE Trans. Plasma Sci.*, 18, 819, 1990.
61. Liu, J. and G. R. Govinda Raju, *J. Phys. D: Appl. Phys.*, 25, 167, 1992.
62. Liu, J. and G. R. Govinda Raju, *J. Phys. D: Appl. Phys.*, 25, 485, 1992.
63. Overton, D. N. and D. E. Davies, *J. Phys. D: Appl. Phys.*, 1, 881, 1968.
64. Cho, M., *J. Phys. D: Appl. Phys.*, 26, 1398, 1993.
65. Raju, G. R. Govinda, *Proc. 7th International Conference on Properties and Applications of Dielectric Materials*, Nagoya, June 1–5, 2003, vol. 3, p. 999. The relationship between α/N and $f(\varepsilon)$ is given by $\dfrac{\alpha}{N} = \left(\dfrac{2e}{m}\right)^{1/2} \dfrac{1}{W} \displaystyle\int_{\varepsilon_i}^{\infty} \varepsilon^{1/2} Q_i(\varepsilon) f(\varepsilon) d\varepsilon.$
66. Pitchford, L. C., S. V. O'Neil, and J. R. Rumble, *Phys. Rev. A*, 23, 294, 1981; www.Siglokinema.com.
67. Raju, G., unpublished, 2003.
68. Watts, M. P. and A. E. D. Heylen, *Int. J. Electron.*, 56, 235, 1984.
69. Brennan, M. J., A. M. Garvie, and L. J. Kelly, *Aust. J. Phys.*, 43, 27, 1990.
70. Dwarakanath K. and G. R. Govinda Raju, Ion thrusters for space electric propulsion, *Indian Space Research Organization Report*, 1978, pp. 1–25.

12 High Frequency Discharges

Radio frequency discharges are a special class of high frequency phenomena because of their immediate use in the microelectronics industry. During the 1940s vigorous efforts were devoted to the study of microwave discharges as part of the war effort. During the past twenty years or so RF discharges have received much attention.[1] Capacitively coupled RF plasma sources are used extensively in the microelectronics industry, for commercial production as well as laboratory research and development. Applications include thin film deposition for circuits, surface catalysis, surface polymerizations, and other forms of surface chemical reactions. Reactive ion-etching plasma reactors are used to etch circuit wafers. Surface studies in plasma chemistry require the production of RF discharges.

Uniformity and reproducibility of plasma etching and deposition processes are key requirements of the industry. However, despite widespread industrial use and significant research efforts, the physics of the process involved in an RF plasma discharge is not well understood yet. The advent of fast computers, associated with developments in numerical techniques, has helped in understanding the discharge process in the application of theoretical concepts to experimental findings. More experimental work in the public domain is needed to keep pace with computational efforts.

In this chapter, the basic discharge mechanism is described and observed characteristics of the discharge, such as the voltage profiles across the gap and charge distributions, are dealt with. Experimental efforts that have been directed towards investigating the physics of capacitively coupled RF discharges are dealt with. The author is constrained to refer to work available in the public domain. The book by Lieberman and Lichtenberg[1] gives a solid foundation to the theoretical concepts, and the review by Gupta and Raju[2] deals with numerical and analytic techniques for studying capacitive RF discharges. A number of interesting articles appear in a special issue of the Institute of Physics.[67]

12.1 BASIC PLASMA PHENOMENA

Let us first consider a parallel plane gap with an applied steady voltage, as shown in Figure 12.1. The gas number density is low and, following the establishment of a self-maintained current, the current continues to build up, with ionization taking place in the interelectrode region. As the positive ions drift towards the cathode the first generation is easily absorbed, but the subsequent generations are slowed down, much like baseball viewers admitted into the stadium through a gate. The buildup of ion population gives rise to a positive potential and a sheath is formed close to the absorbing electrode.

Langmuir[3] first introduced the term "sheath" for the space charge layer. Bohm[4] states that the "characteristic of a plasma is its tendency to maintain itself in a neutral and field-free state in spite of any forces or processes tending to change it." Note that the plasma region is field free and will have a potential V_p. Electrons falling on the edge of the sheath near the anode bounce back into the plasma. Within the sheath the ion density is approximately

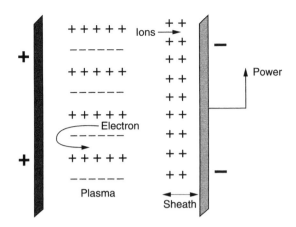

FIGURE 12.1 Formation of the sheath and motion of ions and electrons falling on the sheath edge. Here the powered electrode is the cathode and the plasma is at zero potential.

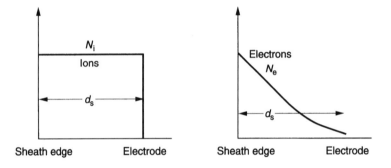

FIGURE 12.2 Ion and electron densities within the sheath. The ion density is constant while the electron density decreases exponentially. Sheath thickness is denoted by d_s.

constant, whereas the electron density falls off exponentially. The decrease of electron density N_e from the sheath edge to the cathode is, according to the Boltzmann law,

$$N_e = N_s \exp - \frac{V(x)}{T_e} \tag{12.1}$$

where N_s is the electron density at the sheath edge, T_e the electron kinetic temperature, and V is the potential at x; $V(0) = 0$ and $V(d_s) = -V_c$. Since the electron temperature is about $5\,\text{eV}$ and $V_c \approx 100\,\text{V}$, one sees that the density of electrons within the sheath falls off rapidly (Figure 12.2).

The plasma has the same number of ions and electrons and therefore there is only a small resultant space charge. The sheath has predominately ions. The region between the plasma and the sheath has therefore approximately the same number of charges, making it a region of semineutrality. The sheath does not have sharp edges as shown in Figure 12.1 because we are dealing with a large number of charge carriers with varying energies. This region between the plasma and the sheath is sometimes referred to as the presheath region. The sheath edge begins where the density of electrons falls to about 1% of the ion density.[5] By varying the potential applied to the cathode one can draw ion or electron current (positive potential is required in the latter case), both of which have a saturation value.

One often encounters plasma enclosed in a tube; the ordinary fluorescent tube is a good example. The electrons that bombard the wall give it a negative potential that is in the logarithmic ratio of electron to ion temperature. The electron temperature is about 1.5 eV and the ion temperature is the same as that of the neutrals (0.025 eV), rendering the wall temperature about seven times the electron kinetic temperature.[6]

The ions leaving the plasma and entering the sheath (see Figure 12.1) must have a critical speed. The condition stated by Bohm[4] is that the potential must be such as to accelerate the ions to a velocity corresponding to half the mean kinetic energy of the electrons. This criterion is mathematically expressed as

$$V_{\mathrm{p}} = \frac{T_{\mathrm{e}}}{2} \tag{12.2}$$

which, in terms of ion velocity, gives

$$v_{\mathrm{i}} \geq \sqrt{\frac{eT_{\mathrm{e}}}{M}} \tag{12.3}$$

This ion velocity, usually referred to as the Bohm velocity (v_{B}), is a function of electron energy and ion mass M, but not ion energy. Figure 12.3 shows the plasma potential and

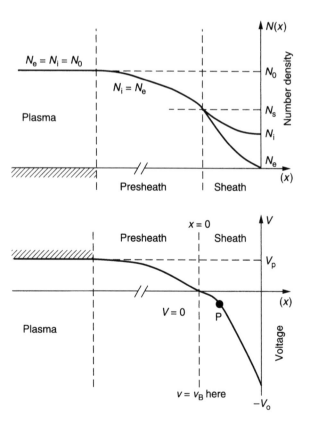

FIGURE 12.3 Potential and charge density in a plasma with cathode at $-V_{\mathrm{c}}$ volts. The ion density in the presheath region is equal to the electron density. The outer edge of the sheath is at zero potential and the velocity of ions here is v_{B}. Note the small decrease in N_{i} in the sheath region.

charge densities within the presheath and sheath regions. The plasma has a potential V_p which penetrates into the presheath region and then begins to fall off. At the outer edge of the sheath the potential and the electric field are zero and the ion velocity is given by v_B (Equation 12.3). The voltage drop within the sheath equals the voltage applied to the cathode. Note that the ion density within the sheath decreases because the ions are accelerated, but the current must remain the same.

The potential of the plasma, small as it is, is maintained nearly constant. Any potential drop that occurs is neutralized by a redistribution of electrons that restores the constant potential condition. However, potential changes that are not too large compared with the mean kinetic energy do exist because these are not strong enough to bring about a redistribution of electrons. As stated previously, the entire potential drop between the plasma and the electrode occurs in the sheath. The sheath thickness d_s is calculated using Child's law

$$d_s = \left[\frac{4}{9} \varepsilon_0 \left(\frac{2e}{M} \right)^{1/2} \frac{V_c^{3/2}}{J} \right]^{1/2} \quad (m) \tag{12.4}$$

where J is the current density (A/m^2), e the electronic charge (C), M the ion mass (kg), and ε_0 is the permittivity of free space (8.854×10^{-12} F/m). Equation 12.4 is derived by making the following assumptions: (1) the outer sheath edge is a sharp boundary and does not change with time; (2) the current is entirely due to ions; (3) the entire potential drop due to the applied voltage occurs across the sheath. The length of the presheath region is 20 to 50 times the thickness of the sheath. In direct voltage discharges the product Nd_s has been observed to be constant.[7]

It is instructive to calculate the sheath thickness for a parallel-plane discharge. For this type of discharge, voltages of the order of 250 V prevail and the sheath thickness for a current density of 500 A/m^2 is obtained as 1.6×10^{-4} m ($\sim$0.2 mm). The electric field within the sheath is $(250/1.6 \times 10^{-4}) \sim 1.5 \times 10^6$ V/m, which is quite high.

12.2 DEBYE LENGTH

As stated previously, any potential applied to a plasma results in a restoration of charge carrier distribution. If one attempts to "stir" the plasma by inserting a probe connected to the positive terminal of a battery, the probe attracts electrons and a layer of negative charge will be built around the probe. However, the layer cannot extend throughout the plasma region because the electrons in the layer repel other electrons approaching the probe. The situation is somewhat similar to passengers entering a suburban morning train in Bombay. Those who are outside push everyone in their way to get on the train, and, once in, push out anyone trying to get in. The length to which the effect of the probe extends, known as the Debye length or Debye shielding distance, is given by

$$\lambda_D = \left(\frac{\varepsilon_e \varepsilon_0}{N_e e} \right)^{1/2} = 7437 \left(\frac{\varepsilon_e}{N_e} \right)^{1/2} \quad (m) \tag{12.5}$$

where N_e is the electron density (m^{-3}) and ε_e is the electron energy (eV). The Debye length decreases as the electron density increases because the repulsive force is stronger. Further, the shielding distance becomes larger with increasing energy, due to increased thermal agitation. In the absence of thermal agitation the charge cloud would simply collapse into an infinitely thin layer.

If one substitutes $\varepsilon_e = 5\,\mathrm{eV}$ and $N_e = 10^{16}\,\mathrm{m}^{-3}$ in Equation 12.5, one gets a shielding distance of $0.17\,\mathrm{mm}$. The physical explanation of Debye length is easy to understand. If there are fluctuations of charge density within the plasma or potential changes in a region of the dimension of λ_D, the charge neutrality of the bulk of the plasma will not be affected and one uses the term "quasi-neutrality" for the plasma. The essential condition for quasi-neutrality is that the dimension of the plasma (L) should be much larger than the shielding distance, that is, $\lambda_D \ll L$. If this condition is satisfied, then $N_e = N_i = N_0$ and one speaks of plasma density, dispensing with the electron and ion densities. Debye shielding also occurs in beams of single charge entity, such as electron streams in magnetrons or a proton beam in a cyclotron.

One distinguishes a plasma from ionized gas by specifying three criteria:

1. The plasma dimensions are much larger than the Debye length.
2. Inside a sphere having the radius of Debye length the number of particles (N_D) is very large, $N_D \gg 1$.
3. An electron within the plasma oscillates many times at the plasma frequency before it makes a collision.

The last aspect is discussed in Section 12.4.

12.3 BOHM SHEATH MODEL

The Bohm criterion for a stable formation of the sheath is that the ions should have a velocity of at least half that corresponding to the mean kinetic energy of the electrons. The derivation of this condition is now considered.

The Poisson's equation in one dimension in the region $x > 0$ is given as

$$\nabla^2 V = \frac{d^2 V}{dx^2} = -\frac{e}{\varepsilon_0}[N_i(x) - N_e(x)] \tag{12.6}$$

where N_i and N_e are ion and electron densities as shown in Figure 12.3. The electron density at any point P with coordinate x is given by Equation 12.1

$$N_e(x) = N_s \exp\left[\frac{V(x)}{\varepsilon_e}\right] \tag{12.7}$$

where ε_e is the electron energy in eV. Note that at $x = 0$ (sheath edge) $V = 0$ and therefore $N_e = N_s$.

We now need an expression for N_i. This is obtained from the ions' kinetic energy translated into ion velocity, after noting that the ion energy at $x = 0$ is eV_P,

$$v_i = \left(\frac{2eV_P}{M}\right)^{1/2} \tag{12.8}$$

The ion current density at $x = 0$

$$J_i = N_s e v_i = N_s e \left[\frac{2e}{M} V_P\right]^{1/2} \tag{12.9}$$

The ion current density at P is

$$J_i(x) = N_i e \left[\frac{2e\{V_P - V(x)\}}{M} \right]^{1/2} \tag{12.10}$$

Since the current is the same at every point the ratio of currents is unity and therefore one gets

$$N_i = N_s \left[\frac{V_P}{V_P - V(x)} \right]^{1/2} \tag{12.11}$$

Equation 12.11 shows that the ion density decreases as x increases. We now substitute Equations 12.7 and 12.11 into 12.6 to obtain

$$\frac{d^2 V}{dx^2} = -\frac{e N_s}{\varepsilon_0} \left\{ \left[\frac{V_P}{V_P - V(x)} \right]^{1/2} - \exp\left[\frac{V(x)}{\varepsilon_e} \right] \right\} \tag{12.12}$$

This equation is known as the plasma sheath equation. After integrating and substituting the boundary conditions

$$\left. \begin{array}{l} V = 0 \quad \text{at } x = 0 \\ \dfrac{dV}{dx} = 0 \quad x = 0 \end{array} \right\} \tag{12.13}$$

one obtains the electric field as[4]

$$E^2 \approx \frac{e N_s [V(x)]^2}{\varepsilon_0} \left[\frac{1}{\varepsilon_e} - \frac{1}{2V_P} \right] \tag{12.14}$$

Obviously the solution exists only for the positive values of the term within the large square brackets or, in other words,

$$V_P \geq \frac{\varepsilon_e}{2} \tag{12.15}$$

The solutions are exponentials for positive values and oscillatory for negative values, conveying that a stable sheath cannot form. Substituting this condition, with the equality sign, into Equation 12.8, one obtains the Bohm velocity

$$v_B = \left(\frac{2eV_P}{M} \right)^{1/2} = \left(\frac{e\varepsilon_e}{M} \right)^{1/2} \tag{12.16}$$

For an average energy of $5\,eV$ for the H_2 ion we have $\varepsilon_e = 5V$, $M = 3.35 \times 10^{-27}\,Kg$, which gives $v_B = 1.5 \times 10^4\,m\,s^{-1}$.

12.4 PLASMA FREQUENCY

We have already stated that the plasma is a region of quasi-neutrality. Let us suppose that a group of electrons in a small region decide to run away with higher speed; the region that they leave will have a positive potential, attracting back the electrons. However, the electrons, because of inertia, overshoot their position and the displacement will now be in the opposite direction. The restoring force will also be in the opposite direction. The frequency with which the electrons oscillate about their mean position till conditions are restored is known as the plasma frequency. The frequency is so high that the ions, due to their heavier mass, do not respond to the oscillating field and remain relatively fixed in position. Without giving the proof, we note that the plasma frequency is given by

$$\omega_p = \left(\frac{N_0 e^2}{m\varepsilon_0}\right)^{1/2} = 56.4(N_0)^{1/2} \quad (\text{rad/s}) \tag{12.17}$$

The plasma frequency increases with plasma density and no other factor; for a plasma density of $10^{16}\,\text{m}^{-3}$ the plasma frequency is 5.6 GHz. If the mean time between collisions is τ, the condition $\omega_p \tau \gg 1$ must be satisfied for a discharge to be classified as plasma. The discharge then possesses a collective interaction with the electric field rather than particle interaction.

Let us consider a plasma upon which electromagnetic radiation of frequency ω_0 is allowed to be incident. The wave will be partly transmitted and partly reflected, depending upon the relative magnitudes of ω_0 and ω_p. The electrons in the plasma will interact with the wave and extract energy from the electric field associated with it. Due to collisions taking place within the plasma, the absorbed energy will attenuate the transmitter electromagnetic wave. If $\omega_0 > \omega_p$ the electromagnetic wave will be transmitted with little attenuation.

Speaking in terms of the plasma density, there is a critical number density of the plasma, N_c, which determines the highest frequency that will be reflected by the plasma. If $N_0 < N_c$ then the highest frequency reflected will be lower than ω_0. In concise form one can state that

$$\left.\begin{array}{lll} N_0 > N_c; & \omega_0 < \omega_p & \text{incident radiation essentially reflected} \\ N_0 < N_c; & \omega_0 > \omega_p & \text{incident radiation largely transmitted} \end{array}\right\} \tag{12.18}$$

At the critical density the plasma frequency will be equal to the incident radiation frequency. This condition will give the relationship between N_c and ω_0 as, from Equation 12.17,

$$N_c = \frac{\omega_0^2 m\varepsilon_0}{e^2} = \frac{4\pi^2 f^2 m\varepsilon_0}{e^2} \quad (\text{m}^{-3}) \tag{12.19}$$

Substituting values for the constants one derives

$$N_c = 1.24 \times 10^{-2} f^2 \quad (\text{m}^{-3})$$

The relation between N_c and f is important from the communications engineering point of view. Reflection of electromagnetic radiation from the ionosphere has received attention from the days of radio transmission round the curved surface of the Earth. The lower part of the ionosphere, called the D-region, has low conductivity and reflects low frequency radio waves. The next higher layer, the E-layer, reflects medium frequency waves. The top F-layer, which has the highest concentration of free electrons, is the most useful for radio transmission. However, this layer does not reflect in the region of 10^8 to 10^{10} Hz (TV signals broadcast at 300 MHz to 30 GHz escape into outer space), which makes it necessary to employ launched satellites (apart from relay stations or coaxial cable) for reflection and retransmission of TV signals.

12.5   PLASMA CONDUCTIVITY

The conductivity of plasma in terms of basic parameters is

$$\sigma = \frac{e^2 N_e}{m\nu} \quad (\text{S/m}) \tag{12.20}$$

where ν is the electron collision frequency (s^{-1}). For $N_e = 2 \times 10^{18}\,\text{m}^{-3}$, $\nu = 3 \times 10^7\,\text{s}^{-1}$, the conductivity is obtained as $1.9 \times 10^3\,\text{S/m}$, which may be compared with the conductivity of copper at room temperature, $10^8\,\text{S/m}$.

12.6   AMBIPOLAR DIFFUSION

In an ionized gas with electron density lower than that in a plasma, the diffusion coefficient of ions (D_i) is much smaller than that of electrons (D_e). However, in a plasma, let us suppose that a bunch of electrons, due to their higher thermal energy, leave the plasma first. The region which they had occupied now has excess positive charge. One can visualize that the positive ions are attracted by the long-range Coulomb force and follow the electrons at the same rate as the electrons. The diffusion that occurs now is called ambipolar diffusion and is expressed by

$$D_a = \frac{\mu_i D_e + \mu_e D_i}{\mu_e + \mu_i} \tag{12.21}$$

Since $\mu_e \gg \mu_i$, Equation 12.21 may be simplified to

$$D_a = D_i + \frac{\mu_i}{\mu_e} D_e \tag{12.22}$$

By substituting the relationships

$$\left. \begin{array}{l} \dfrac{D_e}{\mu_e} = \dfrac{kT_e}{e} \\[2mm] \dfrac{D_i}{\mu_i} = \dfrac{kT_i}{e} \end{array} \right\}$$

one gets

$$D_a = D_i + \frac{T_e}{T_i} \tag{12.23}$$

If the temperatures of both species are the same, that is $T_e = T_i$, then

$$D_a = 2D_i \tag{12.24}$$

The diffusion of ions is increased by a factor of two but the diffusion rate of both species is controlled by the ions.

12.7   RF PLASMA

A major distinction between a direct voltage plasma and RF plasma is that the power transfer in the latter occurs through displacement currents whereas in the dc case power transfer occurs through current flowing through the electrodes.[6] The RF plasma is coupled to the power source either inductively (oscillating magnetic field), capacitively (oscillating electric field), or at microwave frequencies (both fields). Inductively coupled plasma does not have electrodes in the traditional sense, the power being fed through a number of turns outside a tube within which the plasma develops. We confine ourselves to capacitively coupled RF discharges where the electrodes employed are plane parallel in configuration. Power is absorbed by the plasma and converted into themal motion of electrons. Lieberman and Lichtenberg[1] have written an excellent book on RF discharges, and Gupta and Raju[2] have published a review of numerical and analytic techniques for studying RF discharges.

The electrons are the main current carriers as the ions do not respond, due to their inertia, to frequencies above about 1 MHz. The plasma will have high conductivity and the ion sheaths act as fairly small capacitors. The dissipation of RF power will cause fairly large alternating voltages to appear across the sheaths. To keep the electron and ion fluxes equal the dc potential drop increases till it is approximately equal to the RF peak voltage.[8]

Various arrangements of RF power supply and electrode connections are used in industrial plasma processing applications, depending on the actual plasma process and the application. Some common reactor configurations are for plasma etching, reactive ion etching, and symmetric triode etching. Electrode arrangements can be plane parallel, barrel, hexagonal, or ring-coupled cylindrical. In order to introduce some standardization in this area, the American Physical Society's Gaseous Electronics Conference proposed the Gaseous Electronics Conference (GEC) reference cell.[9] Most parallel plate industrial plasma processing reactors are similar to the GEC reference cell but without its standardization as far as the electrode dimensions are concerned. The frequency adopted is mostly 13.56 MHz.

The plasma reactor, in its simplest concept, consists of two plane parallel electrodes with an RF field imposed (Figure 12.4). An equivalent circuit, although variations are possible,[7] is also shown.[12] The plasma is highly conducting and its conductivity corresponds to the resistance R. The plasma sheath, with its low concentration of ions, acts as a capacitor C_s for the RF current. The diodes are included to show that current can flow if the plasma potential becomes negative with respect to the electrodes. This ensures that the lowest potential of the sheath will be zero.

The choice of the RF field depends principally on its ability to bombard insulating surfaces and its efficiency in promoting ionization and sustaining the discharge. The RF discharge can have a steady-state bias above or below ground potential, depending on whether one of the electrodes is grounded or negatively biased. Most capacitive

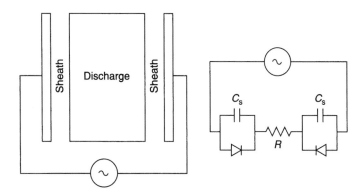

FIGURE 12.4 The RF discharge and an equivalent circuit.

RF plasma reactors in the industry operate without an externally applied magnetic field, though a perpendicular (to **E**) magnetic field has been investigated.[10]

In the parallel plane reactor high energy ions bombard the substrate and often cause considerable damage. The disadvantage may be alleviated by reducing the ion energy without reducing the ion flux. This may be accomplished by placing a magnetic field perpendicular to the electric field, since the effect is to reduce the transverse drift velocity and at the same time increase the ionization efficiency (Chapter 11). The magnetic flux density employed by Bletzinger[10] was approximately 0.06 T. Gases used were argon, CF_4, and SF_6, with number densities ranging from 6×10^{20} to 3×10^{22} m^{-3}.

The plasma is a collection of free charged particles moving in random directions such that on the whole it forms an electrically neutral mass. In a weakly ionized gas the number of electrons generated is of the order of 10^8 and the degree of ionization is typically 10^{-4}. In an electrically driven weakly ionized plasma, electrons oscillate among the heavier, relatively immobile, positive and negative ions. The mobile electrons oscillate in the field, within the space charge cloud of the ions, while the massive ions respond only to time-averaged electric fields. Ions flowing out of the bulk plasma and moving toward the substrate are accelerated by high fields in the sheaths, to high energies and form energetic ion-enhanced products. A typical ion-bombarding energy may be as high as 50% of the applied voltage.

The principal heating mechanisms in a plasma discharge are ohmic heating in the plasma bulk and stochastic heating in the sheaths. The plasma is sustained by the external energy source, which transfers energy to the electrons and indirectly to the ions by various mechanisms. Ohmic heating is caused by the dissipation, through electronic collision processes, of energy gained by the electrons in the electric fields of the plasma.

Stochastic heating is the term used to describe the heating brought about by the net average energy gain over an oscillation period, resulting from electrons impinging on the oscillating sheath edge and being reflected back. Electrons impinging on the sheath edge suffer a change of velocity on being reflected back into the plasma. When the oscillating sheath moves into the bulk plasma, the reflected electrons gain and then lose energy when the sheath moves away. There is, however, a net gain in energy which manifests itself as stochastic heating.

12.7.1 Experimental Studies

A typical plasma reactor consists of a vacuum chamber in which two parallel plane electrodes are mounted (Figure 8.5). In the GEC reference cell referred to earlier,[9] the electrodes are 10 cm in diameter with a spacing of 2.5 cm. One of the electrodes is grounded and the other is connected through a blocking capacitor to an RF generator, commonly 13.56 MHz. There is no net dc current across the system. At a sufficiently high RF voltage a discharge strikes between the electrodes. The luminous part of the discharge is an ionized gas containing electrons which are at a higher energy than the ions. Of course, the ionized column contains excited species and, in the case of attaching gases, various negative ions, depending upon the gas used. On each side, at the electrodes, sheaths depleted of electrons are formed as described above. The sheath at the cathode is a little larger than that at the anode, and the sheaths oscillate as a result of polarity reversal. A computer visual discharge is presented by Pitchford et al.[11]

van Roosmalen et al.[8] have studied the electrical properties of planar RF discharges in O_2 at 7 to 53 Pa gas pressure and measured the sheath thickness as approximately 10 mm at the cathode and ~3 mm at the anode. Thompson and Sawin[12] have studied poly-silicon etching in SF_6 RF discharges and Allen et al.[13] have extended the study of plasma etching of polysilicon to discharges with CF_3Cl + argon mixtures. Thompson et al.[14] have measured the bombardment energy of ions in highly electron attaching gases, SF_6, $CClF_3$ (chloro-trifluoro methane or Freon-13), and CF_3Br (bromo-trifluoro methane), obtaining ion energies as high as 60 eV at an RF power of 80 W. The average ion energy is expressed as

$$\bar{\varepsilon}(eV) \approx 3.5 \times 10^{29} \times \frac{i_0(A)}{\omega(s^{-1})A(m^2)N(m^{-3})} \tag{12.25}$$

where i_0 is the RF current through the plasma, A the area of the electrodes, and N the gas number density. Substituting $i_0 = 1A$, $A = 125\,cm^2$, $\omega = 2\pi \times 13.5 \times 10^6$, $N = 0.65 \times 10^{22}\,m^{-3}$, the mean energy is obtained as 50 eV, which compares well with the measured mean energy.

The influence of addition of electron attaching gases on the impedance of the discharge has been studied by Bletzinger;[10] added attachers were CF_4, C_2F_6, and SF_6. A large increase in the discharge impedance in the high pressure region was observed. Power deposition occurs by a volume process at high pressures and an electrode process through secondary electron emission.

The characteristics of RF discharge in mixtures of $Ar + SF_6$ are shown in Table 12.1. Additional studies have been reported by Butterbaugh et al.,[41a] Beneking,[41b] Godyak et al.,[37] Miller et al.,[41c] Sobolewski,[41d] and Kroesen and de Hoog.[15] A systematic study of the influence of the various parameters on the discharge is, however, incomplete.

A survey of *in situ* diagnostics of elementary processes has been made by Kroesen and de Hoog.[15] A reliable technique for determining electron density is microwave spectroscopy, or, more specifically, cavity resonance detuning, which is especially suitable for the density ranges encountered in RF plasmas.[16] The electron density in argon, and also in electronegative gases such as Freon, has been found to be 1.0 to $10 \times 10^{15}\,m^{-3}$.

Langmuir probe diagnostics are commonly used to measure the electron energy spectrum and various other plasma parameters over a wide range of experimental conditions.[17] Langmuir probe characteristics may be employed to obtain the electron energy distribution function and electron density values, but the procedure for quantitative interpretation is quite involved. Further, the method is specific to discharge geometry and chemistry.

The values for SiH_4 discharges at 13.56 MHz have been reported by Böhm and Perrin,[18] who employed an electrostatic method of using grids to select ions and spatially resolved optical emission spectroscopy. Two different discharge modes were observed, one at high

---

TABLE 12.1
Characteristics of RF Discharge at 14.1 MHz in Mixtures of Ar and SF_6[10]

N (10^{22} m^{-3})	% SF$_6$	Power Deposition (mW/cm^3)	Voltage Drop (V, peak)	d_s (cm)	Peak Electric Field (V/cm) Sheath	Peak Electric Field (V/cm) Volume
0.48	0	1.5–3.7	40	0.76	52	—
0.48	3	3.2–8.0	46	0.18	125	—
0.48	9	4.2–10.5	49	0.11	219	—
0.48	23	4.9–12.3	46	0.076	303	—
3.22	0	6.2	29	0.39	66	4
3.22	3	12.3	18.4	0.12	76	18
3.22	9	12.3	29.4	0.072	204	23
3.22	23	18.5	45.5	0.04	569	27

The voltage drop shown is per electrode; the electric field of the sheath is averaged over the width.

pressure and low voltages, the other at low pressure and high voltages. The mechanisms for the two different modes are based on the mechanism of energy gain by the electrons, from the sheaths and in the plasma.

Information regarding the electron energy distribution may be obtained by time-resolved emission spectroscopy of the atomic lines.[19] Optical emission spectroscopy also can serve as a probe for the estimation of density of neutral species, such as radicals, in the discharge. Infrared (IR) absorption is another tool for quantitative measurements. Apart from density measurements, IR absorption spectroscopy, performed with either high-resolution Fourier-transform spectrometers or tunable diode lasers, can also yield information about rotational and vibrational excitation.[20,21]

The neutral gas temperature effect may be obtained equally effectively by using the Doppler effect in combination with emission spectroscopy, laser-induced fluorescence,[22] or IR absorption spectroscopy. At low pressures the gas temperature is seen to be close to room temperature, as expected. Temperature of the neutrals is subject to two opposing concerns. While higher temperatures ($\sim0.2\,eV$) increase the etching rate they also raise the temperature of the substrate, which is undesirable. Neutral temperatures as high as $0.5\,eV$ have been measured. Electron temperature measurements have been made in various gas mixtures of CF_4 and N_2 at a total pressure of $66\,Pa$.[23]

The positive ion energy distribution at the surface is measured by using a method of retarding grids.[24] An alternative principle, based on energy selection of the ions depending on their trajectories in an alternating field, has been explored.[25] Negative ion density may be measured by mass spectrometric methods in afterglow conditions, but these do not reveal the true plasma composition during discharge.[26] Alternative strategies, using optogalvanic methods[27] or microwave schemes[28] to detect electrons set free from negative ions by the impact of photons, have been investigated. The presence of negative ions is limited to the bulk plasma.

Experimental investigations of an RF plasma discharge in oxygen have been performed using spontaneous luminiscence spectroscopy, a microwave resonance technique using photo detachment,[29] and a laser photo-detachment technique.[30] Nonintrusive diagnostics of electric field in the plasma include laser-induced fluorescence spectroscopy,[31] and a study of optogalvanic signals induced by tunable dye-laser irradiation,[32] as well as plain emission spectroscopy. Laser-induced fluorescence spectroscopy has been used increasingly to study the sheath structure in RF discharges in helium, argon, and hydrogen.[33]

The energies of the bombarding ions have been measured by electrostatic deflection analyzers,[34] cylindrical mirror analyzers,[35] and retarding grid analyzers.[36] Comparison of the relative abundance of different ionic species in the same discharge was made possible in some of these investigations by mass-resolved measurements using quadrupole mass spectrometers.

Electrostatic probe diagnostics have been used effectively in measuring electron energy distribution in low pressure plasmas. However, conventional probe theories for electron and ion currents assume a Maxwellian energy distribution function, and the non-Maxwellian effects are usually ignored on the grounds that a departure of the actual energy distribution from the Maxwellian affects only a small number of electrons. But, in reality, the energy distribution in low pressure discharges is often non-Maxwellian, even in the low energy ranges, and the application of conventional probe techniques may lead to appreciable errors.

Godyak et al.[37] compared the results of applying various probe diagnostic techniques to non-Maxwellian plasmas. Measurements were made in a $13.56\,MHz$ capacitive RF discharge in argon. At low pressures, typically $4\,Pa$ where stochastic electron heating dominates, the energy distribution function is observed to be bi-Maxwellian, indicating cold electron group having a mean temperature of 0.5 eV and a density of $4.2 \times 10^{15} \, \text{m}^{-3}$ together with a hot group having a mean temperature of 3.4 eV and a lower density of $2.4 \times 10^{14} \, \text{m}^{-3}$. On the contrary, at the higher pressure of 40 Pa the energy distribution becomes the Druyvesteyn type,[38] with a mean temperature of 3.4 eV and a density of $2.9 \times 10^{15} \, \text{m}^{-3}$.

Further, the plasma parameters obtained from the energy distribution measurements were compared to those obtained by using conventional procedures assuming a Maxwellian plasma. The methods considered were the Druyvesteyn procedure involving the differentiation of the probe characteristic, and the classical Langmuir procedure applied to the electrode retardation region of the probe characteristics. It was observed that the application of these methods to the probe characteristics obtained in non-equilibrium plasmas may lead to significant error by a factor of 2 to 5. The degree of departure of the energy distribution from the Maxwellian in a particular discharge, and the resulting errors, are not predictable.

Circuit models, in addition to relating electrical measurements to important plasma properties, are also capable of describing the manner in which the geometry of the discharge cell and the electric properties of the external circuitry affect the plasma properties. Models of electrical behavior of RF symmetrical discharges have been developed[39,40] and tested experimentally.[37]

Current and voltage measurements at several points in the discharge have been utilized to determine a circuit model of argon discharges in the asymmetric GEC cell.[41] Electron densities in a nearly identical GEC cell for argon discharges at gas pressures of 13.3 to 133 Pa were measured using microwave interferometry. Scaling laws for the electron density data obtained therein, and the electron temperature reported for discharges in the given pressure range, were utilized by Sobolewski[41] to calculate plasma parameters such as the Debye length and its power-law dependence on current. The spatially dependent ion and electron concentrations of argon discharges in a GEC cell were observed to reach a saddle point maximum in the center of the glow.[42]

Application aspects of plasma-assisted vapor deposition techniques have been reviewed by Schneider et al.[43] Electron-neutral collision data constitute an indispensable part of the data base of chemical and physical properties that are essential to produce a reliable prototype of the discharge model. Huo and Kim[44] and earlier chapters in this volume have attempted to provide such a data base.

12.8 POWER ABSORBED

For practical applications, power absorbed is one of the most important quantities. The complicated processes within the discharge and at the electrodes make it difficult to quantify the power absorbed. As explained already, when an RF potential is applied to a capacitively coupled discharge a large fraction of the voltage can appear across the sheath regions near the electrodes. A smaller fraction of the applied voltage appears across the bulk of the plasma as a longitudinal electric field. The fraction of the voltage that appears across the sheaths depends, of course, upon several operating parameters such as gas pressure, electrode distance, and the frequency. Since the negative ions reside mainly in the bulk, the fraction of voltage appearing across the sheaths decreases as the ratio of the negative ions to electrons increases.[45] In addition, when the electrodes are of unequal size one of the electrodes can develop a large dc bias. Thus the division of the applied voltage depends on (1) voltage across the sheaths, (2) longitudinal electric field in the bulk plasma, and (3) development of the dc bias.

Power dissipation consists of several mechanisms of electron and ion acceleration, classified as follows:[46]

1. Acceleration of plasma electrons by the bulk RF field, considered as Joule heating.
2. Acceleration of positive ions through the sheath, drifting from the plasma, and of secondary electrons released from the cathode.
3. The plasma sheath boundary is not sharp (Figure 12.3) and the plasma jets into the sheath for a short distance, resulting in the acceleration of plasma electrons by the sheath field. This process is called "wave riding" in the literature.[18]
4. Electrons from the plasma are reflected from the sheath during the half cycle that the powered electrode is anodic (Figure 12.1).

van Roosmalen et al.[8] have measured the power absorbed in oxygen in a commercial plasma reactor made of aluminum and operating at 13.56 MHz. Measurements were carried out at 7 to 53 Pa and with 50 to 800 W of RF power. The dc potential developed between the powered electrode and ground was measured with a probe and an oscilloscope. Dark space thickness was measured visually by a traveling microscope.

The power absorbed in an RF discharge is the sum of the power in three regions: the glow region, the sheath–glow boundary, and the sheath itself. A summary is provided below.

12.8.1 GLOW REGION

The glow is usually represented as an equivalent series or parallel resistance. The glow resistivity ρ is given by

$$\rho = \frac{1}{eN_e\mu} \quad (\Omega\,\text{m}) \tag{12.26}$$

where μ is the mobility of the electrons and N_e the density. Considering the glow as a cylinder of cross section A (m^2) and length L (m), the resistance of the glow column is

$$R = \frac{\rho L}{A} \quad (\Omega) \tag{12.27}$$

In a typical discharge $N_e = 10^{14}\,\text{m}^{-3}$, $\mu = 10^3\,\text{V}^{-1}\text{m}^2\text{s}^{-1}$, $L = 0.1\,\text{m}$, and $A = 0.25\,\text{m}^2$. The power calculated using the above two equations for a voltage drop of 50 V along the glow column is 10 W. Since the power absorbed in a typical discharge is 200 to 400 W it is clear that the loss of power in the glow column is only a small fraction of the total. This is also evident from the fact that the voltage drop along the glow column is relatively small.

12.8.2 SHEATH–GLOW BOUNDARY

To calculate the power absorbed by the sheath–glow boundary one needs to calculate the potential there. This is accomplished by using Poisson's equation. The glow–sheath boundary has been analyzed by van Roosmalen et al.[8] and the reasoning is easy to follow, as under.

The electrons approaching the sheath are reflected back and therefore the glow–sheath boundary may be considered as an electron-reflecting electrode during most of the RF cycle. For simplicity, let us assume that the positive ions are distributed uniformly between the electrodes and are relatively immobile. Further, let us assume that the sheath is perfectly reflecting, that is, there are no electrons in the sheath. Under these conditions, the ion charge density in the sheaths is constant, equal to the charge density of electrons in the glow, assuming that the ion charge density in the glow is instantly neutralized by the thermal electrons. The one-dimensional Poisson's equation is

$$\frac{d^2V}{dx^2} = e\frac{N_e}{\varepsilon} \tag{12.28}$$

The instantaneous wall-to-plasma potential is the sum of the dc bias and the alternating voltage,

$$V = V_{dc}(1 + \sin \omega t) \tag{12.29}$$

This potential causes the glow–sheath boundary to oscillate. Integrating Equation 12.28 twice, using Equation 12.29 as a boundary condition, and differentiating x with respect to t gives the instantaneous velocity of the boundary. The root-mean-square value of the velocity over one RF cycle is

$$\bar{v} = \sqrt{\frac{\varepsilon \omega^2 V_{dc}}{2eN_e}} \quad (m/s) \tag{12.30}$$

The electron current in the glow is

$$J_G = 0.25 e N_e v_{th} \quad (A) \tag{12.31}$$

Combining Equations 12.30 and 12.31, Roosmalen et al.[8] derive the power transferred to the glow electrons as

$$P = \frac{J_G}{eA}(2m\bar{v}^2) \quad (W) \tag{12.32}$$

To show the method of application of Equation 12.32 for calculating the power, we first calculate the mean thermal velocity using Equation 1.25:

$$v_{th} = \left(\frac{8kT}{\pi m}\right)^{1/2} \quad (m/s) \tag{1.25}$$

Following Cantin and Gagne,[47] we substitute $T = 15{,}000$ K and $N_e = 5 \times 10^{13}\,m^3$ to get $v_{th} = 7.6 \times 10^5\,ms^{-1}$ and $J_G = 1.5\,A$. The mean-square velocity calculated from Equation 12.30 for the same electron density is 1.6×10^4 m/s. Substituting these values, the power calculated from Equation 12.32 works out to be 40 W, which is about 10% of the actual power absorbed.

12.8.3 SHEATH

The third component of the power absorbed is within the sheath. The positive ion current density in the sheath is given by the Bohm criterion as[48]

$$J_i = 0.4 e N_e v_{th}\left(\frac{m}{M_i}\right)^{1/2} \quad (A/m^2) \tag{12.33}$$

and the corresponding power is

$$P = J_i A V_{dc} \quad \text{(W)} \tag{12.34}$$

where J_i is the current density. Substituting the values, one finds that the calculated power is too low to account for the experimentally observed values.

12.8.4 DISCUSSION

Lieberman and Lichtenberg[1] have developed a theory in power absorbed in nonattaching gases by assuming that the density of charge carriers is nearly constant in the center of the plasma and varies steeply near the edges. At high pressures the density tapers down sharply from the center towards the electrode, with a cosine distribution. An effective gap length, d_{eff}, is defined in terms of the gap length and electron density. The power absorbed is taken as the sum of ohmic heating and stochastic processes. This theory was applied to electronegative oxygen by Raju[49] and the calculated power agreed very well with the measured power. However, it was pointed out by Gupta[50] that the RF frequency was inadvertently taken as too high and necessitated a revised approach.

The influence of a two-temperature electron energy distribution was examined by Gupta and Raju,[51] using the p-set given by Equation 6.33. The energy distribution enters the calculation of power through the rate coefficient, which is defined as

$$K_j = \int_0^\infty F(\varepsilon) \left(\frac{2e}{m} \right)^{0.5} Q_j(\varepsilon) d\varepsilon \tag{12.35}$$

where Q_j is the energy-dependent cross section for a particular reaction j and $F(\varepsilon)$ is the energy distribution function. The energy lost per ion–electron pair created by collision (ε_c) is given by

$$\varepsilon_c = \sum K_j \varepsilon_j = \varepsilon_{ion} + 3 \frac{m}{M} \frac{T_e K_{el}}{K_{ion}} + \varepsilon_{exc} \frac{K_{exc}}{K_{ion}} \tag{12.36}$$

where ε_{ion} and ε_{exc} are the energies lost per ionization and excitation collision respectively. The second term on the right-hand side of the above equation corresponds to the energy lost for polarization scattering in electrons. Vibrational excitation and attachment losses are neglected because of the high value of E/N that exists. Mateev and Zhelyazkov[52] have suggested that the loss to walls of the container is $6T_e$. This loss was also neglected since the discharge was assumed to be confined to the electrode area.

The basis for a bimodal energy distribution is that the electrons are accelerated to high energies by collisions at the sheath and this can produce a species of electrons that have a higher energy than those in the bulk plasma.[53,54] A two-temperature distribution takes the form

$$G(\varepsilon) = [(1 - \gamma)F_c(\varepsilon) + \gamma F_w(\varepsilon)] \tag{12.37}$$

where the subscripts c and w imply cold and warmer components of the plasma. Figure 12.5 shows such a distribution, having two groups with different mean energies and several ratios.

A two-temperature distribution can maintain a discharge at a much lower bulk temperature due to ionization and excitation produced by the warmer component. The ratio of the two temperatures was taken as 7.0 and, replacing $F(\varepsilon)$ with $G(\varepsilon)$, the rates (Equation 12.35) were recalculated numerically and the power absorbed computed.

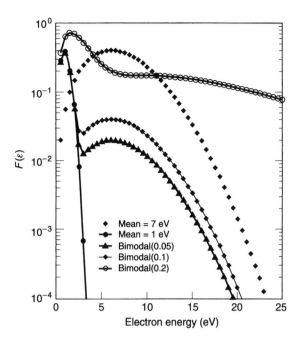

FIGURE 12.5 Bimodal energy distribution of electrons having mean energies of 7.0 and 1.0 eV for various values of γ in Equation 12.37. ($\blacklozenge$) 7 eV, Maxwellian; (—$\bullet$—) 1 eV, Maxwellian; (—$\blacktriangle$—) bimodal, $\gamma = 0.05$; (—$\blacklozenge$—) $\gamma = 0.1$; (—$\circ$—) $\gamma = 0.2$.

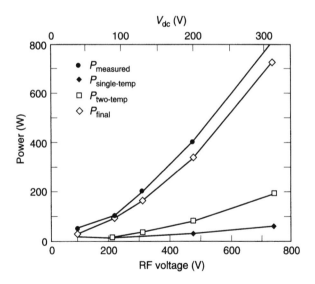

FIGURE 12.6 Power absorbed in RF discharge in molecular oxygen as a function of RF voltage. The measured values are due to van Roosmalen et al.[8]

Figure 12.6 shows the results of computations, in reasonable agreement with the measured values. In oxygen, basically, there are three electronegative regimes, depending on the neutral pressure and applied power. At low pressure the negative density, even at the center of the plasma, is quite small, and the presheath region behaves essentially electropositively. At high pressure and low power, a significant density of negative ions may exist, giving no electropositive sheath edge region. However, for a large intermediate region the electronegative ion density at the center may be quite large, but there will exist a significant edge presheath region in which the plasma is essentially electropositive. This effectively means that it is possible to accept the Bohm sheath criterion in the case of oxygen discharges in this regime. The conditions that exist in other gases must be carefully evaluated before the power absorbed in each gas is computed; the general framework has thus been outlined.

12.9 MICROWAVE BREAKDOWN

Extensive researches were carried out in the 1940s on microwave breakdown in view of its application in defense and industrial applications of technologically advanced products. An excellent volume has been written by MacDonald,[55] with additional resources being available in Hirsch and Oskam[56] and Roth.[6] We provide a brief description for the sake of completeness. The concepts necessary to understand microwave breakdown have already been dealt with.

The breakdown phenomenon is a function of the applied voltage and its frequency provided all other parameters such as the type of gas, gap length, etc. remain constant. One can broadly delineate three frequency regimes, in the order of increasing frequency to distinguish the mechanisms. In the first region the frequency is so low that the processes that occur are similar to those that occur at steady voltage, provided one remembers that the peak of the alternating voltage determines the collision processes. The upper value of frequency in the first regime is determined by the time taken by the ions (slower species) to reach the electrodes.

Typically, if one has a gap length of 1 cm in nitrogen, the reduced mobility of ions is $1.4 \times 10^{-4} \, m^2 V^{-1} s^{-1}$ (Table 1.8), resulting in a drift velocity of $3 \times 10^3 \, ms^{-1}$. The frequency corresponding to this drift time is 100 kHz, meaning that the peak of the alternating voltage at $f > 100$ kHz, at breakdown, is lower than the steady-state breakdown voltage. This condition defines the maximum frequency for no change in breakdown voltage as

$$f_{max} > \frac{\mu_i E}{\pi d} \qquad (12.38)$$

where μ_i is the ion mobility and E the peak ac field.

Though the ions oscillate in the discharge region, the electrons reach the electrode and secondary avalanches build up, leading to breakdown. If, however, one substitutes the electron mobility in Equation 12.38, one obtains the maximum frequency above which the electrons too oscillate within the discharge region. Breakdown occurs by increased ionization in the gas, though the electrons are confined, and this type of breakdown is generally called "diffusion controlled" in contrast with "mobility controlled" breakdown at lower frequencies. The latter involves loss of electrons primarily by their mobility. The loss of electrons in the diffusion-controlled mechanism is entirely due to diffusion and the breakdown condition is that the rate of ionization must be equal to the rate lost by diffusion.

To distinguish clearly between ion drift, and electron drift, the frequency corresponding to the latter is called the "cutoff" frequency (f_c). Thus

$$f_c = \frac{\mu_e E}{\pi d} \qquad (12.39)$$

where μ_e is the electron mobility. For the condition $f_{max} < f < f_c$ the breakdown voltage will be somewhat lower than the dc value (Figure 10.10). Paschen's law is found to hold good for frequencies up to ~ 1 MHz.

As the frequency is increased beyond about ~10 MHz the oscillations of the electrons result in a neutralization of the space charge built up and one has to apply a higher voltage to achieve breakdown. Thus the breakdown voltage plotted as a function of frequency has the shape of a "U" with the minimum occurring at ~1 to 10 MHz. The right limb after the minimum will be higher than the left limb, meaning that the breakdown voltage is higher than even the steady-state voltage. At still higher frequencies of the alternating voltage, or at low N, the electron–molecule collision frequency enters the picture. If $f > \nu_M$ the electron oscillates as though in a box with its sides less than the free path, and no ionization can occur. This condition is referred to as the "collisionless limit." Under these conditions one observes a breakdown voltage higher than the steady-state voltage, as stated previously.

To account for the frequency of the applied voltage, an effective electric field is defined according to

$$E_{\text{eff}}^2 = \frac{\nu_c^2}{\nu_c^2 + \omega^2} \hat{E}^2 \tag{12.40}$$

where ν_c is the electron–molecule collision frequency (s^{-1}), ω the angular frequency of the applied voltage (s^{-1}), and $\hat{E}$ denotes the rms value of the electric field. Compare Equation 12.40 with 11.30 to realize the similarity, in concept, of the effective electric field. High frequency radiation has a wavelength that is dependent on the medium and the free-space wavelength is denoted by λ (m); therefore the mean free path is denoted by L (m) to avoid confusion in the present context.[6] The free-space wavelength accordingly is

$$\lambda = \frac{2\pi c}{\omega} = \frac{c}{\nu} \tag{12.41}$$

A parameter called the diffusion length (Λ, m) is defined to denote the characteristic length over which the particle must diffuse in order to be lost from the discharge. A relatively simple expression for the diffusion length is obtained as follows.

The starting point is the equation for the loss of flux density in terms of the diffusion coefficient, Equation 1.75,

$$J = -D\nabla N_e \tag{12.42}$$

The time-rate of change of electron density due to diffusion is

$$\frac{\partial N_e}{\partial t} = -\mathbf{V}\cdot\mathbf{J} = \nabla^2 D N_e \tag{12.43}$$

The rate of increase of density ($m^{-3} s^{-1}$) due to ionization is

$$\frac{\partial N_e}{\partial t} = \nu_i N_e \tag{12.44}$$

where ν_i is the ionization rate. The breakdown criterion is set as

$$\nu_i N_e + \nabla^2 D N_e = 0 \tag{12.45}$$

Recalling that the diffusion coefficient is assumed to be constant, and treating Equation 12.45 as one dimensional (due to uniform field), one derives

$$\nu_i N_e + D \frac{\partial^2 N_e}{\partial x^2} = 0 \tag{12.46}$$

The ionizing collision frequency is the product of the ionization coefficient (α) and the drift velocity of the electrons (W) and therefore

$$\nu_i = \alpha \mu_e E \tag{12.47}$$

or

$$\nu_i = \eta \mu_e E^2 \tag{12.48}$$

where $\eta \ (= \alpha/E, \ V^{-1})$ is the ionization efficiency. At high frequencies an ionization efficiency due to diffusion is defined in a way analogous to Equation 12.48 as

$$\hat{\eta} = \frac{\nu_i}{DE^2} \tag{12.49}$$

It is convenient to express this equation as

$$E = \sqrt{\frac{\nu_i}{D} \frac{1}{\hat{\eta}}} \tag{12.50}$$

To solve Equation 12.46, the electron density is assumed to vary sinusoidally, having a maximum at the mid gap and zero at either electrode (that is, $x/d = \frac{1}{2}$ or 0), according to

$$N_e = N_m \sin \frac{\pi x}{d} \tag{12.51}$$

The solution is found as

$$E_b = \left(\frac{\pi}{d}\right) \sqrt{\frac{1}{\hat{\eta}}} \tag{12.52}$$

where E_b is the breakdown field. Comparing Equations 12.50 and 12.52, one gets

$$\frac{\pi}{d} = \frac{1}{\Lambda} = \sqrt{\frac{\nu_i}{D}} \tag{12.53}$$

A characteristic diffusion length is defined as

$$\Lambda = \left(\frac{D}{\nu_i}\right)^{1/2} \tag{12.54}$$

This is the microwave equivalent to Townsend's breakdown criterion under steady-state voltages. From the breakdown voltage measurements the quantity $\hat{\eta}$ may be determined in order to appreciate the influence of the microwave frequency on the ionization coefficient. The rms breakdown electric field is also expressed as

$$E_b \approx \frac{C_2}{\Lambda N} \tag{12.55}$$

in which C_2 is a constant. This equation shows that the breakdown field is inversely proportional to the gas number density in the long free path (low N) collisionless regime. The field also decreases with increasing characteristic diffusion length. At higher pressures the rms breakdown field is proportional to

$$E_b = C_1 N \tag{12.56}$$

We give an example of variation of the breakdown voltage as a function of gas number density. Figure 12.7 relates to small additions of Hg in helium (the mixture is called Heg gas), the idea being that every mercury atom would be ionized by the metastable energy level of helium, 19.8 eV—the ionization potential of this pseudo gas is usually taken as 19.8 eV since this is the energy level of helium that causes ionization. The collision frequencies obtained from microwave studies are $7.36 \times 10^{-14} \mathrm{m}^3 \mathrm{s}^1$ in helium[56] and $1.6 \times 10^{-13} \mathrm{m}^3 \mathrm{s}^{-1}$ in air.[57] Additional data are given in Table 12.2.

The correspondence between steady-state and microwave breakdown voltages has been further examined by Heylen and Postoyalko[58] who show that the sparking voltages with dc and microwave frequencies in hydrogen are compatible if a value of $v/N = 9.6 \times 10^{-14} \mathrm{m}^3 \mathrm{s}^{-1}$ is taken. The microwave studies have resulted in a value of $v/N = 1.8 \times 10^{-13} \mathrm{m}^3 \mathrm{s}^{-1}$. Such differences also exist in the collision frequency reported in air. The microwave data yield $v/N = 9.3 \times 10^{-14} \mathrm{m}^3 \mathrm{s}^{-1}$,[59] whereas the magnetic field studies in three different laboratories, namely England, Australia, and India, have yielded $v/N \approx 2.8 \times 10^{-13} \mathrm{m}^3 \mathrm{s}^{-1}$. These differences have remained unexplained.

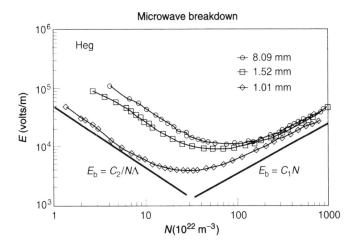

FIGURE 12.7 Microwave breakdown field in Heg gas at a frequency of 2.8 GHz as a function of N for various characteristic diffusion lengths. Note the decreasing field with increasing Λ. The linear dependence of the field in accordance with Equations 12.55 and 12.56 is clearly seen. Symbols are experimental and curves are calculated from the theory. Data taken from MacDonald, A. D., *Microwave Breakdown in Gases*, John Wiley, New York, 1966. With permission.

TABLE 12.2
Momentum Transfer Collision Frequencies in Selected Gases

Gas	$\nu/N\ (\times 10^{-14}\ m^3 s^{-1})$		
	$T = 300\ K$	500 K	1000 K
N_2	0.594	0.959	1.77
O_2	0.286	0.437	0.804
CO_2	10.05	9.68	7.48
H_2O	77.90	56.25	346
He	7.36	10.1	15.0
Ne	0.07	0.118	0.217
Ar	0.22	0.146	0.094
Kr	1.775	1.315	0.713
Xe	5.29	3.81	1.90
Dry air	0.529	0.845	1.550

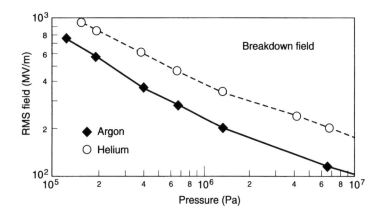

FIGURE 12.8 Laser breakdown field strength in helium and argon as a function of gas number density, determined from the energy in the giant optical pulse from a ruby laser.[60]

12.10 LASER BREAKDOWN

Brief comments on laser-induced breakdown in gases are given for the sake of completeness. Electromagnetic radiation in this form, concentrated into a small area, if containing enough energy, can cause electrical breakdown in gases at high pressures. Meyerand and Haught[60] used a giant pulse derived from a ruby laser system with a peak power of the order of tens of megawatts. The pulse had a duration of 30 ns and the luminosity from the test gap lasted about 50 μs. A typical value of the energy involved is one joule for a duration of 30 ns, giving a peak power of 30 MW. Figure 12.8 shows the field strength necessary to produce breakdown in two gases, argon and helium, as a function of gas number density. In each case the breakdown threshold is observed to decrease with increasing pressure, leveling off at higher pressures. Attempts to explain the breakdown by extending the theory of electron–atom collision to optical frequencies have had very limited success. In the experiments of Meyerand and Haught[60] the electron energy changes by the order of

10^{-3} eV during each cycle, whereas the energy of the ruby laser radiation is 1.7 eV. Since the energy of each quantum is greater than the classical energy change, one must look for a quantum mechanical description of breakdown at optical frequencies.[61]

Morgan et al.[62] have identified four stages in the development of a non-conducting gas into a highly conducting plasma: initiation, growth, plasma development, and finally extinction. The initiatory stage is the arrival of the laser beam at the focal point situated between the electrodes. An electron is needed and it is liberated by multi-photon resonance absorption of the laser light by the atom.[63] The formative stage is the multiplication of the electron, forming electron–ion pairs until sufficient density is built up to be characterized as breakdown. The breakdown criterion is different from the traditional definition of collapse of the applied voltage, because the power source in laser breakdown is electrically isolated from the gap. Of course, a small potential of 100 to 200 V is applied for the collection of charges to make oscilloscopic measurements, and Meyerand and Haught[60] did not observe any change in the energy deposited or the time to breakdown due to this small voltage. This observation confirmed that the breakdown was "purely" induced by the laser beam.

The breakdown condition is arbitrarily defined as the electron density reaching a certain concentration, 10^{21} m^{-3}, or the degree of fractional ionization of atoms in the focal region reaching ~0.1%.[62] The combined time duration for the availability of the initiatory electron and subsequent buildup of the required number density of ion pairs is a few nanoseconds.

Following the creation of the critical density of electrons, a rapid plasma development occurs if the laser intensity continues to increase. During this stage the plasma channel is formed and the hot expanding gas is subject to further absorption and hydrodynamic effects. The final stage, which may be of the order of several microseconds, is the extinction stage. During this phase the plasma gradually decays, the electron loss mechanisms being diffusion, attachment, and recombination. Radiation emission contributes to the loss of energy from the plasma and, Meyerand and Haught[60] measured an interval of 50 μs for the entire process, as against 30 ns which was the duration of the pulse.

The growth stage involves the process known as inverse bremsstrahlung absorption, which may be briefly described as follows. When a high energy electron, moving at a fast speed, encounters an atom, it is rapidly slowed down because of the electric field around the atomic nucleus. The electron radiates X-rays (10^{-9} to 10^{-11} m) which may cover a wide range of frequencies down to a certain minimum value which depends upon the incident energy of the electron. In the inverse bremsstrahlung process an electron can acquire energy from a photon, provided it is in the field of an atom or ion so that momentum is conserved.[61]

An electron in the focal region of a laser beam gains energy from the radiation field by the inverse process mentioned above. The energy of the electron is built up in a series of relatively small increments until it undergoes an exciting collision with a gas atom. The electron loses energy and the whole cycle repeats again. The first excitation levels of argon and helium are 11.55 and 19.8 eV respectively, and energy gain from radiation of 1.7 eV must necessarily involve a series of steps. However, the energy levels above the resonance levels are closely spaced and the atom, once it is excited, becomes rapidly ionized by absorption of three photons in the case of helium and two photons in the case of argon. This process leads to cascading growth of electrons exponentially and may be identified with avalanche growth at lower frequency fields. Essentially one could view this approach as an extension of microwave breakdown theory to optical frequencies.

Elastic collisions cause loss of energy and the free electrons are lost from the focal region owing to diffusion, attachment, and recombination. If the rate of production exceeds the rate of loss the concentration grows, leading to breakdown. Using the continuity equation, Morgan et al.[62] calculated the threshold intensities in helium and argon and compared them with measurements. The collision frequencies used were

$$\left. \begin{array}{ll} \dfrac{\nu}{N} = (9.3 - 1.2) \times 10^{-13} & (\text{m}^3\,\text{s}^{-1}) \quad \text{argon} \\[2mm] \dfrac{\nu}{N} = 7.5 \times 10^{-14} & (\text{m}^3\,\text{s}^{-1}) \quad \text{helium} \end{array} \right\} \tag{12.57}$$

The multiphoton absorption and inverse bremsstrahlung process are the most important electron generation mechanisms. The relative importance of these two processes changes considerably with the gas number density. Multiphoton ionization is expected to become dominant at low gas number densities and short duration laser pulses because the inverse bremsstrahlung mechanism is a three-body process involving an electron, photon, and atom. The process is therefore pressure dependent. Further, if the electron–neutral collision frequency is greater than the laser radiation frequency the inverse bremsstrahlung process becomes efficient (Gamal and Harith[68]). Data on laser breakdown in selected molecular gases (O_2, N_2, CO, CO_2) may be found in Dewhurst[64] and L'Huiller et al.[65]

12.11   CONCLUDING REMARKS

This chapter attempts to provide the necessary foundation to understand the vast scope that is available for further study. Systematic studies of RF discharge phenomena in gases still remain to be completed. The sheath–plasma boundary region is not well understood and evidence exists that there may be more than one point within the plasma where the Bohm criterion is satisfied. The computation of power absorbed is another area in which theory and experiment should give better agreement. In a series of articles, Franklin[66] has highlighted the contemporary issues in theory and experiments to be studied by the gaseous electronic community. This area of research with its various manifestations awaits to be classified as fully explored and is capable of yielding rich dividends, deserving continued efforts.

REFERENCES

1. Lieberman, M. A. and A. J. Lichtenberg, *Principles of Plasma Discharges and Materials Processing*, John Wiley, New York, 1994.
2. Gupta, N. and G. R. Govinda Raju, *IEEE Trans. Dielectr. Electr. Insul.*, 7, 795, 2000.
3. Langmuir, I. *Gen. Elec. Rev.*, 26, 731, 1923.
4. Bohm, D. in *The Characteristics of Electrical Discharges in Magnetic fields*, ed. A. Guthrie and R. K. Wakerling, McGraw-Hill, New York, 1949, p. 77.
5. Franklin, R. N., *J. Phys. D: Appl. Phys.*, 36, R309, 2003.
6. Roth, J. R., *Industrial Plasma Engineering*, Institute of Physics, Bristol, 1995.
7. Bletzinger, P., and M. J. Flemming, *J. Appl. Phys.*, 62, 4688, 1986.
8. van Roosmalen, A., W. G. M. van den Hoek, and H. Kalter, *J. Appl. Phys.*, 58, 653, 1985.
9. Bletzinger, P., A. Garscadden, M. L. Andrews, and D. Cooper, *IEEE Int. Conf. Plasma Science* Virginia, 1991, IEEE Catalog No. 91-CH3037-9, 1991, p. 136.
10. Bletzinger, P. *IEEE Int. Conf. Plasma Science*, 1–3 June 1992, Conference Record, p. 86.
11. Pitchford, L. C., J. P. Boeuf, and J. P. Morgan, 1996. Available at www. siglo-kinema.com.
12. Thompson, B. E. and H. H. Sawin, *J. Electrochem. Soc.*, 133, 1887, 1986.
13. Allen, K. D., H. H. Sawin, M. T. Mocella, and M. W. Jenkins, *J. Electrochem. Soc.*, 133, 2315, 1986.
14. Thompson, B. E, K. D. Allen, A. E. D. Richards, and H. H. Sawin, *J. Appl. Phys.*, 59, 1890, 1986.

15. Kroesen, G. M. W. and F. J. de Hoog, *Appl. Phys.*, A, 56, 479, 1993.
16. Haverlag, M., G. M. W. Kroesen, T. H. J. Bisschops, and F. J. de Hoog, *Plasma. Chem., Plasma Proc.*, 11, 357, 1991.
17. Hershowitz, N., *Plasma Diagnostics*, ed. A. Auciello and D. L. Flamm, Academic, Boston, MA, 1989.
18. Böhm, C., and J. Perrin, *J. Phys. D: Appl. Phys.*, 24, 865, 1991.
19. Köhler, W. E., R. J. Seebock, and F. Rebentrost, *J. Phys. D: Appl. Phys.*, 24, 252, 1991.
20. Knights, J. C., J. P. M. Schmitt, J. Perrin, and G. Guelachvili, *J. Chem. Phys.*, 76, 3414, 1982.
21. Richards, A. D., B. E. Thompson, and H. H. Sawin, *Appl. Phys. Lett.*, 50, 492, 1987.
22. Nakano, T., N. Sadeghi, and R. A. Gottscho, *Appl. Phys. Lett.*, 58, 3414, 1991.
23. Kobayashi, H., I. Ishikawa, and S. Suganomata, *J. Appl. Phys.*, 33, 5979, 1994.
24. Ingram S. G. and N. St J. Braithwaite, *J. Phys. D: Appl. Phys.*, 21, 1496, 1988.
25. Maneneschijn, A., G. C. A. M. Janssen, and F. J. de Hoog, *J. Appl. Phys.*, 69, 1253, 1991.
26. Kono, A., M. Haverlag, G. M. W. Kroesen, and F. J. de Hoog, *J. Appl. Phys.*, 70, 2939, 1991.
27. Gottscho, R. A. and C. E. Gaebe, *IEEE Trans. Plas. Sci.*, 14, 92, 1986.
28. Haverlag, M., A. Kono, D. Passchier, G. M. W. Kroesen, W. Goedheer, and F. J. de Hong, *J. Appl. Phys.*, 70, 3472, 1991.
29. Stoffels, E., W. W. Stoffels, D. Vender, M. Kando, G. M. W. Kroesen, and F. J. de Hoog, *Phys. Rev. E.*, 51, 2425, 1995.
30. Ivanov, V. V., K. S. Klopovsky, D. V. Lopaev, A. T. Rakhimov, and T. V. Rakhimova, *IEEE Trans. Plasma. Sci.*, 27, 1279, 1999.
31. Moore, C. A., G. P. Davis, and R. A. Gottscho, *Phys. Rev. Lett.*, 52, 538, 1984.
32. Ganguly, B. N., and A. Garscadden, *Appl. Phys. Lett.*, 46, 454, 1984.
33. Kim, J. B., K. Kawamura, Y. W. Choi, M. D. Bowden, K. Muraoka, and V. Helbig, *IEEE Trans. Plasma. Sci.*, 26, 1556, 1998.
34. Kuypers, A. D. and H. J. Hopman, *J. Appl. Phys.*, 67, 1229, 1990.
35. Olthoff, J. K., R. J. Van Brunt, S. B. Radovanov, J. A. Rees, and R. Surowiec, *J. Appl. Phys.*, 75, 115, 1994.
36. Flender, U. and K. Wisemann, *J. Phys. D: Appl. Phys.*, 27, 509, 1994.
37. Godyak, V. A., R. B. Piejak, and B. M. Alexandrovich, *J. Appl. Phys.*, 73, 3657, 1993.
38. Heylen, A. E. D., *Proc Phys Soc.*, (London), 79, 284, 1962.
39. Lieberman, M. A., *IEEE Trans. Plasma. Sci.*, 16, 638, 1988.
40. Lieberman, M. A., *IEEE Trans. Plasma. Sci.*, 17, 338, 1989.
41. (a) Butterbaugh, J. W., L. D. Baston and H. H. Sawin, *J. Vac. Sci. Technol.*, 8, 916, 1990. (b) Beneking, C., *J. Appl. Phys.*, 68, 4461, 1990. (c) Miller, P. A., H. Anderson and M. P. Splichal, *J. Appl. Phys.*, 71, 1171, 1992. (d) Sobolewski, M. A., *IEEE Trans. Plasma. Sci.*, 23, 1006, 1995.
42. Overzet, L. J. and M. B. Hopkins, *Appl. Phys. Lett.*, 63, 2484, 1993.
43. Schneider, J. M., S. Rohde, W. D. Sproul, and A. Matthews, *J. Phys. D: Appl. Phys.*, 33, R 173, 2000.
44. Huo, W. M. and Y. K. Kim, *IEEE Trans. Plasma Sci.*, 27, 1225, 1999.
45. Kushner, M. J., *IEEE Trans. Plasma. Sci.*, PS-14, 188, 1986.
46. Belenguer, P. and J. P. Boeuf, *Phys. Rev. A41*, 4447, 1990.
47. Cantin, A. and R. R. J. Gagne, *Appl. Phys. Lett.*, 30, 316, 1977.
48. Chapman, B. N., *Glow Discharge Processes*, John Wiley and Sons, New York, 1980.
49. Govinda Raju, G. R. 8th *Int. Symp. Gaseous Dielectrics*, Virginia Beach, June 2–5, 1998.
50. Gupta, N., Private communication, 1999.
51. Gupta, N. and G. R. Govinda Raju, *J. Phys. D: Appl. Phys.*, 33, 1, 2000.
52. Mateev, E. and I. Zhelyazkov, *J. Appl. Phys.*, 87, 3263, 2000.
53. Surendra, M., D. B. Groves, and I. J. Morey, *Appl. Phys., Lett.*, 56, 1022, 1990.
54. Godyak, V. A., R. B. Piejak, and B. M. Alexandrovich, *J. Appl. Phys.*, 73, 3657, 1993.
55. MacDonald, A. D. *Microwave Breakdown in Gases*, John Wiley, New York, 1966.
56. MacDonald, A. D., and S. J. Tetenbaum, in *Gaseous Electronics, Vol. 1, Electrical Discharges*, ed. M. N. Hirsch and H. J. Oskam, Academic Press, New York, 1978.
57. Brown, S. C., *Introduction to Electrical Discharges in Gases*, John Wiley, New York, 1966.

58. Heylen, A.E.D. and V. Postoyalko, *Int. J. Electron.*, 68, 1113, 1990.
59. MacDonald, A. D. and S. J. Tetenbaum, in *Gaseous Electronics, Vol. 1, Electrical Discharges*, ed. M. N. Hirsch and H. J. Oskam, Academic Press, New York, 1978, p. 193.
60. Meyerand, R. G. and A. F. Haught, *Phys. Rev. Lett.*, 11, 401, 1963.
61. Wright, J. K., *Proc. Phys. Soc.*, 84, 41, 1964.
62. Morgan, F., L. R. Evans, and C. G. Morgan, *J. Phys. D: Appl. Phys.*, 4, 225, 1971.
63. Tozer, B. A., *Phys. Rev. A*, 137, 1665, 1965.
64. Dewhurst, R. J., *J. Phys. D: Appl. Phys.*, 11, L191, 1978.
65. L'Huiller, A., G. Mainfray, and P. M. Johnson, *Chem. Phys. Lett.*, 103, 447, 1984.
66. Franklin, R.N., *J. Phys. D: Appl. Phys.*, 36, Nov. 2003 issue.
67. *J. Phys. D: Appl. Phys.*, Nov. 2003 issue.
68. Gamal, Y.EE-D. and M.A. Harith, *J. Phys. D: Appl. Phys.*, 14, 2209, 1981.

Appendix 1

TABLE A1.1
Fundamental Constants

Quantity	Symbol	Value	Units
Avagadro constant	N_A	6.022×10^{23}	mol^{-1}
Boltzmann constant	k	1.381×10^{-23}	$J\,K^{-1}$
Boltzmann constant in eV	k/e	8.617×10^{-5}	$eV\,K^{-1}$
Electron mass	m	9.109×10^{-31}	kg
Elementary charge	e	1.602×10^{-19}	C
Electron charge-to-mass ratio	e/m	1.759×10^{11}	$C\,kg^{-1}$
Permittivity of free space	ε_0	8.854×10^{-12}	$F\,m^{-1}$
Permeability of free space	μ_0	$4\pi \times 10^{-7}$	A^{-2}
Planck's constant	h	6.626×10^{-34}	$J\,s$
Planck's constant in eV	h	4.136×10^{-15}	$eV\,s$
Bohr radius	a_0	0.529×10^{-10}	m
Proton mass	m_p	1.672×10^{-27}	kg
Proton-to-electron mass ratio	m_p/m_e	1836.152	
Rydberg constant in eV	R	13.606	eV
Speed of light	c	2.998×10^{8}	$m\,s^{-1}$

TABLE A1.2
Selected Conversion Factors

Energy
$1\,eV = 11{,}604\,K = 8065.54\,cm^{-1} = 8.066 \times 10^5\,m^{-1}$
$1\,m^{-1} = 1.2398 \times 10^{-6}\,eV$
$1\,eV = 23.06\,kCal/mol$
$1\,eV = 2.4180 \times 10^{14}\,Hz$
$kT\,(77\,K) = 0.0066\,eV$
$kT\,(300\,K) = 0.026\,eV$
1 rydberg (R) $= 13.606\,eV = 1.097 \times 10^7\,m^{-1}$
Energy (eV) $= 2.843 \times 10^{-12}\,[\text{drift velocity (m/s)}]^2$

Cross section
$a_0{}^2 = 0.28 \times 10^{-20}\,m^2$
$\pi\,a_0{}^2 = 8.791 \times 10^{-21}\,m^2$

Pressure
1 atmosphere $= 760\,torr = 101.31\,kPa = 1.0131\,bar$

(Continued)

TABLE A1.2
Continued

Gas number density (N)

$$N_{P(\text{torr}),\, T(\text{k})} = 3.54 \times 10^{22} \times 273 \frac{p}{T}$$

Reduced electric field
$$1 \text{ Td} = 10^{-17} \text{ V cm}^2 = 10^{-21} \text{ V m}^2$$

$$\frac{E_{\text{V/m}}}{N_{1/\text{m}^3}} = 2.82 \frac{E_{\text{V/cm}}}{p_{\text{Torr}}} \; Td \quad (273\,\text{K})$$

$$\frac{E_{\text{V/m}}}{N_{1/\text{m}^3}} = 3.1 \frac{E_{\text{V/cm}}}{p_{Torr}} \; Td \quad (293\,\text{K})$$

Reduced ionization coefficient

$$\frac{\alpha_{1/\text{m}}}{N_{1/\text{m}^3}} = 2.82 \times 10^{-21} \frac{\alpha_{1/\text{cm}}}{p_{\text{Torr}}} \quad (273\,\text{K})$$

$$\frac{\alpha_{1/\text{m}}}{N_{1/\text{m}^3}} = 3.11 \times 10^{-21} \frac{\alpha_{1/\text{cm}}}{p_{\text{Torr}}} \quad (293\,\text{K})$$

Appendix 2

Term and Electron State Notations

A2   A. ELECTRON QUANTUM NUMBERS

The electron has four quantum numbers, as shown in Table A2.1[1]. The principal quantum number, n, takes integer values $n = 1, 2, 3, \ldots, \infty$. The second quantum number, designated l, is called the orbital quantum number. It is also an integer and takes on values $l = 0, 1, 2, \ldots$ $(n-1)$. It denotes the magnitude of the angular momentum of the electron in units of $h/2\pi$, where h is Planck's constant. For historical reasons, special letters are used to denote the value of l, thus:

l	0	1	2	3	4	5	...
	s	p	d	f	g	h	...

Electrons with $l = 0$ are called s electrons; those with $l = 1$, p electrons; $l = 2$, d electrons; and so on. The *configuration* of an electron is characterized by its quantum numbers n and l. Electrons having the same principal quantum number are said to belong to the same shell. Electrons having the same value of n and l are said to belong to the same subshell. Following Pauli's exclusion principle, that no two electrons in the same atom can have the same four quantum numbers, the maximum numbers of electrons allowed in a subshell are: s, 2; p, 6; d, 10; f, 14. The maximum number of electrons allowed in a shell is $2n^2$.

Electrons having different principal quantum numbers n differ in energy widely whereas electrons with the same value of n but different values of l have less energy difference. Two configurations are different if at least one of the quantum numbers is different (having different n or l). Several electrons with the same n and l are present in an atom (except in atomic hydrogen) and the number of electrons having the same n and l are shown by a superscript. Thus, $3p^6$ means six electrons having quantum numbers $n = 3, l = 1$. More often

TABLE A2.1
Quantum Numbers of an Electron

Symbol	Quantum Number	Allowed Values	
n	Principal	$n = 1, 2, 3, 4, \ldots$	Electron energy is quantized
l	Orbital angular momentum	$l = 0, 1, 2, 3, \ldots (n\text{-}1)$	Orbital angular momentum is quantized
m_l	Magnetic	$m_l = 0, \pm 1, \pm 2, \ldots, \pm l$	Orbital angular momentum along an external magnetic field is quantized
m_s	Spin	$S = \pm \frac{1}{2}$	Angular momentum due to electron spin is quantized

than not, the electron configuration of an atom is given for the outermost electrons only since the inner shells are repetitive. In the absence of a magnetic field, states with the same values of n and l but different values of m_l and m_s have the same energy. The electrons are then said to be "degenerate" with each other.

Each configuration consists of many *energy states*. When transition occurs between states the selection rule $\Delta l = \pm 1$ should be obeyed. This is the reason for showing the various electronic transitions in an energy level diagram of an atom by slanted lines and never by vertical lines. This rule is remembered as "escalators, but not elevators." Transitions that follow the selection rule are called "allowed transitions"; transitions with no change of l or with change of 2, 3, 4, ... etc. are not allowed and such transitions are called "forbidden transitions."

A2   B. TERM NOTATION FOR ATOMS

In atoms with more than one electron there is interaction between electrons, and the individual angular momentum no longer has much meaning. In general, the orbital angular momenta l_1, l_2, l_3, ... of the individual electrons are strongly coupled among themselves and the resultant angular momentum, which is also quantized, is designated by the symbol L. L characterizes the whole atom in place of l which characterizes an individual electron. Capital letters are used to denote the resultant angular momentum, thus:

L	0	1	2	3	4	5	6	...
	S	P	D	F	G	H	I	...

L is the vector sum of the angular momenta of individual atoms, l_i, and the selection rule for electronic transitions is the same as for l; $\Delta L = \pm 1$ although $\Delta L = 0$ can also occur.

The spin quantum number of electrons, m_s, has values of $+1/2$ and $-1/2$. For atoms with many electrons the spins s_1, s_2, s_3, ... are considered as strongly coupled and the total spin of the atom is the arithmetic sum of the individual spins, denoted by S. The total angular momentum is the vector sum of L and S, denoted by J. J has discrete values and therefore L and S cannot be oriented in any arbitrary direction but only in certain directions. This is called "space quantization." The selection rule for J is that $\Delta J = 0, +1, -1$.

For given values of L and S the number of possible values (components) of J is given by $J = 2S + 1$ for $L \geq S$. The number of possible values of J is given by $J = 2L + 1$ for $L < S$. Unfortunately the same symbol is used to denote the quantum number S and the symbol for $L = 0$. A little familiarity with the notations removes the ambiguity. The values of J are restricted to

$$|L - S| \leq J \leq L + S \qquad (A2.1)$$

increasing in steps of $+1$. When an atom contains an odd number of electrons J is a half integer; with an even number of electrons it is an integer. One should not get confused with the number of J values (components of J) and those values themselves.

The total angular momentum denotes the different energy states of the components of singlets and multiplets. For doublets, $J = 1/2$ for S terms, $J = 1/2, 3/2$ for P terms, $J = 3/2, 5/2$ for D terms, and so on. The selection rule for J is that $\Delta J = 0, +1$ or -1. For triplets, $J = 1$

for S terms, $J = 0$, 1, 2 for P terms, $J = 1$, 2, 3 for D terms, and so on. It is important to bear in mind that the S term of the singlet system (^1S) is different and generally lies a little higher than the S term of the triplet (^3S) of the same element. J values for the doublet and triplet terms are shown in Table A2.2. For higher multiplicities the expression A2.1 must be used.

It is easy to see that if transition occurs between ^3D and ^3P levels there will be altogether nine lines, but only six lines are observed experimentally because the selection rule does not allow transitions $^3D_3 \rightarrow {}^3P_1$, $^3D_2 \rightarrow {}^3P_0$, and $^3D_3 \rightarrow {}^3P_0$. For each of these transitions $\Delta J \neq 0, +1$ or -1.

These considerations are built into the term designation suggested by Russell and Saunders.[2] It may be expressed as

$$(s_1, s_2, s_3, \cdots)(l_1, l_2, l_3, \cdots) = (S, L) = J \tag{A2.2}$$

The coupling scheme is based on the assumption that, when there are several electrons in an atom, each with a definite l, and each with $s_i = \pm 1/2$, the individual ls combine vectorially. The resultant vector L differs in energy considerably. For two electrons, L is limited by

$$|l_1 - l_2| \leq L \leq l_1 + l_2 \tag{A2.3}$$

For more electrons one can obtain the possible values for L by first combining two ls and then combining each possible resultant with the third l and so on. The resultant L will always be integral.

Likewise, the individual s_is combine vectorially and the resultant S will differ in energy considerably. As the spin for each electron is $s = 1/2$ the resultant spin, say for x electrons, can have all values up to $x/2$, each differing by unity. If x is odd, $S = 1/2, 3/2, 5/2, \ldots, x/2$; if even, $S = 0, 1, 2, 3, \ldots, x/2$. A given spectrum contains only states with integer or half integer values of S.

In the Russell–Saunders coupling scheme, the left superscript, equal to $2S + 1$, denotes the multiplicity; superscript $1 = $ singlet ($S = 0$), $2 = $ doublet ($S = 1/2$), $3 = $ triplet ($S = 1$), $4 = $ quartet ($S = 3/2$), etc. The capital letters S, P, D, F, G, etc. denote the L value according to $L = 0$, 1, 2, 3, 4, 5 respectively. Thus ^3D means $L = 2$, $S = 1$ (because $2S + 1 = 3$). From Table A2.2, J can have a value 1, 2, or 3. The J value is given as the right subscript. Symbols for the three levels are $^3D_1$, $^3D_2$, $^3D_3$. A second example is ^7S: here $L = 0$, $S = 3$, J has a single value, 3, because of Equation (A2.1). The term is $^7S_3$, which denotes the ground state of chromium (Cr) with 24 electrons or molybdenum (Mo) with 42 electrons. The term for platinum ($Z = 78$) is $5d^9$ ($^2D_{5/2}$) $6s$ $^3D_3$; the nine $5d$ electrons combine to yield the grand

---

TABLE A2.2
J Values for the Singlet, Doublet, and Triplet Systems. The Terms Are Shown in Brackets

Subshell	L	Singlet Terms $S = 0$ J	Doublet Terms $S = \frac{1}{2}$ J	Triplet Terms $S = 1$ J
S	0	0 ($^1S_0$)	1/2 ($^2S_{1/2}$)	1 ($^3S_1$)
P	1	1 ($^1P_1$)	1/2, 3/2 ($^2P_{1/2, 3/2}$)	0, 1, 2 ($^3P_{0, 1, 2}$)
D	2	2 ($^1D_2$)	3/2, 5/2 ($^2D_{3/2, 5/2}$)	1, 2, 3 ($^3D_{1, 2, 3}$)
F	3	3 ($^1F_3$)	5/2, 7/2 ($^2F_{5/2, 7/2}$)	2, 3, 4 ($^3F_{2, 3, 4}$)
G	4	4 ($^1G_4$)	7/2, 9/2 ($^2G_{7/2, 9/2}$)	3, 4, 5 ($^3G_{3, 4, 5}$)
H	5	5 ($^1H_5$)	9/2, 11/2 ($^2H_{9/2, 11/2}$)	4, 5, 6 ($^3H_{4, 5, 6}$)

TABLE A2.3
Electron Configurations and Term for Selected Elements

n	1	2		3			4				5					
Element	1s	2s	2p	3s	3p	3d	4s	4p	4d	4f	5s	5p	5d	5f	5g	Term
1. H	1															$^2S_{1/2}$
2. He	2															$^1S_0$
3. Li	2	1														$^2S_{1/2}$
4. Be	2	2														$^1S_0$
5. B	2	2	1													$^2P_{1/2}^{\ 0}$
6. C	2	2	2													$^3P_0$
7. N	2	2	3													$^4S_{3/2}$
8. O	2	2	4													$^3P_2$
9. F	2	2	5													$^2P_{3/2}^{\ 0}$
10. Ne	2	2	6													$^1S_0$
11. Na	2	2	6	1												$^2S_{1/2}$
12. Mg	2	2	6	2												$^1S_0$
13. Al	2	2	6	2	1											$^2P_{1/2}$
14. Si	2	2	6	2	2											$^3P_0$
15. P	2	2	6	2	3											$^4S_{3/2}$
16. S	2	2	6	2	4											$^3P_2$
17. Cl	2	2	6	2	5											$^2P_{3/2}^{\ 0}$
18. Ar	2	2	6	2	6											$^1S_0$
19. K	2	2	6	2	6		1									$^2S_{1/2}$
20. Ca	2	2	6	2	6		2									$^1S_0$
32. Ge	2	2	6	2	6	10	2	2								$^3P_0$
35. Br	2	2	6	2	6	10	2	5								$^2P_{3/2}^{\ 0}$
36. Kr	2	2	6	2	6	10	2	6								$^1S_0$
53. I	2	2	6	2	6	10	2	6	10		2	5				$^2P_{3/2}$
54. Xe	2	2	6	2	6	10	2	6	10		2	6				$^1S_0$

n	1–2	3	4s	4p	4d	4f	5s	5p	5d	5f	5g	6s	6p	
55. Cs	10	18	2	6	10		2	6				1		$^2S_{1/2}$
80. Hg	10	18	2	6	10	14	2	6	10			2		$^1S_0$
81. Tl	10	18	2	6	10	14	2	6	10			2	1	$^2P_{1/2}^{\ 0}$

parent term $^2D_{5/2}$, and the combination of $5d^9$ and $6s$ electrons gives the term $^3D_3$. The general notation in the scheme is $^{2S+1}L_J$.

For even multiplicities J takes half integral values, 1/2, 3/2, 5/2, etc.; examples are $^2P_{1/2}$, $^2D_{3/2}$. For odd multiplicities J takes integral values, 1, 2, 3, etc.; examples are $^3P_1$, $^3D_3$, etc. For singlets the J value is the same as the L value; examples are $^1S_0$, $^1P_1$, $^1D_2$. The terms are preceded by the principal quantum number, as in $2\,^1S_0$, $4\,^3P_2$, $5\,^3D_3$ etc. Inert gases have singlets and triplets, alternating in the periodic table, column 8. The terms of atoms with an odd number of electrons have even multiplicities. The terms of atoms with an even number of electrons have odd multiplicities. Table A2.3 shows the ground terms for selected elements we have encountered in the text.

Nonequivalent electrons are those having different n or l. Equivalencies for configurations may be found in Bacher and Goudsmit[3] and Gibbs et al.[4]

When the arithmetical sum of all ls of the electrons is even, one obtains even energy levels. In the other case, one obtains odd levels. The symbol for odd levels is a zero sign as the right subscript.[5] An example is $3p\ 3d\ [^3D_2]_0$, meaning one electron with $n=3$, $l=1$; one electron with $n=3$, $l=2$; resultant $L=2$, $S=1$, $J=2$, odd energy level.

The Russell–Saunders scheme is a special case of interaction between the electrons of an atom. If the electrostatic interactions between different electrons are much larger than the spin–orbit interactions, the coupling scheme holds true. This is generally true for atoms with small nuclear charge. With increasing nuclear charge or with highly ionized atoms the spin–orbit interaction increases more strongly than electrostatic interactions and the coupling scheme shows increasing deviations. An alternative scheme, not used in this book, is the jj coupling in which the spin–orbit interaction is stronger than the electrostatic interaction.

Racah[6] proposed another type of vector coupling for excited states in which the total angular momentum of the parent ion, J, couples with the orbital angular momentum l of the outer electron and gives a resultant K. K is then coupled to the spin of the outer electron and their resultant is J. The notation has a round bracket for the term of the parent ion and a square bracket for the value of K. The square bracket has a subscript for J value and o as superscript for odd parity. The two brackets are separated by the configuration of the outer electron; often, the round brackets are omitted for brevity. Example: the transition with threshold energy of 13.08 eV (term value 21648.70 cm^{-1}) in argon, is denoted as $(^2P_{3/2}^{\,o})\ 4p\ [5/2]_3$, meaning that the ion core has the term $^2P_{3/2}^{\,o}$, the outer electron is $4p$, $K=5/2\ (3/2+1)$, $J=3\ (5/2+1/2$ spin), and even parity. Abbreviated Racah notation is simply $4p\ [5/2]_3$.

Resorting to several schemes such as L–S coupling, jj coupling (used mostly in relativistic calculations) of equivalent electrons yields a *term*, and the value of the term is expressed in units of cm^{-1} (wave numbers). The *electronic level* has a lower hierarchy than configuration with the value of J specified. A group of levels which differ only in their value of J is called a *multiplet*. In spectroscopic notation the term is a multiplet and the level is a line. Levels lie in order of their J value. The component of a line is the *electronic state* with all four quantum numbers specified.[7] In the literature, as well as in this volume, one finds the words *term*, *energy level*, and *state* used interchangeably.[8] One also finds the expression *term value* used synonymously with *energy value*. We mostly refer to energy value, in units of eV, unless the expression *term value* is indispensable.

In following the literature on excitation cross sections one needs to know the level designation, the threshold energy, and often the wavelength of the allowed transition. Generally speaking, these notations are encountered: the spectroscopic, the Russell–Saunders coupling scheme, the Paschen notation, and the Racah notation. Fortunately, equivalent states are listed in Bacher and Goudsmit.[3]

Conversion from energy to frequency of radiation (ν, Hz) or emission is made according to the elementary formula

$$\nu = \frac{\varepsilon_1 - \varepsilon_2}{h} \tag{A2.4}$$

Conversion from energy to wavenumber (ν', cm^{-1}) is according to

$$\nu' = \frac{\varepsilon_1 - \varepsilon_2}{hc} \tag{A2.5}$$

where c is the velocity of light. Wavelength (λ, cm) is the reciprocal of wavenumber ($1/\nu'$).

Table A2.4 shows a short list of the more important lowest energy levels of the rare gas and mercury atoms.

TABLE A2.4
Short List of the More Important Lowest Energy Levels of the Rare Gas and Mercury Atoms

Atom	ε_M (Metastable) (eV)	ε_r (Radiative) (eV)	Configuration	Wavenumber (m^{-1})
He	19.80	21.21	$1s^2\ {}^1S_0$	0
			$2s\ {}^3S_1$	15985031.8
			$2s\ {}^1S_0$	16627170.0
			$2p\ {}^3P^0_{2,1,0}$	16908100.0
			$2p\ {}^1P^0_1$	17112914.8
			$3s\ {}^3S_1$	18323108.0
			$3s\ {}^1S_0$	18485906.0
			$3p\ {}^3P^0_{2,1,0}$	18555900.0
Ne	16.62	16.85	${}^1S_0$	0
			$3s\ [3/2]^0_2$	13404379.0
			$3s\ [3/2]^0_1$	13482059.1
Ar	11.55	11.61	${}^1S_0$	0
			$4s\ [3/2]^0_2$	9314380.0
			$4s\ [3/2]^0_1$	9375063.9
			$4s'\ [½]^0_0$	9455370.7
			$4s'\ [½]^0_1$	9539987.0
			$4p[½]1$	10410214.4
Kr	9.91	10.02	${}^1S_0$	0
			$5s\ [3/2]^0_2$	7997253.5
			$5s\ [3/2]^0_1$	8091756.1
			$5s'\ [½]^0_0$	8519241.4
			$5s'\ [½]^0_1$	8584750.1
			$5p[½]1$	9116931.3
Xe	8.32	8.45	${}^1S_0$	0
			$6s\ [3/2]^o_2$	6706804.7
			$6s\ [3/2]^o_1$	6804566.3
			$6s'\ [½]^o_0$	7619729.2
			$6s'\ [½]^o_1$	7718556.0
			$6p\ [½]1$	7726964.9
Hg	4.667	4.886	${}^1S_0$	0
			$6p\ {}^3P^o_0$	3764508.0
			$6p\ {}^3P^o_1$	3941230.0
			$6p\ {}^3P^o_2$	4404298.0
			$6p\ {}^1P^o_1$	5406878.0
			$7s\ {}^3S_1$	6235046.0

A2   C. ELECTRONIC STATES IN MOLECULES

The ideas explained in connection with atomic excitation levels may be extended to molecules.[8] In an atom the electric field due to the nucleus is spherically symmetric, but in a diatomic molecule this symmetry is reduced; there is only symmetry along the axis joining the two nuclei. The orbital angular momentum of the electrons loses its significance whereas its component along the internuclear axis, the magnetic quantum number M_L, remains well defined. Accordingly, it is more meaningful to define the energy states with reference to the value of M_L than with reference to that of L. The component of L in the diatomic molecule is denoted by Λ, according to

$$\Lambda = |M_L| \tag{A2.6}$$

TABLE A2.5
Symbols for Various Values of *L* (Atom) and *Λ* (Molecule)

	Atom				
L	0	1	2	3	4
Symbol	S	P	D	F	G

	Molecule				
$Λ$	0	1	2	3	4
Symbol	Σ	Π	Δ	Φ	Γ

Note that the symbol used on the left hand side of Equation A2.6 is Greek Lambda, corresponding to L in the atom. The vector $Λ$ represents the component of the orbital angular momentum along the internuclear axis and it can take the values $0, 1, 2, 3, \ldots, L$. As in the case of an atom, the various values of $Λ$ are designated symbols as shown in Table A2.5. M_L can have two values, $+Λ$ and $-Λ$, with the same energy. Therefore molecular states Π, Δ, Φ,... etc. are doubly degenerate. Σ states are non-degenerate.

The resultant spin of the molecule, in analogy with the spin of the atom (S), is denoted by the Greek symbol Σ and the values allowed by the quantum theory are

$$Σ = S, S - 1, S - 2, \ldots, - S \tag{A2.7}$$

with a total of $(2S + 1)$ values. In contrast to $Λ$, which is always positive (see Equation A2.6), Σ is allowed to have both positive and negative values. The quantum number $(2S + 1)$ is called the multiplicity of a state and, if the molecule does not rotate, Σ states are singlets. In the term notation for molecules the multiplicity is shown as the left superscript, as for atoms. However, two differences between atoms and molecules arise. For $Λ \neq 0$, full multiplicities occur (in atoms full multiplicities do not always occur for $L \neq 0$) and the multiplets are separated by equal energy increments (in atoms the multiplets are separated by unequal energy increments).

It is unfortunate that the same symbol Σ is used for both the resultant spin quantum number and the state of $Λ = 0$ (see Table A2.5), but a little familiarity with notations removes the ambiguity. It is recalled that one encounters the same situation in the symbols for atoms. S is used to denote both the resultant spin and as a symbol for a term having $L = 0$.

The total angular momentum of the molecule is denoted by the Greek symbol $Ω$ (analogous to J for atoms) and is obtained by adding the vectors $Λ$ and $Σ$ i.e.,

$$Ω = Λ + Σ \tag{A2.8}$$

For reasons too involved to explain here[9] the value of $Λ + Σ$ is used in place of $Ω$ to distinguish multiplet components. In that case

$$Ω = Λ + Σ \tag{A2.9}$$

The value of $Λ + Σ$ is added as a subscript for the term symbol.

The term of a molecule has even or odd multiplicity, just as in atoms. Even multiplicities are indicated by the subscript g (German, gerade) and odd multiplicities are indicated by the subscript u (German, ungerade) in the term of the molecule. The presence of this symbol denotes, mostly but not always, that the two atoms have the same nuclear charge. The term

TABLE A2.6
Molecular Terms Resulting from Combinations of Atoms

States of the Separated Atoms	Molecular States	Comment
$S_g + S_g$	Σ^+	Unlike atoms, even parity ($L=0$)
$S_u + S_u$	Σ^+	Unlike atoms, odd parity ($L=0$)
$S_g + S_u$	Σ^-	Unlike atoms, even and odd parity ($L=0$)
$S_g + P_g$	Σ^-, Π	Unlike atoms, even parity ($L=0, 1$)
$S_u + P_u$	Σ^-, Π	Unlike atoms, odd parity ($L=0, 1$)
$S_g + P_u$	Σ^+, Π	Unlike atoms, even and odd parity ($L=0, 1$)
$S_u + P_g$	Σ^+, Π	Unlike atoms, odd and even parity ($L=0, 1$)
$^1S + ^1S$	$^1\Sigma_g{}^+$	Like atoms, both same parity ($L=0$)
$^2S + ^2S$	$^1\Sigma_g{}^+, {}^3\Sigma_u{}^+$	Like atoms, both same parity ($L=0$)

symbol may also have a superscript $+$ for the Σ state which indicates that the reflected wave from a plane containing the internuclear axis has the same polarity as the incoming wave. If this superscript is $-$, the reflected wave from such a plane has opposite polarity.

The term of a molecule can be built up by first assuming that the combining atoms are at an infinite distance from each other to start with and then approach each other. We recall that equivalent electrons have the same quantum numbers, n and l; unequal electrons have different n or l. Selected examples of terms that result from the combination of atoms are shown in Table A2.6, both for equal and for unequal electrons.

A2 D. ROTATIONAL AND VIBRATIONAL EXCITATION

A comparison of spectra obtained with an atom and a molecule shows that the atomic lines spread into bands in the molecule. This is because, in addition to electronic excitation (energy ε_{ex}), two additional energy loss mechanisms occur during electron collisions with molecules. These are vibrational excitation (energy ε_v) and rotational excitation (energy ε_k). The threshold energies decrease in the order $\varepsilon_{ex} > \varepsilon_v > \varepsilon_k$.

In the vibrational mode, the atoms forming a diatomic molecule vibrate relative to each other along the internuclear axis. A change of vibrational energy may take place concurrently with electronic excitation if the electron energy is sufficiently large. If radiation is emitted it corresponds to a decrease in the vibrational energy of the molecule; if radiation is absorbed the vibrational energy increases. Molecules may rotate as a whole along an axis perpendicular to the internuclear axis and passing through the center of gravity. There is a distribution of angular velocities just as there is a distribution of linear velocities. Temperature affects both velocities.

Both vibrational and rotational energies are quantized. If the rotational energy increases there is absorption of radiation and, correspondingly, there is emission of radiation if the rotational energy is diminished. If rotational energy alone decreases the emission takes place in the far infrared region of the spectrum (wavelength $\sim 100\,\mu m$, $10^{12}\,Hz$). If both vibrational and rotational excitation occur together the spectrum will be in the near infrared region (wavelength $\sim 1\,\mu m$, frequency $10^{14}\,Hz$). If, however, the electron energy exceeds the first electronic excitation threshold the spectrum will have visible and ultraviolet wavelengths.

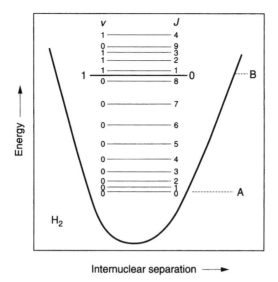

FIGURE A2.1 Rotational and vibrational levels of the hydrogen molecule. The lowest energy levels are associated with the rotational excitation, and the first vibrational excitation occurs above $J = 8$. There are about 30 levels between the levels marked A and B.

In the simplified theoretical models the rotation of a diatomic molecule is approximated to a rigid rotor with an atom at each end of a weightless rigid rod. The solution of the Schrödinger equation exists for eigenvalues of energies in terms of the rotational quantum number J that can take integral values 0, 1, 2, 3 . . . This means that only certain rotational frequencies are possible. The discrete energy levels increase quadratically with increasing J. The quantum number may change by only one unit when transitions occur.

The vibration of the two nuclei of a diatomic molecule can be approximated to the simple harmonic motion of a single molecule of reduced mass M_R, whose potential is assumed to be proportional to the square of the displacement. The solution of the Schrödinger equation exists for eigenvalues of energies in terms of the vibrational quantum number v that can take integral values 0, 1, 2, 3, ... The energy levels are separated by equal amounts and are shown in Figure A2.1. In contrast to the rotational frequency, which is zero at the lowest level $(J = 0)$, the vibrational energy is finite at the lowest level, i.e., at $v = 0$.

In considering the vibrational excitation of diatomic molecules, two different situations arise. If the two atoms are identical, as in H_2, O_2, N_2, etc. the molecule does not possess a dipole moment and is known as nonpolar. If, however, it is made of dissimilar atoms, then the molecule possesses a permanent dipole moment and is known as polar.[10] If the internuclear distance changes, the dipole moment of a polar molecule changes, and it may be assumed that the change in the dipole moment is proportional to the change in r. In an alternating electric field the dipole changes direction with the same frequency as the field and, according to classical electrodynamics, radiation is emitted or absorbed. According to quantum mechanics, transitions can only occur according to the selection rule $\Delta v = \pm 1$. All transitions give rise to the same frequency, in the infrared region. Thus, we have equidistant lines in the spectrum, due to rotational excitation, and a single, relatively strong line due to vibrational excitation.

For a molecule consisting of two like atoms, the permanent dipole moment is zero and therefore no transitions between the different vibrational levels occur as a result of long-range force of a dipole. In certain homonuclear molecules there is a quadrupole moment

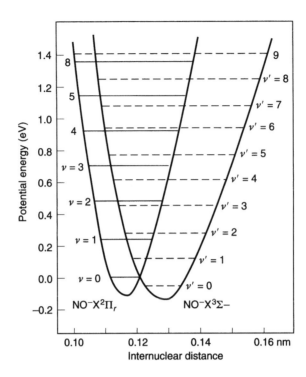

FIGURE A2.2 Potential energy and internuclear distance for vibrational excitation of the nitric oxide (NO) molecule and the NO^- ion.

which causes transitions after electron impact. There is no infrared emission or radiation. Of course, this does not mean that vibrational excitation does not occur, because the electronic polarization still occurs. For example, the vibrational levels in the H_2 molecule, $v = 3$, and $v = 2$, are separated by $3333.3\,cm^{-1}$ (10^{14} Hz) in the ultraviolet region.

We show two examples of low energy inelastic collision in molecules. Figure A2.1 shows the lowest energy levels of the ground electronic state of the hydrogen molecule with both rotational and vibrational levels marked. The lowest energy levels are associated with the rotational excitation, and the first vibrational excitation occurs above $J = 8$.[11] There are about 30 levels between the levels marked A and B.

The second example is the molecule of nitric oxide (NO), which is electron attaching and has a dipole moment (Figure A2.2). The NO molecule has the ground state $X\,^3\Pi_{1/2,\,3/2}$ and NO^- has the ground state $X\,^3\Sigma^-$ (see below) which is the same as that of O_2. Note that $v = 0$ and $v' = 0$ do not coincide with zero energy, indicating a small energy in the ground state. The $v' = 0$ level of the ion is 24 meV below the $v = 0$ level of the molecule.[12] Vibrational excitation between any two levels in the same particle gives rise to radiation of the same wavelength.

Terms of diatomic molecules are often stated with a preceding capital or lower-case letter. X indicates the ground level molecule, such as $X\,^1\Sigma_g^+$ for CO. The higher levels are denoted by capital letters A, B, C, etc. to show their relative position, with the same multiplicity (usually 1) in the pantheon of electronic states. Letter A has lower energy than B, and so on. As an example, the $A\,^1\Pi$ state of the CO molecule has 8.06 eV and $B\,^1\Sigma^+$ has 10.78 eV. Different multiplicities are denoted by a, b, c, etc., as in $a\,^3\Pi$ (6.04 eV), which is a triplet state of CO. For historical reasons, A, B, C, etc. refer to triplet states and a, b, c, etc. to singlet states of N_2. Table A2.7 shows selection rules for some simple molecules and representative values for rotational excitation.[13]

TABLE A2.7
Selection Rules for ΔJ and Total Cross Sections for Rotational Transitions Due to Electron Impact

Interaction	H_2	N_2	CO	H_2O
Nonresonant weak quadrupole	Influence covered by resonance	$\Delta J = 0, \pm 2$ $\leq 1.5 \times 10^{-22}$ m^2		
Weak dipole			$\Delta J = \pm 1;$ $\approx 1.7 \times 10^{-21}$ m^2	
Strong dipole				$\Delta J = \pm 1;$ $\approx 2.5 \times 10^{-20}$ m^2
Shape resonance:	$\Delta J = 0, \pm 2$	$\Delta J = 0, \pm 2, \pm 4$	$\Delta J = 0, \pm 1, \pm 2, \pm 4$	Influence covered by dipole
Attractive	$\approx 8 \times 10^{-21}$ m^2	$\approx 2.\,8 \times 10^{-21}$ m^2	$\geq 10^{-20}$ m^2	
Repulsive	Influence not measurable			$\Delta J = 0, \pm 1$ 3×10^{-21} m^2

Reproduced from Jung, K. et al., *J. Phys. B*, 15, 3535, 1982. With permission IOP, U.K.

REFERENCES

1. Kasap, S. O., *Fundamentals of Electrical Engineering Materials and Devices*, McGraw-Hill, New York, 2000.
2. Russell, H. N. and F. A. Saunders, *Astrophys. J.* 61, 38, 1925.
3. Bacher, R. F. and S. Goudsmit, *Atomic Energy States*, Greenwood Press, New York, 1968, p. 10.
4. Gibbs, R. C., D. T. Wilber, and H. E. White, *Phys. Rev.*, 29, 790, 1927.
5. Bacher, R. F. and S. Goudsmit, *Atomic Energy States*, Greenwood Press, New York, 1968, p. 5.
6. G. Racah, *Phys. Rev.*, 61, 537, 1942.
7. Martin, W. C. and W. L. Wiese, in *Atomic, Molecular and Optical Physics Handbook*, ed. G. W. F. Drake, AIP Press, Woodbury, NY, 1996, Chapter 10.
8. Herzberg, G. *Molecular Spectra and Molecular Structure*, Van Nostrand, Princeton, NJ, 1957, p. 8.
9. Herzberg, G. *Molecular Spectra and Molecular Structure*, Van Nostrand, NJ, Princeton, 1957, p. 216.
10. Raju, Gorur G., *Dielectrics in Electric Fields*, Marcel Dekker, New York, 2003.
11. Levine, I. N., *Molecular Spectroscopy*, Wiley International, New York, 1975, p, 150.
12. Tronc, M. A. Huetz, M. Landau, F. Pichou, and J. Reinhardt, *J. Phys. B: At. Mol. Phys.*, 8, 1160, 1975.
13. Jung, K., Th. Antoni, R. Müller, K.-H. Kochem, and H. Ehrhardt, *J. Phys. B: At. Mol. Phys.*, 15, 3535, 1982.

Appendix 3

TABLE A3.1
Cross Sections of Methane (CH$_4$)

			Methane (CH$_4$)				
Energy (eV) Ref.	Q_T (10^{-20} m^2) 1	Energy (eV)	$Q_V(I)$ (10^{-20} m^2) 2	Energy (eV)	$Q_V(II)$ (10^{-20} m^2) 2	Energy (eV)	Q_i (10^{-20} m^2) 3
1.00	2.3	0.162	0.00	0.374	0.00	15	0.235
1.20	2.9	0.165	0.0427	0.380	0.143	17.5	0.687
1.40	4.0	0.170	0.241	0.400	0.330	20	1.279
1.70	4.9	0.180	0.384	0.450	0.440	22.5	1.879
2.00	6.3	0.200	0.527	0.500	0.495	25	2.042
3.00	10.3	0.230	0.459	0.550	0.442	30	2.544
4.00	15.1	0.300	0.300	0.600	0.360	35	2.766
5.00	19.9	0.400	0.181	0.650	0.330	40	2.964
6.00	23.9	0.500	0.159	0.700	0.308	45	3.177
7.00	26.5	0.600	0.165	0.800	0.273	50	3.279
8.00	27.4	0.700	0.170	1.00	0.235	60	3.435
9.00	26.5	0.800	0.177	2.00	0.235	70	3.521
10.0	25.8	1.00	0.180	3.00	0.235	80	3.528
15.0	23.0	2.00	0.180			90	3.505
20.0	19.6	3.00	0.180			100	3.461
25.0	17.5					110	3.396
30.0	16.1					125	3.283
35.0	14.6					150	3.086
40.0	13.6					175	2.921
50.0	12.3					200	2.774
60.0	11.7					250	2.445
70.0	11.0					300	2.211
80.0	10.3					400	1.841
90.0	9.6					500	1.595
100	9.0					600	1.391
200	6.31					700	1.247
300	4.76					800	1.103
400	3.90					900	1.019
500	3.18					1000	1.006
600	2.71						
700	2.49						
800	2.21						
900	1.98						
1000	1.78						

$Q_V(I)$ relates to cross sections for vibrational excitation bands with threshold energy 0.162 eV and $Q_V(II)$ for bands with threshold energy 0.374 eV.

REFERENCES

1. Zecca, A., G. Karwasz, R. S. Brusa, and C. Szmytkowski, *J. Phys. B: At. Mol. Opt. Phys.*, 24, 2747, 1991. Joint measurements were made in two laboratories (Gdańsk and Trento) in the electron energy ranges from 0.9 to 100 eV and 77.5 to 4000 eV respectively.
2. Schmidt, B., *J. Phys. B: At. Mol. Opt. Phys.*, 24, 4809, 1991.
3. Straub, H. C., D. Lin, B. G. Lindsay, K. A. Smith, and R. F. Stebbings, *J. Chem. Phys.*, 106, 4430, 1997; Tian, C. and C. R. Vidal, *J. Phys. B: At. Mol. Opt. Phys.*, 31, 895, 1998. Average values of the two studies are shown in the table.

Appendix 4

TABLE A4.1
Reduced Ionization Coefficient (α/N) as a Function of Reduced Electric Field (E/N)

E/N (Td)	$\alpha/N\ (10^{-20}\,\text{m}^2)$					
	He	Ne	Ar		Kr	Xe
Ref.	1	2	3	2	2	2
6		2.00 (−4)				
7		3.00 (−4)				
8		5.00 (−4)				
9		7.00 (−4)				
10	5.00 (−4)	8.00 (−4)				
15	2.00 (−3)	2.70 (−3)				
20	4.8 (−3)	5.90 (−3)		3.00 (−4)	2.00 (−4)	
25	1.03 (−2)	1.05 (−2)	9.28 (−4)	9.00 (−4)	7.00 (−4)	1.00 (−4)
30	1.68 (−2)	1.62 (−2)	2.08 (−3)	2.20 (−3)	1.60 (−3)	3.00 (−4)
35	2.28 (−2)	2.26 (−2)	4.04 (−3)	4.20 (−3)	3.10 (−3)	6.00 (−4)
40	2.89 (−2)	2.93 (−2)	6.95 (−3)	7.10 (−3)	5.50 (−3)	1.10 (−3)
50	4.32 (−2)	4.40 (−2)	1.57 (−2)	1.51 (−2)	1.29 (−2)	3.50 (−3)
60	5.91 (−2)	5.98 (−2)	2.77 (−2)	2.57 (−2)	2.36 (−2)	8.20 (−3)
70	7.46 (−2)	7.66 (−2)	4.21 (−2)	3.89 (−2)	3.67 (−2)	1.43 (−2)
80	8.98 (−2)	9.38 (−2)	5.81 (−2)	5.50 (−2)	5.24 (−2)	2.38 (−2)
90	0.1046	0.1113	7.51 (−2)	7.34 (−2)	7.00 (−2)	3.57 (−2)
100	0.1192	0.1290	9.3 (−2)	9.23 (−2)	8.97 (−2)	4.86 (−2)
125			0.141	0.1505		
150	0.1910	0.2158	0.195	0.2029	0.209	0.1443
175			0.255	0.2770		
200	0.2627	0.2957	0.320	0.3417	0.344	0.2653
225			0.390	0.4064		
250	0.3297	0.3739	0.463	0.4719	0.483	0.4107
300	0.3920	0.4449	0.614	0.6057	0.627	0.5633
350	0.4489	0.5088	0.764	0.7346	0.776	0.7112
400	0.4979	0.5692	0.910	0.8598	0.918	0.867
450	0.5361	0.6239	1.050	0.9811	1.058	1.025
500	0.5636	0.6748	1.180	1.102	1.197	1.186
550	0.5832	0.7204	1.300	1.218	1.328	1.347
600	0.5975	0.7606	1.410	1.330	1.453	1.506
650	0.6095	0.7968	1.510	1.436	1.573	1.6610
700	0.6218	0.8308	1.610	1.535	1.688	1.814
750	0.6374	0.8627	1.700	1.628	1.798	1.964
800	0.6589	0.8907	1.780	1.719	1.904	2.110
850	0.6893	0.9138	1.860	1.810	2.008	2.249
900	0.7312	0.9335	1.930	1.896	2.110	2.381
950			1.990	1.974		
1000		0.9685	2.060	2.048	2.302	2.6254
1100		1.0005				
1500			2.500	2.653	3.070	3.713
2000			2.760	3.068	3.609	4.576

$a(b)$ means $a \times 10^{b}$.

TABLE A4.2
Reduced Ionization Coefficient (α/N) as a Function of Reduced Electric Field (E/N) in Hydrocarbon Gases

E/N (Td)	α/N (10^{-20} m^2)				
	Methane		Ethylene	Acetylene	Ethane
Ref.	4	12	4	4	5
70	8.00 (−4)	2.40 (−4)			
80	1.80 (−3)	8.60 (−4)			0.0003
90	3.40 (−3)	4.30 (−3)			0.0009
100	6.90 (−3)	7.60 (−3)			0.0023
125	2.32 (−3)	1.91 (−2)	0.0031	0.0019	0.0122
150	4.91 (−2)	4.33 (−2)	0.0095	0.0078	0.0359
175	8.31 (−2)	7.18 (−2)	0.0202	0.020	0.0681
200	0.121	0.107	0.0359	0.040	0.113
225	0.168	0.161	0.0588	0.067	0.168
250	0.223	0.205	0.0886	0.100	0.222
300	0.336	0.303	0.158	0.190	0.358
350	0.452	0.422	0.247	0.292	0.490
400	0.576	0.497	0.351	0.410	0.622
450	0.693	0.623	0.457	0.524	0.759
500	0.813	0.701	0.576	0.648	0.904
600	1.066	0.904	0.836	0.940	1.18
700	1.315	1.11	1.10	1.20	1.51
800	1.559	1.30	1.35	1.46	1.78
900	1.787	1.47	1.61	1.71	2.03
1000	1.992	1.76	1.88	1.94	2.28
1250	2.441		2.54	2.52	2.88
1500	2.802		3.04	3.06	3.41
2000	3.334		3.65	3.83	4.40
2500	3.746		4.12	4.45	5.09
3000	4.058		4.52	4.98	5.48
4000	4.498		5.14	5.74	6.03
4500					6.53
5000	4.800		5.55	6.26	
6000	5.009		5.83	6.593	
7000	5.162		6.09	6.77	
8000	5.339		6.34	6.94	
9000	5.625		6.57	7.26	

E/N (Td)	α/N (10^{-20} m^2)			
	Propane	Butane	Propylene	Butene-1
Ref.	5	5	6	6
100	0.0003			
125	0.0024	0.0002	0.0012	
150	0.0090	0.0025	0.0054	0.0014
175	0.0255	0.0092	0.013	0.0051
200	0.0513	0.0254	0.026	0.013
225	0.0894	0.0488	0.045	0.024
250	0.134	0.0790	0.070	0.042
300	0.248	0.174	0.132	0.095
350	0.378	0.294	0.216	0.168
400	0.522	0.423	0.318	0.256

(*Continued*)

TABLE A4.2
Continued

E/N (Td)	α/N $(10^{-20}\,\text{m}^2)$			
	Propane	Butane	Propylene	Butene-1
Ref.	5	5	6	6
450	0.658	0.563	0.424	0.357
500	0.792	0.711	0.531	0.461
600	1.10	1.02	0.766	0.684
700	1.40	1.34	1.031	0.947
800	1.68	1.64	1.33	1.23
900	1.96	1.94	1.63	1.50
1000	2.26	2.24	1.91	1.76
1250	2.95	3.00	2.6	2.50
1500	3.55	3.71	3.26	3.24
2000	4.74	4.96	4.14	4.29
2500	6.00	6.30	4.93	5.28
3000	7.13	7.40	5.68	6.17
3500			6.33	6.86
4000	8.12	8.35	6.86	7.40
4500	8.25	8.64	7.26	7.88
5000	8.37	9.14	7.52	8.39
5500	8.64	10.08	7.63	8.98

$a(b)$ means $a \times 10^b$. Measured values are interpolated at the same E/N for economy of printing space.

TABLE A4.3
Reduced Ionization Coefficient (α/N) as a Function of Reduced Electric Field (E/N) in H_2, D_2, N_2, and Hg

E/N (Td)	α/N $(10^{-20}\,\text{m}^2)$					
	H_2		D_2		N_2	Hg
Ref.	7	8	7	9	10	11
45	0.0002					
50	0.0005		0.0012			
60	0.0023		0.0048			
70	0.0058		0.0111			
80	0.0119		0.0199			
90	0.0203		0.0153		0.0003	
100	0.0314		0.0061		0.0007	
125	0.0687		0.085		0.0033	
150	0.116		0.135	0.106	0.0095	
175	0.167	0.142	0.190	0.158	0.0202	
200	0.220	0.167	0.245	0.220	0.036	
225	0.275	0.225	0.297	0.281	0.057	
250	0.328	0.292	0.352	0.335	0.083	
300	0.426	0.409	0.448	0.414	0.145	
350	0.514	0.515	0.530	0.480	0.220	

(Continued)

TABLE A4.3
Continued

E/N (Td)	α/N (10⁻²⁰ m²)					
	H₂		D₂		N₂	Hg
Ref.	7	8	7	9	10	11
400	0.594	0.587	0.616	0.535	0.302	
450	0.665	0.649	0.688	0.580	0.391	
500	0.728	0.677	0.743	0.620	0.483	
600	0.826	0.745	0.843	0.679	0.668	0.947
700	0.912	0.782	0.928	0.710	0.851	1.17
800	0.974	0.812	0.988	0.727	1.03	1.53
900	1.02	0.828	1.033		1.208	1.89
1000	1.07	0.847	1.072		1.37	2.15
1250	1.16	0.872	1.160		1.72	2.82
1500	1.23	0.871	1.236		2.00	3.48
1750					2.23	4.10
2000	1.31	0.821	1.27		2.43	4.45
2500	1.35	0.748			2.69	5.10
3000		0.672			2.84	5.60
3300					2.87	
3500						6.12
4000						6.54
4500						6.92
5000						7.18
6000						7.58
7000						7.73

$a(b)$ means $a \times 10^b$. Measured values are interpolated at the same E/N for economy of printing space.

REFERENCES

1. Chanin, L. M. and G. D. Rork, *Phys. Rev.*, 133, 1005, 1964.
2. Kruithof, A. A., *Physica*, 7, 519, 1940.
3. Phelps, A. V. and Z. Lj. Petrović, *Plasma Sources Sci. Technol.*, 8, R21, 1999.
4. Heylen, A. E. D., *J. Chem. Phys.*, 38, 765, 1963.
5. Heylen, A. E. D., *Int. J. Electron.*, 39, 653, 1975.
6. Heylen, A. E. D., *Int. J. Electron.*, 44, 367, 1978.
7. Rose, D. J., *Phys. Rev.*, 104, 273, 1956.
8. Shallal, M. A. and J. A. Harrison, *J. Phys. D: Appl. Phys.*, 4, 1550, 1971.
9. Cowling, I. R. and J. Fletcher, *J. Phys. B: At. Mol. Phys.*, 6, 665, 1973.
10. Haydon, S. C. and O. M. Williams, *J. Phys. D: Appl. Phys.*, 9, 523, 1976.
11. Overton, G. D. N. and D. E. Davis, *J. Phys. D: Appl. Phys.*, 1, 881, 1968.
12. Davies, D. K., L. E. Kline and W. E. Bies, *J. Appl. Phys.*, 85, 3311, 1989.

Appendix 5

TABLE A5.1
Rate Constant (in $10^{-16}\,\text{m}^3\,\text{s}^{-1}$) of Dissociative Attachment as a Function of Mean Energy in Industrial Gases

Molecule	Mean Energy (eV)						
	0.25	0.5	0.75	1	1.5	2	2.5
CCl_4	700	450	350				
$CHCl_3$	150	195	130	80			
CH_2Cl_2	35	90	120	110			
C_2HCl_3	40	80	95	70			
$1\text{-}1\text{-}1\text{-}C_2H_3Cl_3$	44	50	52	36			
$1\text{-}1\text{-}2\text{-}C_2H_3Cl_3$	16	35	40	30			
CH_3Br	0.48	0.2	0.1	0.022	0.0075	0.0016	
C_2H_5Br	0.19	0.65	0.84	0.63	0.34	0.16	0.075
$n\text{-}C_3H_7Br$	0.23	0.67	1.0	0.65	0.32	0.17	0.085
$iso\text{-}C_4H_9Br$	0.33	0.85	1.2	0.75	0.40	0.18	0.12
$iso\text{-}C_5H_{16}Br$	0.45	1.0	1.4	0.8	0.42		
$iso\text{-}C_6H_{13}Br$	0.48	1.2	1.5	0.9	0.5	0.3	0.2
$cis\text{-}C_4F_6$	320	130	68	40			
$2\text{-}C_4F_6$	370	420	400	300			
$1,3\text{-}C_4F_6$	570	300	190	135			
$cis\text{-}C_4F_8$	390	350	260	170			
$2\text{-}C_4F_8$	300	230	170	120			
$cis\text{-}C_5F_8$	1000	480	310	210			
$cis\text{-}C_6F_{10}$	980	480	300	200			
$cis\text{-}C_6F_{12}$	940	540	330	220			
C_7F_8	1000	540	330	220			
C_8F_{16}	930	540	330	220			
$cis\text{-}C_7F_{14}$	200	150	100	70			
CCl_3F	500	220	100				
CCl_2F_2	15	18	20	14			
$CClF_3$		0.027	0.16	0.3			
CO_2					0.0003	0.0011	0.009
F	130	90	60	45	25	20	16
Br_2	0.65	1.3	1.5				
HBr	4	8.5	7.6	6.2	4.6		

Reproduced from Grigoriev, I. S. and E. Z. Meilikhov (Eds.), *Handbook of Physical Quantities*, CRC Press, Boca Raton, FL, 1999, p. 506. With permission.

Appendix 6

TABLE A6.1
A. Reduced Ionization Coefficient (α/N) as a Function of Reduced Electric Field $(E/N)^*$.
B. Dissociative Electron Attachment as a Function of Electron Energy in SO_2.[6]

E/N (Td)	A. α/N (10^{-20} m^2)			B. SO_2: Q_a (10^{-22} m^2)				
Gas Ref.	Air 1, 2	CO_2 3	CO 4, 5	Energy (eV)	O^-	S^- 6	SO^-	Total
80		0.0022		2.5	0.01	0.0	0.0	0.01
90		0.0040		3.0	0.10	0.013	0.02	0.133
100		0.0075		3.1	0.16	0.019	0.0	0.179
110				3.2	0.22	0.034	0.03	0.284
120	0.0062			3.3	0.36	0.056	0.03	0.446
125	0.0073	0.0201	0.0023	3.4	0.58	0.079	0.10	0.759
150	0.0185	0.0362	0.0069	3.5	0.89	0.122	0.16	1.172
175	0.0306	0.0592	0.0155	3.6	1.34	0.167	0.36	1.867
200	0.0521	0.0944	0.0224	3.7	2.20	0.225	0.54	2.965
250	0.1090	0.170	0.0545	3.8	3.16	0.276	0.84	4.276
300	0.199	0.242	0.101	3.9	4.38	0.298	1.41	6.088
350	0.265	0.409	0.172	4.0	5.58	0.313	2.05	7.943
400	0.368	0.525	0.256	4.1	6.92	0.310	3.27	10.500
450	0.458	0.626	0.376	4.2	7.52	0.295	4.58	12.395
500	0.549	0.744	0.489	4.3	8.08	0.271	6.94	15.291
600	0.760	0.974	0.701	4.4	8.02	0.246	8.51	16.776
700	0.908	1.19	0.906	4.6	6.48	0.191	10.80	17.471
800	1.091	1.43	1.12	4.8	4.69	0.164	10.53	15.384
900	1.33	1.64	1.32	5.0	2.14	0.114	7.76	10.014
1000	1.43	1.82	1.48	5.2	1.90	0.065	4.84	6.805
1100	1.51	1.98	1.64	5.4	1.12	0.044	2.80	3.964
1200	1.71	2.12	1.81	5.6	0.69	0.028	1.21	1.928
1300	1.87	2.28	1.96	5.8	0.45	0.013	0.02	0.483
1400	1.98	2.46	2.08	6.0	0.43	0.015	0.39	0.835
1500	2.06	2.66	2.16	6.2	0.81	0.008	0.33	1.148
1600	2.17	2.84		6.4	1.17	0.017	0.40	1.587
1700	2.29	3.00		6.6	1.59	0.022	0.44	2.052
1800	2.40	3.14		6.8	2.10	0.017	0.58	2.697
1900	2.48	3.27		7.0	2.56	0.027	0.42	3.007
2000	2.54	3.39		7.2	2.41	0.030	0.42	2.860
2200	2.68			7.4	2.41	0.031	0.44	2.881
2500	2.99	4.06		7.6	1.89	0.030	0.43	2.350
2800	3.17			7.8	1.33	0.027	0.28	1.637
3000		4.63		8.0	0.93	0.015	0.21	1.155
3500		4.88		8.2	0.61	0.015	0.11	0.735
3600		4.88		8.4	0.35	0.013	0.0	0.363
				8.6	0.25	0.016		0.266
				8.8	0.09	0.021		0.111
				9.0	0.04	0.027		0.067
				9.2	0.02	0.017		0.037
				9.5	0.01	0.002		0.012

$^*a(b)$ means $a \times 10^b$. Measured values are interpolated at the same E/N for economy of printing space. Coefficients in units of 10^{-20} m^2.

TABLE A6.2
Number Density Reduced Attachment Coefficients in SO_2[7]

E/N (Td)	η/N (10^{-22} m²) Three-Body	E/N (Td)	η/N (10^{-22} m²) Dissociative
3	3.16	30	0.40
6	3.60	40	0.46
9	3.44	50	0.63
12	2.73	60	0.91
15	2.08	70	0.115
18	1.60	80	0.144
21	1.24	90	0.174
24	0.98	100	0.201
27	0.79	125	0.231
30	0.63	150	0.248
		200	0.238
		250	0.220

REFERENCES

1. Moruzzi, J. L. and D. A. Price, *J. Phys. D: Appl. Phys.*, 7, 1434, 1974.
2. Rao, C. Raja and G. R. Govinda Raju, *J. Phys. D: Appl. Phys.*, 4, 494, 1971.
3. Bhalla, M. S. and J. D. Craggs, *Proc. Phys. Soc.* (London), 76, 369, 1961.
4. Bhalla, M. S. and J. D. Craggs, *Proc. Phys. Soc.* (London), 78, 438, 1961.
5. Lakshminarasimha, C. S., J. Lucas, and N. Kontoleon, *J. Phys. D: Appl. Phys.*, 7, 2545, 1974.
6. Grigoriev, I. S. and E. Z. Meilikhov (Eds.), *Handbook of Physical Quantities*, CRC Press, Boca Raton, FL, 1999, p. 505.
7. Moruzzi, J. L. and V. K. Lakdawala, *J. Phys.*, Colloque c7, 11, 1979.

Appendix 7

TABLE A7.1
Ionization Potentials

Gas	Symbol	Ionization Potential (eV)		
		I	II	III
Acetylene	C_2H_2	11.40		
Ammonia	NH_3	10.07		
Argon	Ar	15.759	27.629	40.740
Bromine	Br_2	10.516		
Carbon dioxide	CO_2	13.773	37.2	
Carbon monoxide	CO	14.014		
Carbon tetrafluoride	CF_4	15.61		
Chlorine	Cl_2	11.48		
Deuterium	D_2	15.46		
Disilane	Si_2H_6	9.74		
Ethane	C_2H_6	11.56		
Fluorine	F_2	15.697		
Freon-12	CCl_2F_2	12.05		
Germane	GeH_4	≤ 10.53		
Heavy water	D_2O	12.636		
Helium	He	24.587	54.416	
Hexafluoroethene	C_2F_6	14.48		
Hydrogen	H_2	15.43		
Iodine	I_2	9.28		
Krypton	Kr	13.999	24.359	36.95
Mercury	Hg	10.437	18.756	34.2
Methane	CH_4	12.60		
Neon	Ne	21.564	40.962	63.45
Nitric oxide	NO	9.264		
Nitrogen	N_2	15.60		
Nitrogen dioxide	NO_2	9.586		
Nitrous oxide	N_2O	13.38		
Octafluoropropane	C_3F_8	13.70		
Oxygen	O_2	12.069		
Ozone	O_3	12.43		
Propane	C_3H_8	10.95		
Propene	C_3H_6	10.95		
Silane	SiH_4	11.00		
Sulfur dioxide	SO_2	12.34		
Sulfur hexafluoride	SF_6	15.34		
Water	H_2O	12.617		
Xenon	Xe	12.130	21.21	32.10

Index